TUNNELLING '94

TUNNELLING
'94

Papers presented at the seventh international symposium, 'Tunnelling'94', organized by the Institution of Mining and Metallurgy and the British Tunnelling Society, with the cooperation of the Transport Research Laboratory, Department of Transport, and held in London from 5–7 July, 1994.

Published on behalf of the Institution of Mining and Metallurgy and the British Tunnelling Society

CHAPMAN & HALL

London · Glasgow · Weinheim · New York · Tokyo · Melbourne · Madras

Published by Chapman & Hall, 2-6 Boundary Row, London SE1 8HN

Chapman & Hall, 2-6 Boundary Row, London SE1 8HN, UK

Blackie Academic & Professional, Wester Cleddens Road, Bishopbriggs, Glasgow G64 2NZ, UK

Chapman & Hall GmbH, Pappelallee 3, 69469 Weinheim, Germany

Chapman & Hall USA, One Penn Plaza, 41st Floor, New York, NY10119, USA

Chapman & Hall Japan, ITP - Japan, Kyowa Building, 3F, 2-2-1 Hirakawacho, Chiyoda-ku, Tokyo 102, Japan

Chapman & Hall Australia, Thomas Nelson Australia, 102 Dodds Street, South Melbourne, Victoria 3205, Australia

Chapman & Hall India, R. Seshadri, 32 Second Main Road, CIT East, Madras 600 035, India

First edition 1994
Reprinted 1995

Printed in Great Britain by St Edmundsbury Press, Bury St Edmunds, Suffolk

ISBN 0 412 59860 4

A Catalogue record for this book is available from the British Library

∞ Printed on acid-free paper, manufactured in accordance with ANSI/NISO Z39.48-1992 and ANSI/NISO Z39.48-1984 (Permanence of Paper)

Contents

Session 6: Case histories

Session 7: Tunnel-boring machines

Session 8: Research

Session 9: Case histories

Session 10: Design

Session 11: Case histories

Session 12: Hydropower

Session 13: Tunnel machinery

Organizing Committee

United Kingdom

Dr. John Temporal (*Conference Chairman*)
O. M. Bevan
R. N. Craig
J. P. Doody
I. H. Elliott
Dr. R. J. Fowell
D. J. Hindle
R. F. Jennion
T. I. King
M. C. Knights
C. N. McKenzie
Dr. T. W. Mellors
Dr. M. P. O'Reilly
G. E. Pearse
Dr. C. P. M. Snee
Eric Snowdon
Dr. S. C. A. Walker

Overseas

Dr. S. Babendererde (Germany)
Professor W. E. Bamford (Australia)
S. Bjørgum (Norway)
Professor E. T. Brown (Australia)
Professor R. Ciccu (Italy)
Professor H. Duddeck (Germany)
T. Franzen (Sweden)
Dr. G. Fukuchi (Japan)
Dr. A. Haack (Germany)
R. W. O. Kersten (South Africa)
Professor Liu Baochen (China)
R. P. Lovat (Canada)
S. Nelson (U.S.A.)
M. Panet (France)
Dr. J. F. Raffoux (France)
R. J. Robbins (U.S.A.)
Dr. G. Sauer (Austria)

Contributors

Roald Aabøe, Norwegian Public Roads Administration, Directorate of Public Roads, Road Research Laboratory, Oslo, Norway
Vidar Aarvold, N S B Gardermobanen High Speed Railway Project, Norway (formerly, Veidekke–Selmer Group, Norway)
Y. Abd Al-Jalil, Department of Civil Engineering, The University of Texas at Austin, Austin, Texas, U.S.A.
J. R. Araujo, Capital Engineering and Planning Department, Waterworks Division, Massachusetts Water Resources Authority, Boston, Massachusetts, U.S.A.
L. J. Arthur, Lahmeyer MacDonald Consortium, Maseru, Lesotho
Klaus Axhausen, Alfred Kunz GmbH and Co., Munich, Germany
Manfred Bandmann, Tiefbau-Berufsgenossenschaft, Munich, Germany
D. A. Barratt, Transport Research Laboratory, Crowthorne, England
Jon Baston-Pitt, G B Geotechnics, Ltd., Bottisham, Cambridgeshire, England
J. G. Beatty, U.S. Heavy Division, Perini Corporation, Chicago, Illinois, U.S.A.
Uwe Beckmann, BRP Consult Beckmann, Ruppert + Partner GmbH, Braunschweig, Germany
L. J. Bergen, Sverdrup Civil, Inc., Boston, Massachusetts, U.S.A.
D. Blakey, Sir Robert McAlpine, Hemel Hempstead, England
K. H. Bowers, Transport Research Laboratory, Crowthorne, England
K. T. Burgess, Howard Humphreys and Partners, Ltd., Leatherhead, England
H. P. Caspe, Sverdrup Civil, Inc., Boston, Massachusetts, U.S.A.
P. Catling, Cementation Mining, Ltd., Bentley, Doncaster, England
P. J. Chapman, TML (Balfour Beatty International), Thornton Heath, England
I. L. J. Chudleigh, London Underground, Ltd., London, England
P. M. Claye, Thames Water Utilities, Ltd., Reading, England
B. J. Cole, North West Water Engineering, Ltd., Southport, Merseyside, England
A. W. P. Coutts, Jubilee Line Extension, London Underground, Ltd., London, England
R. S. Curtis, Hatch Associates, Ltd., Mississauga, Canada

A. W. Darby, Tunnels Division, Mott MacDonald, Ltd., Croydon, England
H. R. Davies, Jubilee Line Extension, London Underground, Ltd., London, England
P. G. Davies, Sumber Bumi Knight Piésold, Sdn. Bhd., Kelantan, Malaysia
Graham Daws, Graham Daws Associates, Ltd., Riddings, Derbyshire, England
A. P. Deane, Mott MacDonald, Ltd., Croydon, England
Walter Dietz, Ed. Züblin AG, Stuttgart, Germany
I. W. Farmer, I. W. Farmer and Partners, Newcastle upon Tyne, England
J. P. Farrow, Thames Water Utilities, Ltd., Reading, England
D. F. Fawcett, Babtie Shaw and Morton, Maidstone, England
G. A. Featherstone, Ninham Shand, Inc., Randburg, South Africa, and Lesotho Highlands Tunnel Partnership, Maseru, Lesotho
A. P. Finch, Mott MacDonald, Ltd., Croydon, England
Sigurdur Finnsson, The Robbins Company, Kent, Washington, U.S.A.
R. J. Fowell, Department of Mining and Mineral Engineering, University of Leeds, England
F. J. Fredriksen, Norwegian Public Roads Administration, Directorate of Public Roads, Road Research Laboratory, Oslo, Norway
R. J. Ganey, Continental Conveyor and Equipment Company, Winfield, Alabama, U.S.A.
D. C. Gilchrist, Messrs. Sandberg, London, England
M. A. Ginary, Taylor Woodrow Construction, Ltd., London, England
M. J. Gollick, Dosco Overseas Engineering, Ltd., Tuxford, Nottinghamshire, England
Erik Gutenwik, Skanska Raise Boring Team, Stockholm, Sweden
Richard Harpf, Alfred Kunz GmbH and Co., Munich, Germany
D. I. Harris, Geotechnical Consulting Group, London, England
G. J. Hasle, Norwegian Public Roads Administration, County Roads Office of Vestfold, Tønsberg, Norway
J. E. Hellings, Pell Frischmann Consulting Engineers, Ltd., London, England
D. A. Hobson, Howard Humphreys and Partners, Ltd., Leatherhead, England
Nicola Innaurato, Dipartimento di Georisorse e Territorio—Politecnico, Turin, Italy
Sandor Jancsecz, Wayss & Freytag AG, Frankfurt am Main, Germany
H. A. Janzon, The Robbins Company, Kent, Washington, U.S.A.
C. Kettle, Bachy (UK), Ltd., Godalming, England
J. E. Killingsworth, Continental Conveyor and Equipment Company, Winfield, Alabama, U.S.A.

A. Y. Kim, Sverdrup Civil, Inc., Boston, Massachusetts, U.S.A.
M. C. Knights, Howard Humphreys and Partners, Ltd., Leatherhead, England
H. Kruizenga, Tarmac HBM JV, Chatham, Kent, England
C. Laughton, Department of Civil Engineering, The University of Texas at Austin, Austin, Texas, U.S.A.
Y. Leblais, Simecsol, Paris, France
L. Leblond, Department of Engineering, TML French Site, Calais, France
H. J. Lees, North West Water Engineering, Ltd., Southport, Merseyside, England
R. P. Lovat, Lovat Tunnel Equipment, Inc., Toronto, Canada
J. P. Love, Geotechnical Consulting Group, London, England
Dan MacDonald, Vaughan Engineering, Ltd., Sydney, Nova Scotia, Canada
R. J. Mair, Geotechnical Consulting Group, London, England
B. W. L. Mak, Montgomery Watson, Hong Kong
Yannis Malios, Yannis Malios & Associates, Ltd., Athens, Greece
Renato Mancini, Dipartimento di Georisorse e Territorio—Politecnico, Turin, Italy
P. Marcheselli, Passante Scarl, Milan, Italy
Bruno Marrai, Knight Piésold and Partners, Ashford, England
C. J. H. Martin, Trafalgar House Construction (Tunnelling), Bentley, Doncaster, England
R. Matthews, Taylor Woodrow Construction, Ltd., London, England
Daniel Morfeldt, Mineconsult AB, Stockholm, Sweden
A. D. Murray, Knight Piésold and Partners, Ashford, England
P. P. Nelson, Department of Civil Engineering, The University of Texas at Austin, Austin, Texas, U.S.A.
B. M. New, Transport Research Laboratory, Crowthorne, England
M. P. O'Reilly, Consultant, Camberley, England (formerly, Transport Research Laboratory, Crowthorne, England)
M. A. Oswell, Montgomery Watson, Hong Kong
G. Pakes, Gerard Pakes Consultants, Wirksworth, England
Pierre Jean Pellegri, C. C. Camargo Corrêa S.A., São Paulo, Brazil
J. Y. Peron, Neyrpic Framatome Mécanique, Tunnelling Machine Department, Lyon, France
B. Rafoneke, Planning and Design Division, Lesotho Highlands Development Authority, Maseru, Lesotho
H. Ramisch, Tiefbau-Berufsgenossenschaft, Munich, Germany
R. Remington, Thames Water Utilities, Ltd., Reading, England
G. Richardson, Dosco Overseas Engineering, Ltd., Tuxford, Nottinghamshire, England
Enrico Rondena, ENEL S.p.A., Centro di Ricerca Idraulica e

Strutturale, Milan, Italy
Maurício Santana, C. C. Camargo Corrêa S.A., São Paulo, Brazil
J. Scholey, Howard Humphreys and Partners, Ltd., Leatherhead, England
Walter Steiner, Balzari & Schudel AG, Bern, Switzerland
John Temporal, Transport Research Laboratory, Crowthorne, England
Guilherme Tibau, C. C. Camargo Corrêa S.A., São Paulo, Brazil
R. B. Warren, Taylor Woodrow Construction, Ltd., London, England
N. S. Webb, London Underground, Ltd., London, England
C. R. Weeks, Tunnel Division, Mott MacDonald, Ltd., Croydon, England
T. R. Winterton, C. V. Buchan Ltd., Northwich, Cheshire, England
Attilio Zaninetti, ENEL S.p.A., Centro di Ricerca Idraulica e Strutturale, Milan, Italy

Foreword

It is now more than twenty years since a proposal was first mooted to hold an international tunnelling symposium in Britain. At the time of the first symposium, held in London in 1976, the Channel Tunnel project had just been shelved. Last weekend a charity walk was held in the finished tunnel, which will be open for business later in the year. Tunnels have figured prominently, and at times spectacularly, in the development of national and international links and it is hoped that such links gather pace in the future. It is particularly pleasing that Alastair Biggart of Storebælt has agreed to deliver the twenty-sixth Sir Julius Wernher Memorial Lecture of the Institution of Mining and Metallurgy, entitled 'The changing face of tunnelling', at the start of this event.*

Although almost every edition of the technical journals on tunnelling reports another £1 billion scheme somewhere in the world, it would be unfair of me to suggest that tunnelling is restricted to these prestigious schemes for major transport links. Much of the work that makes modern life possible receives hardly a mention outside the technical press and one suspects that society at large applies the 'out of sight, out of mind' attitude even more readily to underground construction than it does to other forms of engineering. Clearly, there is a continuing need to improve the capacity and performance of our infrastructure, while having a careful regard for the environment. In a world dominated by economic constraints, cost-effectiveness can best be achieved by stimulating innovation and progress in the design, construction, operation and maintenance of all tunnels, whatever their size or purpose, through events such as 'Tunnelling '94'. While doing this, we must also endeavour to raise the profile and public perception of our work within society as a whole.

'Tunnelling '94' is organized by the Institution of Mining and Metallurgy (IMM) and the British Tunnelling Society (BTS), with the cooperation of the Transport Research Laboratory (TRL). All three bodies

*_Trans. Instn Min. Metall. (Sect. A: Min. industry),_ **103,** 1994, in press.

have provided volunteers to serve on the Organizing Committee. In addition, many others—in particular, representatives from overseas—have been coerced into joining the Committee and helping with all the tasks that go on behind the scenes. As this is the first symposium in the series for which I have been Chairman, I am especially indebted to all the members of the Organizing Committee not only for their hard work and dedication but also for their personal support and encouragement.

I am particularly pleased to record my thanks to Martin Knights, who took on the job of chairing the papers committee at short notice, following the sad and untimely death of Professor Barry Whittaker, to whom this volume is dedicated; to Roy Jennion and Dr. Steve Walker for organizing the technical visits that form a traditional part of the 'Tunnelling' series of conferences; to Colin McKenzie and John Doody for their liaison with Brintex for the associated Exhibition; to Michael Jones, IMM Secretary; to Liz Munro and the IMM Conference Office team; to Dr. Myles O'Reilly, who has readily agreed once more to undertake the technical summary at the closing session of the conference; to Geoff Pearse, who, equally readily, has agreed to prepare another of his reports of the discussions at the technical sessions for publication in the *Transactions* of the Institution of Mining and Metallurgy; and to Atlas Copco Rocktech AB for its support of the conference. Finally, the authors of all the papers in this volume are to be thanked for their hard work in putting together and presenting their excellent contributions. Delegates to the conference and readers of this volume must decide the worth of all these efforts: I believe that the papers that appear in *Tunnelling '94* will be found to be both stimulating and informative.

As I have already mentioned, Professor Barry Neil Whittaker died early in 1993, shortly after agreeing to chair the papers committee for this symposium. Barry had been involved in the 'Tunnelling' series for many years and was one of its most enthusiastic supporters. It is most fitting that we should dedicate this volume to the memory of such a well-respected scholar, gentleman and friend.

Dr. John Temporal
Chairman, Organizing Committee
February, 1994

In Memoriam

Professor Barry Neil Whittaker
1936–1993

Many tributes were paid to Professor Barry Whittaker, who died on 17 January, 1993. Given below are those which were published by the Institution of Mining and Metallurgy.

Professor Barry Neil Whittaker was born in Yorkshire on 11 June, 1936. He started work as a miner in the Yorkshire coalfield at the age of 16 and in his distinguished career progressed to become the world's leading expert in the field of soft rock mechanics. It is typical of Barry that he chose the subject of soft rock mechanics, the most capricious area of geotechnical engineering, as his specialism. In the process he graduated with a B.Sc. with first-class honours and the degree of Doctor of Philosophy from the University of Newcastle-upon-Tyne, and that of Doctor of Science from the University of Nottingham. He held the First Class Certificate of Competency under the Coal Mines Act. After graduation he returned to industry before joining the University of Nottingham as a Research Fellow. At Nottingham he established himself as one of the world's leading practitioners in the field of rock mechanics and rose through lecturer and reader to become Professor of Rock Mechanics. Since August, 1992, he was Professor of Mining Engineering at the University of Leeds. He was author or joint author of numerous papers in his field of study and co-authored three internationally recognized textbooks. He was a consultant to many large mining organizations. Starting as an undergraduate he received several prestigious awards, culminating in the Douglas Hay Medal of the Institution of Mining Engineers. Professor Whittaker's approach to research was to grind problems down as finely as possible to gain the maximum amount of knowledge. He was tireless in his efforts and was never satisfied to just intensively study the literature. Many of us believe he 'talked to the rocks' and he found his answers in practice as well as theory.

Many eminent mining engineers around the world have reason to be grateful to him for his meticulous teaching and guidance—and they are! Barry cared for people as well as his work. Many students experience financial hardship during their studies: Barry had a remarkable talent for finding solutions to their problems. Sometimes the solutions defied the laws of gravity (and occasionally those of the Bursar), but no one ever let him down and he had the great satisfaction of seeing those young people graduate because of his help.

Barry was greatly respected for his scholarship, but he was also universally liked for his great humour, friendliness and his ability to work with and get the best out of others. Everyone felt privileged to be his colleague. He was a man who cared, a man of great compassion and a family man. The world will be the poorer for his passing.
Tom Atkinson

Many will indeed be able to vouch for what Tom Atkinson has written about Barry Whittaker. In the months before his death, Barry had been giving me and my Laporte colleagues valuable consultancy advice on a ground support problem at one of our mines in Derbyshire. Towards the end of November he unexpectedly requested an underground visit to monitor progress. While he was with us he took particular trouble to teach us how to do the calculations so that, as far as possible, we could manage without him. Although obviously unwell, he made light of his condition. We did not then know what he had already been told by his doctor—that he had very little time left. His visit was the typical action of a very fine man.

Barry performed many other similar kind and conscientious acts during his last days. I shall miss him for that as I shall remember his valued contribution to the work of the IMM Executive committee.
J. V. Bramley

Professor Whittaker's early training and experience were in underground mining, but he developed a keen interest in civil engineering and he felt strongly that there was much information about design and methods that could usefully be shared between mining and civil engineers. One of his plans at Leeds University was to inaugurate a degree course in tunnelling engineering.

At the 'Tunnelling '88' conference Professor Whittaker commented on the causes of the Abbeystead water scheme disaster and similar incidents

in civil tunnels worldwide in a definitive paper on methane hazards in tunnelling operations. He was invited to join the organizing committee for the next tunnelling-related conference in 1990, to which he contributed two geotechnical papers. He continued on the committee for the 'Tunnelling '91' event and at that he provided outstanding support, including standing in at very short notice for the President, who was indisposed, to thank a Government Minister for his opening address, introduce and respond to the Sir Julius Wernher lecturer and open the exhibition. Later that same day he presented a paper on tunnel stability and, on the following day, he chaired the first session. Needless to say, he was invited to serve on the 'Tunnelling '94' committee. His advice and enthusiasm will now be deeply and sadly missed by all his friends and colleagues.
G. E. Pearse

Although a relative newcomer to the Council of the Institution, having been one of its members since 1989, Barry Whittaker had been active in our affairs since he joined the IMM as a Member in 1971. Not only was he involved with journals and conferences but as the author, with Russell Frith, of *Tunnelling—design, stability and construction*, which the Institution published in 1990. Typical of Barry's generosity was his waiving of any royalty payments from that book in favour of the IMM Benevolent Fund. More recently, he served as a very conscientious Chairman of the Professional Affairs committee—an undertaking that necessitated significant time in dealing, in particular, with the Engineering Council and the accreditation of a variety of university courses throughout the United Kingdom and related matters beyond this country.

In all these tasks Barry showed both the diligence and enthusiasm for the Institution that others have seen elsewhere in his academic and industrial life. His premature death is the more tragic in having come so soon after his move from Nottingham to Leeds.

I shall miss Barry as a personal friend and as a keen supporter of the Institution in so many ways. It is our great loss that we have been robbed of a future President: his contributions, nevertheless, mark him out as a very special person who will long be remembered with affection.
Michael Jones

Session 1: Hydropower

Lesotho Highlands Water Project: design and layout of underground works for 'Muela Hydropower Project

L. J. Arthur
Lahmeyer MacDonald Consortium, Maseru, Lesotho
A. W. Darby
Tunnels Division, Mott MacDonald, Ltd., Croydon, England
B. Rafoneke
Planning and Design Division, Lesotho Highlands Development Authority, Maseru, Lesotho

Abstract
The paper describes the design of the underground works for the 'Muela Hydropower project which is part of the multi-purpose Lesotho Highlands Water Project which diverts water from the Kingdom of Lesotho through a series of reservoirs and tunnels being constructed in a series of phases.

The 'Muela project, for construction of the first major power station in Lesotho, is being incorporated in the first stage of the scheme currently under construction and the paper describes the inter-relationship between power generation and water delivery elements together with the constraints imposed on the design and implementation of the tunnels and underground powerhouse.

Geotechnical aspects specific to the project are reviewed and the testing and investigation programmes carried out are summarised. Although design methods follow largely orthodox lines, specific features of the excavations and linings required to meet the particular conditions of the site are noted.
Keywords: Geotechnical investigations, hydro-electric works, Lesotho, planning and design of underground works, rock tunnelling, tunnels for water.

1. Lesotho Highlands Water Project

The Lesotho Highlands Water Project is being constructed to supplement water supplies in the heavily populated area of South Africa centred around Johannesburg. Water will be transferred from the Katse Reservoir on the Malibamatso river in the Maluti Mountains of Lesotho through a series of tunnels to Clarens on the Ash River in the Orange Free State [Figure 1]. On completion of the present stage (Phase 1A),

18 m³/sec will be delivered into the catchment basin of the Vaal River. In Phase 1B, this will be increased to 33 m³/sec and planned subsequent developments could increase this to a maximum of 70 m³/sec after 5 phases of development. The tunnel system drops around 300 m from an elevation of 1990 m above sea level at Katse to the outfall 100 km to the north [Figure 2]. Around 200 m of this fall occurs at the Nqoe valley, 60 km from Katse Dam, where a regulating reservoir is required between northern and southern parts of the tunnel system and there is an opportunity to locate a hydro-electric plant close to the village of 'Muela. This plant will generate 72 MW (base load) on completion of Phase 1 of the project, with the possibility for the installation of a further 110 MW of generating capacity at a later stage.

Figure 1. Key and Location Plan

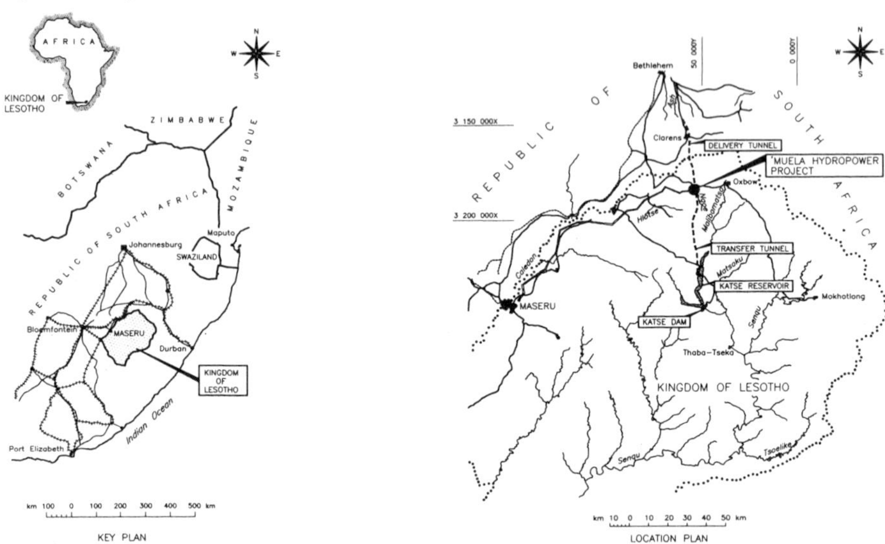

Figure 2. Lesotho Highlands Water Project : Plan and Longitudinal Section

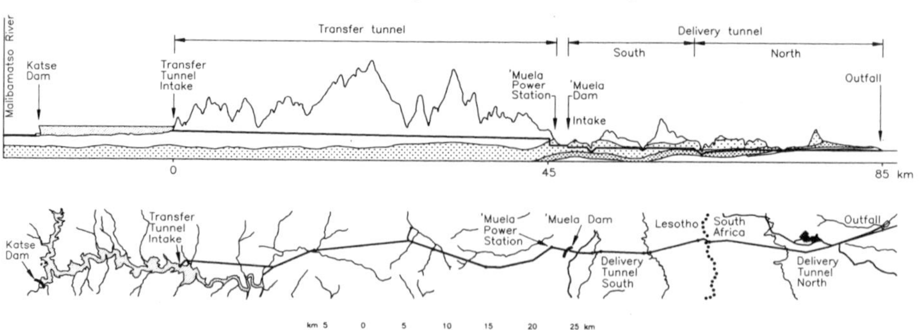

At present, the electricity consumed in Lesotho is provided by the South African power utility ESKOM apart from a small number of mini-hydro stations. The 'Muela Hydropower Project will thus enable both the Lesotho Electricity Corporation to provide its own consumers directly as well as raising the possibility of integrated operation with the adjacent South African network.

The underground works for the 'Muela Hydropower Project described in this paper, together with the associated electrical and mechanical contracts, are scheduled to be commissioned simultaneously with the start of water transfer from Lesotho to South Africa at the end of 1996. At the time of writing, tenders have been received and construction is expected to start during 1994.

The works are being designed and built for the Lesotho Highlands Development Authority (LHDA). The detailed design, preparation of construction drawings and supervision of construction of the 'Muela Hydropower Project is being undertaken for LHDA by the Lahmeyer MacDonald Consortium (Lahmeyer International of Germany and Mott MacDonald Ltd. from the U.K. in association with the Lesotho consulting engineering group, ABCC).

2. 'Muela Hydropower Project

The location of the elements of the hydro plant is governed largely by the alignment of the water transfer system together with the requirement to avoid any construction at the present stage which could interfere with possible developments later [Figure 3].

A number of sites were examined for the dam necessary to create the reservoir which both regulates the flow of water to South Africa and provides the tailpond for the hydro development. The site chosen, in a narrow gorge on the Nqoe river north of 'Muela village, limits the possible locations for the outfall from the tailrace tunnel. The position of this outfall is further constrained as the river carries a high silt load during floods and it is essential that the outfall does not become blocked by any silt accumulation behind the dam over a period of time.

Figure 3. 'Muela Hydropower Project : Plan of Project Area

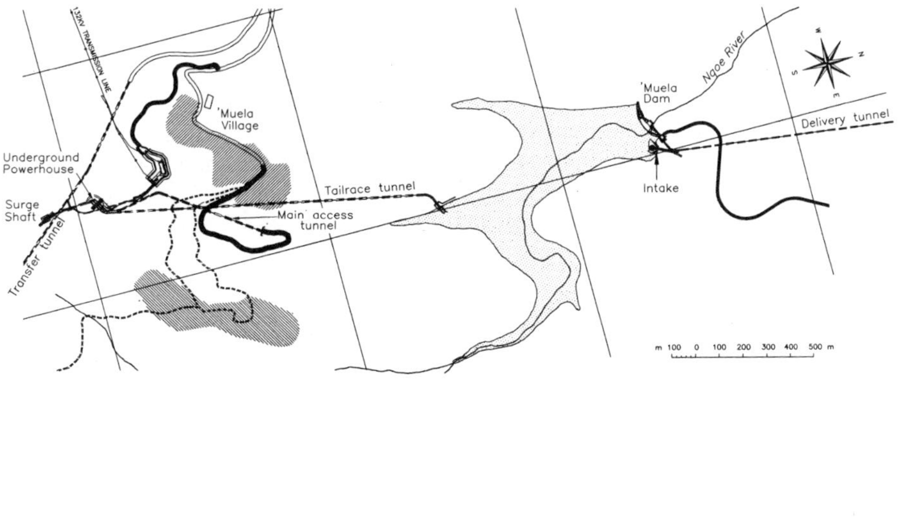

Upstream of the power station, the connection with the 45 km long Transfer Tunnel from Katse Reservoir is governed by the need to locate a surge shaft at a point where the topography is such that sufficient height is available to prevent overflow of water during maximum upsurges.

A number of alternative locations for the powerhouse were investigated between these upstream and downstream limits. The considerable additional length of steel lining associated with a surface station on the edge of the reservoir, and the very deep excavation required for a shaft power station, indicated the choice of a totally underground station entirely on economic grounds, although there are also concomitant environmental advantages.

An Operations Building for the administration and control of the entire Lesotho Highlands Water Project, as well as the hydropower station, will be sited at ground level adjacent to 'Muela village. Originally, it was intended that this would be located directly above the underground powerhouse with an interconnecting lift and cable shaft for ease of access and ventilation. In addition, it was envisaged that a by-pass pipeline would be required at a future date to ensure continued water transfer in the event of extended downtime of the turbines preventing flow through the powerhouse. Provision for a connection to this by-pass was required at the penstock guard valve chamber immediately downstream of the upstream surge shaft.

Optimisation studies were carried out to examine the total cost of the powerhouse, waterway tunnels (including steel linings), access tunnels and roads, provision for the by-pass, and high voltage electrical connections for five alternative configurations. These showed that a saving of around 10% in the capital cost of these works could be achieved by moving both powerhouse and shafts the maximum practicable distance of around 250 m upstream, but with the disadvantage of a marginally less convenient access for operational purposes.

The principal hydropower elements follow orthodox lines, with a steel lined penstock shaft and manifold tunnels, the underground powerhouse and concrete lined tailrace tunnel. Although it was initially proposed to drive a new main access tunnel to the powerhouse, it was finally determined that this could be provided more economically by enlarging the existing exploratory adit. The works are designed to be constructed independently of other parts of the water transfer project [Figures 4 and 5].

A high level access tunnel will connect the Operations Building both to the powerhouse by means of a shaft and to the penstock guard valve chamber which contains the guard valve to isolate the power station from the Transfer Tunnel upstream. In addition to the access underground, the top of the upstream surge shaft will be directly accessible for construction. Although the scheme is physically quite small, the construction programme is tight and the large number of access possibilities will significantly assist in both achieving the construction schedule and facilitating permanent operation of the station.

Figure 4. 'Muela Hydropower Project : Plan

Figure 5. 'Muela Hydropower Project : Longitudinal Section

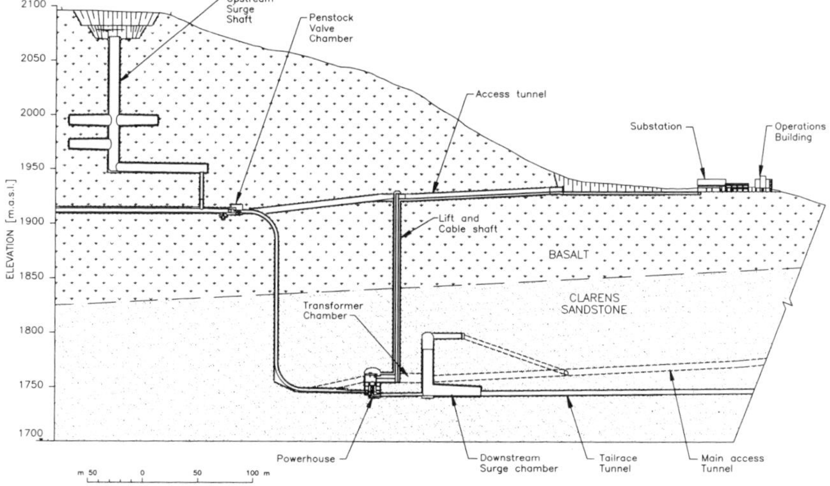

3. Geology

Prior to detailed design, an extensive site investigation programme was undertaken involving over 4000 m of boreholes together with in-situ and laboratory testing. In addition, an exploratory adit 550 m long was excavated to investigate the geology of the future powerhouse site. This programme has since been supplemented by geological mapping of the sites of the principal surface works.

The upstream works will be constructed in basalts of the Lesotho Formation. These are horizontally layered, strong, widely jointed and are amygdaloidal in some areas. Permeability is generally low, although weathering can be present up to depths of around 10 m. Horizontal in-situ stresses are relatively low and a horizontal to vertical stress ratio of 0.5 has been adopted for design purposes. The lower 70 m of the basalts consist of the "Mixed Beds" which lie immediately above the underlying Clarens Sandstone. In this stratum the lithology is variable but dominated by basalt and other rocks of volcanic origin.

The underground powerhouse, low level waterways and the main access tunnel are all located within the Clarens Sandstone. This is lithologically and mineralogically homogenous, strong, fine/medium grained, massive and very widely jointed. Stress relief joints are present near the surface in some areas but there is limited weathering.

The region is intruded extensively by dykes and sills but none have been identified in the area of the underground works except for a dolerite neck (diatreme) which is expected to be encountered by the tailrace tunnel. The areas is tectonically stable and is subject to only low levels of seismic activity.

The exploratory adit excavation demonstrated the competence of the sandstone with few support difficulties and low water inflows, although in certain areas degradation of the sandstone surfaces has been observed over a period of time. Excellent tunnelling conditions have also been observed in the TBM drive for the Delivery Tunnel South, north of the Nqoe River, and in drill-and-blast excavations for the Transfer Tunnel in the basalt at the upstream end of the hydropower development.

The consistency and nature of the formations has enabled the configuration of the underground works to be optimised with a minimum of geological constraints to achieve lowest construction costs. Although it was found more economic to relocate the main cavern some distance upstream from the position determined at the time of the exploratory works, further detailed investigation of the new location has not been considered necessary in advance of construction.

Table 1 Summary of rock properties

		Unconfined compressive strength (MPa)	Modulus (GPa)	Minimum principal stress ratio
Basalt	Lower bound	45	12.5	
	Best estimate	75	20	0.5
Sandstone	Lower bound	20 to 40	12	
	Best estimate	40 to 60	20	0.5

4. Upstream surge shaft

The largest single underground element of the hydropower project will be the upstream surge shaft with a total excavated volume of around 32,000 m^3, which compares to the excavated volume of the powerhouse of approximately 20,000 m^3 [Figure 6]. The necessity for this structure arises from the need to control hydraulic surges induced in the 45 km long Transfer Tunnel waterway system arising from operation of the turbines. It is believed that this will be the longest headrace tunnel of any hydro-electric scheme world wide.

Figure 6. 'Muela Hydropower Project : Upstream Surge Shaft

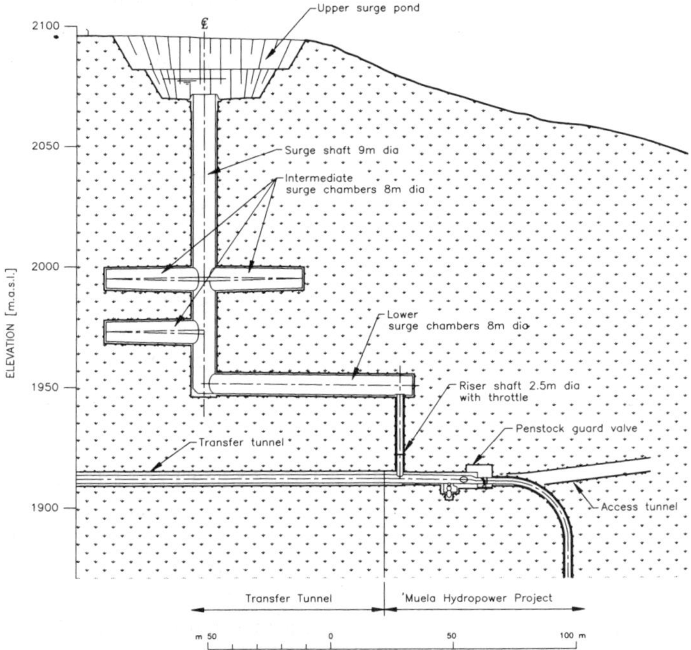

Originally proposed as a simple cylindrical shaft, further consideration of the critical turbine operations resulted in a change to the present layout incorporating expansion chambers. This arrangement arises from consideration of the requirement to achieve the fastest possible acceleration or deceleration of water velocity in the Transfer Tunnel after a turbine load change, together with rapid stabilisation of water levels and attenuation of oscillations. The levels of the chambers were selected to minimise the temporary drop in power generation after load acceptance over the entire turbine operating head range.

The local topography determines the location of the shaft in order to prevent any overflow during maximum upsurges. Orientation of the expansion chambers has been governed by geotechnical considerations together with the need to avoid any works which could impinge on the adjacent "corridor" for subsequent stages of the water transfer scheme.

The lining to the upstream surge shaft must meet the following criteria:

- provide an adequate margin against hydrofracture of the surrounding rock,
- limit leakage due to any unfavourable geological conditions,
- prevent any induced slope instability in the surrounding ground,
- resist stresses induced by internal and external hydrostatic pressures.

Initially, it was considered that it would be necessary to provide the upstream surge shaft with a heavily reinforced concrete lining with a waterproof membrane throughout. However, examination of the in-situ stress measurements showed a margin against hydrofracture in the range 1.35 to 5. From measurements in the boreholes, leakage in the rock appeared to be restricted to fissured zones which could be readily grouted and there is no evidence of discontinuities which could produce kinematically feasible instabilities. Under internal hydrostatic pressures, the lining would crack and the load transferred to the rock. Thus a concrete lining with nominal reinforcement to control cracking could be adopted in conjunction with orthodox contact and fissure grouting procedures.

The additional cost arising from the more complex shape of shaft has been compensated by the simplification of the lining design. A saving has been obtained which is estimated at around 10% of the total capital cost of this element, without taking into account the benefits from improved operation of the power station.

5. Upstream waterways

The upstream waterway system consists of the penstock valve chamber, the penstock shaft and the upstream manifold to the powerhouse. The valve chamber will contain a 2.5 m diameter bi-plane butterfly guard valve and a degraveller, together with provision for man access into the tunnel. Provisions for connection to a future pipeline to bypass water around the entire power station are also included at this point.

Steel linings are required to the upstream waterways where the in-situ stresses in the rock provide insufficient containment against hydrofracture (i.e. where the minimum principal stress < internal design pressure + 20%). In addition, linings are needed to prevent leakage where the hydraulic gradient between the tunnel and an unlined opening such as the powerhouse exceeds the design limits of 3 in the sandstone and 5 in the basalt.

Whilst in theory the central section of the penstock shaft does not require a steel lining, such an arrangement would be impractical and these linings are to be provided throughout the entire upstream waterway system. The linings are designed against full external hydrostatic pressure and for internal pressure as a free standing penstock. Reference to current European design codes showed that there is no advantage to be gained by allowing a contribution from the rock to the support of the linings.

Although the possibility of adopting an inclined penstock shaft was considered, economic considerations favoured a vertical shaft with the distance from the upstream surge shaft to the powerhouse reduced to the least practicable. The increase in costs of the access tunnels incurred by this arrangement was mitigated by the savings in the steel lined waterway and simplified shaft construction. The penstock diameter was determined as 2.5 m following an optimisation exercise which took account of construction costs, turbine efficiency, head losses, effects of water hammer and energy benefits.

6. Powerhouse

The powerhouse is dimensioned to house the three 24 MW Francis turbines, the generators and the associated ancillary plant. An unusual feature is the location of the principal auxiliary equipment at the opposite end of the cavern to the entry of the main access tunnel and the loading bay. This will facilitate incorporation of additional generating capacity at a later stage when it is envisaged that the cavern will be extended [Figure 7].

Figure 7. 'Muela Hydropower Project : Powerhouse : Longitudinal Section

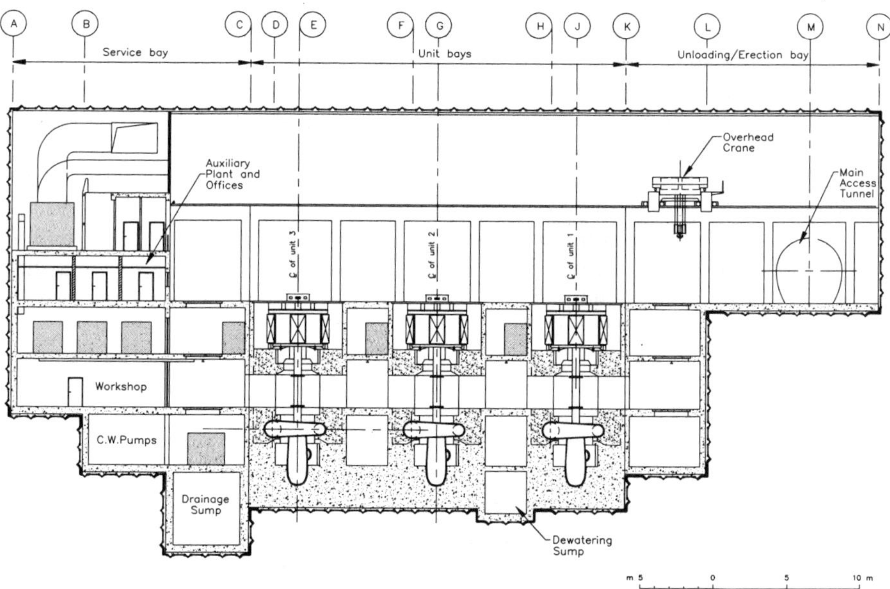

Consideration was given to the provision of rails for the main overhead crane on ledges formed in the rock at the arch springing level, but it was concluded that this could produce difficulties in the event of unfavourably orientated discontinuities at this point of the excavation and the crane is supported directly from reinforced concrete columns.

Control of the flow of water is achieved with inlet valves upstream of each turbine and hydraulically controlled flap gates in the draft tubes. This arrangement allows any one turbine to be dewatered for maintenance whilst allowing the other two to continue generating.

Power from the generators is conveyed from the powerhouse to the Operations Building by means of a lift and cable shaft 6.6 m excavated diameter and around 180 m in depth. Alternative generator transformer locations underground and at the surface were reviewed and it was found to be more economic for the transformers to be housed underground in an enlargement of the access tunnel adjacent to the powerhouse. In addition, the higher voltage of the cables in the cable shaft resulting from this arrangement reduces both energy losses and ventilation requirements.

7. Downstream waterways

Downstream of the powerhouse, the flow from the three draft tubes passes into the 4.1 m diameter tailrace tunnel leading to the outfall structure at 'Muela Reservoir.

A surge shaft with upper and lower chambers is required downstream of the powerhouse due to the length of the tailrace tunnel (1400 m). It has been possible to size the upstream and downstream surge shafts such that common dimensioning of the shafts and chambers could enable use of individual pieces of some of the more complex formwork to be maximised.

The quality of the Clarens sandstone is such that TBM driven tunnels of similar diameter in other parts of the project will remain unlined. However, the degradation of some of the surfaces in the exploratory adit and the need to maintain security of operation of the power station indicated that a minimum thickness concrete lining would be needed throughout, particularly as it is likely that the tailrace tunnel will be excavated by drill and blast methods.

8. Design of underground works

Due to the size of the underground works, it is anticipated that drill and blast excavation methods will be adopted. Although linings will be required in all waterways and in the shafts, it is not anticipated that in situ concrete linings will be required in the access tunnels except immediately adjacent to portals.

Sprayed concrete and/or rock reinforcement will be installed immediately following excavation to suit the ground conditions encountered. Reinforced sprayed concrete will be obligatory in all shafts and in all unlined permanent access tunnels as a safety measure. Rock reinforcement design follows methods based on rock classification/precedent practice with installation to be reviewed as work proceeds in the light of rock conditions encountered.

The risk of discontinuity generated failures is not anticipated to be great as the joints are comparatively widely spaced. However, potential failure possibilities have been identified and appropriate notes prepared for reference by the site supervision staff during construction.

The powerhouse has been orientated in the optimum direction with respect to discontinuity orientations. Rock support has been designed on the basis of assessment of possible failures arising from unfavourably located discontinuities, the results of two-dimensional boundary element analysis and consideration of the Rock Mass Classification value (Q). Nominal support pressures of 50 kN/m^2 in the roof and 25 kN/m^2 in the walls were derived from these, and it is proposed to use 200 kN capacity rock bolts, 4 to 6 m long at appropriate spacing, together with 50 to 100 mm sprayed concrete applied to all permanently exposed faces.

For the tunnels and shafts, an excavation support classification scheme has been prepared indicating a range of possibilities from isolated rock bolts through to full pattern reinforcement with mesh reinforced sprayed concrete. The support at any location will be determined by the Engineer. Rates are included in the Bill of Quantities for excavation classes which are defined by the quantity of support required for any size of tunnel or shaft; the amount to be paid for excavation will depend on the excavation class. Payment will be made separately for the rock reinforcement and sprayed concrete.

Acknowledgements

The authors wish to acknowledge with thanks the permission of both LHDA and the Lahmeyer MacDonald Consortium to publish this paper.

References

1. Wallis S. (1992) **Lesotho Highlands Water Project.** Laserline, UK.
2. Wallis S. (1993) **Lesotho Highlands Water Project, Volume 2.** Laserline, UK.

Design and construction of the cooling water intake tunnel system for Point Aconi generating station, Nova Scotia

Graham Daws
Graham Daws Associates, Ltd., Riddings, Derbyshire, England
Dan MacDonald
Vaughan Engineering, Ltd., Sydney, Nova Scotia, Canada

Abstract
This paper discusses the design of the Cooling Water Intake (CWI) system for Nova Scotia Power Incorporated (NSPI) Point Aconi generating station. The CWI system includes shafts, submarine tunnels and offshore marine works. The project incorporated a unique combination of factors that included undersea tunnelling through coal measure strata; shaft sinking exposed to the North Atlantic environment; offshore blind shaft drilling; construction of a submerged intake velocity control cap subject to arctic ice flows and adherence to very stringent environmental stipulations. This paper addresses all aspects of these factors that had to be considered in the development of the design. Construction methods are also discussed in order to present a complete case study of this project.

1 Introduction

The Nova Scotia Power Corporation is developing a new coal fired generating plant at Point Aconi, Nova Scotia, Canada. It will be designed to accommodate up to three 165 MW generating units. Due to the proximity of the plant to the Atlantic Ocean, the simplest option for cooling of the plant is by the use of sea water.

Environmental stipulations dictated the thermal gradient of the discharge cooling water. To meet this requirement, the water at the intake must be a defined temperature. This temperature for year round operation, required that water be drawn from seabottom at a minimum depth of 10 metres. The profile of the seabed immediately off the shore at Point Aconi is relatively shallow, and as such it was necessary to go offshore a distance of approximately 1 kilometre to obtain 10 metres of water depth. Considering the distance and the environmental stipulations on disturbance of the seabed, NSPI, upon studying the feasibility of various options for obtaining the required cooling water, recommended the construction of a submarine tunnel connected to an offshore intake structure and to the onshore pumphouse forebary.

The principle design consultant was Vaughan Engineering Ltd of Sydney, Nova Scotia. Tunnel construction methodology was subcontracted to Associated Mining Consultants Ltd of Alberta, Canada who engaged Graham Daws Ltd as tunnel designers and geotechnical specialists.

The terms of reference for the design of the CWI system also included a detailed ice study. This was required in order to assess the ice flow conditions on the intake structure cap. The design package included a detailed cost estimate of the proposed system and culminated in the provision of a full tender package to NSPI. There was a requirement to design a system which would require minimal maintenance over a 50 year life.

2 Regulatory Considerations

The design and subsequent construction of the CWI structure, tunnel and shaft has the rare feature that it involves three distinct habitats namely; surface construction, tunnelling and ocean bottom construction. These areas involve considerable number of regulatory bodies, both Provincial and Federal, and require a range of statutory requirements to be adhered to with permits requiring formal application.

The Nova Scotia Department of Labour is the Agency whose mandate is to protect the health and safety of the workforce whether on the surface, underground or on/in water. The Provincial and Federal Departments of the Environment are responsible for maintaining and minimising impact due to industrial activity on the marine and land based environments. The design and construction methods of the CWI structure and tunnel were required to adhere to all Federal and Provincial statutes and guidelines to ensure the safety of the workforce and the environment.

The following is a summary of these regulatory responsibilities.

Environmental

The Pt. Aconi Power Plant is the result of many years of research and development by Nova Scotia Power Corporation to develop a means to burn high sulphur coal reducing the production of SO_2 and NOX emissions which are considered the primary contributors to acid rain. As a continuance of this philosophy the design and construction of the Cooling Water Intake Structure and tunnel had to comply with all Provincial and Federal environmental guidelines and regulations. By incorporating these into the design and construction methodologies, the necessary permits and approvals proceeded in a timely manner. The following agencies were involved in the environmental control of this project.

Nova Scotia Department of Environment responsible as Regulatory Agency for all land based activity at the sit.

Environment Canada responsible for standards on Environmental Codes of Practice for Steam Electric Power Generation.

Federal Department of Fisheries and Oceans. Responsible as Regulatory Agency for any activities in the marine environment.

Transport Canada. Responsible for the Navigable Water Projection Act which is the permitting body for the offshore work.

Construction Safety

The Nova Scotia Department of Labour administers the statutes of the Occupational Health and Safety Act, Statues of Nova Scotia, Acts of 1985, Chapter 3. Various divisions of the Nova Scotia Department of Labour were called upon during the course of this project when a particular question regarding a specific area of expertise was required. These included

Occupational Safety
Occupational Health
Mine Safety
Accident Prevention

Although this site is clearly not a 'Mine' as defined in the 'Metalliferous Mines and Quarries Regulation Act' of Nova Scotia and the 'Coal Mines Regulations Act' several activities on this site such as shaft sinking and tunnel drivage employed machinery and construction methods and dealt with safety concerns commonly encountered in a mining environment. Consequently, certain rules were invoked from the Coal Mines Act and Metalliferous Mines and Quarries Act to supplement corresponding sections of the 'Construction Safety and Industrial Safety' Regulations. These additional requirements were dictated by the Department of Labour (Mines Inspection) to the Contractor prior to commencement of construction. Due to the fact that the shafts/tunnel were to be constructed in coal measures strata and it was anticipated that numerous thin coal seams would be encountered, all electrical equipment used on the site was required to be classified as Intrinsically Safe (i.e. flameproof).

3 Shaft Sinking and Tunnelling

The overall concept of the scheme was to construct a shaft and tunnel network through which could be take ice-free water to cool the Point Aconi power plant. The general arrangement of the tunnels and shafts is shown in Fig. 1. Previous site investigation work had been carried out in this area but the extent and depth had been limited as a previous scheme utilised the concept of a pipeline liad on the ocean floor or within a shallow trench. This concept proved not to be feasible in the aggressive conditions of the Atlantic Ocean off Nova Scotia and consequently the tunnelling option was chosen. A new set of sit investigation data was therefore required.

3.1 Site Investigation

The cooling water intake (CWI) system at Point Aconi was to be constructed within the Sydney Mines formation of the Morien Group. This is a coal measures sequence containing major sandstone bodies composed of individual or multiple channel sandstones approximately 1m to 5m thick. Minor but important thin seams of coal (0.1 to 0.3m thickness) would also be encountered.

It was considered that Section 106 (I) of the Coal Mines Regulation Act (1988) which stipulates that the minimum cover of solid measures from the seabed should be 100 ft (30.8m) should be applied to this project. This part of the Act creates a barrier between undersea workings and the seabed in order to help prevent breakthrough. In this respect, the Act is similar to the 30m of cover required by the Labour Canada Coal Mines (CBDC) Regulations 1990 Section 55k. Consequently, this requirement was taken as the minimum depth of cover at the Intake Shaft location. The tunnel was to have a slope of 0.5% towards the Forebary shaft in order to assist water drainage during construction. This geometry therefore dictated the required depth of the new offshore boreholes. Four offshore holes were planned to supplement the existing data and a borehole was placed on the centre line of the Forebay Shaft.

The results from the site investigation enabled the final position of the tunnel to be chosen. From the original data, a channel sandstone had been predicted and the Forebay Shaft/CWI tunnel elbow was positioned within this sandstone body to aid stability. However, the results from the supplementary site investigation did not confirm the existence of this channel sandstone. In fact the location chosen was found to consist of a very weak mudstone. This resulted in a complete re-alignment of the system and required the tunnel to be placed at a greater depth than originally planned. The final tunnel alignment was carefully considered bearing in mind the need to keep the gradient as close to 0.5% as possible as it was considered that the Contractor would need to use locomotive haulage during construction. It was ultimately possible to place the tunnel at a location which satisfied this demand and also placed the shaft/tunnel elbows in the best possible strata.

The principle rock types encluntered in this area, other than the thin coal seams were as follows

Rock Type	**Rock Mass Rating (RMR)**	**NGI-Q-Value**
Sandstone	42	0.8
Siltstone	39	0.6
Grey Mudstone	37	0.5
Weak Red Mudstone	17	0.05

Both RMR and Q embody the actual rock properties and consequently these are not quoted separately.

Observations of groundwater pressures and levels were made in the two deep on-shore boreholes. Vibrating wire piezometers were installed at 15m and 61.5m depth and initially indicated excess porewater pressure above hydrostatic. However, after three days, the head had fallen to below hydrostatic, indicating the possible presence of connate (perched) water.

A further test hole was drilled on line of the connector tunnel and pumping and drawn down tests carried out. Water pressures above hydrostatic were not found but concern was noted about the possibility of full hydrostatic head when designing the shaft and tunnel linings.

3.2 Design of the Shaft and Tunnel Linings

3.2.1 Shaft Linings

The diameter of the Forebay Shaft was fixed at 8m as it was assumed that the Contractor would require this amount of space to install tunnelling equipment. The internal diameter of the Intake Shaft was set at 2.7m after carrying out flow calculations. In view of the resuslts from the water level monitoring at the Forebay Shaft centre borehole, it was decided to design the shaft lining for full hydrostatic head. There is some debate (Bell, 1982) as to whether, when a hydrostatic load exists, the rock load should be ignored or added to the hydrostatic load. After some consideration, it was decided to add the rock load to the hydrostatic load. The theory used to calculate the stresses on the lining is the one proposed by Coates (1981). This assumes that the ground around the shaft is in a state of plastic equilibrium with a stress relationship governed by Mohr's strength theory. The pressure on the lining, P is then given as follows

$$P = \frac{2\gamma a}{\text{Tan } \varnothing \ (f-2)} \left(\frac{f-2}{f}\right)^{f/2}$$

where
a = Radius of the shaft (m)
γ = Rock Density (MN/m^3)
f = Tan^2 (45 + $\varnothing$/2)
$\varnothing$ = Angle of Friction of the Rock Mass (given from putput to Geomechanics system)

The required thickness of the concrete lining was then calculated by the use of Lame's quation (Bell 1982). This was carried out as follows

$$t = a \left[\sqrt{\frac{fc/FOS}{((fc/FOS) - 2P)}} - 1 \right]$$

where
t = lining thickness (mm)
a = Radius of Shaft (mm)
fc = Concrete Strength (MPa)
FOS = Factor of Safety
P = Pressure on Lining (MPa)

This gave a minimum lining thickness of 315mm for the Forebay Shaft and 75mm for the Intake Shaft.

A check was carried out on these design thicknesses for lining stiffness and both were found to be acceptable. It was considered impractical to construct a lining 75mm thick for the Intake Shaft and consequently this was specified as a minimum. During construction, 150mm wall thickness pre-cast concrete liners were used as described later.

3.2.2 Tunnel Linings

For a shallow tunnel such as this, the Rock Load concept was used. Two systems were used to estimate rock load, namely Terzaghi's system (1946) and the output from the Geomechanics Rock Mass Classification system (Bieniawski 1984). The loads generated by the two systems were in close agreement and it was also decided to allow for full hydrostatic load. The tunnel was to be concrete lined, either by segments as the drive progressed or cast in-situ. The lining thickness was calculated by the use of Lame's theory as proposed by Hoek and Brown (1980). This is given as follows

$$P = 0.5 \quad \sigma_C \left[1 - \frac{(r - t)^2}{r^2} \right]$$

where P = Load on Lining (MPa)
σ_c = U.C.S. of Concrete (MPa)
r = Excavation radius to rock (mm)
t = Lining thickness (mm)

For the worst case ground conditions anticipated along the tunnel drive, this gave a lining thickness of 250mm. During construction the lining was cast in-situ to give a lining thickness of 390mm. This extra lining thickness was placed at the request of the Contractor to ease placement behind the formwork.

The tunnel to connect the Forebay Shaft to the forebay excavation was situated in mudstones/siltstones with an RMR of 40, Q of 0.64. It was assumed that the Contractor would drive this tunnel as a conventional arched shaped mining tunnel and that the support would be rockbolts and mesh before the final circular lining was cast in-situ. The rockbolt support system was therefore designed using current mining practice. This approach used the Geomechanics Rock Mass Classification system to predict likely rock load and the rock mass confinement approach is used to design the rock bolting pattern (Daws, 1991). Using this approach, the principle design equation is given as follows

$$PS^2 + qUS - qUL = 0$$

where P = Rock Load (tonnes/m)
S = Bolt Spacing (m)
q = Tan^2 (45 + Ø/2)
Ø = Angle of Friction of Rock Mass
U = Yield load of rockbolt (tonnes)
L = Bolt Length (m)

For coal measures strata, the rock load, P, is given by the Geomechanics system as follows

$$P = ht \times W \times \gamma \quad \text{(tonnes/m)}$$

where $ht = \left(\frac{100 - RMR}{100}\right) . W$

W = tunnel width (m)

RMR = Rock Mass Rating

As a guide, bolt length, L is given by the following expression (Barton et al 1974)

$$L = 2 + 0.15B/ESR$$

where B = Excavation Width (m)

ESR = Excavation Support Ratio - taken as 2 to 3 in mining application

3.3 Construction Aspects

The contract to construct the CWI system was let to a joint venture company comprising of The Foundation Company of Canada and Harry Vie Inc. of Montreal.

3.3.1 Forebary Shaft

The excavated diameter of the Forebay Shaft was 8.6m and it was sunk to a depth of 63m using the full face blasting method. 1.8m rounds were used and a John Deere 555 side drop loader was used to muck out. A 150 tonne Link Belt 518 crane was used to hoist muck and materials. As sinking continued, blast vibrations were closely monitored in order to comply with environmental constraints. No blast exceeded a ppv of 25mm/sec within the contract limit and air overpressure did not exceed 128 dBA. These limits were given in the environmental code issued by NSPI.

Temporary support to the shaft was afforded by the use of 25mm dia. x 1.8m long resin anchored rockbolts set on a 1.5m grid pattern. Welded wire mesh was also used. In areas identified as potentially unstable, i.e. where the coal seams intersected the shaft, then sprayed concrete was also used. This was generally 100mm thick and applied by the dry process. It was necessary to heat the water due to the extremely low temperatures encountered during the winter period at this exposed location.

The lining of the Forebay Shaft was eventually placed from the bottom up in order to ensure that a monolithic lining was constructed. Steel forms were used and the concrete was placed into two hoppers connected to 150mm diameter pipes located on both sides of the shaft collar. At the bottom of the pipes, a re-mixing chamber was connected to hoses allowing the placement of concrete around the perimeter of the forms. Each form was 7.1m in height and was filled at a rate of 1 metre per hour, requiring 130 to 140m³ of concrete to complete each pour. The finished diameter of the shaft was 8m.

3.3.2 Offshore Shaft

The offshore work was subcontracted out to Pitts International of Toronto and Zeni Drilling Ltd of Morgantown, West Virginia. Mobilisation started on June 15th 1991 at a wharf in Sydney which is 15 kilometres from the offshore site. Pitts equipment consisted of the 'Pitts Olympic' barge on which was mounted a 4600 Manitowoc Crane/ clamshell, the 1000 cub yard Pitts No. 12 scow and several tugs. Work could not begin until after July 15th when the local lobster fishing season had finished. After arriving at the drilling site the precise location was achieved by survey triangulation and maintained by using a laser mounted on shore.

Excavation commenced with an 8 cubic metre clam on the 200 ton crane to excavate the intake cap area and glory hole where the Intake Shaft was to be drilled. The rock proved to be too competent for the clam so a truck mounted drill was brought in and drilled a series of 700mm diameter holes in a circular pattern. The clam was then able to excavate the rock to the required depth. The area was cleaned of any debris by airlifting material to the surface. More than 1400 cubic metres of rock was removed from the ocean bottom during the excavation process.

A prefabricated concrete collar 5 metres in diameter 200mm thick and 4m in height was lowered into the glory hole and concrete was tremied around to hold it in place. The work to prepare the area for Zeni Drilling was completed on August 30th 1991.

On September 3rd Zeni moved onto the glory hole. The Zeni drill was mounted on the 12m x 24m jackup barge (See Fig. 2). The Zeni drill rig had installed horsepower of 650 HP and an installed torque on the rotary table of 75000 lb/ft. A steel drill collar 4.1m in diameter 25mm thick and 25m in height was installed into the concrete collar and located by the use of tremied concrete.

A 650mm pilot hole was drilled to a depth of 39m. The main bore was then drilled a depth of 37m at a diameter of 3.9m. A circular bit with disc cutters and 80 tons of weights was used for boring (Fig. 3). Drill cuttings were airlifted up the drill string to a shaker box and loaded on a barge for disposal on land with the mud recirculated in the drill hole. The method for cuttings disposal prevented environmental damage to ocean floor. The drilling rate for the Zeni drill was 300mm drilled per hour.

When the drilling had been completed 60 tonnes of coarse gravel was deposited in the bottom of the shaft. With the gravel acting as a base the Intake Shaft was ready for the final lining. Four concrete lining sections, each 8m high with an inside diameter of 2.75m and a wall thickness of 150mm had been previously constructed at a nearby wharf. The first precast segment was lowered into the shaft and back grouted using tremied concrete. The remaining segments were lowered into position and also back grouted.

An 8 hour regarder was added to the tremie concrete when it left the batch plant. The mix was unloaded from the batch trucks into 3.5m^3 concrete buckets and loaded on a barge.

The barge was towed to the Intake Shaft location where the concrete was unloaded into concrete trucks located on another barge. The concrete trucks remixed the concrete which had been standing for 3 hours, and discharged into a Schwing 5000 concrete pump. The mix was pumped down a 150mm hose and placed on the ocean floor by divers.

A total of 550m^3 of concrete was used in completing the offshore work. Working conditions were made much more difficult by the stormy weather encountered in the North Atlantic. The date was October 20th and the decision was made to halt the marine work until the following Summer. A temporary steel cap was installed on the Intake Shaft so that underground activities could continue.

Offshore work for the 1992 season commenced on July 23rd. The intake cap area was cleaned of any debris by airlifting. Four hundred and fifty tonnes of crushed stone mattress were deposited with the clam shell and leveled with a screed beam. The 1500 tonnes intake cap structure was towed to the glory hole and anchored into position. Eight 5 tonne tirfors, four on top of the structure and four on the bottom of the structure, were used to position the intake structure pricisely in place. The sinking of the intake structure was started on July 31st at 8.00 am and completed by 2.00 pm (See Fig. 4). The mattress was grouted and six 40mm diameter x 13m long rock anchors were placed to hold the intake cap in position. The intake cap collar was concreted and the sheet piling and superstructure removed.

3.3.3 Tunnel Drive

The contractor had a TBM available and therefore elected to use this machine even though the diameter was slightly larger than specified.

The TBM used was an electro-mechanical Jarva Mark 12 with a bore of 4.5m. The machine had been overhauled in Solon, Ohio by Boretec Ltd and then transported by truck to the job site. The machine was flameproof, approved for use in a gassy environment. Four 125 horsepower motors were used to power the TBM which ad a stroke of 660mm. The TBM transformer stepped down the voltage from 4160 to 480 volts.

The main body of the TBM was lowered down the shaft and installed in a chamber that had been excavated by drilling and blasting. The TBM was advaned 35m until there was sufficient space to install the balance of the TBM trailing package. Two trains, each made up of a loco, six muck cars and a material car were installed on the 600mm wide track.

A three week period from September 27th to November 17th 1991 was required for familiarising the work force with the TBM, for setting up the balance of the trailing package and for initial equipment problems. The TBM had not been in operation since 1986 and initially there were several problems with motor failures and replacement of electric panels. Drivage of the 1000 metre tunnel commenced on November 17th 1991 and was completed on February 20th 1992.

Tunnel support consisted of 100mm x 100mm circular steel sets installed on 1.2m centers with wooden lagging installed in the roof and floor. The four piece steel sets were erected 1.5m behind the tunnel face. The steel sets were installed manually, expanded with a 2000 psi hydraulic ram and held in place by installation of a 'Dutchman'.

A 20 tonne capacity skip bucket hoisted on guides by the 150 ton crane was used for shaft mucking. Three muck cars filled up one skip bucket and two skip buckets were required for one muck train. A single track with a switch at the shaft bottom allowed the second train to travel to the tunnel face while the first train was being unloaded. After two 600mm pushes had been completed a steel set was erected under 2m long 'fingers' that extended back from the top of the TBM head. After the TBM had advanced 2m the steel set was no longer under the fingers and could be expanded against the rock.

A forcing/exhausting system was used to provide ventilation in the tunnel. Two 30 HP fans on surface delivered 500 cubic metres of fresh air to the face via a 1050mm duct. A 25 HP fan mounted on the TBM exhausted 140 cubic m of air through a 600mm duct to surface. A 600mm and a 1050mm cassette maintained ducting for this system.

Three constant methane monitors were installed, one at the bottom of the shaft, one at the TBM transformer and one at the TBM head. The monitor on the TBM head was interlocked to stop the TBM if 0.5% methane was detected and hand held detectors were carried by tunnel personnel and checked on a shift basis.

Horizontal probe holes 75mm in diameter were drilled 45m ahead of the face to check for gasses and water. As the TBM approached to within 8m of the end of previously drilled holes, additional holes were drilled. Later in the drivage a third hole was drilled 3m vertically to give an indication of weak roof strata. A single laser mounted on the wall was used to align the bore of the TBM and normal practice was to move the laser for 250m of advance.

As the TBM approached to within 10m of the Intake Shaft tunnelling was stopped and a series of probe holes were drilled, intersecting the Intake Shaft. A water flow of 150 litres per minute was observed. Drivage was continued to within 5m of the Intake Shaft and another series of holes were drilled at various angles, the last hole resulting in a flow at a rate of 3000 litres per minute. Within 3 hours of the last hole being drilled the Intake Shaft was emptied and a constant flow of 100 litres per minute of water was observed.

TBM excavation was maintained into the Intake Shaft where coarse gravel that had previously been deposited by Zeni Drilling was encountered. Excavation was continued to 0.5m on the north side of the Intake Shaft. It was discovered that the initial low water flow through the probe holes had been caused by a false bottom in the Intake Shaft. During the concreting of the Intake shaft several cubic m of concrete had inadvertently been deposited on top of the coarse gravel. The concrete-gravel mix had solidified and formed a shell at the bottom of the shaft. The initial probe holes had broken through below the false bottom.

The last hole had pierced the shaft above the false bottom, resulting in the draining of the Intake Shaft.

Stripping of the TBM was begun immediately. All of the trailing gear was removed and the cutter brackets and head unbolted. The transformer and the power pack were loaded on trams in front of the TBM to maintain power and the TBM was walked out of the tunnel to the Forebay Shaft. Most of the hydraulic cylinders were blocked off to facilitate faster stroking of the remaining cylinders. After arriving at the shaft the TBM was broken down into several components and hoisted to surface.

The three main phases of tunnel concreting consisted of pouring concrete, cleaning up the invert and resetting the forms. Five piece Everest forms 3.72m in diameter were used to concrete the tunnel which had been bored to 4.5m. A total of 64 linear metres of form with a skin thickness of 6mm was installed in the tunnel. The concrete was poured on day shift, with an average 160m³ of concrete being required to fill 36m of form.

Concrete was delivered to the site in concrete trucks with a 4.3m³ loads. The concrete was dropped down a 200mm pipe into a set of baffles that remixed and poured the concrete into Moran agitator cars of 4.3m³ capacity. Two trains consisting of two agitator cars transported the concrete to the pump 1000m away. The Moran cars fed one into the other and were air powered. As the first train travelled to the end of the tunnel the agitator cars of the second train were loaded.

The concrete was placed with a 150mm slickline going over the top of the form. A schwing 5000 track mounted pump was used to pump the concrete. With this system 3 trains or 26m³ of concrete could be placed in one hour.

After the concrete had been poured for the day, the concrete lines were cleaned and concrete pump pulled to the shaft by the second shift. A Kubota loader with a 0.5m³ bucket was used to clean the invert and load muck into a 4m³ muck box. 30m of invert were cleaned per shift with the railway track and piplines also being removed.

The third shift moved the forms for the next days pour. The form carrier was used to move 9.1m sections of form at one time. The carrier then retreated through the balance of the forms where the 9.1m section of form was erected and bolted in place. This procedure was repeated until 27 or 36m of form had been installed. Steel pins threaded through the forms were used to maintain the position of the forms. The concrete pump was pulled into position and the slickline hooked up for the next 3 days pour. Concrete work on the tunnel was started on April 20th and completed on June 3rd.

With the concrete completed, backfill grouting was required to fill any voids. Grout holes were drilled on 3m centres in the roof and grouted at a pressure of 1.75 bar with six 20 kg bags of grout required per linear metre of tunnel. Some minor patching and finish work was required.

4 Offshore Marine Works

The intake structure for the Cooling Water Intake system as shown in Fig. 5 was an offshore submerged type designed to control the flow of water into the system. To optimise the geometry of the intake structure five alternatives were developed and analysed. The development of these alternatives was progressive in which each alternative was an enhancement of the previous. Alternatives 4 and 5 as shown in Fig. 6 were fabricated and input into the physical model to determine the optimum configuration for the intake cap. The key factor in the development of these alternatives was the performance of the structure in relation to ice, both in resistance to forces generated by the ice, and prevention of ice blockage. Alternative No. 5 was selected from the results of the model study, as it proved to give the best durability.

One of the prime considerations in the overall design of the cooling water intake system was the location of the intake cap as it dictated the alignment and total length of the tunnel which greatly impacts the capital cost of the system.

As stated the intake cap had to draw water from greater than 10m depth. Locating the structure in a seabed depression would be advantageous for defence against ice ridges, but increases the potential for siltation build-up around the intake cap and eventually into the conduit system. Conversely placing on a subsea ridge reduces siltation problems, but increases exposure to ice and ice forces. Also, as the onshore end of the tunnel was fixed, the location of the intake cap dictated the tunnel alignment and as the geology was varied it was important to align the tunnel to follow the most competent rock profile in the area.

Establishing the intake cap position was an interactive process of analysing these requirements. A location was chosen along the 11m contour (Fig. 7) which from review of the siesmic sediment charts had minimum silt deposition and would remain swept clean by wave action. Being on a slope provided some protection from ice ridges and the resulting tunnel alignment was perpendicular to the geological rock dip a benefit to construction of the tunnel.

The design loads for this intake cap were high and are as follows

Ice Load

- Shear at top - 1000 tonnes
- Concentrated - 6.5 tonnes/m^2
- Uplift - 7.8 tonnes/m^2

Wave Loads

Vertical Down	-	3.12 tonnes/m^2
Vertical Up	-	2.82 tonnes/m^2
Horizontal	-	3.12 tonnes/m^2

Due to the harsh environment to which this structure will be exposed, it was essential that the materials selected for construction ensure the integrity of the structure for the designated 50 year life span.

Reinforced concrete was the material selected for the intake cap construction as it presented the most practical construction material for this structure. The permeability of the concrete is the most important characteristic to ensure durability when exposed to sea water. Pozzolain additives which reduces the large pores in the hydrated cement paste, Type 20 cement and tight control of water cement ratio during construction were some of the measures taken to reduce the permeability of the concrete.

Even with tight quality control on the design and construction of the concrete seawater will penetrate the concrete and chemicals will attack the reinforcing steel to protect against this all reinforcing steel in the intake cap was epoxy coated.

The construction of the intake cap involved the following major tasks

- Excavation of a glory hole to sound rock, placement of the drill casing and tremie concreting of the casing in place.
- Fabrication of the 1100 tonne intake cap in a ship building drydock.
- Placement of a removable steel sheet pile cofferdam to increase the freeboard of the intake cap for towing.
- Floating and towing of the intake cap 200 miles in open sea conditions.
- Placement of a 600mm crushed rock mattress on the glory hole for levelling to receive intake cap.
- Sinking of intake cap into position over intake shaft. This was accomplished by placing anchors in the four directions and using eight 5 tonne tirfors, on top of the structure and 4 on the bottom of the structure, lowering the cap precisely in place.
- After position of the intake cap the concrete mattress was grouted to increase shear resistance to ice forces.

Six rock anchors were installed to add additional resistance to ice shearing forces.

Next the false bottom was removed, the tunnel flooded, the temporary lid removed from the riser shaft and the permanent lid placed on the intake cap.

5 Bio Fouling System

To negate marine growth in the tunnel and shafts (musels) it was necessary to design and install a biofouling system. Several systems were studied including chlorine injection and coating systems.
The chlorine injection system was chosen because it was the proven system, the other systems are relatively new and have not been proven over time.

The chlorine injection system involved the injection of 3 ppm into the tunnel and shafts system during the mussel growing season.

To accomplish this chlorine in a solution of 3500 ppm is pumped via two 100mm diameter pipes as shown in Fig. 8 to a diffuser ring in the Intake Shaft just below the intake cap. A complete backup duplicate system was installed.

The piping and the diffuser rings were constructed of reinforced fibreglass piping as this was the only material that was guaranteed to withstand the high concentration of chlorine over the design life of the system.

6 Environmental Site Remediation

The development of NSPI's Pt. Aconi Generating Station was an environmentally sensitive project. To ensure the construction of the CWI system with respect of the impacts, it would have to the surrounding environment. An environmental management plan was developed and input into the design process resulting in many potentially negative situations being avoided altogether, while mitigative measures and contingency planning was implemented to effectively control other site operations.

The planning and design measures adopted were applied to the following key issues

6.1 Drilling and Blasting

All blasts were monitored for noise and vibration levels measured at the boundaries of the project site. The compliance levels for these were

- Air concussion 128 DBA
- Ground vibration 25mm/sec.

6.2 Waste Rock Laydown and Disposal

Construction of the shafts and tunnels intercepted a widely varying geology and the waste rock consisted of sandstones, mudstones, shales, coal and bentonitic clays. These materials generate a low pH from associated runoff waters. To mitigate this impact a prepared base was constructed with a 10^{-7} m/s impermeability rating and a drainage catchment system to direct runoff to a treatment pond.

6.3 Coastline Erosion Control

The close proximity of the Forebay Shaft to the shoreline left it vulnerable to shoreline erosion. To protect the shoreline and the associated beach, geotextile fabric and armour rock protection was installed.

6.4 Drainage Management

Site drainage was generated from three sources, general site runoff, waste rock pile runoff and tunnel dewatering. These sources were channelled through a catchment system to two ponds, the first being a sedimentation pond to allow settlement of suspended particles and the second a treatment pond in which lime was introduced to amend the pH.

The compliance requirement for the treated waters to the receiving environment were

- suspended solids : 25
- pH : 6.5 to 9.5

6.5 On-Site Sewage

Sewage from the construction activities was handled by the installation of a holding tank which was emptied periodically by a licensed contractor.

The geotechnical conditions at the site did not allow for the development of an onsite disposal field and the main plant sewage system was not constructed in time for this project.

6.6 General Construction Activities

The environmental management plan incorporated measures for the following

- storage of explosives
- storage of diesel fuel
- storage of potentially dangerous materials
- dust control

7 Conclusions

The CWI System was

- complted on budget
- completed 7 days ahead of the original schedule
- completed within the stringent environmental stipulations with minimal impact on environment.
- completed with no major claims by the contractor
- complted with an excellent safety record.

8 Acknowledgements

The co-authors wish to express appreciation to Nova Scotia Power Incorporated for their permission to print this paper and also to Kevin Beaton, P. Eng. for information he supplied on the construction of the system.

Any views expressed are those of the authors and not necessarily those of NSPI or any other body.

9 References

[1] Barton, N., Lieu, R and Lunde, J "Analysis of Rock Mass Quality and Support Practice in Tunnelling and a Guide for Estimating Support Requirements" Norwegian Geotechnical Institute Report No. 54206 June 1974.

[2] Bell, M.J. "The Design of Shaft Linings in Coal Measure Rock" Symposium on Strata Mechanics - Newcastle upon tyne 1982 pp 180 - 186.

[3] Bieniawski, Z.T. "Rock Mechanics Design in Mining and Tunnelling" A.A. Balkema, 1984.
[4] Coates, D.F. "Rock Mechanics Principles" Energy Mines and Resources, Canada. Monograph 874 1981.
[5] Daws, G "The Use of the Geomechanics Rock Mass Classification System in the Design of Coal Mine Roofbolting System" Univ. of Nottingham, Dept. of Mining Engineering Magazine. Vol. No. XL111 1991
[6] Hoek, E and Brown, E.T. "Underground Excavations in Rock" IMM 1991
[7] Terzaghi, K "Rock Defects and Load on Tunnel Support" In Tunnelling with Steel Supports, Proctor and White, Commercial Shearing and Stamping Co. Youngstown 1946.

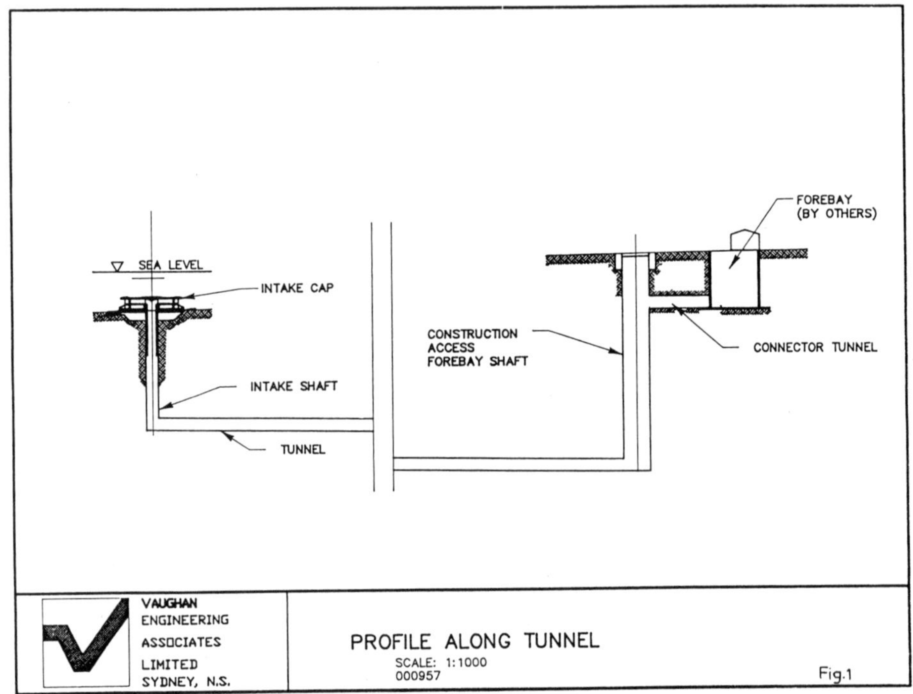

Fig.1

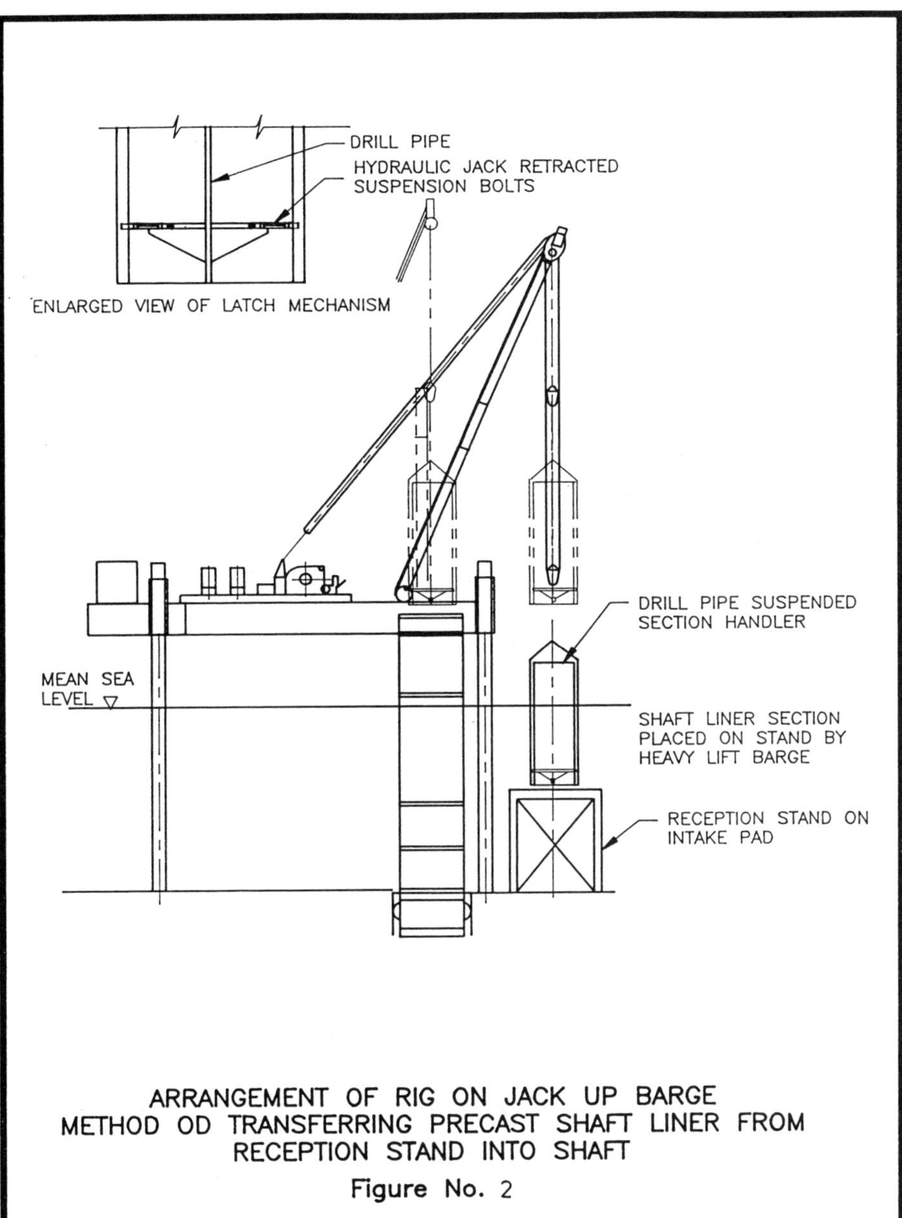

ARRANGEMENT OF RIG ON JACK UP BARGE
METHOD OD TRANSFERRING PRECAST SHAFT LINER FROM
RECEPTION STAND INTO SHAFT

Figure No. 2

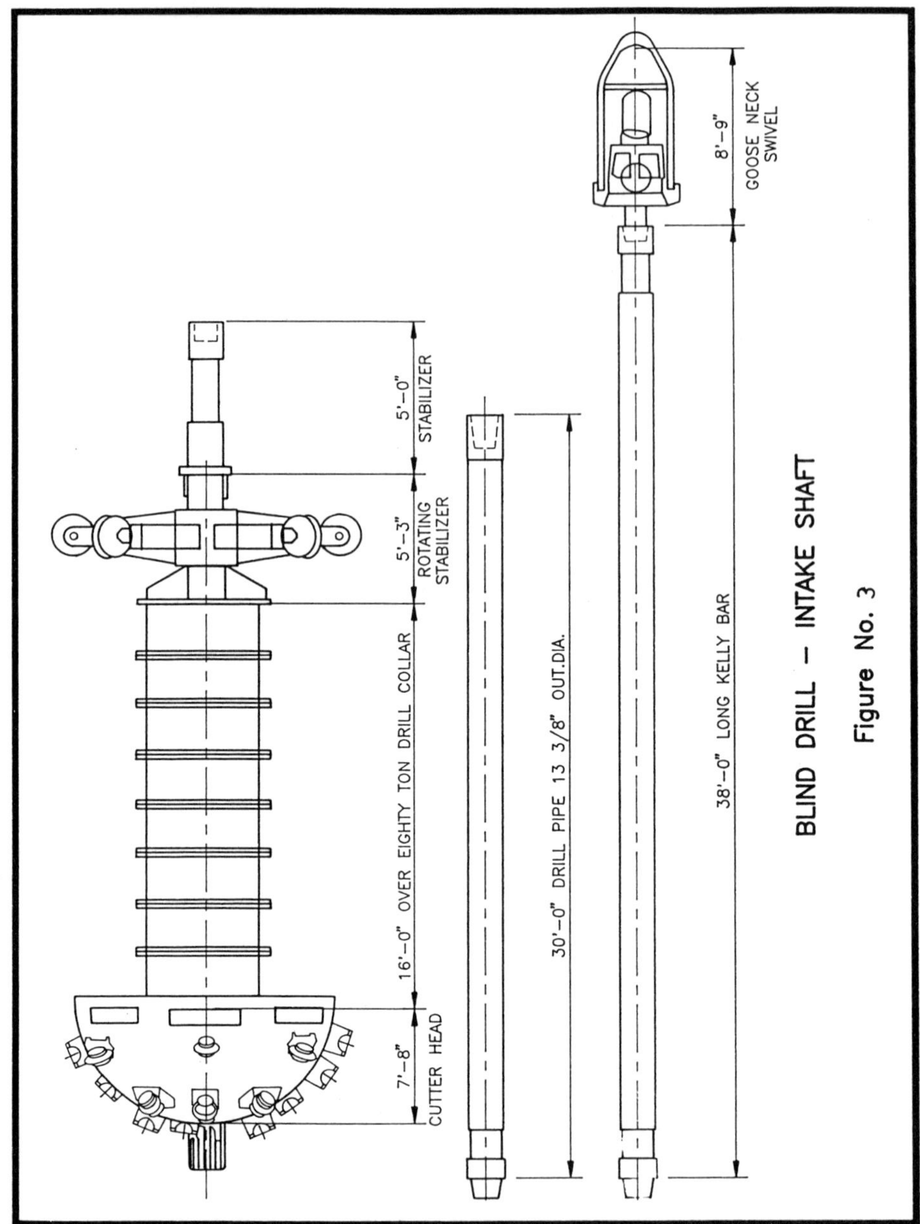

BLIND DRILL – INTAKE SHAFT

Figure No. 3

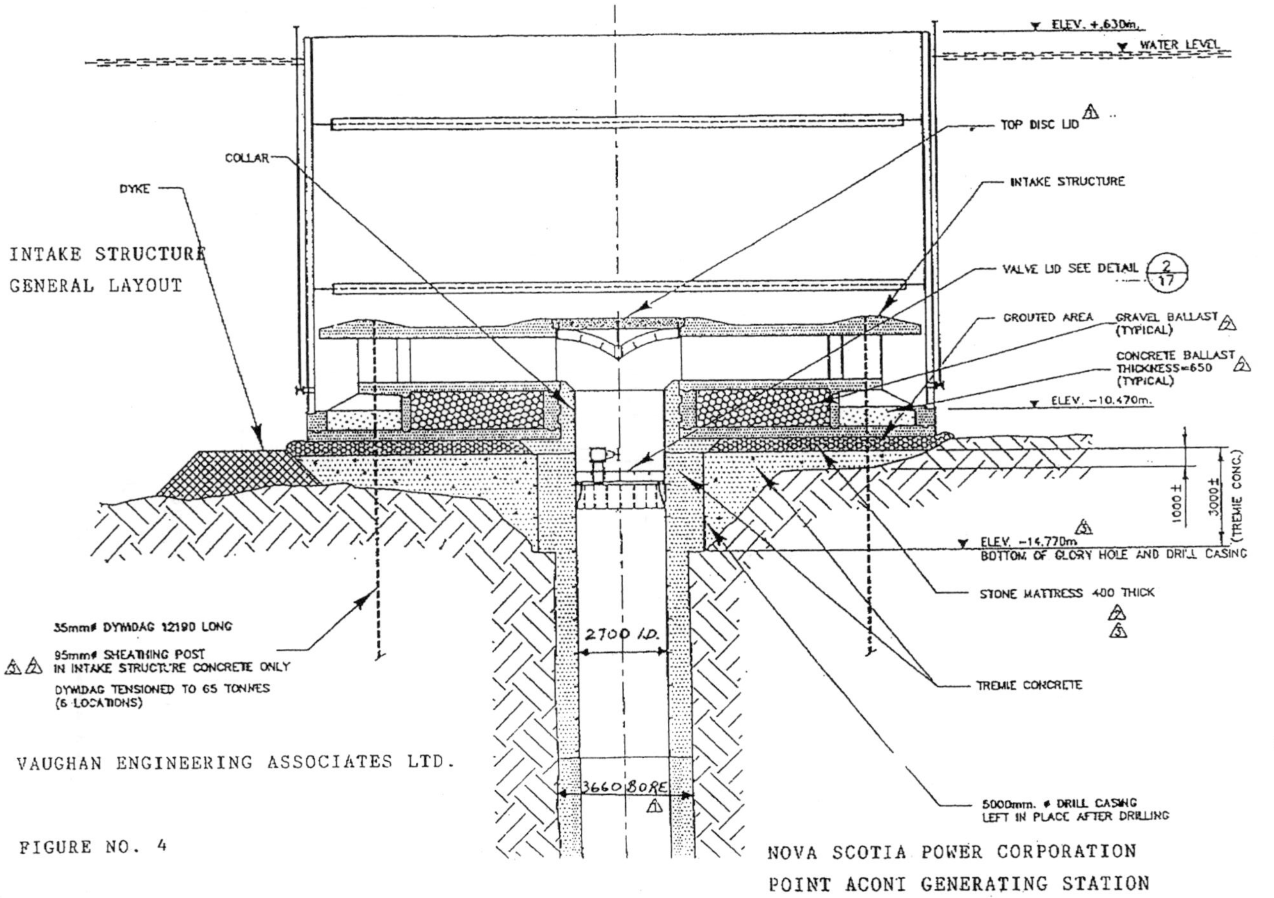

INTAKE STRUCTURE
GENERAL LAYOUT

VAUGHAN ENGINEERING ASSOCIATES LTD.

FIGURE NO. 4

NOVA SCOTIA POWER CORPORATION
POINT ACONI GENERATING STATION

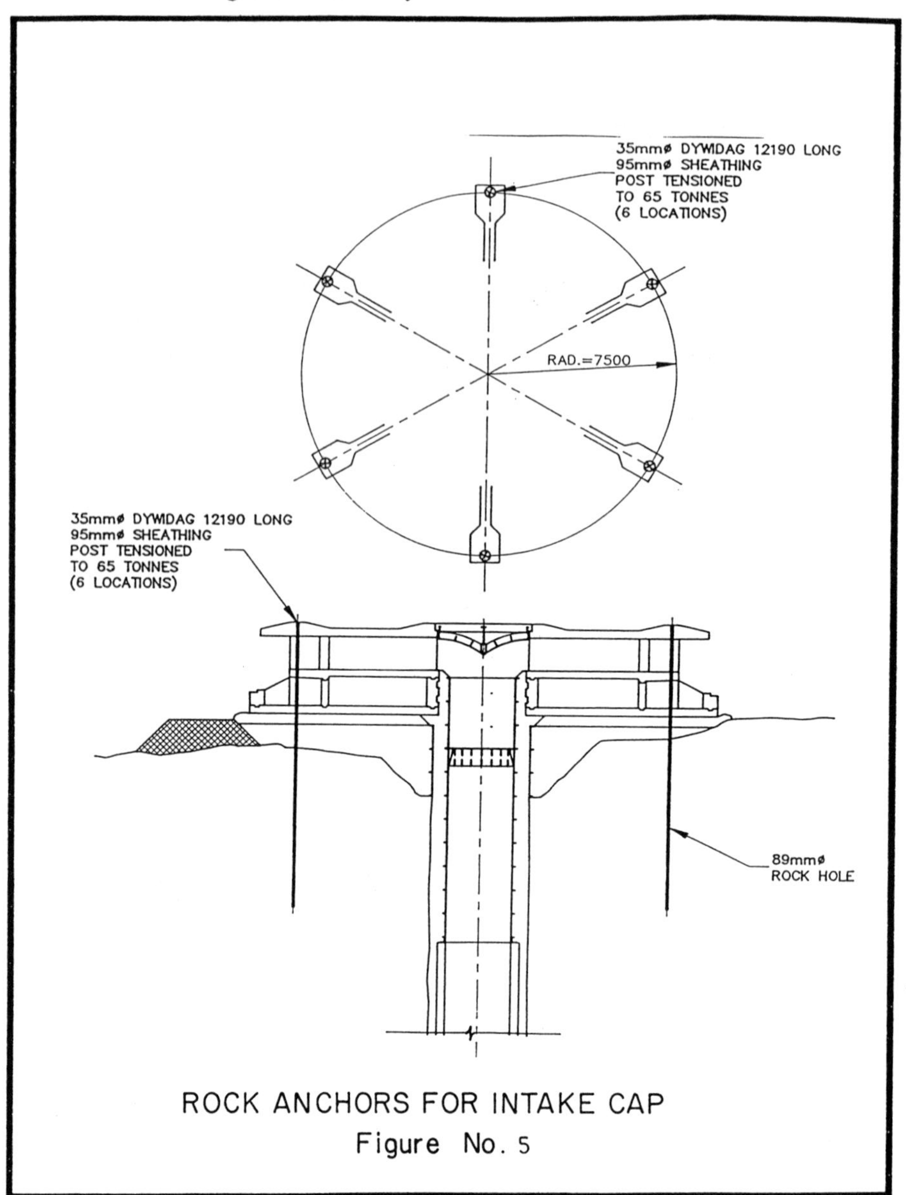

ROCK ANCHORS FOR INTAKE CAP

Figure No. 5

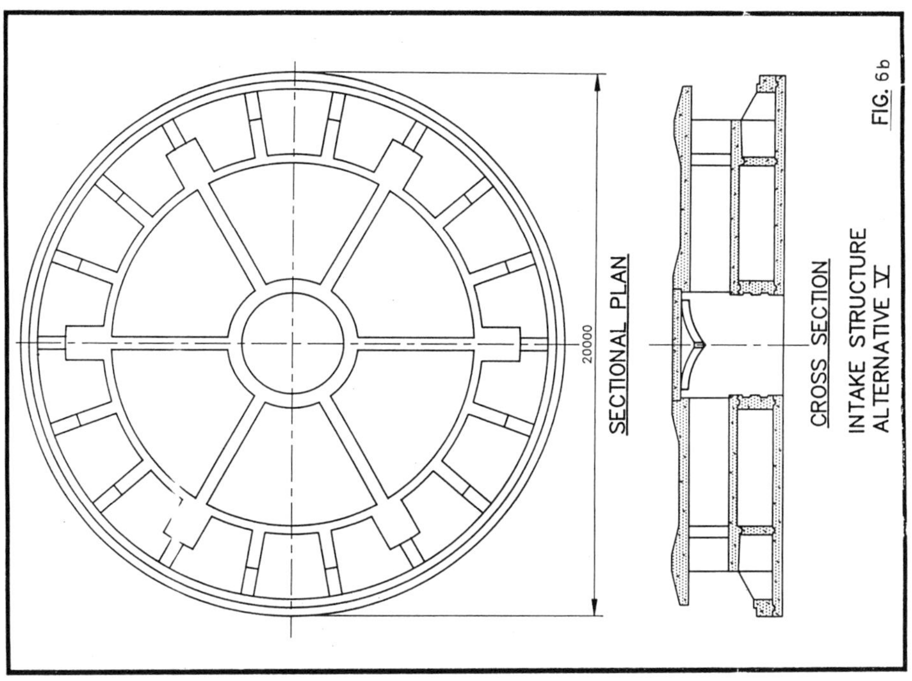
20000
SECTIONAL PLAN
CROSS SECTION
INTAKE STRUCTURE
ALTERNATIVE V
FIG. 6b

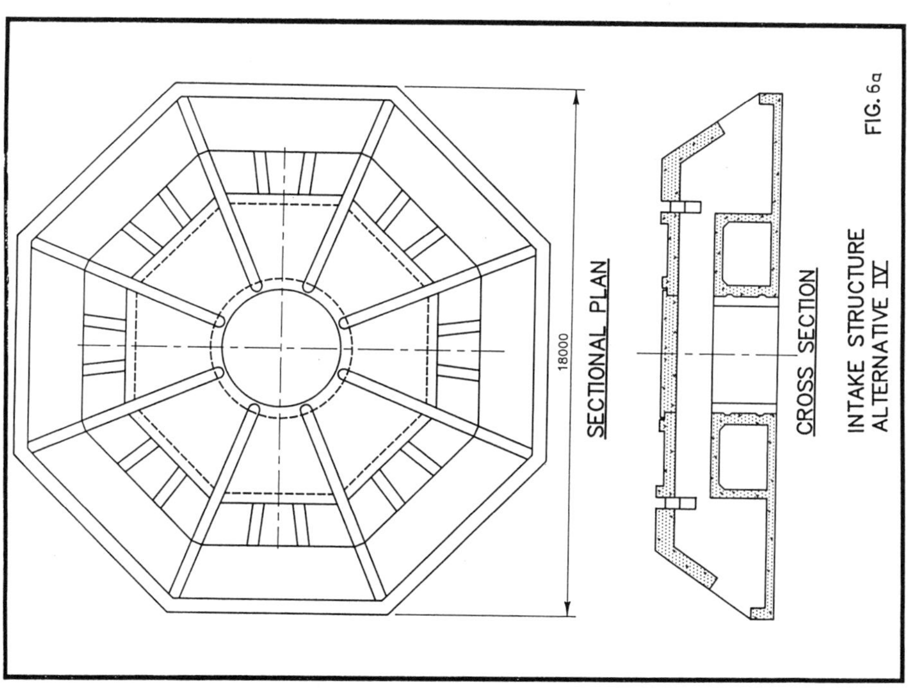
18000
SECTIONAL PLAN
CROSS SECTION
INTAKE STRUCTURE
ALTERNATIVE IV
FIG. 6a

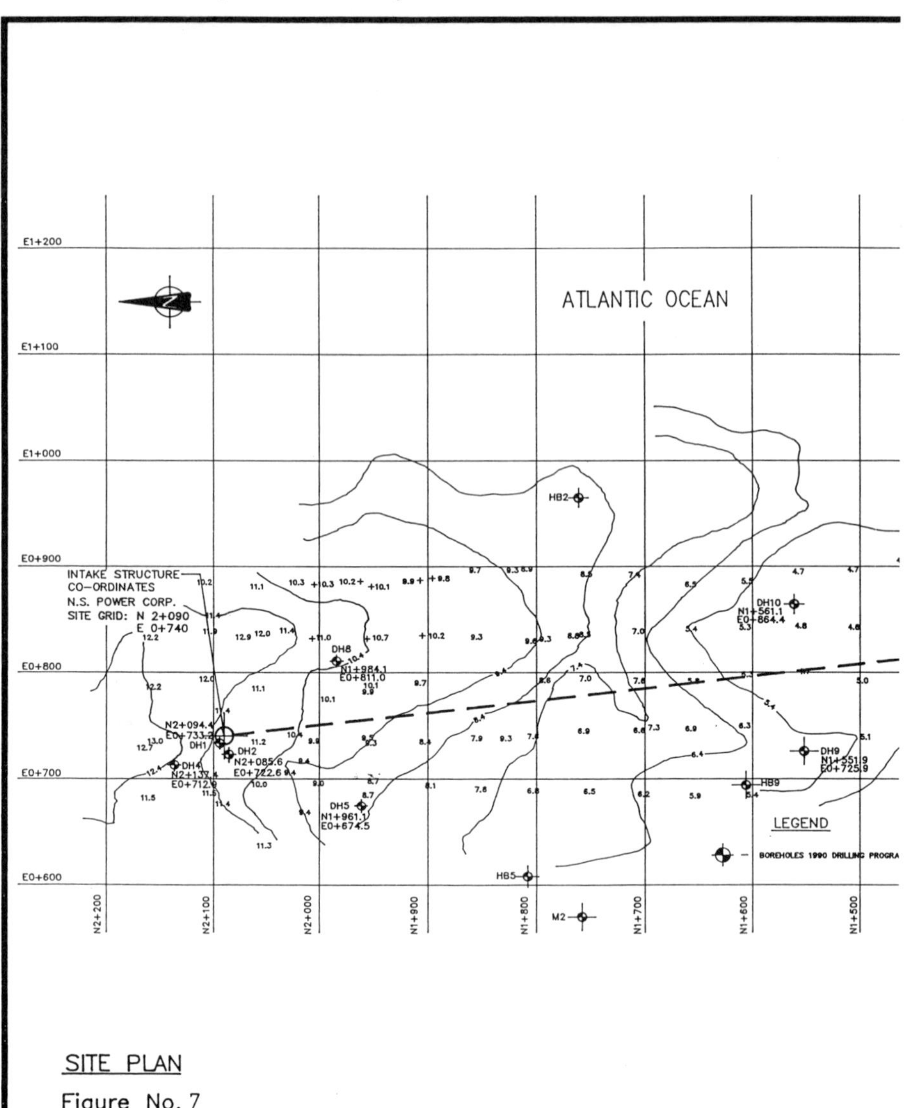

SITE PLAN

Figure No. 7

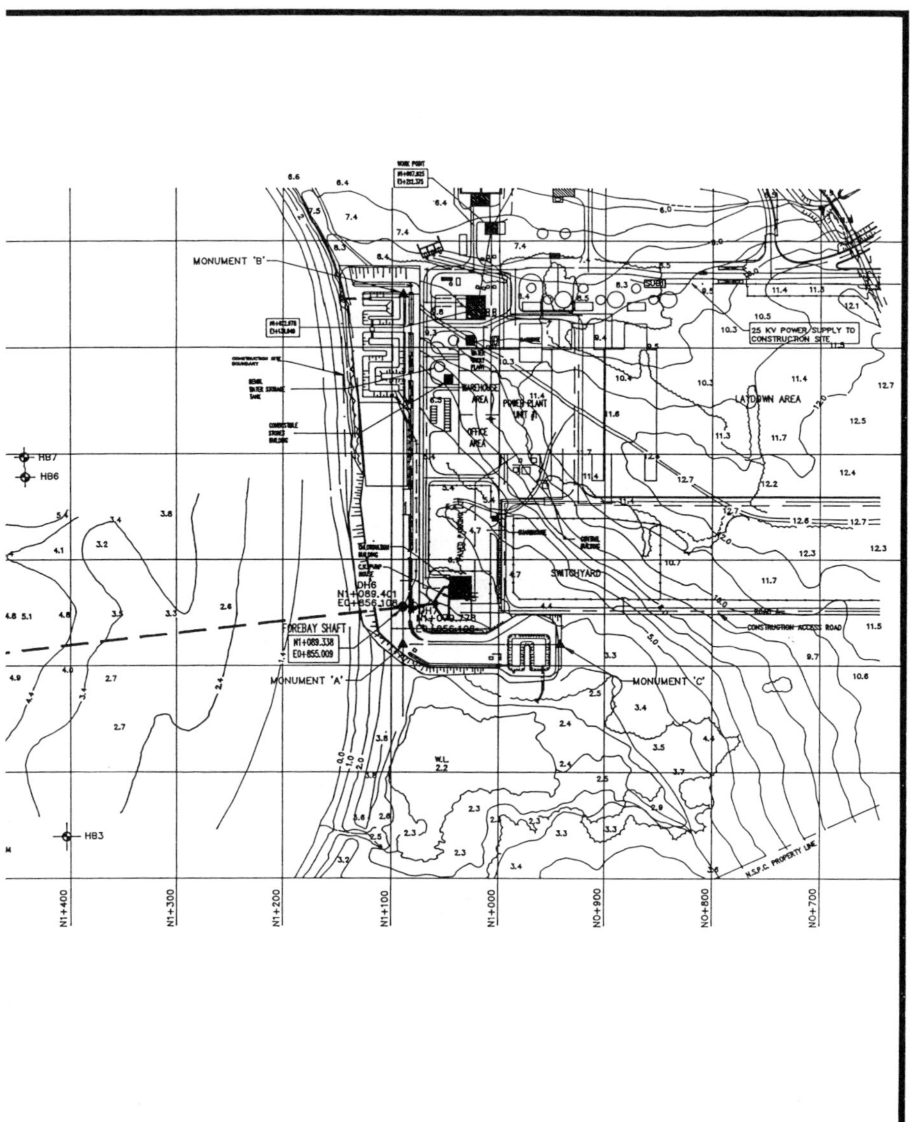
MONUMENT 'B'
MONUMENT 'A'
MONUMENT 'C'
HB7
HB6
HB3
FOREBAY SHAFT
N1+089.338
E0+855.009
DH6
N1+089.401
E0+856.108
25 KV POWER SUPPLY TO
CONSTRUCTION SITE
LAYDOWN AREA
WAREHOUSE
AREA
POWER PLANT
UNIT #1
OFFICE
AREA
SWITCHYARD
CONSTRUCTION ACCESS ROAD
N.S.P.C. PROPERTY LINE
W.L.
2.2
N1+400
N1+300
N1+200
N1+100
N1+000
N0+900
N0+800
N0+700

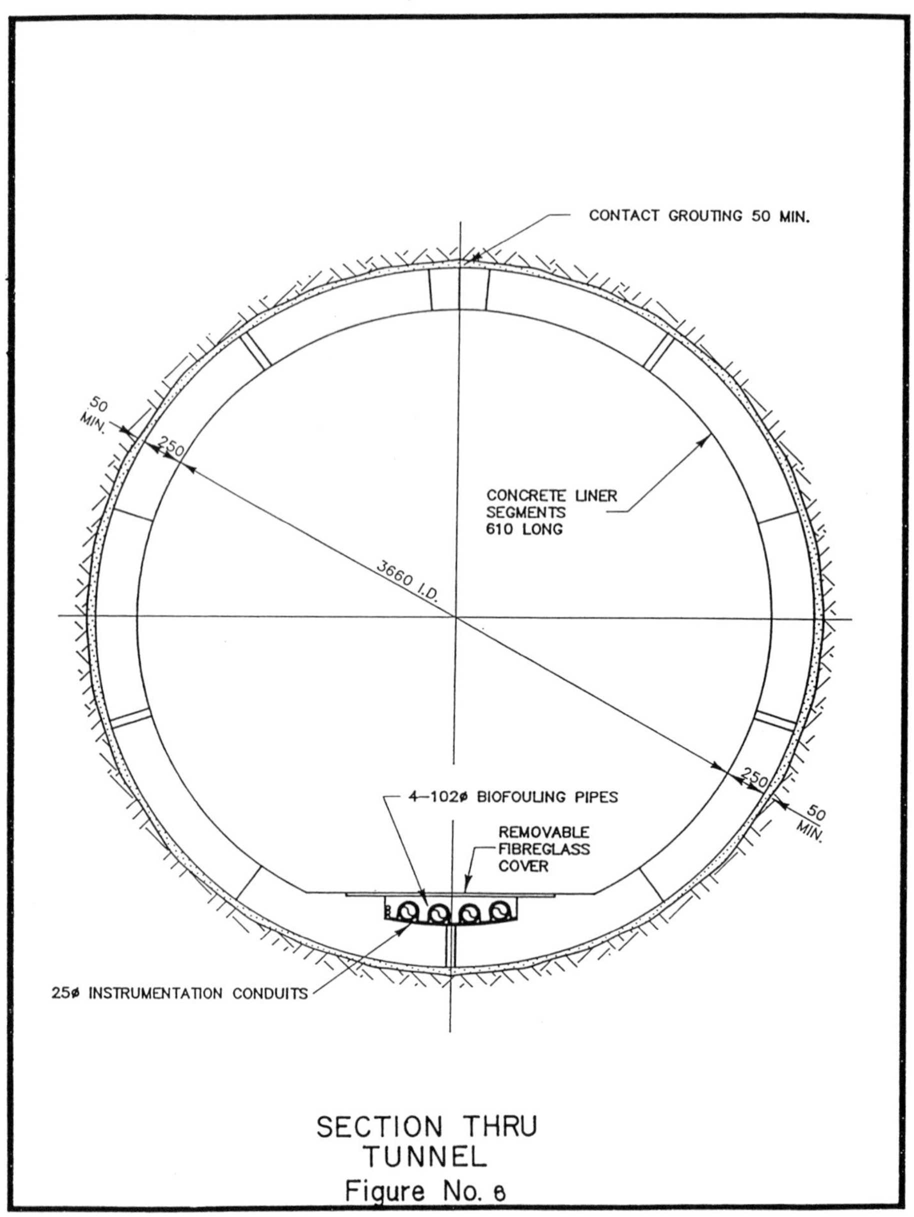

SECTION THRU TUNNEL
Figure No. 6

Comparison of two classification systems as applied to the Alpe Devero tunnel, Italy

Nicola Innaurato
Dipartimento di Georisorse e Territorio—Politecnico, Turin, Italy
Renato Mancini
Dipartimento di Georisorse e Territorio—Politecnico, Turin, Italy
Enrico Rondena
ENEL S.p.A., Centro di Ricerca Idraulica e Strutturale, Milan, Italy
Attilio Zaninetti
ENEL S.p.A., Centro di Ricerca Idraulica e Strutturale, Milan, Italy

Abstract
The case of a tunnel 14 km long, with 3.5 m diameter, driven by TBM in metamorphic rocks, as a part of an hidro-power project in North Italy, is described.

A more detailed analysis is carried out for the central stretch, around 3800 m, where it has been possible to compare different rock classification methods, pertaining to boreability and to stability.

In this stretch the rock classification according to the Austrian norms system and the Wickham Rock Structure Rating were available; based on these data, a boreability classification developed by the authors in previous studies has been checked for reliability, by comparing actual advance rates to the predicted values.

Keywords: Austrian norms, rock classification, TBM tunnelling.

1 Foreword

The use of the TBM's is widespread in Italy both in the construction of tunnels for hydroelectric plants, and other uses, (i.e. the pilot bored tunnels for enlargement in road tunnel construction).

On hydroelectric projects, ENEL (Italian Electricity Board) has encouraged TBM use and undertook with the Dipartimento di Georisorse e Territorio of Politecnico in Torino researches on TBM aimed at supplying the contractors with some basic concepts for the optimum use of machines and forecasting criteria of TBM performances.

From the first uses of TBM's to the present state of the art, problems about machines use and design arose, i.e., the wear of the tools caused by the abrasiveness of some rocks, the difficulty when boring some geological formation, and the problems of tunnel stability, since the machines are also intended to bore poor and difficult ground : more and more, for the users, the main problem is to forecast the time that must be spent at the construction of a certain tunnel in more or less known geological and geotechnical conditions. This time depends basically on two factors: - the boreability of the rock and - the stability of the tunnel.

Fig. 1 Schematic view of the tunnel layout

The case described discusses these factors , dealing with the presence along the tunnel path of rocks of great hardness and abrasiveness and of stretches where some (a few cases very serious) problems of instability occurred.

The tunnel, whose layout is shown in fig. 1, comprises three stretches (Bondolero-Cairasca, Bondolero-Devero and Devero-Agaro) : only the two latter are dealt with here.

2 The Alpe Devero tunnel

This tunnel (see fig. 1), 14800 m long and 3.5 m excavation diameter, will feed, with the waters of two streams (Bondolero and Cairasca), an artificial lake (Agaro lake) representing the upstream reservoir of a hydropower pumping plant now under construction near Domodossola (Novara district, North Italy).

The intake and diversion works are composed essentially of :

1-a large underground room (A) that will serve for water flow distribution and was used as a TBM launching room (size : 58x13.5x9.55 m);

2- the principal diversion tunnel, divided into three main sections : - from Devero site (A) to Agaro lake (D) , 3651 m long; - from Devero to Bondolero (A-E), 3400 m long; - from Bondolero to Cairasca (E-F), 6900 m long;

3- a tunnel for the access to the launching room (B),120 m long, 24 m^2 cross section area;

4- a tunnel at Bondolero site (E), 120 m long , cross section area of 13.7 m^2, that will be used as canal for adducting the waters of the Bondolero stream to the principal tunnel;

5- intakes at Bondolero and Cairasca sites (G,F).

From the end of 1988 to the end of 1992 the tunnel was constructed starting from the launching room towards Agaro lake to complete the first section at the beginning of 1990; then the TBM was dismantled and re-assembled in the launching room, from where it excavated the second section in about 11 months. Finally the work of the machine ended after having excavated about 15 km of tunnel.

The tunnel was excavated using a TBM Mk 12 with the following technical specifications:

Diameter 3.5 m
Thrust 6200 kN
Power 4x150 kW
Stroke length 1.5 m
Disc number. 22+4
Disc diameter : 39.37 cm
r.p.m. 10.6
Weight 1500 kN.

The average TBM performances can be seen in table 1.

Table 1. Average TBM performances in the Devero-Agaro (3651 m)and Devero-Bondolero (3797 m) sections.

	Devero-Agaro	Devero-Bondolero
Advance rate per day	20.9 m/day	19.2 m/day
Net penetration rate	2.12 m/h	2.65 m/h
Mean specific energy	23.5 kWh/m^3	23 kWh/m^3
Mean thrust	4467 kN	3455 kN
Disc wear	0.0032 discs/m^3	0.0037 discs/m^3
Utilisation coefficient(UC)	0.22	0.33

The TBM behaviour appears quite similar in the two sections, with the exception of the utilisation coefficient that in the first stretch was highly influenced by the crossing of a zone of difficult ground which forced the machine to stop at 1504 m chainage for three months (from July to October) to reinforce the tunnel walls and roof. Likewise in the Devero-Bondolero section , starting from around 2800 m chainage from Bondolero, the presence of altered and weathered rock again slowed down the advance rate due to reinforcing operations of the ground. Where the rock was sound, TBM progress was very high, for example, in the second section, in July 90, when 1010 m of tunnel were excavated (average 39 m per day, maximum 48 m per day!).

3 General geology and lithology

Generally speaking, the zone intersected by the tunnel layout belongs to the Pennidic nappes of the Alps, formed under complex structural and metamorphic conditions.

In particular the tunnel crosses rocks of various type and nature as calc-schists (passing sometimes to true quartzites, or possibly to chloritic-talc schists), micaceous schists, metaconglomerates and micaceous marls, sometimes gas bearing.

Because of the high metamorphism of the zone, coupled often with hydrothermal phenomena, the rocks are intensively folded and faulted, frequently with the presence of shear zones, sometimes altered and weathered.The principal systems of discontinuities, besides foliation and stratification (general dip NW), were constituted by fractures with the following features: type 1: 1 set, spacing > 4 m, sound rock; type 2: 2 or more sets, spacing <1 m, open (1-5 mm) and filled; and faults (generally subvertical). Here great instability problems frequently arose because of thick belts of brecciated or milonitized rock. Water flows were very variable from dripping to inflows of up to 100 l/s.

Fig.2 and 3 show the geological profiles of the two stretches Devero-Agaro and Agaro-Bondolero, with a comparison between the results of geological mapping (top) and the actual conditions(bottom).

As shown, a general agreement was found concerning the lithotypes crossed during the tunnel driving; predictions, as expected were not so accurate in details: the location of fault zones was grossly predicted, and the same can be said for the main water inflow chainages.

Also rock variability within the same broad lithological type was found during excavation : i.e., under the general calcschist label, several variations have been recognised: clear type, dark type, talc-schist bearing, garnet bearing and so on.

4 Geomechanical properties of the rocks.

The geomechanical properties of the rocks encountered by the tunnel were studied before the construction on core samples and checked (systematically only for the stretches Devero-Agaro and in some zones of the second stretch) against values from cores taken from tunnel walls.

General agreement was found between the mechanical properties forecast and checked. Table 2 summarizes the properties for the principal lithotypes involved.

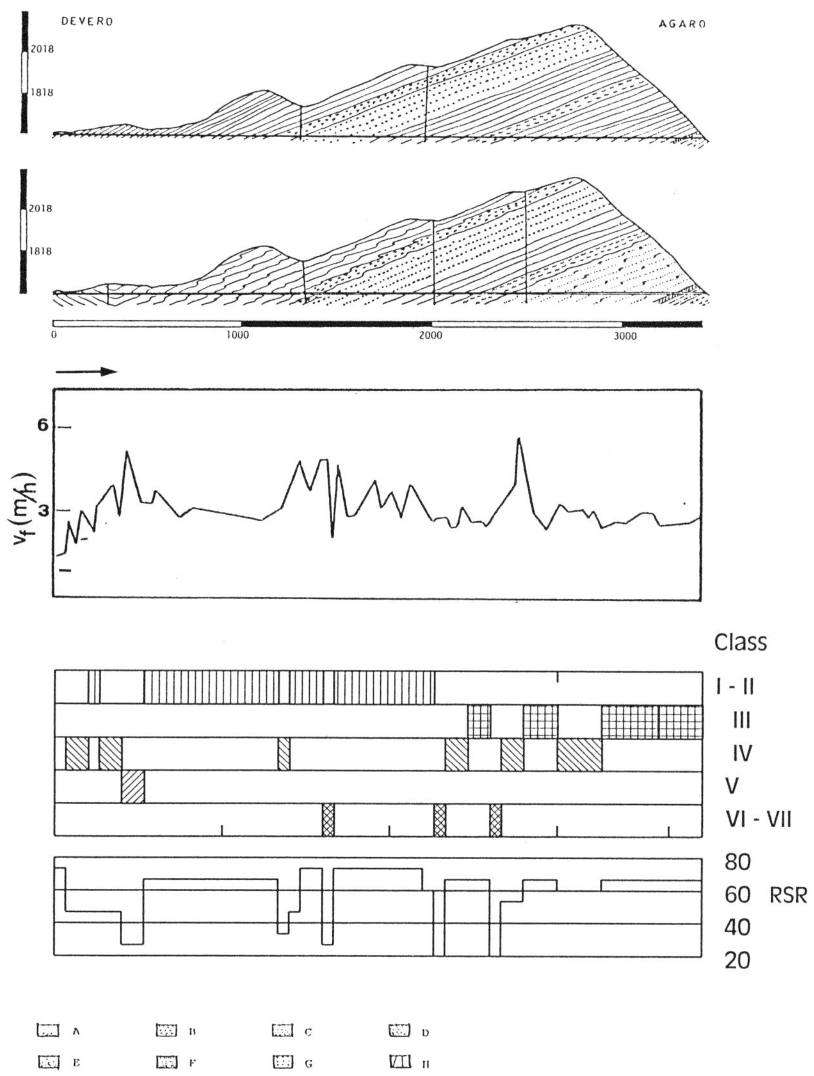

Fig. 2 Geological profile of the Devero-Agaro stretch from the surface survey (top), and from actual underground observations (bottom). Graph of the net penetration rate (Vf), of the RSR factor and partition of the various Austrian classes according to the prevailing class.
A Calc-schist; B Micaceous schist and micaceous limestone; C Metaconglomerate; D Talc-chloritic schist ; E Quartziferous calc-schist and quartzite; F Garnet bearing micaceous limestone; G Orthogneiss; H faults.
The arrow marks the direction of drive.

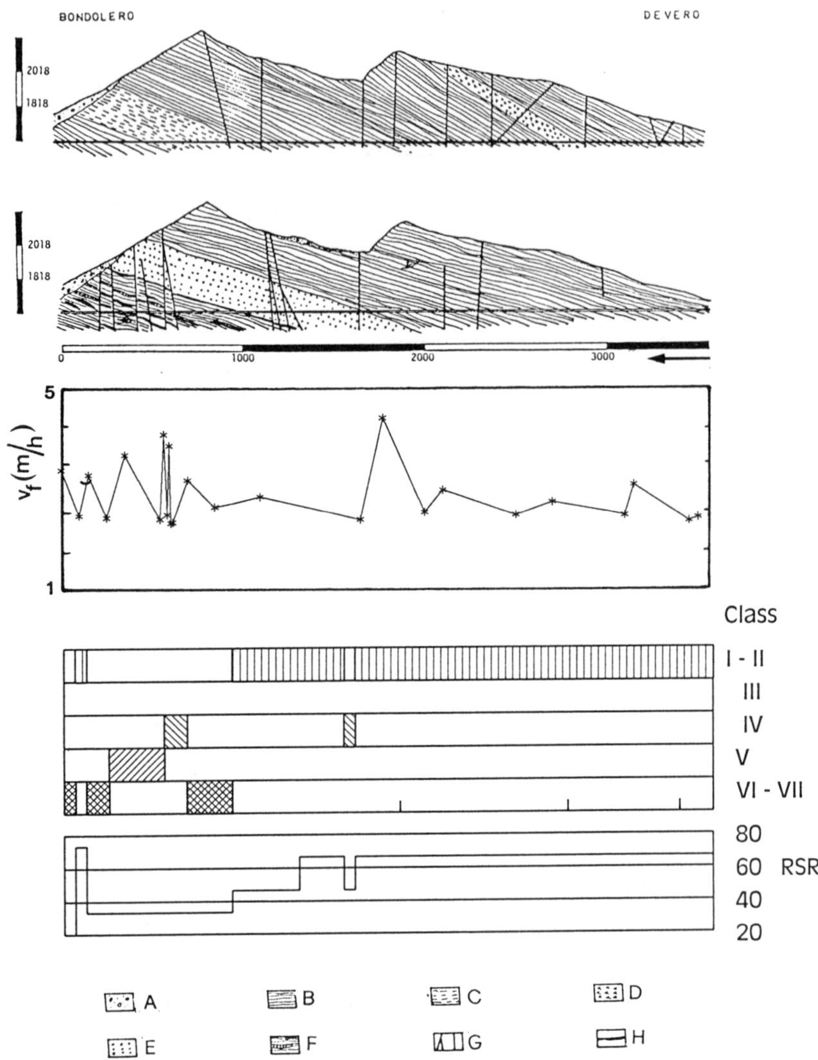

Fig. 3 : Geological profile of the Devero-Bondolero stretch from surface survey (top), and from actual underground observations (bottom). Graph of the net penetration rate (Vf), of the RSR factor and partition of the various Austrian classes according to the prevailing class.
A Clastic burden-moraine; B Micaceous calc-schist; C Gneiss; D Schist; E Metaconglomerate; F Micaceous schist and marls; G Faults; H Tectonic contact. The arrow marks the direction of drive.

Table 2. Mechanical properties of the rock encountered by the tunnel.

Lithotype	σ_c MPa	σ_t MPa	I_{NCB}
Calcschist	60-80	4-7	7-10
Micaceous schist	110	5	5-6
Quartzif. schist	110	3-7	11
Micaceous limestone	60	5	----------
Fine grained gneiss and metaconglomerate	60-130	3-9	11

σ_c and σ_t are respectively the compressive and the tension strengths of the rock and I_{NCB} is the Cone Indenter hardness (National Coal Board Test).

5 Rock mass classifications.

During the works on behalf of the contractor a rock classification was used that relates to the Austrian Norms B2203 (1983).This rock classification system ,that combines NATM type support requirements (and the concepts of Stand-up time/ Unsupported span), with TBM tunnelling,is shown in table 3.

The Austrian classification above quoted is based on informations on excavation and support setting gathered during the work.

On the contrary, other classification systems are based on rock features. Among them we choose, and extensively used in previous studies on TBM performances, the Rock Structure Rating of Wickham [1] , mainly because of its simplicity.

As well known RSR factor depends upon the following parameters:(A) -General area geology (basic rock type; geological structure); (B)-Joint pattern, direction of drive; (C)-ground water , joint conditions. Although the Wickham classification can be considered essentially a guide for the choice of support (and, specifically for the choice of the steel arches) we use this classification because the parameters above mentioned surely affect the TBM rate of advance and, on the other hand, the parameters involved are simple enough to be available without great difficulty . Certainly it can happen that the Wickham parameters related to the direction of drive are not completely correct in order to describe the actual behaviour of the TBM, faced with the direction of the discontinuities, but in this field the knowledge is today still very small.

On this basis the authors found a correlation between the net penetration rate (NPR), the compressive strength of the rock(σ_c) and the Rock Structure Rating of Wickham (RSR), based on the consideration of many cases of TBM works. [2] . Fig. 4 shows the lines of iso-net penetration rate in the plane σ_c -RSR. From the same previous experiences also a gross correlation was found between the Utilisation coefficient (UC) of the machine and the RSR factor (Fig.5).

Table 3 Rock classification chart as used in Alpe Devero tunnel

Rock Class	Rock mass behaviour	Support measures Type	Influence on advance
I	Long term stability	Spot rockbolts	None
II	Local rock fall	Spot rockbolts-shotcrete	None
III	Frequent rockfall in machine area	Rockbolts (up to 2/m)-shotcrete (0.05 m^3/m)-wiremesh	Short hindrances
IV	Frequent rockfall in machine area	rockbolts (4/m)-shotcrete (0.15 m^3/m)-wire mesh	Hindrances after each stroke
V	Frequent rockfall in cutter-head area after each stroke	Rockbolts (up to 8/m)-shotcrete (0.3 m^3/m)-shotcrete facing machine head	Long hindrances after each stroke
VI	Large overbreak in cutter-head area-Self supporting capacity reduced	Rockbolts (up to 8/m)-shotcrete (0.5 m^3/m)-steel ribs- some reinforcements facing machine head	Long hindrances after each partial stroke
VII	No self-supporting capacity	According to the conditions (up to the rock reinforcements)	

The idle time is most frequently spent installing support and/or in consolidation, grouting containing water inflows, as both current literature and the present experience demonstrate. An effort to arrive at a rock classification suitable for TBM work which incorporates the two aspects of stability and boreability is then worth considering.

An attempt was then made to see if, in the case of the Alpe Devero tunnel, RSR correlates with the Austrian classes. Consequently, in some 33 stretches of the Devero-Agaro and Devero - Bondolero sections, where the Austrian classes were known, the RSR factor was also determined. The correlation obtained is shown in fig. 6: it is possible to observe that I-III classes correspond approximately to a large range of RSR values.

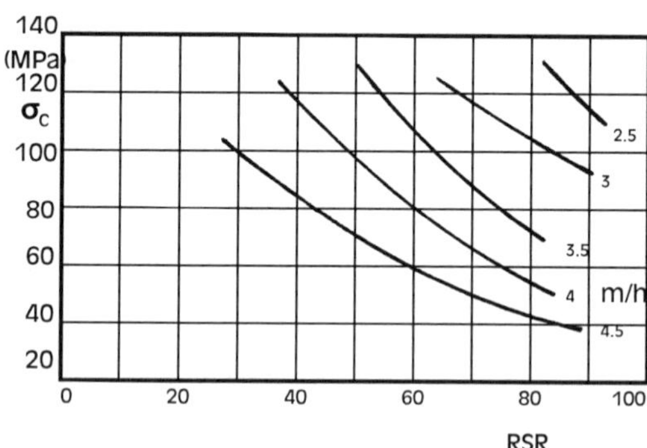

Fig.4 : Iso-penetration rates in the plane σ_C - RSR .

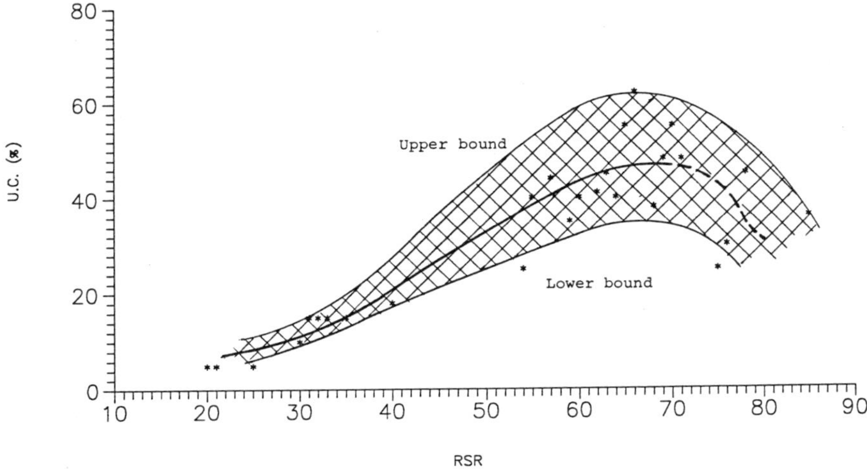

Fig. 5. Utilisation coefficient vs. RSR factor.

Therefore from the sole consideration of RSR in the above presented correlation, I-III could be considered as a single class. However when analysing in detail the significance of the Austrian Norms, it is possible to acknowledge some important differences between I-II and III classes, i.e. the different behaviour of the rock in class III under the gripper pad pressure, and the rock reinforcing that is claimed by the same Norms (cfr.A.N. B2203).

A limit value of the daily gross advance rate can be assigned to each class of the Austrian norms because of the dependence of the Utilisation Coefficient (UC) of the machine on the rock class.

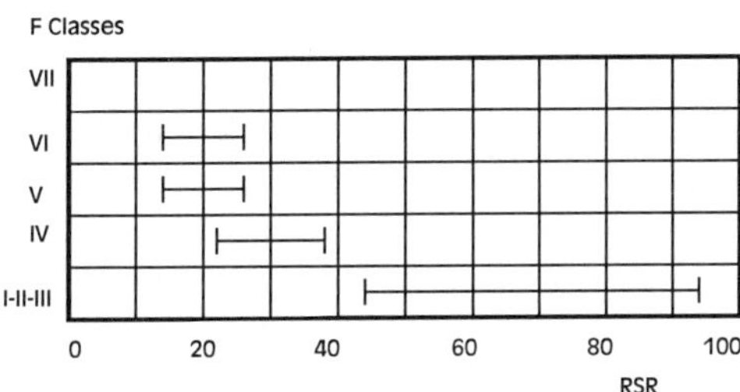

Fig. 6. Ranges of the RSR factor values (horizontal axis) observed within the Alpe Devero stretches, assigned to different Austrian classes (vertical axis).

Recently Lauffer [3] published a rock classification (1988-1990) in which a mathematical relationship between unsupported span,stand-up time and a rock quality index (z) is shown, and found also a correspondence between his classes and the ones of Austrian Norms.Following this concept, Seeber [4] shows an application to a case of a tunnel 3.26 m diameter wide, where the limit value of the daily advance rate and the unsupported span values are defined in connection with the Lauffer and corresponding Austrian classes.

By compounding the RSR/UC correlation (fig. 5), the RSR/σ_c/NPR correlation (fig. 4) and the RSR/Austrian classes correlation (fig.6), a chart can be prepared, tentatively shown in fig. 7, where the gross advance rate is related to RSR or to Austrian classification for different values of the compression strenght of the rock. Taking into account the usual range (60- 80 MPa) of the strenght values of the Alpe Devero rocks, it can be seen that the points representing the performances of the TBM in operations quantitatively agree with the predicted trend.

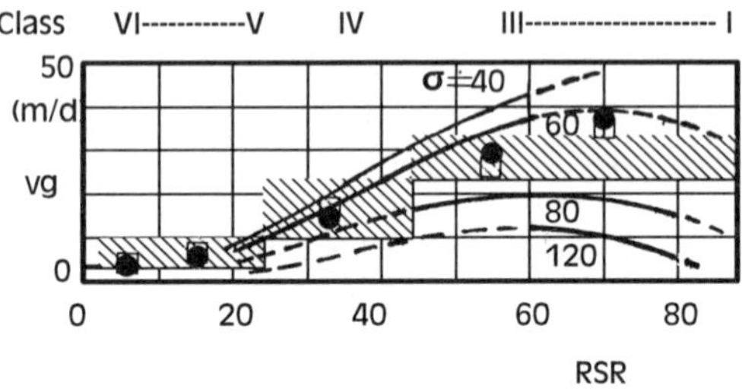

Fig. 7. Gross advance rate vs. RSR (lower horizontal scale)and Austrian classes (upper horizontal scale). Comparison between the Lauffer values (dashed field) and the theoretical values derived from the net penetration rate (fig.4) and the Utilisation coefficient (fig. 5). The points show the actual values of the Devero-Agaro section (●) and of the Devero -Bondolero section (□).

6 Concluding remarks

The Austrian Classification system is, in principle, different from other systems proposed, being founded on the observation of rock behaviour during the works rather than on rock mass properties measured, inferred or simply guessed before excavation. Nevertheless it should be and, at certain extent, can be correlated to other systems. We analysed the Austrian-RSR comparison with reference to a 7 km TBM operation. Correlation observed (fig. 6) is not strict, but still workable. Other classification systems will be checked by the authors on other operations, being admittedly scarce the resolving power of the RSR system at the extreme of the scale (i.e. in poor ground). More important,when attention is paid to progression rate as well as to stability problem, a further parameter has to be added to describe the rock, accounting for the difference in net progression rate observed within the same rock class; the compression strength of rock will cover this need, as shown in fig. 4, and, in order to predict gross progression rate (Vg), a parameter, depending on UC and NPR. A complete picture is obtained by means of Rock class/Vg/σ_c graph, as shown in fig. 7.

7 References

1 Wickham et alii (1974) Ground support prediction model. RSR concept. N.T.I.S Rep.,S.Francisco.
2 Zaninetti A., Rondena E., Innaurato N., Mancini R.(1992) Hydraulic tunnels: review of the TBM excavation features on the basis of Italian Electricity Board experiences, in Proc. Towards New Worlds in Tunnelling, Balkema, Rotterdam.
3 Lauffer H. (1988) Zur Gebirgsklassifizierung bei fraesvortrieben, Felsbau,6,3,137-149.
4 Seeber G. (1988) Die Neue Oesterreichische Tunnelbaumethode und der mechanischer Vortrieb mit Tunnelbohrmaschinen, Felsbau,6,2,63-68.

Session 2: Metro/Rail

Planning and design of tunnels for the Jubilee Line extension, London

A. W. P. Coutts
H. R. Davies
Jubilee Line Extension, London Underground, Ltd., London, England

Abstract
The extension to the London Underground Jubilee Line from Green Park to Stratford is one of the major infrastructure projects currently under construction in the United Kingdom. The tunnelling contracts represent some 30% of the £1.8 billion total cost of the Project. This paper concentrates on the planning and design concepts and the proposed practice. It includes a brief history of the Project, the project management and procurement strategies adopted and the planning and parliamentary processes undertaken and describes aspects of the detailed design of the tunnels and their assumed construction methods together with the current status of construction, including contractors' alternative designs and proposed methods.

1 Introduction

The Jubilee Line Extension (JLE) will be the most substantial addition to London's Underground Railway System since the Victoria Line was constructed in the 1960s.

The Project consists of an 11 station, 16 kilometre extension to the existing Jubilee Line. The JLE will help sustain the regeneration of London's Docklands, relieve congestion on other lines, improve links between existing rail termini and bring parts of inner London on to the Underground system for the first time (Fig. 1).

The underground section between Green Park and the portal at Canning Town incorporates a total of 10.4 route kilometres of twin bore running tunnel and 8 underground stations, 5 of which are wholly or partly constructed by tunnelling methods. The Project also includes significant works to the existing Northern Line station at London Bridge to improve safety and reduce congestion.

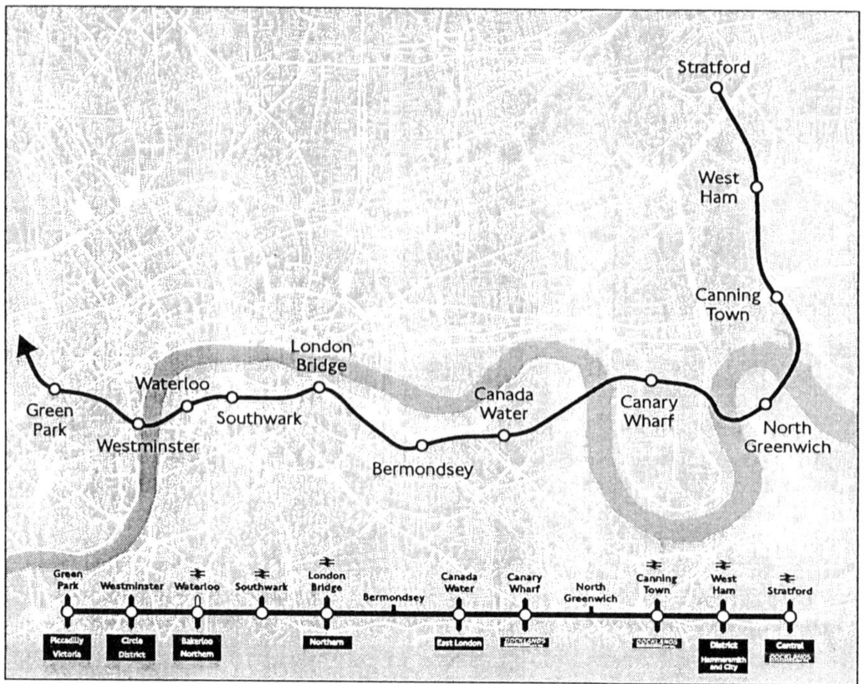

Fig. 1 Route Plan

The main difference between the JLE and existing Tube lines is the provision of enhanced safety measures, including mid station ventilation and escape shafts, tunnel accessways for use by the emergency services and platform edge doors. Exciting station architecture and extensive use of lifts and escalators will also be a feature. The more difficult tunnelling conditions that the route encounters over most of the route south of the River Thames have dictated the use of closed face tunnel boring machines (TBMs) which have only been widely available in the last decade.

The design has been largely completed and the main civil and E&M contracts have been awarded following Secretary of State approval on 29 October 1993, with construction now underway.

The planning and design concepts and the proposed construction methods are described below.

2 Planning

2.1 History

The history of extending the Underground system towards South East London goes back many decades, the most recent proposal being in the 1970s when it was planned that the Fleet Line should pick up one of the branches of the Bakerloo Line at Baker Street and provide a new route parallel to the Bakerloo Line through Central London to Charing Cross and then on an alignment north of the River Thames towards the City before crossing the River into South East London.

In the event, the Fleet Line became the Jubilee Line that we know today with its terminal station at Charing Cross. It was realised in the late 70s/early 80s that, if regeneration of the vast area of disused London docks were to be successful, then much better transport links with the rest of London were essential. The Docklands Light Railway (DLR), connecting the Isle of Dogs to the City and Stratford with extensions planned to South East London and Eastwards to the Royal Docks, was seen to be the answer. This proved to be a catalyst for redevelopment on a much grander scale than had been envisaged when the DLR was initially planned, with the result that it could not provide the necessary capacity for the potential population of the developments being planned, such as Canary Wharf. The Canary Wharf developers, realising this, sought to promote a scheme for an independent underground railway linking Canary Wharf to Waterloo Station. At this point the Government asked London Underground Limited (LUL) to incorporate this proposal into the Underground network.

A number of routes were studied but the general alignment was very much determined by the availability of land, the desire to provide good transport interchange with the British Rail main line termini at Waterloo and London Bridge and other Underground stations and the need to avoid existing obstructions such as building foundations. This dictated the preferred alignment with step-plate junctions with the existing Jubilee Line to the east of Green Park Station, connecting with Westminster Station and crossing the River to Waterloo.

From Waterloo the route follows the alignment of the British Rail viaducts south of the River to stations at Bermondsey and in the Surrey Docks before crossing the River to the Isle of Dogs and Canary Wharf Station. The tunnels then cross under the River to the North Greenwich peninsula and back again to the north side emerging after a short section of cut-and-cover tunnel to the south of Canning Town Station. The railway then runs in the existing British Rail North London Line corridor at grade to the terminal station at Stratford. A maintenance and stabling depot is located to the south of Stratford. This route was the one deposited before Parliament as a Private Bill in November 1989.

2.2 Parliamentary Process

The Project proposal passed successfully through Parliament by means of a number of Bills. Although there were a number of petitioners opposing the scheme, the main debate relating to tunnelling was an alternative alignment between Canary Wharf and Canning Town which kept to the north of the River. The Select Committee found in favour of the preferred alignment with better potential transport links both south of the River and to the Royal Docks.

A separate Bill for the safety and congestion relief works for the Northern Line station at London Bridge was progressing through Parliament at the same time. The necessary statutory powers for JLE were successfully achieved in March 1992.

In order to meet the Project programme, it was necessary to progress the design in parallel with the parliamentary process and this was substantially achieved in that the first two construction contracts, which were required to meet the programme requirements relating to track and line-wide E&M equipment installation from the depot, were ready for award coincident with the enactment of the Bill. The Project was then delayed due to problems with the provision of the promised private sector funding.

2.3 Project Management

The Client has directly employed a substantial in-house project management team responsible for defining the Project, seeking the necessary legislation, procuring the design and contractors and supervising the works to ensure that quality, cost and programme requirements are met.

Architectural and civil engineering consultants have been employed to undertake the detailed design work based on conceptual designs prepared by the Project team. Separate architectural practices have been appointed to design each of the stations and the station architecture reflects the best the profession has to offer.

Civil engineering contractors are generally employed on a construct-only basis which provides the Client with more certainty over the finished product.

Strict cost control measures similar to those being used during construction were applied during the design phase.

2.4 Client Requirements

The Client has required the JLE to be a railway for the next century within the constraints imposed by the existing Underground system, both at the interfaces at interchange stations and as an extension to the existing Jubilee Line with its fixed structure gauge. Almost universally the architects have placed emphasis on creating spacious public areas and carrying natural light to the underground spaces notwithstanding the structural and geometrical challenges posed. Adequate provision for the mobility impaired is being provided.

New design standards have been prepared that take into consideration the lessons learnt following the King's Cross fire, LUL's extensive operating experience, the appropriate principles of the new American National Standard NFPA 130 and recent metro construction overseas. The Client has also required that the main structural elements of the substructure works, including the tunnels, should have an enhanced design life. As far as the tunnelling is concerned, the main effects of these requirements have been a tunnel diameter of 4.35 m, which is greater than the normal 3.81 m found on other lines, to accommodate an accessway at track level to enable emergency services personnel to pass a stalled train. The use of platform edge doors has obviated the need for larger diameter, transition tunnels which would have been necessary to reduce the air velocity in stations caused by the piston effect of the trains.

2.5 Environmental Factors

Although the majority of the JLE is below ground or beside an existing railway line, the environmental impact was fully considered in the planning stage and various aspects were debated during the parliamentary process.

The effects of ground settlement on overlying and adjacent structures is a significant engineering challenge. Advances in construction methodology, including compensation grouting, can be expected to minimise these effects.

Both the effects of construction noise and operational noise, particularly through re-radiated ground-borne vibration, have been fully considered both in the locations of work sites and in terms of the track support with the use of long welded rail, resilient track and floating slab track.

The effect of construction traffic was carefully considered and it is intended that over 50% of the excavated spoil will be removed by barge from riverside tunnelling shafts.

2.6 Geology

A detailed site investigation was carried out to augment and confirm the database of previous information about the geology of this part of London (Fig. 2).

At the depth required for the tunnels to meet the various constraints, including interchange with existing lines and river crossings, the Project can be divided into two sections: to the west of London Bridge the running tunnels are constructed completely in London Clay and conventional tunnelling techniques were assumed in the planning process; to the east of London Bridge the tunnel route passes through variable geological strata and it was assumed that closed face TBMs would be required. As the tunnels approach Bermondsey Station, the London Clay gives way to Woolwich and Reading Beds which can contain water-bearing sand lenses and occasionally limestone bands. From Canada Water to Canary Wharf the tunnels will pass through Thanet Beds, which comprise densely packed silty sands until these sands dip away to Woolwich and Reading Beds towards North Greenwich. The final part of the tunnel is expected to be in London Clay. The shallower works at stations and shafts will encounter made ground and Thames Gravels.

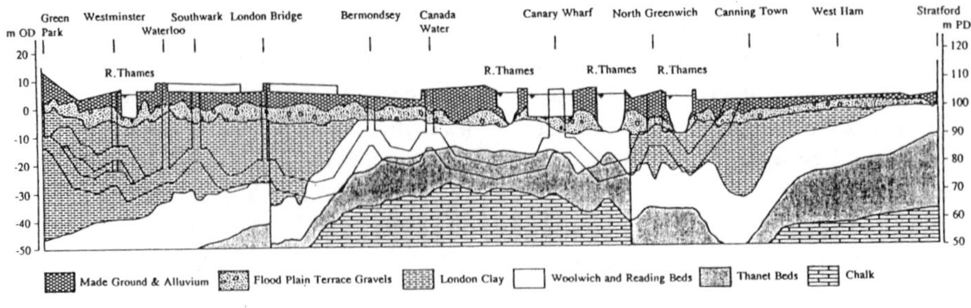

Fig. 2 Geological Section

2.7 Construction Methods

The spatial, environmental, interchange requirements and river crossings mentioned above have automatically dictated that a substantial section of the JLE is underground with virtually all running tunnels and the majority of station tunnels in bored construction. Where space permits, cut-and-cover construction is utilised for the underground stations with the advantages of construction economy, more generous space and ease of construction in difficult ground conditions. Therefore, Canada Water, Canary Wharf and North Greenwich Stations are of cut-and-cover construction.

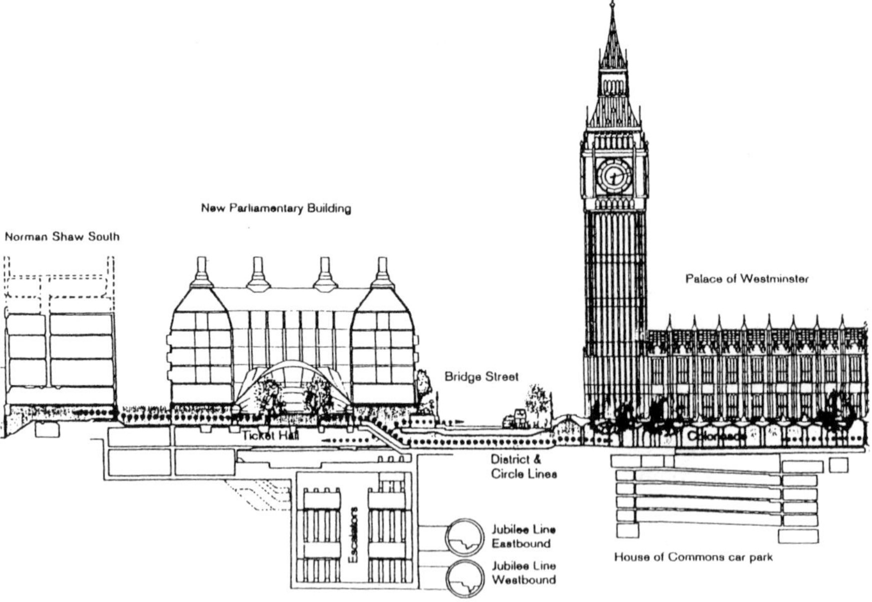

Fig. 3 Westminster Station - Cross-section

Westminster has developed as a station with the 150 m long platform tunnels positioned one above the other with adits to a 30 m deep station box incorporating the existing District and Circle Line station and with a new parliamentary building being constructed over the station box (Fig. 3). Waterloo Station is a more conventional tunnelled station with a booking hall at ground level incorporated within the existing British Rail infrastructure. Southwark Station also has tunnelled platforms but the escalator connection to a subsurface ticket hall is by cut-and-cover construction under the railway viaduct arches. Both the JLE and Northern Line works at London Bridge (Fig. 4) are of more typical LUL bored tunnel construction. Construction of the step-plate junctions around the operational Northern Line running tunnels will require careful construction, particularly the northern one which is located under the River. Bermondsey is a hybrid station, with part of the platforms constructed by tunnelling and part by cut-and-cover method within the main station box which allows for future development in the air space above.

There are a number of other step-plate junctions including the connection with the existing Jubilee Line and crossovers required for operational reasons. Platform tunnels are either 7.0 or 7.7 m in diameter depending on passenger volumes with the largest tunnels being 10.0 m diameter which are required in some escalator shafts and for step-plate junctions.

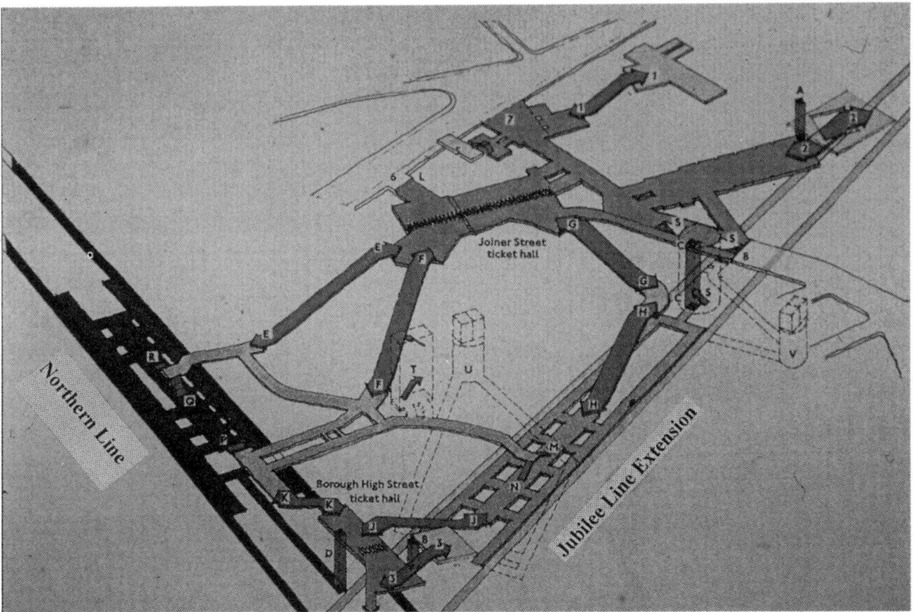

Fig 4 London Bridge Station - Isometric Sketch

2.8 Contract Strategy

The civil engineering and fitting out and builders' works are split on a geographical basis seeking to parcel the work into significant packages to minimise contractual interfaces and to require and attract major contractors, whilst at the same time not over exposing the Project to any one contractor. This objective is very much affected by the availability of adequate tunnelling access shaft sites and the optimal type of construction method. The cut-and-cover stations were considered to be separate contracts, although North Greenwich Station was combined with the adjacent running tunnel contract because of the need for a tunnel access shaft at this location. The extent of the works at London Bridge Station, including the Northern Line, warranted a contract of its own. This resulted in six tunnelling contracts: three in the west in London Clay, namely Contracts 102, 103 and 104, and three in the east in mixed water-bearing strata, Contracts 105, 107, and 110 (Fig. 5). By dividing the Project into six tunnelling contracts, it was considered that the risk of serious programme over-run would be minimised by increasing the resources, materials, plant and labour present on the Project and that any delay in tunnel driving could be more readily recovered due to the relatively short drive distances and the possibility of driving from the other direction.

The five tunnel design contracts were generally split in a similar manner with the express purpose of involving consultants who had not previously worked on major LUL projects and with a view to involvement in future projects such as CrossRail and the Chelsea - Hackney Line.

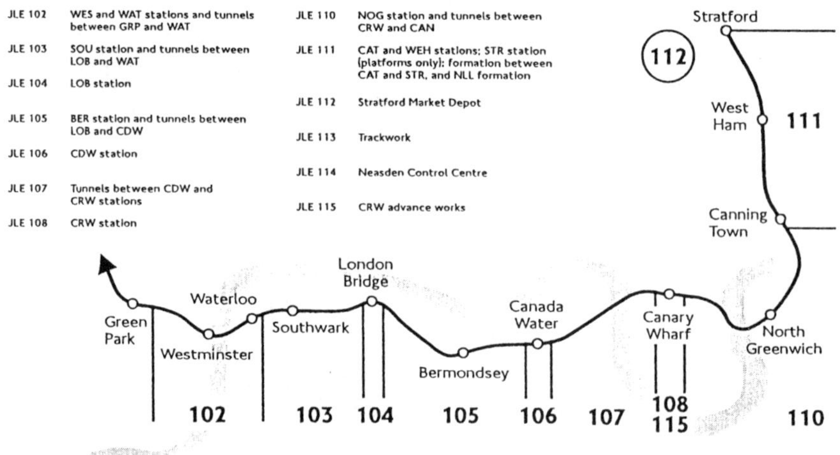

Fig. 5 Civil Engineering Contracts

Both the design contracts and the construction contracts were let through an international prequalification, selected short list, competitive tender process on a strategy of value for money. The design contracts were let on a lump sum basis and the construction contracts are based on contract form of Federation Inernationale Des Ingenieurs - Conseils (FIDIC) origin, incorporating allocation of specific risk, mediation, admeasurement, interim payment schedules against milestone achievement and with provision to instruct acceleration to recover delay. Tenderers for the construction contracts were encouraged to submit alternative designs to suit their plant and expertise.

3 Design

3.1 Parameters

The Client required that substructures constructed for the JLE have an extended design life beyond the 120 years which is normally assumed in modern designs. This requirement reflects the age of some of the existing Underground system which has already exceeded 120 years and is expected to continue in service for the foreseeable future. Last year, for example, heralded the 150th anniversary of Brunel's Thames Tunnel which is still in operational use as part of the East London Line.

In order to achieve the required design life, the existing standards have been raised in particular areas. A good example is the JLE requirement for precast concrete tunnel linings and insitu concrete structures cast against the ground to be tested for permeability to water and diffusivity to chloride and oxygen ions. There is no agreed European standard for water permeability and a detailed test method has been developed with the assistance of specialists in this field at Leeds University and Imperial College, London. It is now believed that there are workable test methods, concrete mixes and curing requirements that will meet the specification and will result in concrete which is 10 times less permeable to water than normal C30 insitu concrete.

The concrete cover to the tunnel segments has also been increased to a minimum of 40 mm, which in combination with C50 concrete takes the durability beyond the range quoted in BS8110.

There has also been a determined policy by the various design consultants to avoid cast-in ferrous items such as bolt pockets, sockets and grout holes. In this way, the durability of tunnel segments should be greatly enhanced. Although the existing ground water along the route has been found to be benign, with a pH of 8 and chloride content of 1 g/l, there has been acknowledgement of the rapid rise in ground water table throughout the Thames Basin which may be significant during the design life of the tunnels.

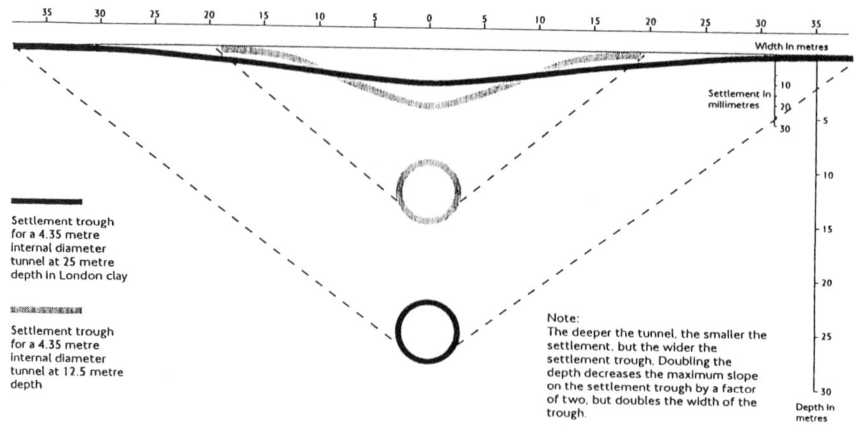

Fig. 6 Typical Settlement Trough

3.2 Settlement

In order to limit the actual settlement of structures with corresponding limits of differential settlement and tensile strain as required by the specification, it was recognised that ground treatment measures must accompany the tunnel excavation.

Normally settlements must not exceed 25 mm. Structures at risk were determined by producing settlement contour maps of the station sites using accepted methods of settlement prediction with a typical Gaussian settlement trough (Fig. 6). Mandatory ground treatment for those structures which come within the 25 mm or greater settlement contour has been specified. In order to demonstrate that compensation grouting could be carried out successfully in London Clay, a trial was carried out at the Redcross Way Site located at London Bridge in early 1992. This trial showed that grout could be injected at a desired location to produce heave and that this heave could be controlled. When this grouting technique is used in conjunction with adequate monitoring, the settlement caused by tunnel excavation will be able to be controlled within tight limits. This was proved more recently during the construction of a new LUL escalator tunnel required in connection with Waterloo International Terminal. The grouting was carried out in the London Clay and its interface with the Thames Gravels to protect the existing Waterloo and City Line tunnels and the Victory Arch. Predicted 'green field' settlements of 80 mm were reduced by compensation grouting to approximately 10 mm.

In order to limit face loss and therefore settlements, the use of compressed air has been specified for the construction of the Northern Line northern step-plate junction under the River. Although the theoretical water head requires the compressors to be capable of providing 2 bars of pressure, due to the 6 m of London Clay cover, a working pressure of below 1 bar is anticipated.

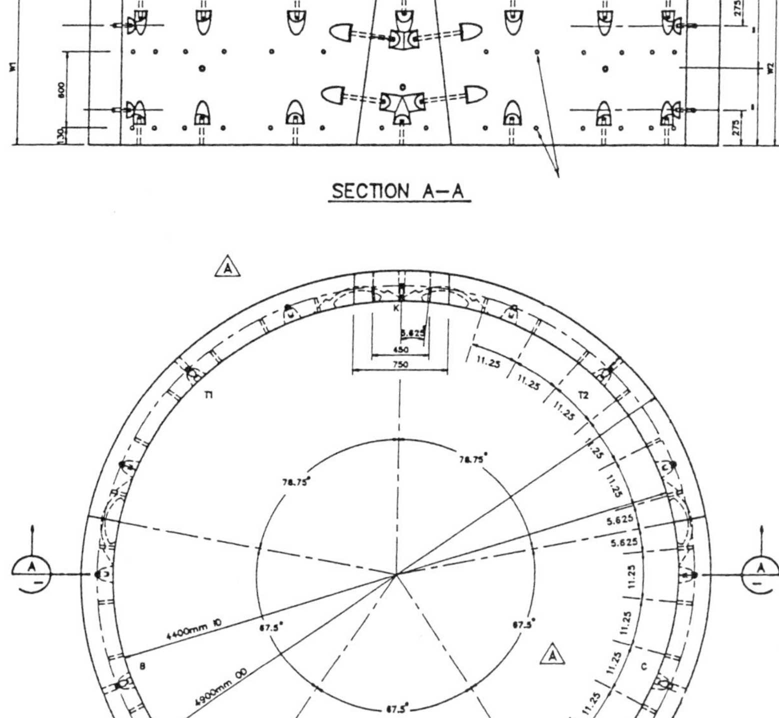

Fig. 7 Typical Section of 4.40 m Internal Diameter Precast Concrete Tunnel Lining

To maximise the clay cover in this area, the use of elliptical Spheroidal Graphite Iron (SGI) linings is proposed to encompass the existing Northern Line running tunnel.

3.3 Linings

The eastern group of contracts in mixed ground all have the same ring design which accounts for approximately 60% of the running tunnels (Fig. 7). This design is based on the assumption that the contractors will use closed face machines to drive the 4.40 m inside diameter tunnels. The ring is 250 mm thick and of robust design with a maximum shove force of 25,000 kN having been assumed in the design. The ring has only 5 segments with a wedge block key and is 1200 mm wide and incorporates curved bolts for both cross and circle joints. The rings are all tapered to avoid having to utilise excessive circle packings to negotiate the many horizontal and vertical curves along the route. The circle packing is limited to 6 mm in order that the sealing gasket, placed close to the extrados, can provide an effective seal against 3 bars of water pressure. The specification calls for a dry tunnel, only damp patches being permitted above axis level. The three contracts were designed by two consultants, one of whom adopted the other's ring design.

For the three western contracts where the running tunnels are in London Clay, two ring designs were produced with the designer for Contract 103 adopting the Contract 104 ring. The rings can be more lightly reinforced than those in the east due to the anticipated reduction in shove forces as it was assumed that the contractors would opt for simple, open faced machines with backhoe excavators. The ring proposed for Contracts 103 and 104 is of 4.35 m inside diameter and 200 mm thick. The ring has 8 segments, with a wedge block key, and is a tapered ring of nominal width 600 mm, with curved bolts again used to connect cross and circle joints. The designer for Contract 102 has adopted a ring design with a 4.35 m inside diameter and 200 mm thick, but with 9 plates and a straight key. The 1000 mm wide ring is straight, as the drive from Green Park to Waterloo is mostly straight or of large radius curvature. The 300 m radii called for at Green Park step-plate junctions will be negotiated by packing, as experience has shown that the watertightness in the London Clay should not be compromised in this area of London.

The additional 50 mm in running tunnel diameter allowed at the east is to provide the contractors with increased tolerances to line and level for closed face shields. Throughout the route, SGI linings have been adopted at openings to facilitate the construction of the junction and the supporting steelwork. The majority of platform and escalator tunnels will have SGI linings.

3.4 Procurement Strategy

The Project has departed from LUL's previous strategy of procuring the tunnel linings in that they will be supplied by the contractors to the Engineer's design. Reasons of this departure from previous practice are in line with the Project philosophy that requires contractors to take responsibility for those areas where it is appropriate. It also provides the opportunity for contractors to propose alternative lining designs in accordance with the Project policy. Also, the provision of approximately 150,000 tunnel segments required within a period of 20 months, 12 months after Project commencement, would have placed an extremely high demand on the chosen manufacturer, particularly if production and quality problems arose.

As a guide to the contractor each Particular Specification detailed the Engineer's assumptions as to how the contract could be completed within the programme period. These assumptions however did not include the number of tunnel machines nor the assumed drive rates. In the east, provision was made for either Earth Pressure Balance Machines (EPBMs) or Slurry Machines but it was left to the contractors to choose the most suitable type of machine for the ground conditions.

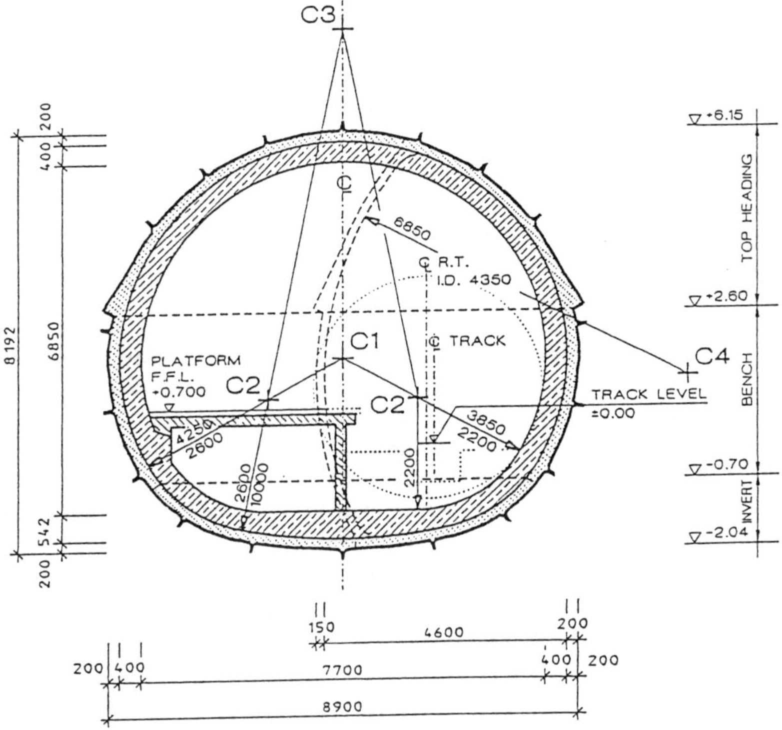

Fig. 8 Typical NATM Station Tunnel Section

It was expected that environmental considerations and space requirements might dictate against slurry machines.

3.5 Alternative Designs

In line with the Project policy of encouraging contractors' alternative designs and construction methods, a New Austrian Tunnelling Method (NATM) Particular Specification was prepared and provided to tenderers who expressed an interest in a NATM alternative based on an assumed typical cross-section (Fig. 8). This specification closely limits the excavation sequences and both the temporary and permanent lining thicknesses. Overall design parameters and loading criteria were also stated thereby ensuring that the tenderers would be adopting a common design approach.

The major advantages anticipated by the use of NATM are:

1. Reduction in surface settlement.
2. Increased flexibility of design due to ability to change the tunnel cross-sections which result in reduced excavation quantities.
3. Reduction in the overall numbers of skilled labour compared with conventional hand-built rings.
4. Reduction in construction cost.

It was a requirement that any tenderer proposing a NATM alternative would also have to carry out a section of trial tunnel using NATM to prove his method and to submit a tender which fully complied with the Engineer's design. This provided a direct cost comparison with the alternative and, in the event of the contractor failing to satisfy the Client and other interested parties of the acceptability of his methods during the trial, a conventionally tunnelled scheme could be adopted. The use of NATM for station construction in London Clay had been seriously considered during the planning and design stage, but had been rejected for the Engineer's design in favour of a conventional ring design, because before the results of the NATM trial tunnels for Heathrow Express were known there were some reservations in certain quarters about its use in a heavily built-up area and in London Clay.

4 Construction

4.1 NATM

Major alternative designs have been accepted, subject to successful trials, involving the use of NATM for the construction of Waterloo and London Bridge Stations.

In order to ensure there is no delay to the programme, the trials will take the form of temporary shafts and adits which are required in any event to provide access to the running tunnels. For Waterloo, the access adit is approximately 100 m long and will be driven, at running tunnel diameter, from the Contract 102 tunnelling access shaft at Jubilee Gardens. At London Bridge, a large cross-section, fully instrumented backshunt will be formed from the access shaft at Redcross Way site. This will demonstrate the contractor's ability to construct the largest 10 m diameter tunnels at London Bridge Station. An access adit will also be driven to intersect with the running tunnels adjacent to the station using NATM methods. The contractors will be required to use the plant and labour which they intend to use for the remainder of their NATM tunnelling works.

The trial at Redcross Way will be a full scale demonstration of NATM, not only for the Project team and its Client, but also for the petitioners with whom there are undertakings to protect their buildings and structures adjacent to the works. The contractor will also be required to carry out compensation grouting in conjunction with this trial tunnel excavation to demonstrate that this method can control ground movements within specified maximum parameters.

4.2 Programme

With regard to other tunnel contracts, the contractors have generally proposed higher drive rates than were assumed in the Engineer's pre-tender programme. It was agreed at tender time that in some cases these increased rates could be accepted and the number of TBMs reduced.

The procurement of the tunnel shield machines is the most important single purchase to be made by the contractor. The shield design must be carefully tailored to suit both the geological strata to be encountered and the design of the tunnel linings to be built. Generally, the assumptions made during the design phase were correct with open face machines adopted in the West and a preference for EPBMs in the east, although slurry machines may be used on one of the contracts.

4.3 Other Contractors' Alternatives

The contractor for Contract 105 is proposing the use of compressed air at below 1 bar to facilitate the construction of the platform tunnels at Bermondsey Station. These tunnels will be excavated and lined by hand by breaking up from a running tunnel pilot. The geology is sandy Woolwich & Reading Beds with overlying Thames Gravels. Protective grouting and ground stabilisation is a mandatory requirement for a number of the adjacent buildings.

Another alternative design involves Contracts 102 and 103 where both contractors have considered the use of expanded linings for the running tunnels. Traditionally, LUL have used expanded linings for the majority of its running tunnels since the 60s. The Victoria Line, the Piccadilly Line extension to Heathrow and the existing Jubilee Line were all constructed using expanded linings and the advantages and disadvantages are well known.

New St Clair River tunnel, Canada–U.S.A.

R. S. Curtis
Hatch Associates, Ltd., Mississauga, Canada
A. P. Finch
Mott MacDonald, Ltd., Croydon, England
R. P. Lovat
Lovat Tunnel Equipment, Inc., Toronto, Canada

Abstract
When completed, the new 1800 metre long, 8.4 metre internal diameter St Clair River Tunnel will accommodate double-stack container trains running between Canada and the USA. The concrete lined tunnel will be driven in soft clay by a full face Earth Pressure Balance TBM. The development of this tunnel project, its relationship to the original nearby tunnel built in 1890 and an overview of the rail transportation considerations will be provided. Details of the field investigations, alignment selection, design of tunnel and portal structures will also be provided together with a discussion of risk sharing between client and contract. The development of the 9.52 metres outside diameter EPBM will be described with details of the provisions incorporated to cater for the expected ground conditions which include soft squeezing clay, limited cover to the river bed, the potential presence of large granitic boulders and the possibility of encountering shale bedrock.

1 Introduction

CN North America, through its subsidiary the St Clair Tunnel Company, is planning to construct and operate a new rail tunnel under the St Clair River, between Sarnia, Ontario and Port Huron, Michigan, as shown in Figure 1. The 1800 metres long, 8.4 metres internal diameter tunnel will be built about 26 metres north of the existing 100 year old St Clair River Tunnel which, although still in use, is too small to allow passage of the taller railcars now in service, particularly double stack container cars. When completed in late 1994 the new tunnel will carry all CN rail traffic in the Sarnia - Port Huron corridor, including the taller railcars presently ferried across the river by barge.

The decision to proceed with the tunnel was taken by CN in late 1991 following a detailed feasibility study undertaken for the railway company in the same year. Several methods of dealing with the bottleneck created by the existing tunnel were studied, including: enlargement of the existing tunnel, construction of a bridge, a deep tunnel, a shallow-cover bored tunnel and an immersed tube tunnel. These options were compared with the "do nothing" option or maintaining the status quo.

Following careful consideration of many factors such as disruption of rail service during construction, environmental impact, construction safety, property requirements and, of course, cost it was concluded that the most feasible option would be the construction of a new shallow-cover bored tunnel immediately north of the existing one at approximately the same crown elevation. It was also concluded that the tunnel should be built by an earth pressure balance (EPB) tunnel boring machine (TBM) designed specially for this project.

An environmental assessment of the project was conducted as part of the feasibility study and culminated in the publication and circulation of an Environmental Study Report. The report concluded that construction of the new tunnel would result in net positive benefits and that all potential negative impacts could be effectively mitigated.

In March 1992, following an international call for proposals, consultants Hatch Associates Limited of Canada in Joint Venture with Mott MacDonald were retained by CN to undertake the final design of the tunnel project and to manage construction.

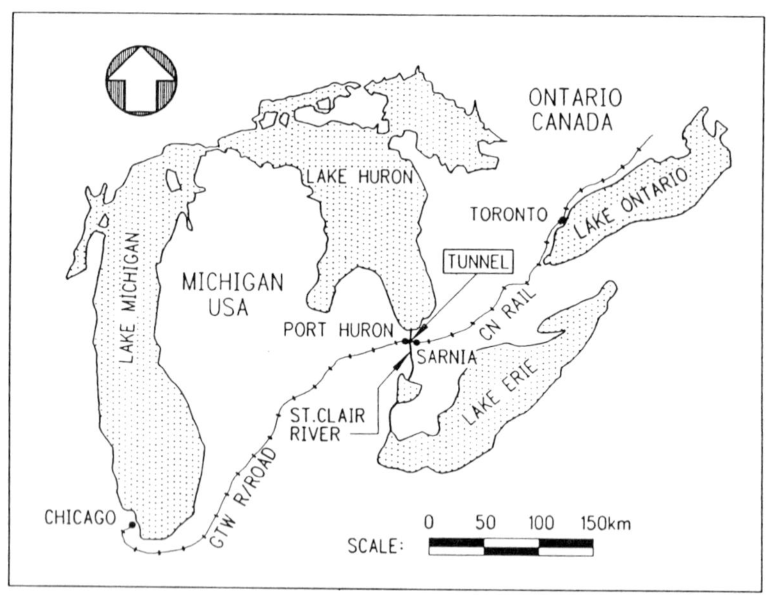

FIG. 1 Location Map

Tunnelling is proceeding from Canada to the USA, resulting in all 130,000 cubic metres of tunnel spoil being evacuated at the Sarnia Portal and disposed of in Ontario. In addition approximately 80,000 and 200,000 cubic metres of material will be excavated from the Sarnia and Port Huron approaches respectively.

2 Geotechnical

Preliminary site investigations were carried out during the feasibility study. In early 1992, following the decision to proceed with the shallow-cover bored tunnel, a more extensive programme of site investigation was conducted with particular attention being paid to the river section. To increase knowledge of the site, further geotechnical investigations were conducted during the detail design phase of the project. In total some 60 boreholes were drilled, including 32 in the river section.

In addition, a significant amount of useful information was obtained from records of both the investigation for, and the construction of, the existing tunnel in the late 1800's.

The general stratigraphy of the site is shown in Figure 2 and can be summarised as typically 30 to 35 metres of superficial deposits overlying the shale bedrock as follows:

- Fill
- Upper Sand
- River Bed Alluvium
- St Clair Till (clay)
- Lower Till
- Shale Bedrock

The fill was encountered locally in several places along the alignment.

The Upper Sand deposit was encountered only on the Port Huron side where deposits typically 3-4 metres thick were found at the surface.

River Bed Alluvium was encountered in the majority of the boreholes drilled in the river. This material is generally a fine to medium grained gravel with some sand but containing a low percentage of fines, evidence of substantial sediment transportation in the fast flowing river. The thickness of Alluvium found ranged up to 2.5 metres in depth but deeper pockets may exist in former scour paths. Attempts were made to map the Alluvium/Till interface by geophysical exploration methods; but these did not give results which could be relied on with confidence.

The St Clair Till, (referred to as clay in Figure 2), the most extensive of the superficial deposits encountered, consists of a predominantly silty clay

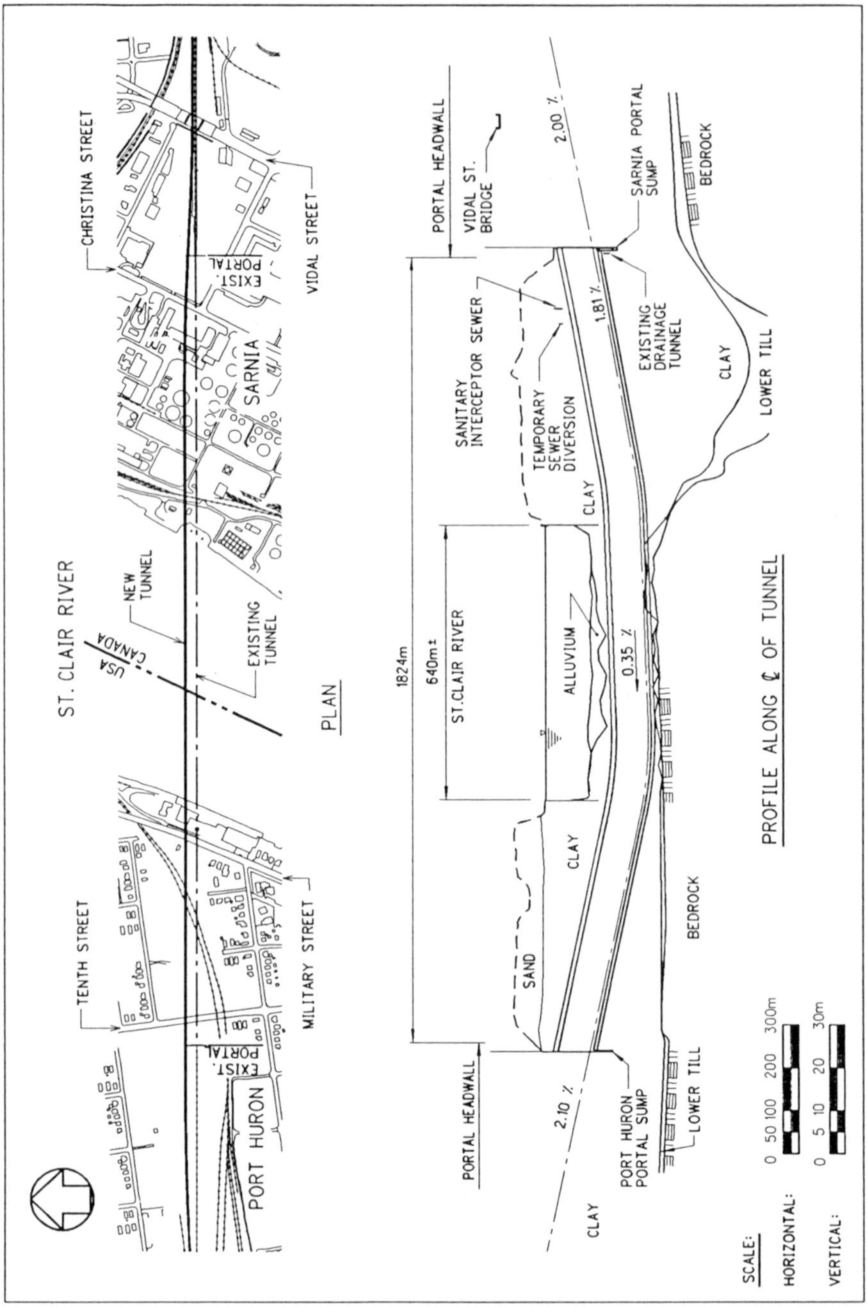

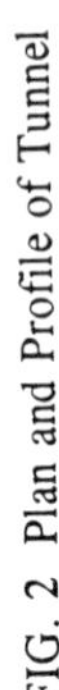

FIG. 2 Plan and Profile of Tunnel

material containing sand, gravel, cobbles and boulders. Consistent with most "till" deposits the St Clair Till includes pockets and lenses of silt, sand and gravel. Boulders are known to be present from local construction experience and are recorded in reports of the construction of the existing tunnel. The lower portion of the St Clair Till is close to normally consolidated, making it unusual among till deposits since it has never been subjected to past over-consolidation from glaciers. The upper portion of the St Clair Till is over-consolidated due to desiccation and weathering. The undrained shear strength of this clay is in the range 40 - 80 kPa. Hydraulic conductivity in the clay ranges between 1×10^{-9} m/s to 4×10^{-11} m/s with calculated ground water velocities less than 9mm per year (Golder Associates 1992).

The Lower Till is an over-consolidated material of extremely variable composition comprising a very dense broadly grained sand and silt mixture containing varying proportions of gravel and clay. This deposit is apparently continuous on the Canadian side of the river but tends to reduce in thickness and become discontinuous towards the west.

The Bedrock consists of a Shale of the Kettle Point Formation which is notable for the presence of strong to very strong limestone concretions appearing in the form of near spherical "kettles" or boulders within the shale. The shale is, typically, moderately strong in the range 25-50 MPa. However, the kettles have unconfined compressive strengths in excess of 100 MPa.

A fresh water aquifer is situated between the lower till and the bedrock. Average hydraulic conductivity in the aquifer is estimated at 5×10^{-6} m/s and groundwater flow within the aquifer has been estimated at 0.57 $metres^3$ per year per unit aquifer width (Intera, 1992). Groundwater pressures in the bedrock are higher than the hydrostatic head from the river by 1 metre equivalent. This indicates that there is a small, but important tendency for water to flow upwards from the lower strata.

3 Railroad Considerations

Approximately 24 trains per day are expected to pass through the new tunnel. The majority of movements will be freight trains up to 2350 metres long, hauled by diesel-electric locomotives. The two daily passenger trains presently in service will be diverted to the new tunnel.

Design operating speeds for the new tunnel will be 100kph for both freight and passenger traffic, whilst the actual speed operating will be about 80kph.

Horizontal curvature of the track and consequently the

tunnel has been limited to a radius of 6985 metres, while vertical curves are derived from CN standards based upon the design operating speed.

Design railroad loading for the tunnel track is Cooper E-100, which consists primarily of a series of 45,000kg axle loads spaced generally 1.5 metres apart. The track will consist of 136lb AREA rails carried on prestressed concrete sleepers set in slag ballast.

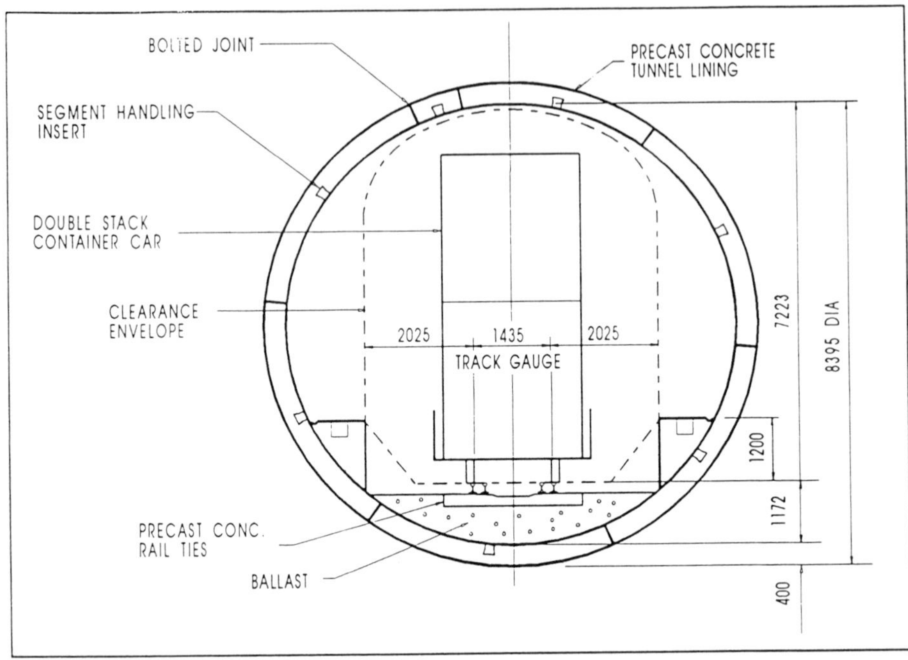

FIG. 3 Tunnel Cross Section

With an internal diameter of 8.395 metres the new tunnel will accommodate double stack container cars where the top of the rail car is 6.27 metres above the rail level. If the railway is ever electrified, the required additional vertical clearance will be obtained by lowering the track and directly fixing rails to a concrete track slab which would be cast in the tunnel invert. Figure 3 which shows a cross section through the tunnel, illustrates the clearances available.

4 Alignment

Several natural and man-made features influenced the selection of tunnel alignment, both horizontally and vertically. In addition, train operating and braking considerations dictated that track gradient should not exceed 2.2%.

The establishment of the horizontal alignment was controlled by the following:

- Alignment of the existing tunnel
- Foundations of Vidal Street vehicular and pedestrian overhead bridges in Sarnia
- Waste disposal and monitoring wells in Sarnia
- Dock structure and sheet piling at the Sarnia shoreline
- Port Huron Dock wall
- Existing track at both the Sarnia and Port Huron tunnel approaches and summits.

Vertical alignment was selected based on consideration of the following:

- Elevations of existing rail track at the two tunnel approaches and track summits
- Location of a major trunk interceptor sewer about 11 metres below grade under Christina Street, Sarnia
- Location and profile of bedrock under the river
- Profile of the river bed
- Existing elevation of 16th Street, in Port Huron and the elevation of the street after it has been lowered the maximum extent possible
- Underside of sheet piles at the Port Huron shoreline

Study of the soil profiles indicated that it would be possible to locate the tunnel alignment within the St Clair Till/Lower Till horizon while still maintaining a minimum of 4 metres cover above the tunnel in the under-river section. The rock profile interpolated from boreholes indicated that it was generally uniform, dipping slightly from east to west. The extrados of the tunnel invert was fixed 900mm above the interpolated rock profile and as a consequence the low point of the tunnel lies close to the USA bank of the river.

Considerations regarding the stability of the cut slopes at the Sarnia approach, together with the desirability of avoiding large scale modifications to the substructure of the busy four lane Vidal Street bridge, dictated that the tunnel be placed as close as possible to the underside of the Christina Street sewer. As a result, the top of the new tunnel passed within about 500mm of this un-reinforced concrete sewer, which handles peak flows of $0.95m^3$/second. To safeguard the sewer, it was temporarily exposed prior to the tunnel drive and externally strengthened to bridge over the settlement zone.

5 Environmental Issues

Environmental considerations played a significant role in the planning and design of the new tunnel, particularly as many of Ontario's petrochemical plants are located in Sarnia, close to the tunnel site.

Between 1958 and 1975, industrial waste from several of these plants was disposed of by deep well injection into the underlying bedrock strata. Subsequently, it is believed, some of these wastes migrated upwards into, and contaminated, the fresh water aquifer which is situated at the bedrock/clay interface. As the tunnel drive will intersect this aquifer under the St Clair River, near the Canadian shoreline, and remain in contact with it for about 200 metres, it was therefore considered necessary to determine the chemical content of the water in the aquifer in order to establish appropriate handling and disposal practices. Water was collected from several boreholes that penetrated the aquifer, and from within the existing tunnel where it is believed that seepage water originates at the aquifer. Chemical testing indicated that water in the aquifer is marginally contaminated with phenols, chlorides and polyaromatic hydrocarbons and that it will require treatment if it is discharged off site. To achieve adequate water quality, a simple carbon treatment system will be installed at the Sarnia portal, to which all tunnel seepage will be directed prior to discharge.

An extensive sampling and chemical testing campaign was also carried out to characterise the soil that will be excavated from both the tunnel and its approaches. As expected, the highly impermeable clay, which will be the predominant excavated material, was found to be uncontaminated when compared to both Michigan and Ontario Standards. Nevertheless, additional sampling and testing is being undertaken during construction to confirm the above findings.

The alluvial river bed sediments above the tunnel also were found to be uncontaminated. This is not surprising given the high current velocity (3.5km/hr average) and high flow volumes (5,100m^3/sec average) in the St Clair River, and the fact that the nearby petrochemical plants are all downstream of the tunnel site.

6 Tunnel Linings

Consideration which influenced the choice of the tunnel lining system included the following:

- The lining must be capable of carrying all the imposed loads, both the long term external pressures and operating loads, and probable construction loads, particularly those from the TBM

- Buoyancy: cast iron and steel segments would require ballasting when tunnelling under the river; concrete segments would not require ballasting
- It must be durable, providing at least the specified 100 year design life of the tunnel, and capable of being made watertight
- It must allow for the rapid rates of advance possible with modern TBMs

After evaluation of available options, a single pass precast concrete segmental lining was selected, with gaskets between the segments to provide the primary seal against water inflow.

Following initial design calculations and discussions with the TBM supplier and representatives of the pre-qualified contractors it was agreed that the lining would be 1.5 metres wide and 400mm thick, and would consist of six plates plus a wedge key per ring. The standard segment weighs 7.25 tonnes, well within the capability of modern hydraulic erectors. The linings will be tapered to provide for alignment adjustment.

In determining the thickness of the segments, both the TBM thrust load (6,000 tonnes) and the extremes of ground and water loading were considered. These effects were found to result in similar stresses. Additionally, live loading resulting from the railway traffic was taken into account, particularly the extremely high axle and dynamic loads from the heavy freight locomotives. Finite element analyses of the ground/lining interaction were carried out using a specially developed computer program which modelled the stages of construction and different ground strata expected.

The high concentrated loads at the radial and circumferential joints result in splitting forces within the concrete which are resisted by welded steel reinforcement mats made in a semi-automated welding plant and formed into cages for segment production.

Bolts, required across the joints in order to erect the lining, are set in pockets formed along the edges of the segments. These pockets have the effect of reducing the lining cross section locally. In order to minimise this loss, sloping bolts, have been provided with cast in sockets on both the radial and circumferential joints - see Figure 4.

Groundwater chloride (4,000 ppm) and sulphate (155 ppm) levels combined with a hydrostatic head of up to 35 metre, dictated that dense high quality concrete be specified for the linings. The concrete specification calls for a minimum strength of 60N/mm^2; but in addition, a low permeability to chloride ions was specified which was monitored by an ionic diffusion test developed by the designers. These requirements demanded much of the segment manufacturers, (Pre Con-Sehulster J.V.), both in the design of the concrete and in the casting and curing

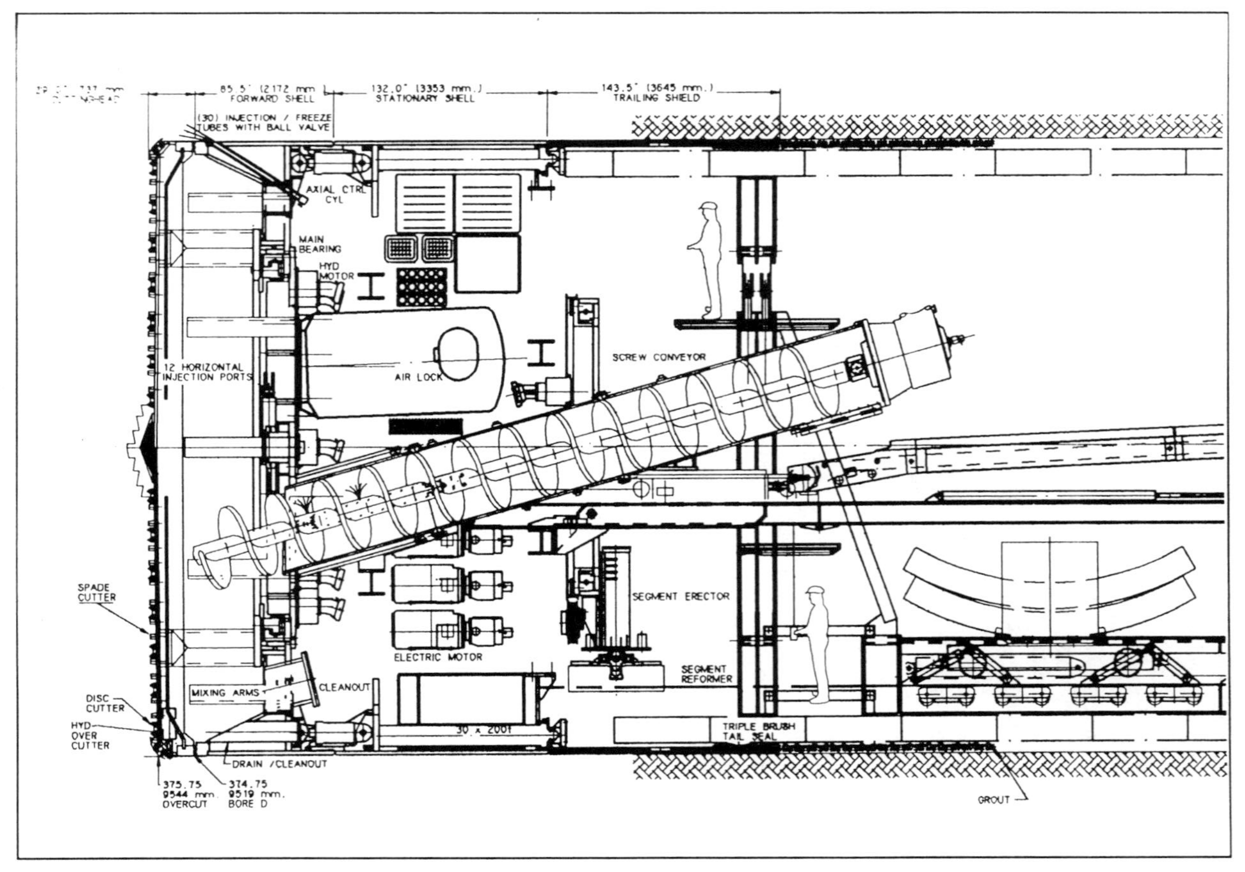

FIG. 4 Section Through Lovat ME 375SE TBM

of the segments.

The reinforcement is epoxy coated using a fluidised bed technique. With this method, a complete reinforcing cage is pre-heated then dipped into a tank full of epoxy powder which is fluidised by air jets. The powder fuses to the preheated steel resulting in a uniform protective layer. While this technique has been used on similar projects in Europe this is believed to be the first occasion that it has been used for coating reinforcement cages in North America.

In order to provide the required dry environment for railway operating equipment, the lining joints are sealed using synthetic rubber gaskets located in grooves cast in the segments. In the event that this does not form a perfect seal, grooves are provided on the front (intrados) face of all of the joints for a caulking medium to be installed.

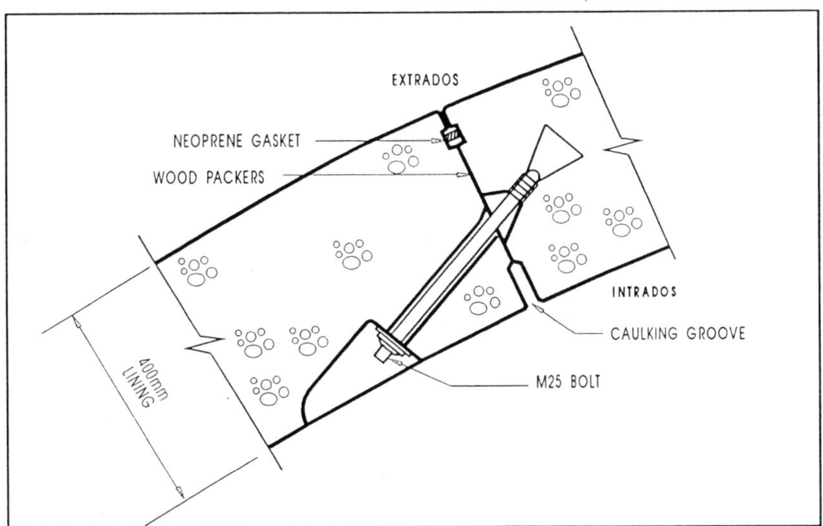

FIG. 5 Radial Joint Detail

7 Portal Structures

Key design considerations for the portal structures, particularly the one in Sarnia from where the TBM was launched, included:

- Adequate and suitable access for the installation of the TBM and associated plant and subsequently, in Port Huron, for its removal.
- Ability to withstand high loads such as the TBM thrust loads and design earth pressures of up to 150 kPa. In addition to the geotechnical investigations past slope failures near the portals were examined

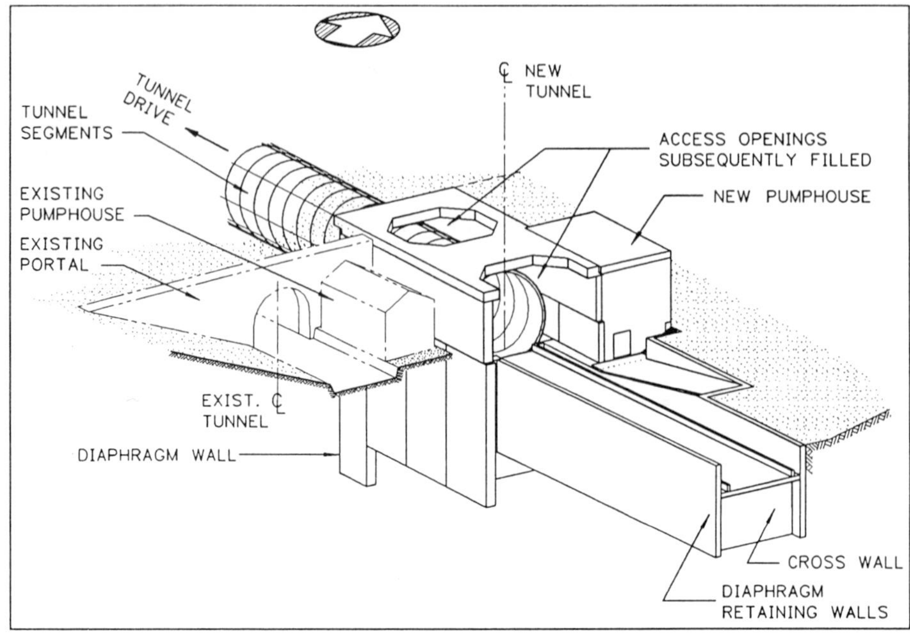

FIG. 6 Sarnia Portal During Construction

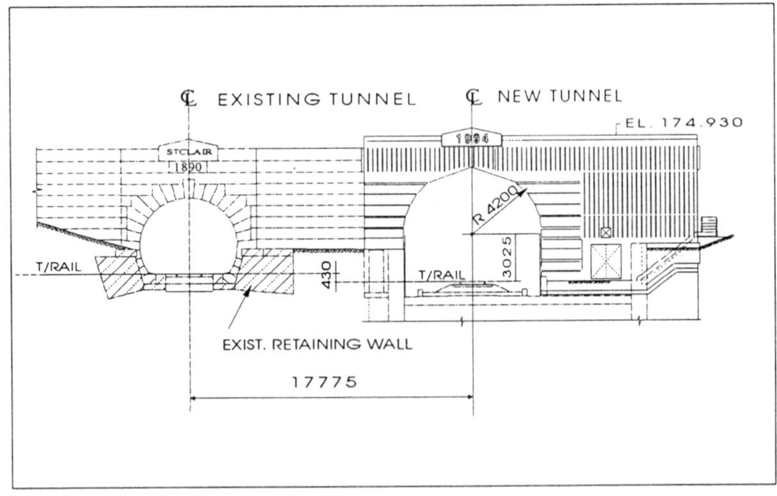

FIG. 7 Sarnia Portal

to determine appropriate loading

- Ability to accommodate the ground heave expected as a result of overburden removal due to the portal excavation.
- Desirability of both minimising physical impact and of being visually compatible with the existing portal structure.

The portal structure contained in the Tender documents consisted of a series of interlocking 1.2 metres diameter secant piles. Following award of the tunnel contract to Traylor and Associates, the Sarnia portal was modified to reflect the diaphragm wall construction preferred by the contractor. Key elements of the Sarnia portal, during construction are shown in Fig. 6. When complete, the structure will be fully faced with precast concrete panels as shown in Figure 7.

Concern about open cut slope stability was not confined to the portal area and as a result, a 125 metres long diaphragm wall retaining structure, as shown in Figure 6, was constructed beyond the portal to retain the side slopes. The one metre thick reinforced concrete walls were constructed using the bentonite slurry trench method and are strutted by a continuous horizontal slab about 3 metres below top of wall and by un-reinforced concrete cross walls every 5 metres. The slurry walls are about 9 metres deep and required temporary strutting at the top until the horizontal slab was placed.

8 Contract Documentation

In preparing the contract documentation, the Owner and the Engineer acknowledged that, despite carrying out extensive geotechnical investigations, it was impossible to foresee all of the conditions that might be encountered during the tunnel drive. Accordingly, it was decided to embrace fully the recommendations for "Avoiding and Resolving Disputes During Construction" prepared by the Underground Technology Research Council (ASCE, 1991).

A Disputes Review Board was set up with the powers to resolve major problems. This Board comprises three members, the owner and contractor each nominating one member and an independent chairman being selected by the first two members.

The Contract documents included provisions for all documentation of the successful bidder's estimate to be placed in escrow, whilst still remaining the property of the contractor. These documents are then accessible, jointly by the owner and the contractor, to assist in resolving any pricing issues on claims or disputes.

Geotechnical information was provided in a Geotechnical Design Summary Report (GDSR). The GDSR outlined the findings of the geotechnical subsurface investigations and discussed the influence which the anticipated subsurface conditions would have on tunnel construction. This document will form the baseline for identification of varied ground conditions. Areas of potential difficulty were identified and carefully reviewed with the aim of determining an equitable sharing of risk between the owner and the contractor. To share this risk, unit prices were included in the tender documents for potential work items which could not be quantified. These items include the removal of various types of obstructions and possible disposal of unknown quantities of contaminated material, should this be encountered.

Additionally, Partnering between the owner, engineer and contractor has been initiated in a further effort to create an environment of trust and teamwork to prevent disputes and facilitate a successful project for all partners.

9 Tunnel Boring Machine

The tunnel will be largely driven through the St Clair Till, which due to its low shear strength, will act as a "soft squeezing" or "rapidly squeezing" clay and which will require to be positively supported at all times. The existing tunnel was driven using two hydraulically powered shields working from either end of the tunnel. During tunnelling in free air, under the land, it is documented that as much as 50% ground loss occurred. Tunnelling under the river was completed successfully by applying up to two atmospheres of compressed air.

The development, over the last decade, of TBM's capable of providing positive support at the face should permit the new St Clair tunnel to be driven without the need to employ compressed air, except when it becomes necessary to enter the Plenum Chamber for routine maintenance or possibly for the removal of obstructions. The feasibility study identified that an Earth Pressure Balance Machine (EPBM) would be the most appropriate type of machine. The use of a Slurry machine was considered but rejected due to the problems with muck handling and separation, particularly if contaminated ground were to be encountered.

In order to fast track the project, CN placed an order for the TBM at the end of the feasibility study, long before the award of the tunnel contract. This was made possible by forming a TBM Technical Review Committee, including representatives of seven pre-qualified world class tunnel construction consortia, which advised CN and agreed upon TBM selection. Only those construction companies which contributed to this TBM Committee were permitted to bid for the tunnelling work. Following an international tender call, the TBM supply contract was awarded to Lovat Tunnel Equipment Inc. of Toronto, Canada.

The design of the machine is based upon that of previous Lovat TBM's with due cognisance of the scale of the project and the anticipated ground conditions. The TBM layout is a conventional triple shell cylindrical construction comprising a steerable forward shell which houses the cutter head, main bearing and drives, a central stationary shell which houses the operators cabin, power units and main thrust jacks, and a rear tail shell in which the permanent tunnel lining is constructed. The pressure bulkhead forms the rear face of the forward shell upon which is mounted the main bearing and the screw conveyor.

The cutter head is attached to the inner ring of the main bearing, the outer ring being fixed to the bulkhead. The cutter head comprises six reinforced spokes in which are fitted the cutter mountings capable of being equipped with either drag bits or roller discs. The edges of the spokes are fitted with spade teeth. The cutterhead is designed for operation in either direction with a variable speed drive, and spoil is fed into the openings between the spokes. Lovat's patented flood doors located between and integral with the spokes provide for sealing of the chamber in an emergency, and when the machine is stationary.

Lengthy discussions took place during the design development phase as to the need to make provision for use of disc cutters. It was considered extremely likely that a number of granitic boulders, and possibly intrusions of shale bedrock, would be encountered during the tunnel drive and it was necessary to develop appropriate, safe methods of overcoming such eventualities. If only drag bits were to be provided then there was a very strong possibility that on several occasions during the drive it would be necessary to stop the tunnel drive and evacuate the plenum chamber to enable man entry into the plenum chamber to be effected in order to remove or break up potential obstructions. Thus it was considered prudent to make provision for the installation of disc cutters in order to be able to, at a minimum, break any boulder encountered into smaller, more manageable pieces.

As originally envisaged the face plating was flush with the front of the spokes; and with the arms being widened to incorporate the disc housings, the area of the flood door openings was reduced to 27% of the face area. The TBM Technical Review Committee considered this to be insufficient, arguing that such a restriction on muck flow would lead to excessive cutter head torque demand and reduced forward penetration rates. Subsequent design development found that by moving the spokes 200mm forward of the face plating it was possible to increase the flood door opening to 32% of the face area.

The main bearing is a large diameter triple roller annular type assembly, lubricated continually by a pressurized oil system. The outer sealing of the bearing is effected by a combination of mechanical seals and double lip seals, and an automatic greasing system is installed, with grease continuously pumped out at pressure to prevent ingress of material from the plenum chamber.

Within the plenum chamber there is facility to inject water, bentonite or other additives to condition the spoil. Substantial mixing arms are provided to ensure that the spoil is adequately mixed to an optimum plasticity before entering the screw conveyor. The plenum chamber pressure is monitored by 8 pressure sensors and 2 pore water pressure gauges with an additional 2 sensors on the screw conveyor.

The screw conveyor is 1200mm diameter x 12.20 metres long, capable of handling boulders or rock fragments up to 450mm in size. The screw helix is end driven by a variable speed, reversible 225kw hydraulic motor. The screw can be retracted for maintenance purposes and then sealed from the plenum chamber by a pair of caliper gates. At the rear end a sliding guillotine gate controls the muck discharge onto a 1200mm wide belt conveyor.

The tunnel lining segments are delivered to the trailing gantries by rail cars. On arrival at the gantry the segments are rotated by 180° and then transported forward to the erector by an hydraulically operated trolley system. Segment erection is carried out within the protection of the rear tail shell using an erector mounted on a rotational ring. The erector is fitted with a rotating clamping device which locates into a ductile iron lifting pot cast in at the centre of gravity of the segment.

In order to minimise the potential for settlement due to ground movement at the tail of the TBM, grout is injected directly into the tail void as it is formed. Grout pipes, grease lines and accelerator pipes (arranged in an array of thirteen locations) have been incorporated into the tail skin of the machine. The grout is injected by 3 Schwing grout pumps each feeding a pair of grout pipes. The rate of flow and pressure of grout injected is constantly recorded, with the rate of injection being automatically coupled to the TBM advance to ensure complete filling of the annular void.

In the event that access is required to the Plenum Chamber, provision is made for compressed air to be applied to the Chamber. Man entry would then be effected via a 2-chamber, 3-man air lock which is permanently fixed to the bulkhead. In order to empty the Plenum

Chamber, in the event of a problem with the main screw conveyor, an auxiliary 400mm diameter screw conveyor has been provided which can be fitted to the bulkhead at a low level.

Equipment for probing ahead is installed within the TBM enabling, as necessary, cased or uncased holes to be drilled ahead of the machine to prove the geology or check for water or gas inflows. Provision has also been made to install grouting tubes should it be necessary to consolidate ahead of the tunnel face.

Technical data and key features of the Lovat ME-375SE series 12500 being constructed for this project are provided below.

- Bore Diameter - 9.52 metres Nominal
- Shield Length - 10.25 metres
- o/a TBM Length including back-up - 105 metres
- Shield Weight - 524 tonnes
- Cutting Head Speed - Minimum 1.5 rpm, Maximum 2.5 rpm
- Cutting Head Power - 1800 kW
- Thrust Cylinders - 30 @ 200 tonnes
- Thrust Cylinder Stroke - 2.25 metres
- Segment Erector Capacity - 7.3 tonnes

10 Construction

Delays in obtaining approval for the project from the US Army Corps of Engineers and the Canadian National Transportation Agency meant that construction work did not start until March 1993.

In order to fast track the project, construction was procured through a number of separate contracts, as follows:

- Supply of Tunnel Lining
- Sarnia Approach Contract
- Port Huron Approach and Portal
- Tunnelling and Sarnia Portal
- Tunnel Enabling Works
- Mechanical and Electrical Services

The above split also allowed the work to be contracted to several Canadian and US firms.

The tunnelling contract was awarded on 1 March 1993 to Traylor Associates, a consortium of Traylor Bros. Inc., The Foundation Company of Canada, Frontier-Kemper Constructors, Inc. and Wayss & Freytag AG. Key dates in this contract are as follows:

- Contract Award March 1, 1993
- Sarnia Portal Construction Completion July 1993
- Install TBM August - September 1993
- Start tunnelling October 1993
- Complete tunnelling (Projected) July 1994

The tunnelling will be carried out on a 24 hr/day, 7 days

per week basis under the sensitive structures and river sections.

11 Progress to Date (at January 1, 1994)

The Sarnia Portal Structure and Approach have been completed as are the Tunnel Enabling Works. The Port Huron Approach and Portal Structure are 50% complete. 252 metres of tunnel has been constructed (total length of 1824 metres). Tunnelling under the sensitive Imperial Oil Research facility has been successfully completed. An extensive compensation grouting installation was provided for settlement protection on this important building. However in the event only a limited quantity of grout (5000 litres) had to be injected to compensate for the effects of tunnelling.

12 Conclusion

The new St Clair River Tunnel represents a major technical challenge which potentially is one of the more difficult tunnelling projects to be carried out in North America in recent years. The selected TBM reflects the "State of the Art" in soft ground tunnelling machines and is one of the largest currently operating in the world. The successful completion of this tunnel project will signify a major step forward in tunnel construction techniques, comparable to the achievement of completing the original tunnel 103 years ago.

Application of NATM to design of underground stations in London Clay

A. P. Deane
Mott MacDonald, Ltd., Croydon, England

Abstract
Three of the underground railway schemes currently under design or construction in London are using NATM for the station tunnels. The paper reviews the reasons for this change from previous practice which was largely based on segmental tunnelling and examines the advantages of the system with reference to particular structures in the three schemes. It discusses design methods for NATM, possible settlement effects and concludes with a few points to be considered when using NATM.
Keywords: NATM, London Clay, stations, design

1 Introduction

The demand for better travel facilities in cities has led to a significant increase in the interest in underground rail systems around the world in the past five years. London is no exception to this trend. After a hiatus following the construction of the first part of the Jubilee Line there are now a number of schemes for tunnelled rail facilities for the capital in various stages of design and construction.

London already has one of the most extensive underground systems in the world. There are many reasons for this but one of the most important is the ground on which the city is built. Most of the central area north of the Thames is constructed over a thick layer of blue/brown over- consolidated alluvium known as London Clay. This is an extremely uniform material with very low permeability, reasonable cohesive strength and few major discontinuities. These parameters come together to provide a near ideal material for soft ground tunnelling that can be excavated by roadheader, back-actor or hand-held pneumatic tools; it has a good stand-up time, can be cut to give a uniform smooth excavation and in most locations the low permeability allows very dry tunnelling.

However providing new rail lines to the existing transport systems is not enough. In London, as elsewhere, operators are demanding more extensive spaces underground. Some projects are using the larger stock and longer trains more normally associated with standard suburban railway systems. These trains produce much higher throughput of passengers along platforms and up and down escalators. Even where London Underground size stock is to be used modern safety requirements and the peak passenger flows, which often extend for several hours, combine with users expectations to produce demands for more space.

Ventilation tunnel sizes have increased to cope with more stringent regulations and travellers requirements for greater comfort. Particularly within Central London, these passages are often required to be constructed in complicated shapes to avoid the existing underground infrastructure and to utilise small and difficult sites to vent to atmosphere.

The vast majority of tunnels built in London have segmental linings of concrete or cast iron but, while highly suited to long tunnels of constant diameter, precast segments are not an ideal solution to the requirements of a modern underground railway station. Complex shapes and wide openings are extremely disruptive to the segmental system resulting in expensive tunnels.

The alternative, which is now planned or under construction for a number of stations on different underground rail schemes in London, is to use a shotcrete lining over mesh and steel arches as the primary support followed by a second permanent lining, usually of insitu concrete. This method of construction is often referred to as the New Austrian Tunnelling Method or NATM, though it is generally accepted that the term NATM implies much more than simply the use of shotcrete for tunnel lining.

Shotcrete has many advantages as a ground support. It can be applied quickly, immediately seals the face against which it is applied to slow desiccation or leaching, can be used in almost any shape of tunnel and, although it has a long term modulus approaching that of concrete, it has a relatively low stiffness at early age. This allows some deformation of the ground around an excavation and assists in generating the composite ground/lining support system which is a central part of the NATM philosophy. This is in contrast to segmental lining which is usually designed to carry significant external load as an isolated structural unit and only relies on the surrounding ground to maintain its stability.

2 NATM Stations in London Clay

The remainder of this paper will review the use of NATM in three specific projects, all urban underground railway schemes to be constructed primarily in London Clay. These are the Heathrow Express rail link, the Jubilee Line extension and the CrossRail project. It will examine the advantages and limitations of the NATM system, appropriate methods of design and how they interact with the construction method.

The first example of the use of the technique is on the Heathrow Express rail link, specifically the main stations and ancillary structures which are currently being constructed in NATM under Heathrow Airport.The initial planning and station layouts for Heathrow Express were based on a conventional segmental approach but during the detail design it became clear that a scheme based on a shotcrete primary lining and insitu concrete permanent support had significant economic advantages over the segmental design.

There are a number of reasons for these advantages and since the majority of them apply to all three projects they are described in some detail.

Firstly the NATM approach overcomes one of the major problems inherent in using segments in a station in that it provides the flexibility to change the shape of the tunnel to suit the users requirements, rather than enclosing everything in a circle. Typically station platform tunnels need greater width than height and, since these large diameter tunnels often represent a significant proportion of the cost of the station, any saving made by using a more cost effective shape can be substantial. Figure 1 shows a typical cross-section of the Heathrow Express platform tunnel.

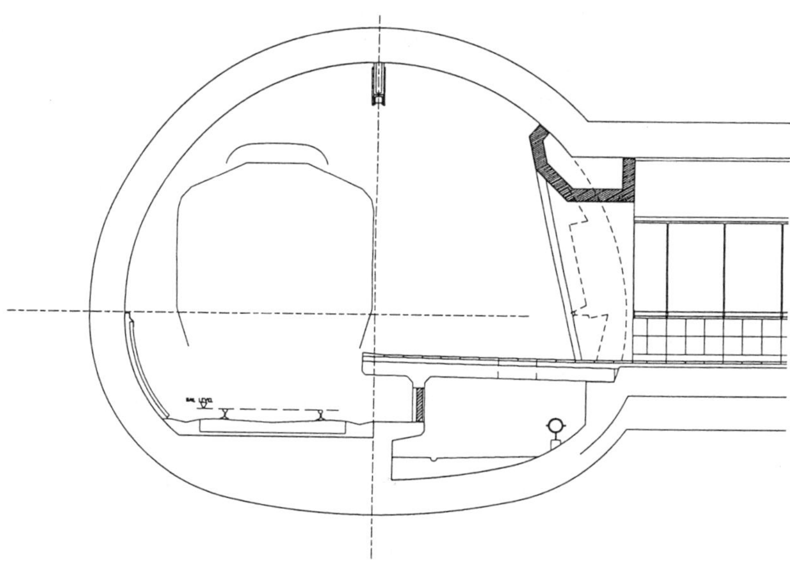

FIGURE 1 - HEATHROW EXPRESS PLATFORM TUNNEL

It is also easier to construct large openings in shotcrete lined tunnels. Despite the bolts between the rings, segmental linings are generally designed to act as a series of individual short cylinders so that if some rings are broken to construct an opening, a substantial structure, usually a heavy steelwork frame, is required to distribute the load from these rings around the opening. The shotcrete and insitu lining of a NATM tunnel acts as a three-dimensional shell structure rather than a series of two-dimensional units allowing load to be transferred in shear to the complete cylinder either side of an opening and reducing the size of structure needed around the junction. Typical openings into the concourse at Heathrow are four metres wide which is significantly larger than the majority of openings on the existing LUL network.

Stations frequently consist of many different sized tunnels, all relatively short, which are not suited to segments. The time taken to build shield chambers, and erect and dismantle shields may equal or exceed the driving time. Advance rates on short shield-driven tunnels are often not much greater than those using shotcrete lining because of the learning curve required to get to full production rates in segmental construction. Even allowing for the subsequent permanent lining, which may or may not be off the critical path, the programme for complicated groups of tunnels can usually be significantly shortened using NATM.

The programme advantage is further improved by the short mobilisation time required for the system. Most of the excavation and shotcrete plant is standard rather than job specific and productive work can start much more quickly than with a segmental tunnel, particularly if a shield is required.

The NATM system is therefore inherently flexible allowing the numerous requirements of a station to be addressed. As an example of the complexity of modern systems. the CrossRail station at Tottenham Court Road referred to later has, in addition to two exits to new ticket halls, connections to the existing LUL Central and Northern Lines and a provision for the future Chelsea-Hackney line. This results in 27 different main tunnel cross-sections and 109 tunnel junctions. Variety of this type would be totally uneconomic in segments.

At Heathrow, one specific example of the flexibility of the NATM system are the cross-overs to be constructed south of the central terminal area. These will allow an extension of the system to Terminal 5 to be made at a future date. In traditional tunnelling these would have been step plates utilising several different diameters of segmental lining, probably in cast iron. The tunnels would have been constructed by hand, taken a considerable time and, unless great care was taken with the temporary ground support, would have generated significant settlement at the ground surface.

The solution adopted is a continuously varying NATM cross section with a primary shotcrete lining followed with a permanent lining of insitu concrete. The excavated space is reduced to only that which the operator requires and a potential bottle neck in the construction programme is minimised (Figure 2).

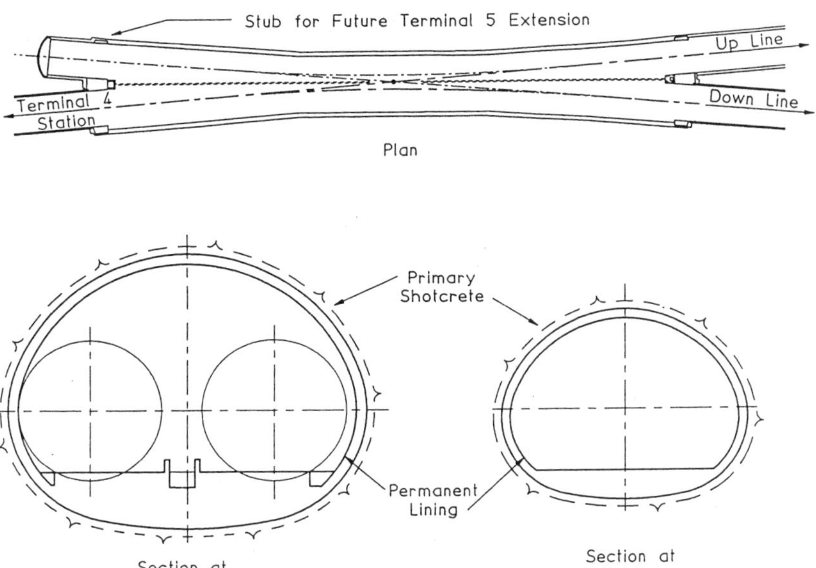

FIGURE 2 :- HEATHROW EXPRESS CROSSOVER TUNNEL

For the Jubilee Line Extension the planning was also done for segmental station tunnels and this was carried through to the detail design, although the client realised that alternative NATM solutions could be more economic and these were encouraged in the tender documents. In the event, for most of the tunnelled stations the NATM approach was found to be up to 15% cheaper and for the main stations nearest to central London NATM alternatives to the Engineers' designs have been accepted.

During a limited tender period it is impossible to make major changes to a station layout. Circular tunnels can be replaced by non-circular NATM structures which satisfy the individual space requirements of specific cross sections but the implicit decisions made during the station planning, which could have been driven by the limitations of segmental lining system, cannot be modified without the background knowledge that has been accumulated over the outline and detail design phases by the station designer. Although there is some saving in excavation, this is relatively small. The heavy junction steelwork required by segments can be reduced and the time-saving on the construction programme described above may be achieved. However, the alternative design does not take advantage of the flexibility of NATM that allows structures which cannot normally be considered when segmental construction is to be used.

By accepting an alternative NATM design the client is obtaining a more economic solution but the full advantage of the system can only be fully realised by designing the station from the ground down for NATM construction.

The CrossRail project aims to realise the potential advantages of NATM more effectively. As with the previous schemes the outline planning for the stations was carried out using a segmental lining approach but because the programme is some few years behind that of Heathrow Express and the Jubilee Line Extension, more familiarity with the NATM approach existed in the U.K. so that at the start of detail design the decision was taken to review the station layouts assuming that NATM would be used. This has allowed the architects considerably more freedom to produce interesting and efficient spaces underground. Two examples from the extension to Tottenham Court Road underground station for CrossRail are the platform tunnels which have been designed as an egg-shaped cross-section to provide a large space over the platform area but not duplicating this over the trains (Figure 3) and a beehive shaped circulation area where five different tunnels meet (Figure 4). Although these structures could theoretically be built by other means NATM is the only practical and economic way of producing the shapes that the architect required.

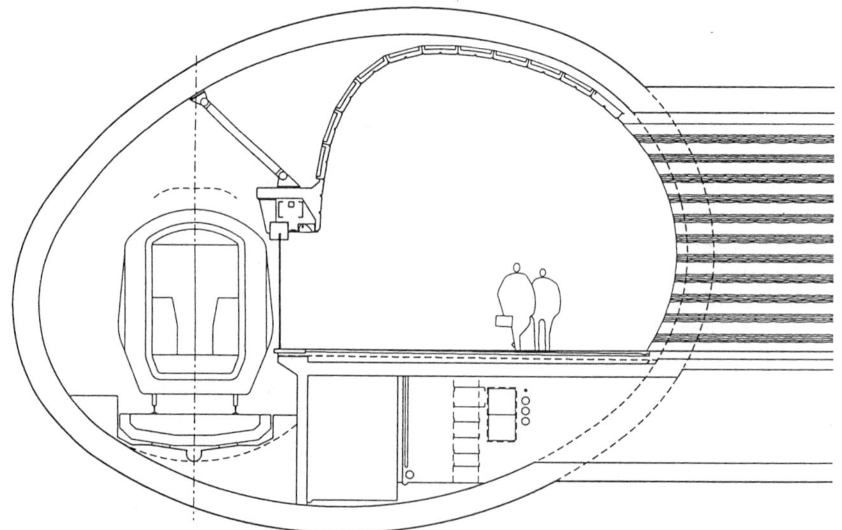

FIGURE 3 - CROSSRAIL PLATFORM TUNNEL

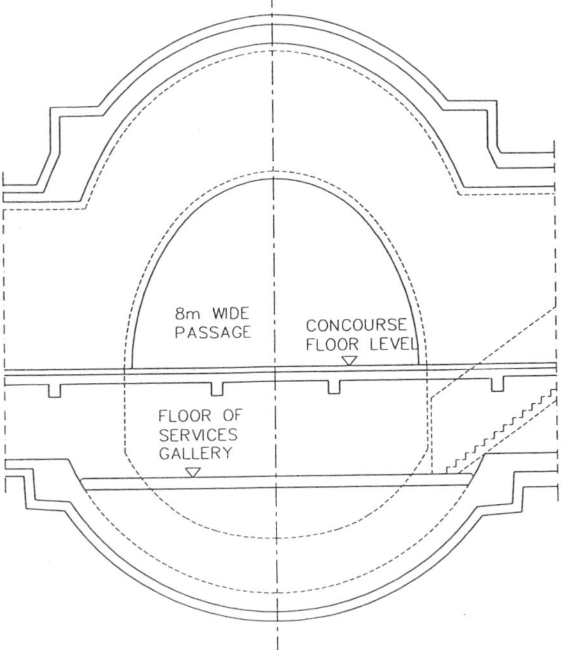

FIGURE 4 – CROSSRAIL 12m DIAMETER CIRCULATION HALL

3 NATM Design

From the large number of tunnels already constructed in London, a body of information exists both about the clay and also about the behaviour of tunnels constructed within it. Clearly the data on existing tunnels should be of great value, but in practice it is difficult to interpret directly for NATM design.

The two principal pieces of information that a designer requires from previous tunnels are the loads in the lining and the movements of the lining. Ground stresses and ground displacements are interesting and help to build an overall design methodology but the designer's first priority is to provide a stable, economic tunnel and this relates directly to the lining system.When viewed in this light the information available from the existing tunnels in London Clay is somewhat limited. Displacements of the tunnel lining have seldom been measured except where they have been observed to be exceptionally large. Where they have been measured they do not agree well with the measured ground parameters.

Most London Clay is over consolidated with the horizontal insitu stress between 1.5 and 3 times the vertical stress. This is a consistent figure measured over many site investigations but there is virtually no evidence of this stress ratio in the existing tunnels. The vast majority of these have shown a shortening of the vertical diameter which would suggest, if anything, K_0 figures of less than 1. There are only a limited number of measurements of stresses in existing linings. As is to be expected they show considerable scatter but the general trend is that between 40% to 60% of the overburden load is carried in the lining 28 days after installation and that there is a further slow increase in this load (Reference 1). The few records that exist suggest that after several years the load in the lining is a very high proportion of full overburden. For urban tunnels this apparently high loading has not been a major design problem since precast linings of minimum practical thicknesses will carry full overburden loads at depths up to about 30 metres, where most of the metro tunnels in London have previously been constructed.

A similar situation arises with NATM tunnels. For near circular tunnels the design of NATM in London Clay is relatively straightforward.As the tunnel profile becomes less circular then the loading applied by and particularly the restraint offered by the clay become more significant.

The method chosen to model the ground has a major effect on the design.Analysis based on "full overburden" loads, as used for many of the segmental tunnel designs, together with lower bound modulus and cohesion figures for the clay, results in high bending stresses being calculated in the lining which in turn dictate relatively high percentages of reinforcement.

Similarly if this loading is used to design the large openings in the tunnels now being demanded by operators and users the resultant stress concentrations near the opening require complicated reinforcement in the insitu lining.

To produce a more economic design it is necessary to develop a new methodology which takes account of both the known behaviour of the clay and the high stresses measured in some segmental tunnels under London.Until more empirical data is available from NATM tunnels in the clay any new design approach also needs to be relatively conservative.

A number of models have been used for the design of prismatic tunnels in NATM. These are often finite element or finite difference calculations, where the three-dimensional nature of the tunnel construction is represented by a modification of the known parameters in two dimensions. Methods used include assuming a different (hypothetical) modulus for the shotcrete or by reducing the modulus of the ground within the tunnel.

These approaches are based on back-analysis of available data of previous NATM tunnels, which, as noted elsewhere, is in short supply for London Clay. With the rapid reduction in the cost of computing power, it has been possible to investigate non-linear design approaches which, even five years ago, would have been the province of major research establishments.

A method which has been developed by Mott MacDonald is to use the model of London Clay as a strain softening material (Reference 2).

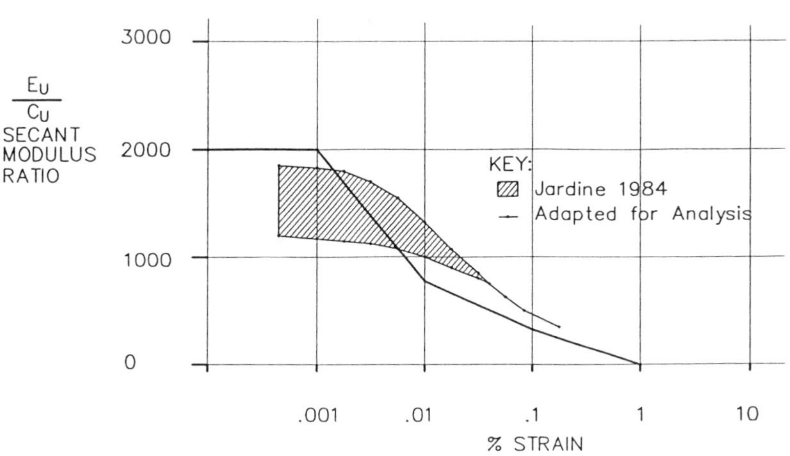

FIGURE 5 NON-LINEAR SOIL STIFFNESS

Figure 5 shows a typical relationship between modulus and strain derived from small test samples of the clay. The tunnel lining is modelled within this non-linear continuum using two and three dimensional finite element analysis.The concrete strength and any reinforcement required is then calculated in a conventional manner using the stresses computed in the analysis.

The method is currently used with relatively high load and material factors to allow for the substantial unknowns involved but as more information is obtained from the NATM schemes now under construction it will be possible to carry out a more rigorous stochastic analysis of the data and to refine the load factors and model to be used in future designs.

4 Settlement

Settlement is one of the major considerations to be taken into account when constructing tunnels in urban areas. Modern underground stations consist of larger and more numerous tunnels than those constructed previously. This exacerbates the problems of settlement with its potential to damage buildings and buried services.

From the large amount of data available concerning ground movements above tunnels it has been possible to establish empirical formulae to predict settlement caused by a segmental tunnel .Claims have been made for NATM that it results in less settlement than segmental tunnelling. Most of the available evidence, however, is for tunnels constructed in ground substantially less plastic than London Clay. Where NATM has been utilised in soft ground, such as in Frankfurt, the settlement measurements can be difficult to interpret because of the concurrent dewatering which causes its own ground movements.

At the time of writing the best directly applicable NATM data is for the Heathrow Express trial tunnel, but it is to be hoped that more settlement information will become available during 1994 with the construction of the main Heathrow Express works and the Jubilee Line Extension. The trial tunnel and its results are fully discussed in a separate paper at this conference but in summary it was found that, with conservative construction techniques, the resulting settlements were similar to or slightly less than those that might be expected by a good standard of conventional tunnelling.

The case for substantially reduced settlements with NATM is, therefore, not proven in London Clay but the system, if constructed properly, is probably no worse than segmental tunnelling, and for locations where large diameter hand tunnelling might be required such as step plates, the NATM approach will almost certainly cause less ground movement than the conventional approach.

5 Flexibility

NATM not only allows the designer a flexibility in form but it also presents the possibility of change during the construction process. In some circumstances the ability to modify the excavation and lining system as construction advances, usually as a result of the observed deformations and stresses of the constructed section of tunnel, can be seen as one of the primary advantages of the system.

Whether this flexibility is a major advantage for urban underground stations is less certain. Except for the platform tunnels, most drives in stations are short and have frequent disruptions, such as openings, along their length. Any variation to the construction method-for example longer individual advances-could result in changes to the ground movements around the tunnel and therefore the settlement above. Until a data base of experience is built up about the effects of various changes in the method it is likely that NATM tunnels will be constructed without major modifications to the original design.

6 Points for Special Attention

Even with all its advantages there are some points which require careful consideration if NATM is to be used successfully. The system is not a total panacea for urban tunnelling.

As noted above soft ground NATM requires sophisticated design techniques to provide economic solutions, particularly at junctions. Possibly more importantly it requires considerably more technical input and supervision during construction than would be the case with segments. The flexibility of the method means that there are more opportunities for errors during construction which can cause major problems.

Significant monitoring is usually installed to confirm the acceptable behaviour of the tunnel during construction. This, together with procedures necessary to implement the system, can require a different contractual emphasis from that normally found in the U.K.

It is also necessary to consider the type of tunnel to be constructed. A segmental tunnel, although it requires a major start-up investment in time and money, may under the right circumstances be advanced more quickly than the two pass NATM system. For the foreseeable future it is unlikely that running tunnels, which are usually a constant cross-section with few interruptions such as junctions, will be driven using NATM but for station tunnels with their multiple openings the system would appear to be well suited to London Clay.

7 Conclusion

The current schemes for Heathrow Express, Jubilee Line Extension and CrossRail will demonstrate NATM's viability and economy even more clearly and the experience gained should allow NATM techniques to be used for much of the future tunnelling under the capital and assist in designing tunnels for other cities requiring metro systems in poorer ground than London Clay.

8 References

1. Barrat and Tyler (1975) "Measurements of Lining Behaviour on the London Underground at Regents Park" TRRL - Report 684.

2. Jardine et al (1986) "Non-linear Stress-Strain Characteristics in Soil-Structure Interaction "- Geotechnique Vol. 36.

Session 3: Water/Waste

Underground works on Third Nairobi Water Supply Project

K. T. Burgess
M. C. Knights
Howard Humphreys and Partners, Ltd., Leatherhead, England

ABSTRACT

This paper describes the Third Nairobi Water Supply Project and concentrates on the underground works on the two contracts (C207 Transmission Tunnels; C208 Dam) which were based at N'Dakaini in the foothills of the Aberdare Mountains, some 80 km by road from Nairobi, Kenya.

The Project involves the damming of the Thika River at N'Dakaini, the passing of water through the Thika Intake, along the 4.2km Transmission Tunnels to the Chania River. Increasing the capacity of the Mwagu Intake to supply an enlarged Ngethu treatment works. After treatment the water would be piped over 40km to holding reservoirs in the outskirts of Nairobi. The water will then be passed into an improved distribution network on to the inhabitants of Nairobi.

Keywords: Underground Works, Water Supply, Kenya, Dam

1. Introduction

Nairobi is the capital city of the Republic of Kenya, which is a World Bank member country and is being assisted by the Bank with development of its water supplies.

Future water demands have been estimated from past experience and projected growth as follows:

Year	1984	1995	2010
Population	983 000	1 853 000	3 667 000
Pre-capita Domestic Demand (lcd)	72	90	103
Demands (Mld)			
Domestic	71	168	378
Commercial/Public	32	65	156
Industrial	9	29	64
Total	112	262	598
UAW (as percentage of water put into the supply)	41	20	20
Total Water Supplied (Mld)	190	327	747

The Third Nairobi Water Supply Project constitutes a further stage in the development of the north-western rivers, turning this time to the Thika River which is being dammed at N'Dakaini, 40km upstream of Thika Town. There is potential for further expansion of Nairobi supply by tapping yet more rivers north of the Thika, but such schemes will not be needed until the 21st century.

At the turn of the century poor quality water was drawn from the Nairobi river in the Athi catchment, but this source was later abandoned in favour of the development of the Kikuyu Springs which served the city until 1938. Water has since been drawn from the Ruiru, Sasumua and Chania rivers which today produce about 240,000m^3 a day, just about enough to satisfy demand.

Source	Year	Yield (Mld)	Cumulative Yield (Mld)
Kikuyu Springs	± 1900	4.5	4.5
Ruiru Dam	1950	21.5	26.0
Sasumua Dam	1956	57.0	73.0
Sasumua Dam Raising Chania I Chania II	1960-84	111.0	184.0

A central feature of the $150M Third Nairobi Water Supply Project is the Thika Dam, set in an area renowned for the resplendence of its flame trees. Indeed, the dam is the mainstay of the scheme - an earthfill embankment 70m high, holding back 70 million cubic metres of water, located 2,000 metres above sea level in the foothills of Kenya's Aberdare mountains. The project area lies just south of the equator and covers around 7,500 km^2 of land, including the city of Nairobi. It extends from the Aberdare mountains, rising to 4,000m, 75km north of the Athi plains in the south. Mean annual rainfall ranges from 2,000mm in the north-west to about 300mm in the south. The deeply incised valleys which drain the forested eastern slopes of the Aberdare range provide many good dam sites but only small reservoir capacities.

Formed from 2.7 million cubic metres of material transported from within the reservoir area, the dam will provide a secure supply to the city via Transmission Tunnels with a capacity of 450,000 cubic metres per day.

Designed by Howard Humphreys for the Nairobi City Council's Water and Sewerage Department and the Government of Kenya, and supervised by Howard Humphreys (Kenya), the project is the latest stage in the development of the regional water resources.

The scheme, which is due to be completed by the end of 1994, will increase source and treatment capacity, and transmission and distribution systems, to meet the domestic, industrial, commercial, public service and agricultural demands of the region's three-and-a-half million inhabitants.

Also included in the scheme are 4km of transmission tunnels and 43km of transmission pipelines; extensions to the Ngethu treatment works, tripling its capacity to process 460,000 cubic metres of water per day; construction of concrete service reservoir with a combined capacity of 46,000 cubic metres; 35km of treated water trunk distribution pipelines; and provision for hydroelectric power generation.

During construction of the dam, the natural flow of the river has been diverted through a 3m-diameter tunnel, 306m in length through the left abutment. Once the dam is complete this will be used for downstream releases and compensation water. The main contractor for the dam and tunnel contracts (C207 and C208) is Strabag-Lima JV.

Water will be transferred to Nairobi via three intake adits to an 80m-deep intake shaft and an underground powerhouse designed to generate 2MW of hydropower utilising the available reservoir head - a useful additional source of power for the local residents and contribute to the national grid.

Funding for the project - which is split into 12 independent civil, mechanical, electrical and instrumentation contracts - is being provided by the World Bank through IDA, OECF of Japan, African Development Bank and Fund, European Investment Bank and the Nairobi City Council.

2. Summary of the Project

The following table summarises the underground works for the two contracts based at N'Dakaini:

THIRD NAIROBI WATER SUPPLY PROJECT UNDERGROUND WORKS		
Location	Length/ Depth (m)	Description
THIKA INTAKE		
Upper Adit	35.23	2.5m I.D. horseshoe
U A Valve Chamber	11.52	7.5m NATM lozenge
Middle Adit	100.75	2.5m I.D. horseshoe
M A Valve Chamber	11.52	7.5m NATM lozenge
Lower Adit	162.45	2.5m I.D. horseshoe
LA Valve Chamber	10.30	7.5m rectangle
Intake Shaft	67.0	5.5m I.D. cylinder
Powerhouse	14.5	18.75mx14m I.D.
TRANSMISSION TUNNELS		
Thika/Kiama Tunnel	954.4	2.5m I.D. horseshoe
Kiama Shaft		4mx3m I.D. cylinder
Kiama/Chania Tunnel	3366.7	2.5m I.D. horseshoe
Kimakia Shaft		8mx3m cylinder/square
DIVERSION		
Diversion Tunnel	303.845	3.0m I.D. horseshoe
Shaft	29.5	5.5m I.D. cylinder
SPILLWAY		
Spillway Tunnel	169.5	5.5m I.D. circular
Shaft	30.0	5.5m I.D.+ 2m fins
DRAINAGE ADITS		
Left Abutment D.A.	417.5	2.5m I.D. horseshoe
Right Abutment D.A.	280.0	2.5m I.D. horseshoe
RADA Access Adit	49.9	2.5m I.D. horseshoe

3. General - Geology

The geology of the N'Dakaini site of the Third Nairobi Water Supply Project is one of a series of pyroclastic flows associated with the volcanic activity of the East African Rift Valley. The terrain is made up of a series of tuffs and ash flows of varying thicknesses.

The Thika Reservoir lies mainly within deeply incised valleys of a classic "V" shape. This changes as the dam site is approached. The valley broadens and a distinct flat area is present on top of a sequence of similar lapilli tuffs.

The widespread nature of this lapilli tuff sequence was of great benefit to the project. Many of the features of the project at N'Dakaini were founded on or driven through it. The Transmission Tunnels and Powerhouse were excavated in it. The Dam, Stilling Basin and Secondary Intake Shaft were founded on it.

Another lapilli tuff above the first was also of significance to the project. Within the hill of the left abutment the Secondary Intake Shaft was excavated and supported through this rock layer and it formed the foundation to the bend and thrust block of the Spillway Shaft/Tunnel.

Between these two major lapilli tuff sequences were a series of other tuffs of various weathering grades within which the tunnels associated with the Thika Dam were excavated.

Unusual geology seen on the project included:

In the Chania/Kiama Tunnel between Ch 1+390 and Ch 1+375 a fine grained tuff dipped into tunnel level of welded lapilli tuff. Being less welded and more porous, it deposited a large quantity of water into the mainly dry tunnel. The section at Ch 0+053 to 0+066 was interpreted as the rolled end of a flow. There were significant quantities of manganese oxide (70%) within large open joints (100-200mm). There were 3 other major joints (up to 200mm wide) which required special treatment (dental concrete). These were located at Ch 3+341, Ch 3+067 and Ch 2+896.

In the Left Abutment Drainage Adit at Ch 60 a mud filled fumurole (adit diameter) was intersected. The mud fell into the adit immediately on excavation.

In the Right Abutment Drainage Adit the excavation encountered a full size tree (still partially fibrous) lying perpendicular to the tunnel alignment at Ch 150 some 50m below the surface.

The Diversion tunnel was excavated in the lower of the two residual soils between the two strong lapilli tuff sequences. This material ranged from hard to very hard and was the only ground to exhibit some form of stress relief, with cracks appearing in 100mm layers after a few hours.

The Spillway Tunnel excavation showed how variable volcanic terrain can be, with 5 units visible at the portal. Each unit changed in thickness and the upper lapilli tuff varied in weathering grade from II/III to IV/V, with large saprolitic sections. This highlights the problems of drilling in trying to define precisely what would be encountered in a tunnel excavated in volcanic terrain.

4. Construction

4.1 Adits and Tunnels

4.1.1 Design

The Transmission Tunnels, Intake Adits, Drainage Adits and Diversion Tunnel were all designed to the horseshoe cross-section.

Based on the NGI "Q" value method four Classes of Support were designed to suit the expected site conditions. The Classes were based on the use of shotcrete and/or steel ribs as Permanent Support. The concrete lining of the tunnels was not required to have any structural component except in the Diversion Tunnel. The following table lists the Support Classes and their criteria:

Class of Support	Q Value	Comments
1	>1 grade III/II or better >4 grade III or worse	Spot rockbolting and patches of shotcrete of 50mm nominal thickness
2	0.1 to 4 grade III/II or better 4 to 10 grade III or worse	Complete shotcrete coverage of 50mm nominal thickness
3	0.1 to 1.0 all grades	100mm shotcrete with reinforcing mesh
4A/B	< 0.1 all grades	50mm shotcrete shell braced by 150mm/23.4kg steel set. 'A' with invert strut.

In the section of the Chania/Kiama Tunnel where Class 4 support was used between Ch 120 and Ch 600, the Contractor opted to thicken the shotcrete support to 100mm with a mesh reinforcement midway.

4.1.2 Excavation and Support

The majority of the Chania/Kiama Tunnel (3367m) was excavated from Chania with an EIKHOFF roadheader, breaking through at Ch 460 with a drill and blast/back hoe operation coming from Kiama. The roadheader had also excavated the first 117m of Kiama/Chania Tunnel from Kiama.

The majority of the Thika Intake (Adits and Powerhouse), the whole of the Thika/Kiama tunnel (954m), and part of the Chania/Kiama tunnel were excavated using drill-and-blast techniques.

Behind the roadheader in the Chania to Kiama drive, a railway mucking system was used. This used diesel locomotives and a train of Hagglund cars (maximum of 4). Little room was left between the Hagglunds and the side walls, due to the tubing (ventilation, water extraction etc) which ran down the tunnel.

The drill-and-blast technique used a "burn-cut" method. In general advances in sound rock were 2.4m with up to 60 parallel holes in the face. A blasting ratio of over 2.0 kg gelignite per cubic metre was used, with blasting gelignite of 80% strength. The blast was delayed from close to the centre spiralling outwards to the periphery. A centre relief hole was not always used.

Where ground softened, excavation reverted to Shaeffer back-hoe. This equipment was also used to muck the drill-and-blast operation. To support the back-hoe, diesel tunnel dumpers were used in all drill-and-blast operations and in the roadheader operation from Kiama. Where tunnel dumpers were not available a CAT910 or small ZEPPELIN front-end-loader was used.

The Contractor adopted blow-in ventilation in all adits/tunnels, with the 600mm/up to 1000mm/fans being stationed at the portal.

Within the roadheader section of the transmission tunnels, major problems were encountered when dry rock was excavated due to the blow-in ventilation (4 fans). Dust was carried back down the main tunnel section resulting in difficult conditions for carrying out ancillary works in the tunnel away from the "working zone" around the roadheader.

The principal means of support used in the underground works was shotcrete. The specification required shotcrete to reach 30 MPa at 28 days. Shotcrete was mixed at one of three batching plants with minus 20mm aggregate (in practice minus 10mm aggregate). Initially 400 kg cement per cubic metre was used, but this was raised to 440 kg early in the project, eventually to be reduced again at a later stage. Kenya Standard cement (15% pozzolanic) was used initially but was changed to BS12 early in the job and finally to sulphate resisting BS 4027 (which was a finer grind).

The moisture content of the shotcrete mix was below 2-3% and the shotcrete was transported by concrete mixer from the batching plant to the tunnel portals. It was then transferred into the tunnel dumper/Hagglund/CAT910 bucket and transported to the shotcrete delivery machine near the face.

Accelerator was added at the shotcrete pump (4% liquid SIGUNIT by dispenser; or 6% powdered SIGUNIT by trowel) and water at the nozzle. In general batches of 2.5m or perhaps 5.0m would be used. In certain circumstances 1.0 and 2.0m batches would be used.

In the Chania/Kiama tunnel where shotcrete was not required, regular spot-bolting was undertaken using tensioned 2.4m long spun resin galvanised rock-bolts.

Within Class 4 support, the design required 150mm cross-section HEB ribs. As the contract progressed these "I" section ribs were changed to a thicker rib of lesser cross-section and later ALWAG TH21 ribs were substituted. The designed support method braces the shotcrete support from a large number (16) of blocking points off the rib. ALWAG ribs are fully enclosed in shotcrete. An invert strut was to be used if required. No invert struts were installed in ALWAG ribs.

Convergence arrays were installed in a variety of ground conditions. In general they were placed at 100m intervals in the Transmission Tunnels, only a limited number were installed elsewhere. Little or no movement was recorded.

Extensometers were installed in the powerhouse roof and at four locations in the Transmission Tunnels.

Little or no movement was recorded from the Instrumentation

4.2 Valve Chambers

4.2.1 Design

There are three valve chambers in the Thika Intake, associated with the junction between the Intake Adits and the Shaft. Considering the shaft centreline as Ch 0, the entrance to the Valve Chamber is at Ch 3.25 and the end of the Valve Chamber at Ch 14.75 (Upper and Middle Adit) or Ch 13.55 (Lower Adit).

The design concerned the concrete structure, leaving the ground support for construction. This design concept was changed early in the contract with the shape of the Upper and Middle Adit excavation changed to a "NATM" shape and the ground support fully specified. The Lower Adit remained as originally designed with only the junction with the Powerhouse changed.

A four part lattice beam was designed to enclose the 6m finished diameter approx ellipsoid excavation at 1m centres approx.

4.2.3 Excavation and Support

Upper Adit Valve Chamber was excavated by drill-and-blast using a heading and bench method. The whole of the heading was excavated and supported in advances rarely greater than 1.2m. from adit to shaft followed by the bench excavation and support.

The Middle Adit Valve Chamber was excavated by heading and bench with the full heading supported prior to commencing excavation of the bench. Excavation of the heading was by backhoe/jackhammer, whereas the bulk of the bench was excavated using the CAT910 front-end-loader with profiling by jackhammer/backhoe.

The Upper Adit Valve Chamber was supported to Class 4b. This involved installing 200mm shotcrete with lattice beams on 1m approx. centres above springing line and 200mm shotcrete below springing line through the invert. The sequence of support involved 50mm shotcrete applied to the ground, a layer of mesh reinforcement was applied to the shotcrete followed by fixing of the lattice beam (with foot niche supports and plated junctions). The lattice beam would be encased in shotcrete. A further mesh layer placed and then a further layer of shotcrete applied. The lattice beam at Ch 6.0 was applied full circle (i.e. Class 4a).

The Middle Adit Valve Chamber support was similar to the Upper Adit Valve chamber except that it was all Class 4b and 150mm of shotcrete was applied below the springing line.

The Lower Adit Valve Chamber had no shotcrete applied, except for a small patch near the junction with the adit proper and the section closest to the Powerhouse where three lattice beams were placed between Ch 5.25 and Ch 3.25 where 250mm of shotcrete was placed in the crown and lattice beam niches.

4.2.4 Lattice Beams

The lattice beam was constructed from 6 lengths of T25 rebar arranged 2 to each vertex of a triangular cross-section. The T25 bars were held in position by 8mm rebar spot-welded in a herringbone pattern across the lattice beam. Cross section approximates to 108mm.

4.3 Shafts on C207

4.3.1 Design

There are three shafts within the C207 contract, the Thika Intake Shaft, the Kiama Intake Shaft and the Kimakia Intake Shaft.

The Thika Intake Shaft forms the connection between the Thika Intake Adits and the Transmission Tunnels. Some 68m in height and 6.5m in diameter, it carries the pipework from the Valve Chambers to the Powerhouse and provides access to the Powerhouse from the Control House. The Thika Intake Shaft has a 500mm reinforced concrete lining.

The Kiama Intake Shaft is a short 4m drop shaft from the side of the Kiama Intake. This excavation was regarded as an extension of the tunnel with a similar diameter.

The Kimakia Intake Shaft is 8m in length and transfers water from the Kimakia River into the Transmission Tunnel. It is a 3.5m diameter excavation with a 500mm reinforced concrete lining.

4.3.2 Excavation and Support

Within the Thika Intake Shaft the majority of the ground was in weathering grade IV/V to VI. This material was sufficiently cohesive and stiff that it could be successfully excavated by a Menzi mucker directly. This four legged spider-like backhoe, sat in the centre of the shaft and excavated all around itself. It discharged into mucking "kibbles" which were drawn up the shaft by crane. Where the material hardened making direct backhoe excavation impossible the ground was drill-and-blasted.

Approximately one third of the shaft was excavated via a "glory" hole which had been excavated from the Powerhouse and mucked out using the CAT910 front-end-loader via the Lower Adit.

The Kiama Intake Shaft was excavated from below by the roadheader and from above by drill-and-blast.

The top 1m or so of the Kimakia Intake Shaft was excavated to a broad platform by surface equipment. Below this was excavated entirely by jack-hammer through strong rock, due to the proximity of the Kimakia River and the danger to the whole excavation that blasting could have been.

The support used depended on Class but corresponded to that used elsewhere in C207. The specified shotcrete strength of 30 MPa was achieved with difficulty.

4.4 Powerhouse

4.4.1 Design

The underground powerhouse lies at the base of the Thika Intake Shaft, it was designed as a 12m wide by 20m long by 9m high excavation. The roof was set on a 15m radius curve, with all other walls vertical except for the powerhouse/shaft/lower adit intersection where a semi-circular section of shaft lies outside and adjacent to the powerhouse.

The originally designed support consisted of patterned 5m tensioned rock bolts on a 1.5m diamond, installed perpendicular to the roof, with a series of 3m tensioned rockbolts installed above and below the roof/wall junction. The roof was to be covered by 100mm shotcrete with mesh reinforcement and the walls coated with 50mm of shotcrete.

This original design could not be used when it was found that the roof was too weak to form a compressed arch by pattern rockbolting. The design was changed so that the roof could be supported by lattice beams. The roof radius was changed to 12m, the excavation widened to 13m (14m at the roof niches) and the end crown/wall modified with horizontal and vertical curves.

The revised design involved short-term support to the roof transferred by lattice beams into roof niches. The long-term support was provided directly from the concrete walls through a rebar connection. The excavation was sprayed with 50mm of shotcrete followed by mesh, the lattice beam was erected in two pieces, connected in the centre by a plated junction. The niche footing was shotcreted in and then the lattice beam shotcreted over until it was completely enclosed. Finally a mesh layer would be attached and shotcreted to a depth of 50mm. This resulted in a 250mm thick shotcrete layer.

4.4.2 Excavation and Support

The powerhouse area was approached from the Lower Adit, skirting around the powerhouse edge of the shaft to the approximate longitudinal centreline of the powerhouse.

A drill-and-blast excavation was then advanced to the far end of the powerhouse leaving the excavation approximately 2m from the end wall and 2m from the roof.

After completion of the shaft a programme of exploratory drilling revealed that a redesign was required.

Excavation commenced with the roof arch, including niches. The excavation started at the shaft end of the powerhouse, but became problematical. Excavation moved to the far end of the powerhouse and worked back to the shaft in 1.2m full width blocks with completed lattice arches prior to further advance.

Excavation was by drill-and-blast with trimming of the last 1m to the walls, to minimise blast damage to the support niches and the final excavation outline. The roof excavation was taken down to niche level with a trench excavated above the concrete wall line to accommodate the starter bars to the support wall, which were attached to the lattice beams.

The excavation was advanced to the edge of the shaft at roof level. The bench was then taken out to general floor level by larger scale blasts, with jackhammer trimming to final profile. Following this the end wall was trimmed to its final shape. The last full width lattice arch (close to the shaft) forms the top end of a series of arches perpendicular to it. This final section of roof was excavated prior to excavation of the various sumps and the discharge valve in the invert.

Following concern for drainage pressures, strips of drainage (first corrugated-sheeting, secondly half rounds) were placed on 2m centres vertically on the side walls, the walls were then shotcreted to 100mm. Similar drainage was applied to the end wall with 200mm of shotcrete applied.

4.4.3 Lattice Beams

The lattice beams used in the Powerhouse roof and the Lower Adit Valve Chamber/Powerhouse intersection were of similar type although of different radii.

Constructed from 6 x T25 bars in a triangular cross-section with 2 bars at each vertex. Using 8mm bar in herringbone pattern between the vertices. Erected in two pieces with plated junction in the crown and foot plate at each end. Each lattice beam had starter bars to the concrete walls spot welded in place.

4.5 Spillway Tunnel

4.5.1 Design

The design of the Spillway Tunnel was based on the premise that the ground support was constructional and the permanent support would be provided by a designed reinforced lining.

In Class 4A and 4B the permanent support consisted of a 500mm thick reinforced concrete lining to C30 which fully enclosed any steel ribs which formed part of the constructional support. In all other Classes this lining was reduced to 400mm.

In order to improve the design, the approach was changed to see whether it was possible to incorporate the constructional support within the whole lining. Thus the constructional support and the permanent lining could be combined as primary and secondary linings to produce a long term structural support.

The ground conditions were modelled using the FEA program CRISP. After extensive analysis, the secondary lining was reduced to 300mm of C30 reinforced concrete which combined with a minimum of 200mm of 30MPa shotcrete for the primary lining (230mm in Class 4).

The design required the use of 23.4 kg/m run 150mm HEB ribs in Class 4 to be placed on 1500mm or 750mm centres as necessary. Provision was made to install a partial rib with legs as Class 4B or a full rib as Class 4A. The substitution of ALWAG TH21 ribs was accepted but the revised design required these to be increased to ALWAG TH25/58 ribs.

Ribs were installed at 500mm centres between the portal and the first aeration groove at Ch 97 and at 1000mm centres from there to the grout curtain (Ch 52). Upstream of the grout curtain steel ribs were no longer required by the design, and were only necessary if required as constructional support.

4.5.2 Excavation and Support

The Spillway Tunnel was excavated from the downstream portal (Ch 180) using the heading-and-bench method. The excavation was advanced by a mixture of drill-and-blast and jack hammering. Initially the bench lagged some 4m behind the heading. This was soon modified when a ramp was built from the bench to the heading with the bench excavation stopped at Ch 146. The heading was advanced to the start of the Spillway Bend at Ch 10.75 and then the bench was excavated back towards the portal. A channel was excavated down the centre of the tunnel bench for drainage. The remainder of the bench was removed by jack hammer.

The majority of the excavation was performed by drill-and-blast either as a full scale blast or as crack blasting. Soft ground excavation was done by back-hoe. Due to the requirement that the excavation line was relatively precisely controlled the final shaping of the excavation was generally performed by jack hammer. Mucking was generally performed by back-hoe and tunnel dumpers.

After excavation, 50mm of shotcrete was applied immediately to the ground. This was followed by a mesh layer which in Class 4 was held in place by an ALWAG TH25. The whole was then shotcreted in. The shotcrete was then built up to provide cover to the ALWAG by a nominal 10mm. The "ALWAG" shotcrete layer reaches a thickness of 130mm. A further layer of mesh was then applied and covered with 50mm of shotcrete to complete the support. In other support Classes the "ALWAG" layer was reduced to 100mm.

Each rib was provided with a 500mm niche plate so that the upper ribs would be supported during the bench excavation.

Regular installation of convergence arrays and extensometers revealed little movement. Long term monitoring of the four extensometer arrays installed in the Spillway Tunnel was planned, results are not yet available.

4.6 Spillway Shaft

The underground part of the bellmouth Spillway from el.2038.5 to el.2008 was designed as a structure stiffened by 7x2m by 1m fins evenly spaced around the periphery of the 6.5m o.d. shaft. The design did not include constructional support design.

Within the Spillway Shaft the majority of the ground was in stiff to hard grade V to VI dry tuff. Between el. 2031 and el. 2016 there were many small boulders of phonolite giving the ground a saprolitic texture. Some boulders required crack blasting but otherwise the material was successfully excavated by a Menzi mucker as in the Thika Intake Shaft.

The 2m deep by 1m wide fins were excavated relatively simply with the use of the back-hoe and jack-hammering. The stability of the excavation was significantly compromised by these fins.

4.7 Diversion Tunnel

The fully lined Diversion Tunnel (u/s portal at Ch 90.205, d/s portal at Ch 396.05) provided both a route for the diversion of the Thika river during construction and for the outlet for the Secondary Intake Shaft including compensation water downstream. The finished diameter was larger than the Transmission Tunnels, being a 3m i.d. horseshoe.

The tunnel was excavated from both portals using drill-and-blast and direct back-hoe excavation. The majority of the tunnel was excavated in hard grade V/VI which was easily excavated although requiring blasting. This overconsolidated clay showed signs of stress relief after excavation, and for this reason required support. The excavation was mostly dry. Blinding concrete in the invert was advanced with the excavation. Over the upstream portal six by six metre long T25 passive dowels were installed to improve the stability.

4.8 Secondary Intake Shaft

The Secondary Intake Shaft is a free standing structure which extends from the underground part (el. 2000.5 to el. 1970.5) up to 2045 (dam crest level).

The original design involved the stiffening of the shaft by fins. As the upper lapilli tuff appeared to be a massive rock horizon the fins were deleted. When excavated it was found that the large boulders of solid rock were surrounded by seams of grade IV/V lapilli tuff. Due to the discontinuous nature of the hard rock, four rows of 24 anchorages (6m long made from T32 rebar), angled downwards at 20 degrees were installed into 75mm holes to prevent rotation and minimise lateral movement of the shaft. These rows were set at el. 1993, el. 1991.5, el. 1990 and el. 1988.5. Grout (30 MPa at 28 days, 1:1 by weight) was tremied into the holes.

The shaft was excavated in a similar manner to the other shafts on the project.

4.9 Drainage Adits

4.9.1 Design

The two Drainage Adits were designed to a similar cross-section as the Transmission Tunnels (2.5m i.d.) (see Section 4.1). As no investigation was performed prior to construction, the design was based on 100% Class 4A. The tunnel would be fully lined and a drainage membrane placed circumferencially, piped through into the adits and with the wells, drained via the invert.

4.9.2 Excavation and Support

The Drainage Adits were excavated using both back-hoe and drill-and-blast techniques. The adits were generally dry and blinding concrete was periodically advanced in the invert to improve running conditions. As well as tunnel dumpers a small ZEPPELIN front-end-loader was used to muck out.

Commencing at the portal of the Left Abutment Drainage Adit (Ch 25) Class 4B was used. Due to the proximity of two temporary access roads and the Spillway Cutting the Class 4B support was extended to Ch 90. Here a thin fine grained tuff had risen in the excavation to the crown. This tuff remained in or just above the roof for the majority of the entire length of the adit.

The majority of the Right Abutment Drainage Adit was supported Class 2 except for near the portal of the Access Adit (Ch -54 to Ch -40) and beyond Ch 160, where a strong massive grade II/III lapilli tuff dominated the excavation. There Class 1 was used with patches of shotcrete along the sides in the early part.

4.9.3 Drainage Membrane

In both Drainage Adits a drainage membrane was specified (DELTA MS 20). This "eggbox" was to be placed over the entire surface area of the adits above the invert. A permeable pipe was to be laid horizontally at the base of the drainage membrane. This was connected to an inclined pipe which passed through the concrete lining draining to the invert. The shotcrete support skin was to be broken out in numerous holes to allow drainage from the ground.

5. Update on Progress

At the time of preparing this paper the Thika River dam was 95% complete with completion expected by the end of February 1994. Thereafter impounding the Thika River would commence and finish in approximately 12 months. The Intake Tunnels/Shafts, Transmission Tunnels, Spillway and Diversion Tunnels were all completed by 1993. The pipeline was installed in the Intake Tunnel and Shaft and the valve plugs should be installed by the end of February 1994.

The powerhouse is complete but the client has not yet decided to proceed with the installation of turbine equipment for hydro power generation.

6. Conclusions

As in any large project some aspects of the design and construction did not run as smoothly as planned. The common problem with tunnel schemes are generally ground and logistics related and the Nairobi project was no exception. The various problems related to excavation and support are addressed in Section 4.

As an example, the design of the powerhouse roof was based on the expection that the roof could be sited in strong grade II lapilli tuff thereby allowing a `flat' arch with 5m long resin rock bolts to reinforce the rock. During the design period the location of the cavern was moved 20m. It was assessed from the site investigation information that the ground conditions would still apply for the revised position. However, on excavation of the lower part of the intake shaft near the junction with the cavern roof, the ground conditions were found to be significantly different, with moderatelly weak grade III/IV lapilli tuff.

This being the case a revised roof support design was required. Following discussions with the contractor a lattice arch and shotcrete roof was designed with integral connections with the r.c. cavern wall. Preformed drain holes were provided in the shotcrete to reduce the possibility of water pressure build up behind the shotcrete. PVC channels are located under the roof to collect and direct seepage to the cavern drainage system.

The specification allowed the contractor to select his method of excavation based on his interpretation of the ground conditions. Naturally he chose to excavate short lengths of tunnels/adits in hard ground using drill and blast and in soft ground using backacter machinery. The contractor chose to excavate the 4.2km Transmission Tunnel by roadheader which was reasonable providing the infrastructure and logistics to support this method were methodically adhered to. Generally this was not achieved as there were frequent muck train rail derailments; the abrasive rock put high demand on the ventilation system and pick wear was high. Progress of 15m/day was planned but approximately half that anticipated progress rate was met. Drainage in the Transmission Tunnel also added to the problem. The relatively remote location also added to the problem of back up supply when additional plant/equipment was needed to sort out an unexpected problem. There was a particular problem if specialist items were required from Europe.

However, the contractor is experienced in working in Central and East Africa and in time the problems were attended to and were satisfactorily overcome to complete this phase of the water supply requirement for Nairobi.

7. References

1. BARTON,, LIEN R. and LUNDE J. Engineering classification of rock masses for the design of tunnel support. Rock Mechanics, Volume 6, No. 4, 1974 pages 189-236
2. Hoek E. and Brown E.T. Underground excavations in rock. The Institution of Mining and Metallurgy, London, 1980.

8. Acknowledgements

The authors are very grateful for the cooperation and assistance of the Nairobi City Council, particularly for Mr W J Odhiambo, General Manager, Water and Sewerage Dept., in allowing this paper to be published.

We are also grateful to Mr. Laurence Attewill the `Engineer' and Mr. Tony Bruggemann for ideas and data in assembling the paper.

Design and construction of caverns for an underground sewage treatment plant, Hong Kong

M. A. Oswell
Montgomery Watson, Hong Kong
I. W. Farmer
I. W. Farmer and Partners, Newcastle upon Tyne, England
B. W. L. Mak
Montgomery Watson, Hong Kong

Abstract
80,000m^3 of underground caverns to house a sewage treatment plant were constructed at depths up to 60m in typical Hong Kong granites comprising strong to very strong massive rock with widely spaced joints. The objective of the cavern design was an unlined cavern roof and unlined walls above the level of the tanks in the process area. The paper describes the geotechnical site investigation, the cavern support and design philosophy and construction experience.
Keywords: Underground caverns, Rock excavation, Tunnel design, Site investigation, Roof support, Underground construction

1 Introduction

The Stanley Underground Sewage Treatment Works site is in granite at the northern end of the Stanley peninsula in Hong Kong Island, adjacent to Stanley Prison. It is above sea level with a road access in an area with no habitation and a direct sea access to a beach in the Prison security area. The factors determining the choice of an underground option for the works - the lack of a suitable surface site and its relative cheapness compared with other options - have been discussed by Oswell et al (1989).

The basic arrangement of the treatment works required three general areas (Figure 1) - services, main treatment and disinfection and an access tunnel. A total underground volume of about 80,000m^3 was needed to contain works to service a residential population of 27,000 covering Red Hill, Tai Tam, Ma Hang, Chung Hom Kok and Stanley. The main caverns had maximum dimensions of 17m span and 17m high. Spans were increased at junctions. The service area about 130m long included inlet works, flow distribution chambers, screens, sludge dewatering plant, liquid and dewatered sludge handling systems, air blowers, hypochlorite storage, transformers, parking and control room and operator facilities. The treatment caverns were about 100m long. The main treatment works comprised two aeration tanks about 70m long and 10m high, pumping station

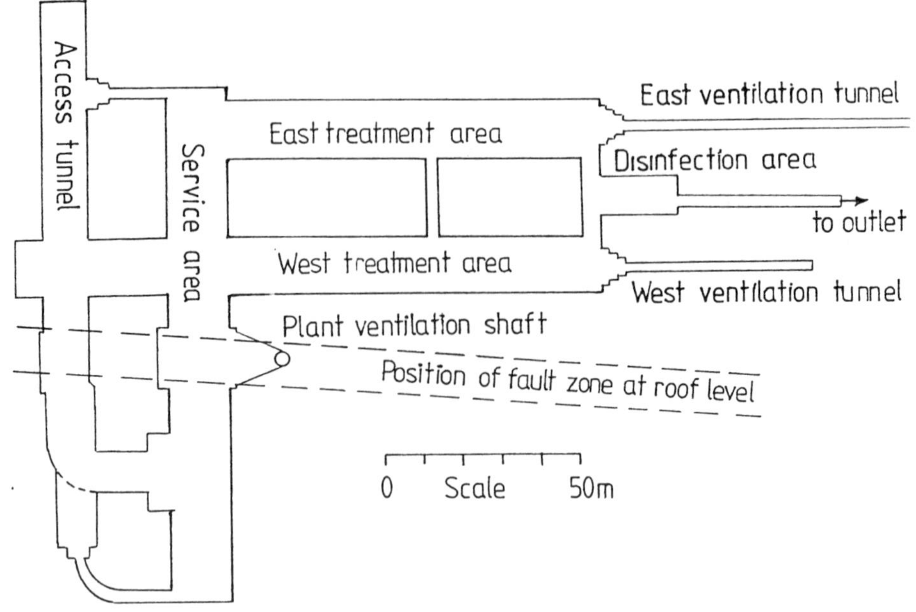

Figure 1 Layout of the underground treatment works

and final settling tanks about 30m long. The disinfection area where the treated flow is disinfected before discharge to sea was about 23m long, but of limited volume. The access tunnel had an 11m span and was 6m high.

2 Geology

The solid geology of Hong Kong Island and adjacent mainland areas comprises mainly Middle Jurassic volcanic tuffs through which have been intruded granites of Early Cretaceous age. Sedimentary rocks are not an important feature. The geology of Hong Kong Island and Kowloon is covered in detail by Strange and Shaw (1986) and McFeat-Smith et al (1989).

The granites are generally coarse grained and widely jointed (0.5-2m) with sub-vertical joints following the dominant regional trends, NW, NE and ENE. Sheeted structures are often present near the surface, giving the rock mass an occasional stratified appearance. Volcanics are generally fine grained and closely jointed (50-200mm).

Most of the rocks are faulted and sheared and a particular feature comprises recent basalt dyke swarms of varying width up to 2m which usually follow the ENE trend. Faults and dykes are predominantly vertical or subvertical.

Information on regional residual stress directions is limited, but horizontal principal stresses in ENE, NNW direction can be inferred. There is no reason to assume that magnitudes are extreme.

Aerial or sub-aerial weathering has particularly affected granites and basalts. Surface granites have been weathered to considerable depths - usually 10-20m and are often associated with colluvial deposits. Fault and dyke zones can be weathered to quite large depths. The hydraulic conductivity of unweathered massive granites and tuffs is relatively low (<10^{-8}m/s) and flow is through fissures.

The Stanley peninsula is wholly granite with a colluvium cover, and weathered to depths up to 15m. Basalt and aplite dykes occur frequently.

3 Site Investigation

The site (Figure 2) is located beneath a saddle rising to a level of 84mPD. The surface is covered with boulder colluvium with occasional outcrops of rock. The site is bounded by photogeological lineaments to the NE and NW - the latter a probable fault zone - which effectively constrained the initial location of the underground works. The site investigation, which was designed to provide information on the overall site structure with especial emphasis on the cavern roof horizons, comprised 8 vertical and 7 inclined boreholes with a total length of 1050m. The boreholes can be grouped as follows:

S3, S9	which passed vertically through the Service Area Cavern
S4	which passed vertically between the treatment caverns
S7	which was inclined at 11.5° and reached roof level between the Treatment Caverns
S7A	which was inclined at 17° and passed through the roof of the intersection of the Service Area and East Treatment Cavern
S17N	which was inclined at 16° and was close to the roof of the East Treatment Cavern
S5	which passed vertically close to the Inspection Area Cavern
S2, S6, S8A, S21	which were vertical and outside the cavern area
S8, S17W	which were inclined respectively at 6° and 27° and intersected the fault zone to the south of the caverns
S18	which was inclined at 15° and passed above the caverns
S20	which was inclined at 19.5° and passed over the Service Area and West Treatment Cavern, and intersected the fault zone

The fault zone is positioned on Figure 1 at the cavern roof level. The surface location of the fault zone can also be estimated from the contours and is clearly visible at the site.

Information collected from the site investigation included, for each borehole:

- complete core photographs
- complete impression packer discontinuity logs with orientations and dip directions
- pressurized packer tests over 10m intervals
- water level records
- total and solid core recovery percentages
- fracture frequencies and RQD percentages
- estimates of degrees of weathering from unweathered (Grade I) to decomposed (Grade VI)
- point load tests at 1m intervals
- uniaxial compressive strength and modulus tests at 2m intervals

Additional information from selected boreholes included:

- Goodman Jack tests in boreholes S2, S3, S5, S7, S7, S14, S17N and S20
- Insitu stress measurements in boreholes S17W and S20
- Triaxial test data from boreholes S2-S6, S9 and S18

A review of the site investigation data confirmed that the predominant rock at the site was a typical Hong Kong granite which comprised strong to very strong massive equigranular medium coarse crystalline rock with widely spaced joints. The granite bedrock at the Stanley site was overlain by decomposed granite varying in thickness from about 1m to about 15m.

Three main discontinuity sets following the regional pattern were identified; two were steep, striking NW and NE with dips to either side of the strike line. There was also a well defined, sheeted, subhorizontal discontinuity set. There was variation over the site, but except in the fault zone, the discontinuities were invariably narrow (EN grade) and discoloured but not altered (Grade II weathering). The RQD was mainly in excess of 90% and the average number of discontinuities was 1 or 2 per metre. Apart from the major fault/dyke zone, there did not appear to be any significant secondary faults or lineaments. There were aplite dykes which were extremely strong and did not represent a weakness, and there were occasional zones of weaker and sheared granite. It was thought that these would not significantly affect the behaviour of the rock mass.

The rock uniaxial compressive strength ranged from 50-150MN/m^2 with a predominance of higher values. The point load IS50 index ranged from 2-8MN/m^2 with a predominance of

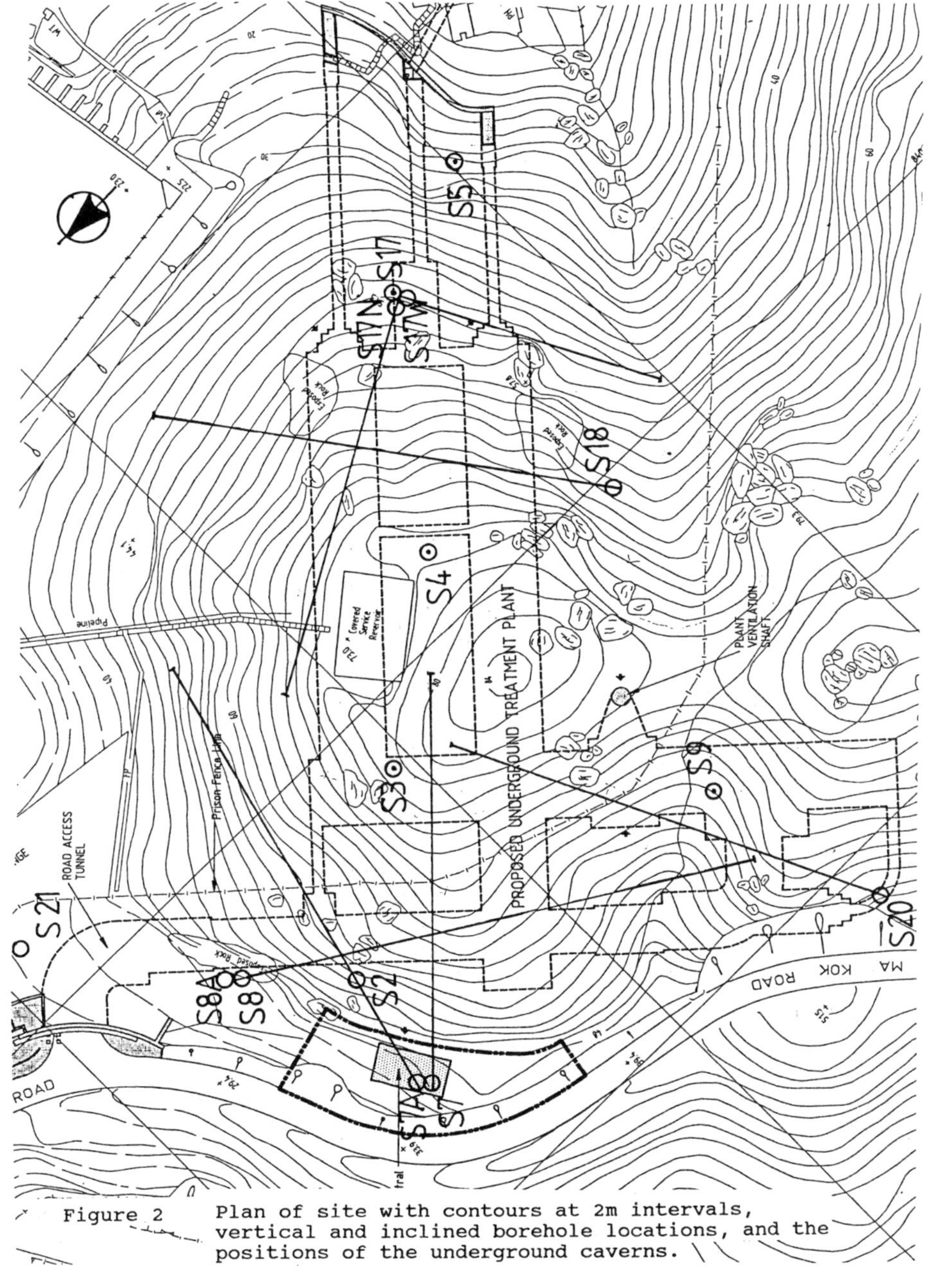

Figure 2 Plan of site with contours at 2m intervals, vertical and inclined borehole locations, and the positions of the underground caverns.

values over 6. In both these tests and particularly the point load test, lower values should be treated with caution, if not discounted, and it is reasonable to describe the intact rock as strong to very strong, with a strength of about 2-3 times that of good structural concrete.

There was an interesting variation between modulus measured in the laboratory (invariably in the region of $50GN/m^2$) and in-situ, using the Goodman Jack (about $15GN/m^2$). This agrees with previous observations (Hueze and Amadei, 1985) that the device returns unacceptably low values of modulus. FitzSimmons (FitzSimmons, 1992) suggests that the main problem lies with incomplete seating of the jack, which is difficult to overcome. In-situ stress measurements were also unreliable.

The major geotechnical problem at the site was perceived as the fault/dyke zone. This can be seen at the surface as a major topographical feature (Figure 2), where basalt in the fault zone has been eroded by sea/salt water/air contact to a silty consistency. The zone is approximately 12m wide and dips at about 85° to the NE. The fault was parallel to and about 15m from the floor of the proposed West Treatment Cavern. It would, however, affect the Service Area Cavern and the Access Tunnel which both intersected the fault normally.

It was not possible from the site investigation to postulate the effect of the fault zone on the support of these structures. The reason for this was that Borehole S8 which intersected the fault zone at 32mPD (about the level of the Service Cavern roof) showed 3.3m of lost core and 7.80m of poor quality but supportable rock. Borehole S20 which intersected the fault at about 40mPD (about 7m above the roof) had 4m of lost core and 3m of poor quality rock. Borehole S17W which intersected the fault zone at a shallow depth showed more lost core. It was considered probable that at the cavern roof level, the excavation would be supportable.

The existence of open discontinuities in the upper part of the granite bedrock, close to the weathering contact was demonstrated by the pump test results, which showed quite high flows in these zones. There was also evidence of high groundwater levels in some of the boreholes, particularly S4. This was investigated on site and a major water emission was found near S4. It was thought to be a leak from the prison reservoir immediately above the proposed cavern. During construction, it was found that the source of water was the overflow from the reservoir, which was discharged into the ground through broken drainage channels. The main body of unweathered rock was, however, of low conductivity in the range 10^{-7}-10^{-8}m/s.

4 Design

Design of large caverns in rock - particularly at shallow depths where weathered and open joints will probably be present - is a difficult process. There is an increasing tendency to base everything on "precedence practice", through comparison with previous practice using case history experience (Sharp, 1989) or formalized rock quality bands (Barton, 1989). While this approach is important in comparing support options, it is not a satisfactory substitute for engineering design, which is essentially a mechanical process.

In the case of large underground caverns, this process may be summarized as estimating the disturbance caused by the construction of a cavern or caverns on the host rock mass. This is in turn conditioned by the geology and geohydrology of the area and the geological structure and geomechanics of the rock mass.

In such an analytical approach, the first requirement is an estimate of insitu stress. This is almost impossible to measure with accuracy at shallow depths (see Irvin et al, 1987) - principally because the measured stress levels are low compared with factors such as boundary yield and rock strength. For instance, published data from China (Carlsson et al, 1989) and Hong Kong (GEO, 1992) indicates horizontal major principal stresses of different relative magnitudes and at right angles to each other in the same structural regime. Best estimates suggest a NW-SE major horizontal principal stress direction with a magnitude of around 1.5 times the intermediate and minor horizontal and vertical principal stresses which have a magnitude equal to overburden stress. If this was the case the major principal stress would be parallel to the treatment cavern axes (Figure 1) and normal to the service cavern and access tunnel axes. In this case the effect of the redistributed compressive stresses following excavation would be to close fissures in the roof and sidewalls, and the magnitude of the redistributed stresses - even at the corners of the excavations would be insufficient to cause fracturing of the rock. This was confirmed by BEM modelling using various assumptions for rock mass modulus and strength and discontinuity stiffness and shear resistance, and using sensitivity analysis and probabilistic methods to determine the effects of inherent unreliability.

Specific roof conditions were considered with reference to the site investigation. The roof level of the East Treatment Cavern was about 28mPD. At its deepest point, near and below the service reservoir (Figure 2) it had a cover of about 50m of which about 40m was crystalline granite. At its shallowest (the southeast end) this cover was reduced to 25m of which only 12-15m was crystalline granite. Inclined borehole S18 (at 70-80m) passed through this area at about 8m above the cavern soffit where the granite has an RQD of 100 and a uniaxial compressive strength of 80-100MN/m². The same area was intersected by inclined borehole S17N (at 30m, and 12m above soffit level) where RQD was 96% and strengths about 100MN/m². This borehole followed the line of the cavern, approaching the level of the roof at 90m and throughout its length passed through crystalline granite, although zones of increased fracturing, possibly indicating shears, reduced RQD to 60-70% in one or two places. Boreholes S3, S4 showed with minor variations, uniformly high rock quality (90-100% RQD) and high uniaxial compressive strength (80-150MN/m²) in the 25m above the cavern soffit. The axis azimuth of the East Treatment Cavern was approximately 315°. There appeared from impression packer tests and from borehole and surface mapping to be 3 main joints sets in the roof rocks with approximate dip directions and dips: 160/60; 220/65 and sub-horizontal, all with extremely narrow apertures, planar and irregular. This configuration is relatively stable to potential roof falls with well spaced and narrow discontinuities. Thus the roof rocks appeared strong to very strong, continuous and relatively competent and would present few difficulties in excavation or support.

The West Treatment Cavern was similar to the East. Additional factors were the closer proximity of the fault zone and the additional information from borehole S7 which confirmed the higher quality of the roof rock, but indicated an additional joint set at about 310/70. The cover varied from 60m at the centre and West to 23m at the East with up to 12m of weathered decomposed surface granite. The minimum unweathered granite cover was about 10m. The horizontal distance between the fault zone (up to 12m wide, but of variable quality and thickness) and the cavern was about 15m and interaction was considered unlikely.

The roof level of the Service Area Cavern was approximately 32mPD. The cover varied from a maximum of 45m to a minimum of 35m, with an upper 10-12m of decomposed granite. The maximum cover of 35m was at the centre of the tunnel where it crossed the steeply inclined fault/dyke zone up to 12m thick which comprised zones of weak, fractured, crushed and shattered rock. The fault/dyke zone was identified in the vicinity of the cavern by boreholes S8 and S20, and dipped to the North-East at about 85°, so that the footwall of the zone occurred on the Southwest side of the cavern. Borehole S20 intersected the fault/dyke about 7m above the roof and S8 at about the level of the cavern roof. In S8 there was about 3m of basalt dyke where no core was recovered, and which was described as highly decomposed. There was an additional 3m of moderately strong fault/dyke. In S20 there was about 4m of missing core described as moderately weak basalt and clayey silt and sandy silt. There was a further 2m of weak basalt. Based on these boreholes it was expected that there would be a tunnel length of 3 to 4m in very weak ground - possibly with very low stand up time, even in a small section tunnel.

On either side of the fault/dyke zone and also in boreholes S3 and S9 the roof rock was of high quality. RQD was 100% in S3 and S9 at and above roof level and roof uniaxial compressive strengths were 150MN/m^2 (S3) and 85MN/m^2 (S9). In S8 and S20 RQD was 100% on each side of the fault/dyke zone. In S3, S8, S9 and S20 point load test results all indicated very strong rock.

The investigation results also suggested a probability that discontinuities parallel to the fault (220/70-85) would daylight in the southwest sidewall of both the Treatment Caverns and a possibility that secondary discontinuities (315/70) would daylight in the sidewalls of the Service Cavern.

The basic conclusion from this analysis was that the caverns would be stable and would require minimal temporary support during construction and limited permanent support after construction, apart from the fault zone and one or two isolated fracture zones. Some care might be needed in cavern sidewall support in the fault vicinity.

5 Previous Experience and Support Specification

Determination of temporary and permanent support in a form suitable for contract conditions can be improved by looking at recent experience in similar rocks. There are three relevant case-histories in Hong Kong and similar granites - the Tai Koo MTR Station (Sharp, 1989) and the Quarry Bay station (Matson, 1989) in Hong Kong and the Guangzhou pumped storage power house (Carlsson et al, 1989) in China. Some available data from these are summarized in Table 1, together with data from Stanley. In each case there were few problems during construction apart from care needed in faulted, sheared or weathered zones. The supports used cover the range of available supports for rock caverns. Although final massive concrete arch linings may ultimately have been specified, in most cases the caverns were inherently stable, and relatively limited supports were required to guarantee long term safety and stability. Suitable long term support systems, depending on rock conditions were:

(a) no support or spot bolting in very good rocks
(b) pattern bolting (equal spacings, with length equal to more than twice the spacing) in good rocks. Bolts tensioned or untensioned depending on the quality of the rock, and mesh used in fractured rocks
(c) pattern bolting with tensioned bolts and shotcrete, unreinforced or reinforced, depending on thickness and rock quality, in poor quality rocks

Table 1 Data from previous excavations in Hong Kong and similar granite compared with Stanley Caverns

Dimensions (m)	**Tai Koo Station (Sharp, 1989)**	**Quarry Bay Station (Matson, 1989)**	**Guangzhou Power House (Carlsson et al, 1989)**	**Stanley Cavern**
Span	24	15 (max)	22	17 (max)
Height	16	6.5	45	17 (max)
Length	250	20	146	130
Depth to roof	20-70	170	350	20-60
Direction	EW	EW	EW	NW, NE
Rock Structure				
Joint/Fault	Sub vertical	Sub vertical	Sub vertical	Sub vertical
Strike	NE, NW	NE, NW	NE, NW	NE, NW
Directions	Sub horizontal	NNW	NNW (tension sub horizontal	Sub horizontal
Joint Character	smooth planar unweathered dry	smooth planar unweathered	smooth planar	smooth planar
Joint frequency m^4 RQD%	0.5-2	0.3-1	0.3-1 90	0.5-1 mainly + 90
Rock Quality	Poor to fair	Q=20-50	RMR-80	Q mainly + 20
		Good-very good	Good-very good	Good-very good

There are several historical guides to estimation of rock bolt support, usually specifying support density in terms of notional rock quality. None of these are entirely satisfactory, but Barton's (1989) Q value, which had been modified for Hong Kong conditions, was used as a working guide to support design. Apart from the fault zone and the short transition zone between weathered granite and rock at the portals, the average predicted Q value of the rock in the cavern was above 20 confirming good to very good quality rock. Table 2 gives the design support allocated to corresponding Q values for the contract.

6 Construction

Due to the hilly terrain in the whole Stanley Peninsula, space was scarce and a separate works area was not available to the contractor. This required the creation of working space by cutting steep slopes varying from 60° to 75° in weathered granite at the portal entrance. The temporary slope formed was stabilized by soil nails from 7m to 10m in length at 1m spacing. A 50mm to 75mm thick mesh reinforced shotcrete layer was then sprayed on the slope after the nails had been grouted to prevent surface infiltration and erosion of the slope surface. The portal was established in weathered granite beneath horizontal fully grouted bolts.

The total excavation quantity of 80,000m^3 had to be completed within 12 months in order to avoid delays to the project and it was necessary to work round-the-clock and at two to three faces at the same time. Excavation was carried out by conventional drill-and-blast techniques using jumbo mounted hydraulic drills in a three stage upper heading, and a lower bench excavation. The pull was maximized by using a burn cut pattern in the headings, because it was possible to drill large diameter relief holes easily. Smooth blasting methods were used for the final tunnel profile. A typical main face drill-charge-blast-scale-support-muck cycle for an approximate quantity of 150m^3, which was equivalent to a face area of 6m by 7m and 3.5m pull, is given in Table 3. The excessive mucking out time was caused by restricted truck access. A total of 80,000 kg of explosive was used, giving an average production rate of 1m^3 of rock per kilogramme of explosive. The caverns were above the water table and there was no water ingress into the tunnel apart from minor seepage at isolated locations, mainly through sheared joints.

Except in the fault zone, where support was designed to interact with construction, support was based primarily on the estimates of rock quality in Table 2, with occasional subjective modifications at more sensitive areas such as junctions. Details of roof supports densities are given in

Table 2 - Guide to support specification based on previous practice (after Barton 1989)

Q Value	Cavern (Eq dimension 11.5m)	Other Tunnels (Eq dimension 7-10m)
>40	Spot bolting only	Spot bolting only
10 - 40	3m untensioned bolts at 1.5m spacing	3m untensioned bolts at 1.5m spacing
4 - 10	3m tensioned bolts at 1m spacing	3m untensioned bolts at 1.m spacing; and 30mm shotcrete to be decided
1 - 4	3m tensioned bolts at 1m spacing with 75mm reinforced shotcrete	3m untensioned bolts at 1m spacing with 50mm reinforced shotcrete
0.4 - 1	3m tensioned bolts at 1m spacing with 100m reinforced shotcrete	3m tensioned bolts at 1m spacing with 75mm reinforced shotcrete
0.1 - 0.4	3m tensioned bolts at 1m spacing with 100-200mm reinforced shotcrete	3m tensioned bolts at 1m spacing with 75-200mm reinforced shotcrete
0.1 - 0.1	3m tensioned bolts at 1m spacing with 200-750mm reinforced shotcrete	3m tensioned bolts at 1m spacing with 200-250m reinforced shotcrete

Table 3 - Typical tunnelling production rates in main headings

Description of works	Time (hour)	Labour/No		Plant/No	
Preparation work and setting out	1½	Surveyor Chainman	1 1	Jumbo	1
Drilling shotholes: Depth of hole = 4.5m No of holes = 80	2½	Foreman Operator Miner	1 2 5	Jumbo	1
Temporary support using spot rockbolts	1	Operator Miner	1 2	Jumbo	1
Loading and firing 190 Kg explosive and 80 detonators	1	Foreman Shotfirer Miner Operator	1 2 5 2	Jumbo	1
Ventilation and scaling	½	Miner	4		
Mucking out	4	Operator Driver	1 3	Loader Dumper	1 3

Figure 3. A photograph of the supported roof in Figure 4 give indications of exposed roof qualities and support density.

From the site investigation, it was evident that a major fault zone would cut across the access road tunnel and the service tunnel near the ventilation shaft. When the fault was approached, a pilot tunnel was driven to allow detailed inspection and support design. The fault consisted of a greyish green completely decomposed basalt and Grade V and IV granite, highly metamorphosed with clayey and sandy silts with fine gravels (C/HDG) striking at about 310°N with dip angle 85° cutting across the site from NW to SE. The materials were similar to those identified in the site investigation. No unexpected material was found and, as predicted, the material was supportable. The thickness of the fault varied from 9m to 12m, close to the predicted width of 12m.

Excavation of the fault zone was by conventional multiple heading and benching. A small pilot heading tunnel with 1m

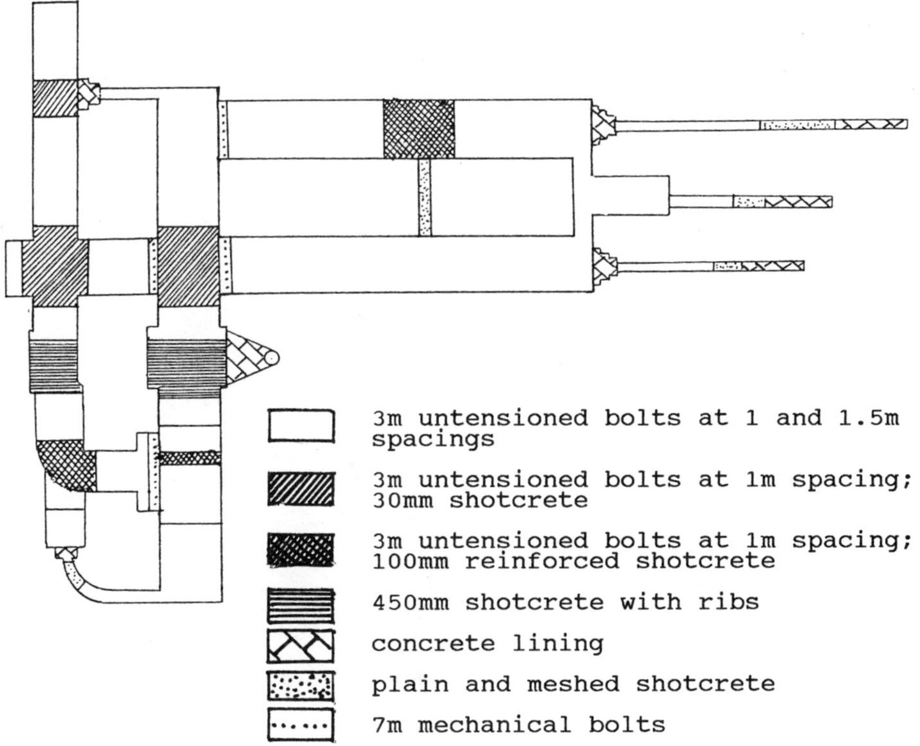

Figure 3 Roof supports installed in the completed cavern excavation

Figure 4 Excavated roof profile in the East Treatment cavern showing the final roof support pattern of 3m untensioned bolts at 1.5m centers

advance rate was first driven. The roof was stabilized initially by shotcrete with steel fibre to improve its strength and steel ribs were erected. The next step was to widen the pilot heading. A second heading was driven and stabilized by shotcrete with steel fibre. Steel ribs were joined to those erected during the pilot tunnel excavation to form a continuous arch at the roof. The sequence was repeated until the crown was fully excavated. At this stage, the roof load was taken by the steel ribs and was transferred to the abutment by an extension of the crown excavation over the sidewall. The shotcrete layer was then increased to the design thickness so that the composite action of the steel ribs and the shotcrete lining could effectively transfer the vertical roof load to the abutment. The bench was then excavated and the sidewall shotcreted to complete a fully supported section.

7 Environmental precautions

The main environmental concern at the site was the close proximity of underground blasting to nearby sensitive structures including the high security Stanley Prison, the sports centre at St Stephen's College Primary School Section with a swimming pool across Wong Ma Kok Road and classrooms on the other side, and a water supply reservoir 50m immediately above the crown of the East Treatment Cavern. The Water Supplies Department had imposed a restriction that the maximum peak particle velocity (ppv) measured at the reservoir site should not exceed 13mm/s. Blast vibration control was therefore an important aspect of the blast design.

Vibration monitoring was carried out using vibrographs to measure ppv at critical structures such as at the reservoir walls, the perimeter wall of the prison, the swimming pool and the site access road to the resident engineer's office. From the monitoring results, it was observed that ppv at these locations was below 9mm/s most of the time. On rare occasions this was exceeded and a maximum ppv of 14.6mm/s was recorded at the swimming pool. This good control was achieved by maximizing the use of deci-second delay detonators instead of using half second or milli-second delays. Milli-second delays have a higher tendency to reinforce shock waves from individual intervals. They were mainly used for the initial cut holes. Half-second delays have the advantage of allowing successive pulls to develop more fully but can result in the "fronts" being blown off adjacent later firing blastholes. The use of deci-second delays allowed optimum blasting between the two extremes. They allowed sufficient reinforcement of shock waves for rock fracturing and yet limited the vibration to below acceptable levels. There was enough time for the broken rock produced by the preceding blastholes to be ejected from the cut before the succeeding shots fired. This gave a better balance between fragmentation and throw and hence a more efficient blast design.

Measured vibration data are plotted in Figure 5 using the cube root scaling method recommended by Dowding (1985). This is less conservative for distances around 30m than square root scaling. W is the charge detonating within an 8ms time period. The data were compared with the square root vibration attenuation curve for granite suggested by Holmberg and Persson (1979) which is plotted for 1kg of explosive. This was used for preliminary site vibration estimates and there was good agreement, despite the divergence at greater distances. The upper envelope is perhaps overly conservative and the care needed in using any arbitrary prediction method for a particular site is

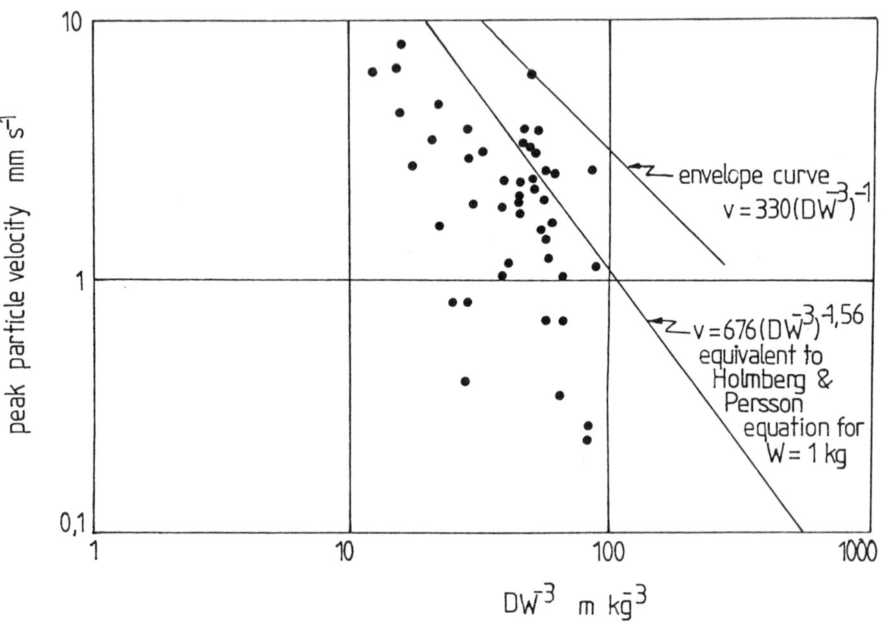

Figure 5 Vibration data collected during cavern blasting

emphasized.

Apart from blast vibrations, noise generated from blasts and also from plant used on site had an important effect on the environment and caused disturbance to nearby residents - in particular noise created at night when the ambient noise level was low. Close monitoring of noise at night was carried out. It was found that the noise level due to blasting, at St Stephen's School, which is the closest residence to the site, varied from 60dBA to 70dBA for the majority of readings. The noise generated by other plant was between 50dBA to 60dBA.

8 Instrumentation and Performance

Although this was not the first major underground cavern project in Hong Kong, it was the first that dispensed with the need for secondary concrete lining except at the fault zone. It is intended that the roof and sidewalls above the tank level will remain unlined. It is normal practice to monitor the long term performance of large span caverns. It was desirable to do so at Stanley as the use of underground space is being explored in Hong Kong and a database record of cavern performance to allow the design of future caverns is desirable. However because of the nature of the rock - stiff with relatively tight discontinuities, very little deformation was expected or observed.

During construction, temporary monitoring of roof lowering was carried out by inverse levelling to an accuracy of 0.2mm to centre roof stations installed immediately before blasting of the centre heading. Although some stations were dislodged by blasting, it was possible to detect some increases in roof deformation following blasting of the centre heading, widening of the upper heading and removal of the bench. The total observed lowering was of the order of 1 to 1.5mm, about twice the estimated elastic roof deformation, and of the order of that measured at Dinorwic by Douglas and Wilson, (1979).

Inverse levelling has been maintained for long term monitoring together with tape extensometer deformation measurement stations to determine lateral deformation of the upper caverns above the treatment tanks. Convergence meters with an accuracy of .025mm were installed at the junctions between the Service Cavern and each of the Treatment Caverns to measure long term roof to floor convergence.

Long term monitoring of movement in the roof will be carried out by the three borehole extensometers and 40 instrumented rockbolts. The borehole extensometers were installed at the mid-point of the East Treatment Cavern, mid-point of the West Treatment Cavern and the junction of the Service Cavern and West Treatment Cavern. Each borehole extensometer comprised three anchors together with head assembly, terminal box, interconnecting cable and readout unit to allow remote monitoring. For a competent rock like massive granite, the plastic zone will be very small and as very little movement was expected, the first anchor was set as close to the head as possible. The three anchors were set at 0.10m, 0.50m and 5m. No differential movement has been observed.

The other more extensively used long term monitoring instrument was the instrumented rockbolt. This comprised a miniature vibrating wire strain gauge embedded and clamped in a hole drilled along the central axis of a standard rockbolt. The gauge monitors the strain and hence the load in the bolt after installation. It is then linked to the control station to allow remote monitoring. After six months, no change in bolt loading had been observed.

9 Conclusion

The project will be commissioned in mid 1994, by which time the only visible site area will be the portal along the Wong Ma Kok Road. The success of this project is a major step forward in the utilization of underground space in Hong Kong.

Acknowledgements

The authors acknowledge the permission of the Director of Drainage Services Department, Hong Kong Government for the publication of this paper.

References

Oswell M A, Guilfoyle J J, Thomas R L (1989) Stanley underground sewage treatment works - Hong Kong. Rock Cavern - Hong Kong, IMM, London pp 75-82

Strange P J and Shaw R (1986) Geology of Hong Kong Island and Kowloon. Hong Kong Geological Survey, Memoir No 4, Government Information Services Department, Hong Kong

McFeat-Smith I, Workman D R, Burnett A D & Chuau P Y (1989) Geology of Hong Kong. Bull. Assoc. Eng. Geol. Vol 26 No 1 pp 17-107

Hueze F E & Amadei B (1985) The NC borehole jack; a lesson in trials and errors. Int. Jl. Rock Mechs. Min. Sci. & Geomech. Abstr. Vol 22 pp 105-112

FitzSimmons C (1992) Review and analysis of Goodman Jack. Internal Report, University of Arizona.

Sharp J C (1989) Design of major caverns in granitic rock based on precedence practice in Hong Kong and elsewhere. Rock Cavern -Hong Kong, IMM, London pp 247-270

Barton N (1989) Cavern design for Hong Kong rocks. Rock Cavern -Hong Kong, IMM, London pp 179-202 & Geomech.

Irvin R A, Garritty P, Farmer I W (1987) The effect of boundary yield on in-situ stress measurements using overcoring techniques. Int. Jl. Rock Mechs. Min. Sci. Abstr. Vol 24 pp 89-93

Carlsson A, Olsson T, Richards L & Zhu Y (1989) Engineering geological aspects of underground construction, Guangzhou pumped storage power station. Rock Cavern - Hong Kong, IMM, London pp 309-320.

Geotechnical Engineering Office (1992) Guide to cavern engineering, Geoguide 4, Government Information Services Department, Hong Kong

Matson C.R. (1989) Rock conditions art Quarry Bay station applied to cavern construction in Hong Kong. Rock Cavern-Hong Kong, IMM, London, pp271-292

Dowding C H (1985) Blast Vibration Monitoring and Control, Prentice-Hall, NJ

Holmberg R & Persson P A (1979) Design of tunnel perimeter blasthole patterns to prevent rock damage, Tunnelling '79 pp 280-283, IMM, London

Douglas T H and Wilson W (1979) Dinorwic Power Station. Movements recorded at completion of cavern excavation. Proc. 4th Int. Cong. ISRM, Montreux, Vol 3, pp 312-315

The Snake: a rock tunnel for storing combined sewage water, Stockholm

Daniel Morfeldt
Mineconsult AB, Stockholm, Sweden

Abstract
The fullface-driven rock tunnel "The Snake" has been bored in a geological environment consisting of hard rocks, such as granite and gneiss, beneath the city of Stockholm. Previous lowering of groundwater levels have recurrently caused structural damage to old buildings and other structures on the surface. To minimise this damage, the Snake Project was designed with a very high limitation on the acceptable inflow of groundwater, allowing only 2 litres per minute for every 100 metres of tunnel length (9.6 m^2). The importance of a professional and competent engineering geological approach to the lay-out and planning of the tunnel route is described, in addition to the in situ control programmes and documentation routines practised. Settlement surveys in the project area and control of other ongoing underground projects are mentioned as tools to understand the hydrogeological situation. Grouting routines and mechanical equipment for driving TBM tunnels are described, and the importance of a control programme is underlined, together with a presentation of the volume of grout consumed and the different qualities of grout used.

The conclusion that can be derived from the Snake Projects is the significance and necessity of an engineering-geological approach to tunnelling in urban areas and the indisputable advantages of using a TBM in combination with pregrouting equipment directly at the face.

1 Introduction

The bedrock in the Stockholm region is part of the Svekofennian ridge, formed 18-1900 million years ago. All the east of Sweden and western parts of Finland are from this geological period.

The oldest rocks from this era are those that settled in a water-bearing environment, namely schists and sandstones in combination with volcanic rocks. Different magmas have thus intruded these fine crystalline rocks, primarily during two main periods of geological activity, and the granites subsequently crystallized in a deep environment over a long period of time. As a result of this, the formation of mineral crystals could achieve considerable size and dimensions. The last generation of these granites, designated "Stockholm granite", comprises many different types with slightly different characteristics, all of which are around 1800 million years old.

During the cooling of the bedrock after magmatic intrusions, all the cracks and fissures which are to be found today were formed. Some of these fissures have been active during several different periods of geological activity and many larger zones of movement have occurred, caused by existing stresses in the rock mass.

The granites in this area normally show major irregularities in the fissure configuration and most of the rock surfaces are extremely cracked. In recent years (10 000 years ago from a geological viewpoint), the ice age scraped clean these rock surfaces and shaped the landscape as we can see it today. Depressions in the surface strata are filled with loose materials such as clay, gravel and sand. Eskers are the best evidence from this period of geological activity.

How to best use this environment in the construction of systems of tunnels and other underground excavations and to establish a better way of living on the surface, has been a major challenge to many engineers in Sweden over the past 100 years. This paper highlights the Snake Project in Stockholm; a rock tunnel driven with TBM technology in highly competent, hard granite rock formations.

2 The Snake

Rain water, which falls on the city of Stockholm, is led to Henriksdal's waste water treatment works. The network and the treatment plant did not have sufficient capacity to cope during periods of heavy precipitation. To avoid flooding, the mixture of rain/waste water was allowed to overflow into the nearest water course. To reduce this overflow, a waste water tunnel has been built where the "surplus water" can be temporarily stored, until the pressure on pipelines and the waste water treatment works is reduced. The Snake has a storage capacity of approx. 35 000 m^3.

To achieve optimal interaction between the tunnel and the main drainage system, the water level is regulated and monitored with the help of computers.

A transport tunnel about 300 metres long and a 200 metre long loading station have been blasted and built. A Tunnel Boring Machine (TBM) was installed for excavation of a 2.8 km long tunnel beneath the Vasastaden and Östermalm residential and business districts of Stockholm.

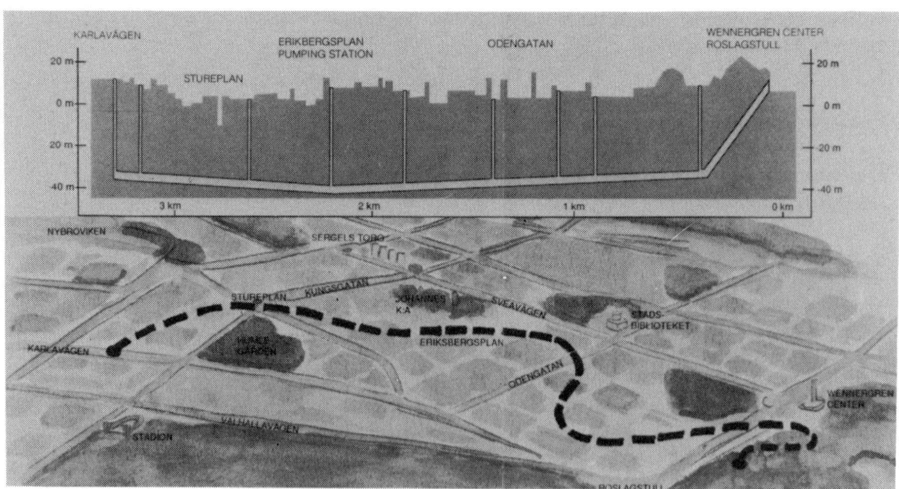

Fig. 1. Profile and view of the Snake Project.

The tunnel meanders like a serpent, hence its name. The TBM FORO 900 is the same machine that was used to drive the Saltsjö Tunnel, with only a few minor modifications. A new generation of hard rock TBMs, with a short body which made it possible to position the probe and grout drilling machines very close to the drilling head, was developed by Atlas Copco. The machine type FORO 900 is only 7 m long from the head to the grippers, and the groutdrills are capable of rotating around the towing bridge, thereby allowing for the positioning of holes around the entire periphery.

The FORO 900 with a cutterhead of 3.5 m diameter and rotating at a speed of 12.3 rpm has a bore stroke of 1.5 m. The head is equipped with 27 disc cutters penetrating the granite. Hydraulic grippers anchore the rear end of the machine to the rock with a force of up to 2 000 tonnes, thus providing a steady base for the three thrust cylinders which are pushing the head against the face with a force of 700 tonnes. A track to carry the 122 m long back-up rig is laid under the 20 metre long towing bridge. The TBM operator is sitting to one side of the forward end of the back-up rig. From this position he is monitoring the operation of the TBM on his control console. All fixed installations (hydraulic power pack, electric cubicles, transformers, compressors and dust exhaust equipment are in the rear end. The conveyor belt for transporting rock materials is running from the head over to the top of the back-up rig to the rear of the TBM and dumped into mine cars for transportation out of the tunnel. This TBM equipment rig can manage a curve radius of down to 200 m.

Behind the TBM, other equipment was used for exploratory drilling. Two Atlas Copco COP 1238 hydraulic rock drills drilled 51 mm diameter exploratory holes down to a depth of 26 metres, at an angel of 6°. Support equipment behind the TBM can be moved forward on a steel frame so that it can be used separately, without interfering with tunnelling operations. The exploratory holes were used for pregrouting. Since this was carried out ahead of the tunnel face, any penetration of groundwater into the tunnel was prevented.

The total length of the tunnel, including the transport tunnel to Eriksbergsplan, is 3.7 km. The transfer station is located at a depth of approx. 40 metres and includes a service workshop for daily inspections of equipment.

The completed tunnel has an area of 9.6 m^2, whereas the downward access tunnels have an area of 25-40 m^2. The tunnel slopes from both directions down towards the pumping station at Eriksbergsplan, approximately 50 metres below street level.

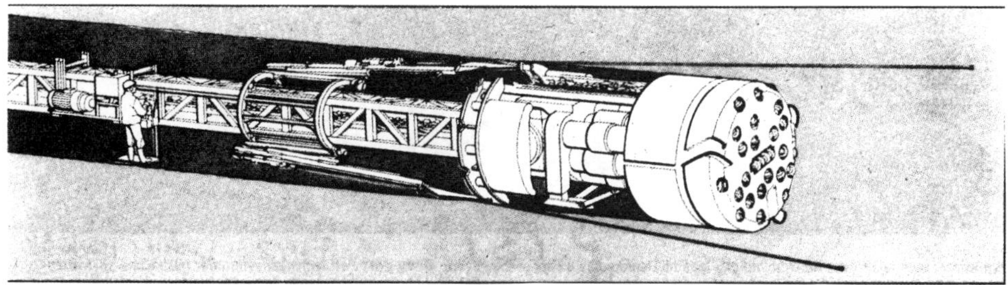

Fig. 2. Drilling of probe and grout holes directly behind the TBM head.

Combined water is led down from eight locations on the combined sewer network via movable weirs, controlled by a computerised control system. From these weirs, the combined sewer is led to the storage tunnel via vertical holes, each with a diameter of between 0.6 and 1.2 m, drilled through the rock. When the flow in the sewer network drops, waste water is pumped from the pumping station 55 metres below Eriksbergsplan up to a network, where it is then transported to Henrikdal's waste water treatment plant. A rock chamber with a storage capacity of about 5 000 m^3 has been blasted and constructed for the pumping station.

3 Why TBM?

3.1 Groundwater environment grouting

The "Snake" project enabled the use of modern TBM technology, thereby reducing the degree of intervention and disturbance to a major city to the minimum. No disturbances therefore arose from noise or blasting operations. The project also enabled monitoring and control of the disturbance effect of vibration as well as raising the degree of information obtained, thus achieving a better tolerance level among the real estate owners affected.

Planning of the Snake Project was carried out by Stockholm Vatten AB with some help from a few consultants. Mineconsult AB was responsible for the engineering geological criteria analyses and the implementation of geological control programmes and documentation routines during the tunnelling operations.

The reservoir is located in an extremely variable geological environment. The region is already well known from earlier rock works for its extensive fissured and weak zones, as well as highly variable types of soil strata and clay-filled depressions in the underlying rock surface. The region is also intersected by the large Brunkebergs Esker (remains).

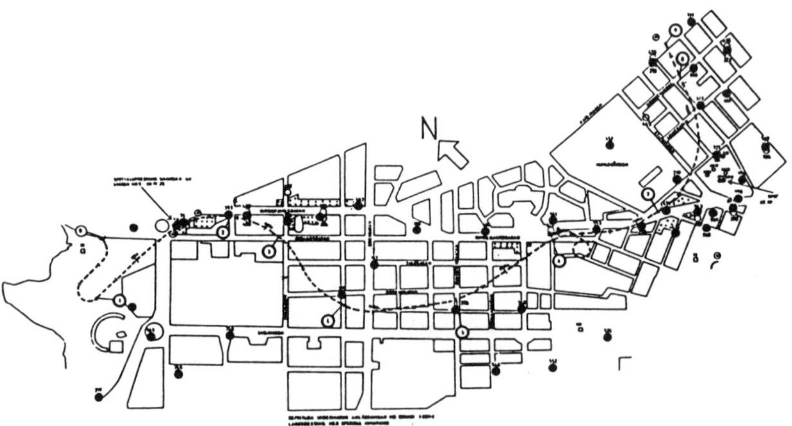

Fig. 3. Location of the observation wells for control of groundwater levels.

Buildings within the region comprise different types of premises supported on different foundation techniques (wooden-steel-concrete piles, foundation walls in variable base materials, etc). An overlapping risk analysis in consideration of the geological environment was therefore conducted, to enable a survey program to be planned with regard to the prevailing groundwater situation in and surrounding the tunnelling area (groundwater model).

3.2 Groundwater observation holes

The municipal authorities for the city of Stockholm have had a functional observation monotoring system that has covered the major part of this worksite area for some considerable time. This was complemented with a small number of newly installed observation holes in order to achieve total coverage of the area. Accurate and well-documented observation readings conducted over a long period, both before and after tunnelling, are of great significance for understanding the "groundwater model" surrounding the project.

A complementary follow-up programme was compiled, so that during the preparation of the tender documents, it was possible to determine specific "risk zones" that would require extra checks and monitoring of the groundwater balance to be conducted as the project progressed.

Extra planning for the collection area was necessary in order to achieve optimal use of the reservoir. Among other things, it was necessary to build a main drainage system from the down discharge shaft by pipe jacking as far as Sveavägen, etc.

This working plan with the planning based on "geological terms", subsequently proved to be well in agreement with the forecasts given as the project progressed, in addition to being a useful and accurate guide for the special measures taken with grouting, etc. Tender documents were thus compiled on the basis of:

(a) Geology
(b) Hydrogeology
(c) Foundation
(d) Soil types disposition

and required that the tunnel was to be driven in full-face sections (minimal zone damage).

Grouting throughout the full-face tunnelling process was continuously followed up by intensive monitoring from the previously mentioned observation holes. In certain sensitive environments, it was necessary to carry out monitoring and measurement as a rule on a weekly basis, and sometimes daily. Through applying this system of monitoring control, a flow model evolved whereby the influence and effect from both seasonal variations and other rock construction activities in the area, could be analysed and interpreted. Regular summaries of this expanded control and monitoring/analytic routine have formed the basis for determining how the grouting work has been carried out.

The main contractor too carried out control measures in the tunnel with the aid of monitoring dams, which in turn allowed continuous measurement of the amount of water leaking. When necessary, this provided the opportunity to temporarily halt tunnelling in order to perform secondary grouting.

Fig. 4. Dam in bottom area of a TBM tunnel for measurement of groundwater inflow.

4 Leakage limit value: TBM technique

Complemented by an extensive pregrouting and follow-up control programme, it was possible to stipulate an approved limit of the leakage volume to a mere 2 litres per minute per 100 tunnel metres. The TBM technique makes this possible since it is easy to spot the smallest leakage inside the tunnel. The smoothly polished surfaces of the tunnel walls allow highly detailed studies of even the smallest leakage path. A remedial programme with secondary grouting, particularly in the marked "risk zones", proved to be sufficient for keeping within the highly restricted leakage limit.

In older city or town centres, as with new constructions in sensitive hydrogeological regions, there exists the risk of damage and lowering of the groundwater table when driving tunnels or when carrying out different forms of underground constructions beneath the existing groundwater level. Drainage of the "lower" groundwater reservoir will eventually result in drainage of the "upper" groundwater reservoir.

The grouting operations performed on the Snake Project have therefore been guided by measurement and control of primarily the "lower" groundwater reservoir. The identified, sensitive risk zones mainly comprise those premises supported on wooden pile foundations that have previously been affected. Over a period of time, before full-face boring commenced, extra settlement and subsidence measurements were conducted on these buildings, and the findings subsequently documented after the tunnel face has bypassed these premises. Efficient control and advance inspection were also carried out in consideration of the blasting operations that occurred during the excavation of the down discharge shaft and at the access tunnels at Roslagstull and Snickarbacken (pump station). An additional inspection of premises, mainly with regard to the status of basement floors and interior walls at basement level, were also well documented, both before and after tunnelling operations. In addition, ground subsidence checks were conducted along the entire length of the tunnel and accompained by extensive photographic documention and records.

5 Result of settlement documentation and measurements

The buildings within the identified risk zones did not at any time show changes in the settlement records which could be connected to the tunnelling work. The foundations show the normal settlement records of around 15 mm per year. The tunnel operation had a good control of the leakage of groundwater and with this very restricted allowance (2 litres/100 m of tunnel and minute) of groundwater disturbance the identified risk zones were kept under control. The measured settlements in old buildings on weak foundations will continue and the foundations will in the future be re-piled due to other circumstances as old underground works and the uplift of the landscape after the last glacial period in Sweden. Our photo documentation of the surface along the tunnel route did not show any drastical changes which could be connected to the tunnelling work.

6 Grouting model

Tender documents for the Snake Project specify the special geological circumstances that apply, and the primary demand for an impervious tunnel, to eliminate the risk of lowering and damage to the groundwater table. The control programme that was described earlier was also included in the tender documents; and which also emphasised the importance of controlled pregrouting and the need for secondary grouting in those sections where the leakage limit could not be attained. The main contractor is responsible for the grouting work and control thereof.

Sealing of the tunnel was mainly achieved through continuous pregrouting cycles. Following minor modification of the grouting-drilling equipment, the grout team began by drilling four collar holes at equal spaces around the tunnel periphery. Following inspection and measurement in the tunnel, and additional measures in the form of extensive secondary grouting, it was decided to increase the number of collar holes to eight. Following this, subsequent secondary grouting was only necessary on an occasional basis.

The length of the grouting holes was chosen in order to allow grouting to be mainly carried out at night-time, thus enabling the TBM equipment to be used to maximum effect, as boring with the TBM was prohibited between the hours 10 pm and 6 am.

While the grouting equipment was installed on the "back-up" part of the TBM, it remained close to the tunnel face at all times and has functioned well with pregrouting as part of the daily work cycle.

Probing and grout drilling were carried out with hole lengths up to 30 metres. The choice of hole lengths was partly governed by curves of the tunnel in the horizontal plane, and partly by scheduled TBM drilling for the following day. This drilling was carried out using two Atlas Copco COP 1238 hydraulic rock drills, positioned immediately behind the TBM cutter head in order to minimise the quantity of "unnecessary" drill metres.

A total of 44 000 m of probe and grout holes were drilled. This means 16 m/m of tunnel. At the tender stage of the project, it was estimated that it would be necessary to drill 29 000 m, or 11 m/m of tunnel. One explanation for the big difference between the initial estimate and actual reality is that it was decided at an early stage to use grouting patterns throughout with eight holes.

The choice of the grout medium was determined during water loss measurement. With water losses of up to 5 litres per minute per hole, Microcement alufix was used; and with water losses of between 5 and 20 l/min, Microcement W650. Where water losses exceeded 20 l/min, Cemcil was used, though instances of this were few and far between. Consumption of grouting materials totalled 164 tons, or 60 kg/m tunnel, which coincides well with the estimate quoted at the tender stage.

As pregrouting proved so successful, secondary grouting was only necessary on a small scale and in just a few sections of the tunnel. The results following grouting have been extremely satisfactory, which is largely due to the fact that the tunnel has been bored full-face. This is particularly noticeable when tackling leakage on the tunnel floor. It is extremely easy to spot and follow a fissure formation around the entire periphery of the tunnel.

Altogether, a total of 2 300 metres was drilled for secondary grouting and 10 tons of grout were consumed for this purpose. The type of grout used was exclusively Microcement, mainly W650. Cement was used exclusively throughout grouting operations. No chemical or water glass grouting agents were used whatsoever. Microcement consists of a normal water-cement mixture, but the cement is grinded some extra times in order to achieve very deep penetration value even in very "small" fissures.

Cemcil is an accelarator additive to a microcement mixture, used only when large waterleakage appears.

Summary: Water loss measurement - Grouting agents

Water loss	*Grout mix (type)*
≤ 5 l/min	Microcement (alufix)
5-20 l/min	Microcement W650
> 20 l/min	Cemcil

Total consumption 164 tons = 60 kg/m TBM tunnel
Total grout hole m = 44 000 = 16 m/m TBM tunnel

Grouting of the tunnel has been described as a complete success. It can therefore be claimed that the use of the right technique (TBM) in combination with a well proven grouting routine, can drastically reduce the risk and likelihood of "the swirling mists of a black art".

7 Instructional model

The geological and hydrogeological circumstances surrounding Humlegården have been the case for an in-depth study, following the completion of a full-face boring project through this "disturbed" area. Results from mapping and leaking measurements have been visually reproduced in a scale model (1:400), that has been used for control and guidance with regard to secondary grouting and additional reinforcement measures. In a couple of tunnel sections, an additional covering of shotcrete was recommended, mainly for the purpose of reducing future maintenance needs (erosion wear caused by water flows).

The engineering geological information have been digitalized and the documentation methods are presented on the following pages (Fig. 5 and Fig. 6).

8 Summary

The full-face driven rock tunnel reservoir, The Snake, was constructed in a geological environment that demanded extensive measures with regard to pilot studies and control programmes. The technique chosen, combined with a controlled and well proven pregrouting programme along the entire length of the tunnel, resulted in a substantially lower risk of possible disturbance to surrounding surface buildings and other underground installations. The control and documentation programmes carried out have resulted in the client, Stockholm Vatten AB, obtaining a solid platform of invaluable information with regard to the geological and hydrogeological parameters of this rock construction project. Operational management of the tunnel reservoir too has been made easier, and in the future it will be possible to conduct rock inspections and controls of the groundwater status and situation within the region, using these follow-ups as a basis.

Experiences from this project have provided us with reasons for conducting more overlapping studies of different underground projects, and their effect and influence on the groundwater balance in specific regions. Even with observations of seasonal variations and isostatic uplift, etc, one will still wonder whether Stockholm is in possession of a well documented groundwater status situation. How will the limit for permissible leakage be implemented on the next project in this region? The answers to questions like these can not be regarded as clear cut, and this, together with the project described above, prove the great significance and importance of engineering geology in the expansion of a city's or town's infrastructure.

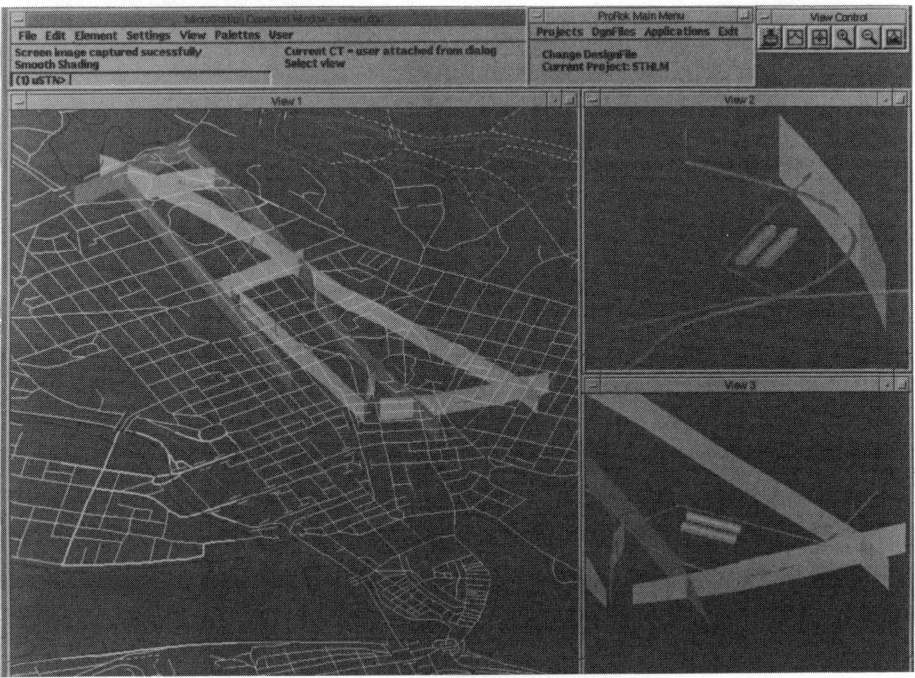

Fig. 5. Computerized picture of the engineering geological situation around the Snake in Central Stockholm.

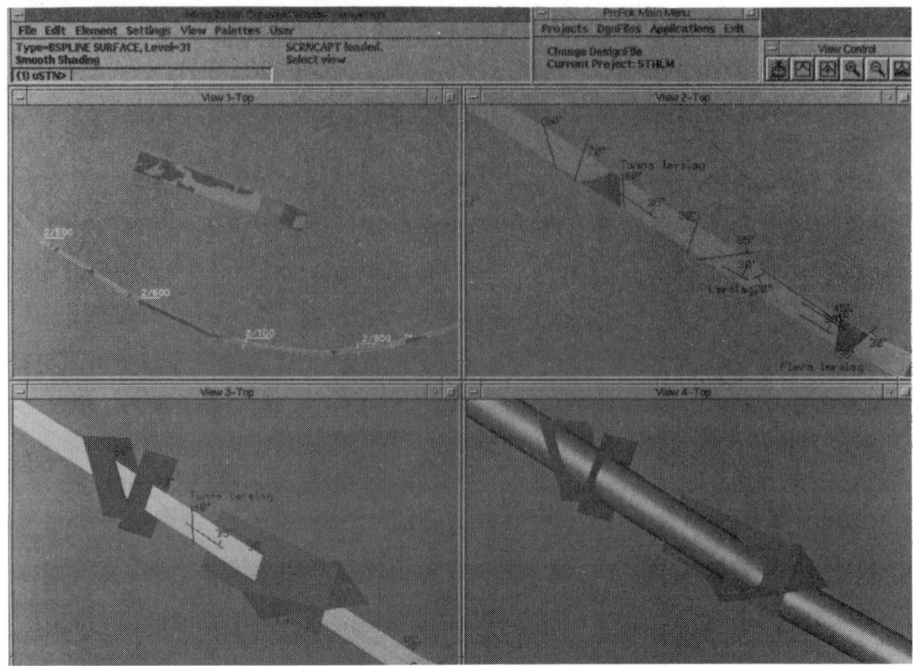

Fig. 6. Computerized documentation of fissures crossing the Snake tunnel.

Fig. 7. TBM tunnel in hard rock.

Tunnelling on the Penzance and St. Ives sewerage scheme, Cornwall, England

C. J. H. Martin
Trafalgar House Construction (Tunnelling), Bentley, Doncaster, England
G. Pakes
Gerard Pakes Consultants, Wirksworth, England

SYNOPSIS

This paper describes the design and construction of tunnelling works on the Penzance & St.Ives Sewerage Scheme and Sewerage Treatment in south west Cornwall.
The scheme includes two major tunnels:

1. A sea outfall from Gwithian adjacent to Godrevy Point in St.Ives Bay approximately 2.65km seaward.
2. An interceptor sewer and storage tunnel parallel to the coastline at Penzance.

The paper describes the initial concept and detailed design of the two tunnels and illustrates how the design and construction elements have been inter-related to the benefit of both.

1. PENZANCE INTERCEPTOR AND STORAGE TUNNEL

1.1 Historical Background

The design and construct contract awarded in 1992 to Trafalgar House Water Projects by South West Water Services Limited, included interceptor sewers in Penzance and an outfall from Chyandour for storm water. As an alternative to this configuration, Trafalgar House proposed the construction of a storage tunnel to be combined with the interceptor sewer in order to eliminate the need for an outfall. Storm flows originally intended to be discharged directly from the Chyandour outfall, would then be stored in the tunnel and pumped from the Chyandour Pumping Station for treatment at Hayle, before further transmission to the outfall at Gwithian on the north coast. The advantages of this scheme were many, and included the following.

1. It was the least cost option.
2. It caused minimum disruption during construction.
3. It promised minimum environmental nuisance during operation.
4. It avoided discharge into Mounts Bay at frequencies greater than a 1 in 30 year storm.

1.2 Tunnel Geometry

The geometry of the interceptor sewer is determined by the necessity to provide:

1. A conduit for foul and surface water transfer from Newlyn to a pumping station at Chyandour adjacent to Penzance Railway station.
2. Connections for five sewers at fixed points between West Penzance and Chyandour.
3. Approximately 16,000m^3 of stormwater storage in the system.

A detailed study of the likely ground conditions resulted in a division into three lengths with different construction methods.

1. From Newlyn to West Penzance, at Bolitho Gardens in Wherrytown, - an opencut pipeline.
2. From Bolitho Gardens, via Wherrytown promenade, - approximately 800m of 1.83m diameter interceptor sewer tunnel.
3. From the end of the 1.8m diameter tunnel, approximately 1600m of 3.2m diameter combined storage and interceptor sewer tunnel to Chyandour, via Battery Rocks and Albert Pier.

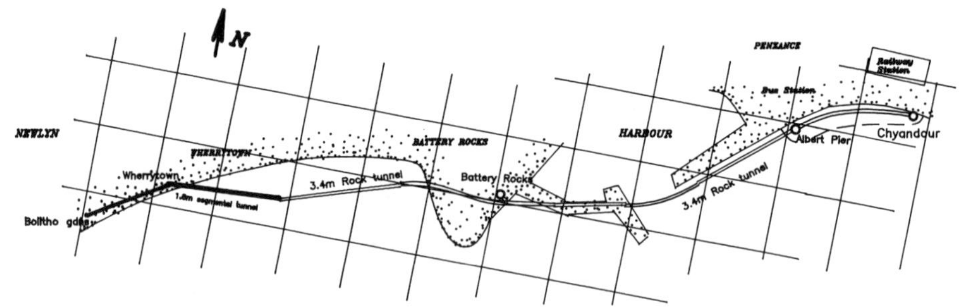

Fig 1 Penzance Tunnel

The tunnels are at a depth of between 24 and 30 metres below ordnance datum and for much of the length are located parallel to the shoreline and under the sea.

1.3 Geotechnical Assessment

The south west region of Cornwall is dominated by the Lands End Granite which occupies about 60% of the area. The remainder consists of Devonian sediments with intrusions of igneous rocks. The coastal area in the vicinity of Penzance is generally a Mylor Slate formation of Devonian age which is a grey/green or dark grey slate. The Lands End Granite intrusion and other igneous intrusions have produced some contact metamorphism of the slate on the interface. The Mylor Slate is also intruded by a second phase of dykes and sills of dolerite and the induration of the Devonian sediments has produced a "greenstone". In addition to the dolerites, intrusions of quartz- feldspar-porphyry, locally known as "elvan", are present in the area. These dykes trend east/northeast to west/southwest and vary from 1m to 20m in width, but are usually about 10m wide. They dip to the north or northwest and some can be traced for distances of up to 2 km.

The Mylor Slates are regarded as weak to moderately strong (ie. from 5 MPa to 50 MPa) while the dolerites are classed as very strong to extremely strong (100 to 200 MPa) and the elvans, strong to very strong (50 to 200 MPa). The implacement of the Lands End granite and other orogenies has lead to a complex history of faulting in the area and the line of the tunnel is crossed by many faults. All rocks are of generally low permeability, but the extensive faulting results in a high rock mass permeability, with the possibility of considerable water ingress into the tunnel during construction.

During the past four hundred years, there has been considerable metaliferous mining in the area, including a major tin mine offshore from Wherrytown exploiting the mineralised areas on the boundaries of an elvan dyke. Although the exact positions of the workings are unknown, the tunnel route had to avoid the postulated offshore shaft positions.

A number of exploratory boreholes were drilled along the tunnel line, which combined with surface mapping, confirmed the general geological assessment and allowed the choice of construction methods. The borehole data was used as a basis for detailed design of the tunnels and shafts.

The borehole investigation indicated that a significant number of dolerite/slate interfaces occurred in the region between Wherrytown and the eastern end of the tunnel section. This area also appeared to be highly faulted and weathered, even at depth. To the east of Wherrytown, an Elvan dyke was found to run in the same direction as the tunnel but about 150m offshore. From the Battery Rocks Shaft eastwards to Chyandour, the tunnel line was occupied mainly by Mylor Slates and greenstones, containing several large faults, including a major one underlying the harbour entrance.

1.4 Penzance Tunnel Detailed Design

Despite the intrusions of dolerite in the Wherrytown area, the generally weathered, faulted slates at the western end of the tunnel appeared to have low stability ratings and short stand-up times, requiring immediate support. A lining of 1.8m diameter, smooth-bore, bolted segments was therefore chosen for this length. The segmentally lined section extended some 400m each side of the working shaft at Wherrytown, the western length running under to the promenade to the terminal shaft, and the eastern leg angled obliquely out to sea in a direction to intersect the offshore elvan dyke.

Geological mapping of the major elvan outcrop parallel to the coast indicated that it dipped between 60° and 45° and was sufficiently wide to be utilised for part of the tunnel alignment. The positions of all major intersecting faults were determined by surface mapping both at low tide on the foreshore and from surface land features. This indicated that the elvan dyke probably lay under the western breakwater of the harbour allowing it to be followed for a total of about 800 metres. East of the dyke it would be necessary to cross the main fault under the harbour entrance but from there onwards through Albert Pier shaft to Chyandour the rock should be broken, but generally competent Mylor Slates and Greenstones.

Consideration of the rock mass properties indicated that the most economic tunnelling method was drill and blast excavation, with primary support using rockbolts and shotcrete. The support systems were designed using the Norwegian Geotechnical Institute method of rock quality evaluation. The elvan was predicted to have a Q value of greater than 10, normal slates and greenstones between 1 and 10 and faulted areas as low as 0.05. Four different standard primary support classes were therefore established, requiring the tunnel face to be mapped after each blast, the Q value determined, and the appropriate support class selected and implemented.

To fulfil the requirements for a smooth lining for the storage tunnel, a secondary lining of insitu concrete was designed. As some lengths of the tunnel would be located in fault breccia and other gouge filled joints, the lining was designed to resist full hydrostatic and geostatic loading. Preliminary calculations indicated a 250mm thick mass concrete lining would be sufficient for a tunnel 3.2m finished diameter.

The design life of the project is 100 years, and in order to maintain maximum durability, reinforcement steel was avoided wherever possible. The concrete mix, a nominal C35, and the aggregates, were selected to resist chemical attack from the stored dilute sewage and stormwater, as well as the saline groundwater. In order to reduce the potentially acid atmosphere in an empty storage tunnel as far as possible, the dry weather sewage flow was restricted to a low flow pipe, cast into the lower corner of the lining.

At the Wherrytown shaft site, the detailed ground ıvestigation indicated weathered and broken rock tending to below tunnel invert level. A 4.57m ameter, caisson-sunk, bolted segmental shaft with an situ concrete lining was chosen. At Albert Pier, and yandour, the shafts were to be constructed in exposed positions on the marine side of the sea wall on reclaimed bunds. The design took into account the method of working which involved keying-in 6.1m diameter segments to the rock at low tide and building a collar upwards as fill was placed. Once completed to a safe level above high tide underpinning of the segmental shafts could be commenced. Although the original intention was for primary supported rock shafts with insitu linings below weathered rock level, the ground in both locations was so poor that segments had to be carried to full depth. At Battery Rocks, where a connection was to be made, a vertical drop pipe and a man-access were accomodated in a shaft, raised from the completed tunnel.

After construction had started it was decided that the Chyandour shaft should be located closer to Albert Pier, thus reducing the total length of the 3.2m diameterrock tunnel by some 200m. To compensate for the decrease in the storage volume, the finished diameter of the remaining tunnel was raised to 3.4m. A redesign of the permanent lining for the increased diameter indicated it would be necessary to raise the nominal thickness to 275mm of class C35 concrete. The additional 25mm was accommodated within the excavation profile.

1.5 Construction

Construction of the Penzance Tunnel took place from two separate working shafts at Albert Pier and Wherrytown some 1.6km apart.

The 1.8m segmental tunnel was driven simultaneously in both directions from the main Wherrytown Shaft. Because of the comparatively short lengths of the drives and the known variability of the strata it was decided that hand excavation techniques, supplemented as required by blasting, would be the most economical means of tunnel advance. Ground conditions varied from clay-like weathered Mylor Slates to very strong dolerite intrusions.

Shields were installed in both East and West drives to provide immediate support in the weaker ground and to assist in building "Bucline" segmental linings. In harder, more competent rock, particularly where blasting was required, immediate support was not required and the shields were allowed to lag some 3m behind the excavation face.

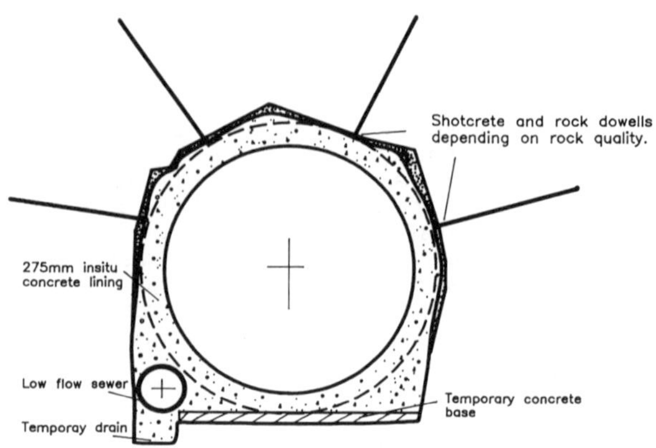

Fig 2 Penzance Tunnel Cross Section

Although probe holes were kept ahead of the face at all times it was decided not to pre-grout the small diameter tunnel unless absolutely necessary. The preferred technique was to accept the water inflows at the face and to seal them off behind the face by pressure grouting against the segmental lining. This method of water control proved generally successful with a final pressure grouting operation being required on completion of tunnelling and prior to caulking of joints.

Once the tunnel was dry, and had been caulked, a 525m diameter half-round concrete pipe was installed in the invert and benched to form a "low-flow" channel.

The second working shaft was located at Albert Pier whence an approximately 4m horseshoe rock tunnel was driven 180m east to Chyandour Pump Station shaft and 1200m west to intercept the segmental tunnel described above. Whilst strata conditions along the length of this tunnel were known to vary from very strong dolerite to weak Mylor slates, these were not expected to be as weathered as those at Wherrytown. The tunnel was therefore planned as a conventional drill-and-blast drive.

The short, easterly section was equipped with a single-boom, crawler mounted electro-hydraulic drill rig, whilst in the long west drive a two-boom, Tamrock Minimatic rig was used. Blasted rock was removed from both tunnels to the shaft bottom initially by diesel powered 2m³ G.H.H. load-haul-dump unit, supplemented by a 4m³ G.H.H. unit as tunnelling progressed. At the shaft bottom spoil was loaded into 4m³ capacity skips and hauled up the shaft by crane. Probe holes 30m long were drilled after every 25m of advance and, if water was intercepted, a 12 hole pattern was drilled and cement grouted ahead of the advancing face. Initial support requirements were determined using the NG1 Q-system following face mapping, and was provided by 2.4m long "swellex" rock bolts and steel fibre-reinforced shotcrete dry batched on site and applied using an Aliva 260 shotcrete machine.

Tunnelling eastwards proceeded very much according to plan for a distance of 140m from the shaft. At this point, immediately after passing through a 20m wide dolerite intrusion, the strata abrubtly reverted to Mylor slates and significant quantities of water were encountered issuing from what appeared to be an open channel to the sea bed in the roof. Probe holes drilled ahead of the face encountered further large quantities of water indicating a band of highly fractured and disturbed ground about 10m wide. After fully supporting the roof and face, several grout covers were drilled and injected. As the Chyandour reception shaft had, by this time, been sunk to full depth, the decision taken to tunnel west from Chyandour for 38m in order to approach the water-bearing area from the other side. Further grout covers were then drilled and grouted from both sides before the final 10m of tunnel was carefully holed through by heading and bench, supported by shotcrete and lattice arches.

Although the cause of this highly disturbed zone has not been established, there are strong suspicions that the old Cornish tin and copper miners had been active in the area before mining records began.

The drive westward from Albert Pier also started as planned in Mylor slates. As described Section 1.4 the plan was to drive this tunnel beneath Penzance harbour through the slates and intercept a known elvan dyke near the Battery Rocks shaft. The tunnel would then follow the line of the dyke, taking advantage of the anticipated competent igneous rock parallel to the shore until meeting up with the 1.8m segmental driven eastwards from the Wherrytown site.

Two events occurred which caused this plan to be changed. On first contact with the dyke, the elvan proved to be as competent as anticipated and a good tunnelling medium. Further into the centre of the dyke, however, the joints became more open and rock blockier. Whilst still a strong rock of high quality, overbreak became a problem, and the quantity of shotcrete used in strata support increased dramatically. This had detrimental effects on progress and costs. However, when the tunnel diverted temporarily out of the dyke in order to intercept the Battery Rocks connection, it was found that the indurated shale in the dyke margins was a more competent and better tunnelling medium than the dyke itself, and the problems of blocky overbreak disappeared.

Following further investigation by core drilling from the tunnel face, it was decided to take advantage of the improved conditions in the dyke margin and the tunnel was diverted out of the elvan and was completed in the baked shales on a parallel alignment.

Although ground treatment by grouting during tunnel construction had controlled water inflows to allow efficient face advance, there remained on completion about 150 litres/second of saline ground-water leaking into the tunnel from various locations along its length. This water had to be controlled and channelled in order to allow the placing of the insitu concrete lining. A "Visqueen" membrane was fastened to the excavation roof and walls by shot nails and was brought down the walls to a drain encased in no-fines concrete running either side of the tunnel. Once the concrete lining was completed the drains could be sealed off and back grouted.

The secondary concrete lining was cast using travelling formwork consisting of five, fully-circular sections each 6m long with four additional invert sections. The low flow sewer pipe was installed in the tunnel floor prior to concreting the invert. The daily cycle consisted of advancing, building and concreting four top sections then advancing, building and concreting four inverts ahead of the completed top sections, producing 24m of full lining daily, with the invert leading by one day.

Concrete was pumped from the surface at the access shafts at Battery Rocks and Albert Pier directly into the shutter, leaving the tunnel clear for membrane fixing, drain laying and other ancillary work ahead of the advancing lining.

2. STONES OUTFALL TUNNEL

2.1 Background

After treatment at Hayle, effluent is to be discharged through a long sea outfall at Gwithian on the eastern side of St.Ives Bay on the north coast adjacent to Godrevy Point. A rising main some 15km long connects the works at Hayle with the outfall. The outfall is 2.65km in length and terminates offshore of a rock outcrop known locally as the "Stones".

As originally designed, this was to be a seabed outfall installed in an excavated trench in the foreshore. The tunnel outfall is an alternative proposed by Trafalgar House Construction (Tunnelling) in view of difficulty in forming a suitable trench through both the drift deposits and the rock in the surf zone. The outfall pipe, laid within the tunnel, is 900mm in diameter and terminates in four sea bed diffusers each containing four ports of 140mm diameter.

2.2 Initial Considerations

It was appreciated at the planning stage that the outfall tunnel would encounter a large variety of rock conditions along its route. Five site investigation boreholes and detailed examination of the exposed cliffs along the shoreline indicated extensively folded sedimentary rocks varying from strong competent sandstones to weak weathered shales. Fault fissures, some tight and dry, others broken, gouge-filled and very wet could be expected at approximately 30m intervals. In addition, quartz veins and stringers, in places several metres thick, would be encountered.

The tunnelling method adopted would need sufficient flexibility to handle all the expected conditions as well as having the facility for advance probe hole drilling and ground treatment. The tunnel profile, however, was not in this case a constricting factor. As the requirement for the outfall was a 900mm diameter conduit, this could be formed by laying a 900mm pipe in a tunnel of the most economic cross section. Thus the tunnel profile could be designed to fit the construction equipment chosen rather than vice-versa.

In the early stages of planning consideration was given to the use of a tunnel boring machine (TBM). It was recognised, however, that a normal open faced rock TBM would not be able to cope with the water bearing fissures without ground treatment. Cover drilling and grouting ahead of a full-face TBM, especially of small diameter, is an awkward and time consuming operation.

The development of fully shielded rock TBMs has progressed over the last few years however and this type of machine was considered. A system of this type would need to provide:

2.2.1 A machine and necessary tunnel lining designed to withstand a 60m hydrostatic head in order to eliminate the need for ground treatment.

2.2.2 The finished tunnel should not be larger than approximately 2.5m diameter or the cost of the machine and the lining becomes prohibitive.

2.2.3 The cutter-head would need to penetrate the hard sandstones and quartz intrusions and also handle the softer and highly fractured zones known to exist along the route. Continual changes between roller cutters and picks on this size of enclosed TBM was not considered a practical option.

2.2.4 A TBM would almost certainly need to be procured second hand as the design and construction of a new machine was unlikely to meet the tight programme.

Following discussions with suppliers of both new and secondhand machines it was concluded that a machine to meet all the above constraints did not exist and, whilst theoretically feasible, was unlikely to be available in the time required. The decision was therefore taken to pursue a "drill-and-blast" tunnel design. Whilst a drill and blast tunnelling method is generally slower than a TBM, it had the following advantages for the Stones Outfall:

2.2.5 The necessary equipment is standard. Tried and tested plant was readily available.

2.2.6 The blast hole drilling rig can also be used for probe drilling and cover drilling and allows ready access to the face. This greatly simplifies the treatment of water-bearing fissures and faults ahead of the advancing face.

2.2.7 Support types can be varied to suit conditions encountered.

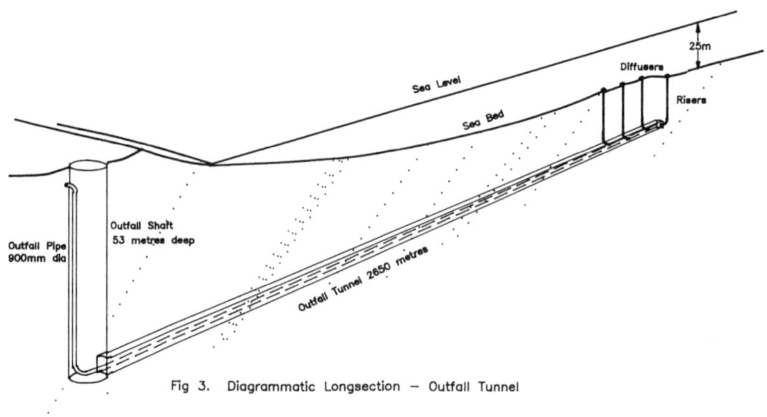

Fig 3. Diagrammatic Longsection – Outfall Tunnel

2.3 Tunnel Geometry

The tunnel is 2.65km in length and driven on an uphill grade from a 53m deep shaft in the old Gwithian Sandworks area. It terminates in four risers containing the diffuser upstands.

Hägglund shuttle-train equipment was chosen for spoil removal from the tunnel, and this determined the cross-section. The minimum size, allowing for ventilation ducts and pumping ranges, is 2.5m wide by 2.8m high. Englargements are made for passing places and marshalling areas.

The diffusers are designed to be drilled from an offshore jack-up barge down to the level tunnel in advance of the tunnel construction. The prefabricated 400mm diffuser units are then grouted into the drilled holes. Blank flanges on the diffuser ports maintain these watertight until the tunnel is completed, the connections made, and the system commissioned.

The outfall pipe laid within the tunnel is to be a 900mm GRP pipe with a concrete surround for protection against rock falls etc.

2.4 Geological Assessment

The tunnel is located mainly in the Devonian flysch deposits consisting of the Mylor Slates and Gramscatho Beds which form the coastline on this particular part of North Cornwall. They consist mainly of indurated mudstones and sandstones. The beds are highly faulted and fractured and although general rock mass permeability can be quite low the water ingress to a tunnel in the fault areas can be very high. The beds lie generally perpendicular to the tunnel line and are dipping away so that the natural tendency of the rock was to break to a square profile. This property was utilised in the choice of tunnel cross-section.

Extensive metaliferous mining has occurred in the area of the foreshore but few records exist. An inspection of the remains of mines on the cliff face indicated that they would probably be shallow and would not affect the tunnel.

2.5 Tunnel Detail Design

The tunnel primary support system has been designed in accordance with Norwegian Geotechnical Institute Q System. As the final conduit is the outfall pipe laid within the tunnel, it was considered unnecessary to provide a secondary tunnel lining. The primary support is therefore the final tunnel support. A design ESR value of 1.6 was chosen, classifying the tunnel as a permanent water tunnel, rather than temporary works.

A total of seven standard support systems were derived for various Q values extending from the minimum expected of 0.001 up to a maximum expected of about 10. Five of the support systems utilise combinations of rockbolts and shotcrete, while two are steel arch girder systems for use in very poor, highly permeable, rock conditions, where lagging and extensive grouting could be required.

The outfall has a design life of 100 years and the rockbolting systems are intended to maintain the stability of the tunnel for this period. Once completed and commissioned it is intended to allow the tunnel to flood, thereby reducing all loads on the support systems. Only in the improbable event of it being necessary to gain access to the tunnel to repair faults in the outfall pipe, will it be necessary to pump the tunnel dry, and a re-entry protocol has been formulated in order to minimise the stress on the support systems and to allow safe access in this eventuality.

The rockbolts used are all 1.8 metre long fully grouted galvanised solid steel dowels. As the rock permeability is low, it is considered that once the tunnel is flooded, water flow and corrosion will be minimised. The epoxy resin bolt grout gives secondary protection against corrosion.

The driveshaft on land was designed as a conventional segmental shaft for a depth of 22m until competent rock was reached and thereafter a rockbolt and shotcrete supported excavation. On completion of the shaft an insitu concrete lining 200mm thick was placed for the full depth.

2.6 Diffuser Design

The diffuser upstands consist of 400mm polyethylene pipe connected to a manifold on the end of the outfall pipe in the tunnel. The polyethylene pipe is enclosed within a steel tube which is grouted into a hole drilled into the rock located about 2m offset from the tunnel. The diffuser sea bed head units are in concrete, with polyethylene ducts terminating in flanges on which rubber "duck bill" valves can be mounted. During construction, these flanges are blanked off, in order to prevent sea ingress into the tunnel workings. The polyhedral design of the concrete diffuser head is intended to reduce the possibility of snagged fishing nets, or anchor flukes.

2.7 Construction

The drill and blast equipment combination chosen consisted of a rebuilt, two-boom, Tamrock, electro-hydraulic, rail mounted, drill rig, together with a Hägglund 8HR loader and three 140 Hägglund shuttle cars pulled by a SIG 15 tonne battery operated locomotive. A fourth 140 Hägglund car was introduced into the train at the halfway point to allow the length of round drilled to be increased and to improve efficiency.

Full column resin anchored rockbolts 1.8m long were installed by hand held drilling machines and fibre reinforced shotcrete was dry batched on site, transported to the face in $4m^3$ Mulhauser cars and applied by a SBS shotcrete machine mounted on a purpose made train.

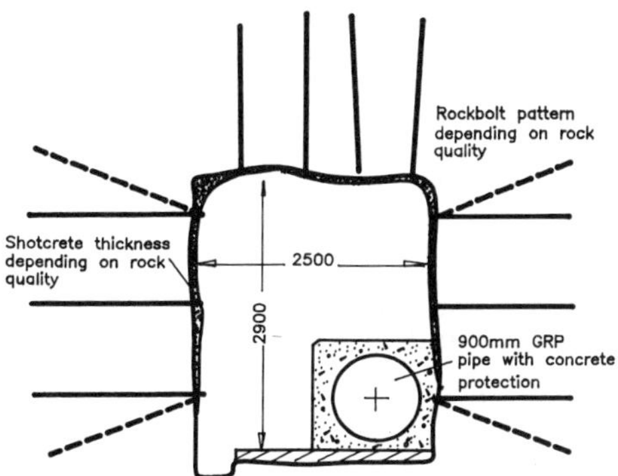

Fig 4 Outfall Tunnel Cross-Section

Advance probe holes were drilled at 30m intervals and, if more than 0.75 L/S of water was intercepted, an 8 hole cover was drilled and grouted. Some 70% of the tunnel drive has required ground treatment to date and, on several occasions, the 8 hole cover had to be increased to 16 holes before the water inflow could be contained sufficiently to allow tunnelling to proceed.

On completion of the tunnel, it is calculated that despite all ground treatment, it will be necessary to pump about 150 l/s,or nearly 3 million gallons per day, from the tunnel.

Bulkhead doors, to protect the workforce and the works in the event of catastrophic water ingress have been constructed at 1000m intervals.

To date the equipment has performed to plan with advances of 70m plus per week being achieved in good ground but reduced to 20-30m per week in soft broken areas requiring considerable support and ground treatment.

At the time of writing this paper, about 70% of the tunnel has been driven, and the support requirements reflect the predictions made in the initial design.

	Class	1	1A	2	3	4	5
Predicted		80%	0	12%	6%	1%	1%
Actual		75%	13.5%	10%	1.1%	0.4%	

On completion of tunnelling the bottom of the diffusers will be located, with short side headings, connections made and the 900mm i.d. outfall pipe laid in the tunnel and up the shaft. The pipe will be surrounded in concrete for protection and, after successful commissioning of the system, the pumps will be removed and the tunnel allowed to flood.

3. COMMENTS ON DESIGN AND CONSTRUCT METHOD

The tunnels described in this paper have been constructed on the basis of an initial concept design, with the detailed design completed while construction progressed. Both operations were under the direct control of the tunnelling Contractor.

This overall control has allowed an ambitious programme to be achieved. When the contract was awarded to Trafalgar House in early March 1992 the tunnels were no more than a concept. By June 1992 the site investigation was completed and detailed design well under way. By the end of July 1992 work had started on site. Tunnelling at Penzance was completed in March 1994 and is expected to complete in May 1994 at the Stones Outfall.

Once the concept design is complete, design-and-construct projects change the basic philosophy on site, from design-led operations to construction orientated. The detailed designer has to react to the Contractors requirements rather than vice-versa. This results in a very efficient system, where amendments can easily be made to a design in order to save costs on the much more expensive construction phase. Variations, both those initiated by the Client, and those forced upon the Contractor by ground conditions can be readily accommodated, as the final detailed configuration of the works need not be decided until immediately before construction.

The initial concept design must, however, be retained in a project such as the Penzance Scheme which was tendered as a fixed price, and the designers and Contractors have only that framework in which to move. The question of contract form for design and construct tunnelling works, where unforseen conditions are common, should perhaps be reconsidered.

Work is now well under way for the commissioning of the whole scheme on programme in early 1995. Meanwhile relationships between the Client, his Consultant, the Main Contractor and the Tunnel Contractor remain excellent and the possibility of any contractual claim is slim. It is improbable that the above could have been achieved under any other form of contract than design and construct.

ACKNOWLEDGEMENTS

The authors wish to thank the client, South West Water, and their consultants, Pell Frishman Water, and also the Main Contractor, Trafalgar House Water Projects for their permission to publish this paper.

Session 4: Tunnel-boring Machines

Two tunnel-boring machines for Lesotho: a design and case history

Sigurdur Finnsson
The Robbins Company, Kent, Washington, U.S.A.

Abstract
This paper discusses the Muela and Hlotse tunnels being driven by two new Robbins Tunnel Boring Machines. The paper is divided into four sections. In the first part, a short overview of the scope of the project is presented. Secondly, the geological conditions along the two tunnel drives are briefly introduced. The third part discusses the various design criteria and the resulting machine specifications. The paper concludes with a case history describing the Muela and Hlotse tunnelling operations from start-up to the time of the writing of this paper.
Keywords: TBM Design Criteria, TBM specifications, TBM operation and performance.

1 Introduction

1.1 The Lesotho Highlands Water Project

This project is one of the largest construction projects underway in the world today. The principal project design involves transferring water from the Kingdom of Lesotho to South Africa and simultaneously utilizing the head difference for power generation.

Currently, the first phase of the project is under construction and is scheduled for completion in the year 1996. Phase 1A includes the construction of two dams, an underground power station, as well as about 82km of tunnel, about 5m in diameter.

The Katse dam, approximately 180m high, will create a reservoir with a surface area of about 36km^2 and a total storage capacity of 1950 million m^3. Further to the north, the Katse intake adit is the start of the transfer tunnel. The transfer tunnel ends at the surge shaft of the underground Muela hydro power complex where the power station will have a 70MW generating capacity. The water is discharged into the Muela tailpond, formed by the 55m high Muela concrete gravity dam.

The tunnelling operations provide three main tunnels, the above mentioned transfer tunnel, the delivery tunnel south (DTS) and the delivery tunnel north

(DTN). The transfer tunnel and the DTS were contracted to Lesotho Highlands Project Contractors (LHPC) which is a joint venture of Spie Batignolles, LTA, Ed Züblin, Balfour Beatty and Campenon Bernard. The Hochtief/Marti-Inter/Concor (HMC) joint venture was awarded the DTN.

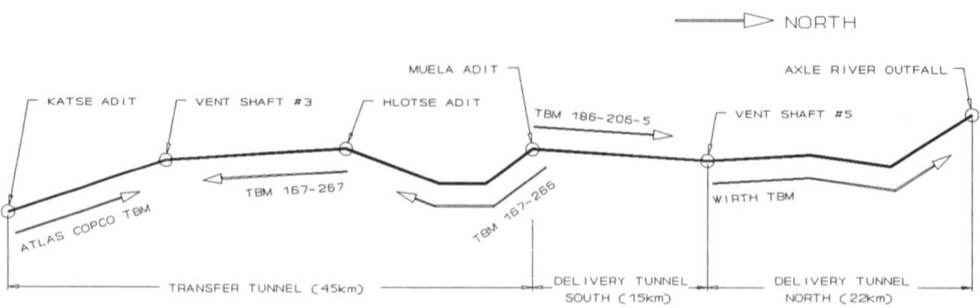

Fig 1. Schematic Overview of the Lesotho Highlands Tunnels

The transfer tunnel is 44.6km long and with a bored diameter of 5m. The tunnel starts at the Katse adit and ends at the Muela adit, from which one of the new Robbins TBMs began boring. This tunnel is being excavated by three open hard rock TBMs, two of which are the main topic of this paper. The third TBM, a Jarva Mk15 boring from the Katse portal, was supplied by Atlas Copco prior to its acquisition of Robbins.

The DTS, starts at the Muela tailpond reservoir and ends just south of the Caledon river at ventilation shaft no 5. This tunnel is 15.1km in length and 5.2m in diameter. The TBM boring this tunnel is a refurbished open hard rock Robbins TBM, model 186-206-5.

The DTN, is being excavated by a 5.4m diameter shielded TBM manufactured by Wirth. This tunnel begins where the DTS leaves off and ends at the outfall into the Axle River. As the tunnel goes under the Caledon River, it will cross the border into South Africa.

The geology along this tunnel is considered the most difficult of the three. Of the 22km being driven by this TBM, just over 19km will be fully lined with a reinforced concrete segmental precast lining.

2 Geology

The geology of the project area is mainly characterized by the Clarens sandstone and Lesotho/Drakensberg basaltic formations.

The basalt forms the mountains of the highlands, overlying the sedimentary formations of the lowlands. The sedimentary rocks include the Elliot and Molteno formations, which consist of silt, sand and claystones respectively. The latter two formations will be encountered in the DTN with the final stretch being in the Beaufort group which consists mainly of claystones.

The transfer tunnel runs exclusively in the basaltic lava. The lava consist of numerous horizontal flows, ranging from 5cm to over 70m in thickness. At the tunnel horizon, the average lava flows are expected to be on the order of 5 to 6m thick. The occasional occurrence of hard doleritic feeder dikes is also expected. The UCS values for the basalt are expected to range from about 80 to 190MPa with values as high as 300MPa being recorded for the dolerite dikes.

Generally speaking, the rock along the transfer tunnel is favorable for TBM excavation and few support problems were expected to occur. The machines are however most likely to encounter blocky ground conditions in the area of faults and within some of the dikes, requiring additional support.

At upstream end of the DTS there is also a drastic change in the geology. The tunnel traverses the massive Clarens sandstone formation. The sandstone is mostly competent and core samples had high RQD ratings. Compressive strength tests yielded values between 20 and 180MPa with the majority being in the 50 to 120MPa range.

3 TBM Design Criteria

3.1 TBM Selection

Based on the geological conditions expected to occur along the transfer tunnel, it was
clear that the machines chosen for the Muela and Hlotse drives would be of the open hard rock type. These machines are well suited for good ground conditions, where support will be occasional and consists mainly of ring beams, rock bolts or wire mesh. Due to the open design, access to the walls and crown is relatively easy, allowing for the placement of supports as required.

The Robbins open hard rock TBM consists of three major components, the Cutterhead/Cutterhead support, Gripper and Main Beam assemblies.

The cutterhead support houses the main bearing, main drive gear, seals and the cutterhead drive trains. These consist of an electrical motor, gear reducer and hydraulic clutch. The cutterhead support assembly also includes the front, side and roof supports. These supports firmly stabilize the cutterhead during boring and ensure precise cutter tracking.

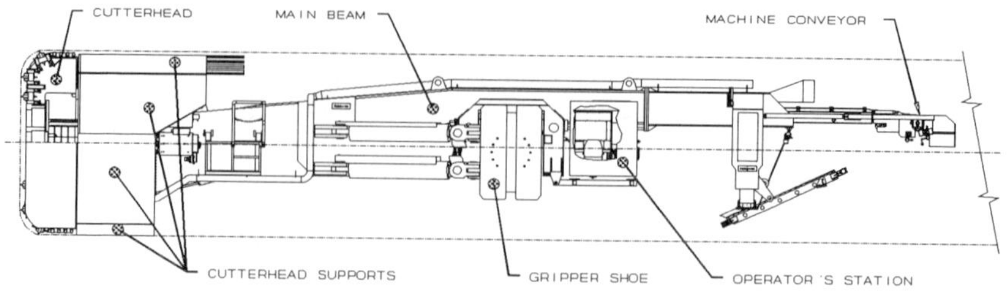

Fig. 2. TBM Model 167-266/267

The main beam assembly consists of the main beam and rear support which is connected to the main beam by a bolted flange. The forward end of the main beam assembly is bolted to the cutterhead support. The aft end of the main beam is supported by the gripper assembly during the boring cycle. The gripper assembly consists of the gripper cylinder, gripper carrier and gripper shoes. This assembly provides reaction to the forward thrust of the machine. The main beam can be moved up, down or sideways relative to the tunnel axis throughout the full working stroke. This is how the TBM is steered. The function of the rear support is to support the weight of the machine during the regrip cycle, when the gripper carrier is being moved forward to begin the next boring stroke. Additionally, the back-up towing system is connected to the rear section support legs.

3.2 Cutter Size

In the initial proposal stage, due to the doleritic dikes, it was suggested that the machines be of the High Performance (HP) class, utilizing the large 483mm diameter cutters. These cutters have load capacities of 311kN compared with the 222kN capacity of the 432mm cutter. As the percentage of tunnel expected to be within the dikes was small, it was later decided to revert to the 432mm cutters. It should however be pointed out that presently two types of 432mm cutters are available, load rated at 222kN and 266kN respectively.

The decision was made to design and construct the TBMs such that at a future date they could be upgraded to HP machines. This is reflected in the design of various components of the machines, as will be discussed in more detail below.

The selection of cutter size to be used on a TBM for given geological conditions is a function of both the strength and abrasive characteristics of the rock. In hard rock formations, large diameter cutters are selected for their high loading capabilities. This means higher penetration and less overall project costs. Brief specifications of the machines for both versions are presented in Table 1.

Table 1. TBM specifications for -266/267 and HP version.

	166 - 266/267	HP TBM Version
Diameter	5.03m	5.03m
Cutter Size	17" (432mm)	19" (483mm)
Load Capacity	222kN	311kN
Number of Cutters	35	34
Cutterhead Thrust	7770kN	10608kN
Cutterhead Power	1575kW (5x315kW)	2205kw (7x315kW)
Cutterhead Torque	1505kNm	2107kNm
Cutterhead RPM	10	10

In the hardest formations, the high load cutters often are the only way to excavate the rock economically, since smaller diameter cutters may not be able to sustain the loads required to break the rock efficiently.

In abrasive formations, the increased wear volume due to larger circumference and cross sectional area results in longer cutter life. This is because more ring material can be worn before the disc will have to be changed. This in turn leads to fewer cutter changes which improves the utilization of the machine.

3.3 Cutterhead Design

The cutterhead has a diameter of 5.03m with new cutters. Due to the likelihood of encountering some blocky ground, it was decided to use recessed back loading cutters. The smooth profile, flat design of the cutterhead helps to support the tunnel face in blocky conditions. This minimizes the possibility of blocks falling from the face and potentially jamming the cutterhead.

Backloading cutters offer a very important safety feature for the tunnel workers. Under unstable ground conditions, the workers can safely change the cutters from within the confines of the head and are therefore never exposed to unstable conditions at the face.

However, should the ground be stable, as was expected to be the case for most of the transfer tunnel, all but the outermost gauge cutters could also be changed from the front. This would result in less cutter change time, since this method is easier and less time consuming.

In order to upgrade the machines to the HP version, a new cutterhead with 483mm cutters would have to be mounted.

3.4 Main Bearing and Drive System

The main bearing is a three axis roller bearing. This is the same design used on the five Robbins HP TBMs built to date. The bearing was chosen due to its capability of supporting the higher loads realized under HP operation.

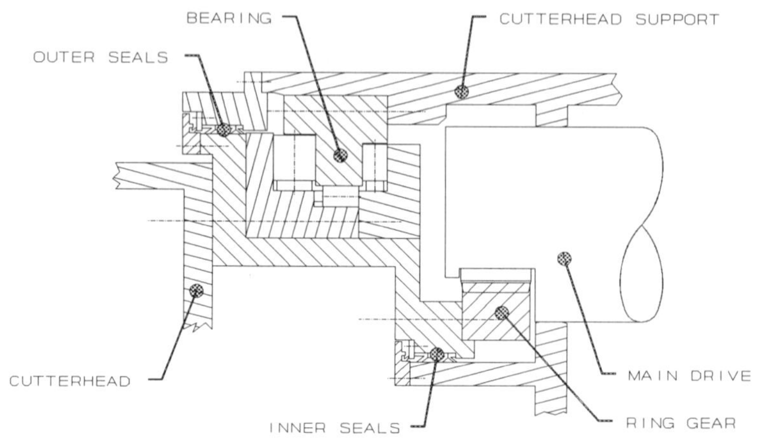

Fig. 3. Main Bearing Concept

The three axis bearing design is very well suited for sustaining the eccentric loads which can occur in blocky or mixed face conditions.

Each drive train consists of a single speed electrical motor, gearbox and a hydraulically activated friction clutch. The drive trains all mesh on a common main drive gear. This provides a rugged system, capable of delivering very high break out torque as well as sustaining a high running torque.

The anticipated tunnel conditions are very well suited to this system, since most of the tunnel is expected to be in fairly uniform conditions.

Should the cutterhead get stuck in blocky conditions, the high break out torque can be very useful. By momentarily applying approximately two times the running torque, the head can in many cases be broken free. This is done by bringing the motors up to full speed and then taking them off line. The clutches are engaged and the rotational inertia of the motors is transferred to the cutterhead.

The machines currently have 5 drive trains, each rated at 315kW. Total installed cutterhead power is therefore 1575kW or 2111Hp.

The cutterhead support has 7 drive train mounts, allowing the cutterhead power to be increased to 2205kW by adding two motors for the HP version. The transformers have adequate capacity for the added electrical requirements.

The RPM of the head is 10. This provides a cutterhead peripheral speed of about 160 m/min which has been found to be a safe operating speed.

3.5 Gripper and Thrust Systems

The hydraulic system of the TBMs is designed to provide the required thrust for the larger 483mm cutters. This includes sizing all the cylinders to be able to operate in the HP mode without having to revert to higher hydraulic pressures. As a result, the machines are currently operating below the gripper and thrust design pressures. This further increases the safety factor already designed into the hydraulic system. Hence, the system wear factors are decreased.

The area of the gripper shoes was also chosen with the upgrade in mind. The gripper load is a linear function of the thrust load. Therefore, the gripper shoes have to be sized for acceptable ground loads at the higher thrust levels. The shoes in the current TBM configuration subject the tunnel wall to lower surface loads, resulting in lower rock stresses.

In the case of ring beams being installed with close spacings, the gripper shoes are slotted. This allows the shoes to straddle the beams without interfering with the operation of the TBM.

3.6 Auxiliary Systems

The TBMs are equipped with an invert conveyor mounted between the rear support legs. This allows the invert to be cleaned by depositing muck from the invert onto the back-up transfer conveyor.

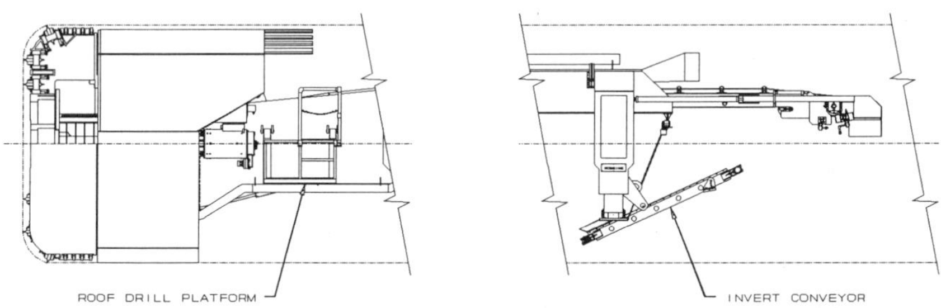

Fig. 4. Auxiliary Systems

Near the front of the TBMs, behind the drive motors, the machines have a drilling platform mounted on either side of the main beam. These platforms are primarily intended for the mounting of rock drills for installation of roof bolts. The platforms can also be used for the installation of ring beams if required.

A hydraulic probe drill is mounted on the first platform of the back-up system trailing each of the machines. The drill allows probe holes to be bored on one side of the machine up to 30m in front of the cutterhead. The angle of the drill allows the ends of the holes to be within 4m of the tunnel perimeter.

3.7 Back-up System

The back-up system was designed and built by Emil Lechner AG, a Swiss company. The system is approximately 190m long. The first gantry of the back-up system is a skid mounted working platform where the probe drill rig is mounted. The remainder of the system consists of 31 decks, rolling on the tunnel rails. The first deck is for the unloading of invert segments with the two following decks providing access and room for additional equipment. Then, five decks provide room for the TBM hydraulic system, the TBM and back-up electrical systems, the scrubber system and lunch room. Three decks constitute a rail switch leading onto a California switch which is located on twelve decks, allowing for muck car loading on one side and parking of the empty muck train on the other. One of the decks on the California switch, houses a car mover. Behind the California switch, another rail switch is mounted on two decks. The following two decks incorporate a second car mover and a safety stop which prevents the train from rolling off the platforms. Towards the rear of the system a single deck, houses the high voltage cable reel, with three other decks supplying mountings for secondary ventilation, duct cassette unit and a working platform for maintenance. The final two decks constitute the ramp, which slides on the tunnel rails.

The system has four conveyors, the bridge conveyor, two loading conveyors and a discharge conveyor. The bridge conveyor transfers the muck from the TBM conveyor and up onto the superstructure of the back-up decks. The loading conveyors transfer the muck back along the back-up system while the discharge conveyor loads the muck cars.

4 TBM Operation

4.1 Manufacture

Lesotho Highlands Project Contractors (LHPC) entered into a contract with the Robbins Company for the supply of two 5m diameter open hard rock TBMs in early April, 1991. Project financing required LHPC to source equipment from the UK, Germany and South Africa. As a result, Robbins decided to manufacture the two machines in the UK.

In joint venturing with Markham and Company, Ltd for the construction of the two TBMs for the Channel Running Tunnels on the English side, Robbins had identified a very capable manufacturing source. Markham of course has been involved in the design and manufacture of soft ground TBMs for quite some time.

In the case of the Lesotho TBMs, the manufacturing of the structural components of the TBMs was subcontracted to Markham. The design work was all done by Robbins in the US.

The hydraulic and electrical power packs, cylinders, main bearing and so forth were Robbins supply and were shipped directly from the various sources to Markham's facility in Chesterfield.

The first drawings where released in late April of 1991, and the two machines shipped on February 19 1992, just over 10 months after the contract signature.

The TBMs left Liverpool on March 2, and arrived in Durban, South Africa on March 22.

4.2 Site Assembly

Overall, the assembly of the 167-266 machine at the Muela site went well. Most of the pieces of the machine were on site by the end of March and assembly commenced. A truck transporting the gripper carrier overturned, causing damage to the carrier and various hoses and brackets. The operator platform sustained extensive damage and the carrier had to be repaired. The machine was assembled and ready for transport into the starter tunnel on May 5, some five weeks after assembly began. The back-up system had yet to be assembled. The starter tunnel rail, however, had not been fully laid. Since the TBM was to be transported on dollies to the face, this was delayed somewhat. At the end of May, the machine was in the starting chamber. By mid June the back-up had been assembled outside the portal and was moved into the tunnel. The next two weeks were spent on final assembly and various testing. The TBM bored it's first stroke on June 24th. The start station for the machine was at chainage 677m.

The assembly of the -267 machine did not go as smoothly. The site experienced labour strikes which affected the assembly to some extent. The machine components arrived on the site at the end of March. The TBM was assembled and ready to move into the tunnel by May 12, but as on the other site, the tunnel rail was not quite complete. So it was not until a month later, middle of June, that the TBM was in the starting chamber. The back-up was assembled and had been moved into the tunnel by early July and the machine bored the first stroke on July 11, after about 2 weeks of final assembly and testing, at station 1194m.

4.3 TBM Performance

After the Muela machine started on June 24, only 4.7m were bored until months end. This was mainly due to minor adjustments that had to be made to various systems of the machine. Table 2 on the following page lists the monthly advance rates of both machines through July 1993.

Table 2. Monthly advance rates for the TBMs

	Muela TBM (m)	Hlotse TBM (m)
June 1992	5	
July	305	130
August	410	458
September	483	867
October	440	896
November	589	240
December	359	274
January 1993	729	647
February	824	483
March	1247	573
April	1343	973
May	1101	783
June	458	1165
July	230	563

Figure 5 depicts the information contained in Table 2. As can be seen in some cases, there are significant differences in performance from one month to the other. Some of the reasons for these variations which include ground conditions and mechanical difficulties will be discussed in the following sections.

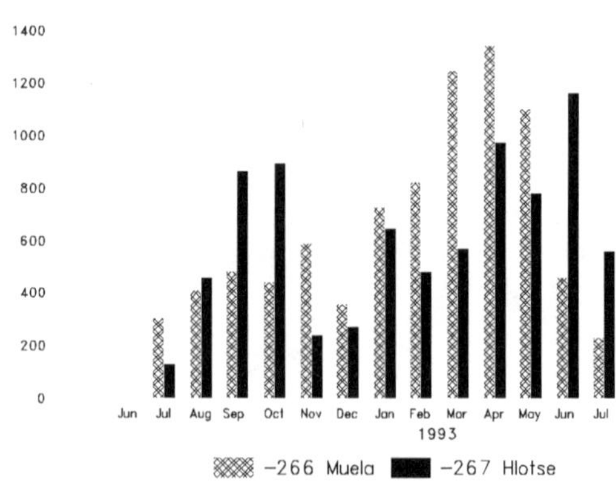

Fig. 5. Monthly Advance Rates

Figure 6 shows the monthly average advance rates, in meters per hour only taking into account actual boring hours.

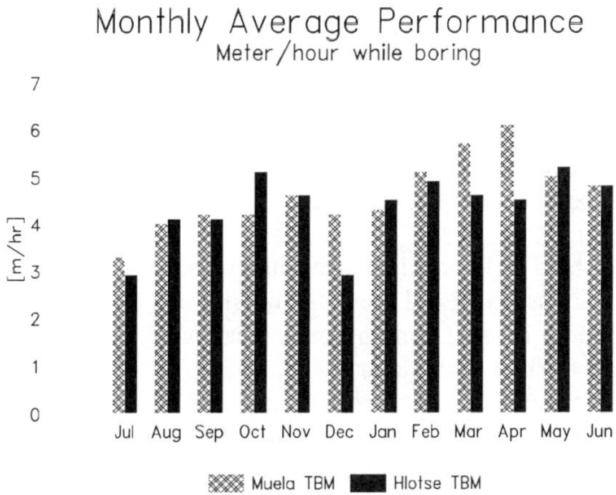

Fig. 6. Monthly Average Performance

Through July of 1993, the Muela machine had bored over 8520m which equates to 655m per month. Corresponding figures for the Hlotse machine are a little over 8050m and 620m per month.

Currently the Muela machine holds the project record at a best day of 83m, a best week of 399m and best month of 1343m.

4.4 TBM Downtime

Overall the machines have functioned quite well to date. There have been the expected minor maintenance requirements. The main problems have been with the seal system of the TBMs and the wheels on the back-up system decks. These will be described in further detail below.

In early November 1992 the rear section on the Hlotse machine was seriously damaged. The bolts on one of the flanges attaching the rear section to the main beam had all been broken. There was also some evident damage to the main beam itself. The repairs were completed in about a week. The exact reason for this mishap is unknown, but it was most likely due to the fact that the wheels on the back-up system had slowly been seezing up. This was assumed to be due to ingress of muck and fines into the wheel bearings. The tow cylinders for the back-up attach to the rear support legs of the machines. As the wheels started seizing, the towing force had to be increased, eventually overloading the flange joint. It is also suspected that the rear support legs were dragging on the invert when the incident occurred further loading the joint.

The wheel bearings are of the bushing type which has been used successfully on other back-up systems. Over a period of time, with much effort, the bearings were changed on all the wheels and an o-ring seal was added. This worked for a while, but the problem has re-occurred. This solution was also implemented on the Muela TBM, but at this point, it is not clear why the problem is worse on the Hlotse machine. In order to assist in moving the back-up, the main thrust cylinders have been used, via a connection from the gripper carrier to the back-up. Air winches, mounted on the back-up and connected to bolts in the tunnel wall, have also been used for moving of the back-up. This has resulted in much increased time for regrip.

The seal system on the TBMs consists of an inner and outer seal arrangement. Each seal arrangement consists (listed from the outside in) of a v-ring seal, two wandering lip seals and an oil seal running on a wear band. The seals are lubricated with oil and flushed with water.

In the course of three months, between February and May 1993, some of the outer seals on both machines failed twice, four failures in total. In three of the four cases, the most likely cause was a combination of packing of muck between the cutterhead and cutterhead support and damage to seal water flushing components. The material caught between the head and the cutterhead support was ground up and eventually found its way to the seals. In one case, the lube lines to the seals had been sheared, most likely by a rock or piece of bucket lip. It is surmised that the loss of lubrication was responsible in that case. In all four cases, all of the outer seals were replaced. This involved removing the cutterhead to gain access. This was done by bolting the cutterhead to the tunnel face and pulling the machine back. On the average, it took about one week to replace the seals and get the machine working again.

The muck build-up between the cutterhead and cutterhead support was apparently the main reason for the seal failures. As a result, the modification focused on improving the muck pick-up on the cutterhead, i.e. trying to minimize the amount of material potentially packing up and secondly, further isolating the seals. Modifications to the head included the installation of new bucket lips, installation of new back-scoop lips and removal of one rear grill bar from each bucket. In addition, the seal flushing components were relocated and further protected.

The seal isolating procedure involved the addition of a cone ring and a guard ring. These now cover the upper 180 degrees of the cutterhead support.

These modifications on the Muela machine were done in the month of August, which coincided with a TBM stoppage due to shaft work.

The same modifications have been accomplished on the Hlotse TBM and neither machine has had any problems since.

4.5 TBM Utilization

Each working day consists of 3 eight hour shifts. The utilization below is defined as actual boring hours per the 24 hour period. The following charts depict the average utilization of the machines during the first year of operation. It should be noted that the "other" category includes events such as electrical outage, muck train absence, shift change, dealing with water inflows, etc.

The following charts depict the utilization for the Muela machine through the third week in May and through June for the Hlotse machine.

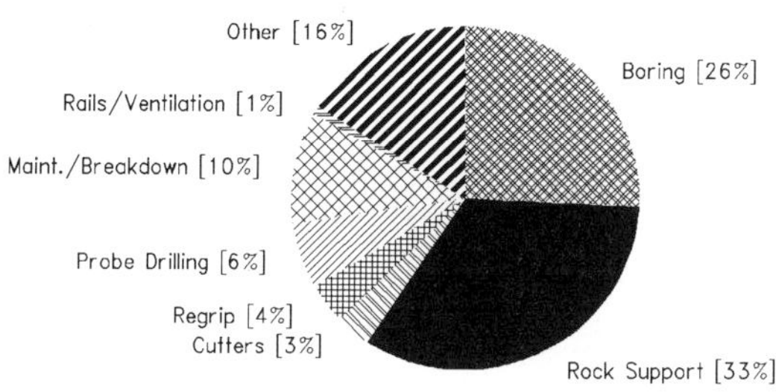

Fig. 7. Utilization, Muela TBM

It is noteworthy how much time was dedicated to ground support on the Muela project, but it should also be noted that there was a sharp decrease in this activity after the beginning of the year. At this point the Muela machine had encountered much better ground, with drastically reduced ground support requirements, while conversely the Hlotse machine was entering slightly worse ground. As an example, of the first six months on the Muela drive, an average of 43 percent of the time was devoted to ground support. During the next six months, through June of 1993, a corresponding figure was only 12 percent. Similarly at the Hlotse drive, the figures are more even at 13 percent and 17 percent respectively.

When the penetration rate of a machine increases, so does the portion of time spent regripping. As a result, one would expect the Hlotse machine to have better performance, since the regrip percentage is higher. This is not the case. One of the main reasons for the difference were the problems with the wheels on the back-up system. This significantly increased the time required for regripping.

Figure 8 on the following page shows the utilization for the Hlotse TBM.

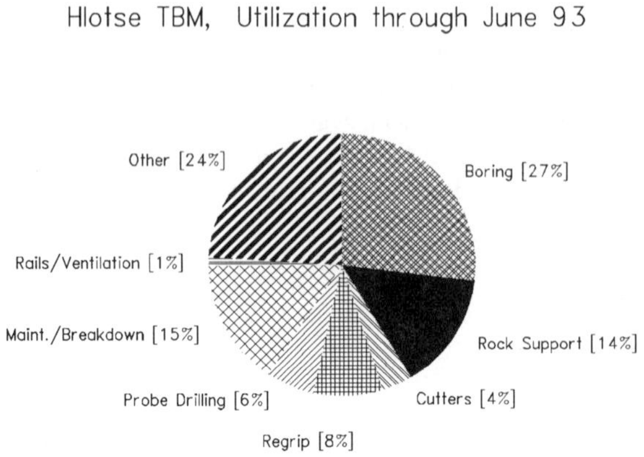

Hlotse TBM, Utilization through June 93

Other [24%]
Boring [27%]
Rails/Ventilation [1%]
Maint./Breakdown [15%]
Rock Support [14%]
Probe Drilling [6%]
Cutters [4%]
Regrip [8%]

Fig. 8. Utilization, Hlotse TBM

4.6 Ground Support

The ground support so far has consisted solely of rock bolting and wire mesh. Much of the bolting has been required for the sidewalls, typically between 7 and 10 o'clock on one side, 2 and 5 on the other. For the bolting work handheld stopers are used, and bolting for the crown takes place from the platforms behind the cutterhead support, while wall bolting is sometimes accomplished from the invert. The amount of bolted support has varied from eight bolts per meter to occasional bolting as required.

The ground along the Muela Drive during the first three kilometers or so was worse than anticipated and this resulted in a large amount of time being spent on ground support. The ground has improved greatly as evidenced by higher monthly advance rates. The Hlotse site has seen a more even situation in terms of ground condition, with slightly less favorable conditions in the second half of the drive.

5 Summary

The Lesotho Highlands Water Project is one of the most ambitious tunnelling projects undertaken in recent years. Five tunnel boring machines will excavate more than 80 km of tunnel. Through the first year of TBM operations, progress has been fairly good. Both the Hlotse and Muela machines have had very good months, in excess of 1100m, and have proven their capabilities. The machines have had problems with the main bearing seals, but recent modifications seem to be working very well. Ground conditions overall have been worse than anticipated, but have improved.

It is expected that the TBMs will continue to advance in a similar fashion and at the end of the project, tunnel boring machines will once again have proven their viability and superior performance over conventional methods on long tunnel drives.

References

Brockway, J. Dollinger, G. and Finnsson, S. (1992) Hard Rock Tunnel Boring Machine Technical Definition in **TUNCON '92: Design and Construction of Tunnels** South African National Council on Tunnelling, Yeoville, R.S.A., pp. 39-44

Davies, A.G. (1992) An Introductory Description of the Lesotho Highlands Water Project with Emphasis on the Implementation of Phase IA in **TUNCON '92: Design and Construction of Tunnels** South African National Council on Tunnelling, Yeoville, R.S.A., pp. 9-16

Wallis, S. (1992) **Lesotho Highlands Water Project**, Project Brochure, S. Wallis, U.K.

Open TBM tunnelling in complex geology for railroad in Sweden

H. A. Janzon
The Robbins Company, Kent, Washington, U.S.A.

Abstract
In competent, hard rock Tunnel Boring Machines of so called open design are generally preferred. In weak, unstable formations, shield machines must be used to create immediate support for the ground which further to the rear comes to rest against the lining of the tunnel.

Both types of machines are well suited to the particular set of conditions they are designed for. It is the area in between – where zones of poor ground will alternate with hard, solid rock – which causes difficulties, if tunnelling is done with either a typical hard rock TBM or a shield machine.

Twin railroad tunnels presently being driven in the SW part of Sweden through the Halland Ridge run in mainly gneissic formations which show extensive fracturization and frequent, clay filled weathering zones.

The measures taken by the supplier of the TBM for the project to adapt the machine to the particular set of geological conditions are reviewed in this paper. They have resulted in a very flexible design which permits probing, grouting, rock bolting and ring beam erection to be done with equipment incorporated from the start.
Keywords: TBM, Design, Adaptation, Flexibility, Hard rock, Mixed ground.

1 Introduction

Early Tunnel Boring Machines (TBM) were only suitable for work in very soft rock formations. This holds true for both the “ancient” TBMs developed in the second half of the last century – example: Col. Beaumont and English’s machine of 1881 – and the first “modern” TBM built by James S. Robbins in 1952.

The manifest advantages of boring a tunnel instead of excavating it by drilling and blasting have, over the last four decades, always focused machine development efforts on making TBMs better suited to tackle ever harder rock formations with acceptable economy.

Combined mechanical and metallurgical developments created cutters that could utilize increased cutterhead thrust and rotational power to achieve the remarkable levels of performance of today’s TBMs with regard to rock hardness.

The many tunnels that have been bored successfully in the mainly igneous formations of the Scandinavian shield and elsewhere – in very hard and abrasive rock – indicate that the hardness of the rock per se no longer forms a problem. What worries engineers now are the zones of poor ground which often occur in the path of a tunnel to be driven in predominantly competent, hard rock.

This paper reviews design considerations for a TBM which has to operate under those conditions. To exemplify various points, the author refers to the layout of the JARVA Mk 27 TBM, presently used to drive twin railroad tunnels in Sweden.

2 The Halland Ridge railroad tunnel

The Halland Ridge – some 90 km north from the city of Malmö in south Sweden, is an up to 150 m high ridge which runs east – west, separating the flatter regions to the north and south of it. The present railway line along the coast from Malmö to the town of Gothenburg follows a rather tortuous valley intersecting the ridge somewhat below its top.

2.1 Project history

In 1985 the Swedish State Railways began an investigation of the general aspects of building a tunnel just under 9 km long through the ridge on a more direct route, as part of the overall upgrading of the line between the two main west coast cities. In 1989 a pre-study of the technical issues involved in tunnelling under the Halland Ridge was commissioned which showed the project to be feasible.

Construction plans and bid documents were completed early 1991 and requests for bids went out to 25 pre-qualified construction companies in July, 1991. During the original bidding period of two and a half months, the point was raised that a double track tunnel might be a suitable alternative to the twin single track tunnels specified in the bidding documents. Consequently, the bid end date was extended to the end of November, '91. A total of nine bids were received in the end.

The two alternatives – a large section single tunnel and smaller twin tunnels – were evaluated against the criteria of investment, stability, safety and operating conditions. This proved to be rather complicated and answers to further queries had to be obtained from several bidders. The questions of safety and stability in particular ultimately tipped the scales in favour of the twin tunnels alternative.

While original plans specified conventionally driven twin tunnels – **drill and blast** – a bid was put in by SVENSKA KRAFTBYGGARNA ENTREPRENAD AB (SKE) for excavation by TBM as well. This turned out to be 6 % lower than their bid for conventional excavation. After intensive negotiations during February, 1992, SKE was awarded the contract on March 5, 1992 and had just started driving the tunnels with a TBM at the time this paper was prepared for submission to this Conference.

2.2 Geology

2.2.1 Macro-geological data

The Halland Ridge is a horst – an **upthrown block of ground** – at the very southern edge of the gneissic strata in SW Sweden, which border on the granites found along the coast and inland further to the north, and the younger sediments to the south. It was probably formed during the younger Cretaceous Period.

The ridge itself consists of a flat dipping (N - NW) red-grey gneiss with parallel-lying amphibolitic inclusions, some centimetres to several tens of metres thick. Several diabase intrusions – which in places have a width of up to 100 m and dip steeply SW – intersect the ridge.

Some 1.5 - 2 km from the tunnel's southern end, a clay-filled crush zone occurs which is believed may have a width of up to 200 m.

2.2.2 Rock compressive strength

Core samples of fresh gneiss taken from bore holes along the tunnel route showed an Unconfined Compressive Strength (UCS) of 195.3 ± 43.7 MPa. Most of the cores showed a quite high joint frequency: approximately 70 % of the samples examined had a joint distance of 10 to 20 cm. The remaining 30 % displayed a joint distance of 2 - 10 cm.

Fresh amphibolite samples exhibited a UCS of 207.1 ± 8.2 MPa.

2.2.3 Faults and water inflows

The ridge is criss-crossed by various fault systems which manifest themselves in the form of crush zones and tension cracks. In the central and northern parts of the tunnel route these cracks generally are open and in the southern part they are most often clay-filled.

As the annual rainfall in the area is 1'000 mm – 550 mm of which finds its way into the ground – the open cracks are expected to lead to considerable water inflows into the tunnel, under fairly high pressure.

To make matters worse, a small lake is situated fairly close to the tunnel line in the northern part. The owner of the fish nursery located in the lake would prefer not to see his brood disappear into the tunnel. It seems safe to assume that the tunnelling crew is similarly inclined!

2.3 Tunnel engineering aspects

While the tunnel is being driven, various types of ground reinforcement will have to be installed, partly to secure the tunnel during the drivage, partly to meet the long range – 100 years! – safety requirements for the end-product. A schedule of eight reinforcement classes has therefore been established, linked to the rock quality (Q-index) intervals.

2.3.1 Rock quality

Of the total distance to be tunnelled, 80 % is estimated to fall in the Q value 1 - 10 range: poor to rather good ground, where a TBM will experience less problems than Drill and Blast excavation would.

Some 15 % of the total length is expected to have a Q value of 0.5 - 5. This represents poor ground, where more than just rock bolting will definitely be required.

Then there is another 5 % with a Q value < 0.1: extremely poor ground, mostly within the southernmost 2 km of the tunnel route.

2.3.2 Reinforcement

The owner has stipulated that the tunnels must be shotcreted over their full length, also in the case of high Q-values. This ties in with the condition that the inflow of water may not exceed 12 l/min per 100 m for the two tunnel bores together. Ring beams combined with shotcrete will be used as a means of support in weaker zones.

The local Water Authority furthermore has decreed that the tunnel must be pre-grouted to control the ground water table in the zone with open cracks – this should take care of the problem of the wandering fish! In zones where the gneisses and amphibolites are converted into clayish material, a decision regarding the need of pre-grouting will be based on the result of probe holes to be drilled as the tunnel is advanced.

2.4 Drivage

2.4.1 Tunnel layout

In the final concept, each of the twin bores will be 8.6 km long and will have an excavated diameter of 9.1 m. The bores will be located 32 m apart C/C, except for the portals, where they will run closer together in cast in situ concrete cover structures.

The bores are to be interconnected at approx. 500 m intervals by 4.5 m square, conventionally driven crosscuts which will house auxiliary equipment for the operation of the tunnel and serve as safety exits.

2.4.2 Driving sequence

The twin tunnels are to be driven in the following sequence:

1. Starting from the north portal, the eastern bore will be driven in a generally SW direction first.
2. After breakthrough at the south portal, the TBM will be turned around and shifted sideways in preparation for the return bore.
3. Starting from the south portal, the western bore will then be driven parallel to the eastern bore to breakthrough at the north portal.

3 Type of TBM selected

From the geological information presented above it was clear from the outset, that a TBM for the Halland Ridge tunnel had to have pronounced hard rock capabilities. Otherwise, it would not be able to tackle the gneiss formations. For this type of ground, one will normally select a machine of so called "open design", due to the excellent accessibility it provides of both the cutterhead – important with a view to cutter changes, likely to occur often in hard, abrasive rock – and of the rock surrounding the machine, for placing the occasional rock bolt.

On the other hand, it was obvious that the machine would also have to deal with conditions of very poor ground which – had they prevailed over the full length of the tunnel – undoubtedly would have resulted in the choice of a tunnelling shield. That type of machines provides immediate support for the ground which – further to the rear – comes to rest against the permanent lining installed under such conditions. Tunnelling shields find purchase against the front end of the lining to develop the thrust on the cutterhead.

In the case of the Halland Ridge tunnel, such a lining was not to be employed, however, and for that reason alone a pure shield machine could not be considered. It thus became a matter of finding a hybrid solution, ie of adapting a hard rock machine in such a way that it would be able to also cope with the stretches of poor ground and at the same time meet the demands of probing and pre-grouting made by the Authorities, as described above.

Fortunately, a good deal of information on the ground to be traversed was available before the design of a machine had to be finalized. As further details became known after the machine had been ordered, the opportunity existed to – while it still was in the engineering phase – adapt the machine's layout thereto in most instances.

3.1 TBM specifications

Having already taken delivery of a 6.5 m diam JARVA Mk 27 TBM from ATLAS COPCO – which by then had started operations in hard ground on the Klippen hydro-electric power project in the northern part of Sweden – it was not illogical for SKE to opt for another MK 27 for the Halland Ridge project.

Because the Mk 27 in its basic layout is designed to cover such a wide range of tunnel diameters – from 6.4 to 12.0 m (10 - 12 m in soft rock only) – many of its main components and sub-assemblies are common, ie diameter-independent.

A realistic chance therefore existed to modify the Klippen machine – should it finish that 10'800 m long project in time – to Halland Ridge specifications to be able to use the thus rebuilt machine for the second of the twin Halland Ridge bores to gain time on that project, if needed.

The main data finally agreed upon for the Halland Ridge Mk 27 are presented in the specification appearing below.

3.2 Mk 27 - Halland Ridge – Main specifications

Boring diameter	
• general range for machine type	6.4 - 12.0 m
• nominal for this project	9.1 m
• overbore capacity	up to 9.4 m
Cutterhead	
• installed capacity	3'360 kW
• rotational speed	0 - 5.77 rpm (at full torque)
	5.77 - 10.0 rpm (at constant power)
• torque	5'560 kNm (0 - 5.77 rpm)
• thrust	18'550 kN maximum
• cutters, standard configuration (9.1 m)	4 centre cutters, 16¼ in. ϕ
	52 face and gauge cutters, 20 in. ϕ, water-cooled
overbore configuration (9.4 m)	4 overbore cutters, 16¼ in. ϕ, on swing out arms
• boring stroke	2'050 mm
Gripping	
• max total gripping force	55'600 kN maximum
• number of grippers	4, arranged in front and rear pairs
• specific ground pressure	2.34 MPa maximum
Drive train	
• main motors, number/type	6 only/TEWC, 560 kW, 660 V/50 Hz

• reduction, planetary gears	43.714 : 1
pinion/bull gear	5.889 : 1
	257.427 : 1 total reduction
• jog/start of cutterhead	electrically, with frequency conversion
• connection motor/planetary gear	torque limiting coupling with adjustable maximum torque transfer

Muck conveyor

• width of belt	1'200 mm
• belt speed	0 - 2.5 m/sec
• capacity	corresponding to ROP 6.0 m/hr
• conveyor retract distance	5.1 m

Electrical system

• installed capacity - TBM	4'000 kVA
- other consumers	360 kVA
	4'360 kVA total

Control system, operator's station

• PLC based, manual operation from control panel with key switch, switches, push buttons and tell tale lights, placed in airconditioned and silenced cabin; closed circuit TV monitoring.

Overall dimensions and weight

• TBM length	21.0 m overall
	17.1 m until rear end of drive train
	3.9 m conveyor overhang
• equipment weight	~ 950 ton metric, TBM proper
	~ 25 ton metric, components on Back Up

4 TBM design for hard rock

The most crucial issues in building TBMs for large diameter tunnels to be driven in hard rock will be examined below. At the same time, we will assess how some design problems were solved for the Mk 27 TBM in general – while further down a few of the special provisions for coping with the poor ground of the Halland Ridge will be reviewed.

4.1 Cutters

Over the years, the demands for high rates of penetration, also in hard rock, have been met by steadily increasing the force applied per cutter, and by rotating the cutterhead faster. This was made possible by enlarging and improving the disc cutter, from a diameter of 12 in. (1980) to the 15½, 16¼ and 17 in. cutters which became the industry standard during the later 1980s.

The increase in diameter permitted the use of larger bearings with higher load carrying capacities. The 12 in. cutters were only rated for a load of some 120 kN. Since 1985, the larger cutters can take continuous loads of up to 250 kN and rolling speeds of up to 2.5 m/sec.

At the same time, metallurgical developments had made available cutter rings of steadily improved steel qualities, so that acceptable ring life now is reached also in hard, abrasive ground, eg solid granite and quartzite.

In the late 1980s, ROBBINS introduced a 19 in. diameter cutter with still larger bearings. It is able to take a 312 kN continuous load.

As cutter loads and running speeds are increased, one meets a problem which was not so pronounced in the earlier days, ie that of **high cutter running temperatures**. A large proportion of the power supplied to the cutterhead of a TBM is converted into heat – at the interface between the cutter and the rock. Part of this heat migrates into the rock and disappears with the muck. Because rock has a low thermal conductivity, the greater part flows into the steel cutter ring and from there back into the cutter body. This heat has a negative influence on the life of almost all cutter components and lowers the quality of the oil with which the cutters are filled to lubricate the bearings.

Some three years ago – before THE ROBBINS COMPANY became a member of the worldwide ATLAS COPCO group in the spring of 1993 – ATLAS COPCO started development of a still more powerful cutter, with a specified design load of 350 kN. Fully aware of the heat problem, the engineers decided to equip the cutter with **water cooling**. This 20 in. diameter cutter became available in time for use on the original Mk 27 supplied to SKE by the end of 1991 for the Klippen project mentioned above.

Though the new cutter showed some initial problems in the first run of the machine – which completed the 2.8 km long bored outlet tunnel section of the project – cutter life and penetration rates met the targets set for a specified design load of over 300 kN. As the cause of the teething problems was known, "dentistry" executed prior to the start of the machine on the outfall tunnel remained fairly simple and proved satisfactory.

Also two Mk 27 machines supplied since for driving highway pilot tunnels under Seoul, Korea, and the Mk 27 for the Halland Ridge project are equipped with this new 20 in. cutter.

4.2 Cutterhead design

Cutters can only be run at high loads, if they are mounted on a cutterhead designed with the necessary **stiffness**. This is a prerequisite in order to evenly spread the thrust over **all** the cutters on the head, not allowing any of them to spring back upon meeting resistance from the rock: they must all penetrate the rock to the same depth at the same time, if the head is to work with optimal efficiency.

When boring hard rock, cutter consumption will be relatively high and cutter changes will occur fairly frequently. To facilitate such changes, easy access to the front of the head must be provided. Cutters are heavy, cumbersome pieces of hardware and become quite a problem to handle in narrow, confined spaces. The head should therefore have a so called **open design** and be rather shallow (short), resulting in wide and short passageways for men and materials to and from the front. A shallow head also has other advantages:

- it will have less "overhang" in front of the main bearing and will be lighter as well, causing less of an "overturning moment" on the main bearing;
- the shorter the head, the easier the muck will flow through it and into the buckets;
- in blocky ground, a short head is less likely to get caught along its periphery and will suffer less friction losses;

- it makes it possible to gain "early" access to the tunnel roof for rock support work, ie at a relatively short distance behind the tunnel face.

4.3 Cutterhead stability

As the cutterhead is rotated and pressed against the tunnel face, the cutters crush the rock in the bottom of their kerfs, penetrating a distance of from 5 to maybe 12 mm - per pass, depending on the rock hardness and the cutter load applied. Under influence of the stress induced by the thrust on the cutter, cracks develop from the kerf. These cracks will propagate in a radial pattern and those closest to the free surface will develop furthest from their point of origin, see Fig. 1.

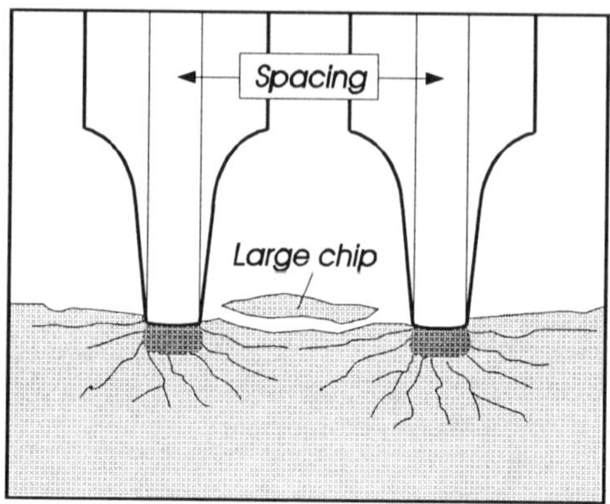

Fig. 1 Chip formation

When cracks from two adjacent kerfs meet, larger rock chips spall off the face. Thus the surface between kerfs is lowered, without actually being touched by the cutters. It is logical that the distance which the cracks will travel must depend on the thrust exerted on the cutters and on the resistance the rock presents to the process.

It is also obvious that a larger spacing can be selected for a given type of rock, if it is possible to increase the cutter load.

If a TBM is to work with optimal cutting efficiency and minimal cutter wear – ie with the best possible economy – it is of utmost importance that the cutterhead be held against the tunnel face in a very **stable** manner during boring, so that the cutters will exactly track the same kerf each pass.

If a cutter is allowed to wander sideways, it will have to start cutting a partially new kerf. This means that it will once more have to penetrate the rock by crushing it along a new track and developing a new system of cracks underneath its edge. In other words: it must repeat part of the work done during previous passes. This of course means reduced efficiency, high side loads and increased wear!

To ensure the best possible cutterhead stability, JARVA type hard rock TBMs have from the very beginning always been equipped with **double gripper pairs** – at the front and at the rear of the main body, spaced as wide apart as practically possible – to anchor the machine in the bore in an absolutely solid manner – see

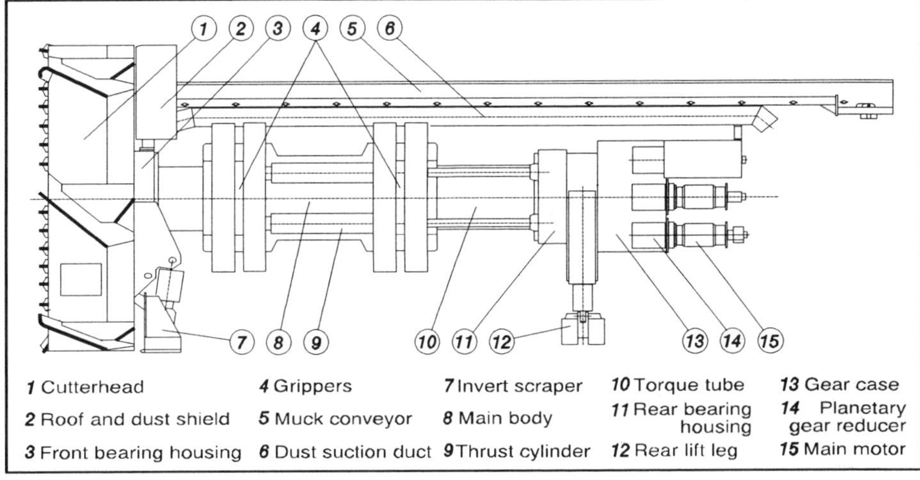

Fig. 2 Basic layout of a JARVA Mk 27 TBM

Fig. 2 which illustrates the overall Mk 27 layout in the 9.1 m diam version.

The torque tube – which, via the front main bearing, carries the cutterhead and slides through the main body during a boring stroke – is supported on and guided by large, **readily adjustable** bronze slide bearings equipped with an automatic greasing system to reduce friction losses and wear. The normal running clearance – of only a few tenths of a mm – thus is easily maintained during the life of the TBM. These two design features together – **double gripping and adjustable slide bearings** – ensure perfect cutterhead stability.

4.4 Main bearing layout and design

A TBM main bearing carries the cutterhead and is without doubt one of the most crucial components of the machine. Especially in machines designed to bore hard rock, it is subject to very large axial forces and will also have to take radial and – sometimes considerable – eccentric loads, of hard to predict magnitudes, particularly so in mixed ground.

For smaller machines – say up to approx. 4.5 m in diameter – standard double conical roller bearings are generally used. This type of bearing is readily available, but has a limited capacity for axial forces. Larger TBMs, with a thrust of say over 800 ton, thus are usually equipped with specially designed bearings of various types.

The Mk 27 TBM offers an interesting example of modern bearing design, in that the two types of loads acting on the cutterhead are not only divided over **different types** of bearings but also taken at **separate points** on the machine. This is shown in Fig. 3 on the next page.

The weight of – and the radial loads acting on – the cutterhead are taken by a single spherical roller bearing in the front bearing housing directly behind the cutterhead. The axial load – which is caused by the machine's thrust – is transmitted to the large (1.9 m diam) **hollow** drive shaft and taken by the front one of twin spherical

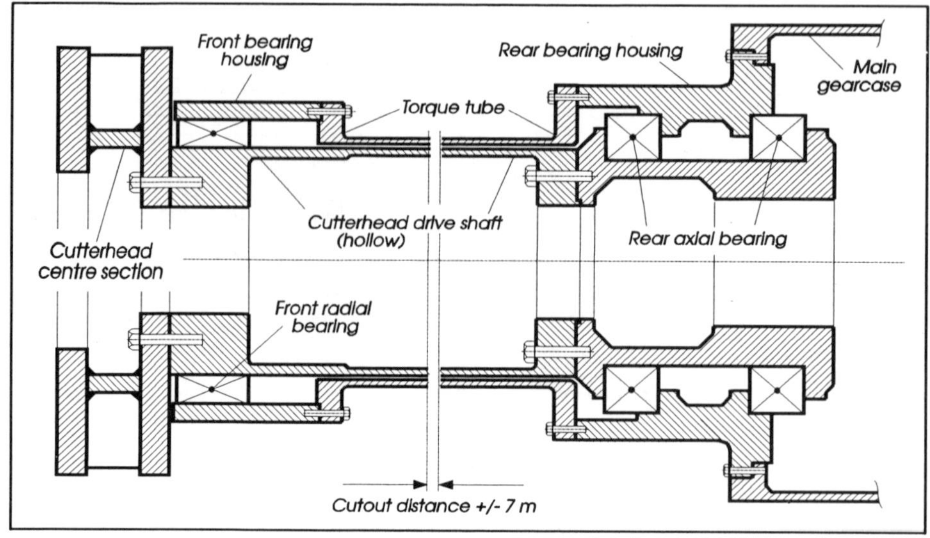

Fig. 3 Mk 27 main bearing layout

axial bearings at the rear end of the machine.

As the twin bearings are identical and because the rear one is not subject to axial loads at all, they can change places after a while to double the life of the package.

Whichever type is used, the **lubrication** system for the main bearing and **sealing it off** against outside dirt – in the form of water, fine dust and abrasive rock particles, together forming a very dangerous grinding paste – are just as important as the pure load carrying capacity.

Main bearing lubrication can be achieved in various manners and by either oil or grease as a lubricant:

- **Grease** with an automatic and continuous supply of grease under pressure; this is less suitable for heavily loaded bearings.
- **Oil in a splash system**; rather risky: when the sealing system fails, all oil can run out in a short time without being noticed and a bearing failure then is likely to result.
- **Oil circulation**; provides excellent opportunities for:
    - temperature control (cooling or heating);
    - filtration;
    - automatic cutterhead-stop in case of a system malfunction.

Oil circulation systems today are practised on all but the smallest of the JARVA type TBMs and have been found well suited for even the most heavily loaded bearings.

4.5 Optimal use of a TBM's capacity

As reliable electronic systems became available in the early 1980s, it was logical to start exploiting the opportunities they offered to optimize the use of all the power built into a modern TBM. Ever since 1984 all JARVA TBMs therefore have been furnished with Automatic Thrust Control (ATC). Likewise, the company has also been the forerunner in applying electrical Variable Frequency Drive (VFD) systems – to vary the speed of the cutterhead – for rock boring purposes.

4.5.1 Automatic Thrust Control

On older TBMs, the operator would set the pressure in the thrust circuit by hand, aiming at as high as possible a Rate of Penetration (ROP) without pushing the main motors' ammeter needles above the red line which marks the nominal capacity current draw. Adjustments made more frequently than once during a full stroke were rare, however, and many operators would run the machine with the same setting for a whole shift. This means, of course, that those machines were not run in the most efficient manner. On the average, they would operate at **no more than some 60 % of their built-in capacity**.

The reason for this is that rock disintegration is not a smooth and stable process which goes on at a constant power level. It is very much a discontinuous process with rapidly and frequently changing power demands, even if the rock would be perfectly homogeneous. In actual fact, no rock has absolutely constant characteristics and this causes further fluctuation of the power required.

Instead of letting a human being try to manually adjust the thrust pressure so as to reach optimal performance, in a manner which at the same time is safe for the machine too, it obviously is far better to entrust this task to an electronic device that can be trusted to accurately read incoming signals and make adjustments at great speed and without ever tiring.

Basically, ATC consists of a regulator which controls the pressure in the thrust circuit **based on the main motor power consumption**. Changes in thrust pressure will have effect in tenths of a second and will be made as often as the rock demands it. The outcome of ATC is that the power taken out is not allowed to sink all that much, before the pressure is adjusted upward to raise the power take-out to a value close to its permitted maximum again. At the same time, the thrust pressure will quickly be dropped, if the power take-out would rise beyond the permitted maximum so that the motors are prevented from ever running overloaded.

With ATC a TBM will – as an average – utilize at least 90 % of its built-in capacity, **which presents a 50 % improvement over the manual control mode**.

4.5.2 Variable Frequency Drive

Variable Frequency Drive (VFD) or the possibility to govern the rpm of the cutterhead motors by means of frequency conversion, is now offered on all JARVA and ROBBINS type TBMs as an option. The main motors have - with a view to their electrical specification and cooling capacity - from the start been laid out to cope with the somewhat higher heat generation VFD causes. The VFD system has been developed in close cooperation with ASEA BROWN BOVERI – who also supply the hardware.

The possibility to vary the cutterhead speed is particularly valuable for operation under widely varying ground conditions: from hard to soft – from solid to broken or blocky. Soft ground will require more torque for the same cutter loading than hard ground. In hard rock, where less torque is needed, it may be difficult to fully utilize the machine's installed capacity, unless the speed can be increased.

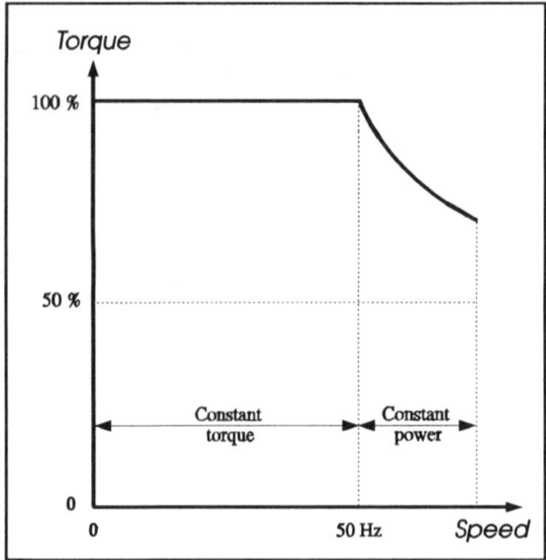

Fig. 4 Torque versus speed with VFD

In blocky ground, it may be advisable to drop the speed, so as not to rip blocks of rock out of the face or the roof of the tunnel and risk damaging the cutterhead. All these various requirements can be met by VFD operation, as is shown in Fig. 4.

The graph shows that an electric motor run with VFD in the range of 0 - 50 Hz will operate at full nominal torque. This is called the **constant torque range**.

The range above 50 Hz is the **constant power range**, where torque falls off as speed increases, as the product of torque and speed – ie power – obviously is constant.

In blocky ground, the cutterhead speed can be reduced to below what corresponds to 50 Hz for the drive motors, while retaining full nominal torque.

The feature of having full torque available at lower speeds, even from start-up – when speed = 0 – is very valuable, particularly in squeezing ground which tends to bind the cutterhead.

A final advantage of VFD is that the nominal speed of the cutterhead can readily be adapted if the machine is to be rebuilt to a different diameter for its next job. On a machine without VFD, one would have to arrive at the new cutterhead speed by altering the mechanical speed reduction between the motors and the head, ie by changing the reduction ratio of the planetary gear reducers or of the ring gear and pinions. Especially the latter is a cumbersome operation, as it also means changing the pitch circle diameter of the drive unit's mounting.

4.5.3 PLC control system

It goes without saying that a rather elaborate control system must be provided for a TBM which is equipped with so many self-regulating features and so much flexibility in its modes of operation. However, the more complex a machine becomes, the greater the need of automation and of interlocks to guard against human errors.

To provide the necessary control and at the same time create the flexibility required to be able to cope with a wide range of conditions in the field, a PLC – Programmable Logic Controller – based control system was introduced on all JARVA TBMs some six years ago. Such a system provides "operative commands" to electrical and hydraulic components, a **sequencing** of operations and the necessary **interlocks,** not in the traditional manner via relays (hardware), but via the program (software) of the controller instead. Thus it facilitates making quick changes in the operation of the machine when conditions so require: no need to rewire – or install new – relays, just rewrite part of the controller's program.

5. Provisions for traversing poor ground

Amongst the engineering aspects of the Halland Ridge project reviewed above, the following points were mentioned regarding measures required by the authorities or the tunnel owner as well as with a view to the planned rock reinforcement methods:

- the drilling of holes for
  - probing ahead
  - grouting
  - rock bolts
- shotcreting
- the use of ring beams
- combinations of all these

We will now review the design features incorporated into the Mk 27 to cope with these project-specific conditions.

5.1 Drilling equipment

5.1.1 Probing and grouting holes

Drilling such holes in combination with tunnel excavation by TBM is not a new technique. In the past, it was customary to pull the TBM back from the face, bring all the drilling equipment required in front of the cutterhead, rig the equipment and drill the holes into the face. When the holes were finished, all the equipment then had to be returned to the rear, before the TBM could start boring again.

While all this can be done, of course, the approach has a number of drawbacks:

- it is cumbersome and time consuming, thus slowing the whole excavation process down to a high degree;
- it forces the drilling crew to work in front of the head, where they are either exposed to rock falls from a possibly unsafe tunnel roof or will be forced to spend even more time securing the roof before they can get on with the actual drilling task;
- if drill steel in holes drilled into the face breaks and the part of the string furthest in cannot be retrieved, it will create havoc for the cutters when the TBM bores into it later on.

In connection with a sewage outfall tunnel bored under the City of Stockholm in Sweden in 1987 in fissured and water logged gneisses and for some 3 km routed directly underneath a lake – where probing and grouting was not only prescribed but also very much a necessity to be able to put the tunnel through at all – ATLAS COPCO therefore developed a new, patented method for drilling such holes from a point further back and **not** into the face. This is illustrated in Fig. 5 on the following page.

Probing holes – which can be used for grouting, too – are drilled in a cone pattern past the TBM. For probing alone, one or two holes will suffice in most cases.

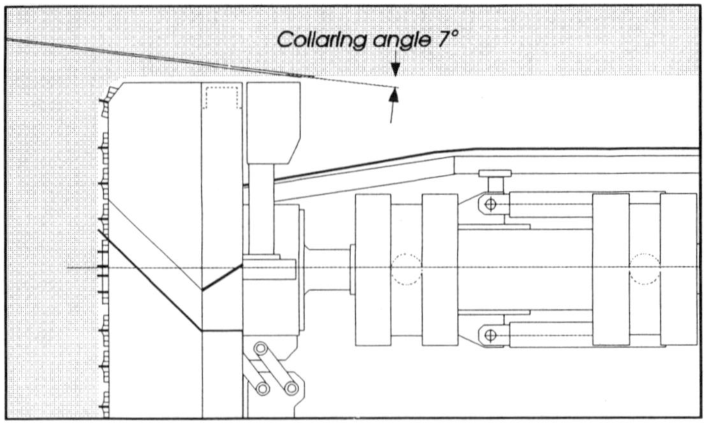

Fig. 5 Probing past the cutterhead

For the purpose of grouting the number of holes will depend on the cover desired. If it is a question of just sealing off **minor** fissures **above** the tunnel, one will normally only drill holes in the upper region of the tunnel.

If the ground is badly fissured and carries water under high pressure, it will be necessary to drill the grouting holes **all around** the tunnel cross section.

The problem in drilling holes in this manner lies in getting the holes started. The flat cone requires a small collaring angle. With the special collaring equipment developed by ATLAS COPCO, it is possible to come down to an angle of only 7°, as indicated in Fig. 5.

To gain time – by not letting the drilling activity interfere with the boring operation – it is desirable to mount the drilling equipment on some kind of a sliding mechanism so that it can remain stationary in the tunnel for as long as it takes to complete the drilling of a hole, while the TBM can advance in the meantime.

How far to the rear the holes will be collared mainly depends on the TBM diameter: on a small diameter machine – which almost completely fills the bore with its own hardware – the drilling equipment will have to be located **behind** the TBM, eg on a working platform between the TBM and the Back Up Equipment.

On larger diameter TBMs – such as the Mk 27 – there will be room for the drilling equipment on the machine itself, and holes can be collared just behind the roof shield, see also Fig. 9.

In the case of the Mk 27 for the Halland Ridge project, four hydraulic rock drills for long hole drilling are carried on rota-extension booms with the boom pedestals mounted on sliding platforms – travel distance in the longitudinal direction of the tunnel: 3.0 m – that are attached to the TBM main body, one in each of the four quadrants of the tunnel cross section, see Fig.6 on the next page.

With this set-up, 40 m long holes are to be drilled, with the hole bottoms at the most 5.0 m outside the bore-to-be. Three of these will be drilled upwards and to the sides as probe holes in first instance.

As a minimum, a further 11 holes will then be drilled, spaced between the first

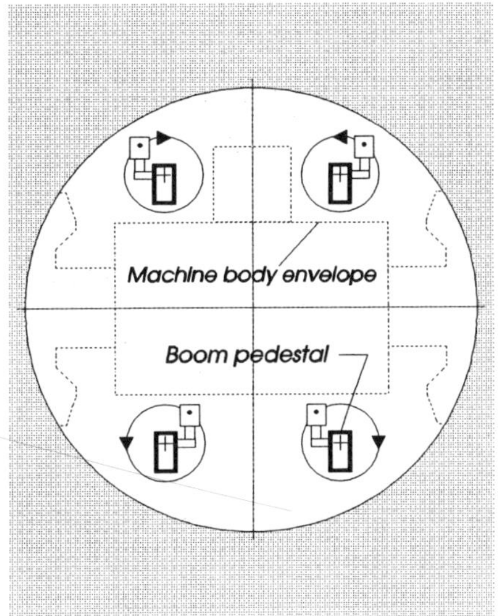

Fig. 6 Drill mountings on Mk 27

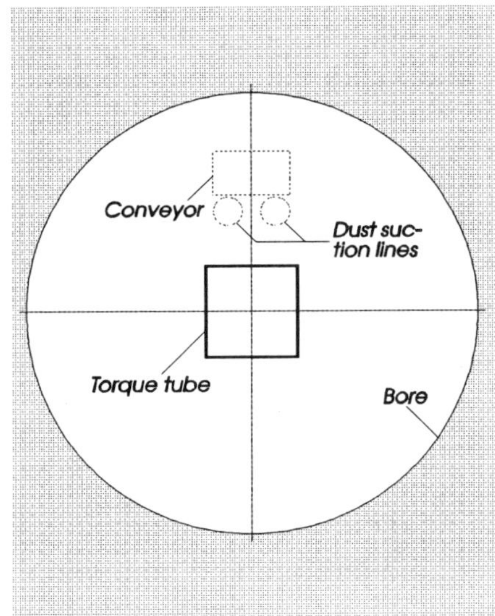

Fig. 7 Maximizing free space

three to form a complete cone, so that all 14 or more can be used for grouting, if the probe holes show the necessity thereof or when grouting is prescribed per se.

5.1.2 Rock bolt holes

These can be drilled with the same equipment as has been described for the probing and grouting holes above. This is possible in such a large diameter tunnel, where the free space allows off-angling the boom heads to such an extent, that the feed beams can be positioned in a plane at right angles to the tunnel longitudinal axis. Thus an almost unlimited number of 4 m long holes for rock bolts can be drilled in a roughly radial pattern.

To create a maximum of free space to drill rock bolt holes, **the conveyor and dust suction lines can be retracted** during drilling. This leaves the 2 x 2 m torque tube section as the only obstruction for the drills in the area immediately behind the dust and roof shields, thus providing the opportunity for "early" placement of rock bolts, very close behind the tunnel face, see Fig. 7 and 9.

5.2 Overboring equipment

It has already been mentioned that ring beams, possibly shotcreted over, will be placed in weak ground sections of the Halland Ridge tunnels.

If such support is placed **inside** the nominal bore diameter, the diameter of the free space in the tunnel decreases and the cutterhead can no longer pass through the reinforced zone. Therefore, it is only possible to place the support **behind** the head as the machine advances.

If - for whatever reason - the machine has to be taken back from the face, it can only be backed so far that the head reaches the forward end of the reinforced zone.

Under certain conditions this may turn out to be a severe limitation.

It may also happen that the ground is so weak, that reinforcement has to be placed **immediately** behind the face, ie when the ground will not stand until the machine has made its way forward sufficiently far for the reinforcement to be placed

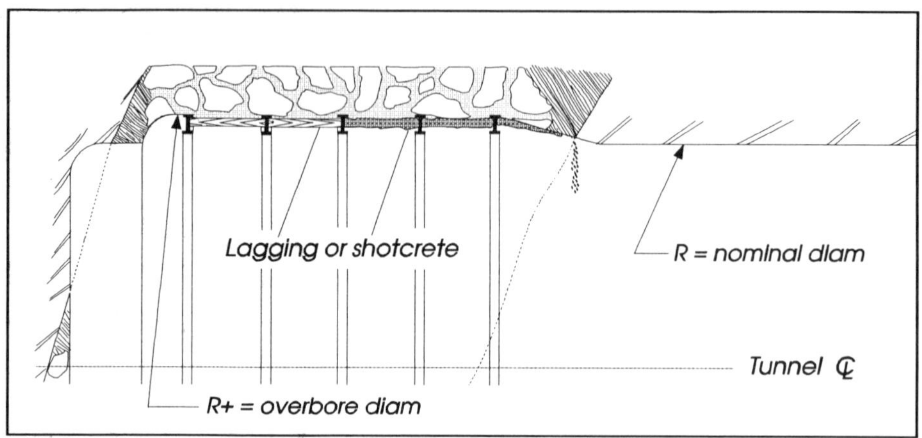

Fig. 8 The ATLAS COPCO overboring system

behind the cutterhead. But any rock support placed **inside** the nominal bore diameter, **in front** of the cutterhead, would be in the way for the head as the machine advances again.

The purpose of the overbore system – see Fig 8 – with which JARVA TBMs can be equipped, is to overcome these problems by temporarily boring the tunnel with a larger than the nominal diameter - **overbore** - so that rock support can be installed **outside** of the nominal diameter, close to the tunnel face and without obstructing the machine in its advance or when it has to be pulled back.

Four so called **overbore cutters** – in the form of 16¼ in. gauge cutters – are each mounted on hydraulically actuated arms which can be swung out when the need arises to increase the nominal bore diameter by some 0.30 m. The arms are controlled by means of valves mounted at the rear end of the machine. Provisions have also been made which allow the cutterhead to still clean the invert in the overbore, where the rock is at a lower elevation than in the nominal bore.

When the ground firms up again and can be secured by means of just rock bolts or a thinner layer of shotcrete applied further back, the overbore cutters are swung back in and the machine continues boring at the nominal diameter.

The overboring technique and hardware was developed during the late 1980s and used for the first time in 1990, on a JARVA Mk 18 TBM operating in Sicily. That machine bored a water collection tunnel – look upon it as a slightly upward sloping, large diameter well – in the tuffaceous strata of the slope of Mount Etna – to tap a valuable source of potable groundwater for the town of Catania. Based on the results reached there, the contractor claimed that 'with the overbore system, one can cope with any kind of rock'.

It goes without saying that the availability of Variable Frequency Drive on the Mk 27 is of great benefit when overboring has to be resorted to. It allows the operator to take things "nice and easy" while the cutters are being swung out and to keep the cutterhead rpm down during overboring, so as not to subject the overbore cutters and their arms to high speed impact loads in broken ground.

Automatic Thrust Control is another great help under those conditions: as it can be set to extremely low thrust levels, it will automatically safeguard the overbore cutters when an already bored section of tunnel is "reamed" later on and the overbore cutters then are the only ones taking any thrust. Without it, there would be a severe risk of overloading those cutters in such an operation.

5.3 Cutterhead layout

On a TBM which will operate in hard, competent ground only, the cutterhead normally is of a so called **open design**. Generally speaking, this will provide the best and simplest access to parts which have to be changed on a regular basis – eg the cutters – and makes the head easy to work on for maintenance in general.

However, if it is known beforehand, that the machine will have to work also in poor ground – be it either blocky, squeezing or otherwise incompetent, as is the case at the Halland Ridge – the head is normally laid out in such a manner, that its outer surfaces – which are in "almost-contact" with the rock in the tunnel face and bore – are as uninterrupted and flush as possible. In that way, blocks of rock will catch against protrusions and jam the head a lot less – ie shock loads in the drive train are reduced – and it will require less effort to keep the head turning smoothly and to start it from standstill.

On the Halland Ridge Mk 27, the cutterhead therefore is equipped with what is generally known as a **false face**. This consists of plate work filling out the spaces in between individual cutters, to provide an even, almost closed front. Large chunks of rock thus are prevented from impacting the cutter saddles and jamming against them. At the same time, the false face in a way even helps to support the tunnel face, if it should have a tendency of caving.

Also along the periphery of the head, openings between spokes and buckets are closed off as much as possible by curved plate work – the so called **rolling can** – to arrive at an almost uninterrupted cylindrical outer surface. During normal operation, the windows through which the overbore cutters and their arms will be swung out when they are needed, are closed off by curved hatches which can easily be removed when the time arrives to start overboring.

5.4 Beam erector

Ring beams consist of a number of curved steel profile sections which are bolted or keyed together to form a complete ring. The difficulty in erecting a ring lies in bringing the sections over the top of the machine, holding them in position while joining them together and then pushing them up against the tunnel roof in the final assembly stage, when the bottom section has to be added to complete the ring.

To **partly** mechanize this operation, the Mk 27 is equipped with a simple beam erector, located immediately behind the dust shield – so that ring beams can be set **as far forward as possible** – in a manner which takes the hard, manual work out of erecting a complete ring. The erector can handle both rings set in the overbore space and rings erected within the nominal bore. It is the intention to avoid using the latter type altogether, of course, but one never knows .....

5.5 Conveyor retraction

When the drilling equipment for rock bolt holes was reviewed above, mention was already made of the fact that the TBM conveyor – together with the dust suction lines – can be retracted to obtain a virtually unobstructed working space immediately behind the machine's roof and dust shield.

All TBMs are equipped with a conveyor retract mechanism of some kind. However, on most of them the retract stroke is limited to what is needed to pull the conveyor out from under the hopper for cleaning and maintenance purposes. On the Halland Ridge Mk 27, the retract distance is as much as 5.1 m, so that the whole area behind the dust shield can be freed up, even when the head has been stroked forward all the way.

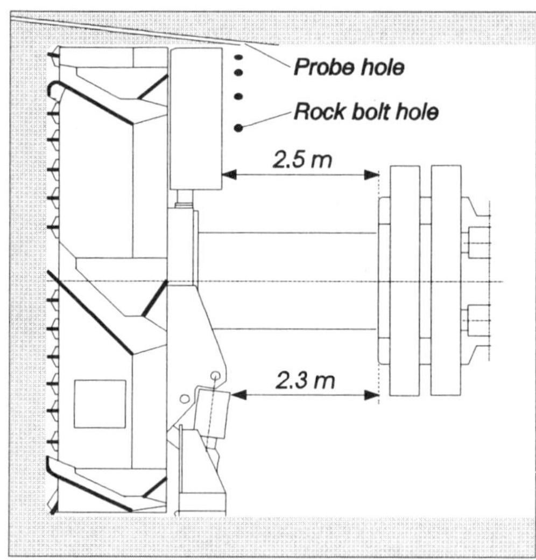

Fig. 9 Free space up front

How much space will be available with the conveyor fully retracted and with the head stroked all the way forward, is shown in Fig. 9: the roof and the walls of the tunnel are unobstructed over 2.5 m. Near the bottom the free space is slightly less: 2.3 m.

This space would not be as free as it is with the MK 27 layout, if the machine had had the main motors, couplings and the planetary gear reducers mounted directly behind the front bearing housing, as is common on ROBBINS type TBMs.

This is one of the reasons why the JARVA layout, with the drive train components at the rear end of the machine, is to be preferred under certain conditions.

The other reasons have to do with the accessibility of the drive train components and the overall balance of the machine. With all the drive components immediately behind the cutterhead, most of the weight of the machine is at the front, making it somewhat nose-heavy. Splitting the heavy components – cutterhead and bearing housing up front, drive train at the rear – as is common for all JARVA TBMs, the machine will be perfectly balanced, no matter where in the stroke the cutterhead is from one moment to the next.

6. Conclusion

This paper has shown that design requirements for large TBMs with typical hard rock capabilities can be met without sacrificing the machines' chances to also operate satisfactorily in stretches of poor ground. The additional features which must be incorporated in the design to make the machines that versatile can be accommodated in a hard rock TBM layout in a fairly simple manner, **provided they are known in advance**.

Installing the necessary auxiliary equipment **after the fact**, when the machine has already become bogged down because of poor ground conditions only discovered some way into the tunnel, would present great difficulties – if it is possible at all.

The features of a machine for poor ground have been reviewed against the background of an actual tunnelling project. In this particular case, a fairly large amount of geological information was available from the start – which allowed tailoring the TBM to the local conditions to a high degree. Of course, the project may run into some problems yet, but we feel confident that the Mk 27 for the Halland Ridge has been engineered with sufficient basic flexibility to cope with them in an acceptable manner.

– o –

Use of sedimentary rock impact indices in evaluation of tunnel-boring machine performance

C. Laughton
P. P. Nelson
Y. Abd Al-Jalil
Department of Civil Engineering, The University of Texas at Austin, Austin, Texas, U.S.A.

Abstract
This paper outlines the analytical methods being used to develop Rock Impact Indices (RIIs), a suite of parameters which are used to link Tunnel Boring Machine (TBM) performance to in situ geologic conditions in sedimentary rocks. RIIs will be developed to describe the rock mass by reference to intact rock strength, water inflow and jointing parameters. However, only the derivation of the RII associated with jointing is described in this paper. Discrete values of this index are obtained along the tunnel and linked to local TBM performance measures, penetration rate and utilization. Emphasis is placed on establishing a reasonable representation of the jointing along the length of the tunnel using the RII, derived from tunnel mapping. The relationships linking the RII to TBM performance parameters, as identified from the back-analysis of case history data, are used in the prediction of TBM performance in a simulation program.
Keywords: jointing, penetration rate, sedimentary rocks, Tunnel Boring Machine, utilization.

1 Introduction

The paper presents some results from a cooperative research program currently involving The University of Texas at Austin, Massachusetts Institute of Technology and The University of Missouri-Rolla. The full research program, entitled "Manufacturing Underground Space", regards the TBM mining operations as being essentially those of a manufacturing process, where the "end product" (supported tunnel) remains stationary while the point of manufacture (the TBM heading) moves through the rock mass [1]. One of the main objectives of the research is to develop an interactive program capable of modeling mining production by simulating the impact of both the underground equipment and rock mass characteristics on TBM production. The program being developed to model TBM operations is modular in structure using embedded deterministic and probabilistic relationships to simulate the performance of mining systems within a rock mass.

One module of the simulation package deals specifically with the equipment aspects of the TBM operations, looking at the reliability of the underground equipment which consists of a series of interdependent sub-systems (mechanical, electrical, hydraulic etc.).

Probability Density Functions (PDFs) are fitted to case history data, modeling both the "time between failures" and "time to repair" of underground system components to simulate equipment performance over time [2].

The jointing RII, based on a joint intensity parameter, is being derived from case history tunnel mapping. However, the final geotechnical module will permit simulation at the feasibility and pre-bid stages of a project using limited amounts of site specific data. The final module will incorporate three RIIs (joint intensity, intact rock strength and water inflow) and will model their combined impact upon TBM performance. Tunnel realizations will be generated to yield a sequence of tunnel lengths, classed according to the combination of RIIs. This RII-based classification will be used to predict TBM performance levels in terms of penetration rate and utilization.

This paper elucidates the basic approach being used to identify the impact of jointing intensity upon TBM performance by reference to a set of four case histories.

2 Overview of the Modeling Procedure

Three detailed sets of data are used to establish the TBM performance modeling relationships. Data sets are required to provide a description of the in situ rock mass conditions (Set 1), the TBM performance achieved (Set 2), and the rock mass conditions, as interpreted from site investigation data (Set 3). These three Sets and the modeling Links required to create the predictive capability of the module are shown in Figure 1.

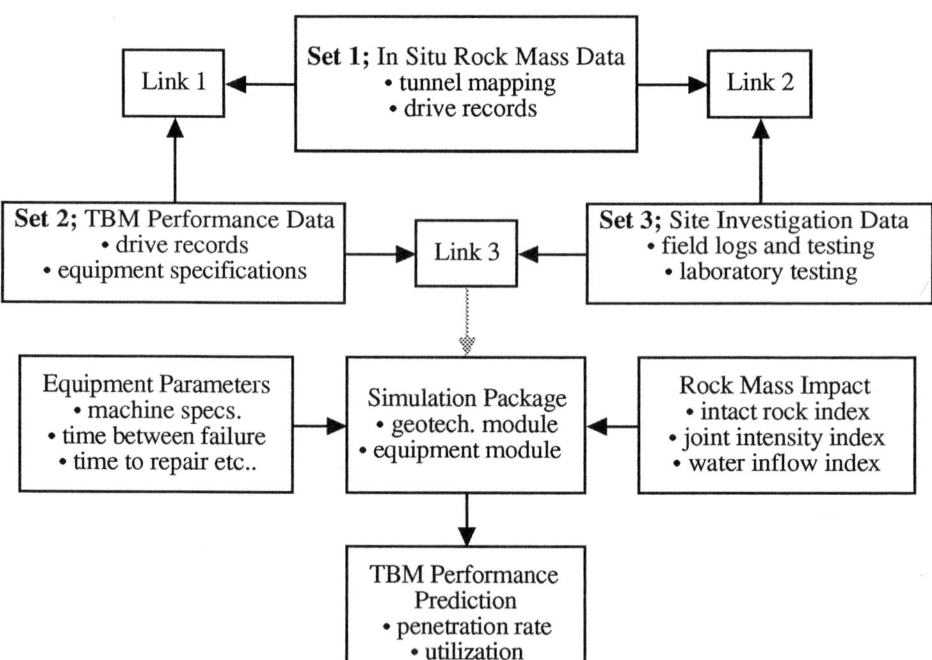

Fig. 1 Steps in the analysis of the detailed case history data sets

The modeling is based on links made between the three identified data sets. Link 1 involves the development of parameters to describe in situ rock mass conditions and the establishment of relationships between these parameters and TBM performance records, expressed in terms of penetration rate and utilization. Link 2 ties rock mass parameters obtained from site investigation with those observed in the mapped tunnel. Link 3 directly ties together site investigation and TBM performance parameters to provide for prediction of mining performance. Throughout this linking process, great importance has been attached to being able to identify and quantify both the average and range of jointing by the use of tunnel map data.

3 Case History Description

3.1 General

The main project, machine, rock and support characteristics of the four selected tunnel drives are summarized below in Tables 1 through 4. The drives studied in this paper were the Niagara Frontier Transit Authority tunnels located in Buffalo, New York [3].

The same rock unit was tunneled in each of the four drives using very similar machine and underground equipment configurations. The alignment was set within the Bertie Formation, a sub-horizontally bedded limestone of Silurian age. The rock unit had been dewatered prior to drivage of the tunnels and the variability in tunneling conditions, observed along the length of the tunnel, was brought about primarily by variation in the extent and condition of jointing.

3.2 Tunnel Layouts

The excavated tunnels were all approximately 5.65 m in diameter. All tunnel access was provided by vertical shaft. The tunnels were driven at shallow depths below rockhead. Vertical alignments were such that gradients never exceeded 2.5%. and drive azimuth varied little over the length of the drive. Drives 1 and 2 (the "C11" Contract) were driven in parallel, as were Drives 3 and 4 (the "C31" Contract), and spacing between tunnels was between 10 to 30 m. A brief overview of the TBM drive layouts is given in Table 1.

Table 1. Parametric description of the four TBM drives

Parameter	Drive 1	Drive 2	Drive 3	Drive 4
Contract Identification	C11 Inbound	C11 Outbound	C31 Inbound	C31 Outbound
Tunnel Length (km)	3.1	3.1	2.2	2.2
Minimum Cover (m)	5	5	5	5
Maximum Cover (m)	15	15	10	10
Tunnel Alignment	NE-SW	NE-SW	NE-SW	NE-SW
Max. Uphill gradient (%)	1.91	1.91	0.56	0.56
Max. Downhill gradient (%)	1.56	1.56	2.46	2.46
Start Year	1980	1980	1980	1980

3.3 Mining Equipment Configurations

The tunnels studied in this paper were each driven using conventional, unshielded Robbins machines and rail based mucking and supply systems. A brief description of the operating features of the four TBMs used on the four tunnel drives is given in Table 2.

Table 2. Basic TBM equipment specifications

Machine Specifications	Drive 1	Drive 2	Drive 3	Drive 4
Manufacturer	Robbins	Robbins	Robbins	Robbins
Condition	New	New	Refurbished	Refurbished
Shield	Finger (roof)	Finger (roof)	Finger (roof)	Finger (roof)
Diameter (m)	5.6	5.6	5.6	5.7
Length (m)	15.9	15.9	14.6	15.9
Stroke (m)	1.5	1.5	1.2	1.5
Cutterhead Rotation (RPM)	5.9, 2.9	5.9, 2.9	7.5	5.1, 2.6
Motors	6	6	6	6
Motor Rating (kW @ amps)	149 @ 216	149 @ 216	149 @ 216	112 @ 165
Maximum Thrust (MN)	8.3	8.3	9.6	8.5
Number of Single Discs	38	38	39	39
Number of Double Discs	2	2	2	2

The similarities among TBM operating characteristics were considerable, the notable exceptions being the higher rotation rate of the Drive 3 TBM cutterhead and the lower power rating of the Drive 4 TBM torque motors. The machines used on Drives 1 and 2 were new, while those used on Drives 3 and 4 were refurbished. Given the limited number of machine differences and the single rock unit excavated, this set of four drives provided a good opportunity to focus on the impact of joint variability upon TBM performance.

3.4 Temporary Rock Support Installed at Tunnel Headings

Table 3 summarizes the support installed for each of the four drives. Three basic levels of rock support were employed to support the rock along these four tunnel drives. Level 1 support was used in the best rock mass conditions; here spot or pattern rock bolting was employed. Level 2 support was used under "intermediate" rock mass conditions and consisted of pattern bolting with internal profile support provided by straps or channeling. Level 3 support was used in the poorest rock mass conditions and consisted of full-ring steel sets with wooden lagging added in the more weathered sections.

The amount and type of rock support installed at the heading varied considerably between the C11 and C31 contracts. Straps and steel sets constituted over 50% of the tunneled length for the C11 drives, and less than 10% in the C31 drives. For all the drives, extensively jointed and weathered zones prompted the use of level 3 support. The range of rock mass conditions, as surmised from site investigation and tunnel mapping, were similar for all four drives.

Table 3 Support level and percentage of total drive length over which it was applied

Level of Heading Support	Drive 1	Drive 2	Drive 3	Drive 4
1 Spot or Pattern Bolts	36.8%	47.4%	96.1%	90.9%
2 Channels, Straps & Bolts	55.2%	43.0%	1.3%	5.8%
3 Full Circle Steel Sets	8.0%	9.6%	2.6%	3.3%

3.5 Rock mass description based on site investigation data

Selected rock mass parameter values are given in Table 4. The values were taken from the bid documents, as assembled to describe in situ rock mass conditions for the bidding contractors [4].

Table 4 Rock Mass Parameter Set for the Bertie Formation Rock

Value	Mean	Value Distribution	Coeff of Variation
Uniaxial Compressive Strength (MPa)	139	normal	74%
Point Load Index (MPa)	2.6	normal	50%
Rock Quality Designation (%)	79	n.a.	n.a.

Values quoted were obtained from vertical NX (54 mm diameter) core. The mean RQD values were determined per three meters of core run and an average value reported. The Point Load and Uniaxial Compressive Strength (UCS) value sets are described by the use of normal distribution parameters. The high values of the coefficients of variation (standard deviation ÷ mean value, reported as a percentage) for the Point Load and UCS data sets indicate a relatively high degree of strength variability associated primarily with variations in weathering within the rock unit.

Each tunnel drive was completed in well-bedded and generally flat-lying sedimentary rock. Bedding thickness varied between approximately 0.5 and 1.0 m. The rock mass had two major orthogonal joint sets. Both joint sets dipped at a sub-vertical angle and the strike of each set lay oblique to the tunnel axis. Joint surface strength conditions ranged from fair to poor, corresponding to an approximate Rock Mass Rating range of values of 12 through 6 [5]. Heavily jointed zones, associated with local areas of extensive weathering, were present at tunnel depth. Variation in the extent and condition of jointing was responsible for the majority of variation in tunneling conditions experienced along the length of the drive.

4 Rock mass description based on tunnel mapping

4.1 General

Considerable efforts were made to develop a rock mass parameter that could be used to describe the variation in tunnel conditions within the mined rock unit. Initially rock mass classification systems (Rock Mass Rating, "Q-System", Rock Structure Rating) were tried. However, it soon became evident that, within a rock unit, the level of logged and mapped detail was inadequate to support consistent derivation of all the requisite classification system parameters. Given the dominant impact of jointing on TBM performance, for these and many other drives in sedimentary rock, a reasonable assessment of the local rock mass conditions, within a rock unit, could be obtained by reference to the extent of jointing.

The joint geometry was thoroughly described for the full length of the tunnel and emphasis was placed on the development of a single joint parameter that would provide a meaningful measure of the local jointing. To assess the jointing variability of the rock mass within the Bertie Formation, the tunnel drive was divided into discrete lengths of tunnel referred to as "tunnel unit cells". A schematic unit cell layout is shown in Figure 2. Unit cell lengths of two tunnel boring machine diameters were systematically adopted for all case history analyses as the length over which average joint intensity and TBM performance values were determined and compared.

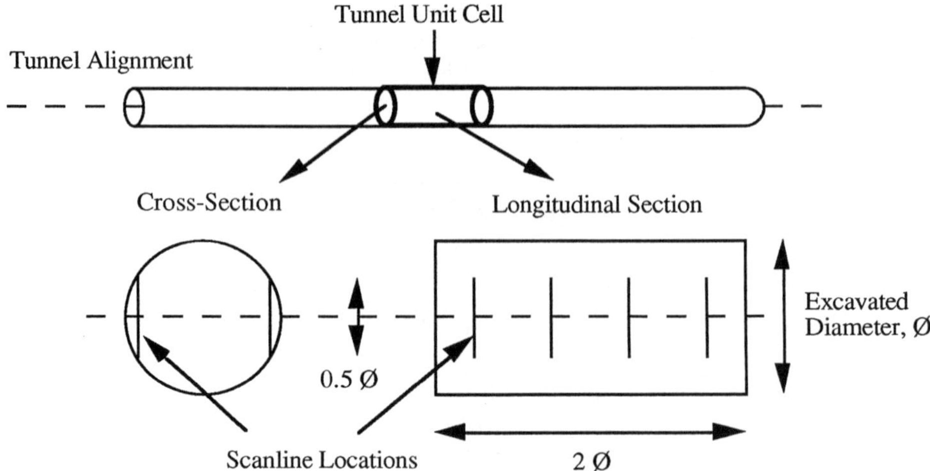

Figure 2 Tunnel Unit Cell Dimensions in Cross and Longitudinal Section

A unit cell length of two bored diameters proved to be an effective normalizing length over which to compare TBM performance and rock mass parameters. For the majority of mapped case histories within the data base each discrete set of machine and joint intensity values corresponded to a driven distance of between 10 and 15 meters.

4.2 Tunnel unit cell joint intensity values

Within the unit cell eight vertical scanlines, a radius in length, are used to quantify the cell's joint intensity. The total number of joint intercepts are counted for these eight scanlines and an average joint intensity value reported as the number of joint intercepts per bored diameter of scanline.

The scanlines are located across the mid-section of the tunnel, where levels of joint observation were shown to be most consistent. Figure 3 shows schematically the basic format used by the majority of tunnel mappers to record the joint exposures on the excavated rock surface and the scanline locations. The cylindrical periphery of the tunnel is "unwrapped" to give a planar projection of the joint traces. Initial analysis of the maps showed that there was considerable bias in the mapping of joints around the periphery of the excavation. The base of the excavation was often obscured by muck and joint trace mapping was not possible on a consistent basis. In addition a significant decrease in the number of mapped joints was observed in the crown of the excavation. Given the reduced "detectability" of jointing in the crown and invert sections of the tunnel, only the sidewall mapping was used in the assessment of joint intensity.

To obtain a representative joint intensity value for each unit cell along the tunnel two approximations were introduced to the basic joint counting procedure. These approximations were introduced to account for units with minimally or heavily fractured zones. Where fractured or weathered zones were identified, without mapping of individual fractures, a nominal fracture count corresponding to a density of fracturing of four fractures per tenth of a diameter, was added to the total unit cell fracture count. Where no fractures were intercepted by the scanlines within a cell, a nominal joint intensity value (0.125 fractures per bored diameter) was allocated to the cell.

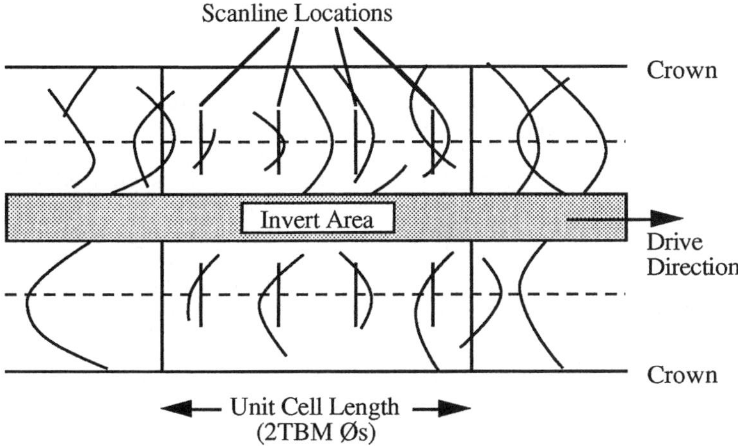

Fig. 3 Schematic Diagram of an "Unwrapped" Tunnel Map Showing Joint Exposures

5 Data Analysis

5.1 General

Once average joint intensity values have been derived for each unit cell of tunnel, corresponding average values for TBM performance, in terms of penetration rate and utilization, are calculated. The average TBM performance parameters for each unit cell are obtained by reference to drive stationing and production shift data, as provided by the contractors. In this research the penetration rate is calculated as the distance mined per unit of cutting or machine operating time. Machine operating time is obtained by reference to the "machine clock", as activated by the start of cutterhead rotation under the application of a minimum torque and thrust. Utilization is defined as the percentage of production shift time during which the TBM is operating. The advance rate is considered to be a derived parameter, the product of penetration rate and utilization.

5.2 Joint Impact Indexing

After average TBM performance values have been allocated to each unit cell, the performance values are grouped into sub sets based on joint intensity limits, each sub set of joint intensity was ranked according to a Joint Impact Index (JII) number. JII numbers were allocated to correspond to joint intensity steps of 1.5 joints per diameter. JII 1 groups unit cell performance parameters with joint intensities of up to 1.5 joints per bored diameter, JII 2 groups unit cells parameters within joint intensity limits of 1.5 up to 3 joints per diameter, etc.. In assessing the impact of rock mass variability upon the mining process this index is used to group and separate data in order to identify distinct sets of penetration rate and utilization data, by use of a simple statistical test. Finer separation of the joint intensity parameter proved unwarranted given the large amount of random or "unexplained" variability that was present within each performance data set and the limited range of TBM system operation. The distributions of penetration rate and utilization values associated with these distinct TBM operating ranges have previously been shown to have normally distributed truncated PDFs [6]. It is these PDFs, defined in terms of their mean and standard deviation, that are used in the simulation package to model the TBM advance through a series of indexed unit cells.

5.2 Variation in penetration rate with joint impact index

To identify the different operating ranges of the TBMs in the jointed rock mass, the unit cell penetration rate and utilization distributions for individual and grouped JII are compared to identify statistically different sub-populations. Statistical testing of the grouped utilization and penetration rate JII values allowed the identification of three distinct machine operating ranges for these four drives. Operating range 1 corresponded to a grouping of JII 1 and 2 values, range 2 corresponded to grouped JII 3 and 4 values and range 3 to a grouping of JII 5, 6 and 7 values. The significance level of testing to identify ranges of differing penetration rate and utilization response with differing joint intensities was set at 15%, using Kolmogorov-Smirnov testing to compare data set value distributions.

In this paper, the impact of joint intensity on penetration rate is illustrated by reference to Drive 3 as, unlike the other drives, the Drive 3 TBM was known to have operated at a constant cutterhead speed. All four drives showed the same trends in average penetration rate with increasing JII to those described below. For drive 3, increases in average penetration rate of some 15% are observed passing from operating range 1 to operating range 2. For further increases in jointing, corresponding to an increase in jointing intensity into operating range 3, significant reductions in average penetration rate are observed. The average penetration rate drops off to some 70% of the maximum registered. The value sets for drive 3 are modeled by the use of normal distribution parameters, as shown in Figure 4.

The initial increase in penetration rate, with increasing joint intensity, can be explained by the reduced need for new fracture creation to achieve penetration. However, as the jointing becomes more extensive, the face itself becomes more unstable leading to "blocky conditions" at the face, resulting in a net reduction in penetration rate. Given the steeply dipping joint intercept angle of the two major joint sets and the presence of weathered joint surfaces, the rock mass conditions were conducive to block instability at the face .

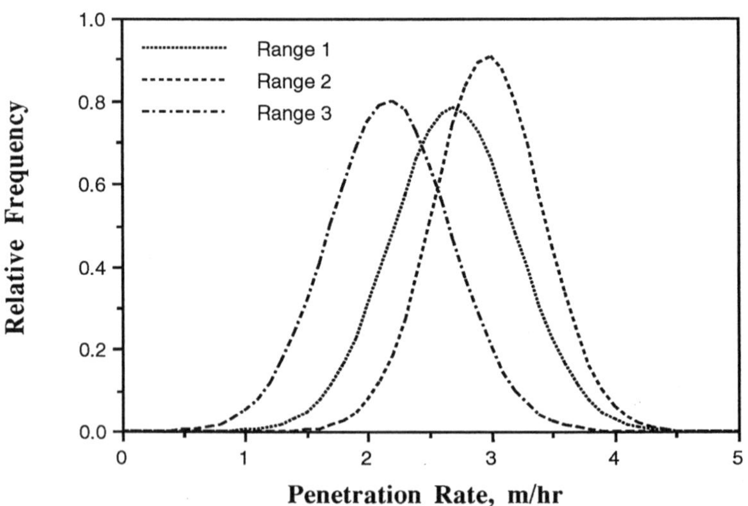

Fig. 4 Variation in Unit Cell Average Penetration Rate Distribution with JII for Drive 3

5.3 Variation in utilization with joint impact index

As might be expected, there is a reduction in average utilization with an increase in joint intensity. The severity of the rock mass impact upon machine utilization is considerable. The average utilization in the poorest rock mass conditions is only 40% of that observed in the best rock mass conditions.

Variations in average unit cell utilization were strongly influenced by both the JII and the level of support installed. As is shown in Figure 5, both bolts and bolts and straps were commonly used to support the rock in the less jointed rock conditions, identified as the TBM operating range 1. In operating range 1 the addition of straps had a relatively small impact upon the mean utilization, an average 5% reduction being observed for the four drives. In the intermediate ground conditions, range 2, all three support mechanisms were employed to effect ground support. Again bolting provided for a slightly higher utilization than for bolts and straps used in combination. However, the use of steel sets had a significant negative impact upon utilization, a mean unit cell utilization value of less than 30% was calculated. In the poorest ground conditions, represented by range 3 operating conditions, only bolts and straps and steel sets were used to effect primary support. Steel set unit cell utilizations averaged-out at just under 20% where as the average utilization using bolts and straps was 17.5%.

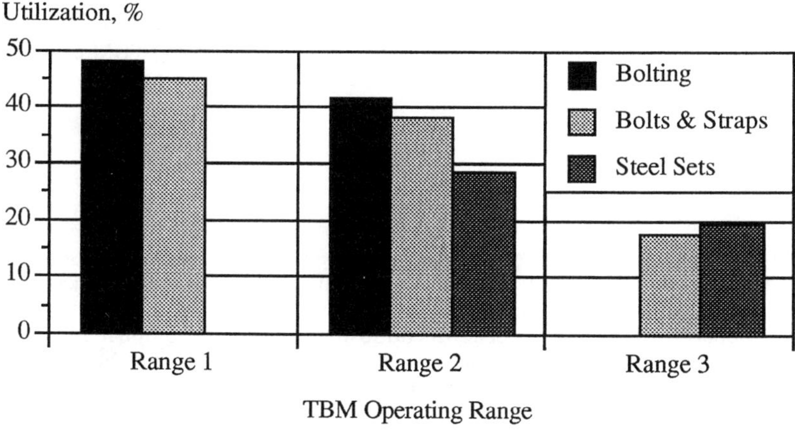

Figure 5; Average machine utilizations for unit cells in different operating ranges

6 Conclusions

For the detailed case histories described in this paper quantitative, in situ descriptions of the rock mass were consistently derived from tunnel maps. These maps have proven to form an effective basis for assessing local rock mass conditions, where jointing is the main source of variation within the rock unit. The analyses show that the derived joint intensity parameter can be indexed and for a given rock unit this index can consistently be used to identify ranges of distinctly different TBM operation using a simple, non-parametric statistical test.

So far, only a limited number of detailed data sets have been acquired from which to derive indexed relationships between jointing conditions and TBM performance. Most of the tunnels analyzed have been driven in relatively undeformed, sedimentary rock with bored diameters between 5 to 7 meters. For these tunnels the set of unit cell values, for each rock type, has provided a consistent mechanism to describe the variability of jointing and this description of variability has allowed for an accurate definition of both the average and extreme mining conditions. An assessment of extreme rock mass conditions is particularly critical to the successful driving of a TBM given the relative inflexibility of the installed heading equipment to cope with significant changes in tunneling conditions.

For the sedimentary rock mass excavated in these four case histories, both the utilization and penetration parameters proved to have a very significant impact upon overall TBM advance rate. The average penetration rate observed under operating range 1 was increased by some 15% in operating range 2 and reduced by some 20% in operating range 3 due to the onset of block instability at the face. The average utilization values dropped off consistently with increasing JII and were reduced by some 60% in the most extensively jointed rock mass, associated with operating range 3.

Further study of a wider variety of case histories, will allow for a more comprehensive indexing of a given rock mass, by including strength and water indices. Further attention will also be paid to gathering and analyzing site investigation data to facilitate the quantification of joint intensity. The use of a representative set of RII values, based on site investigation information, will result in soundly based predictions for the performance of the TBM. The final goal of the modeling work is to provide clients, designers and contractors with a set of practical tools capable of providing feasibility and design feedback on the use of equipment within a rock mass by use of a simply applied simulation package. The relationships identified between rock mass jointing, support and TBM performance parameters, described in this paper, will form a solid basis for mining predictions within the framework of the simulation module.

7 Acknowledgments

The research program has been funded by the National Science Foundation (United States) under the Division of Design and Manufacturing Systems and Geomechanical, Geotechnical and Geo-Environmental Systems. This research has also been supported by the U.S. Department of the Interior's Mineral Institutes program administered by the Bureau of Mines under allotment grant number G1134248 to the Texas Mining and Mineral Resources Research Institute. The support provided is greatly appreciated. The authors would also like to thank those within the international tunneling community for their support in providing the case history data on which the modeling work is based. As always, the realism of both mechanical and geological tunnel modeling will only be as good as the data on which they are based. The quantity and quality of all data supplied is greatly appreciated.

8 References

1. Nelson, P.P., Petersen C.R., Einstien H.H., Hood M. "Manufacturing Underground Space", Proceedings of the International Congress Towards New Worlds in Tunnelling, Acapulco, May 1992 pp. 29-36.

2. Abd Al-Jalil Y., Nelson, P.P., Laughton C. "TBM Performance Analysis and Rock Mass Impacts", Proceedings of the 2nd Symposium on Mine Mechanization and Automation, University of Lulea, Sweden, June 1993, pp. 201-209.

3. Nelson, P.P., "Tunnel Boring Machine Performance in Sedimentary Rock", Ph.D. Thesis, Cornell, New York 1983.

4. Goldberg Zoino Associates, "Geotechnical Data Report Tunnel Section Station 2190+00 to Station 2375+00" prepared for Niagara Frontier Transportation Authority, Metro Construction Division, Buffalo, New York, 1 August 1978.

5. Bienawski, Z.T. "Rock Mass Classification in Rock Engineering". Proceedings of the Symposium on Exploration for Rock Engineering", Johannesburg, November 1976, Vol. 1 pp 97-106.

6. Abd Al-Jalil Y., Nelson, P.P., Laughton C. "Analysis of Performance Measures of Tunnel Boring Machines" Proceedings of the Eurock '92 Symposium, Chester, September, 1992. Vol. 1 pp. 408-413.

Development and performance of tunnel-boring machines on Phase II of the London Water Ring Main

R. Remington
Thames Water Utilities, Ltd., Reading, England

Abstract

In March 1986 construction commenced on the London Water Ring Main, a major infrastructure project to be carried out by Thames Water to ensure the efficient distribution of drinking water to London and secure its supply capacity well into the 21st century.

The major civil engineering element was the construction of 80km of tunnel of 2.5m average diameter, with associated construction shafts and pumping stations. The project was to be carried out in two phases and be complete by the spring of 1996.

During the course of the project significant advances have been made in the development of both pre-cast concrete tunnel linings and tunnelling systems for use in soft ground tunnelling. These developments have resulted in a significant increase in safety and in tunnel driveage rates, saving both time and money, as a direct result of which the project has been completed two years early and within budget.

Keywords: Construction, Water Supply, Tunnel, Tunnelling Machine, Softground, London Water Ring Main.

1 Introduction

The London Water Ring Main comprises some 80 kilometres of tunnel, 2.5 metres in diameter and 2.9 metres in diameter, with twelve pump out shafts, four input shafts and a further number of shafts which were used for construction and will in future be used for access for tunnel inspection and repair (see Fig. 1).

The Southern Tunnel Main, which was completed in 1974 and was constructed principally from 100 inch wedge block lining, enabled the bulk transfer of water by gravity from Hampton and Ashford Common Water Treatment Works to Merton Pump Out Shaft in South London and provided key service information for the future Ring Main. Following a strategic study for "London's water supply into the next century" completed in 1984, design on the project was started in 1985 and construction commenced on Phase I of the London Water Ring Main strategy on 6th March 1986 and included the construction and uprating of certain stored water transfer tunnels, uprating of the major water treatment works at Coppermills and Hampton, implementation of a control scheme, and some 22km of ring main tunnel extending on from the section already completed to Merton.

The physical properties of the London Clay, which generally underlays the thin

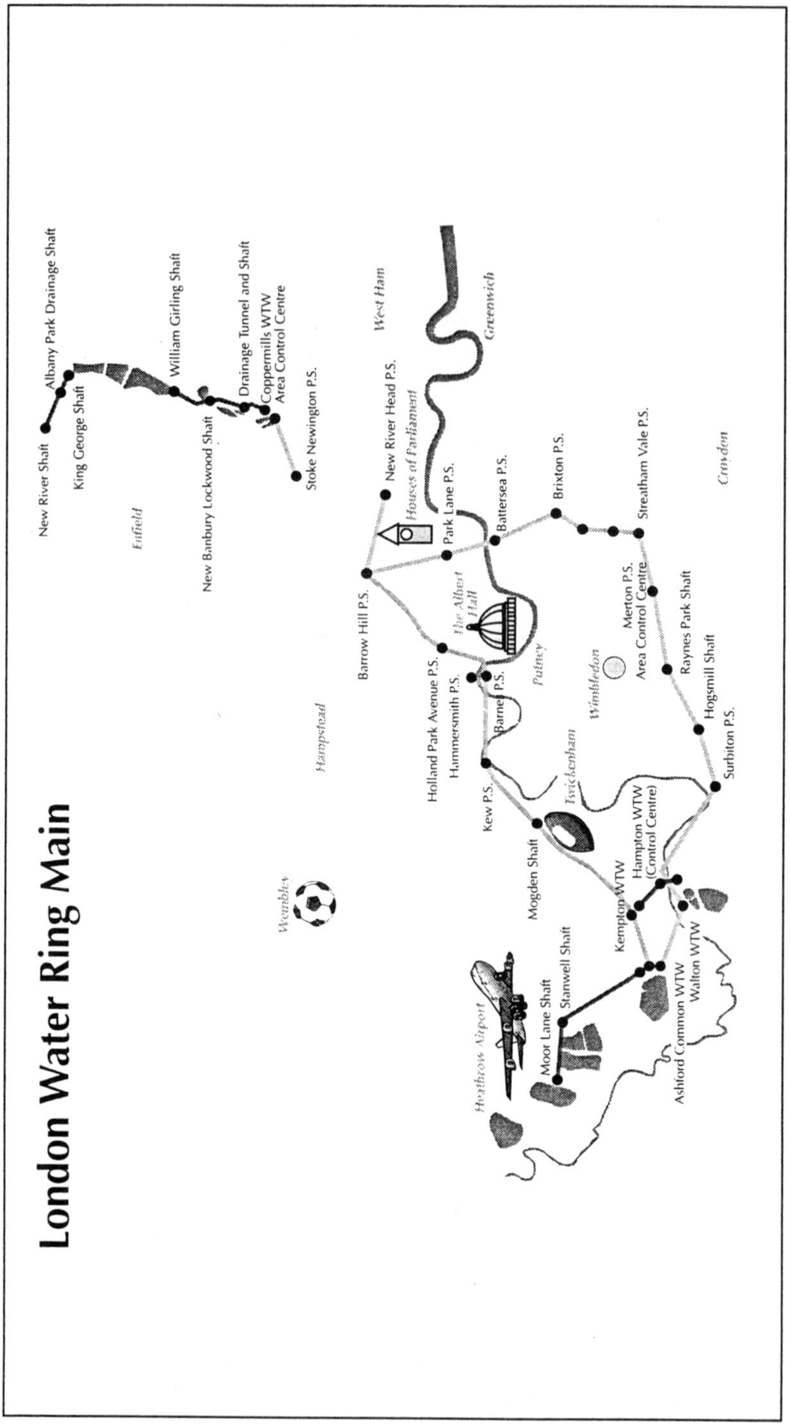

Fig. 1 London Water Ring Main

layer of sands and gravels throughout the London area, permits tunnel construction using expanded wedge block rings. The wedgeblock tunnel lining system was developed by the Metropolitan Water Board in the 1950s and 60s. The main advantage of this lining system is that there is no grouting, no secondary lining, rapid progress and it can carry water pressure equivalent to 1.5 times the overburden pressure. The wedge block expanded lining comprises un-reinforced pre-cast concrete segments, which are expanded against the smooth bored stiff London Clay when the tapered key segment is pushed into position.

Phase I tunnels were divided into four contracts. The ground conditions being extremely variable, a number of concrete segmental lining methods were employed; 100 inch wedge block, bolted smooth bore, bolted segmental and a trapezoidal bolted lining. All bolted linings requiring a secondary cast *in situ* reinforced concrete lining to contain the hoop stresses generated by the water under pressure within the tunnel, when in service.

A variety of machine types was used to construct Phase I tunnels, open face machines with backacter shovels, full face earth pressure balance machines and a conventional full face machine. The transport systems all utilised Clayton type 2.5 to 3.5 tonne locomotives and the unit train concept, with the exception of one contract which used slightly larger locomotives of Italian manufacture. All major civil engineering contracts within Phase I were to ICE 5th edition conditions of contract, production rates and cost of production being contained within the tenders. Handling systems were variable, with some contractors using two cranes in the shaft. However, most contractors used one large crane both to lower into the shaft at one time the segments necessary for one complete ring and the empty muck skip. Mucking out from the shaft normally utilised a large muck skip of one unit train capacity. Installed in the same shaft were often 'Alimak' or similar proprietary temporary construction site shaft lift systems. Analysis of transport and shaft systems was generally to a low level of detail, systems being of basic design. Cycle time analysis was basic and pinch points were rarely removed from the systems. Few drives on Phase I were actually constructed of wedgeblock throughout, and whilst some notable peaks of production were achieved of 36 and 55 rings per shift, on these drives the overall long average for wedgeblock construction was 115m per week.

The commencement of planning of Phase II coincided with the early days of Thames Water as a privatised utility and a reorganisation of Engineering within Thames Water where positive performance was encouraged. The construction of Phase II tunnels, total length 25km of London Water Ring Main and 8km of stored water tunnels, commenced in the spring of 1991. The contract and method of construction used was approached in a radically different manner, which has led to improved production performance and reduced costs. This paper briefly outlines the approach of Phase II design, specification and construction.

2 Phase II: A Radical Approach

During tunnel construction on Phase I, three significant problems were encountered which incurred considerable delays and additional costs. These were:

i) A geological anomaly in the London Clay known as a scour hollow, where the open face tunnelling machine entered a pocket of water-bearing sands and silts, necessitating the use of compressed air at pressures up to 2.4 bar.

ii) Deep bored piles over 50m deep for a four storey residential block were encountered and the tunnel machine had to be diverted around them.

iii) A tunnel machine was inundated with water at 4 bar pressure from the underlying Thanet Sands, and subsequently a new EPBM was installed to complete a 1.2km section of the Ring Main through the Thanet Sands.

Apart from the delays and increased costs, these problems served to highlight the variable nature of the ground despite extensive site investigations, and it became obvious that with 33km of tunnelling still to be completed in Phase II, a complete review of the knowledge gained in Phase I was necessary with a view to minimising future costs and saving time on the project.

Early indications were that to complete the project within the original timescales, carrying out the work in the same way as Phase I, the remaining 33km of tunnel would possibly require at least six tunnelling machines and three contractors, and design requirements would dictate that an increased size of tunnel from 2.5m to 2.95m was required for 50% of the new tunnel. During the summer of 1990 a number of studies were carried out, one of which was into tunnel transportation, power and ventilation requirements.

Tunnels of 2.5m internal diameter have in recent times generally been constructed using unit train transport systems. Each train comprises segment transport vehicles having the capacity for one complete ring of segments and muck skips having adequate capacity for the excavation debris for one complete ring width advance by the tunnelling machine, including allowing for the high bulking factor experienced with London Clay. Locomotive size is limited by the tunnel diameter, particularly with reference to passing places. Passing places have traditionally been installed at 1km intervals. Each train progresses into the next empty passing place and awaits the train coming out loaded with muck from the tunnelling machine to pass it. The cycle is then repeated until finally the tunnelling machine is reached, new segments are unloaded and spoil is loaded at the TBM. The train then proceeds directly to the pit bottom unloading area. Theoretical analysis of locomotive power, train acceleration times, travelling times and real data from existing contracts indicated a maximum capacity of approximately three to four rings per hour at 1km intervals. Tunnel lengths on Phase II vary from 2km to a maximum of 6.1km, thus at 6km lengths and 1km spacing of passing places, a minimum of six locomotives in total were required, including one for shunting. Further analysis indicated that for sustainable performance at four to five rings per hour, passing bay centres needed to be reduced to 800m, and to ensure six rings per hour, passing bay intervals might have to be reduced to 600m. This would have a significant increase in the total number of unit trains and drivers required.

The wedge block rings used on Phase I of the London Water Ring Main were designed by the Metropolitan Water Board in the 1950s. The ring comprised of twelve segments, a ring width giving 685mm of advance. Central to this design originally was the requirement to build rings by hand; each segment was thus limited to 140kg weight, necessitating twelve segments per ring. Clearly, if the ring width was increased and if the unit train system could still be feasible, this would allow an increased advance per hour for the same number of trains in the system. Feasibility studies on tunnel segment design indicated a 1m ring width was feasible and the existing three-and-a-half tonne locomotive would, with reduced acceleration, still be capable of performing adequately with the larger ring and larger muck train capacity. An optimum target of four rings per hour throughout the system was thought to give the best overall economic performance and this formed the specification for the tunnel boring machine transport systems and the surface and shaft handling equipment.

At the same time as reviewing the tunnelling system it was also decided to review the contractual practices adopted on Phase I. Recent development by Thames Water of a Target Cost Formula for use with the Institution of Chemical Engineers' Cost Reimbursable Form of Contract (the IChemE Green Book), requiring a greater element of risk sharing by Thames Water, had proved both flexible in its approach and also successful in permitting a greater degree of client involvement and a teamwork approach to construction. These benefits would be vital with the introduction of client imposed requirements for a new segment lining and improved machine performance. Because of the anticipated enhanced production and the contract flexibility, Thames Water decided to design and manufacture the tunnel rings and also to purchase its own tunnelling machines to its own specified design and free issue these both to the successful contractors to capitalise on the expected savings in time and money.

In order to ensure appropriate machine specification, in addition to the studies of tunnelling systems, many discussions were held with contractors and machine manufacturers. This improved understanding of machine specification anticipated performance requirements and the many support and infrastructure systems required by the contractors to optimise tunnel costs.

3 System Development

The use of mechanised lining erection equipment is widespread within the tunnelling industry. Machine built 100 inch wedge block rings were generally slightly slower than the fastest hand built rings. The availability of mechanised erection, however, enables radical changes to ring design. Reducing the number of segments in each ring has the advantage of less handling actions throughout the system. Mechanised erection systems are easily capable of handling heavier segments and generally far safer than manual lifting systems. A significant design consideration, however, with the proposed increase of ring width and reduction in the number of segments was the larger size of each segment and the bending stress increases this would generate. Design feasibility studies had indicated that the number of segments in an un-reinforced concrete ring could be reduced to a minimum of eight without very significant increase in the bending movement within the segments. The design, development and testing of the new ring is the subject of a separate paper given at this conference. *[Ref. 1]*

The intention was to significantly improve the rates of tunnelling performance, without introducing new risks by changing existing machine parameters but by capitalising on known machine strengths. The intention was that the new system would require four key performance requirements for the new tunnel boring system:

- Safety
- Reliability
- Flexibility, allowing security and progress in poor ground conditions and improved performance in good ground
- Maximum tunnelling performance in London Clay

TBM performance analysis from previous tunnel drives indicated that the new TBMs of a full face type could, over significant distances, have availability greater than 95%.

It was decided to purchase full face tunnelling machines with capability for earth pressure balance working. This would allow the machine to continue to make progress, albeit at a slower rate, in mixed wet soil conditions should they arise.

The machines were also specified to work in an open mode without a rear tailskin whilst erecting wedge block lining, but also to have provided (but not fitted when working in London Clays) a tail skin and earth pressure balanced screw conveyor, to work firstly in sand conditions and secondly in pressurised ground water conditions. In addition, physical and programming constraints did not permit all drives to end at a shaft from which the machines could be removed, so the machines were designed to be disassembled underground and be brought out through the tunnel they had constructed where necessary.

The performance requirement specified was to be able to excavate in London Clays a 1m tunnel advance in five minutes. Additionally, to prove in factory trials that a wedge block segmental ring could be erected in five minutes. Also, that the segment loading and handling into the magazine storage device could be accomplished with the excavation cycle, therefore providing a sustainable 15 minute total machine cycle time.

A tender document was prepared and issued for the purchase of three tunnel boring machines, including recommended spares for the early stages of tunnelling, backup spares in the event of machine breakdown and common spares between machines wherever possible. A product support arrangement whereby inspection and machine refurbishment would be performed in partnership with the machine manufacturer was also included. After careful analysis of the tenders received, the contract was duly awarded to Lovat Tunnel Equipment Inc. of Toronto, Canada for three dual mode TBMs, two 2.9m OD and one 3.3m OD.

In order to ensure that optimum productivity could be achieved when considering all plant and resources from the surface to the machine underground, an analysis of the system logistics, transport to site, surface handling, shaft transport, pit bottom changeover and tunnel transport systems of a number of tunnelling contracts, including those on Phase I of the Ring Main, was carried out. This highlighted that it was often the performance of these backup systems which was the limiting factor and that changes to these systems were likely to be low cost and low risk but could give substantially improved benefits. A key area was the handling of the segments immediately behind the TBM (see Fig. 2), and the releasing of the transport trains as soon as possible to permit machine digging to commence.

The TBM design for segment handling is novel and has exceeded its required performance criteria. Segments are transported along the tunnel on four rotatable segment bogies for each ring (this is necessary because of the size of the segments relative to the internal diameter of the erected lining). The segment bogies are pushed into the rear of the TBM backup system by the locomotive. On reaching the end stop, the segment handling operator engages hydraulic controls which, through hydraulic cylinders, actuate the simultaneous rotation of all four segment bogie decks. This rotates each segment into the correct orientation for insertion into the segment magazine and thence the erecting system. The total turn round time to enter the rear of the TBM and unload segments and to allow the unit train to proceed to conveyor discharge to commence mucking out, takes 20 seconds. During the TBM excavating/shove cycle, the segments are manipulated from unloading position into order in the magazine to allow quick feed to the erector arm. On completion of excavation, the lower shove cylinders are first retracted and the invert segment placed in the invert. This provides a platform onto which the magazine then loads each tunnel lining segment sequentially to match the requirements of the erector arm. The erection cycle then commences whilst the remainder of the shove cylinders are retracted, ring erection taking approximately four to five minutes.

Fig. 2 Segment Handling System Photo

The significance of all elements of the segment handling system cannot be overstressed. Each segment weighs approximately 600kg, and if a segment were to arrive at the TBM in the wrong order, the wrong orientation or damaged, it would be extremely difficult to remove in the confined working area and cause considerable delay.

During the TBM design and development period, the purpose designed segment bogie prototypes were designed and manufactured by Lovat. It was identified at this stage that the tunnelling programme would require a total of 68 vehicles and these would be procured in the UK. The specification was prepared by Thames Water and competitive tenders were invited, with the contract being awarded to Isometal. The design has been very satisfactory, proven by performance.

Improved tunnelling performance would increase transport movements at the shaft head and the surface site layouts had to accommodate the increased delivery of tunnel segments, 10 to 12 loads per day, and the increased number of lorries for muck removal which peaked at over 100 a day. All significant road haulage to and from site was limited to between 7am and 6pm, although tunnelling operations continued 24 hours per day.

The shafts diameters suit their final application and are generally 7m to 12m diameter; clearly most significant problems with shaft transport capacity were met in the smaller shafts of 7m diameter. Two cranes are used in each shaft, with one high power tower gantry or crawler mounted crane of low noise emission being used for muck haulage from the shaft; typically cranes of 80 to 100 tonnes nominal capacity are used for this duty, hauling approximately 20 tonnes payload up the shaft. In slow speed this single lift can take several minutes. A second site crane of lower capacity is used to lower segments down the shaft, move segments in the pit bottom if necessary, unload segment delivery lorries on the surface and to lower and marshall locomotive batteries. Normally one battery change per locomotive mid-shift and one at the end of the shift are required.

An 'Alimak' or similar proprietary lift is installed at each shaft for man access. Segments are delivered to site standing on edge, and are generally unloaded either by crane or fork lift truck. Segments have two lifting points, when lifting by crane, a lifting frame lifts the seven main segments simultaneously; the smaller size keys are handled separately. Segments are lowered down the shaft onto a loading and turning arrangement which loads the segments as required onto the four segment bogies in each unit train. A single track of 600mm nominal gauge runs through the middle of each shaft with the segment handling facility to one side and the shaft muck hoist skip on the other (see Fig. 3).

Fig. 3 Typical Pit Bottom Layout

Unit train muck cars have a rolling side discharge facility, allowing them to be emptied into the shaft muck skip. Shaft muck skips run on rigid guides, normally of rolled steel channel or RSJ type. To facilitate marshalling and loading and unloading at the shaft, each shaft required a back shunt of approximately 30m. Opposite the back shunt in the first part of the tunnel, once drivage has advanced somewhat, a passing and marshalling bay is installed, allowing a unit train ready to travel in-by to pass a full unit train coming out-by (see Fig. 4). Because the segments when loaded on the bogie are slightly over 1m width including retaining facilities, to accommodate passing, the tunnel rails at passing bays need to be significantly above the tunnel invert. Purpose-made structures support the raised rail section, rail type being 15kg per metre. Open mesh steel deck flooring is installed throughout the tunnel length to make walking easier and reduce the likelihood of accidents.

In addition to the normal tunnel passing bays at fixed intervals, a final passing bay approximately 20m behind the TBM's trailing gantry is towed by the machine across the top of the rail track, thus permitting minimum turnaround times for the trains leaving the TBM, having delivered segments and loaded up with spoil. New rails are added to the track within the sliding rail system in the trailing gantry as the machine progresses forward.

Fresh air is fed to the tunnel heading via a 450mm or 600mm flexible ventilation ducting from forced ventilating fans located at the shaft side on the surface. In the unlikely event of dust occurring at the machine, a dedicated ventilation system provided on the TBM backup gantry can be utilised to withdraw it. In addition, all persons entering a shaft or tunnel were required to have with them at all times an emergency breathing set, and gas and oxygen levels were continuously monitored throughout the tunnel during construction.

Fig. 4 Loco and Train in Passing Bay

Signalling systems are used in the shafts with the crane driver, pit top and pit bottom being linked through a shaft signalling system. The depth of shafts, generally 40m to 60m deep, means that the pit top banksman could also observe shaft bottom operations. Certain contractors preferred the use of a closed circuit television system, whereby the crane driver can see a monitor in his cab linked to a camera looking at the shaft bottom operations. In any event, specific safety systems were in operation throughout all lifting operations.

Communication along the tunnel was provided for by telephone at the TBM and pit bottom and surface, and the Tannoy systems, a Tannoy being provided at the pit top and pit bottom, at each pass-by and at the TBM, whereby locomotive drivers can stay in contact with tunnel rail marshalling operations.

4 Construction Performance

The first of the three TBMs was delivered in July 1991 and commenced work at Ashford Common on a Stored Water Tunnel on 12th July 1991. To ensure that performance of the system was closely monitored, construction inspectors collected data, as part of their routine duties, which was entered onto a computer monitoring system developed by Thames Water for the Tooting Bec contract from work carried out on the Channel Tunnel project. Additional detail has been added to the system to enable comparisons to be made on all three drives. *[Ref 2.]* The second machine commenced work at Ashford on 21st August 1991 on a short section of tunnel, 400m long. Prior to commencing work on the drive from Barnes to Barrow Hill the third and largest machine commenced working at Kempton on 5th September 1991. Working shafts for the smaller machines were nominally 10.3m ID, whilst all working shafts for the larger machine were 7m ID, as these shafts are tunnel maintenance shafts and not pump out shafts.

Thames Water significantly benefitted from having three sites running in parallel with three different contractors operating the system. Site data was entered into the computer programme and a range of reports produced which identified performance pinch points. This enabled the systems to be evaluated and proposals made to improve performance, the object at all times to attain optimum system performance and reduce tunnelling costs. This was equally attractive to the contractors, as the form of contract used shared the benefits of any savings between the tunnelling contractors and Thames Water. As would be expected, the first drive at Ashford Common, carried out by Costain Civil Engineering, was very much the pilot for all the drives and some initial difficulties were experienced. The two significant problems were:

i) The over-consolidated clay expanding behind the shield producing a smaller build area for the concrete ring
ii) Achieving the smooth rotation of the turntables on the segment-carrying bogies.

The latter problem, although more annoying for the miners, was overcome with minor modifications to the segment bogies and highlighted the need for good routine maintenance of the bogies system.

The potential problem of clay expanding behind the shield had been identified during the design of the system. Expanded lining systems in the London Clay rely on the minimum machine overbreak so that, when expanding the ring, the range of movement provided by the key segment is sufficient to ensure that the ring is firmly expanded against the clay. This is monitored by noting the key ram pressure required to drive home the wedge. The long shield length of a full face TBM and the inability to easily change the bead on the leading edge make selection of the size and profile of the bead critical. Other factors such as the way the machine is driven, head ram positions, etc., also significantly affect overall performance. The early problems meant that until such time as the correct bead was introduced, some of the wedge key segments had to be trimmed in length to enable them to be driven home. An indication of the complexity of the factors which affected the performance of the system, including the variable nature of the ground, was that the head arrangement on the three machines was always different although similar performance was achieved.

As would be expected with complex full face tunnelling machines, some breakdowns were experienced but with the number of spares that had been purchased with the machines, all problems were manageable within the tunnels. All contractors were required to carry out routine maintenance during driving, supervised by a specialist Thames Water fitter/supervisor. Upon completion of each drive a full maintenance check and service of the machines was carried out under Lovat supervision prior to commencing the next drive. Two bearings were replaced during these routine inspections; one was found to be faulty and suffering excess wear, and this was replaced under warranty. One was changed as a result of fair wear and tear erring on the safe side, having completed over 14km of tunnel. Such was the performance of the machines that machine availability was on average always well over 90% with availability on some drives in excess of 98% (see Fig. 5).

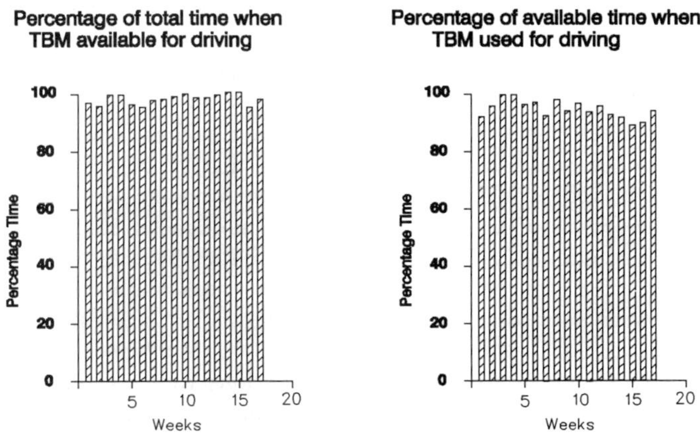

Fig. 5 Machine Availability

The machines were always used in open mode, the ground at no time requiring the use of EPBM operation. This obviously aided the success of the project but full contingency plans and all necessary equipment was available should this have proved necessary.

As has been indicated, TBM performance was extremely good. The original design for the segment handling gear at the rear of the machine has proved extremely reliable and capable of achieving a total ring cycle time of better than 10 minutes at peak performance. The transport system, although intended for a cycle of four rings an hour, in fact had to operate well in excess of the proposed rate to keep pace with the production rates being set. Additional time was required for track installation extension, pipework extension, cable extension, telemetry systems extension, ventilation ducting and maintenance. Very few problems were experienced with the transport systems; a testament to the quality of the kit installed.

The best performance exceeded six rings per hour, with each contractor progressively setting new records for production as they became familiar and more confident in the system, culminating in the statistics in Table 1 below.

TABLE 1

Best Performances Achieved	
Single Shift	56 metres
2 Day Shift	112 metres
10 Shift Week	501 metres
Month (4 Weeks)	1,542 metres

However, whilst peaks of performance are always satisfying, the true test of any system is its ability to sustain high levels of performance over the duration of the project. Our target overall performance for Phase II was a sustained "long average" of 200m per week which in itself was nearly double the long average for wedgeblock on Phase I of 115m per week. The overall average for Phase II was in fact 237m per week, more than twice the average for Phase I, with the best drive average of 324m per week. It is also worth noting that the best week's production of 501m was achieved at a distance of over 5km from the working shaft. The longest single drive was 6.1km, working from a 7m diameter shaft.

As previously mentioned, a large number of activities have to go on to keep the machine moving forward and the link between peak daily performance and sustained high weekly or long term averages is dependant upon these constraints. Machine cable extensions were carried out every 200m of advance, thus at the rates of progress being achieved, mid-week cable extensions were required and a further development of the system was carried out to introduce a degree of mechanisation into the process. This significantly reduced the time and resources to carry out this operation. Further opportunities for developing other parts of the system have been identified, particularly when using the system with bolted, grouted rings.

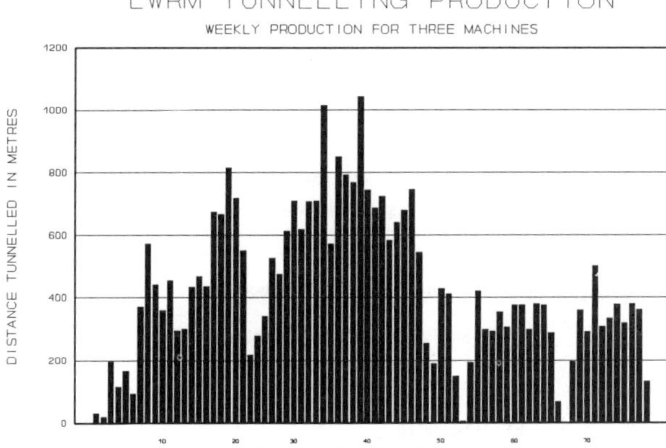

Fig. 6 Overall Production Performance

5 Conclusion

The core business of Thames Water Utilities is the resourcing, storage, treatment and supply of drinking water, and the collection and treatment of domestic sewerage and trade effluent. The use of new contract forms, the purchase of tunnel boring machines and the design and manufacture of tunnel segments is often cited as a significant departure from that core business. However, Thames Water as a private utility company has a significant volume of capital works to implement in order to improve plant capacity and renew infrastructure works. A key element of these works, both for water and sewerage, is the construction of new tunnels of a wide range of sizes.

During the life of the Ring Main project, Thames Water has developed many activities: project management and control, standardisation of design and new quality and safety systems. None, however, have possibly had the profile and degree of success that the tunnelling strategy adopted for Phase II in 1990 has, which itself has set a benchmark for others to try to achieve in the future. A key part of the strategy was to improve safety in tunnelling, particularly necessary when introducing greater degrees of mechanisation into confined spaces. Performance will always need to be sacrificed to ensure safety but Thames Water managed to maintain both at extremely high levels in parallel throughout the project.

The actual performance results achieved speak for themselves. During the "worst" times towards the end of Phase I with the problems at Tooting Bec, it seemed that the Ring Main would be yet another tunnelling project to be late and over budget. The success of tunnelling on Phase II has helped turn this around, ensuring completion of the project two years early and within the £250M budget. Thames Water have managed to maximise the process and minimise costs, ensuring that future work of a similar nature will be carried out with a form of contract that encourages teamwork between the client and the contractor and where both sides share in the successes and failures, if any.

Already one other Thames Water project has benefitted from the use of the system; a storm sewer in north London, and currently the system is being used on a cable tunnel in south London.

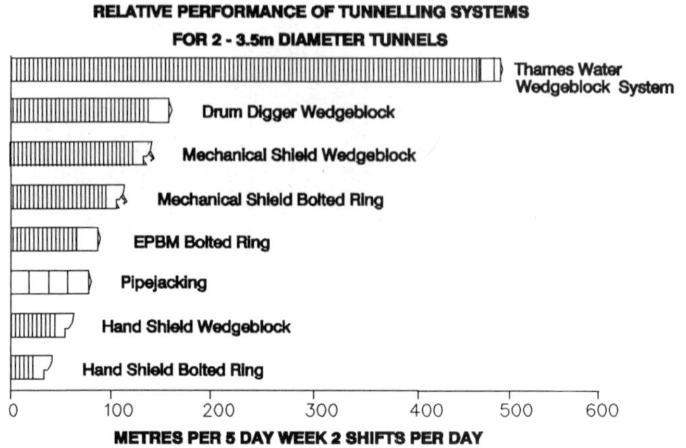

Fig. 7 Tunnelling System Comparisons

Whether it is future improvements to the present system or improvements to other tunnelling systems, Thames Water engineers are committed to achieving the best product possible, safely and at a minimum cost, with the co-operation of its working partners.

References

1. Farrow, J.P., Claye, P.H., Warren, R.B., Ginary, M.A. and Matthews, R. (1994) Tunnel Ring Design, Development, Testing & Manufacture for the London Water Ring Main. **Tunnelling 94 UK Conference Paper**

2. Fawcett, D.F., Remington, R. and Chapman, P. (1994) TBM Drive Management Using Computerised Systems. **Tunnelling 94 UK Conference Paper**

Principal Contractors

- Lovat Tunnel Equipment Co
- Amec Tunnelling
- J Murphy & Sons Ltd
- Costain Civil Engineering
- Taylor Woodrow Civil Engineering
- Isometal

Session 5: Environmental Issues

Environmental planning for the Massachusetts Water Resources Authority's MetroWest water supply tunnel

H. P. Caspe
A. Y. Kim
L. J. Bergen
Sverdrup Civil, Inc., Boston, Massachusetts, U.S.A.
J. R. Araujo
Capital Engineering and Planning Department, Waterworks Division, Massachusetts Water Resources Authority, Boston, Massachusetts, U.S.A.

Abstract
The Massachusetts Water Resources Authority (MWRA) is proposing to construct a 17.4 mile (28 km.) deep-rock tunnel, the MetroWest Tunnel, to provide adequate redundancy for the primary water transmission line serving the metropolitan Boston area. Project planning studies considered an original concept to provide adequate service reliability— reconstruction of an existing aqueduct and addition of connecting tunnels. This aqueduct reconstruction was found to involve substantially greater environmental impacts than the tunnel alternative and provided only partial system redundancy. The tunnel can be constructed as a larger diameter conduit and is being designed to provide full system redundancy. The shafts, construction staging areas, and muck removal portals along the MetroWest Tunnel have been sited to minimize disruption to residential areas. Construction-related environmental issues addressed during final design include treatment of tunnel dewatering discharges, private well impacts, and tunnel muck disposition. The project has been divided into three major contracts for construction of the tunnel and related surface facilities. Project planning has included an extensive geotechnical exploration and testing program to determine boreability characteristics of bedrock along the tunnel alignment. During final design, cost-sharing initiatives, such as value engineering, and contract practices necessary to avoid disputes during construction have been adopted.
Keywords: aqueduct, environmental, MetroWest Tunnel, MWRA, planning, water supply

1 Introduction

The Massachusetts Water Resources Authority (MWRA) is proposing improvements to provide adequate redundancy for its primary water transmission system, which supplies 46 communities and over 2.5 million people in eastern Massachusetts. A key component of the primary water transmission system is the Hultman Aqueduct, a pressure conduit which conveys water 17 miles (27.35 km.) into the Boston area. The Hultman Aqueduct is part of a system of aqueducts and tunnels which transports water 65 miles (104.59 km.) east from the Quabbin-Ware-Wachusett watersheds into Boston (Figure 1).

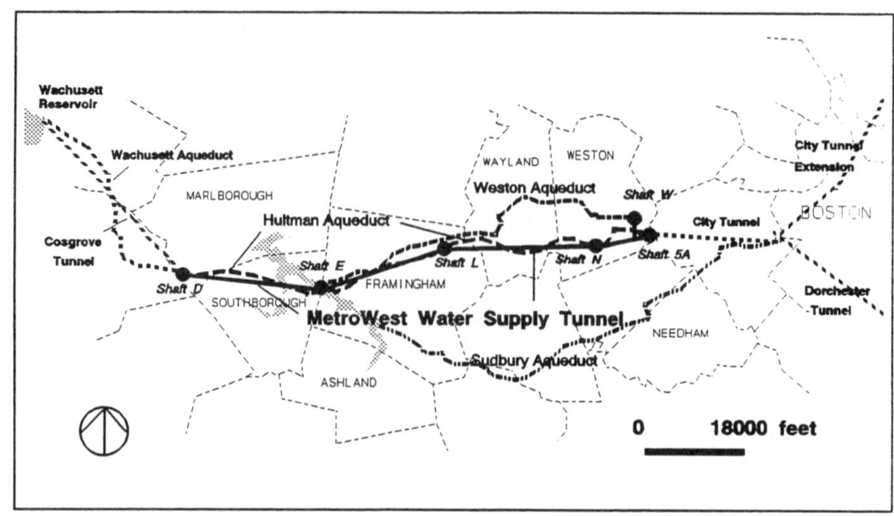

Fig. 1 Existing transmission system and proposed MetroWest Water Supply Tunnel

The Hultman Aqueduct feeds the high service system (areas situated at elevations between 70 and 170 feet (21.3 and 51.82 m.) above sea level) and supplies and operates in tandem with the Weston Aqueduct. The Weston Aqueduct is an older gravity aqueduct supplying the low service system (areas situated at elevations below 70 feet (21.3 m.)). The downstream flow split between the Hultman and Weston Aqueducts is 85%/15%. Portions of the low service area are also supplied from the Hultman Aqueduct directly through pressure reduction.

The Hultman Aqueduct is over 50 years old and currently requires significant repairs. There are 24 identified leaks on the aqueduct, with estimated losses in excess of 20,000 gallons (75.71 m.^{3}) per day from a single leak. As the sole supply line in this portion of the high-service transmission system, the aqueduct cannot be removed from service for a period of more than a day to allow repairs to be effected.

A third unused gravity aqueduct, the Sudbury Aqueduct, has inadequate capacity to substitute for the flows carried by the Hultman Aqueduct. Even with both the Weston and Sudbury Aqueducts in operation, the service area would experience severely reduced service pressures and some areas would go entirely without water if the Hultman were removed from service for more than a day. The Sudbury Aqueduct also relies on reservoirs which do not meet current drinking water standards and would require boiling prior to drinking.

The MWRA plans to construct a supplemental pressure aqueduct (the MetroWest Water Supply Tunnel) adjacent to the Hultman as part of its ongoing programme of water system improvements. The purpose of the project is to provide adequate redundancy for the Hultman Aqueduct. Once constructed, the proposed tunnel would enable the Hultman Aqueduct to be removed from service for inspection, maintenance, and repairs. Under normal operating conditions, the proposed aqueduct would be used to meet roughly half of the system demand to reduce the flow through the Hultman Aqueduct.

In the spring of 1990, the Massachusetts Water Resources Authority selected Sverdrup Corporation to provide services necessary to evaluate, design, and construct a supplemental transmission line. The consultant team working on the project includes Jacobs Associates of San Francisco, Fay, Spofford, & Thorndike, Inc. Engineering, GZA GeoEnvironmental, Inc., Delon-Hampton & Associates, Boston Affiliates, and Judith Nitsch Engineering, Inc. all of Boston.

The first phase of the contract included conducting an environmental assessment and conceptual/preliminary design, culminating in preparation of an Environmental Impact Report to meet state requirements (Figure 2). The difficulties associated with implementing major infrastructure improvements serving and affecting urbanized areas in the metropolitan Boston area became apparent at the onset of the study.

2 Aqueduct reconstruction alternative

The original concept for a supplemental pressure aqueduct involved reconstruction of the Sudbury Aqueduct and addition of new tunnels extending to the existing transmission system. These planned improvements were based on recommendations from a previous study of the MWRA transmission system. This project was originally conceived as the first of a series of phased improvements to provide adequate system redundancy.

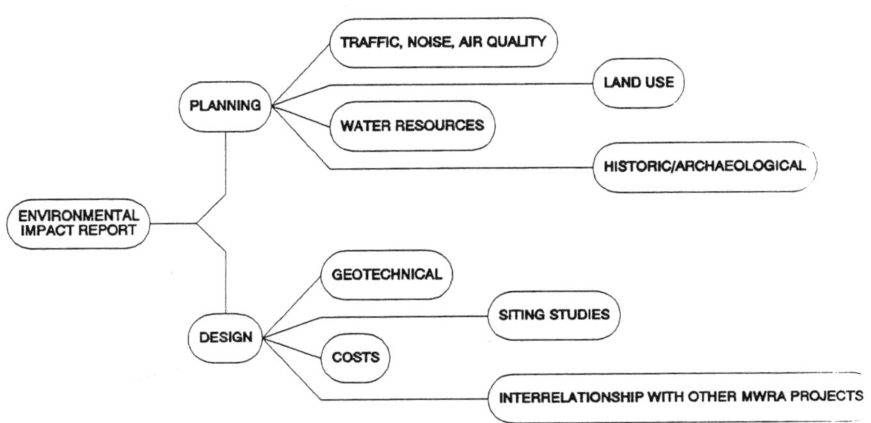

Fig. 2 Planning/design process

The Sudbury Aqueduct is 114 years old and is currently held on emergency standby service. This aqueduct is horseshoe-shaped in cross-section and consists of brick and masonry construction (Figure 3). The Sudbury Aqueduct and related structures are considered historic and are included in a listing on the National Register of Historic Places.

The Sudbury Aqueduct extends 17.4 miles (28 km.) from the Framingham Reservoir system to the eastern end of the City Tunnel, 5 miles (8 km.) downstream from the terminus of the Hultman Aqueduct. In order to integrate this aqueduct into the existing transmission system, construction of 7.5 miles (12 km.) of tunnels was originally envisaged (Figure 4). These tunnels were required to connect the headworks of the Sudbury Aqueduct with the transmission line supplying the Hultman Aqueduct.

In order to pressurize the aqueduct and increase its water-carrying capacity, the inside diameter of the aqueduct was proposed to be enlarged, from an average diameter of 8.5 feet (2.59 m.) to 10 feet (3.05 m.). This reconstruction would require use of cut and cover construction techniques and a construction corridor 60 to 190 feet (18.29 to 57.91 m.) in width.

Evaluations of existing conditions along the aqueduct route identified areas of the existing aqueduct corridor where particularly difficult construction conditions would be encountered, due to the presence of hazardous waste contamination, major transportation routes, and more densely developed urbanized areas. Approximately 7 miles (11.26 km.) of tunnels were proposed to avoid surface construction in these areas.

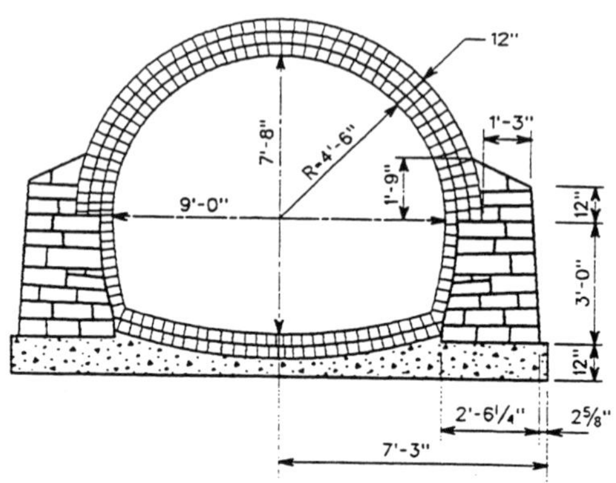

Fig. 3 Typical cross-section of Sudbury Aqueduct

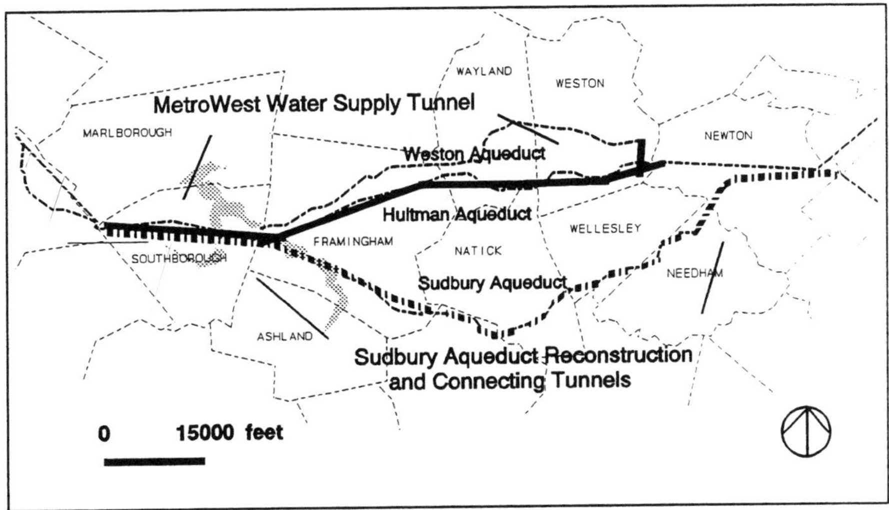

Fig. 4 Project alternatives

Reconstruction was considered to be more feasible along approximately 10 miles (16.09 km.) of the aqueduct corridor, extending through suburban to rural areas. This portion of the aqueduct right-of-way extends through residential neighborhoods in four communities. This reconstruction would have involved disruptions to numerous private and public school properties, two college campuses, town conservation lands, several golf courses, a National Register Historic District, and some of the most affluent areas of metropolitan Boston. Aqueduct reconstruction would have affected hundreds of homes, over 50 roadway crossings, a number of stream crossings, and wetlands adjoining over 9,000 linear feet (2743.2 l.m.) of the aqueduct.

The planning process for this project involved preparation of documents required under the Massachusetts Environmental Policy Act (MEPA) (Figure 5). An Environmental Notification Form outlining the Sudbury Aqueduct Reconstruction approach was prepared and submitted to initiate the state environmental review process. A series of community meetings were held in affected communities to review the project proposal and obtain input into the development and design of the project.

3 Tunneling alternative

The shortcomings identified with the aqueduct reconstruction approach and community concerns led to further development and evaluation of other project alternatives during the planning process. The alternative eventually selected consists of a deep-rock tunnel, situated approximately 200 to 500 feet (60.96 to 152.4 m.) below the ground surface.

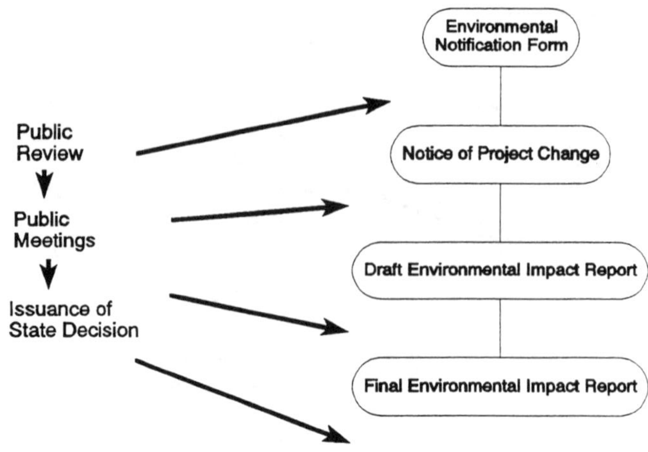

Fig. 5 Massachusetts Environmental Policy Act process (conceptual/preliminary design phase)

The advantages identified with the tunnel alternative included the ability to follow a more direct alignment paralleling the Hultman Aqueduct. This alignment is shorter than the aqueduct reconstruction alternative (17 miles (27.35 km.) vs. 25 miles (40.23 km.)). This tunnel route would also allow ease of connections to the Hultman Aqueduct and to communities drawing water directly from the Hultman Aqueduct.

One of the primary advantages to the tunneling approach is increased hydraulic capacity which can be provided with an underground conduit. The inside diameter of the aqueduct could be increased from 10 to 14 feet (3.05 to 4.27 m.), effectively doubling its water carrying capacity. This design modification would incur only incremental increases in costs, without significantly increasing environmental impacts. The costs of the aqueduct reconstruction approach and all-tunnel alternative were found to be comparable.

The additional capacity afforded by the larger (14 foot (4.27 m.) diameter) tunnel would provide full redundancy for both the Hultman and Weston Aqueducts, enabling the Weston Aqueduct to be placed on emergency standby service. The Weston Aqueduct, as a gravity aqueduct, is subject to groundwater infiltration. This aqueduct and an open reservoir along its length, Weston Reservoir, do not comply with the 1987 U.S. Safe Drinking Water Act Amendments, requiring covered storage of treated water. The Weston Aqueduct was scheduled to be reconstructed as a pressure aqueduct, and covered storage would need to be provided at Weston Reservoir in the near future, at substantial costs. Instead, the tunnel has been sized to allow the Weston Aqueduct to be taken out of service, and provisions to supply the low service area have been incorporated into the project.

In addition to the hydraulic operating advantages posed by the tunnel alternative, underground construction would minimize the extent of surface disruption required along the length of the project. Surface construction would be restricted largely to six vertical shaft sites.

Planning for the proposed tunnel, the MetroWest Water Supply Tunnel, has included preparation of requisite state environmental documentation (Notice of Project Change and Draft and Final Environmental Impact Reports (D/FEIR)). This planning process has been accompanied by and influenced by input received during meetings with the general public and with officials, affected landowners, and residents. Issues addressed in the EIR included impacts associated with tunnel excavation (disturbances to private wells and construction water discharges), surface construction impacts (traffic, noise, air quality), construction staging impacts (land uses and wetlands), and muck disposition. State approvals on the FEIR were granted in April of 1992.

4 Surface facility siting considerations

Many surface-related issues were addressed in the selection of the horizontal alignment for the tunnel. The terminal points for the tunnel were driven by the need to interconnect with the major transmission facilities which supply the Hultman Aqueduct (Cosgrove Tunnel) or receive flow from the Hultman (City Tunnel on high service system and Weston Aqueduct on the low service system). The intermediate shafts were selected to provide good construction access and cross-connections to the Hultman Aqueduct, with considerations given to minimizing community and environmental disruption.

Considerations in selecting shaft sites and in establishing the horizontal and vertical alignment included the following:

- Proximity to existing MWRA transmission facilities at the surface;
- Maintaining a maximum distance of 6 miles (9.65 km.) between each shaft, for economical construction;
- Minimizing the length of tunnel;
- Minimizing community and environmental impacts, considering both direct and indirect impacts, such as construction traffic;
- Proximity to seven existing community pumping stations, which currently draw water from the Hultman Aqueduct.

The new 17.4 mile (28 km.) main tunnel includes six shafts. Five shafts are proposed to be located along the Hultman Aqueduct, with a 0.5 mile (0.8 km.) branch tunnel and a sixth shaft to connect to the end of the Weston Aqueduct. Facilities proposed at the latter shaft include a series of pressure-reducing valves to reduce the hydraulic gradeline to supply low service areas currently supplied by the Weston Aqueduct and a 20 million gallon (0.076 million $m.^3$) covered storage facility to replace the functions of Weston Reservoir.

Of the six proposed shaft sites, three contain construction shafts, and three include raisebored shafts. Five of the six proposed shaft sites involve use of state-owned waterworks property, housing existing transmission facilities. Construction shafts were sited in industrialized areas or areas removed from residential neighborhoods, with proximity to major highways. At project sites closer to residential areas, raisebored shafts were proposed to be constructed to minimize the amount of hauling and truck traffic required through local neighborhoods.

Construction required at three shaft sites (Shafts D, E, and N) would be situated entirely within existing waterworks property. Proposed Shaft D, at the upstream end of the MetroWest Tunnel, is proposed to be constructed as a raisebored shaft and would receive flow from the Cosgrove Tunnel. Proposed Shaft E would be built at the point where the Hultman Aqueduct supplies the Weston Aqueduct. This shaft would be used as a construction shaft and fronts on a state highway. Two raisebored shafts (Shaft N site) would be constructed at Norumbega Reservoir, the distribution reservoir on the existing high service system.

The largest shaft site, at the midpoint of the tunnel, is located on privately owned property which houses a sand and gravel operation. The availability of a large parcel of land currently in industrial use presents an opportunity for on-site storage of a substantial quantity of the material excavated from the tunnel, which would considerably reduce the trucking and traffic impacts during the construction period. This site (Shaft L) is proposed to be used as a major construction shaft and storage area for materials excavated from 12 miles (19.31 km.) of the tunnel. Another advantage to the use of this site is its proximity to major interstate highways.

A fifth shaft (Shaft 5A) on the MetroWest Tunnel would be located at the downstream end of the Hultman Aqueduct and would supply the City Tunnel. This shaft would be built as a construction shaft. Its location within the interchange area of two interstate highways provides ideal construction access to accommodate the proposed staging activities at this site. The construction at this location would entail use of state highway land and the relocation of existing salt shed facilities on the property.

The sixth shaft, Shaft W, is located at the downstream end of the Weston Aqueduct, enabling connections to a series of distribution mains supplied by the Hultman and Weston Aqueducts. These distribution mains serve Boston and areas north and west of Boston. This raisebored shaft is the site for the proposed 20 million gallon (0.076 million $m.^3$) covered storage facility on the low service system. Construction of this facility would require use of privately owned undeveloped land, as well as acquisition of town conservation land. The proposed storage facility was originally proposed to be sited entirely on waterworks property and has been redesigned as a series of smaller tanks to minimize the amount of rock excavation required.

To reduce hauling on residential roads, the raisebored shaft at the site would be used as a drop shaft for materials excavated at the site. This material would be transported through a 0.5 mile (0.8 km.) tunnel segment for removal within the interchange area at Shaft 5A.

The siting and design has proceeded in close cooperation with affected towns. In the two communities most affected by the project, local working groups, consisting of officials from various town boards and local residents, have been formed. These groups have been actively involved in formulating design plans for the two project sites involving the most aboveground construction, Shafts L and W. At Shaft L in Framingham, the primary issue has been the layout for storage of tunnel muck and, at Shaft W in Weston, the focus has been the size and design of the covered storage facility.

In order to address concerns remaining following publication of the FEIR, supplemental environmental documentation will be prepared in the final design phase, currently underway, on issues related to the sizing of the covered storage facility and the storage and disposition of excavated materials.

5 Final design

As discussed above, the project presently incorporates construction of the main transmission line (the tunnel), six major shafts areas, and major surface pipelines at each shaft to interconnect with existing transmission facilities at the surface. Construction of the project must be staged to assure that the service area is adequately supplied with water throughout construction, particularly when existing transmission facilities are taken off-line to allow interconnections to be made.

The system is designed for a maximum internal pressure of up to 270 pounds per square inch (1.86×10^6 newton/meter^2) at tunnel depth. However, the extreme tunnel depth ensures that the tunnel has sufficient rock cover to contain the maximum possible internal pressures. At the surface, the high head combined with the large diameter piping posed design challenges to contain maximum thrust forces within the valve chambers and surface piping. Cross-connections and controls will be facilitated by over twenty valve chambers housing approximately forty 10 foot (3.05 m.) diameter butterfly valves.

The envisaged work will be separated into three distinct contract packages. The scope of each of these packages is as described below.

Tunnel Contract 1 is envisaged to be TBM driven from construction Shaft E westward to raisebored Shaft D, a distance of 4.6 miles (7.4 km.). This would be constructed as a 14 foot (4.27 m.) inside diameter tunnel. At the western end, a surface contract will connect the tunnel to the existing Hultman Aqueduct and the Cosgrove Tunnel, via 1,000 linear feet (l.f.) (304.8 l.m.) of 10 foot (3.05 m.) diameter pipelines. At the eastern end of the contract, 1700 l.f. (518.2 l.m.) of 10 foot (3.05 m.) diameter pipe would connect the tunnel to the existing Hultman and Weston Aqueducts. A small diameter riser from the tunnel will provide a connection to one community pumping station. Connections to this pumping station and two other community pumping stations would be made via 8,100 l.f. (2469 l.m.) of surface pipelines, 2 feet (0.61 m.) in diameter.

Tunnel Contract 2 is the largest contract and will be a 14 foot (4.27 m.) diameter tunnel, again envisaged to be TBM driven both east and west from construction Shaft L. The tunnel segment from Shaft L westward to Shaft E will be 5.2 miles (8.37 km.) in length. The segment from Shaft L eastward to Tunnel Contract 3 will be 6.7 miles (10.78 km.) long. Surface piping (10 feet (3.05 m.) in diameter and 100 feet (30.48

m.) in length) at Shaft L will be included in this contract. At the eastern end of this tunnel segment, two raisebored shafts (Shafts N_W and N_E) will be constructed to permit a connection to Norumbega Reservoir, the distribution reservoir on the high service system. Surface connections between Shaft N and existing transmission facilities would require 2500 l.f. (762 l.m.) of 16 inch to 10 foot (0.41 to 3.05 m.) diameter pipelines. Contract 2 will contain four risers from the tunnel and 400 l.f. (121.9 l.m.) of 2 ft. (0.61 m.) diameter surface pipelines for connections to four community pump stations.

Tunnel Contract 3 will give the contractor the option to excavate the 12 foot (3.66 m.) diameter tunnel using either conventional drill and blast methods or a TBM to construct the 0.8 miles (1.29 km.) of tunnel. The tunneling construction would be staged from construction Shaft 5A. The tunnel will be driven westward to connect to Tunnel Contract 2 and then continue northward to Shaft W. Connections to the City Tunnel at Shaft 5A will be made by 300 l.f. (91.4 l.m.) of 10 foot (3.05 m.) diameter pipelines, with 350 l.f. (106.7 l.m.) of 1 to 2 foot (0.30 to 0.61 m.) diameter pipes to provide a connection to a community system. Also included in the contract is construction of a 20 million gallon (0.076 million m.^{3}) covered storage facility and installation of 2450 l.f. (746.8 l.m.) of 3.5 to 10 foot (1.07 to 3.05 m.) diameter surface piping at Shaft W.

6 Environmental planning

Major environmental issues to be addressed during the final design phase of the project include disposition of tunnel muck, issues related to dewatering of the tunnel during construction, and the sizing and design of the surface storage facility.

The project will involve excavation of approximately 1.5 million cubic yards (1.15 million m^3) of materials, comprised predominantly of rock of igneous, metamorphic, and metasedimentary composition. The materials excavated for this project are considered to have commercial value and would be suitable for use as roadway subbase or other fill material. Approximately half of the tunnel muck generated by the project is proposed to be excavated at the existing sand and gravel property at Shaft L. Negotiations with the property owner have been underway since the preliminary design phase. As currently planned, this material would become the property of the land owner, to be used for commercial purposes.

The sand and gravel property abuts town land which houses three unused municipal wells, which are proposed for reactivation. Issues raised during the early environmental investigations included potential impacts to these wells from leachate generated from the proposed rock storage piles. Plans for on-site storage and groundwater monitoring around the perimeter of the storage areas have been formulated, in consultation with the Massachusetts Department of Environmental Protection and the town.

The remainder of the tunnel muck to be excavated from two other construction shafts would either be used as on-site fill material or would be available for commercial use by interested parties. During final design, interest has been publicly solicited from potential users of the materials. The planning for the disposition of tunnel muck would consider the environmental sensitivity of alternative disposal/storage locations, as well as potential traffic, air quality, and noise impacts associated with the transportation of materials. Once the final destinations for these materials have been identified, supplementary environmental documentation would be prepared and submitted to the state, addressing concerns raised during the environmental review with respect to project transportation routes to be used. The contract specifications would identify proposed locations for muck disposition/storage, incorporating requirements identified in the supplementary environmental documentation.

The discharge of groundwater inflows into the tunnel would be directed into inland surface water bodies. Preliminary estimates of the rate of infiltration into the tunnel anticipated during construction have been made based upon the results of geotechnical and hydraulic jacking tests performed at boreholes along the project alignment. Appropriate conditions for monitoring and treatment of these flows would be incorporated into contract specifications, to ensure that these discharges do not degrade the water quality in the receiving waters.

The dewatering of the rock mass during tunnel excavation is anticipated to affect the recharge area for private bedrock wells in the vicinity of the project. The tunnel alignment extends through several areas which are not currently supplied by the municipal system and which rely on deep rock wells. Based on the geotechnical testing performed to date, preliminary estimates of the width of the zone of potential impact have been made. Preliminary surveys have identified wells within the zone of potential impact, which extend to depths of 120 to 1500 feet (36.58 to 457.2 m.). Surveys of property owners in the vicinity of the project have been conducted during final design, and requirements will be incorporated into contract specifications for more comprehensive water quality and locational surveys of wells along the tunnel alignment prior to construction.

Water supply contingency plans, identifying means of supplying potentially affected properties, will be developed during final design, for use by the contractor. Contract specifications will include requirements for immediate actions to be taken to provide temporary water supply and to restore permanent water service to each potentially affected site.

Lastly, the size of the proposed 20 million gallon (0.076 million m.^{3}) storage facility on the low service system was an issue of contention during environmental reviews conducted during the initial phase of the project. This facility would provide emergency and distribution storage for Boston and other low-lying communities on the eastern and downstream end of the service area.

At the request of representatives from the affected community, the MWRA agreed to abide by the results of an independent review by a panel of experts on the need for and sizing requirements for the proposed storage facility. In consultation with local representatives, a procedure for selection of waterworks experts to serve on the independent panel and for performing the review was established.

The MWRA's proposal for the 20 million gallons (0.076 million m.^{3}) is based on the need to provide at least 4 to 6 hours of emergency storage on the low service system, in the event of a failure of the proposed pressure-reducing valves or other equipment malfunction upstream. This is the time estimated to effect repairs necessary in the event of a failure or other emergency situation. Community representatives have asserted that a smaller volume of storage, 5 million gallons (0.02 million m.^{3}), would be sufficient at this location.

The majority of the independent review panel issued a determination supporting the need to provide 20 million gallons (0.076 million m.^{3}) of storage along the MetroWest Tunnel. Issues related to acquisition of public land for and design of the proposed storage facility are now being addressed, in consultation with community representatives, as previously discussed.

7 Bedrock and surficial geology

The tunnel will traverse structurally complex regions of Precambrian to Paleozoic rocks. The tunnel will cross a major terrane boundary, the Bloody Bluff Fault zone, which separates the Nashoba Zone, a mid-Paleozoic sequence of metasediments from Precambrian Avalonian basement (Figure 6 and Table 1). The Avalonian Terrane has been interpreted as originating as a volcanic arc sequence of basalt, andesite, rhyolite flows and tuffs intercalated with quartz rich sand and mud. The island arc was crushed, possibly during the course of continental collisions, resulting in both ductile and brittle deformation, granitic intrusions, varying degrees of metamorphism, and complex geometric relationships among the original rock units. The terrane was broken up by continental rifting, with fragments of Avalonian basement identified in the British Isles, Newfoundland, Western Africa, and the eastern U.S..

The westernmost portion of the tunnel extends through the Nashoba Zone, with the majority of the tunnel alignment extending through Avalonian basement. In the project area, the Bloody Bluff Fault zone includes the main trace of the fault and an eastern splay, which encompasses a lens approximately 3 miles (4.83 km.) in width. This section is referred to as the Wolfpen Lens.

East of the Wolfpen Lens, five other major rock packages have been identified within the Avalonian basement terrane during the geotechnical explorations for this project. From west to east, they include the Milford Granite Antiform; "Greenstone Complex", previously mapped as Cherry Brook Fm. (Nelson, 1974); a zone of metasediments, previously mapped as Westboro Quartzite and Rice Gneiss (Nelson, 1974); Dedham Granite; and Boston Basin Group.

On the easternmost end of the tunnel, a crossing of another significant fault zone, the Basin Fault, will be required. This fault separates the Dedham Granite from argillites and felsites of the Boston Basin.

Overburden deposits along the alignment include glacial till (consisting of heterogeneous, typically well graded, very dense deposits of clay, silt, sand, gravel, cobbles, and boulders); alluvial and outwash plain deposits (consisting of stratified granular soils); and lacustrine deposits (consisting of stratified fine sands, silts, and clayey silts to silty clays). At most of the shaft sites, the recent alluvium consists primarily of artificial fill or disturbed soils.

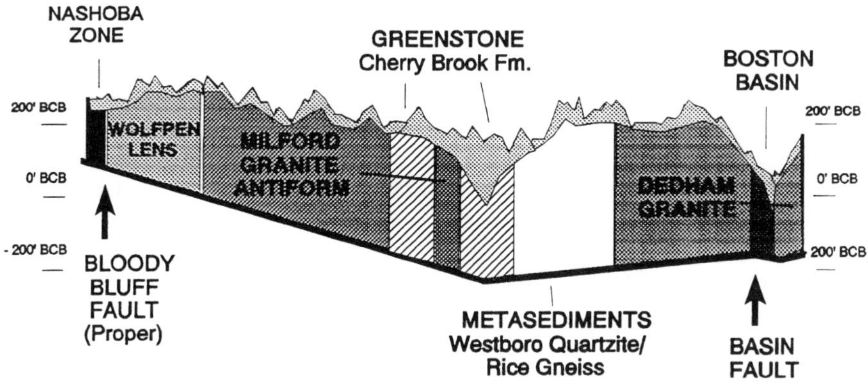

Fig. 6 Generalized geologic cross-section

Table 1. Legend to geologic cross-section

Major Rock Packages	Predominant Lithologies
Nashoba Zone	Biotite Schists and Gneisses
Bloody Bluff Fault	Mylonite believed to be derived from granite
Wolfpen Lens	Biotite Schist/Gneiss and Granite; highly variable distribution with some mylonitization
Milford Granite Antiform	Granite, Granite Gneiss, Mafic intrusives altered to Schist, some mylonitization, minor quartzite
Greenstone (Cherry Brook Fm.)	Mylonitic Chlorite and Epidote rich rocks
Metasedimentary Zone (Westboro Quartzite/Rice Gneiss)	Quartzite, Biotite/Felsic Gneiss, Abundant mafic dikes
Dedham Granite	Granite, Abundant mafic dikes
Basin Fault	Granite to Felsite and Argillite
Boston Basin	Argillite, Mafic Dike Rock, Some Sandstone and Felsites

Soil cover is thin (less than 5) at Shafts N and W. Overburden at Shaft D consists of a layer of silt over glacial till, extending to a depth of about 30 feet (9.14 m.). Glacial sands and gravels at Shaft E extend to a depth of 15 feet (4.57 m.). Shaft excavation through soil at these locations would employ typical construction methods in dewatering the area of excavation.

Two shafts (Shafts L and 5A) are situated close to glacial valleys, with stratified deposits of sands and silts overlying a 15 to 25 foot (4.57 to 7.62 m.) veneer of glacial till. The depths to top of rock at Shafts L and 5A are 90 and 130 feet (27.43 and 39.62 m.), respectively. The shaft construction at these locations will require use of groundwater cutoff techniques, such as slurry wall construction.

8 Geotechnical exploration programme

The geotechnical investigations conducted for the project included surface geologic mapping; geophysical surveys, including seismic refraction; downhole geophysics; VLF surveys; and an extensive boring programme. To date, 43 deep bore holes have been drilled to a maximum depth of 700 feet (213.36 m.). Additional boreholes are presently being drilled to further characterize the rock in specific areas of concern.

As the tunnel will transport water with hydraulic heads up to 630 feet (192.02 m.) above the invert of the tunnel and 250 feet (76.2 m.) above the existing ground surface, the geotechnical programme also addressed the in-situ "tightness" of the rock mass. Over 300 hydraulic jacking tests were undertaken to ascertain the pressures at which existing fractures and discontinuities would open, thus permitting the dissipation of pressure and water loss.

Typically, laboratory testing of samples were also undertaken. Such tests consisted of unconfined compressive strength, point load tests, and index property determination, among others. Many of the unconfined compressive tests resulted in low strength values as a result of failure along pre-existing planes of weakness and foliations. As the tunnel is planned to be excavated using Tunnel-Boring Machines (TBM), high strength and hardness of the rock mass will be a factor in the boreability of the tunnel. An array of hardness and boreability testing was undertaken by TBM manufacturers. Additionally, total hardness determinations were performed by the University of Illinois at Urbana, and Cutter Life Index (CLI) and Drilling Rate Index (DRI) tests were undertaken by the Norwegian Institute of Technology in Trondheim.

Results of the tests are summarized in Table 2.

9 Contract practices

Within the three tunneling contracts, the MWRA is anticipating establishing provisions to implement recommendations of the Underground Technical Research Council's (UTRC) report on "Avoiding and Resolving Disputes in Underground Construction". As such, the contract documents are anticipated to contain the following features:

Table 2
Summary of rock properties, hardness, and boreability test data

Rock Type	Approximate Quartz Content (%)	Unconfined Strength (Robbins) (KSI)	Total Hardness	SINTEF CLI	SINTEF DRI
Biotite Gneiss	15	14.3-29.6	170	9	20
Intrusive Schist	22	25.9-29.7	170+	13	17 & 35
Granite	40	9.4-32.4, 13.3 ave.	200+	6	40
Biotite Schist	56	7.8-12.2	80	15	38
"Greenstone" (Schist)	15	4.2-13.7	90	42	39
Quartzite	95	10.7-14.2	190	4	40
Felsite	36	4.8-14.9 (GZA Data)	No Data	No Data	No Data
Argillite	85	7.6-18.2 (GZA Data)	(50) by others	No Data	No Data

- Full disclosure of all geotechnical data along with a Geotechnical Design Summary Report (GDSR);
- Establishment of a Disputes Review Board;
- Placement of the Contractor's Bid Documents in escrow.

Based on prior tunneling contracts awarded by the Authority, the GDSR will present the geotechnical bases for the design of the tunnel and shafts and define the anticipated influence of geologic and geotechnical conditions upon construction. This establishes the geotechnical baseline for all subsurface and physical conditions anticipated to be encountered and provides the contractor with a basis for his bid. If actual conditions differ from those indicated, and cause an increase in work or change in construction methods required, an adjustment (increase or decrease) in the contract price and/or schedule would be made.

Establishment of a Disputes Review Board provides special expertise to facilitate timely and equitable resolution of disputes, claims, and controversies. A Board of three members is established by having the MWRA and contractor each choose one member who, collectively, would then choose a third member to act as Chairman. As with other MWRA tunnel projects, the Review Board's recommendations would not be binding on either party. However, it would be unlikely that their recommendations would not be accepted.

Consistent with UTRC's recommendations, it is anticipated that the successful low bidder will be required to place his Bid Documents in escrow. The documents remain the property of the contractor, but may be used to assist in negotiations of process adjustments, change orders or settlement of disputes and claims. Access to the documents would be limited to joint review by the MWRA and the contractor.

In addition to implementing the UTRC recommendations, the MWRA has, in the past, instituted Cost Sharing Incentives, such as Value Engineering for the purpose of refining the contract documents so as to contribute to design cost effectiveness or significantly improve the quality of the complete structures. To qualify as a Value Engineering Proposal, the contractor must propose changes that would result in net savings to the Authority without extension of the overall construction schedule. Proposals must also result in savings without impacting any essential operation, functions, or characteristics of the facility.

10 Construction schedule and cost

The final design for the first MetroWest Water Supply Tunnel contract will be completed and awarded in 1995. Final design of the last contract is scheduled to be completed in 1995. The project will be divided into three major tunnel contracts which will include major surface facilities within the contract limits, one contract for pre-purchase materials, and two smaller contracts for surface facilities. These contracts will be advertised for construction on a staggered basis with the last award not expected until mid-1996. By the year 2001, the new tunnel, surface valving, and piping will have been constructed, and pressure tested and ready for operation. At that time, the new project will be cross-connected to the existing transmission system, the Hultman and Weston Aqueducts and Cosgrove and City Tunnels.

The preliminary estimate for project construction is 350 million dollars at present day dollars, which includes 280 million dollars for construction of the tunnels and 70 million dollars for the remaining surface facility work.

This article represents the opinions and conclusions of the authors and not necessarily those of the MWRA.

Palais Bernheimer in Munich: tunnelling for parking space under a historic building

Richard Harpf
Klaus Axhausen
Alfred Kunz GmbH and Co., Munich, Germany

Abstract
This paper describes the construction of four levels of underground parking space underneath a historical building in Munich, Germany, while restoration is under way above ground. The construction elements and their sequence are described.
Keywords: Tunnelling, Underground Work, Car Park

1 Introduction

Times are hard and work is scarce, all over Europe. For the tunnelling division of KUNZ, this meant taking on a very unusual job:

The Palais Bernheimer is a prestigious shopping centre dating back to the 1880's located in the centre of Munich, Germany, next to the Munich Stock Exchange. It is to be restored to its former glory, while being fitted out for the next century. This project includes

- the renovation of the above-ground building to provide a shopping mall and prestige offices, and
- the addition of four levels of underground space directly underneath the building for car parks, as required by the prospective users as well as by building regulations, as well as storerooms and services.

The renovation of the above-ground building will not concern us here. In this paper, we will concentrate on the design and construction of the underground structures.

2 General information

The plan drawing (fig.1) shows the block consisting of nine partly interconnected buildings, arranged around a series of central courtyards. The southern five buildings make up the project. The project area has a ground surface of roughly 65 m by 80 m. The other four buildings are not part of this project and have to be kept operational at all times.

As the sketch shows, the worksite is quite restricted. Access to the interior was initially by a portal in the east facade. In a

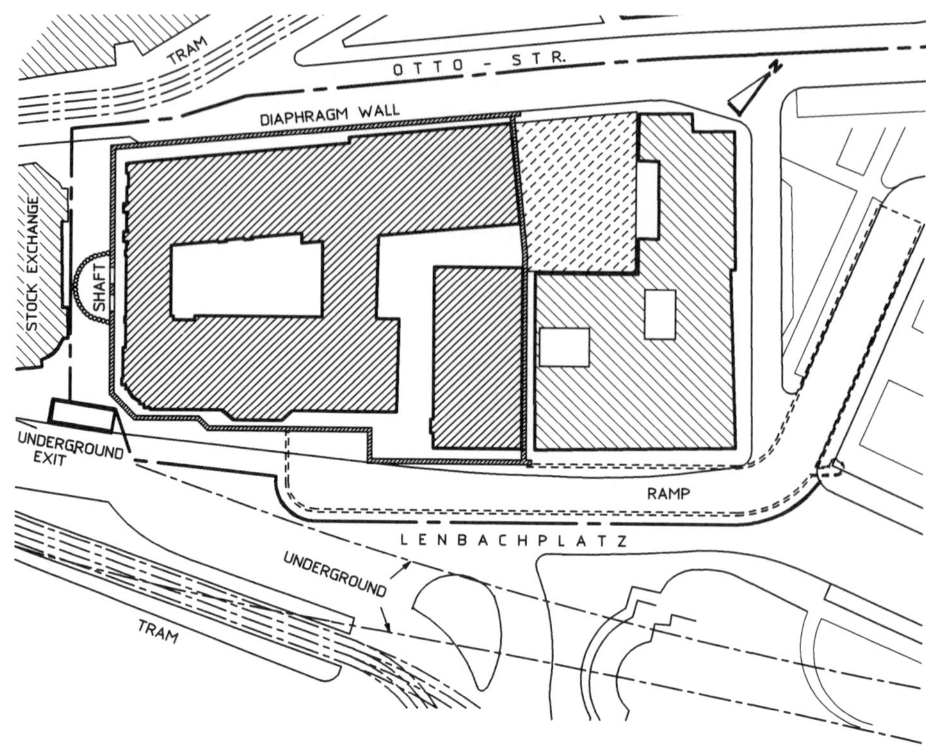

Fig.1. Plan view of Palais Bernheimer

later stage, a vertical shaft situated at the south end in a narrow lane was used. Concurrently, a ramp along the east facade was constructed under a different contract; in the final phase, site traffic will shift to this ramp.

As section (fig.2) shows, roofs generally reach a height of about 29 m above ground. The principal structural materials used are masonry, concrete and steel. The structural loads were originally transmitted by masonry or concrete footings to the subsoil at average depths of 5 m below ground.

The soil profile shows fill over quarternary gravel and layers of tertiary sands and marls. Groundwater is concentrated in the gravel and sand layers. The marl consists of stiff silty clay and is practically impermeable.

3 Underground construction sequence

The building is classified as a "monument", the outward appearance of which must not be changed. Thus, the obvious possibility of taking down the entire building and constructing the parking garages in open cut was ruled out. KUNZ won the contract on the basis of an alternative proposal:

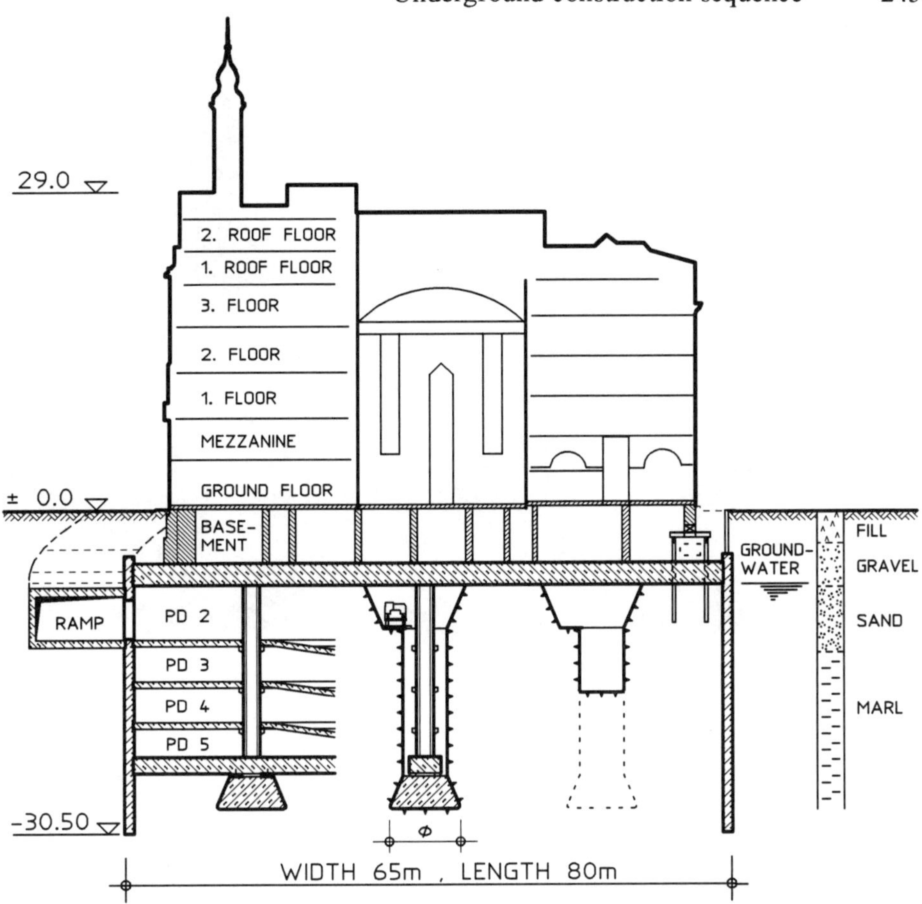

Fig.2. Vertical section with construction sequence

- The entire project area is surrounded by vertical cut-off walls.
- In the existing basement, temporary minipiles are driven
- Column loads are transferred to these minipiles.
- Under the existing foundations, a massive RC "intermediate slab" is constructed, encasing the minipiles and taking over the loads.
- Under the intermediate slab, access tunnels are driven.
- From these access tunnels, 21 shafts are sunk.
- In these shafts, reinforced concrete columns are constructed to carry the loads from the intermediate slab and the parking decks.
- Finally, the remaining levels are excavated and the parking decks are constructed.

Thus, aboveground activities are almost totally divorced from the underground works.

Fig.3. Diaphragm wall construction

4 Preparatory works

Retaining walls had to be constructed to a depth of some 31 m below ground, that is to the final foundation depth and into the impermeable marl layers. They will be part of the final structure. Three sides of the building were secured by conventional diaphragm walls (fig.3), topped by steel H-beams and timber strutting (to be removed later).

Recesses were prepared in the diaphragm walls for the intermediate slab, and Lenton couplers were fixed to the reinforcement cages for connections to the reinforcement of the parking decks. Rigid Quality Assurance procedures helped avoid problems in later stages.

At the north side, within the building complex, access was severely restricted; here, the retaining wall construction was more complicated:

- The invert of the existing basement was lowered, with some shotcrete support for the open face. Thus, a working height of 5 m was achieved.
- From this level of roughly -5 m, jet grouting techniques were used to construct retaining walls to a depth of about 14 m, i.e. into the marl.

- At a later stage, from the level below the intermediate slab with a headroom of 5 m, diaphragm walls were constructed to the foundation level of 30.5 m below ground.
- Finally, a reinforced concrete shell was placed inside the jet-grouted wall, structurally connecting the diaphragm wall below with the intermediate slab above.

Along the south side of the block, adjacent to the Stock Exchange building, the main access shaft for the underground activities was excavated within a bulge in the retaining wall (fig.1). One portal crane and one tower crane were installed, the first for heavy plant and spoil, the second for reinforcement and lighter loads.

In the existing basement, interior walls were largely demolished to provide access to the foundations. Minipiles were driven to an average depth of 7 m below basement level .

Finally, the column loads were transferred by underpinning techniques to the minipiles. This involved the following steps:

- construction of steel frames linking the minipiles to the columns or walls,
- transferring the loads by hydraulic flat jacks.

Fig.4. Excavation between minipiles

For the purpose of determining the required loads in the jacks, theoretical structural analysis proved unreliable. An elaborate monitoring system was therefore put in place to safeguard the existing building. Forces were increased until a maximum lift of 1 mm in the structure was observed.

5 Intermediate slab

The delicate job of underpinning was followed by a more massive undertaking: the construction of an intermediate slab of reinforced concrete of an average 2.0 m thickness. This slab serves to isolate the existing structure from the underground activities. It was built up of rectangular sub-slabs of up to 1.400 m3 volume.

The soil was excavated to a level of roughly 7 m below ground level (fig.4), a blinding layer was poured, reinforcement was installed and the concrete was placed. By their rough surface, the minipiles were safely bonded into the concrete; loads could now be transferred to the intermediate slab.

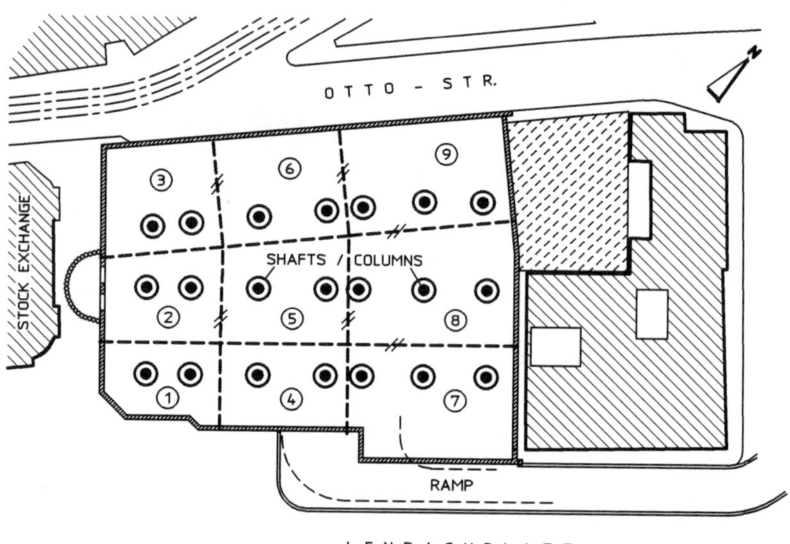

Fig.5. Column layout

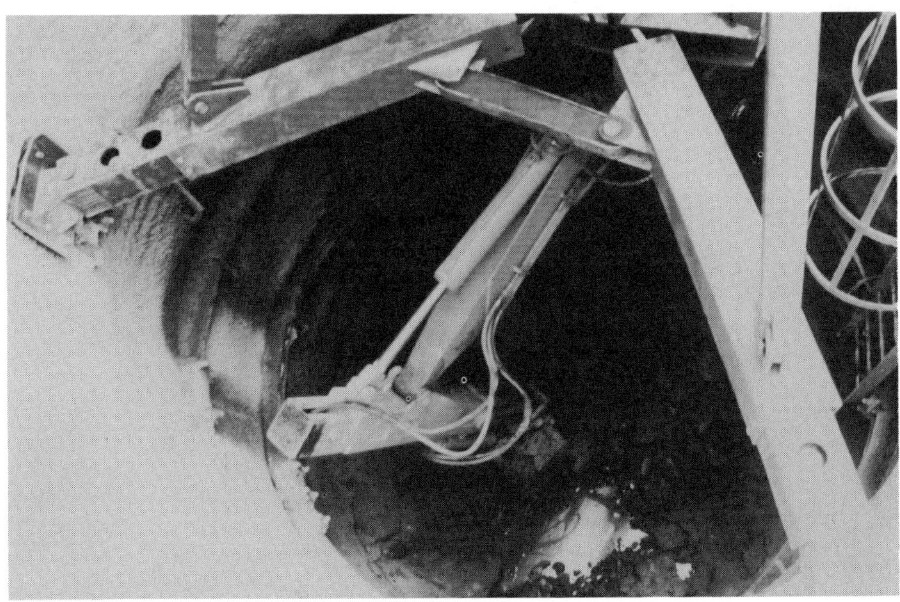

Fig.6. Shaft sinking rig

6 Tunnels and shafts

In the level immediately below the intermediate slab, access tunnels were then driven from the access shaft along the south side of the building and, branching from this, along the three main column axes (fig.5).

These tunnels had an average height of 4 m; their width at the invert was roughly 7 m. The sides were left standing at an angle of about 60 degrees. The walls were supported by 100 mm shotcrete and one layer of steel mesh: the invert had 150 mm shotcrete and two layers of mesh. Minipiles protruding below the intermediate slab were demolished.

This was followed by the sinking of the 21 shafts. Starting level was approximately 11 m below ground level; the excavated diameter was 4.3 m, belling-out to 7.5 m for the at roughly 29 m below ground level (fig.2).

Excavation was performed by a special rig (fig.6) which is shown here during tests. Support consisted of (nominally) 100 mm shotcrete, reinforced by one layer of steel mesh and steel rings at 1.400 m centres (fig.7).

Once the final depth was reached, the permanent main columns were constructed (fig.2): First the enlarged foundations were reinforced and concreted; their upper surfaces were prepared as joints for the future prestressing. Massive RC base were constructed on top. The columns proper were constructed level by level.

Fig.7. Shaft support

Great care was taken during the final pour to achieve full-face contact between the column and the intermediate slab (fig.8). Four 50 mm diameter ventilation pipes were installed in the intermediate slab to ensure no air was trapped at the column head.

As a last step, eight hydraulic jacks, each of 500 tonnes capacity, were installed between foundation and base to transfer the loads into the columns. This operation was controlled by movements observed in the intermediate slab, with a permissible maximum of 2.5 mm. Recorded settlements of the foundations were up to 30 mm.

7 Basement levels

The skeleton of the underground structure was now in place: the retaining walls, the intermediate slab and the columns. This made possible the top-down construction of the underground levels, i.e. alternating excavation and construction of the floor slabs in reinforced concrete. This phase is still under construction; the final stage will hopefully be completed in May, 1995.

8 Conclusion

This exciting project demands a high quality of workmanship, but also - due to the restrictions of time, access and space, the requirements

Fig.8. Contact column/intermediate slab

on noise and dust abatement, and quality assurance - it demands an extremely high standard of management. It is a showpiece for the qualities of the tunnelling industry, in unusual surroundings.

Limiting the damage to historic buildings due to tunnelling: experience at the Mansion House, London

J. E. Hellings
Pell Frischmann Consulting Engineers, Ltd., London, England

Abstract
This paper describes the effects of previous tunnelling on the fabric and foundations of The Mansion House, an 18th century Grade I Listed Building in the City of London. In mid-1988, a tunnel was constructed approximately 15m below the building. This tunnel was to be the first of a number of additional tunnels to be built near the Mansion House, associated with the extension of the Docklands Light Railway to Bank Station. The effect of the first tunnel is described together with predicted and measured movements. The predicted effects of the other tunnels are also outlined. Measures proposed for protecting the Mansion House are described, along with the system which was adopted. A sophisticated system to monitor building and ground movements during and after the construction of the tunnels has been installed. This paper is based on another by Frischmann et al (1994).
Keywords: Historic Building, Settlement, Monitoring.

1 Introduction

The Mansion House (Figure 1) is a mid-18th century Grade I Listed Building, approximately 65m x 30m in plan, located about 400m north of the River Thames in the City of London. It is the official residence of the Lord Mayor and is of considerable historic interest.

London Regional Transport (LRT) proposed the construction of a number of tunnels and underground structures in the vicinity of the Mansion House in order that the Docklands Light Railway (DLR) could be extended to Bank Underground Station. The proposed tunnels (Figure 2) were:

(a) a 4.0m diameter passageway to the Waterloo & City Line, later changed to 3.05m diameter in order to reduce settlement, about 15m below the Mansion House, running approximately east west;

(b) a 4m to 5.75m diameter passageway to the Central Line, about 15m below ground level, approximately 6m in plan distance from the north east corner of the Mansion House at its closest point, but subsequently re-aligned out of the zone of influence of the Mansion House;

(c) a 5m diameter overrun tunnel for the DLR, about 40m below ground level, approximately 13m in plan distance from the north east

Fig. 1 The Mansion House, London

corner of the Mansion House, and a step-plate junction, about 40m below ground level, approximately 20m east of the Mansion House.

In order to distinguish between the lengths of tunnel whose construction might affect the Mansion House a 'zone of influence' was defined as the space bounded on four sides by planes originating from the foundations at 45° to the horizontal or 20m laterally from the building, whichever was greater. The lengths of tunnels within this area are indicated in Figure 2.

In a Parliamentary Undertaking (1985-1986), LRT agreed not to work within this zone of influence without the prior approval of the Corporation of London.

2 The Buildings

2.1 Structure

The Mansion House was built as a typical 18th century town house. During its lifetime the building has been subjected to various alterations, some of which have been detrimental to the overall rigidity of the structure.

At present the building has five storeys, a basement under the front two-thirds of the building and four other levels. The main walls are constructed in load bearing brickwork of London Stock bricks and lime mortar with stone cladding. The floors - with the exception of the basement floor and part of the first floor - are suspended timber floor with floorboards fixed to joists, supported on timber beams, spanning between the load bearing walls on wall plates. The roof is of timber and slate.

The largest rooms are the Egyptian Hall to the south, running the width of the building and some 20m wide, through three storeys to the roof. The other large room is the Ballroom which is situated near the front of the building also extending across the full width of the Mansion House. The Ballroom is about 7m wide and runs through two storeys to the roof. The basement area consists of heavy longitudinal and cross walls which carry brick vaults.

In 1985, a survey was carried out by Pell Frischmann Consulting Engineers Ltd to assess the overall condition of the building. It was found that the structure was generally sound, with the exception of some isolated areas of cracked brick walls. No immediate repairs were considered necessary, though periodic monitoring and maintenance were recommended.

2.2 Foundations and Geotechnical Aspects

It is reported that the course of the River Walbrook ran north-south to the west of the Mansion House (Barton 1962, Grimes 1968), resulting in some soft deposits in the upper soil near the building. Historical information indicates that the Mansion House is underlain by Alluvium, associated with the Walbrook and overlying Flood Plain Gravel, associated with the Thames. Present ground level is about 6m higher than ground level in Roman times, indicating the depth of fill overlying the Alluvium. London Clay underlies the flood plain gravel at about 7m below ground level.

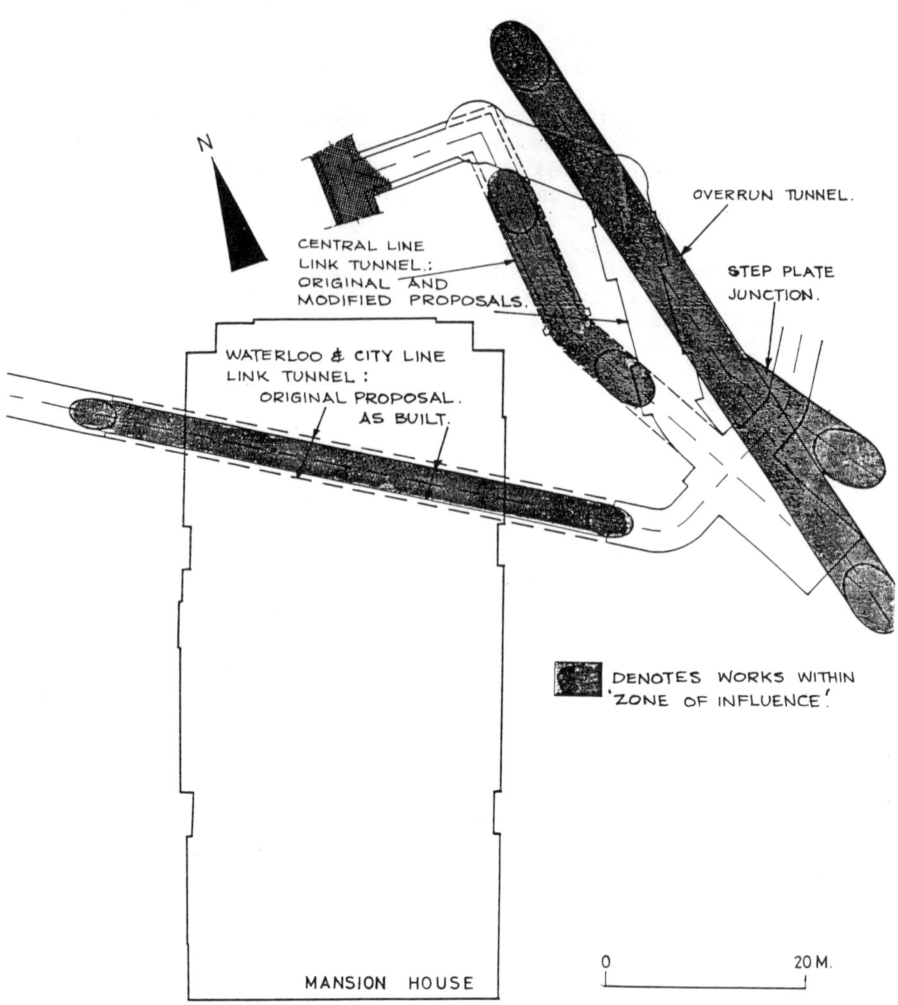

FIG 2. LOCATION OF THE MANSION HOUSE IN RELATION TO PROPOSED TUNNELS

The original foundations (1738-9) were of timber piles, with planking to support the walls. These foundations were underpinned or strengthened on several occasions: in 1868 the timber piles were reinforced with concrete; in 1901 underpinning was required, probably due to the construction of the nearby Central Line underground tunnel; in 1931 further underpinning was required; and foundation strengthening was part of general reconditioning programme carried out following the Second World War after bombing at the adjacent Bank Station.

The 1985 condition survey included an investigation of the foundations by trial pits. These revealed no signs of foundation distress with the previous underpinning generally taken down to the Flood Plain Gravel. No evidence of timber piles or planking was found.

A level survey carried out on a single masonry course indicated that there is a drop in level of about 200mm from the south east corner to the north west corner. If it is assumed that the masonry was level when placed, this indicates a settlement of 200mm of the north west corner, which is the corner nearest the line of the Central Line tunnel.

2.3 Effects of Previous Tunnelling

The Central Line tunnels at Bank Station were constructed around the turn of the century. Originally the west ends of the platform tunnels were in front of the Mansion House with the westbound running tunnel very close but not directly beneath the north portico. As part of the lengthening of the Central Line platforms in the 1930's the platform tunnels at Bank Station were extended westwards by enlarging the running tunnels and realigning the westbound track southwards towards the Mansion House.

No records have been found of settlement due to any of these works but levels of the Ordnance Survey bench mark on the corner of the Bank of England facing the Mansion House and settlement predictions (Frischmann et al 1994) confirm that large settlements took place in the late 1890's and 1930's corresponding to the Central Line tunnelling. At both times tunnelling is known to have caused damage which would not be acceptable nowadays to a building as important as the Mansion House. Major repairs were required and evidence of the damage is still clearly visible, most notably over the ballroom window in the west wall where the joint at the keystone opened by about 25mm.

3 Effects of the First DLR Tunnel

3.1 Predicted and Measured Movements

It is generally considered that, in the absence of any other effects, the displaced ground profile above a tunnel takes the form of an inverted normal probability curve, with its centre above the centre line of the tunnel (Peck 1969, Attewell and Yeates 1984). The total ground settlement which occurs can be divided into two components: initial settlement, which is related to loss of ground, caused mainly by relaxation and over-excavation; and long term settlement, which is

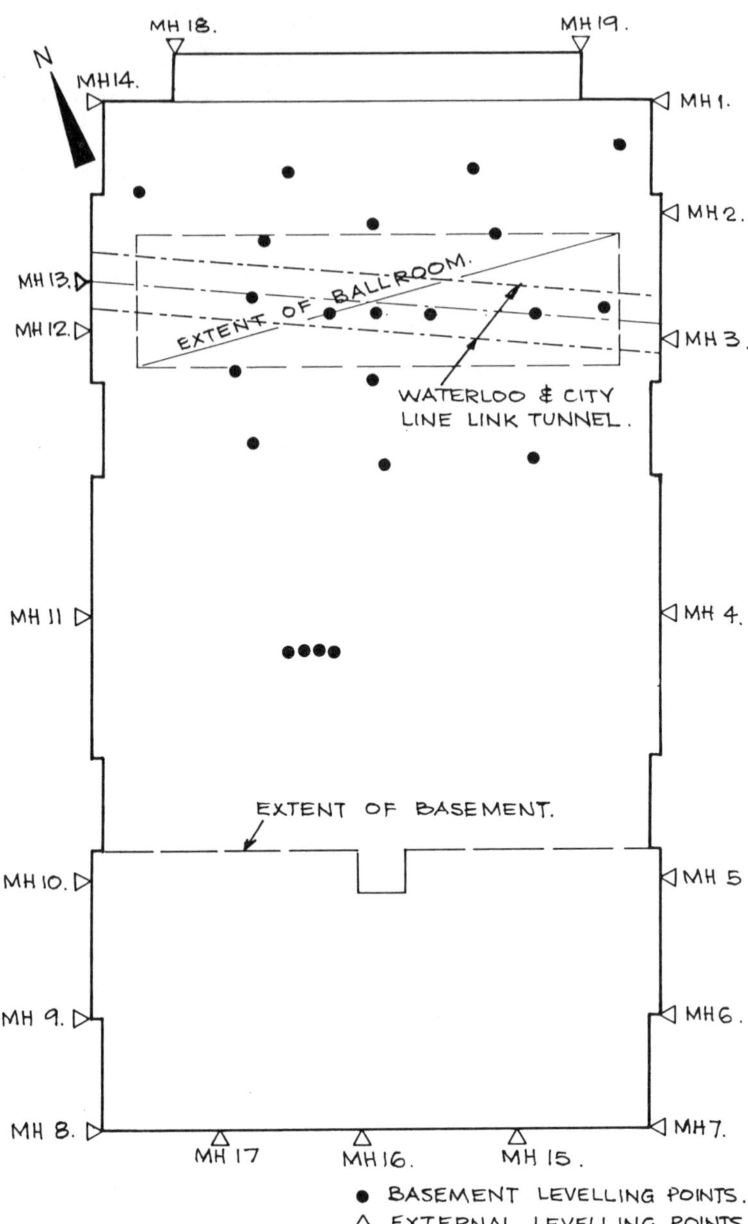

FIG 3. PRECISE LEVELLING POINTS AND LOCATION OF BALLROOM

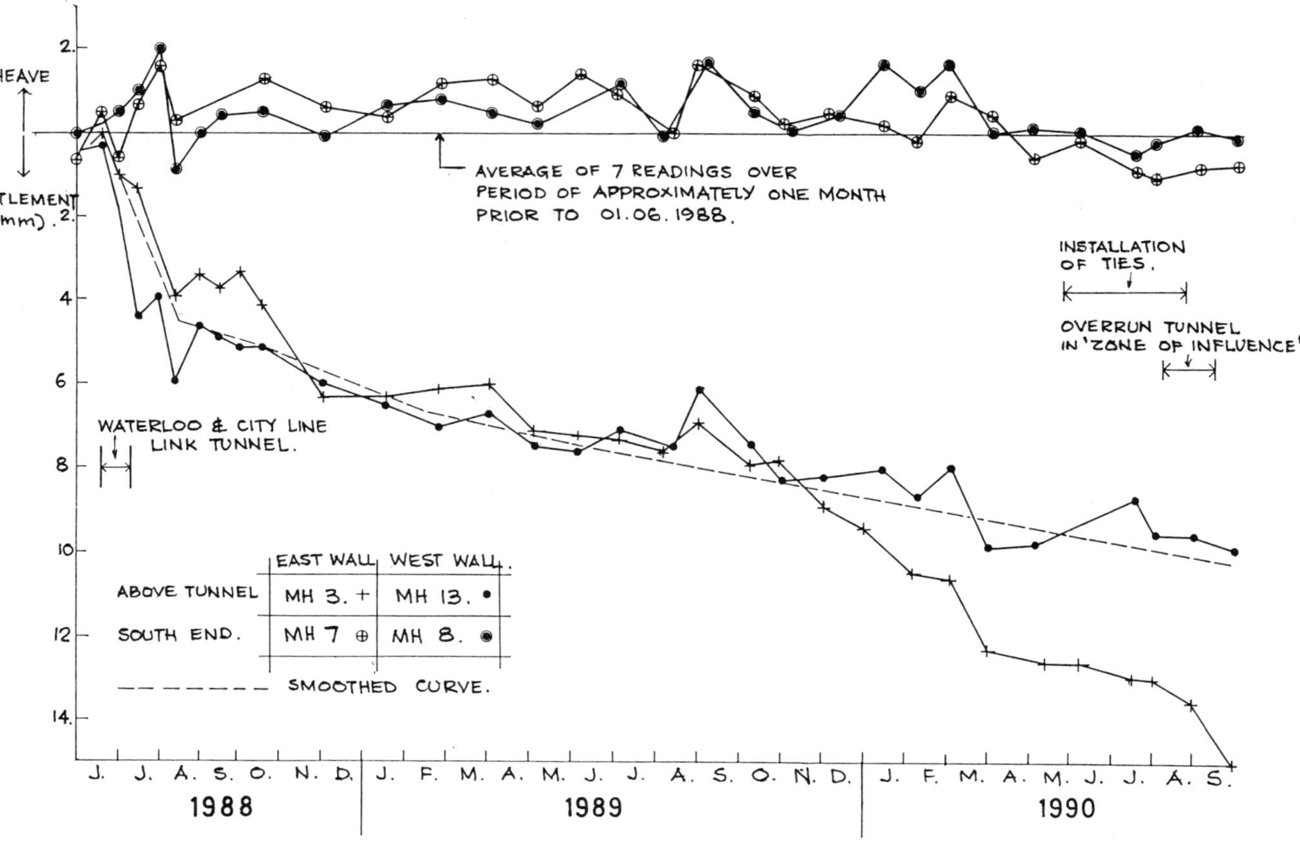

FIG 4. SETTLEMENT AGAINST TIME FOR POINTS ON EAST AND WEST WALLS

the settlement caused by consolidation due to changes in the ground water regime. Initial settlement is typically complete within a few weeks, and can normally be identified with a steep portion of the settlement time curve; the remaining long term settlement may take a number of years before effective completion. Horizontal ground movements towards the tunnel axis are also associated with initial and long term settlements. For tunnels in London Clay, case records (Attewell and Yeates 1984, Barratt and Tyler 1975) indicate that initial and long term settlements are approximately equal.

The Waterloo & City Line Link tunnel was bored under the Mansion House in the middle of 1988. It was decided that, in order to limit any resulting ground movements, the tunnelling techniques and workmanship would be of the highest standard possible for example by limiting the overbreak, grouting soon after placing the linings and working continuously.

The diameter of the tunnel was originally planned to be 4.0m. However, assuming a ground loss of 1.75%, predicted distortions due to hogging near the edges of the settlement trough were calculated to be such as to cause 'substantial' damage according to Burland and Wroth's graphs (Burland and Wroth 1975, The Institution of Structural Engineers 1984). The tunnel diameter was therefore reduced to 3.05m.

Regular precise levelling surveys of the building were carried out at the points indicated in Figure 3 both during the construction of the Waterloo and City Line Link tunnel and thereafter. The exterior and interior points gave mutually consistent results, showing, as would be expected, maximum movements over the tunnel and little apparent movement at the southern end of the building.

Figure 4 shows the results taken from two external points, MH3 and MH13, over the Waterloo and City Line Link tunnel together with the two external points furthest from the tunnel, MH7 and MH8. It can be seen that there is a relatively rapid settlement of approximately 4mm in the first 8 weeks following the start of construction, after which the rate of settlement reduces. This has been taken as the end of initial settlement. The southern point MK7 may have settled up to about 3mm.

Settlement profiles along the east and west walls are shown in Figure 5 from which the localised effect of the settlement trough is evident. The initial settlement profiles can be taken as the profiles of 16 July 1988 for the west wall, and 18 August 1988 for the east wall (the tunnel was driven from west to east). The areas under these curves indicate a ground loss of 1.5% of the tunnel cross-sectional area, which is consistent with good tunnelling practice.

One year after tunnel construction, total settlement had reached 8mm and was increasing. This caused some concern, particularly in regard to the tunnels still to be constructed, as this was already equal to about twice the measured initial settlement. Long term settlement, possibly due to consolidation through drainage to the tunnel lining, was becoming greater than had been anticipated. There was an acceleration in the rate of settlement of the east wall above the tunnel centre line around November 1989 (Figure 4), probably due to other tunnelling works not within the 'zone of influence'. This indicates the arbitrary nature of the definition of the 'zone of influence' as described in Section 1. Mair et al (1993) show that the

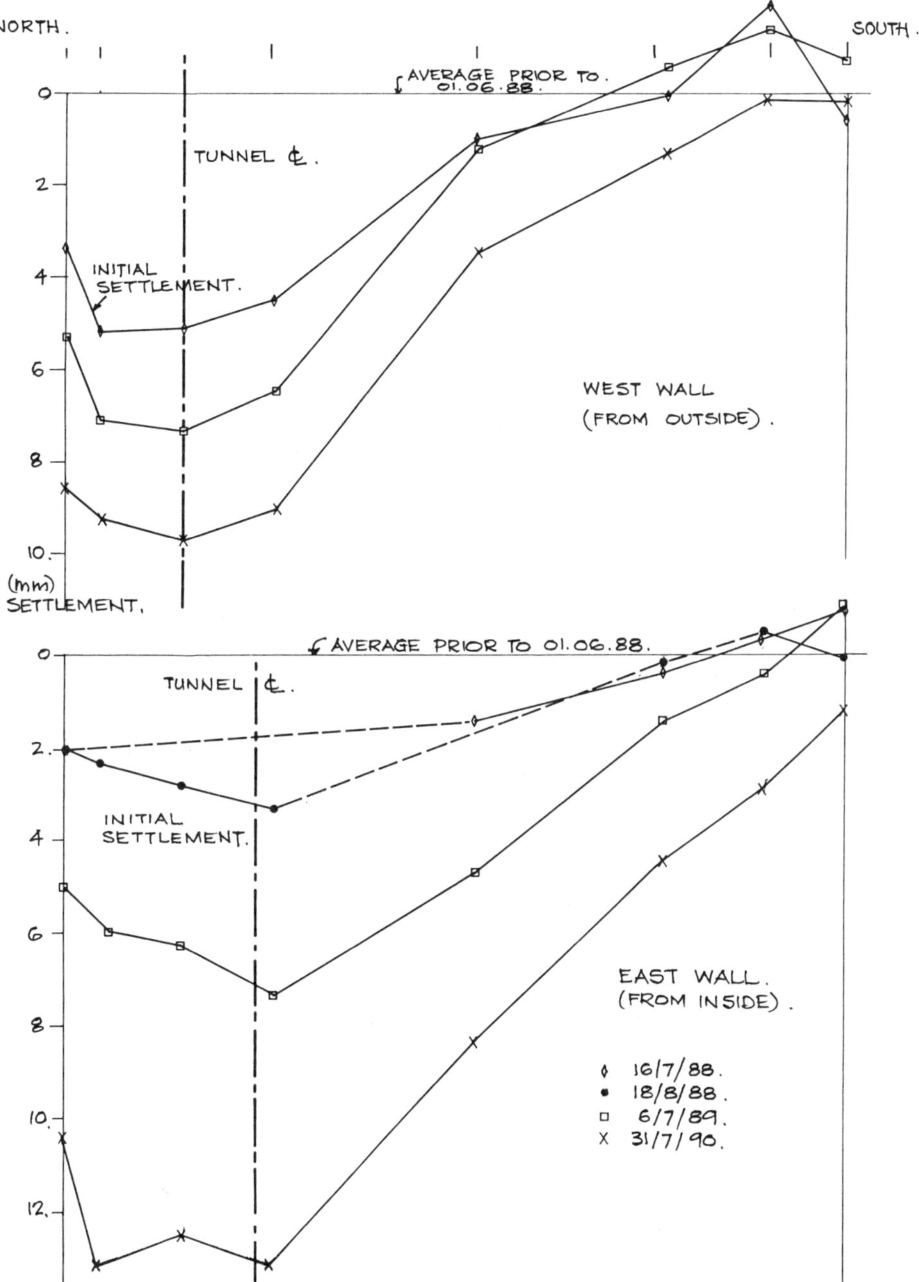

FIG 5. SETTLEMENT PROFILES FOR EAST AND WEST WALLS.

zone of settlement below the ground surface may be wider than has been commonly assumed.

3.2 Calibration of Soil-Structure Interaction Model

The results of monitoring the first tunnel confirmed that the volume of the settlement trough was as predicted for a green field site but the shape of the trough was different. A finite element analysis was therefore undertaken by Sir William Halcrow & Partners Ltd to establish the influence of the structural stiffness of the Mansion House. The results would improve quantifying the influence of soil-structure interaction on settlements when predicting the effects of further tunnelling works.

Figure 6 shows the predicted green field settlement profile, the actual settlement profile of the west wall due to the first tunnel and the best fitting profile from the finite element analysis

In applying the results of this analysis to predict the effects of the later tunnelling works, differences between sagging and hogging behaviour were anticipated and a series of sensitivity tests were carried out for the east wall using different heights for the wall to model the reduced stiffness of the superstructure in the hogging mode.

3.3 Crack Patterns

Further to the 1985 survey, additional crack surveys were carried out prior to construction of the Waterloo and City Line Link tunnel, in February and April 1988, and 12-15 months afterwards, in June, July and October 1989, when occupants reported that cracks had suddenly developed. Demec studs had been fixed to external walls in January 1989 and over selected cracks in the ballroom in June 1989 (Price et al 1994). Displacement transducers were also installed over some cracks in the ballroom.

The 1989 surveys found that cracking in the ballroom area had considerably increased since the 1988 surveys. The ballroom would have been particularly susceptible to cracking because of its location directly above the centre line of the tunnel and because of extremely large windows, taking up about 25% of the area of the external walls.

These crack movements and patterns were consistent with a horizontal movement of the north east ballroom corner towards the tunnel excavation and of a settlement of the south wall. In the north ballroom wall, a number of cracks wider than 3mm or 4mm were found. These would classify the building as moderately damaged according to the scheme adopted by the Institution of Structural Engineers (1989).

4 Predicted Effects of Other Tunnels

4.1 Ground Movements

Analyses were carried out to predict ground movements likely to be caused by the remaining tunnelling works described previously. Their predictions were based on an initial settlement derived from ground losses of 2%-2.5%, depending on the tunnel considered, which were then doubled to give a 'worst credible' condition. Contours of

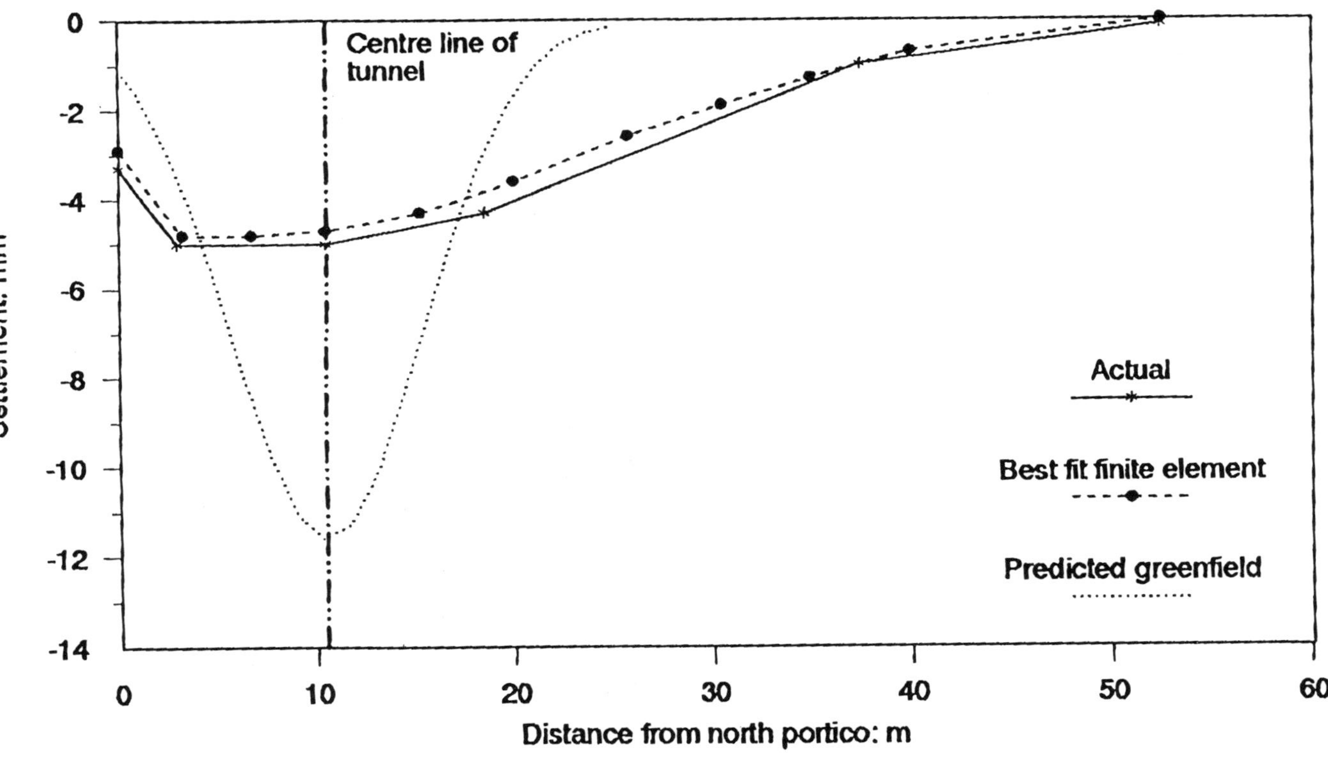

FIG 6. COMPARISON OF ACTUAL AND PREDICTED SETTLEMENTS AT WEST WALL (DUE TO 3.05m. I.D. TUNNEL) (REFERENCE FRISCHMANN ET AL 1994)

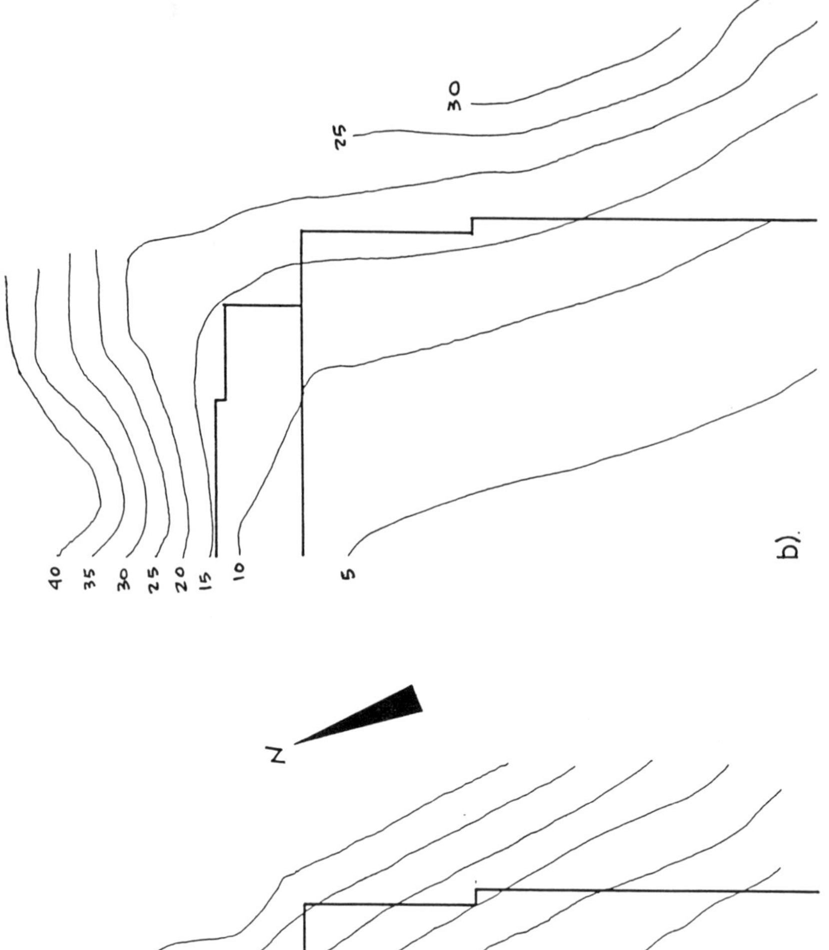

FIG 7. CONTOURS OF PREDICTED SETTLEMENT a) BEFORE RE-ALIGNMENT OF CENTRAL LINE LINK; b) AFTER RE-ALIGNMENT OF CENTRAL LINE LINK

predicted settlement were produced (Figure 7a) which indicated a maximum settlement under the Mansion House of 36mm at the north east corner. Because of the magnitude of the observed movements due to the Waterloo & City Line link tunnel, and the importance of the Mansion House itself, these 'worst credible' conditions were used to investigate the effects of tunnelling on the building structure.

4.2 Effects on Structure

The construction of the Waterloo & City Line link tunnel resulted in settlements which were greatest within the lengths of the east and west walls, causing these walls to deform in a sagging mode. However, the predicted effect of the remaining tunnels was that maximum settlement would occur at the north east corner, resulting in a hogging mode of deformation. It is well recognised that unreinforced walls are much more likely to be damaged by hogging than by sagging (Burland and Wroth 1975).

In order to examine the possible effects, several analyses of the north wall and the north ballroom wall were carried out (Frischmann et al 1994). These walls were considered critical, as the distortions were greatest for these walls.

Depending on the assumptions involved, these analyses indicated that tensile bending strains of around 0.02% to 0.05% could develop in the walls. The published information would suggest that cracks generally start at strains of around 0.02% to 0.03% (Littlejohn 1985, Mainstone 1975), while the core samples taken from the Mansion House brickwork and tested in tension failed at axial strains of between 0.001% and 0.01%. In addition to the effects of vertical movements, tensile forces would be generated in the structure due to horizontal movements.

It is possible that soil-structure interaction might reduce distortions imposed by the 'green-field' settlements adopted for the above analyses. However, finite element analyses described previously illustrated that the interaction depends significantly on the assumption as to the extent of existing cracking, ie, the effective height of wall resisting tensions in the hogging mode of deformation.

Given that the building was already moderately damaged, it was considered that the risk of severe damage caused by ground movements due to the remaining tunnelling would be unacceptable unless some works were carried out to limit the damage.

5 Proposals for Limiting Damage

Initially, two alternative schemes for limiting damage to the Mansion House were proposed: a piled cut-off wall, intended to limit ground movements; and an underpinning system, intended to limit relative building movements.

5.1 Piled Cut-Off Wall

This proposal involved the construction of two bored pile cut-off walls along the north and east sides of the Mansion House. It was intended that the piles should be closely spaced, say 900mm diameter at 1050mm centres, topped with a capping beam, and taken to a depth of around 44m below ground level, about 4m deeper than the deepest tunnel.

The design philosophy was that, by working in friction, the piles would resist downwards movement of the soil on either side of the walls. The compressive forces generated in the piles would be resisted by the length of pile embedded below the zone of significant downward soil movements. Compressive forces generated in the piles would be likely to be relatively small in relation to the strength of concrete.

In order to assess the efficiency of this technique in reducing settlements, two plane strain finite element analyses of a typical cross-section were performed. All materials were assumed to be linearly elastic. Undrained conditions were assumed for the London Clay. Both analyses were identical except that, in one analysis, the wall elements were given soil properties in order to represent the conditions where no wall was constructed. In both analyses, the effect of excavating the tunnels was modelled by removing the appropriate soil elements.

Despite many simplifying assumptions, it was considered that these analyses, while not necessarily giving realistic absolute values of settlement, would give a reasonable estimate of the degree to which settlements would be reduced by the wall. Both analyses showed a much wider distribution of surface settlements than would be expected in practice. This was due to the use of a linear elastic model for soil behaviour and is commonly seen in this type of analysis for tunnels. It was found that close to the wall, settlement is reduced to about one third of its value with no wall present. This effect reduces with distance from the wall until, at around 10m, there is no apparent effect.

It can be seen therefore that it would be necessary to construct the wall as close as possible to the Mansion House. This, together with the building location, could create a number of construction difficulties: various services for and accesses to the London Underground would need to be diverted; great care would need to be taken to ensure that no ground was lost when boring the piles, particularly when going through the gravel; accurate setting out, surveying and construction would be required towards the west end of the north wall where there is limited clearance between the existing Central Line tunnel and an electric cable reserve; and the facade of the Mansion House would need to be protected against damage by the piling rig.

5.2 Underpinning

In association with John Pryke & Partners, a number of possible underpinning options were investigated to control movements of the Mansion House. Conventional underpinning was considered to be unsuitable since the movements caused by tunnelling are deep seated, and so a three stage proposal involving the use of techniques more commonly associated with mining subsidence was put forward. The stages are illustrated in Figure 8.

The first stage would involve decoupling the building from its foundations. 1m wide sections of walls were to be cut away just above foundation level, and Pynford stools mounted on hydraulic jacks and

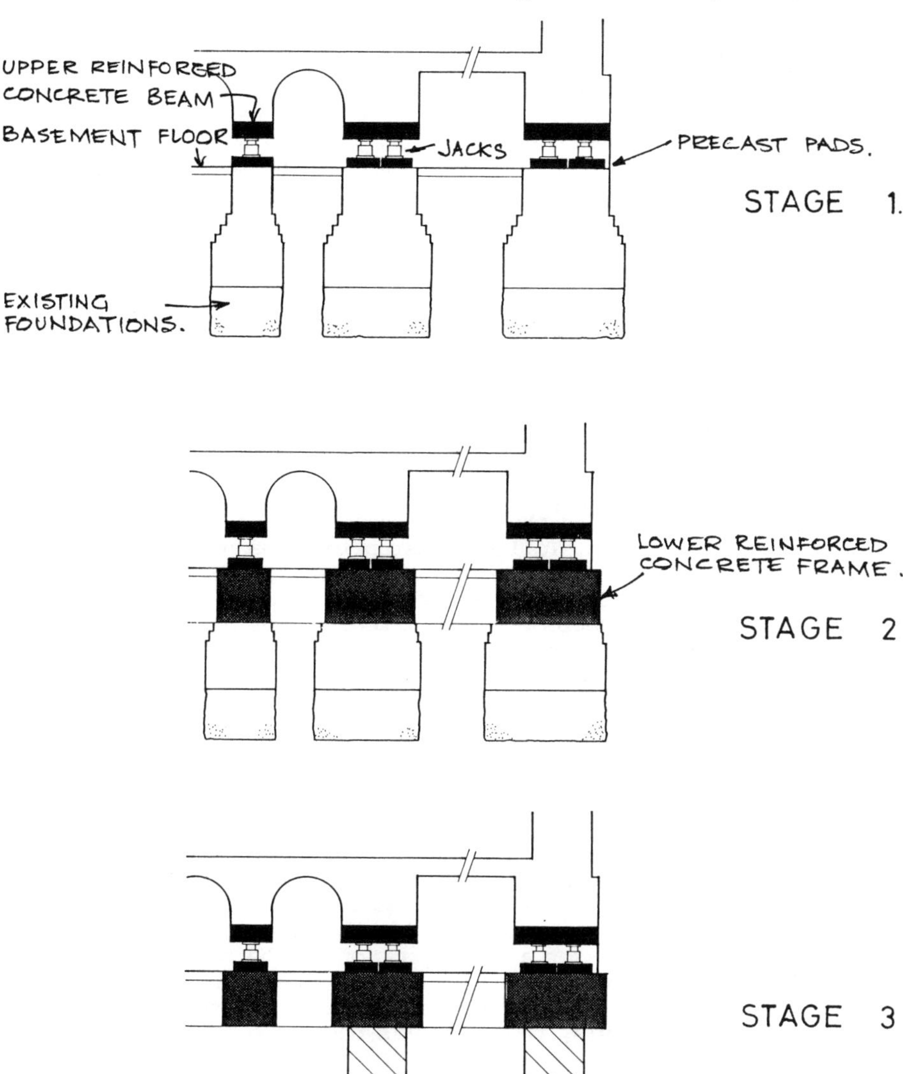

FIG 8. UNDERPINNING PROPOSAL, STAGES 1–3

pads positioned between foundations and walls and packed into place. Once a suitable length of wall was completed, reinforcement would be threaded through the stools, and a section of beam concreted above the jacks. At door openings, linking frames would need to be constructed to maintain an opening and to create a continuous concrete beam. By continuously monitoring building levels and pressure in the jacks, the building could be maintained at its level prior to the construction of the remaining tunnels by adjusting the jacks.

Stage 2 was intended to restore the integrity of the foundations by constructing a framework of reinforced concrete beams beneath the jacks and beams of Stage 1. Initially, drains and services would have to be diverted, then trenches dug adjacent to walls. The concrete beams could then be constructed by deepening the slots created for Stage 1, installing further Pynford stools beneath the jacks, threading reinforcement through the stools and concreting.

In Stage 3, the lower reinforced concrete frame was to be supported by piles if this was found to be necessary. It was considered that hand dug piles, founded below the settlement zone, and designed to accommodate the ground movements due to tunnelling, would be most appropriate. To limit disruption, the number of piles constructed was to be kept to a minimum: around thirty 2m diameter under-reamed piles taken to a depth of 30m-35m were proposed.

In the underpinning option, it was intended that Stages 1 and 2 would provide reasonably long term protection to the Mansion House, with Stage 3 being implemented if considered necessary as a result of movements due to tunnelling works, or if the Corporation of London required permanent protection.

6 Scheme Adopted

In early 1990, the proposed tunnelling scheme was altered as the DLR promoters were reluctant that the underpinning scheme be carried out. The Central Line link was moved to the east of its original position, so that most of the new route lay outside the 'zone of influence' of the Mansion House except for a small portion of the north of the Mansion House at the Central Line platforms. As a result, predicted settlements were now about one half of the original predictions (Figure 7b). On this basis, the proposals described above were no longer deemed to be justified and so a scheme involving safeguarding of the building structure with steel ties was adopted.

6.1 Design

A system of 28 internal stainless steel ties, nominally stressed, was adopted. The locations of the ties were selected so that they would have minimum impact on the appearance of the building, in most cases being located under floor boards or behind plaster work. As a typical example, the tie rod arrangement at third floor level is shown in Figure 9 together with a typical anchoring arrangement.

The ties were intended to increase the rigidity of the structure and would therefore have two main beneficial effects. Firstly, soil-structure interaction would be increased, resulting in a

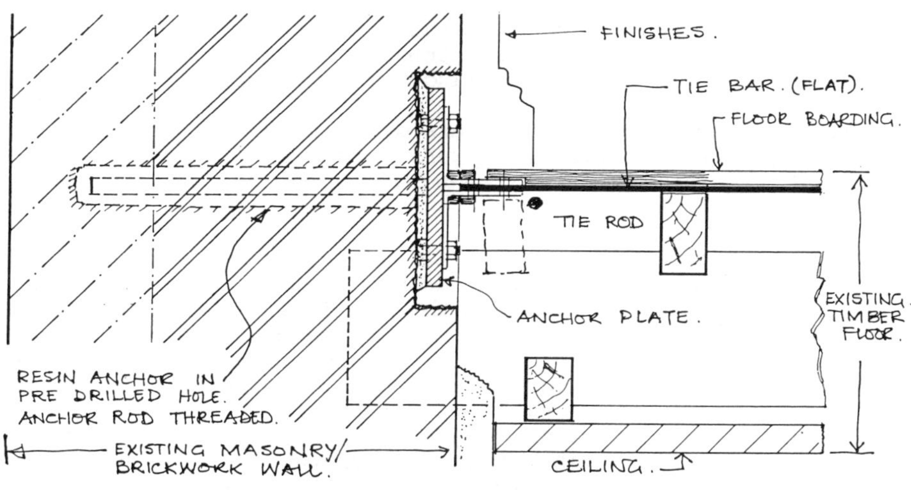

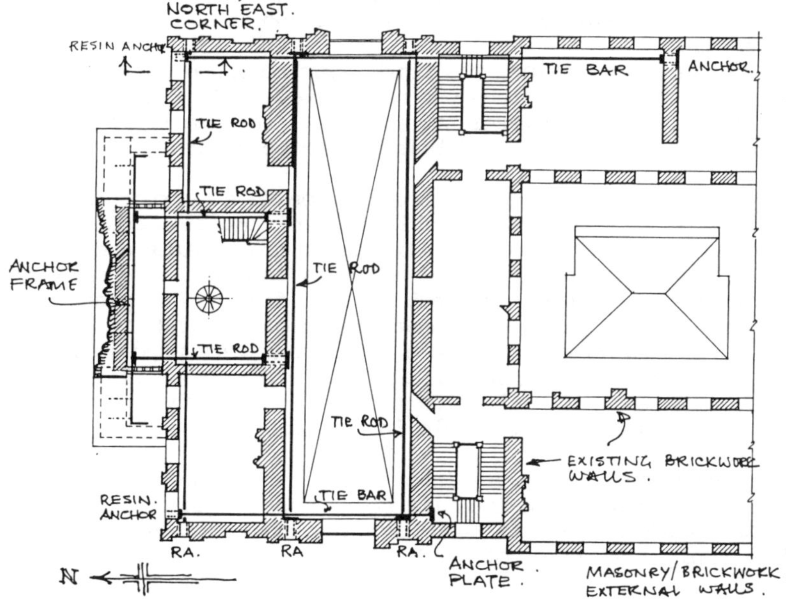

FIG. 9 TIE ARRANGEMENT AT THIRD FLOOR LEVEL AND TYPICAL EXTERNAL WALL ANCHOR DETAIL

re-distribution of settlement, so that the building would tend to deform as a block, reducing the settlement concentration at the north east corner. Secondly, the resulting crack movements would be reduced.

To check the strength of the bars selected, further frame analyses of the type described previously were carried out, using the same material properties but revised settlement predictions (Frischmann et al 1994).

6.2 Construction

A contract was negotiated with Fondedile Foundations Ltd and, in May 1990 work commenced on installing the ties in the upper levels which were deemed to be most critical for a hogging mode of deformation of the walls. The work was carried out in three stages: the first stage comprising 13 ties at third and fourth floor levels, and the second comprising 7 ties and the third stage comprising 8 ties at second and first floor levels. The ties for stages 1 and 2 were installed prior to tunnelling works within the 'zone of influence', while tunnelling was in progress during the installation of the stage 3 ties. The final tie was installed towards the end of August 1990 while tunnelling works for the 5m diameter overrun tunnel entered the 'zone of influence' at the beginning of August 1990 and left in the middle of September 1990.

Throughout the installation, the Lord Mayor remained in residence, and a full programme of functions was maintained. This required that the works were carried out overnight, and that, in some rooms and for certain occasions, any resulting debris had to be cleared away on a daily basis.

Because of the complex nature of the Mansion House structure and services, the precise routes followed by most of the ties could only be finalised on site, following the lifting of floorboards and opening up of walls. The discovery of asbestos in some rooms delayed the work temporarily, however the installation finished on programme, thus causing minimum delay to the construction of the DLR.

In general, 32mm diameter stainless steel rods were used in wall finishes and cornices, while 100mm x 10mm stainless steel flats were used in floors. On the internal walls the ties were anchored by passing them through an oversize hole in the wall into a plate on the far side.

Vibrating wire load cells were incorporated where possible into the rods by Building Research Establishment, most of which were initially tensioned to loads of 10kN-25kN in order to take out any slack and to apply a small proof load.

7 Measured Movements and Forces

In order to be able to fully monitor the movements of the Mansion House during tunnelling works, a further two sets of instrumentation were installed by the Building Research Establishment. Strings of electro-level tilt sensors and a water gauge levelling system were installed. In addition, two invar wires were installed across the diagonals of the ballroom.

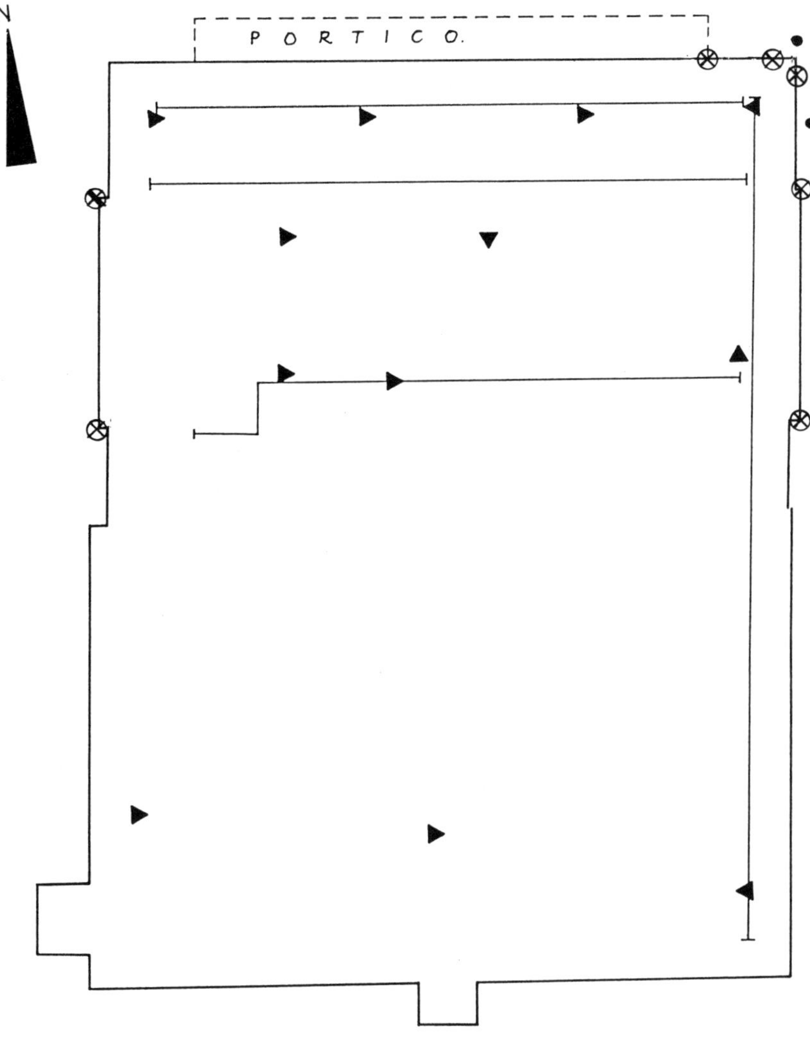

FIG 10. LOCATION OF ELECTRO-LEVEL STRINGS AND WATER LEVEL GAUGES.

The installation and performance of this instrumentation is described by Price et al (1994). The measured movements are described by Forbes, Bassett and Latham (1994).

By measuring the tilt of the electro-levels at various points along a line, it is possible, by integrating the tilt over the distance between sensors, to obtain an estimate of the relative movement of one sensor to the other and, if a datum point is assumed, absolute movements at each point can be calculated. Four strings of such devices to measure vertical movements were located in the basement with the datum taken as a point in the south-east corner. To measure horizontal movements of the walls, seven vertical strings of electro-levels were installed using a point at the base of the wall as datum. Four further strings were installed in two boreholes to measure lateral ground movements, while three horizontal strings measure vertical ground movements in the vicinity of the north-east corner of the Mansion House. The location of these devices is shown in Figure 10.

In addition to the electro-levels, a water gauge levelling system was also installed. This comprised a network of linked water pots with floats, the movements of which were measured by displacement transducers. By measuring the change in water levels at various points in relation to the change in water level at a datum point, relative movements could be obtained. A network of water points (Figure 10) were set up in the basement using the pot nearest the south-west corner as a datum.

Precise level surveys of the building at the points shown in Figure 3 were carried out every week. The monitoring systems therefore gave three independent measurements of building movement. The Demec points and transducers described above were also monitored, while two invar wires were installed across the diagonals of the ballroom in order to measure the horizontal distortion which appeared to be occurring from the results of the original Demec points and transducers. In addition, temperature sensors were placed internally in the Ballroom and the basement and externally on the west, north and east walls. Relative humidity meters were also installed outside and inside the Ballroom.

The load cells, electro-levels, water level gauges, displacement transducers, invar wires, temperature and relative humidity sensors were set up so that readings were taken by an automatic data logging system which could be accessed remotely.

The downward movements of the north-east, north-west and south-east corners of the Mansion House measured by electro-levels, precise levelling and water pots are shown in Figure 11, with readings on 5 July 1990 taken as an arbitrary zero. The conversion of electro-level data to displacements uses the results from all of the electro-levels between the point under consideration and the assumed datum. This means that if one electro-level gives incorrect readings, all of the calculated displacements thereafter are affected. This is evidenced by the scatter in the displacements calculated for the north-west corner, which involves the largest number of electro-level readings of any point in the basement. The south-east corner is the assumed datum from the electro-levels and hence there is no electro-level data for this point.

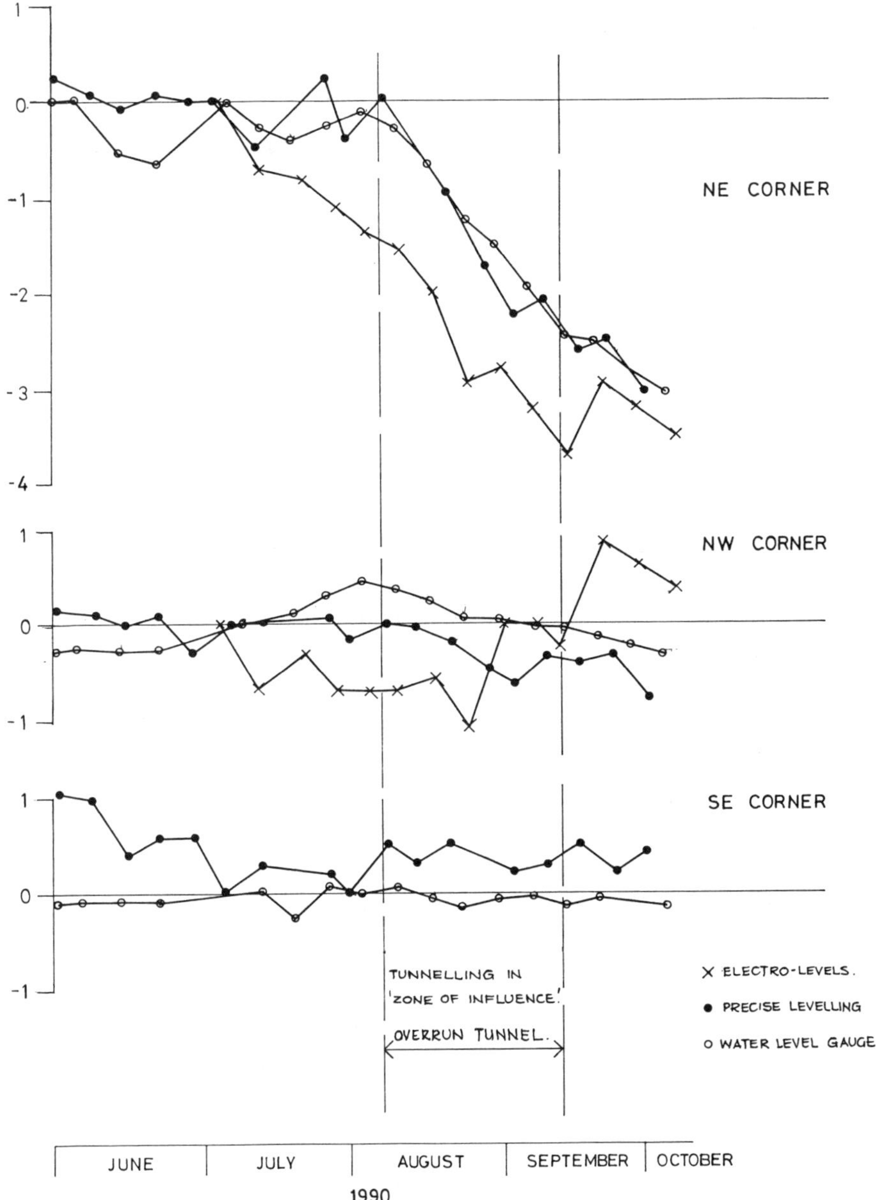

FIG 11. DISPLACEMENT AGAINST TIME FOR THREE CORNERS OF THE MANSION HOUSE AS MEASURED BY ELECTRO-LEVELS, WATER LEVEL GAUGES AND PRECISE LEVELLING

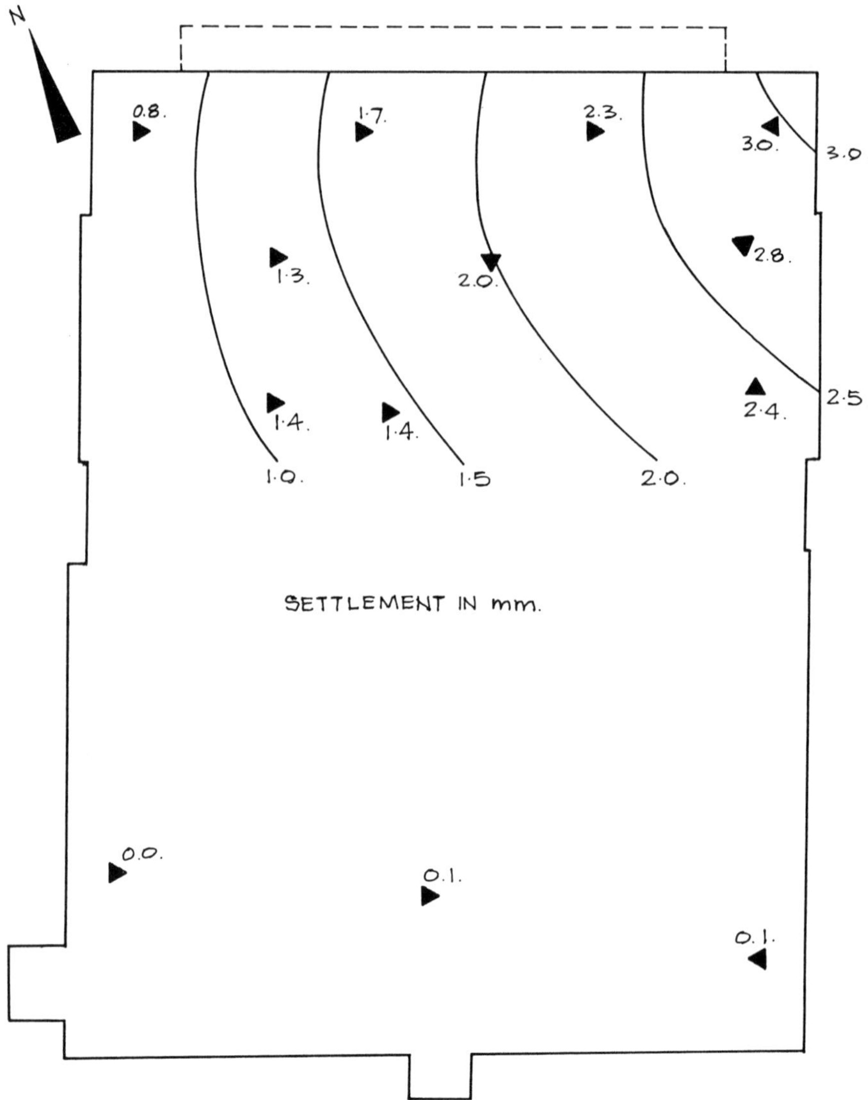

FIG. 12. ACTUAL INITIAL SETTLEMENT CONTOURS FOR THE NORTH END OF THE MANSION HOUSE DUE TO CONSTRUCTION OF THE OVERRUN TUNNEL

The precise levelling and water gauges show that the start of downwards movement of the north-east corner coincides with tunnelling entering the 'zone of influence'. A slight movement of the north-west corner is detectable, while the south-east corner does not appear to have moved measurably.

The data obtained from the water level gauges were used to plot contours of displacement for the north end of the building. Contours of the change in level from 7 August 1990 to 10 October 1990 corresponding to 'initial' settlement associated with the construction of the overrun tunnel are shown in Figure 12. The contours are not particularly curved, relatively parallel and evenly spaced, indicating that this part of the building is generally rotating rather than distorting.

Over the same period, electro-levels in a vertical borehole at the north-east corner showed around 3mm horizontal ground movement towards the tunnel excavation at foundation level. Perhaps associated with this movement there was some renewed opening of cracks on the east end of the south wall of the Ballroom, though there was little, if any, movement of the remaining cracks or of the invar wires across the Ballroom diagonals.

The general pattern of movement of the Mansion House has been one of a continuing general tilt of the building towards the north-east corner. The maximum settlement (at this corner) by February 1991 was 20mm and by February 1993, was estimated to have been 27mm. A settlement of 30mm is predicted by the year 2000 of which 7.5mm has been ascribed as 'construction' or 'immediate' settlement (measured immediately after construction) (Forbes et al 1994). It is of interest to compare this with the settlements predicted in Section 4 - a 'worst credible' settlement of about 20mm

Immediate settlements have been low, reflecting the low ground loss due to good tunnelling practice. Continuing longer term settlement has, however, been rather high in comparison.

Generally, the loads measured in the tie rods dropped initially, probably due to the combined effects of creep and subsequent tensioning of other ties, and then slowly increased. Tensile forces in the tie bars have fluctuated with time, but have generally not exceeded 30kN, about twice the proof load.

8 Conclusions

It is generally agreed that the Mansion House is one of the most important historic buildings in London. Construction of the first part of the DLR Link to Bank station at the Mansion House, the pedestrian passageway, led to the cracking of the building which caused alarm to both occupants and owners. The effects of this passageway were carefully monitored and enabled predictions to be made of likely movement and damage resulting from construction of the remainder of the DLR Link. The estimated scale of damage was deemed to be unacceptable to the building's owners.

Measures for preventing or limiting this damage were investigated and detailed proposals were put forward for underpinning the building. However, the promoters of the scheme decided to alter their tunnel layout in such a way that the predicted movement and resulting damage would be substantially reduced.

A system of tie bars was designed to stiffen the Mansion House and was installed with a minimum of disruption. In addition a comprehensive system to monitor ground and building movements was designed and installed, including a water levelling system, an arrangement of electro-levels on walls and in boreholes and a system of invar wire extensometers in addition to the precise levelling and crack monitoring with displacement transducers. This enabled movements to be monitored with considerable accuracy and for the building owner and tunnel promoter to have the assurance which this gave. It also enabled the tunnelling operations to be controlled in order to safeguard the building.

Significant loads were not carried by the tie bars, probably due to the building deforming generally by tilting. It may also have been that the bars contributed, even with these low loads, to a stiffening of the structure, which would have encouraged a tilting mechanism.

A combination of changes in design, preventative measures, a comprehensive monitoring system and good tunnelling practice contributed to minimising damage to the Grade I Listed building ahead of, and during, an extensive refurbishment programme.

Several new tunnels are planned for construction in London over the next few years, most significant of which will be the Jubilee Line Extension and the CrossRail schemes. Experiences with the Mansion House has shown that there is no need to accept the inevitability of unlimited damage to buildings due to tunnelling. With careful analysis and monitoring it is possible to predict the effects of tunnelling and for the owner to decide if they are acceptable. Should the damage predicted be unacceptable, then appropriate action can be considered to limit the damage.

9 Acknowledgements

Acknowledgement is given to the Corporation of the City of London for permission to publish this paper.

10 References

Attewell, P. B. and Yeates, J. (1984) Tunnelling in soil, in **Ground Movements and Their Effects on Structures** (eds P. B. Attewell and R. K. Taylor), Surrey University Press, London, 132-215.

Attewell, P.B. and Farmer, I.W. (1974) Ground disturbance caused by tunnelling in a stiff, overconsolidated clay. **Engineering Geology, 8,** 361-381.

Barratt, D.A. and Tyler, R.G. (1975) **Measurements of ground movements and lining behaviour on the London Underground at Regents Park.** Report LR684, Transport & Road Research Laboratory, Crowthorne.

Barton, N.J. (1962) **The Lost Rivers of London.** Phoenix House & Leicester University Press, London, 21-25.

Burland, J.B. and Wroth, C.P. (1975) Settlement of buildings and associated damage, **Proceedings of Conference Settlement of**

Structures, Cambridge, Pentech Press, London, 611-654.
Forbes, J. Bassett, R.H. and Latham, M.S. (1994) The Monitoring and Interpretation of Movement of the Mansion House due to Tunnelling, **Proceedings - Institution of Civil Engineers, Geotechnical Engineering**

Frischmann, W.W. Hellings, J.E. Gittoes, S and Snowden, C. (1994) Protection of the Mansion House against damage caused by Ground Movements due to the Docklands Light Railway Extension, **Proceedings - Institution of Civil Engineers, Geotechnical Engineering.**

Grimes, W.F. (1968) **The Excavation of Roman and Mediaeval London.** Routledge & Kegan Paul, London, 92-98.

The Institution of Structural Engineers (1989) **Soil - structure interaction**, I Struct E, London.

Littlejohn, G.S. (1985-86) Discussion, Examples of damage, **Proceedings of Conference on The settlement of structures, Cambridge**, Pentech Press, London, 764-767.

Mainstone, R.J. (1975) Discussion, Examples of damage, **Proceedings of Conference on The settlement of structures, Cambridge**, Pentech Press, London, 771-772.

Mair, R.J. Taylor, R.N. and Bracegirdle, A. (1993) Subsurface settlement profiles above tunnelling clays. **Geotechnique**, Technical Note 315-320.

Parliament Session (1985-86) **London Docklands Railway (City Extension) Act.**

Peck, R.B. (1969) State-of-the-art report. Deep excavations and tunnelling in soft ground, **Proceedings of the 7th International Conference on Soil Mechanics and Foundation Engineering**, Mexico City, 225-290.

Price, G. Longworth, T.I. and Sullivan, P.J.E. (1994) Installation and Performance of Monitoring Systems at the Mansion House, **Proceedings - Institution of Civil Engineers, Geotechnical Engineering**

Compensation grouting to limit settlements during tunnelling at Waterloo Station, London

R. J. Mair
Geotechnical Consulting Group, London, England
D. I. Harris
Geotechnical Consulting Group, London, England
J. P. Love
Geotechnical Consulting Group, London, England
D. Blakey
Sir Robert McAlpine, Hemel Hempstead, England
C. Kettle
Bachy (UK), Ltd., Godalming, England

Abstract
A new escalator tunnel and associated passageways have been constructed at Waterloo Station in London. The tunnels pass through Thames Gravels and London Clay only a few metres beneath two important masonry structures: the Victory Arch and the Waterloo and City Line railway tunnels. In the absence of protective measures, the settlements of these structures would have been large and would have caused significant damage. Compensation grouting was therefore undertaken in the ground between the tunnels and the overlying structures to limit settlements to acceptably small levels. An essential feature of the technique is the carefully controlled injection of grout during tunnelling in response to detailed observations of ground and structure movements. Settlements of the structures due to tunnelling were restricted to less than 15 mm. The relatively novel technique of compensation grouting has proved to be very successful in limiting settlements of structures during tunnel construction.
Keywords: Tunnelling, grouting, settlement, case history, field instrumentation, soil/structure interaction

1 Introduction

London Underground Ltd (LUL) have commissioned the construction of an enlarged ticket hall and a new escalator with associated passageways linking to the Northern Line at Waterloo Station. This improvement of the efficiency of the underground station at Waterloo is part of LUL's Congestion Relief Project. Increased capacity was also required in connection with British Rail's development of Waterloo International Terminal for the Channel Tunnel, and with the Jubilee Line Extension. The project has been designed by LUL and full details of the scheme, which has been recently constructed by Sir Robert McAlpine, are given in a companion paper to this conference [1]. A plan and long section through the tunnels are shown in Figures 1 and 2 respectively. The inclined escalator tunnel is of 7.5 m internal diameter,

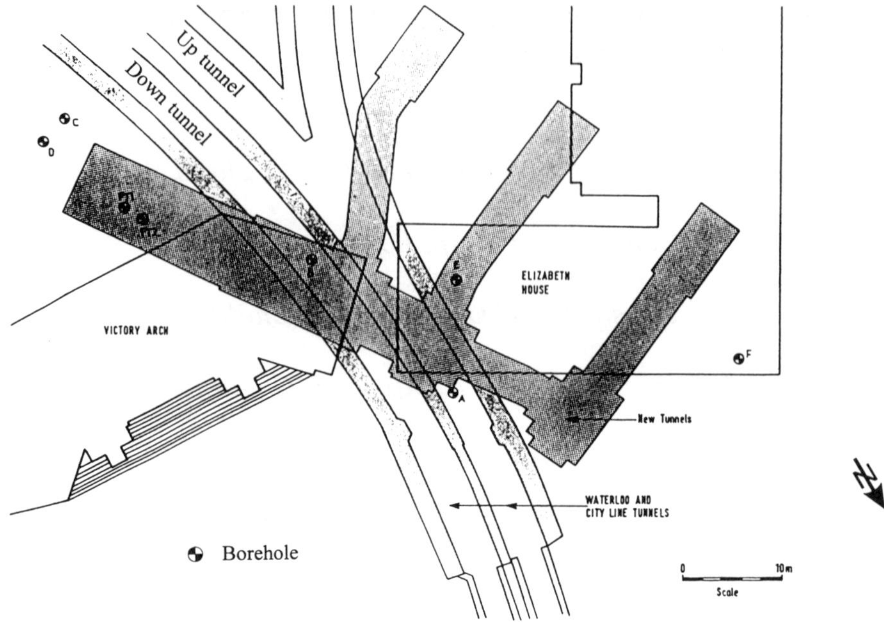

Figure 1 Plan location of new tunnels and existing structures

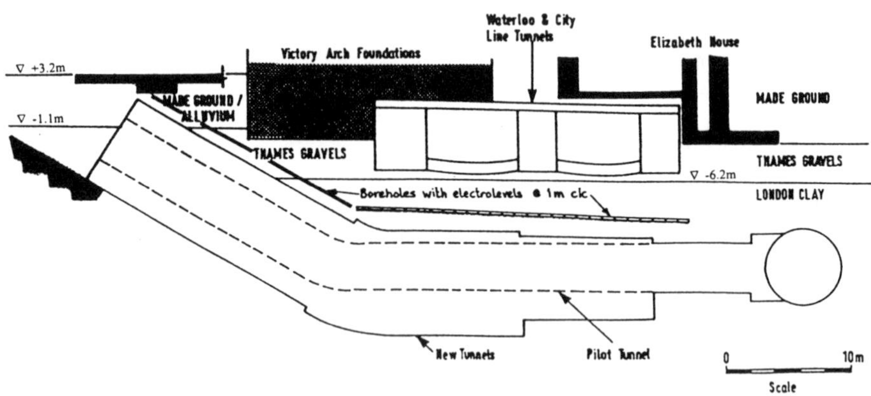

Figure 2 Long section through new tunnels

increasing to 8.25 m in the lower machine chamber at the bottom of the inclined shaft. These tunnels, together with 6.5 m and 3.85 m ID passageway tunnels are lined with spheroidal graphite (SG) cast iron bolted segments. Hand construction methods were used.

As shown on Figures 1 and 2, the tunnels pass beneath three important structures: the Victory Arch (the main entrance to Waterloo Station), the Waterloo and City Line tunnels and Elizabeth House (a 15 storey tower block). The crown of the 7.5 m ID inclined tunnel passes only 2 m beneath the foundations of the Victory Arch, and the crown of the lower machine chamber tunnel is 5 m below the Waterloo and City Line tunnels. Of great concern was the potential settlement of these masonry structures resulting from the tunnel construction. This Paper describes the ground conditions, the potential settlements due to tunnelling in the absence of protective measures, and the application of the technique of compensation grouting which was selected in order to limit the settlements.

2 Ground conditions

The ground profile determined from the boreholes shown in Figure 1 is depicted on Figure 2. Two of the boreholes (BH's A and B) were undertaken from within the Waterloo and City Line tunnels. Made ground and alluvium overlies about 4 m of Thames Gravels which are underlain by stiff London Clay. Bulk excavation in the Thames Gravels for the nearby Waterloo International Terminal project showed them to be generally silty sands and gravels with occasional bands of clay. In-situ permeability tests undertaken in the Thames Gravels indicated permeabilities in the range 10^{-3} m/sec to 10^{-5} m/sec. The ground water table lies at the top of the Thames Gravels.

3 The structures

The Victory Arch, shown in Plate 1, is a complex structure which has developed into its present form in a number of phases. The original structure on the site was built in the 1840's, but only a small part of this building remains. The major redevelopment commenced in 1913 and the Victory Arch facade was added in the 1920's. Part of the structure is founded on the roof of the Waterloo and City Line tunnels, of which a 22 m length was constructed beneath the Victory Arch foundations in the 1880's using underpinning techniques. Sub-basements were constructed as part of this operation which have subsequently been backfilled with mass concrete. Trial pits and coring have been undertaken to investigate the mass concrete and brick footings which are generally founded at the top of the Thames Gravels.

The remaining length of the Waterloo and City Line tunnels of relevance to this project was constructed as a cut-and-cover operation. The tunnels, one of which is

Plate 1 Victory Arch at Waterloo Station

Plate 2 Waterloo and City Line tunnel

Thames Gravels about 4 m below the water table. The invert and roof of the tunnels are shallow brick arches, except beneath the Victory Arch where the roof comprises steel beams with concrete in-fill.

Elizabeth House is a 15 storey tower block of reinforced concrete frame construction, built in the 1960's. The building has a single 6 m deep basement and is supported on a raft founded within the Thames Gravels. Part of the building was designed to cantilever over the Waterloo and City Line tunnels without imposing any load on the tunnel structures, as shown in Figure 2; this required the provision of a very stiff sub-structure.

Of these three structures, the Victory Arch and the Waterloo and City Line tunnels are particularly sensitive to settlement and associated distortion. The Victory Arch is a heavy, predominantly masonry, structure with finishes that are sensitive to settlement. As the main entrance to Waterloo Station, it contains large glass doors and in addition it is directly connected to the glass structures of the new Waterloo International Terminal. The Waterloo and City Line tunnel walls are of masonry construction and are located in water-bearing gravels. The tunnels carry passenger trains from Waterloo to Bank and any damage or cracking to the walls would cause flooding and severe disruption of services.

4 Potential settlements

The potential settlements associated with the tunnelling were assessed by undertaking calculations using empirical methods based on field observations [2]. A maximum settlement in excess of 100 mm was indicated based on a 2% volume loss. Methods to reduce volume loss were adopted including requirements for 24 hour working, tight controls on tunnelling methodology and face support, and the use of a pilot tunnel. Recent measurements of ground movements above and around similar tunnels excavated in London Clay at Angel [3] and for the Docklands Light Railway have demonstrated that construction of a pilot tunnel is beneficial in reducing total settlement. The 7.5 m, 8.25 m and 6.5 m ID tunnels were all required to be constructed by means of a 3.85 m ID pilot tunnel, followed by enlargement to full size. On this basis a volume loss of 2% was assumed for the pilot tunnels and 4 m diameter passageway tunnels, and a volume loss of 1% was assumed for the enlargement operations. For normal tunnelling conditions both these figures are probably conservative, since the available measurements suggest that where a pilot tunnel is constructed the resulting volume loss is as low as 0.5% for the enlargement phase [4]. However the ground beneath the part of the Waterloo and City Line supporting the Victory Arch is relatively highly stressed due to the imposed loadings of the structures, and this potentially could result in larger volume losses and settlements.

The contours of potential settlements based on these assumptions are shown in Figure 3. These were calculated assuming 'green-field site' conditions, i.e. any influence of the stiffness of the structures in modifying the settlement pattern was ignored. Settlements of up to 75 mm were predicted. Even if some allowance for the stiffness of the structures was included, the relative rotations, distortions and associated tensile strains [5,6] induced in the Victory Arch and Waterloo and City Line tunnels by these settlements would lead to unacceptable cracking and damage to the masonry. There was therefore a high probability that the potential settlements would cause significant damage to the Victory Arch and Waterloo and City Line tunnels. In view of the importance of these structures and the potentially serious implications of any damage on the running of both the Waterloo main line station and the Waterloo and City Line trains, it was concluded that protective measures were required.

Of the options examined, compensation grouting to limit the settlements was selected as the most appropriate for these particular circumstances. Analysis of the distortions of the Waterloo and City Line tunnel walls, based on the conservative assumption that the structure would not increase the width of the settlement trough, suggested that a maximum settlement of 10 mm was consistent with inducing negligible damage. The maximum rotation associated with this mode of deformation was 1/1000.

5 Chemical grouting

In addition to the requirement for compensation grouting as a protective measure against settlement, there was a need to ensure safe excavation through the water-bearing Thames Gravels. Permeation grouting of the gravels was selected in order to achieve two aims. The first was to reduce the potential groundwater flow in the gravels, by rendering them as impermeable as possible, and hence to stabilize the ground during excavation. The second aim was to provide a strengthened tunnel face and arch over the tunnel crown to reduce potential settlements of the Victory Arch foundations. The observations of the bulk excavation on the adjacent Waterloo International Terminal site, together with gradings and in-situ permeability measurements for the gravels, were such that chemical grouting was considered necessary to achieve the required grout penetration through the soil. This was carried out throughout the thickness of the gravels in the zone shown on Figure 4 as an advance contract, which was let to Keller Colcrete.

The Contract Specification required that the permeability of the grouted gravels be reduced to a maximum of 10^{-7} m/s and that a minimum unconfined compressive strength of 0.5 N/mm^2 be achieved. Grouting was required to be carried out using tubes à manchettes (TAMs) at a maximum of 1 m centres. Keller Colcrete opted to install 101 vertical and 88 inclined plastic TAMs on a triangular grid at a spacing of 0.8 m. The maximum inclination of holes was at 48° to the vertical. The grout injected was 55% silicate, 8% Stabgel hardener and 37% water.

The sleeves along each TAM were spaced at 0.33 m. The total number of sleeves injected was 3310, of which 850 sleeves were injected a second time and 20 received a third injection. The grouting was controlled on site by close monitoring of the volumes injected. A maximum volume of 100 litres was set for most of the injections at pressures of between 5 and 15 bar, with flow rates typically of 180 to 900 l/hr. A

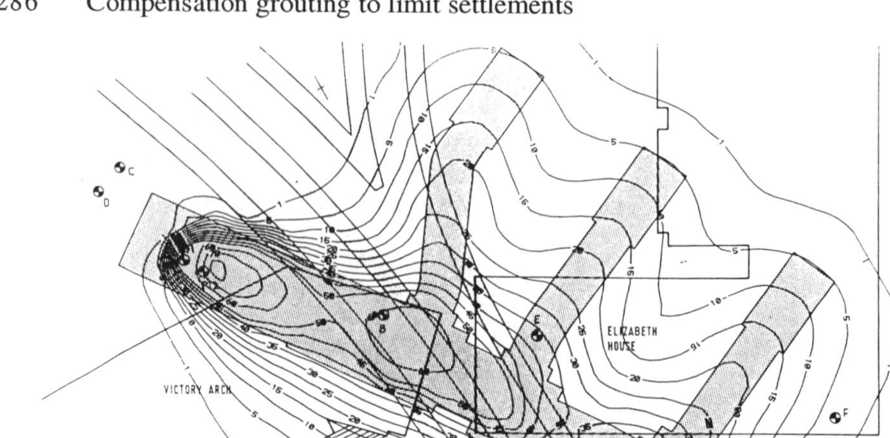

Figure 3 Predicted 'green field site' potential settlement contours (mm)

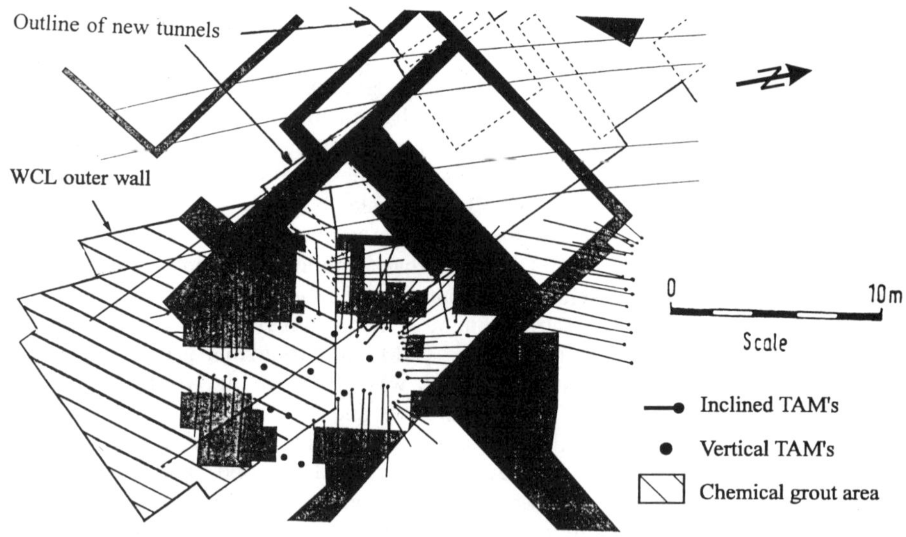

Figure 4 Schematic plan projection of inclined TAMs in the Victory Arch basement and area of chemical grouting

target grout volume of 25% of the total volume of gravels to be treated was set, based on an assessment of the volume of voids which could be effectively permeated. The total volume of grout injected was 219 m^3, which is equivalent to 23%. Validation testing, consisting of rising and falling head permeability tests, was carried out at the end of the treatment. This demonstrated that the permeability limit defined in the Specification had been achieved.

6 Compensation grouting - specification and preparatory works

As a tunnel is excavated the ground undergoes stress relief, resulting in ground loss towards the face with associated ground movements developing at higher levels. Compensation grouting involves the injection of grout between the tunnel and an overlying structure during tunnelling to *compensate* for this stress relief and ground loss. Compensation grouting may also involve the injection of a fluid grout into voids formed beneath foundations to prevent settlement. The aim of compensation grouting during tunnelling is to limit settlement to a specified amount; it is not a question of allowing a structure to settle during tunnelling and then afterwards jacking it up as a corrective measure. Key elements of the compensation grouting technique are instrumentation and monitoring. The grouting operations are undertaken in a strictly controlled manner in response to observed ground and structure movements.

Compensation grouting was selected to limit settlement of the Victory Arch and of the Waterloo and City Line tunnels. (Analysis of the likely performance of Elizabeth House, with its stiff substructure, indicated that the settlement, distortion and tilt would be acceptably small; therefore compensation grouting for this structure was not specified.) The three main techniques available to achieve compensation grouting are compaction grouting, intrusion grouting and fracture grouting [7]. Beneath the Victory Arch, the technique specified was compaction grouting - the injection of stiff mortar paste to form 'bulbs' of grout - in the Thames Gravels (where they had not previously been chemically treated). The Specification required injection positions to be at a maximum spacing in plan of 1.5 m. Beneath the Waterloo and City Line tunnels, the gravel thickness is limited and it was therefore envisaged that the technique would be fracture grouting in the London Clay or intrusion grouting at the London Clay/Thames Gravel interface. The Specification required grout to be injected through tubes à manchettes (TAM's), at a maximum spacing in plan of 1.5 m, beneath each of the three walls of the Waterloo and City Line tunnels over a length of about 45 m.

The Specification defined trigger levels at which compensation grouting had to be undertaken. If the instrumentation detected movements of 1 mm of the structures or 5 mm of the ground relative to the structures (as measured by extensometers) since the previous grouting operation, the Contractor had to grout and would be paid for doing so. Grouting was also permitted before the trigger levels were exceeded, but in practice this only occurred rarely. In addition, overall performance criteria were specified which stated that total settlement and rotation of the structures should not exceed 10 mm and 1/1000 respectively.

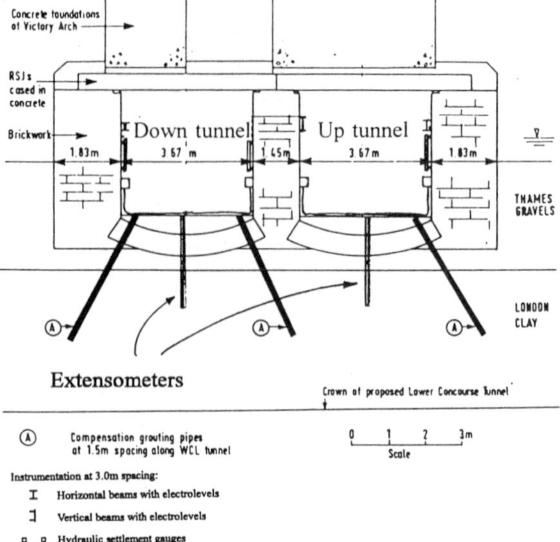

Figure 5 Cross section through Waterloo & City Line tunnels showing location of TAMs and instrumentation

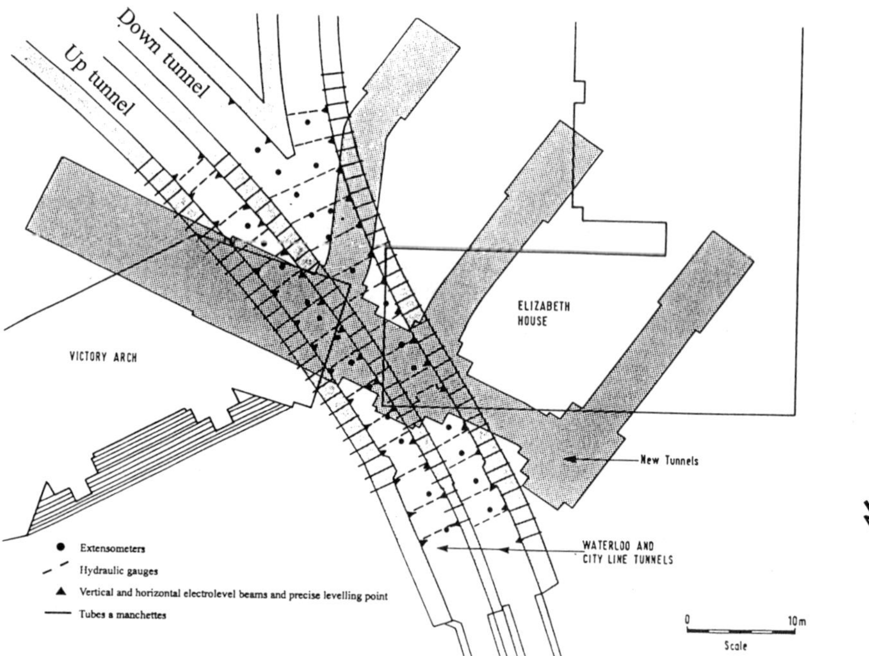

Figure 6 Plan of Waterloo & City Line tunnels showing instrumentation and TAM locations

It was a requirement of the Specification that all grouting tubes and all of the instrumentation (with the exception of electrolevels in a borehole above the lower concourse) were to be installed at least two weeks prior to commencement of tunnelling. The two week period after commissioning of the instruments was stipulated to allow datum readings to be established and the reliability of the instruments to be demonstrated. Access to the Waterloo and City Line tunnels for these preparatory works (and for subsequent compensation grouting operations) was only possible at night and at weekends. The total preparatory works period was three months.

The sub-contractor for the grouting works, Bachy (UK) Ltd, proposed an alternative technique for compensation grouting in the Thames Gravels beneath the Victory Arch. Instead of compaction grouting, it was proposed that a form of intrusion grouting be used; TAM's would be installed (as specified for the Waterloo and City Line) and an unstable, initially fluid, low water-to-solids ratio grout injected. An *unstable grout* is stable when pressurized and contained in the grout hose, but after injection into granular soils the process of pressure filtration ensures that water is rapidly squeezed out of the mix. The grout then remains local to the point of injection as a result of its increased viscosity. By water bleeding from the grout into the gravels as soon as it was injected, grout would be concentrated around the injection point and a similar effect as the originally specified stiff mortar paste 'bulb' achieved. The principal advantage of this alternative was that the TAM's could be re-used for successive compensation grouting operations without the need for equipment to be kept on site to re-drill the grout tubes (which would have been necessary if compaction grouting had been adopted).

A grouting trial to demonstrate the viability of the proposed alternative was undertaken in a zone of gravel on the site which was subsequently excavated as part of the preparatory works for the tunnelling. Various grout mixes were injected, each with a different colour dye. Careful excavation around each TAM allowed examination of the grout in the ground, and it was ascertained that a significant proportion of the grout was clustered around the point of injection, even though there was also evidence of limited fracture with not all of the injected volume being accounted for. Approval was given to the proposed alternative and, as part of the preparatory works, 89 steel TAM's were installed beneath the Victory Arch foundations, as shown in Figure 4. This involved drilling both vertical and inclined holes through masonry from ground level and from within basements.

At night time a total of 92 steel TAM's were installed in the Waterloo and City Line tunnels during the preparatory works period. A rail track mounted rig was used to drill inclined holes, 5 m in length, at 1.5 m centres through and beneath each of the three 2 m thick walls of the tunnels, as shown in Figure 5. Drilling through the masonry walls into the water bearing gravels was a carefully controlled two stage operation to prevent the risk of flooding. The approximately 45 m lengths of Waterloo and City Line tunnel walls beneath which the TAM's were installed are shown in Figure 6.

The preparatory works also included a limited amount of chemical grouting of the gravels beneath a 15 m length of one of the outer walls of the Waterloo and City Line, as shown on Figure 4. This was undertaken to ensure that the zone of grouted gravel extended to at least 2.5 m beyond the periphery of the tunnel to be excavated.

The final stage of the preparatory works involved grouting from approximately 50% of the injection points in the Thames Gravels to 'condition' or 'tighten up' the gravels. This pre-treatment, which was proposed by the Contractor and accepted by London Underground Ltd, was considered to be of benefit because ground movements subsequently occurring during tunnelling could be compensated for more effectively if the larger voids in the gravel had already been filled. The aim of this pretreatment was to grout until the instrumentation on the structures responded, thus enabling a rapid response to be obtained from subsequent compensation grouting operations. Careful monitoring of the instrumentation was necessary during this pre-treatment.

7 Instrumentation

A comprehensive array of instrumentation and survey points was specified to be installed in the ground and on the structures. All instrumentation was installed by the Building Research Establishment, as a sub-contractor to McAlpine. The following instrumentation was installed for detection of ground movements:

(i) Electrolevel sensors installed in an inclined 20 m long borehole and in a horizontal 30 m long borehole, both about 1 m above the enlarged tunnel crown along the tunnel centre-line (as shown on Figure 2).

(ii) 29 sleeved rod extensometers, installed through the invert of the Waterloo and City Line tunnels at 3 m centres (Figures 5 and 6).

(iii) 8 magnetic extensometers installed in the Victory Arch area.

The inclined electrolevel string, the sleeved rod extensometers and the magnetic extensometers were all installed prior to commencement of tunnelling. The horizontal electrolevel string was installed from within the inclined pilot tunnel, the progress of which was halted near the bottom of the inclined section to allow installation.

Within the Waterloo and City Line tunnels, at 3 m intervals along each of the walls, the following instrumentation was installed to monitor movement of the structure, as illustrated in Figures 5 and 6:

(i) Horizontal beams with electrolevels (connected to form a continuous string over a length of 50 m along each wall to measure settlement of the walls).

(ii) Vertical beams with electrolevels (to measure out-of-plane tilt of the walls).

(iii) Hydraulic manometer gauges installed between adjacent walls (to measure differential settlement between the walls).

The hostile environment within the Waterloo and City tunnels meant that certain problems with the instrumentation were encountered. These were, *inter alia*:

(i) Interference from electric currents associated with the railway.

(ii) Effects of humidity.

(iii) Pressure waves associated with passing trains.

These problems were all successfully overcome.

Steel rulers were installed for precise levelling, which was undertaken at regular intervals to provide data on absolute settlement from a remote datum. Electrolevels and levelling markers were also installed on the Victory Arch and Elizabeth House buildings, including a system of connected electrolevel beams within the Victory Arch basement.

All of the instrumentation (with the exception of the magnetic extensometers) was connected to data logging equipment so that measurements could be taken on a 24 hour basis. In total 277 separate sensors were monitored automatically. Readings were generally taken at 30 minute intervals but with increased frequency as required during compensation grouting operations. Data was processed on computers in the site offices of the Contractor and of London Underground. As a precaution, an alarm system was installed which would be triggered if the hydraulic manometer gauges indicated differential settlement of the walls of the Waterloo and City Line exceeding 4 mm; this criterion was based on the maximum allowable rate of change of cant of the rails as defined by British Rail.

8 Compensation grouting - methodology

Tunnelling was undertaken on a 24 hour basis, 7 days per week, and a system was operated whereby the requirement, if any, for compensation grouting was reviewed daily. Every morning, the Contractor produced a report on the results of the automatic monitoring over the past 24 hours, together with data from precise levelling which had frequently been obtained during the previous night. This information was reviewed in detail at an early site meeting between the Contractor and LUL's geotechnical consultant, GCG. Most of these meetings were also attended by British Rail's consulting engineers Charles Haswell and Partners, as well as LUL's site staff. Interpretation of the data was discussed and a consensus reached as to whether trigger levels had been exceeded, or, on the basis of the trends in the data, were likely to be exceeded prior to the next night shift grouting operations.

When trigger levels were identified, grouting requirements were discussed and a detailed grouting plan was subsequently proposed by the Contractor and reviewed at a meeting later in the day. Once the proposal was approved, compensation grouting operations were carried out during the next night. Grouting from within the Waterloo and City Line was restricted to night time possessions; however, the Contractor

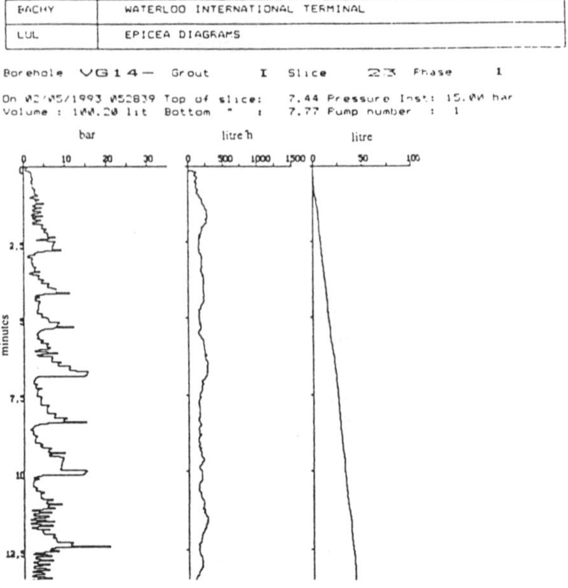

Figure 7 EPICEA characteristic injection plot

elected to carry out all grouting operations at night, including those in the Victory Arch area.

The grouting equipment was located in the basement area of the Victory Arch, from which a grout supply pipe was fed into the Waterloo and City Line tunnels through a single 150 mm hole drilled in the tunnel roof. In practice two grout pumps were operating at any one time, although six grout pumps were available on site which could all be operated simultaneously, if required. The computer-controlled EPICEA system allows injection parameters, such as maximum grout volume, pressure, and flow rate, to be set for each injection position. If any of these are exceeded during the grouting operation, the pump is automatically switched off and an alarm sounds. In order to promote the pressure filtration of the grout in-situ, the operator would manually adjust the injection flow rate. The EPICEA system produces a printed record of the parameters relevant to the injection and allows immediate graphical presentation of the results, as shown in Figure 7. During grouting, the operator was in direct communication with a person monitoring the instrumentation recording movements of the ground and structures.

Periodically the injection sleeves were 'cracked' with water to ensure that a build up of grout did not prevent sleeves from being re-injected. By the end of the compensation grouting programme, some sleeves had been grouted up to 20 times. The majority of the grouting was undertaken beneath the Waterloo and City Line through sleeves at or near the interface between the Thames Gravels and London Clay. The distribution of the total volumes of grout injected into sleeves below the centre wall of the WCL are shown in Figure 8 (a).

The grout used was the same for all injections and comprised a 13:1 PFA:cement mix, with 0.4% bentonite and a water:solids ratio of 0.4. The volume of grout injected as tunnelling proceeded is shown in Figure 8(b). The final volume of grout injected corresponds to just over 4% of the volume of tunnels excavated, with the total volume of solids within the grout corresponding to about 2%. Typical details of grouting are as follows:

	WCL: (Gravels or gravel/clay interface)	Victory Arch: Gravels
Maximum volume per injection (l)	100	100
Maximum volume per TAM (l)	200	400
Maximum pressure (bar)	20	35
Average flow rate (l/hr)	300	300

Excavation commenced through the chemically treated gravels and the effectiveness of the advance contract chemical grouting was confirmed, since only very minor seepages of water were encountered. Compensation grouting was not required until two weeks after commencement of tunnelling, by which time 24 rings (0.6 m each) of the inclined pilot tunnel had been constructed. During later advancement of the pilot tunnel and construction of the passageways, compensation grouting was undertaken typically at a frequency of twice a week. During tunnel enlargement more extensive grouting programmes were proposed which were carried out over more than one night shift. On average grouting was undertaken every other night during this period. This frequency was maintained during construction of the openings and remaining passageways.

9 Monitoring during tunnelling

The observed settlement of the Victory Arch, obtained from the series of connected electrolevel beams and from the precise levelling, are shown in Figure 9. In general reasonably close agreement was found between the two systems of measurement. By the time the pilot tunnel had been completed, the maximum settlement of the structures had reached 4 mm. On completion of the enlargements to the 7.5 m, 8.25 m and 6.5 m diameter tunnels, the total settlement had reached about 7-8 mm. Construction of the passages and the openings (described in detail in [1]) resulted in the maximum total settlement of the Victory Arch increasing to about 12 mm, as shown in Figure 9.

Concurrent with the tunnelling operations, there were other construction activities associated with the Waterloo International Terminal in progress which caused

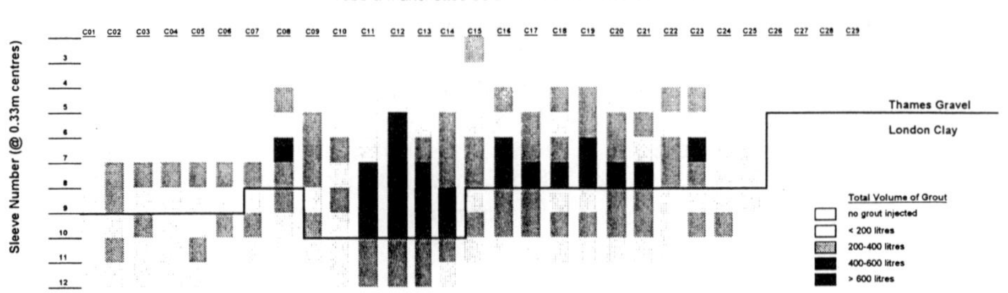

Figure 8a Distribution of total volume of grout injected

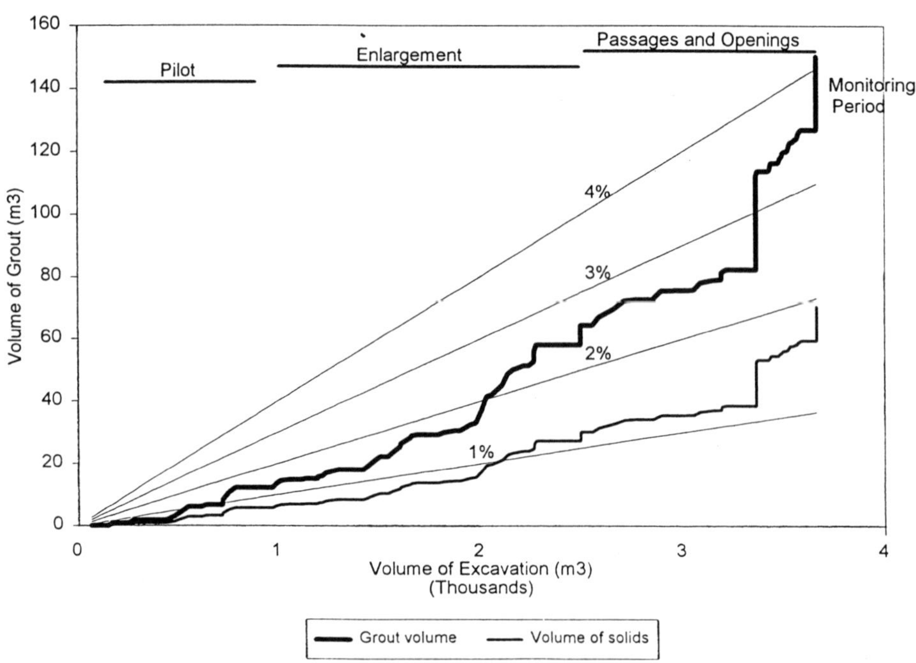

Figure 8b Total grout volume and volume of solids injected against volume of excavation

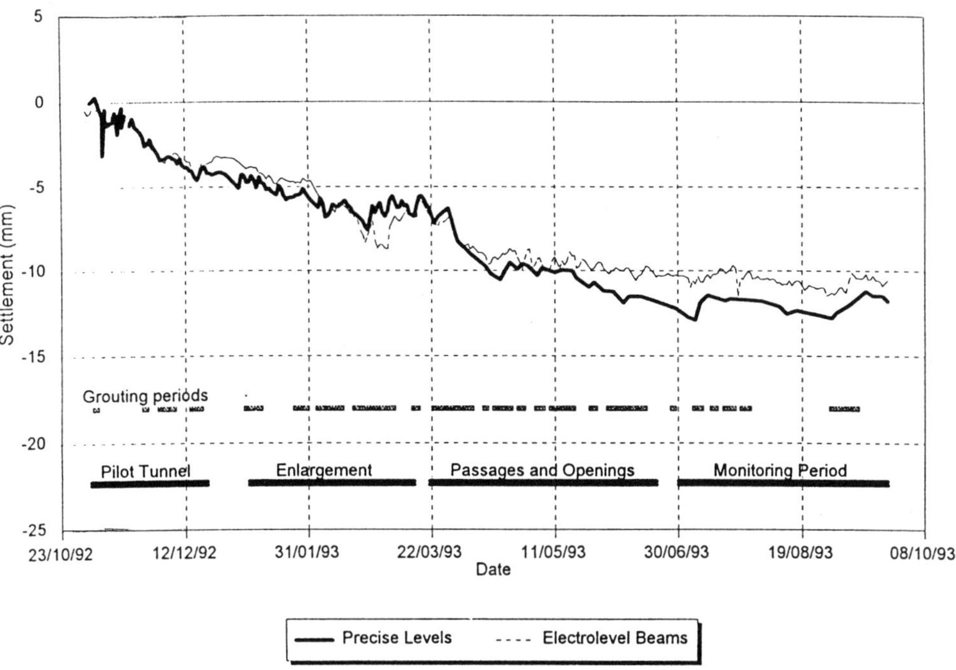

Figure 9 Settlement with time in Victory Arch basement from precise levelling and electrolevel beam system

settlements of the Victory Arch and the Waterloo and City Line. In particular, the construction of the access road behind Elizabeth House required the placement of 3-4 m of fill directly above the Waterloo and City Line tunnels, causing them to settle by up to 3 mm, and the Victory Arch (which is partially supported by the Waterloo and City Line) to settle by 1 to 2 mm. In total it is estimated that up to 5 mm of settlement has been produced within the zone of influence of the tunnelling operations by the Waterloo International Terminal development.

The profiles of total settlement along the three Waterloo and City Line tunnel walls are shown in Figure 10(a). Allowing for the other construction activities referred to above, the profiles corresponding to tunnelling alone are shown in Figure 10(b). It can be seen that the maximum settlement of the 'Up' tunnel outer wall amounted to 14 mm, thereby exceeding the specified 10 mm criterion, but the maximum rotation was 1/1250 which was below the specified value of 1/1000. No cracking of the walls of the Waterloo and City Line was recorded and analysis of the measured deflected shapes demonstrated that the deflection ratios, and associated tensile strains, were below the limiting criteria used in design to determine the specified limits on movements.

The settlement of Elizabeth House was found to correspond closely with that of the Waterloo and City Line (WCL) 'Up' tunnel outer wall directly beneath it. It was

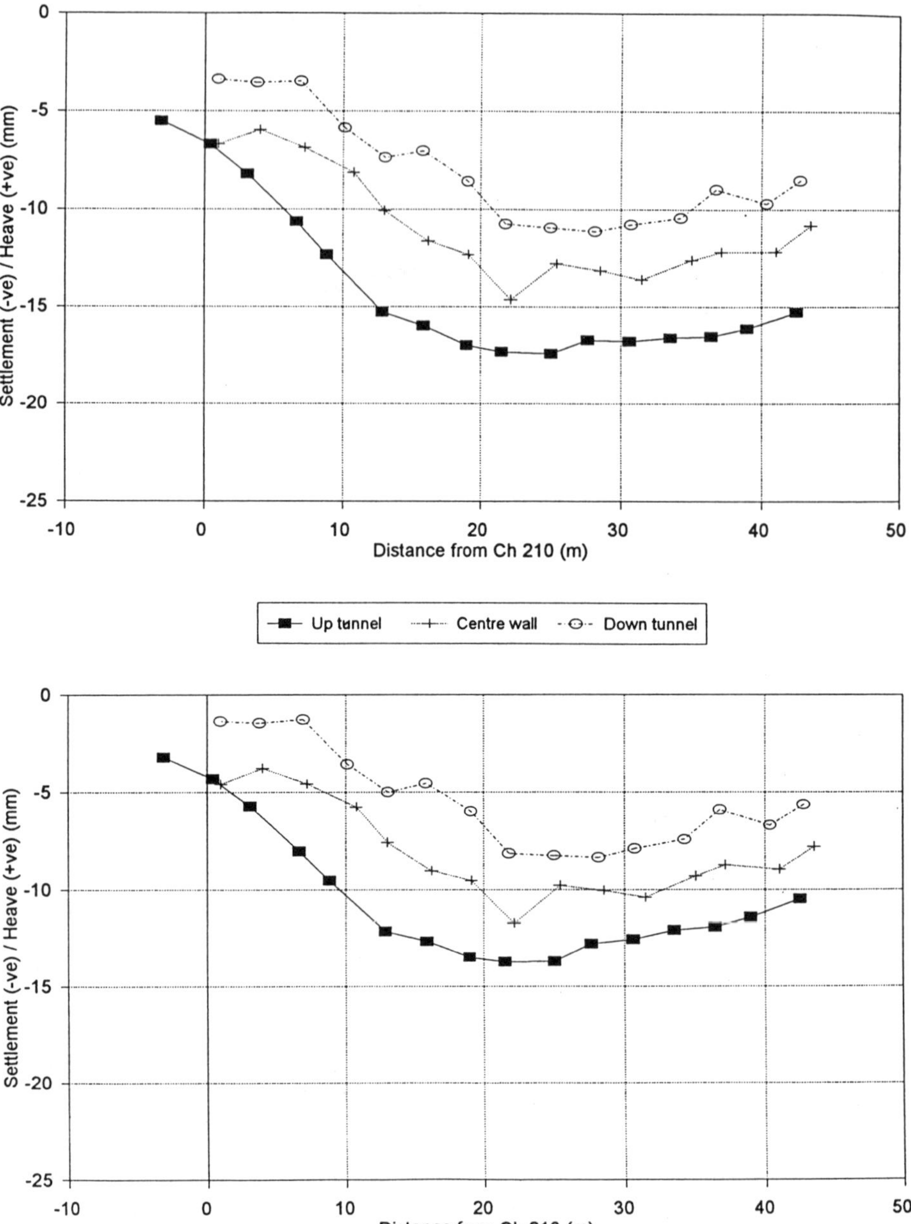

Figure 10 (a) Profiles of total settlement along Waterloo and City Line walls
(b) Profiles of settlement due to tunnelling along Waterloo and City Line walls

concluded from this that Elizabeth House must in fact be in contact with the WCL, rather than cantilevering over it as had been intended by its designer. This is believed to be largely the reason why the WCL 'Up' tunnel outer wall settled significantly more than the other two walls. Nevertheless the maximum rotation was maintained below the specified value, and it was decided that it would be unproductive to attempt to induce heave of the WCL 'Up' wall and Elizabeth House.

The electrolevels in the inclined and horizontal boreholes proved to be invaluable in monitoring the movement of the ground. This is illustrated in Figure 11(a), in which the development of settlement at a level of between 1 m and 1.5 m above the crown of the pilot tunnel is shown as the tunnel face advances. The monitoring data has been corrected to allow for incremental movements caused by compensation grouting operations. A maximum settlement of about 18 mm occurred above the pilot tunnel, and 32 mm developed during enlargement at the same level above both the 8.25 m and 6.5 m tunnels. The benefit of the pilot tunnel acting as a 'dowel' in reducing total settlements is evident, although somewhat less marked than has been observed on other projects [4].

The electrolevels in the inclined and horizontal boreholes were also valuable in monitoring detailed behaviour of the ground. Figure 11(b) illustrates an example, in which the change of slope of a particular electrolevel is shown developing with time as the pilot tunnel face advanced. The rotations measured are consistent with those which would be expected as the 'S' curve ahead of the tunnel face passes a particular point. The cycle of construction of rings is clearly reflected in the 'steps' within the trace, and it can be seen that a production rate of just over 4 rings / 24 hours was being achieved. The effect of some adjacent compensation grouting can also be seen (taking place on 5/12/92). The grouting operations and their effects could be carefully monitored in this way.

The final settlements of the Victory Arch and WCL caused by tunnelling alone at the end of construction are shown on Figure 12. These contours can be compared with the predicted contours of potential settlement (in the absence of compensation grouting) shown on Figure 3. The corresponding settlements above the centre-line of the main tunnel is shown on Figure 13, compared with the predicted potential settlements in the absence of compensation grouting. It can be seen that the compensation grouting was extremely effective in limiting the settlements.

10 Conclusions

Tunnels with excavated diameters of up to 8.8 m (8.25 m ID) have been successfully constructed very close beneath the Victory Arch and the Waterloo and City Line tunnels, each structure being particularly sensitive to settlement. In the absence of protective measures, potential settlements of these structures up to 75 mm were predicted. Appreciable cracking would have occurred, and in the case of the Waterloo and City Line, there was a significant risk of water entry from the surrounding Thames Gravels.

The relatively novel technique of compensation grouting has proved to be very

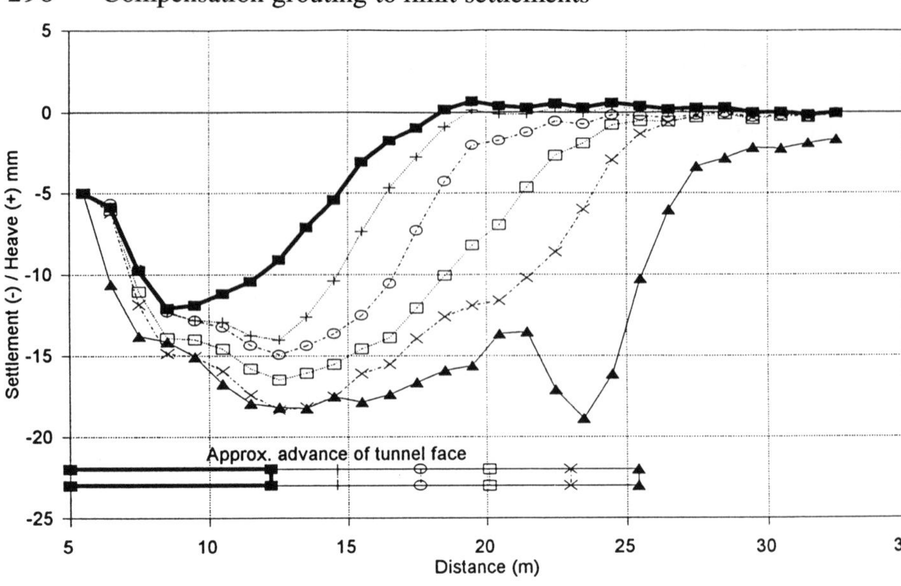

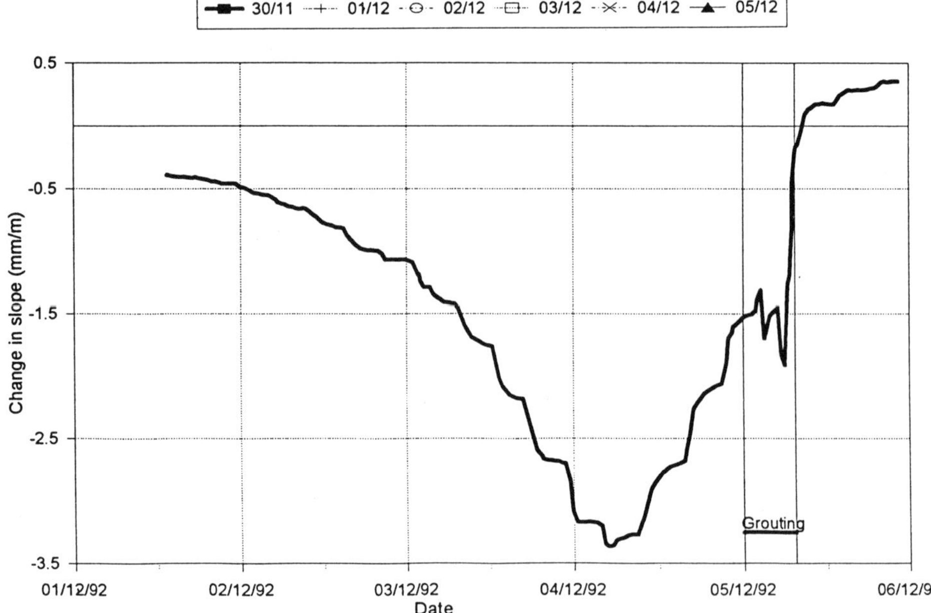

Figure 11 (a) Development of settlement profile above the crown of the pilot tunnel from electrolevels in the horizontal borehole

(b) Change in slope indicated by a single electrolevel in the horizontal borehole as the tunnel advances beneath

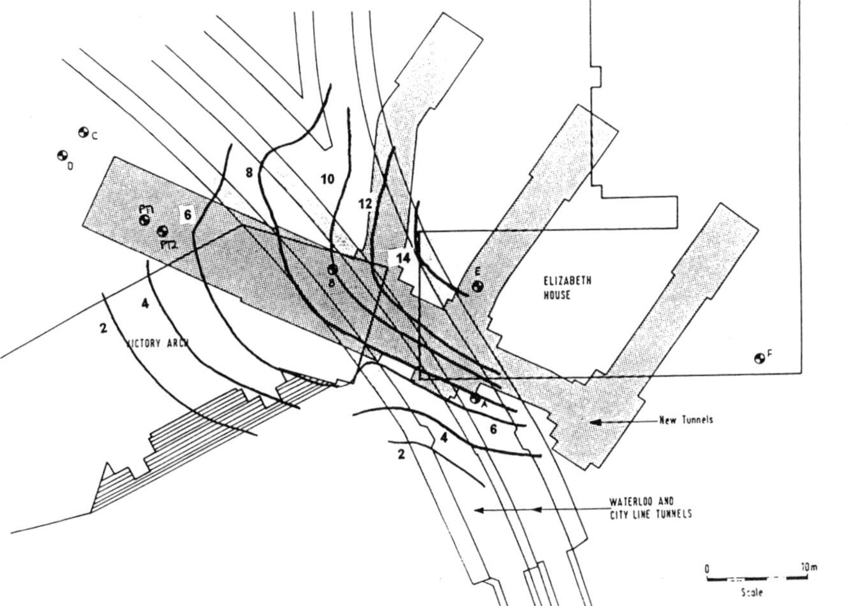

Figure 12 Contours of settlement due to tunnelling (mm)

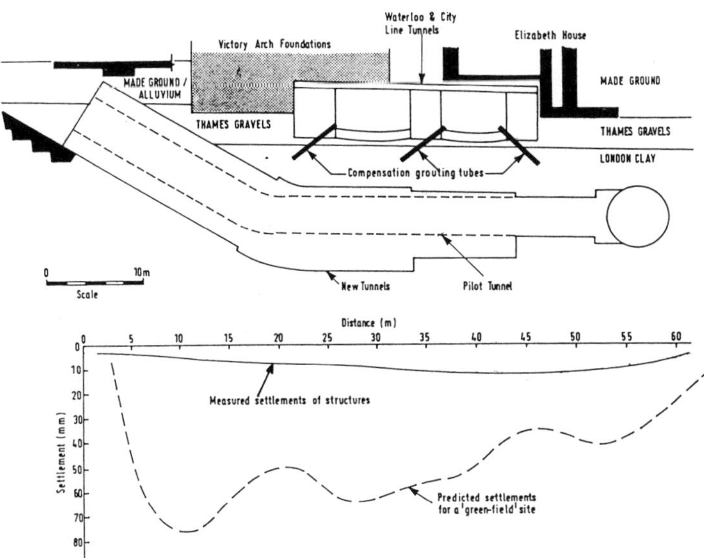

Figure 13 Comparison of measured settlements above the centreline of the new tunnels with the 'green field site' predictions

successful in limiting settlements and, more importantly, distortions of both the Victory Arch and the Waterloo and City Line tunnels. The maximum settlement due to tunnelling alone on completion of tunnel construction was 10 mm in the case of the Victory Arch and 14 mm in the case of the Waterloo and City Line, with corresponding rotations of 1/2250 and 1/1250. The tunnelling operations have been completed with no damage caused to overlying structures and no disruption to the main line BR station or the Waterloo and City Line services.

Key elements of the success of the compensation grouting were installation of comprehensive instrumentation combined with detailed monitoring and interpretation of the data. These ensured that the grouting operations were undertaken in a controlled manner in response to observed movements of the ground and the structures.

Acknowledgements

The authors from GCG acknowledge the valuable contributions from their colleagues Tom Henderson (formerly with GCG), Dr David Hight, Rob Nyren and Dr Neil Taylor. They are also grateful to Dr David Greenwood for his excellent advice during the development of the concept of compensation grouting. Sir Alexander Gibb and Partners provided valuable information about the existing infrastructure at Waterloo. The authors are grateful to London Underground Ltd for permission to publish this paper.

References

1. Chudleigh I. L. J. and Webb N. S., 1994. Construction of new escalator shaft and passageways at London Underground's Waterloo Station. Tunnelling '94.
2. O'Reilly, M. P. and New, B.M., 1982. Settlements above tunnels in the United Kingdom - their magnitude and prediction. Tunnelling '82. Ed. Jones, M.J., pp. 173-181. London, Institution of Mining and Metallurgy.
3. Moriarty, J. M. and Cooper, M. L., 1991. Reconstruction of London Underground's Angel Station, Tunnelling '91, pp. 233-248. London, Institution of Mining and Metallurgy.
4. Mair, R. J., 1993. Unwin Memorial Lecture 1992. Developments in geotechnical engineering research: application to tunnels and deep excavations. Proc. Instn. Civ. Engrs. Civ. Engng., 1993, 93, Feb., pp. 27-41.
5. Burland, J. B., Broms, B. B . and de Mello, V. F. B., 1977. Behaviour of foundations and structures. State-of-the-art report, Session 2, Proc. 9th Int. Conf. on Soil Mechanics and Foundation Engineering, Tokyo. Vol. 3, pp. 495-546.
6. Boscardin, M. D. and Cording, E. J., 1989. Building response to excavation induced settlement. Journal of Geotechnical Engineering, ASCE, Vol. 155, No. 1, pp. 1-21.
7. Mair, R.J., 1994. Report on Session 4: Displacement Grouting, Proc. Conf. on Grouting in the Ground, ICE, November 1992, Thomas Telford.

Ground movement model validation at the Heathrow Express trial tunnel

B. M. New
K. H. Bowers
Transport Research Laboratory, Crowthorne, England

Abstract
Empirically based ground movement modelling techniques are now being widely applied in the design of major tunnelling projects making their validation increasingly important. The Heathrow Express Trial Tunnel was the first excavation to use the New Austrian Tunnelling Method (NATM) in the London Clay and provided an excellent opportunity both to test the ground movement prediction methods and evaluate a variety of different NATM construction sequences.

The Transport Research Laboratory has developed a full range of models for use in the prediction of the pattern and magnitude of ground movements around tunnels and associated works. Earlier work has been extended by the development of a three dimensional model (TUNSUB) which allows consideration of complex excavation geometries.

In particular this paper describes a new technique which models ground movements in the region close to the tunnel. This provides a powerful tool for predicting deformations experienced at adjacent subsurface structures such as other tunnels, services and deep foundations.

The surface and subsurface around the trial tunnel were instrumented and surface measurements of ground movements due to shaft sinking were also recorded. Surface instrumentation consisted of high precision digital levels, tape extensometers and Geomensor EDM equipment. Boreholes around the tunnel contained magnetic ring extensometers with high precision micrometer reading heads, inclinometers, soil stress cells and piezometers. Throughout the shaft sinking and tunnelling works the ground response was monitored and correlated with the construction activities. Maximum settlements, percentage volume losses and settlement trough shapes were established for each section and then related to the modelled outputs.

The field results showed excellent agreement with those modelled and, with the modified assumption of ground loss distribution, the ground movements in the vicinity of the tunnel were also well represented.
Keywords: Ground movement modelling, Heathrow Express, instrumentation, New Austrian Tunnelling Method, settlement prediction, shafts, soil stresses around tunnels, tunnelling induced settlement.

1 Introduction

Tunnelling beneath urban areas is currently assuming renewed significance as environmental pressures and the need for an improved transportation infrastructure continue to increase. Ground settlements induced by tunnelling in soft ground can have a major impact on the overlying environment and place significant restraints on the preferred line and level of the tunnels. Such restraints can result in more devious routeing to avoid vulnerable structures and deepening of the works to ameliorate surface effects. Increases may be incurred in the costs of both construction and operation of tunnels, and the reduced rate of return on investment may thereby threaten the viability of the project.

Considerable research effort has been devoted to the study of ground movements caused by driving tunnels through soft ground. Much of this work in the United Kingdom has been carried out either directly by, or under commission from, the Transport Research Laboratory (TRL, formerly the Transport and Road Research Laboratory). It has involved extensive field monitoring by the Laboratory, universities and consulting engineers as well as a programme of centrifuge modelling and analytical studies. Early work was reviewed in Tunnelling '82 by O'Reilly and New (1982) [1] and later extended and updated by New and O'Reilly (1991) [2]. It led to the development of analytical and numerical models for the prediction of tunnelling induced ground movements, strains and angular distortions under various ground conditions which were validated by extensive site measurements. These models are developments of earlier work by Knothe (1959) [3], Martos (1958) [4], Schmidt (1969) [5], Peck (1969) [6] and others and form the basis for the prediction of ground movements for many major soft ground tunnelling works in the United Kingdom including CrossRail, the Jubilee Line Extension and the Union Railway (Channel Tunnel Rail Link).

The radial flow assumption proposed by O'Reilly and New (1982) allows the calculation of surface and subsurface movements at distances greater than about one diameter from a tunnel using suitable values of a trough width parameter and an estimate of ground volume loss. This has now been extended by the development of a three dimensional numerical model (TUNSUB) which allows consideration of a variety of configurations including sloping and curving tunnels, changes of diameter and cross connecting adits. It is often at these complex locations (for example around stations) that the most critical ground movements occur.

In the region near to the tunnels (where the simple radial flow model is not applicable) new assumptions concerning ground loss have been developed to permit improved modelling of the ground movements. These provide powerful tools for determining the deformations which are experienced at adjacent subsurface structures such as other tunnels, pipelines, basements or foundations. These modelling techniques are now being applied in the design of a variety of major tunnelling projects making their rigorous validation increasingly important.

The Heathrow Express rail link is currently under construction and will provide a fast main line rail link between Heathrow Airport and central London. Prior to construction a trial tunnel was built using the New Austrian Tunnelling Method

(NATM) which promised cost savings for the construction of the stations required for the airport terminals. The trial tunnel was the first excavation to use the NATM in the London Clay and provided an excellent opportunity both to test the prediction models and evaluate a variety of different NATM construction sequences. It should be noted that the Department of Transport also has considerable current interest in tunnelling works using sequential support systems. These include the A20 Round Hill Tunnels, the A27 Southwick Tunnels and proposed works at Worthing (A27) and Hindhead (A3).

The surface and the ground around the Heathrow Express Trial Tunnel were instrumented by TRL. Surface instrumentation consisted of high precision digital levels, tape extensometers and Geomensor EDM equipment. Boreholes around the tunnel contained magnetic ring extensometers with micrometer reading heads, inclinometers, soil stress cells and piezometers. Data were collected and processed on the site using specially developed software, and automated systems where possible, to enable the presentation of key results in graphical form immediately after acquisition.

Throughout the shaft sinking and tunnelling works the ground response was monitored and correlated with the construction activities. Maximum settlements, percentage volume losses and settlement trough shapes were established for each section and then related to the predictions. The measured ground movements were compared with results based on existing and improved modelling techniques, with particular emphasis on subsurface movements in the vicinity of the excavation.

The instruments at the Heathrow site have been regularly monitored since the completion of the excavations in May 1992 and results up to thirteen months after the end of tunnelling are presented. Additionally, near surface ground movements caused by the excavation of the access shaft were measured and used as a basis for a simple empirical model.

The site at Heathrow provided full and uninterrupted access to a complete array of ground monitoring equipment situated in uniform strata of London Clay overlain by a thin layer of gravel. The conditions were therefore ideal in providing the precise and reliable measurements of ground movements necessary for the validation of the models. The results may be of value to other projects and are particularly relevant to tunnelling work in stiff fissured clays.

2 The Heathrow Express Trial Tunnel

The New Austrian Tunnelling Method (NATM) has been proposed for the excavation of the underground stations for the Heathrow Express railway project which will provide a mainline rail link to London's Heathrow Airport. The station tunnels are large and complex openings. Preliminary investigations by the Client, BAA plc, suggested that NATM might provide considerable cost savings and greater construction flexibility, particularly where tunnels with variable cross sections are involved. The application of the NATM to soft ground conditions was a new departure in the UK and the consequences of NATM with regard to ground movements in the London Clay were uncertain. It had been suggested that the immediate ground support might reduce the ground movements to less than those expected from conventional shield tunnelling. This would be advantageous at Heathrow as the stations lie beneath numerous structures in the vicinity of the main terminal areas.

BAA decided to invest in construction of a trial tunnel to enable evaluation of three variations of the New Austrian Tunnelling Method in the London Clay at the Heathrow site and to measure the ground movements which resulted. The intention was that the results of the trial would be used to assist design of the stations for the proposed rail link. It was also hoped that the experience gained during the trial would give confidence in the methods and therefore produce benefits through cost savings on bid prices for the future works. In addition the study yielded valuable data against which settlement prediction methods may be verified and subsequently applied to future road, rail and other tunnelling projects.

The trial works consisted of an 11m diameter, 26m deep access shaft, a 10m length of transition tunnel, and then three test sections each of approximately 30m length. The tunnel had a face area of about 59m^2 and an axial depth below the surface of approximately 21m. Construction took place between January and May 1992. The three test sections were each constructed by a different NATM tunnelling sequence using a Liebherr 902 back-hoe excavator. The three trial designs tested were based on excavation sequences successfully used elsewhere in Europe. The first section used twin sidewall drifts, the second used a single, larger sidewall drift while the third section used a crown, bench and invert sequence with a temporary invert formed with pretzels. The designs are illustrated in Figure 1. Figure 2 illustrates the construction of the single sidewall drift used in the second section.

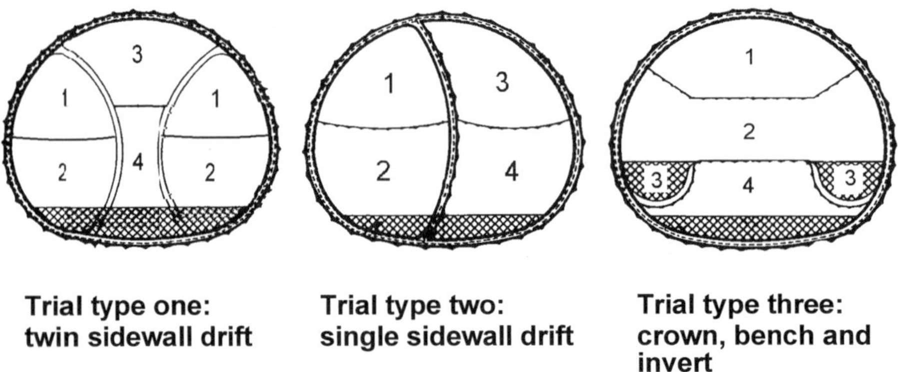

Figure 1: Heathrow Express Trial Tunnel NATM construction sequences

Figure 2: Single sidewall drift being excavated (trial type 2)

The geological profile at the site comprised typically half a metre of made ground over one to three metres of Taplow Terrace Gravel overlying London Clay. The Woolwich and Reading Beds were at depths in excess of 60m putting the trial tunnel entirely within the London Clay. The clay was generally uniform in character with only occasional bands of claystones. Some water was present in the gravel just above the gravel clay interface.

The near homogeneous ground together with excellent accessibility to an area of level car park over the trial tunnel combined to make this a near perfect site for a ground movement monitoring exercise of this type.

3 Instrumentation - Design Considerations

TRL designed, installed and operated the trial tunnel ground movement instrumentation system. The instrumentation measured both surface and subsurface ground movements caused by the excavation and also provided supplementary measurements of soil stress and porewater pressures to assist long-term assessment of ground/tunnel stability. Vertical and horizontal ground movements were measured to allow calculation of the crucial factors of angular distortion and horizontal strain which are essential for structural damage predictions. The procedures used were based on those previously developed and used by TRL at many other tunnels and construction excavation sites in the UK . Monitoring of convergence and lining stress changes within the tunnel were undertaken as part of the main contract as is commonly the case in NATM works.

The instrumentation system layout was carefully targeted to provide cost effective monitoring of the ground movements. Initial ground movement predictions were made and individual instruments were then selected and located to best monitor the ground movements, concentrating on the region where most significant movement was predicted. Figure 3 shows a plan of the layout of instrumentation in relation to the tunnel.

Instrumentation line T ran directly above the tunnel axis while lines A,B and C were located perpendicular to the tunnel axis orientation at the centre of trial sections one, two and three respectively. Line S monitored movements associated with shaft sinking on the side of the shaft remote from the tunnel heading. The subsurface instrumentation was distributed in boreholes principally along lines B and C. Figure 4 shows the distribution of subsurface instruments in cross section along line B. The distribution along line C was similar.

The instrumentation was installed in the spring of 1991 to allow stable datum readings to be obtained prior to the start of tunnelling. Measurements were taken intensively during the construction phase and the following month and thereafter at a reduced frequency to determine the nature of the longer term movements. During the intensive phase all instruments on the site were monitored at least once a week. Those instruments in, or near to, areas where tunnelling was causing the most significant ground movements were monitored more frequently as judged necessary by the engineering and instrumentation teams. In almost every case a reading frequency of once per day, or less, was found completely adequate to record the response of the London Clay to tunnel excavation.

A target accuracy of ± 1mm was considered appropriate for all measuring systems as structures are not likely to be effected by smaller variations. In practice significantly higher precisions were achieved by most of the systems installed (eg. levelling 0.12mm). This high level of precision was of great assistance in the analysis and interpretation of the data.

4 Instrumentation Methodology

4.1 Near-surface displacement

Measuring stations were constructed in small holes excavated through the existing pavement. The station design consisted of a concrete pillar concreted into the gravels underlying the site. The upper part of the pillar was isolated from the pavement because movements of the pavement might not truly reflect those of the underlying ground. A length of reinforcement bar with a machined stainless steel head welded to its top ran through the pillar and projected a short distance from the top of the concrete. This head was tapered and threaded to provide an accurate location point for instruments. In a few areas access was not available to allow the construction of full surface movement stations and therefore levelling pins were bonded into drilled holes

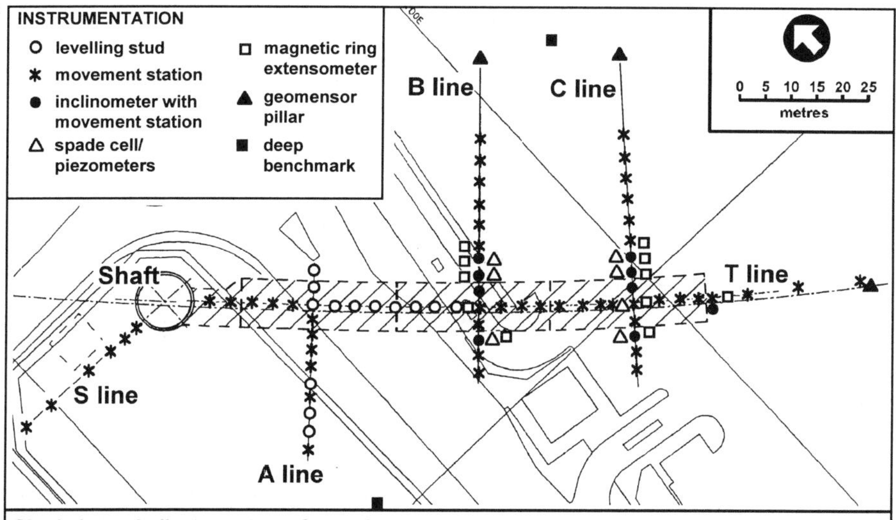

INSTRUMENTATION
levelling stud
movement station
inclinometer with movement station
spade cell/ piezometers
magnetic ring extensometer
geomensor pillar
deep benchmark
0 5 10 15 20 25
metres
B line
C line
T line
Shaft
S line
A line
Shaded area indicates extent of tunnel.
The A line crosses the centre of trial section one, the B line crosses the centre of trial section two and the C line crosses trial section three.
The T line is over the tunnel axis and the S line is affected by the shaft but remote from the tunnel.

Figure 3: Layout of instrumentation

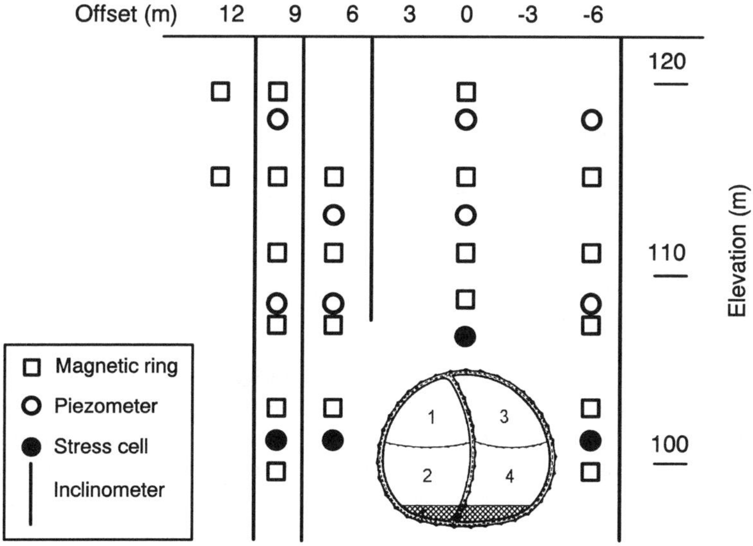

Offset (m) 12 9 6 3 0 -3 -6
120
110
100
Elevation (m)
Magnetic ring
Piezometer
Stress cell
Inclinometer
1
3
2
4

Figure 4: Layout of subsurface instruments in transverse profile B

in the pavement. These permitted surface levelling but no measurements of lateral movements.

Determination of the settlement of surface and near-surface reference stations was carried out using conventional precision levelling techniques. A Wild NA3000 high-precision digital level was used in conjunction with its special bar-coded invar levelling staff. To further enhance the levelling repeatability the level was mounted on concrete pillars founded at a depth of about two metres. This also avoided particular problems of wind induced tripod vibration at this site. Measurements were referred to temporary benchmarks remote from the influence of construction.

Horizontal movements perpendicular to and along the proposed tunnel line were measured using a combination of Geomensor (electronic distance meter) and tape extensometer instruments. Geomensor measurements were taken with the instrument mounted on a tribrach located on reinforced concrete pillars positioned in line with each of the test sections. These pillars were founded in boreholes of 4.5m depth and projected 1.5m above the ground surface. The instrument was sighted on a reflector which was located on a steel pillar which screwed onto the surface movement stations.

The tape extensometer measurements were taken with the instrument suspended between similar steel pillars which screwed onto the surface movement stations. To minimise errors particular extensometers and pillars were dedicated to particular groups of stations. Temperature was recorded at the time of each reading and a correction for thermal expansion of the instrument was made during data processing. The Geomensor was used to control a network of key stations and thus provide an independent reference check on the taping results.

4.2 Subsurface displacement

Vertical movements were measured using magnetic ring extensometers fitted with permanently installed reed switch sensors and a micrometer measuring head. Ring magnets were located at various depths in the extensometer boreholes and held against the walls of the hole by spring loaded arms. The holes were backfilled with a weak bentonite grout which left the magnets free to move vertically with the surrounding ground. Reed switches were suspended on a permanently installed steel tape at the approximate level of each magnet. The tapes were then moved up and down using the micrometer until signals from the reed switch indicated that it was passing the ring. The top of each extensometer borehole was located by levelling and the result of the levelling incorporated in the calculated result to locate fully the ring magnets in the vertical direction.

Horizontal movements were measured using half metre biaxial inclinometers with automatic loggers using methods developed at TRL. Originally aluminium tubes were used but due to a corrosion problem these were replaced with extruded ABS plastic prior to tunnelling. The holes were backfilled with a bentonite/cement grout which was designed to transmit movements of the surrounding ground to the tube. Movement of the inclinometer tube top was determined by a combination of tape extensometer and Geomensor survey and combined with the inclinometer data to locate fully the tubes in the horizontal direction.

4.3 Stress cells and piezometers

Pneumatically operated push-in pressure cells (spade cells) were inserted at the base of the boreholes close to the tunnel at axis level and just above crown level. The cells were orientated to measure horizontal stresses in a direction perpendicular to the tunnel axis. The pressure cells were pushed half a metre into the ground at the bottom of the boreholes, and the boreholes were backfilled with a bentonite grout. These cells required approximately 6 to 8 weeks to equilibrate after insertion.

Porewater pressure at each spade cell was monitored by an incorporated piezometer. Porewater pressures at other levels were measured using pneumatic piezometers installed in sand cells between bentonite plugs within the same boreholes.

4.4 Calibration and backup systems

After completion of the installation four complete sets of readings were taken on each instrument. These served as a check on the correct functioning of all the equipment and provided datum values to which later readings could be referred. Where appropriate the instruments used were regularly checked for calibration errors using well established test ranges and boreholes at TRL. On site a systematic backup system operated. Datum readings were established for the whole site using every instrument. Each instrument was then designated as being the primary instrument for certain survey lines or boreholes and a secondary instrument for others. In case of a problem with a particular primary unit an appropriate secondary was substituted.

4.5 Data capture and processing

The Client required key data to be presented in graphical form on the day on which the measurements were taken. To achieve this and to cope with the large quantity of data gathered automatic logging systems were used for all levelling and inclinometer measurements. All data, whether collected manually or automatically were then downloaded, processed, stored and presented as required on site using a spreadsheet based software system developed by TRL.

5 Results

5.1 Surface movement

Figure 5 shows the variation in centreline settlements along the length of the tunnel immediately after the completion of key stages of the excavation. The figure shows that the greatest short term settlements occurred over tunnel type three (centred at line C) and the lowest over tunnel type two (centred at line B). Most of the large ground movements took place within two weeks of a heading passing under a given point on the surface. This is illustrated in Figure 6 which shows the settlement time histories for points over the centre of each tunnel type. In the plots for tunnel types one and two steps are apparent showing that after rapid settlement associated with the passage of the first sidewalls the movement almost stopped but was then reactivated by the later headings. Over type three much of the settlement occurred after excavation of the crown and bench but prior to excavation and closure of the tunnel invert.

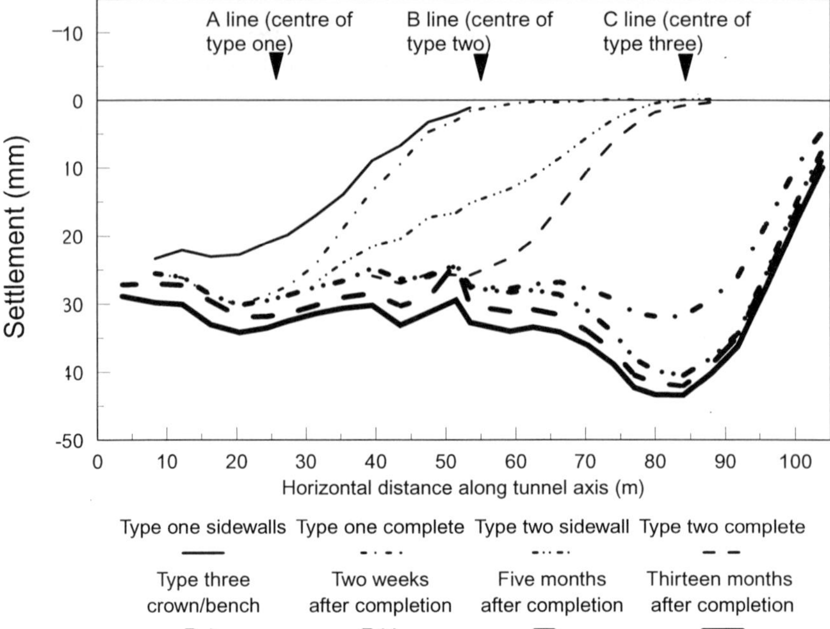

Figure 5: Levelling profile over tunnel centreline (T) at key stages

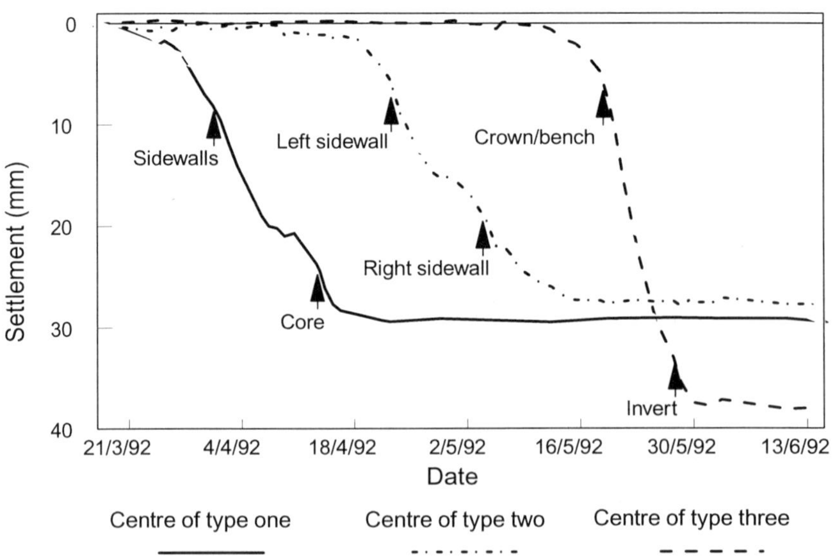

Figure 6: Surface settlement time histories

Figure 7 shows the three transverse profiles approximately two weeks after the completion of tunnelling. This comparison shows that the troughs generated by the type one and type two works have a similar form. However, the deeper trough resulting from type three excavation was relatively narrow and resulted in lower settlements at some points on the flanks. This will have given rise to significantly greater horizontal ground strains over the type three tunnel.

Longer term monitoring has shown both deepening and some widening of the settlement trough. This is illustrated in Figure 8 which shows the change in form of the settlement trough over the type three tunnel in the period between two weeks and thirteen months after the end of tunnelling. In the same period the centreline settlement above the tunnel increased by a mean of 15 per cent (4.3mm), although the maximum increase was 25 per cent (6.7mm). (See figure 5).

The surface instrumentation systems have continued to function as planned throughout the period of the works with the digital levelling system proving to be particularly well suited to this kind of application. Some problems were experienced with the operation of the geomensor in hot weather over large areas of pavement but these were largely overcome by the use of backup systems.

The movements due to shaft excavation were monitored along two radial lines (S and T). Line T was over the centreline of the tunnel and was therefore also affected by the initial adit which was commenced in the later stages of the shaft sinking process. Similar maximum settlements of between ten and fifteen millimetres were recorded on both lines at a distance of approximately 3m from the shaft prior to the tunnel excavation. These movements are discussed in detail below.

5.2 Subsurface movement

The maximum observed horizontal movements were towards the tunnel at a level slightly above the tunnel axis. In the London Clay the magnitude of horizontal ground movements reduced with increasing vertical and horizontal distance from the tunnel. Some larger movements were, however, apparent in the overlying gravels. The vertical movements over the tunnel crown showed an increase of approximately fifty per cent from surface level to three metres above the tunnel crown. The maximum vertical displacements in the boreholes passing to the sides of the tunnel closely approached the values recorded immediately over the crown. The combined horizontal and vertical subsurface displacements over tunnel types two and three are presented below (incorporated within Figure 10). As at the surface subsequent monitoring has indicated that some ground movements are continuing. Again the instrumentation systems have continued to operate successfully throughout the period of the works although particular care was necessary in the use and maintenance of the inclinometers to obtain the required precision.

5.3 Ground loss and settlement trough width

Settlement troughs over tunnels may be described by a Gaussian (normal probability) distribution of the form:-

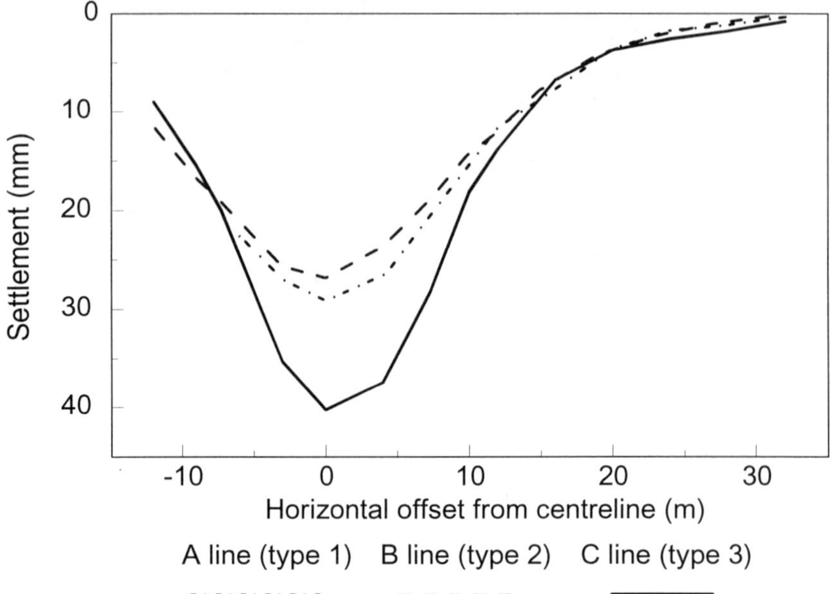

Figure 7 : Transverse levelling profiles at completion of tunnelling

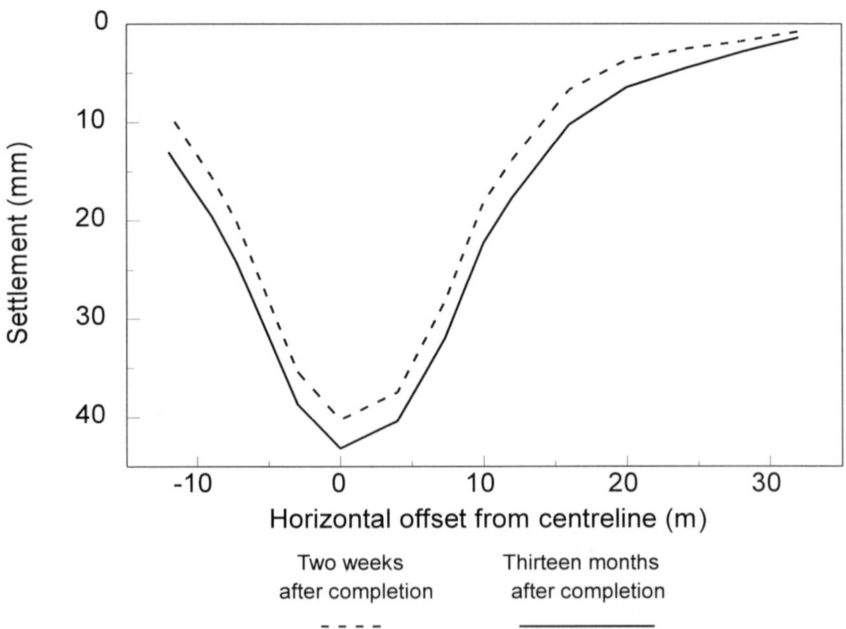

Figure 8: Variation in C line settlement profile over time

$$S=S_{max}\exp(-y^2/2i^2) \tag{1}$$

where S is the surface settlement at a transverse distance y from the tunnel centre line, S_{max} is the maximum settlement (at $y = 0$) and i is the standard deviation of the curve. The value of i provides a means of defining the width of the trough and corresponds to the value of y at the point of inflexion of the curve. Note that $i=Kz$ where K is a parameter dependent on ground conditions and z is the vertical distance above the axis of the tunnel (O'Reilly and New, 1982).

Curves were fitted to the surface settlement data points on the transverse sections for a number of key stages in the work. These related to the completion of each of the three excavation sequences and additionally to the stable states achieved after the excavation of the sidewalls in the first two tunnel types. A personal computer based non-linear regression analysis routine was used to obtain the best fit curves by varying the parameters of trough width and volume loss. For this analysis the calculations were based on the assumption of tunnels of circular cross-section equivalent to the respective headings in the trial. For types one and two excellent fits were obtained confirming that the assumption of the Gaussian form was appropriate as expected. The type three trough was rather narrower and slightly asymmetric. The curve fit was good over much of this trough but did tend to underestimate the small settlements in the wings of the trough. A numerical assessment of the volume loss for type three construction indicated a slightly larger volume loss and this difference has increased in the longer term. The values of volume loss and trough width parameter derived from the analysis, together with the actual maximum recorded settlements, are summarised in Table 1 while Table 2 gives the values based on measurements taken thirteen months after the tunnelling had been completed.

Once the equation for the curve has been obtained values for ground surface slope, horizontal displacement and strain at varying distances from the tunnel may be calculated. These data are essential for the assessment of the likely effects of the ground movements on overlying structures.

Table 1 Summary of settlement troughs two weeks after end of tunnelling

Tunnel section	Max. settlement (mm) S_{MAX}	Volume loss (%) V_S	Trough width parameter (K)
Type one, sidewalls only	20.6	1.17	0.43
Type one complete	27.9	1.15	0.45
Type two, sidewall only	14.5	1.24	0.48
Type two complete	26.8	1.05	0.45
Type three complete	40.3	1.26 (1.37*)	0.35

* by numerical analysis, see text

Table 2 Summary of settlement troughs thirteen months after end of tunnelling

Tunnel section	Max. settlement (mm) S_{MAX}	Volume loss (%) V_S	Trough width parameter (K)
Type one complete	33.1	1.38	0.48
Type two complete	31.9	1.41	0.50
Type three complete	43.2	1.33 (1.53*)	0.35

* by numerical analysis, see text

Excavation of one sidewall drift of the type two tunnel resulted in a settlement trough offset from the centreline of the trial. To determine the extent of this offset a new parameter (β) has been introduced into the equation.

$$S = S_{max} \exp(-(y-\beta)^2/2i^2) \quad \textbf{(2)}$$

This allows freedom to offset the fitted curve in the analysis, with the value of β being equal to the offset distance. The same form of the settlement equation was later applied to the analysis of the type three tunnel results where an offset of about one metre was observed between the centre of the excavation and the centre of the settlement trough. The reason for the latter offset is not clear although it is possible that it relates to asymmetric excavation within the headings in this area.

5.4 Ground stress and porewater pressure

Figures 9a-d show horizontal stress, pore water pressure and lateral ground displacements at locations immediately adjacent to the tunnel. The spade cells indicated a rapid (approximately one week) drop in horizontal stress perpendicular to the tunnel axis as each heading passed the instruments. This was followed by a small recovery toward the original stress over a period of a few months.

Lateral ground movements of up to about 25mm toward the tunnel were recorded by the inclinometers and these movements were always very well correlated with the measured changes in ground stress.

Figures 9e and f show horizontal stress and porewater pressures above the crown of the tunnel and indicate similar but lesser effects to those recorded to the sides of the tunnel. Stress and pressure readings both indicate very clearly the passage of the sidewall and completion of excavations for type two construction.

The piezometers generally showed a rise in pressure immediately prior to the passage of a heading past an instrument location. This was probably attributable to load coming onto the overlying ground ahead of the tunnel face. As the heading went past all but the most remote instruments indicated a rapid drop in the readings. In one case (piezometer ST2P) the reading fell to an apparent zero for a short period, indicating that the true value was probably slightly negative. After the initial fall the piezometers went into a period of recovery in which they generally showed greater increases in pressure than the horizontal stress increases recorded by the spade cells at

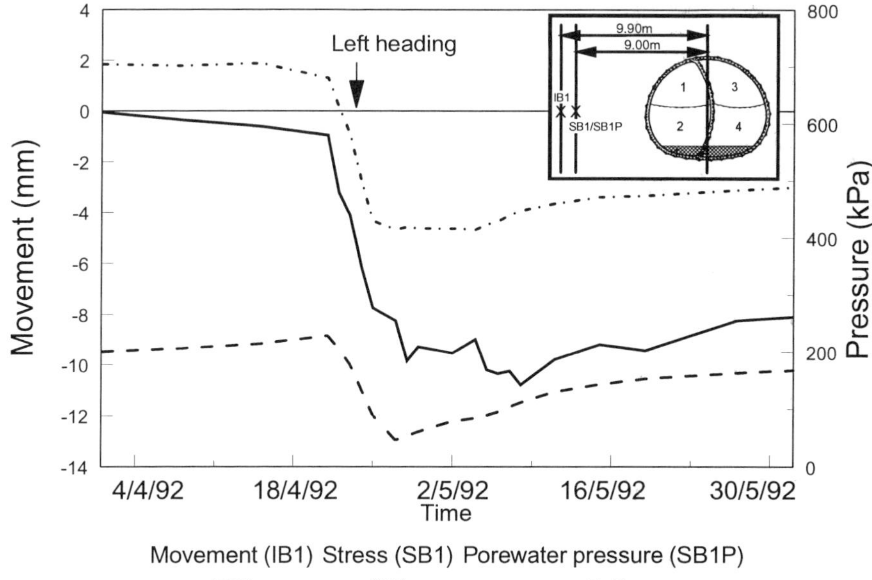

Figure 9a: Type two movement, stress and porewater pressure time history

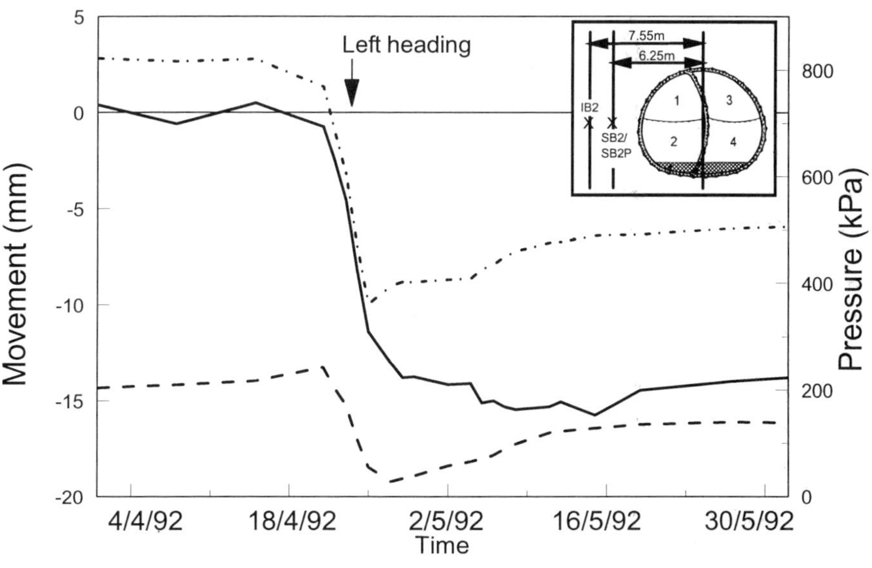

Figure 9b: Type two movement, stress and porewater pressure time history

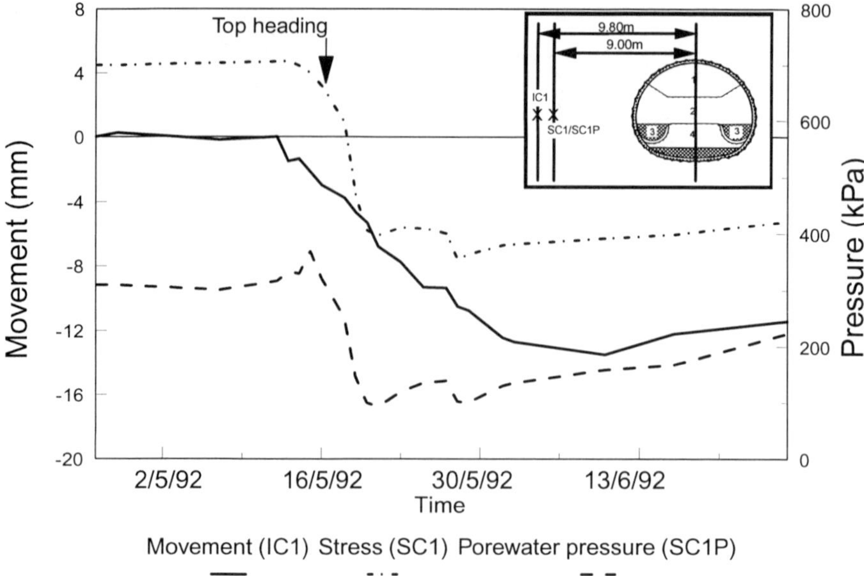

Figure 9c: Type three movement, stress and porewater pressure time history

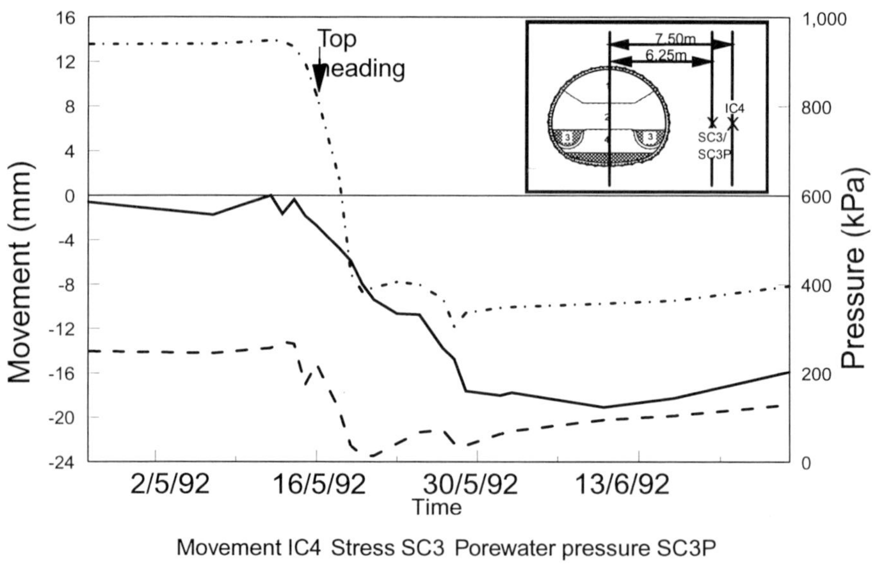

Figure 9d: Type three movement, stress and porewater pressure time history

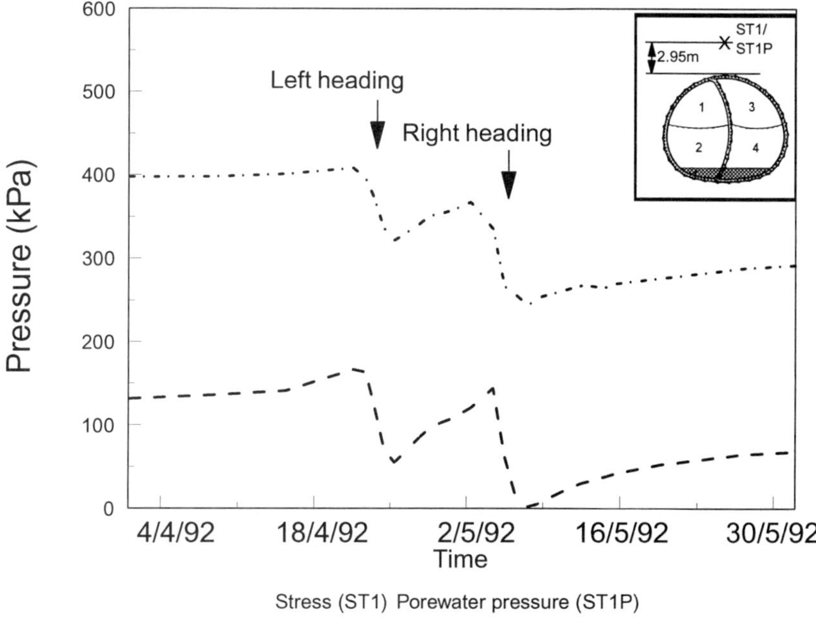

Figure 9e: Type two stress and porewater pressure above crown

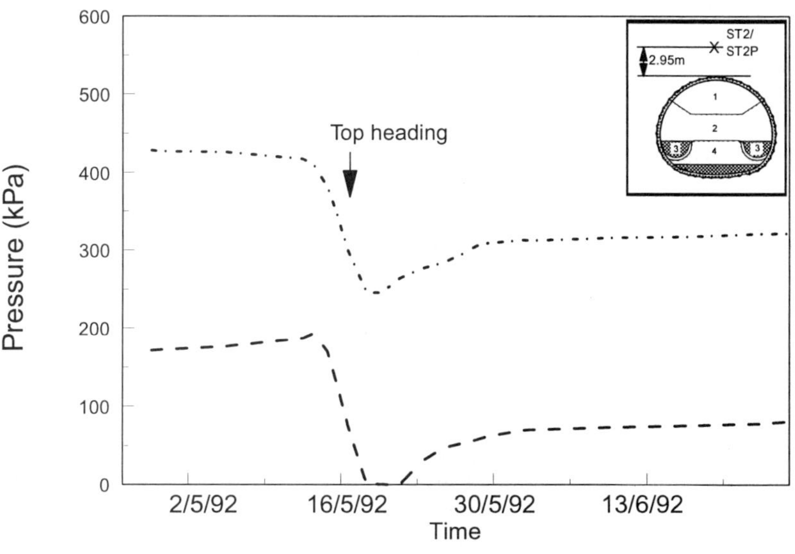

Figure 9f: Type three stress and porewater pressure above crown

the same time. Given the assumptions of an undrained condition and no change in the horizontal axial stress this would imply an increase in the vertical stress in the ground close to the tunnel possibly due to a redistribution of the overburden loads.

6 Ground movement modelling and validation

6.1 Existing empirical model

Models for the prediction of ground movements caused by tunnelling have made significant advances during the past 25 years and have been reviewed elsewhere (New and O'Reilly 1991, Attewell, Yeates and Selby, 1986 [7]). A wide range of field measurements during this period has confirmed the strength of the fundamental assumption that the transverse settlement curve profile takes a Gaussian form as defined in Equation 1. It follows that the longitudinal profile may take a cumulative probability form and this has also become accepted as a reasonable approximation to field experience (Attewell and Woodman,1982 [8]).

For stiff cohesive ground conditions it has been found acceptable to extend the analysis to subsurface movements by assuming constant volume deformation and that all particle displacements flow radially toward a point sink at the tunnel axis (O'Reilly & New 1982). They pointed out that this assumption breaks down in a region close to the tunnel where a point sink is clearly not likely to adequately describe the zone of ground loss.

TRL have implemented this approach for the prediction of ground movements in a suite of computer programmes (TUNSUB) which calculate and plot surface and subsurface ground displacements, strains and angular distortions in three dimensions.

Figures 10a and b show the TUNSUB modelled ground displacement vectors and the measured displacements for the type two and three excavations at Heathrow. Note that beyond a distance of about one tunnel diameter from the periphery of the tunnel the modelled movements agreed very well with those measured. However, as the tunnel is approached the actual displacements became significantly different from those indicated by the model. In particular the model over-predicts centreline settlement, under-predicts trough width, and fails to indicate vertical and horizontal movements at excavation level. These variations are an expected and inevitable consequence of the simple assumption of a point sink located at the tunnel axis.

6.2 Improved model for subsurface displacements

Actual construction works commonly comprise a mixture of intersecting excavations which may change dimension and level and it is at these locations (eg stations) where ground movements can be at their most significant. It may also be necessary to calculate ground movements at various stages during the construction process because, depending on the disposition of the overlying structures, it can be at this time when the most damaging ground movements occur. Figure 11 shows an example TUNSUB output of settlement contours for an example NATM construction sequence from the trial tunnel. The situation modelled is illustrative of two sidewall headings at different stages of advance with the centre excavations just beginning from the transition tunnel.

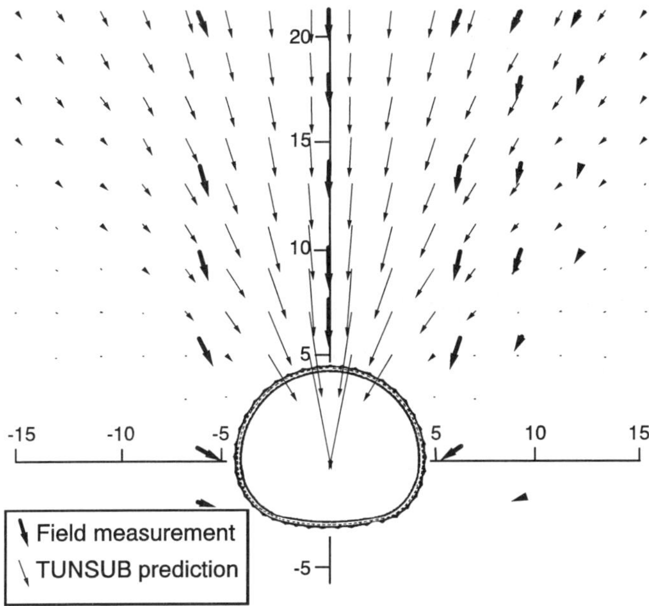

Figure 10a: TUNSUB output for the point sink model of tunnel type two

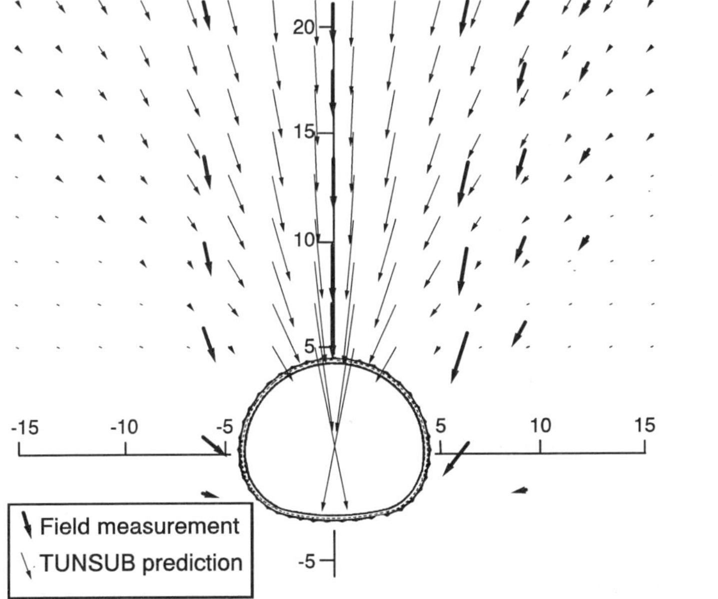

Figure 10b: TUNSUB output for the point sink model of tunnel type three

TRL GROUND ENGINEERING - PROGRAM TUNCON7R
Data File c:\hrow\hrc006.tc4

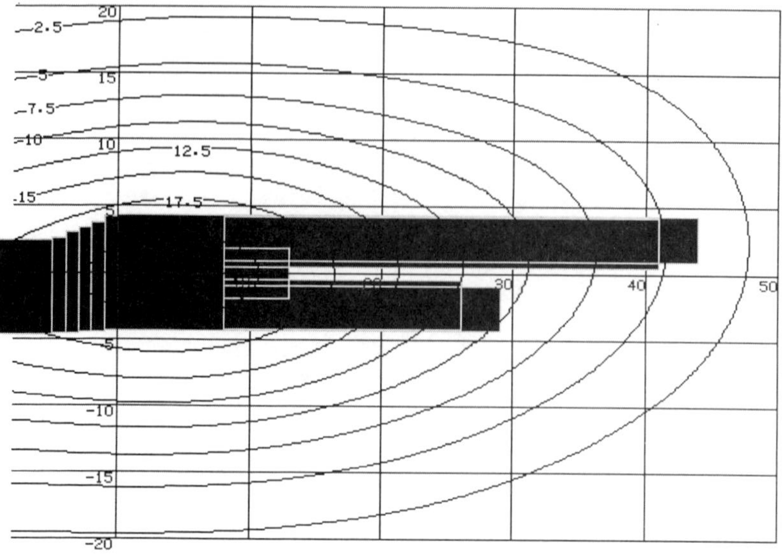

Figure 11: TUNSUB settlement contours for partially constructed type one

The need to improve the prediction of ground movements at depth and in the vicinity of excavations has been prompted by requirements for major projects currently in design. It was therefore decided to develop the model by representing the distribution of ground loss more realistically and modify TUNSUB accordingly. The work was timely because the quality of the field data from Heathrow provided an immediate means to test new assumptions of ground loss geometry.

Three different geometrical forms of ground loss were investigated and appropriate algorithms created as TUNSUB subroutines. In the three cases the ground losses were assumed to be distributed uniformly:-

i) over the cross-section of the tunnel (TUNFILL)

ii) around the periphery of the tunnel (TUNCIRC)

and iii) over a horizontal plane, equal in width to the tunnel, at invert level (TUNRIBBON)

Another option combining i) and ii) in various proportions was also evaluated.

Assumptions i) and ii) required TUNSUB to calculate the movements using a numerical summation of many small sinks distributed over the assumed tunnel area or periphery. This was because the integration of the equations describing the distributions proved rather impracticable.

Assumption iii) allowed an analytical solution as the integration of the loss along the "ribbon" sink proved tractable.

Similar plots to those shown in Figures 10a and b (for the point sink) were produced for all the options and each provided significantly improved correlation between modelled and measured movements in the vicinity of the tunnel. In general it was found that assumption iii) provided the best correlation and this option was also preferred because implementation was more straightforward due to the analytical rather than numerical nature of the solution.

Figures 12a and b show the modelled (option iii TUNRIBBON) and measured ground movements for both sections B and C.

Note a) the actual ground movements are very similar to those given by the model at all levels above the tunnel soffit.

b) whilst significant improvement has occurred to movements to the sides (below soffit) the actual particle trajectories tend to contain a slightly larger horizontal component than modelled. This could be explained, at least in part, by the presence of high horizontal stresses ($K_0 \simeq 2$) in the London Clay which are relieved by the excavation.

The derivation of the ground movement equations are given as an Appendix. Equations A6, A7, and A8 give the three dimensional displacements caused by a tunnel excavation of limited extent whilst Equations A12, A13 and A14 give displacements in the plane normal to the direction of an excavation which may be approximated as being of infinite length.

Calculations of ground movements using these equations may now be readily carried out on most microcomputers with industry standard spreadsheet software. Many of the calculations required during tunnel design can be achieved using this approach. However, openings on a gradient, curved openings and other excavations with complex geometry will, in practice, require the use of dedicated software such as TUNSUB.

6.3 Ground movements due to shaft sinking

Few, if any, published accounts of ground movements in London Clay exist for shaft excavations. Whilst this cause of ground movement is generally of less concern than tunnelling it is of interest because of the extensive requirements for shafts close to structures during construction of current railway projects. In particular arrangements such as the numerous access and ventilation shafts in the Heathrow terminal areas may give rise to ground movements the influence of which needed to be considered during design. Accordingly, additional surface measuring stations (line S on Figure 3) were placed in line away from the shaft in an area likely to remain unaffected by subsequent tunnelling. Also those stations in line T close to the shaft were regularly monitored during shaft sinking works.

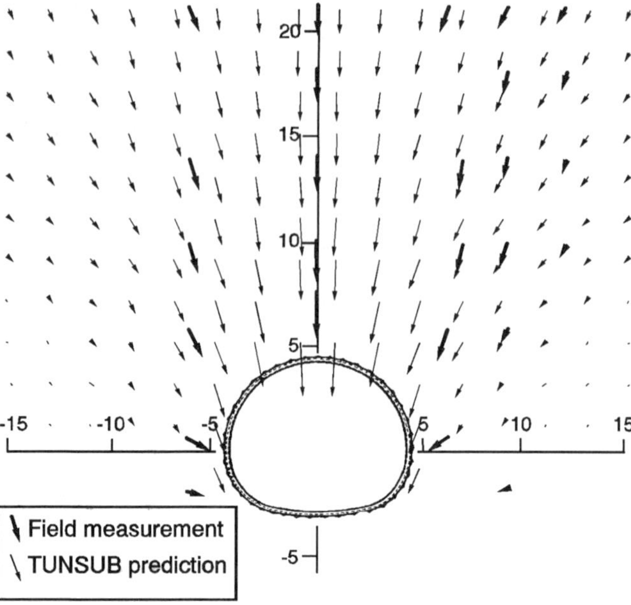

Figure 12a: TUNSUB output for TUNRIBBON model of tunnel type two

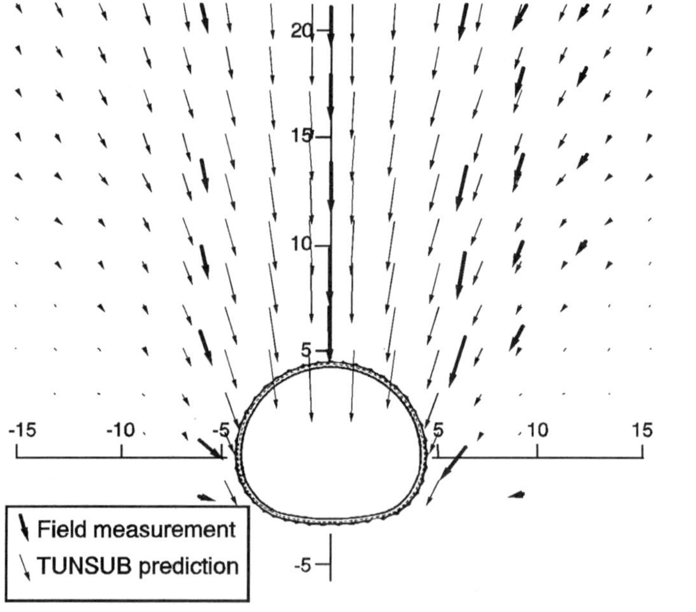

Figure 12b: TUNSUB output for TUNRIBBON model of tunnel type three

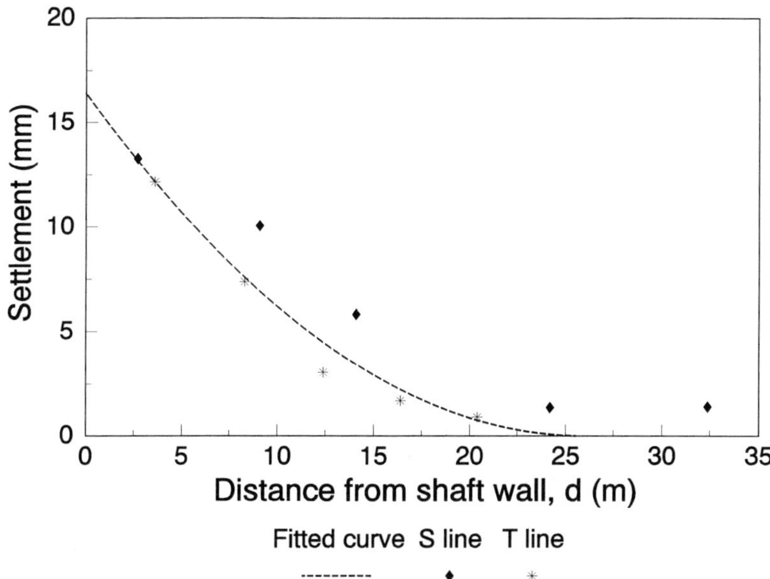

Figure 13: Settlement induced by shaft sinking

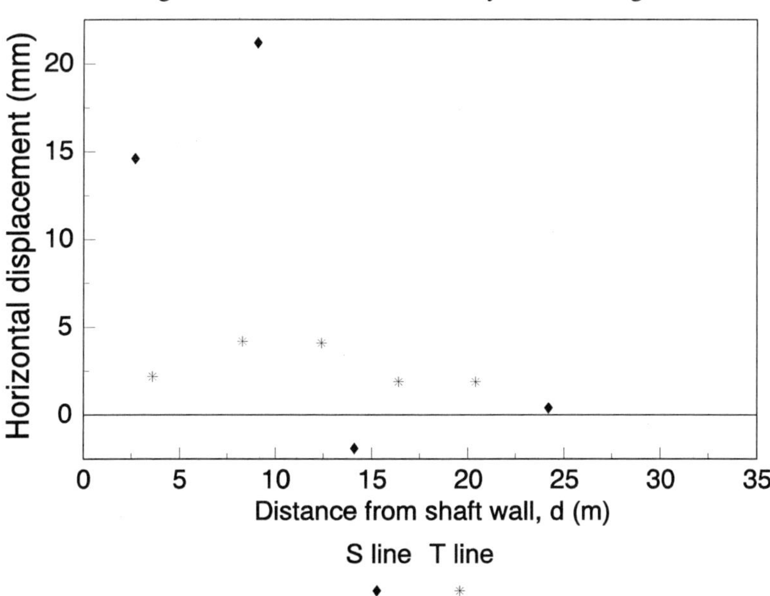

Figure 14: Horizontal displacement induced by shaft sinking

Figures 13 and 14 show the surface settlement and horizontal ground movement toward the shaft when the shaft was approaching full depth (26m) and the tunnelling works were just commencing. The measuring stations on line T remained relatively undisturbed by general site activities and the pattern of ground movements, as the shaft deepened, indicated that the results were likely to be reliable. However, the line S measuring stations were almost continuously subjected to the effect of heavy plant movements and those at 2.7m and 9.1m from the shaft were very close to some shallow excavations for temporary works. These works caused significant horizontal movements during the early stages of shaft sinking and as a consequence it is believed that less than half the observed horizontal movements (at 2.7m and 9.1m) were actually caused by the shaft excavations. Neglecting these locations, all other observed horizontal movements were less than 5mm.

The patterns of ground settlements on both lines T and S were reasonably consistent and it was considered appropriate to fit a simple curve to the data. The curve was chosen such that the settlement becomes zero at a distance from the shaft edge equal to the shaft depth (H) and increases as the square of the distance from the undisturbed ground (at d=H) toward the shaft wall.

The settlement at the shaft wall is taken as αH where α is an empirical constant dependent on ground conditions and construction method. This follows practice for estimating settlements close to retaining walls and similar structures.

The equation takes the form:-

$$S_d = \alpha(H-d)^2/H$$

where S_d is the settlement at a distance d from the shaft wall. Note that when $d = H$, $S_d = O$ and when $d = 0$, $S_d = \alpha H$. This equation provides a good fit with the Heathrow data and may provide some guidance when considering similar shafts in stiff fissured clays. However, its lack of dependence on shaft diameter indicates that it may not be appropriate for shafts of significantly different diameters.

The best fit value for α of 0.0006 (settlement at shaft wall = 0.06% of excavation depth) is consistent with field data for propped retaining structures in stiff fissured clays.

7 Conclusions

The data obtained at the Heathrow Express Trial Tunnel provided an excellent basis for the validation of existing ground movement models and has allowed their further refinement for application in the subsurface region close to excavations. BAA is to be commended its willingness to invest in a trial of this type.

The assumption of radial ground flow toward a point sink at the tunnel axis was found appropriate at the surface, and also at depth for the region outside about a tunnel diameter from the tunnel. However, in the vicinity of the tunnel the model underpredicted the trough width as expected.

More realistic revised assumptions for the zone of ground loss provided considerable improvement to the predicted distributions of ground loss near the tunnel. In particular the assumption of a ribbon shaped sink at tunnel invert level provided not only a greatly improved ground loss distribution but is also convenient in that the equations are analytically tractable and numerical methods are unnecessary. This means that it is relatively straightforward to implement the model on industry standard spreadsheet software and this approach may now be readily applied using the equations provided in the Appendix to this paper.

The ground losses measured for the three different NATM construction sequences were similar to those observed at other sites tunnelled in London Clay. That is, the volume of the surface settlement trough was between 1 per cent and 1.5 per cent of the excavated volume in each case. For the types 1 and 2 excavations the trough width parameter K was in the range 0.43 - 0.50 which again is completely consistent with other field data. However, for type three construction the maximum settlement was about 35 per cent greater and the trough width parameter significantly narrower (K = 0.35).

In the thirteen months following the completion of tunnelling the ground losses increased by 20 per cent, 34 per cent and between 6 and 12 per cent for types 1, 2 and 3 respectively. Thus at the end of this period the total volume losses over each trial section were very similar. Over the same period the surface trough width increased marginally above types 1 and 2 and remained constant for type three.

Near tunnel measurements of porewater pressure, lateral ground movement and stress were very well correlated and the observed movements and changes in pressure could be linked with the detailed log of the excavation sequences. These observations are of significance to the assessment of ground stability in future tunnel lining design.

The ground movements induced by the construction of the 11m diameter shaft were generally small (less than 14mm, except at two locations effected by heavy plant). The movements became negligible at a distance equal to the shaft depth (26m) from the shaft. A simple empirical curve has been fitted to the shaft settlement data.

The instrumentation systems performed well throughout with a reading frequency of once per day or less being found sufficient to describe the movements in the clay. The use of the digital levelling system was found to be particularly well suited to this type of work. Some difficulties were experienced with the use of the Geomensor over certain surfaces during hot weather, and with corrosion of aluminium inclinometer tubes during the stabilisation period before the works began. These problems were overcome by the use of backup procedures and replacement of the aluminium tubes with plastic. Considerable care in the method of use, calibration and maintenance of the inclinometers was required to achieve the required precision.

Automatic logging of levelling and inclinometer data greatly assisted the daily availability of processed ground movements in graphical form. The data handling software was specially developed for the project and proved to be essential for the rapid and effective management and presentation of the large quantities of data obtained.

8 Acknowledgements

The work described in this paper forms part of the programme of the Transport Research Laboratory and is published by permission of the Chief Executive.

The measurements taken at the Heathrow Express Trial Tunnel were collaboratively sponsored by BAA plc Heathrow Airport Limited and Bridges Engineering Division of the Department of Transport.

The authors are grateful to their colleagues in Ground Engineering Division who carried out the extensive field works and also to Dr R D Andrews for his major contribution to the computer modelling studies.

9 References

1. O'Reilly, M.P. and New, B.M. (1982) Settlements above tunnels in the United Kingdom - their magnitude and prediction. Proceedings Tunnelling '82 Symposium. Institution of Mining and Metallurgy, London, pp 173-181.

2. New, B.M. and O'Reilly, M.P. (1991) Tunnelling induced ground movements; predicting their magnitude and effects. Invited review paper to 4th International Conference on Ground Movements and Structures, at University of Wales, Cardiff (Pentech Press 1992).

3. Knothe, S. (1959) The displacements of the surface under the influence of mining extraction and their theoretical interpretation. Europ. Cong. Ground Movement and Colliery Engrg., 36, pp. 24-29.

4. Martos, F. (1958) Concerning an approximate equation of subsidence trough and its time factors. In International strata control congress, Leipzig (Berlin: Deutsche Akademie der Wissenschaften zu Berlin, Sektion für Bergbau, 1958), pp.191-205.

5. Schmidt, B. (1969) Settlements and ground movements associated with tunnelling in soil. PhD. thesis, University of Illinois.

6. Peck, R.B. (1969) Deep excavations and tunnelling in soft ground. Proceedings 7th International conference on soil mechanics and foundation engineering, Mexico, state of the art volume (Mexico: Sociedad Mexicana de Mecánica de Suelos, 1969), pp.225-90.

7. Attewell, P.B., Yeates, J. and Selby, A.R. (1986) Soil movements induced by tunnelling and their effects on pipelines and structures (Blackie, London).

8. Attewell, P.B. and Woodman, J.P. (1982) Predicting the dynamics of ground settlement and its derivatives caused by tunnelling in soil. Ground Engineering, 15, pp13-22.

APPENDIX

Calculation of ground displacement (Ribbon Sink assumptions)

It follows from the assumption that the transverse settlement profile (Y-direction) takes a Gaussian (normal probability) form that the longitudinal profile (X-direction) should take a cumulative probability form; this has been reasonably validated by examining field study reports (Attewell and Woodman, 1982). If i is the standard deviation of the transverse probability curve for a tunnel with centre line depth z, then the trough width parameter K, which is assumed constant for a particular ground material, is defined by the equation

$$i = K z \qquad \text{A1}$$

Here it is assumed that the ground loss occurs at a linearly translating point at the invert of the excavation at a constant rate of V_s per unit length of tunnel, and that all ground deformation takes place at constant volume. The vertical and horizontal displacements S, H_x, H_y at a point (x,y,z) are given by (New and O'Reilly, 1991)

$$S = \frac{V_s}{2Kz\sqrt{2\pi}} \exp\left(-\frac{y^2}{2K^2z^2}\right)\left[erf\left(\frac{x-X_f}{Kz\sqrt{2}}\right) - erf\left(\frac{x-X_s}{Kz\sqrt{2}}\right)\right] \qquad \text{A2}$$

$$H_x = \frac{V_s}{2\pi z}\left[\exp\left(-\frac{(x-X_s)^2+y^2}{2K^2z^2}\right) - \exp\left(-\frac{(x-X_f)^2+y^2}{2K^2z^2}\right)\right] \qquad \text{A3}$$

$$H_y = \frac{yS}{z} \qquad \text{A4}$$

where X_s and X_f are respectively the starting and final locations of the tunnel face, exp(x) is the exponential function and erf(x) is the error function, defined as

$$erf(z) = \frac{2}{\sqrt{\pi}} \int_0^z e^{-t^2} dt \qquad \text{A5}$$

This function is available as a standard function in many spreadsheets and computer programming languages.

Consider a ribbon-shaped zone of ground loss at elevation z=0 running between $x=X_s$ and $x=X_f$ and of width w, lying between y=-w/2 and y=w/2 (see Figure A1, and note that in this analysis z is the depth to tunnel invert rather than axis level). Consider an elementary narrow strip of width δt at y=t. The linear sink strength of the elementary strip is given by $V_s.\delta t/w$. Then the displacements (s,h_x,h_y) at the point (x,y,z) due to the excavation of the elementary strip are given by the following expressions :-

$$s = \frac{V_s}{2wKz\sqrt{2\pi}} \exp\left(-\frac{(y-t)^2}{2K^2z^2}\right)\left[erf\left(\frac{x-X_f}{Kz\sqrt{2}}\right) - erf\left(\frac{x-X_s}{Kz\sqrt{2}}\right)\right] . \delta t \qquad \text{A6}$$

$$h_x = \frac{V_s}{2w\pi z}\left[\exp\left(-\frac{(x-X_s)^2+(y-t)^2}{2K^2z^2}\right) - \exp\left(-\frac{(x-X_f)^2+(y-t)^2}{2K^2z^2}\right)\right] . \delta t \qquad \text{A7}$$

$$h_y = \frac{(y-t)s}{z} \qquad \text{A8}$$

which can be integrated with respect to t from t=-w/2 to t=w/2 to give the total displacements due to the excavation of the ribbon-shaped void (S,H_x,H_y), thus :-

$$S = -\frac{V_s}{4w}\left[erf\left(\frac{x-X_f}{Kz\sqrt{2}}\right) - erf\left(\frac{x-X_s}{Kz\sqrt{2}}\right)\right]\left[erf\left(\frac{2y-w}{2Kz\sqrt{2}}\right) - erf\left(\frac{2y+w}{2Kz\sqrt{2}}\right)\right] \qquad \text{A9}$$

$$H_x = \frac{V_sK}{2w\sqrt{2\pi}}\left[\exp\left(\frac{-(x-X_f)^2}{2K^2z^2}\right) - \exp\left(\frac{-(x-X_s)^2}{2K^2z^2}\right)\right]\left[erf\left(\frac{2y-w}{2Kz\sqrt{2}}\right) - erf\left(\frac{2y+w}{2Kz\sqrt{2}}\right)\right] \qquad \text{A10}$$

$$H_y = -\frac{V_sK}{2w\sqrt{2\pi}}\left[erf\left(\frac{x-X_f}{Kz\sqrt{2}}\right) - erf\left(\frac{x-X_s}{Kz\sqrt{2}}\right)\right]\left[\exp\left(\frac{-(w+2y)^2}{8K^2z^2}\right) - \exp\left(\frac{-(2y-w)^2}{8K^2z^2}\right)\right] \qquad \text{A11}$$

In the case of an infinitely long ribbon, $X_s \rightarrow -\infty$ and $X_f \rightarrow +\infty$, and the displacements take the following forms :-

$$S = \frac{V_s}{2w}\left[erf\left(\frac{2y-w}{2Kz\sqrt{2}}\right) - erf\left(\frac{2y+w}{2Kz\sqrt{2}}\right)\right] \qquad \text{A12}$$

$$H_x = 0 \qquad \text{A13}$$

$$H_y = -\frac{V_sK}{w\sqrt{2\pi}}\left[\exp\left(\frac{-(w+2y)^2}{8K^2z^2}\right) - \exp\left(\frac{-(2y-w)^2}{8K^2z^2}\right)\right] \qquad \text{A14}$$

Finally, in the limiting case as w→0

$$S = -\frac{V_s}{Kz\sqrt{2\pi}} \exp\left(-\frac{y^2}{2K^2z^2}\right) \qquad \text{A15}$$

$$H_x = 0 \qquad \text{A16}$$

$$H_y = H_z\frac{y}{z} \qquad \text{A17}$$

which is precisely the result for the classical infinitely long line sink.

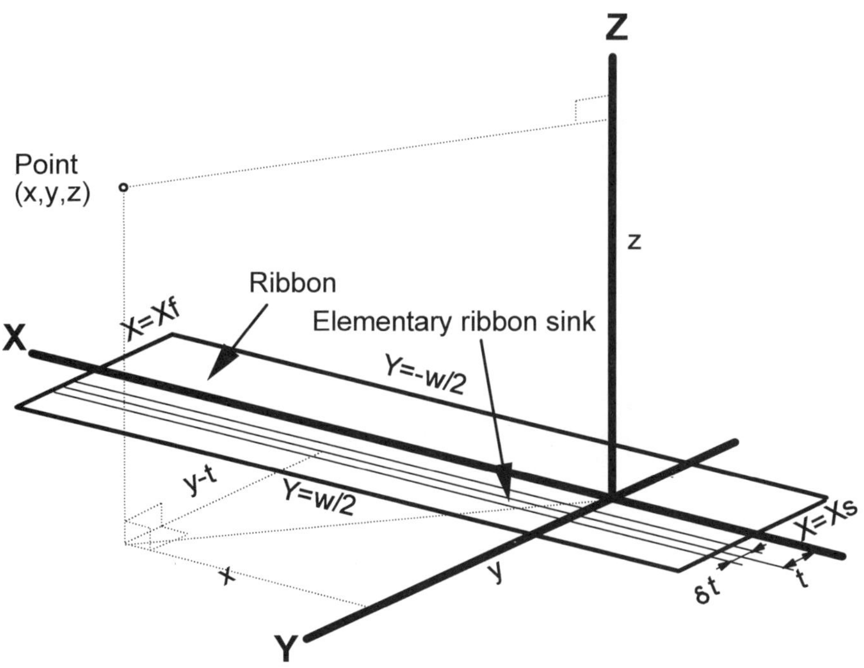

Figure A1: Scheme of notation for ribbon representation

Session 6: Case Histories

Gjøvik Olympic Mountain Hall, Norway

Vidar Aarvold
N S B Gardermobanen High Speed Railway Project, Norway
(formerly, Veidekke–Selmer Group, Norway)

1 Introduction

The construction of the World's largest in-mountain Colo-sseum took place in the city of Gjøvik, Norway from spring 1991 until spring 1993. The construction was based on a close co-operation between leading Norwegian consultants and contractors. By using a combination of Norwegian hard rock technology and highly skilled construction work it was possible to gain the World record span of 61 m.

The Norwegian experience with low cost tunnelling has also made it possible to build the arena to be used for ice-hockey at the XVII Olympic Games at Lillehammer in 1994, within a strict economic budget.

2 World Record

The Gjøvik Olympic Mountain Hall will be used as an ice-hockey rink at the winter Olympics at Lillehammer 1994. The dimensions of the hall are width 61 m, length 91 m, height 24 m.

The hall will be the World's widest rock-cavern with a capacity of 5400 spectators. The main span of the hall is 61 m. The largest span used in rock construction for the same type of purpose was previously 38 meters.

3 The Veidekke-Selmer Group (VS-Group)

The project was carried out as a turn-key project by the Veidekke-Selmer Group, a joint venture between A/S Veidekke and Selmer A/S, the two largest construction companies in Norway. Each company has a yearly turnover of approx. NOK 3 billion or 400 mill. US Dollars with 3000 employees.

The Veidekke-Selmer Group was responsible for the

Gjøvik Olympic Mountain Hall

design as well as the construction works involved in the project.

The basis for the co-operation between A/S Veidekke and Selmer A.S on this project, was better utilization of capacity and resources of the two companies. Another reason for co-operation was the spreading of financial risks involved in a turnkey project of this type. The VS-Group has completed several major engineering projects in Norway, such as the construction of various hydroelectric power plants and military installations.

The two companies in 1993 joined together with a third construction company, Statkraft Anlegg A/S and formed the Nocon Group. The Nocon Group will from this year on, undertake the heavy construction activities of the three firms outside Norway. The firms of the Nocon Group have several major civil engineering works under construction in different parts of the World such as Greenland, Iceland and several countries in Africa.

4 The construction of the Colosseum

The construction works in the field started in April 1991. Prior to that, a systematic pre-investigation and design study had been carried out by the firms Noteby, NGI, SINTEF and Fortifikasjon. These firms are regarded to be among Norways most experienced consultants within hard rock tunnelling.

The site is situated within the city of Gjøvik. This means that the heavy construction works had to be carried out under strict regulations controlling vibration, noise and traffic. Before the works started, the VS-Group in co-operation with the client, Gjøvik Olympiske Anlegg, promoted an information programme to the neighbourhood covering the environmental impact. This information programme was an important factor in the successful accomplishment of the project.

The blasting works started with the driving of two separate access tunnels of 20 and 45 m^2 cross-section towards the future Colosseum. One of the tunnels entered the Colosseum at the very top level.

From this tunnel the 10 m wide Centre Top Heading was driven through the full length of the future Colosseum. This heading was also used as an exploratory tunnel. Several deformation and stress measuring instruments were installed from this tunnel. These instruments together with six MPBX installed from the surface prior to construction, were used for logging the development of rock deformation and settlements during the construction works.

When the Centre Top Heading was completed, the Sloping Lateral Tunnels were blasted.
(See Plan and Cross-section of the Work Schedule Figures).

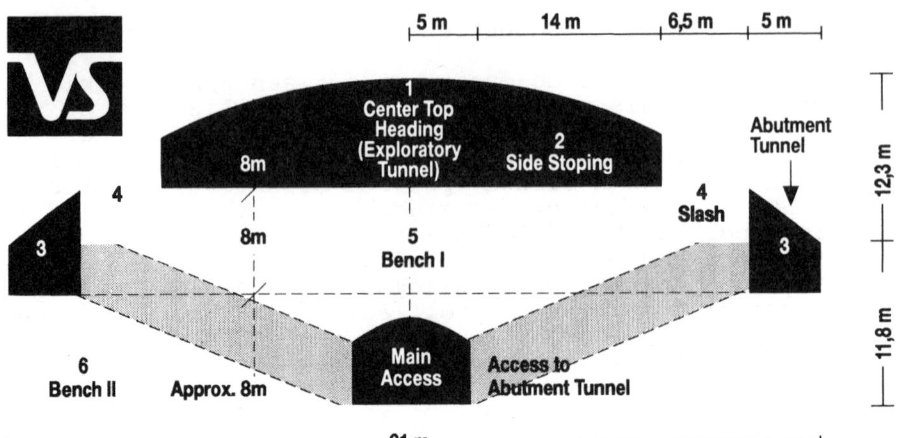

Gjøvik Olympic Mountain Hall
Work Schedule - Excavation

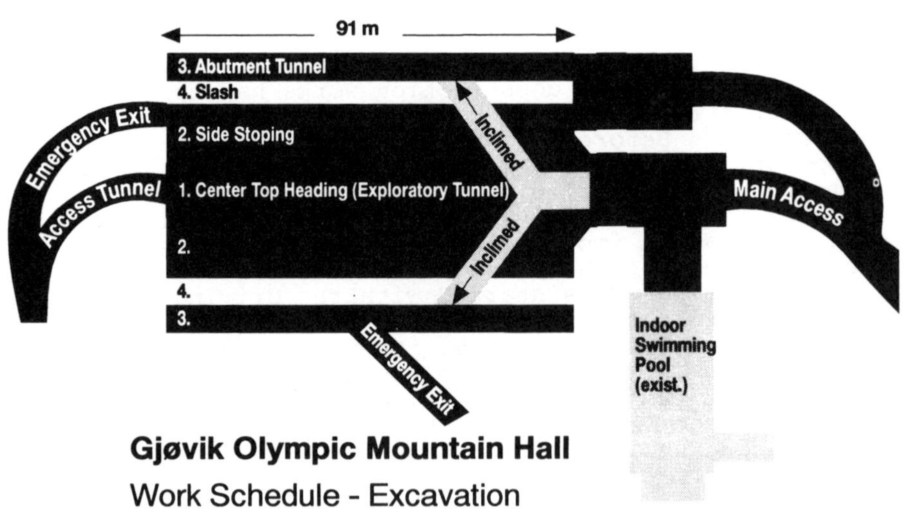

Gjøvik Olympic Mountain Hall
Work Schedule - Excavation

Those sections were each 14 m wide increasing the total span at this stage to 38 m. Parallel to the blasting of the Sloping Lateral Tunnels, the permanent rock support works were started. The support works consist of two layers of wet process steel fibre reinforced shotcrete in combination with 6 m long bolts of diam. 25 mm in a 2,5 x 2,5 m pattern and 12 m long twin strand cables in a 5 x 5 m pattern.

The bolts and cables were untensioned and fully grouted. The shotcrete had a thickness of 10 cm.

While the works at the top of the Colosseum took place, the Main Access Tunnel was driven separately towards the bottom level of the Colosseum. From this tunnel two access tunnels were driven at a maximum slope up to the Abutment Tunnels on each side of the Colosseum. The last part of the work as far as the establishment of the span was concerned, was to blast the remaining rock pillars between the Sloping Lateral Tunnels and the Abutment Tunnels. (See Plan and Cross-section of the Work Schedule Figures).

This was a very exciting part of the construction. The reading from the instruments were monitored daily and inspections of the roof and the walls of the top part of the cavern were carefully carried out. However, nothing indicated that the actual stability conditions and measured values diverted from the predicted values of deformation and settlement. The maximum settlement of the roof of the cavern was measured to be 8 mm. Theory had predicted this value to be between 5 and 9 mm.

The rest of the permanent rock support works for the roof were then completed before the final parts of the Colosseum were blasted. Those lower parts were constructed by ordinary bench blasting techniques with a maximum bench hight of 12 meters. Finally, the support works for the walls were completed. Those works also consisted of steel fibre reinforced shotcrete and systematic bolting, with 3 m and 10 m long untensioned, grouted bolts. (See figure showing the principles of Rock Support Works).

5 Duration of construction period

The construction works started in April 1991 and the cavern and its surrounding tunnels were finished 8 months later, ie. in December of the same year. During that time a total volume of 140.000 cubic metres of rock had been blasted and removed. A total amount of 1300 cubic metres of shotcrete had been placed and 3000 bolts and cables installed; approx 2700 of 6 m length and 300 of 10 and 12 m length. In addition, four vertical ventilation shafts of total length 120 metres for had been drilled.

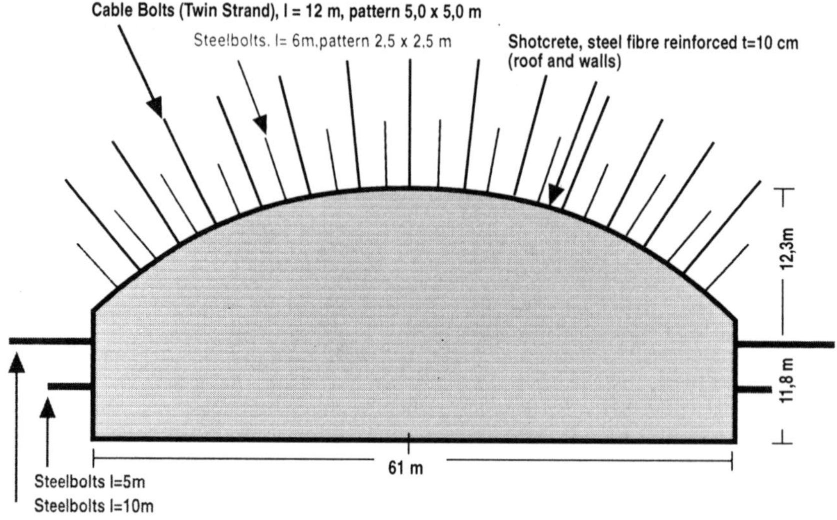

The works were carried out without any major accidents.

The first months of 1992 were used for the construction of foundations and other concrete works in the floor of the Colosseum. Parallel to those works the installation of an artificial ceiling was done. In April 1992 the erection of precast concrete elements started and in June the same year the installation of the technical equipment started. The complete project was finished on the first of April 1993, four months ahead of schedule.

6 The Research-Program

The construction of the Gjøvik Mountain Colosseum Project was supported by a major research programme. Results from the programme were used in the actual design of the cavern and the technical installations. In addition the results will form a general basis for the improved pre-investigations, classification and construction and use of large rock caverns.

In addition to investigation of the rock mechanic aspects of such projects, research was also done on fire- and escape safety, on different ventilation methods and on psychological aspects of underground use. The research work was undertaken by Norways leading research institution SINTEF/NTH in Trondheim and Norwegian Geotechnical Institute in Oslo.

The programme has caught great international interest. Some of the results are now being applied in South-Korea and Japan.

Construction of new escalator shaft and passageways at London Underground's Waterloo Station

I. L. J. Chudleigh
N. S. Webb
London Underground, Ltd., London, England

Abstract
As part of London Underground Limited's Congestion Relief Programme, Waterloo was identified as a priority station, particularly in view of British Rail's plans to construct the Channel Tunnel International Terminal at their main line station and London Underground's plans in respect of the new Jubilee Line connection. A feasibility study was commissioned, and extension of the existing London Underground ticket hall was found to be necessary, together with a dedicated link incorporating three escalators and three connecting passageways to London Underground's Northern Line platforms. A provision to connect with the proposed Jubilee Line Extension was also included.

The alignment of the new escalator tunnels was constrained by the presence of existing British Rail structures. The only suitable alignment resulted in the new tunnels passing in close proximity to the Victory Arch foundations and Waterloo & City Line tunnels and also part of Elizabeth House, a 15 storey building.

Due to the short lengths of drives construction was by hand methods throughout. The tunnels were lined with spheroidal graphite cast iron bolted segmental linings, ranging in size from 3.85m to 8.25m internal diameter. The approximate total length of new tunnels was 150 metres.

The tunnel construction is mainly in the Thames Gravel and London Clay. The Gravel required chemical treatment due to the high water table present. In addition, compensation grouting was used to prevent settlement of structures over the tunnels; comprehensive monitoring of the surface and subsurface structures was carried out by manual and electronic techniques.

In order to minimise settlement and to expedite construction, works continued 24 hours a day, 7 days per week for the whole of the construction period. Works on the permanent connection to the station and the three new overbridges could only be carried out in short night occupations of the track. A method of supporting the station tunnel externally when constructing the overbridges was devised. This avoided the use of a propping system within the existing tunnels, and thus disruption to the Northern Line service.

A working site was negotiated with British Rail but the areas allowed were severely restricted due to the extensive building works under way in the vicinity which included construction of the British Rail International Terminal. Considerable liaison between the various parties was required to ensure the works progressed satisfactorily.

1 Introduction

This paper describes the need to improve interchange facilities at London's Waterloo Underground station and the description of the tunnelling work involved. A detailed account of the ground treatment and compensation grouting is given in a companion paper to be presented at this Conference.

Waterloo has become one of the busiest stations on the Underground Network and with the anticipated increase in passengers due to British Rail's Waterloo International Terminal (WIT) and the proposed Jubilee Line Extension (JLE) it was realised that construction work was necessary to enlarge the station. The new work involved construction of lined tunnels to form a new escalator shaft, lower concourse, cross-passages and connection into the existing station tunnels. The WIT development was commissioned by British Rail (BR) to accommodate passengers from the Channel Tunnel link whose train services are due to commence in July 1994. London Underground Limited (LUL) identified that an increase in passengers using Waterloo Underground station as a consequence of the WIT development would rise by 20%. The Underground station improvements and benefits from the civil engineering works include an additional three new escalators increasing the total number at the station to 10 and a new extension to the Waterloo ticket hall. A provision has also been made to form a connection to the proposed Jubilee Line extension from the new lower concourse.

Details are given of the structural openings formed between the lower concourse and the cross passages which connect to the Northern Line platforms. Construction of overbridges to the Northern Line was carried avoiding disruption to the Underground train service. Stringent safety precautions were taken and a comprehensive monitoring procedure installed to give early warning in the event of construction problems occurring.

2 Background of the Project

London Underground's Waterloo station at present serves the Bakerloo and Northern Lines and provides passenger interchange with both BR Main Line station and the Waterloo and City Line station. It will also serve the new JLE when construction is completed.

There has been a main line station on the Waterloo site since 1848. The Bakerloo Line and the Waterloo and City Line (W&CL) were constructed around the turn of the century. The Bakerloo Line station opened in 1906 and provided a lift service as a means of access. In 1919 a bank of 3 escalators was installed providing direct interchange with the Main Line station. The Northern Line station was added as part of the Kennington Loop works in the mid 1920's. The Underground station was further improved with the provision of additional escalators in 1928 and again in 1967, the latter replacing the original lift connection to the York Road entrance.

The decision to construct the WIT on the site of the former Windsor Line Platforms which were located at the Western side of the existing Main Line station dictated the position of a new interchange facility. Advantage was taken by LUL to improve their station's efficiency which was a requirement of the Underground's congestion relief programme.

3 Description of the Project

The existing LUL Ticket Hall is partly under the BR Main Line station and partly under the new terminal building. As part of their works to construct the International Terminal, BR and their construction manager Bovis, demolished the existing brick arches beneath the old Windsor Line platforms and constructed new slabs for the terminal.

To accommodate the new extension to the LUL ticket hall, BR provided an area under the International departures slab and when Bovis had completed their works, LUL's main contractor was left with a structural shell within which the new ticket hall was to be constructed. Phase 1 of this work was completed and opened to the public on 2 July 1993. Phase 2 of the works which involves further refurbishment of the existing station areas to bring them up to the same standard as the new ticket hall extension will be completed by early 1994.

Access to the new escalators will be from the north west corner of the new ticket hall. Construction of the tunnels started from a diaphragm box structure constructed by Bovis, which will house the upper machine chamber. The box was constructed prior to the tunnelling works, with three walls formed in reinforced concrete, the fourth wall i.e. the north wall, being formed by means of an unreinforced bentonite and cement mix to provide a weaker material for tunnelling through (Fig.1).

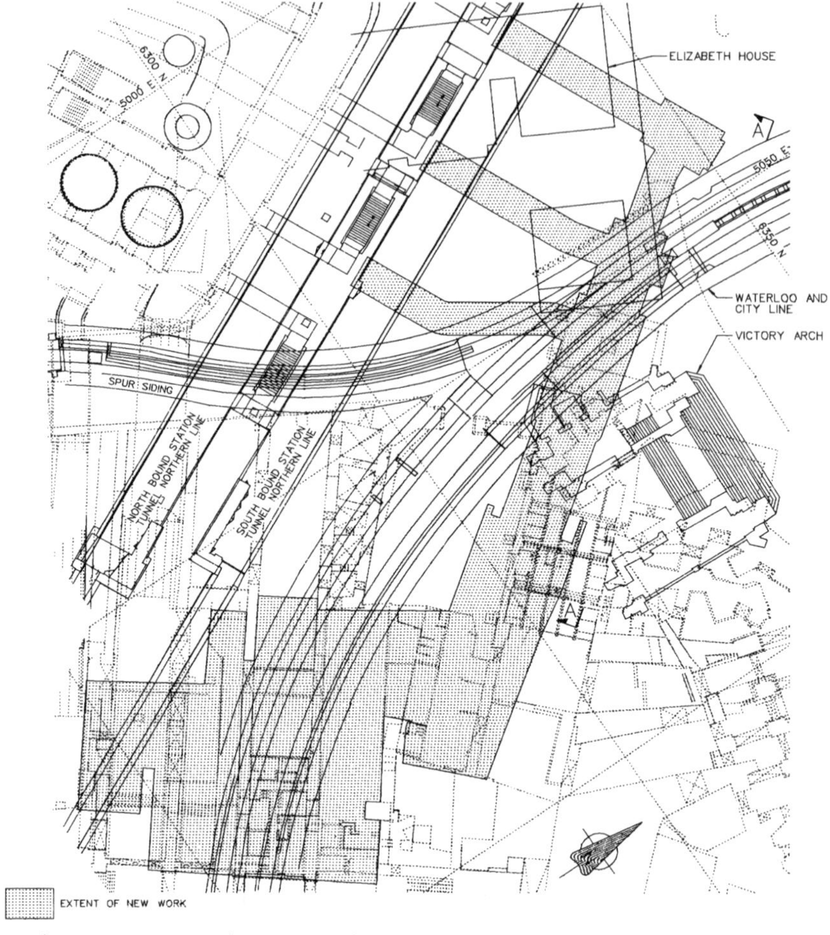

Fig.1 Plan view showing extent of tunnelling works and extended ticket hall

The inclined 7.5m internal diameter (i.d.) escalator tunnel was driven downhill from the box structure by first constructing a 3.85m i.d. tunnel. Enlargement commenced following the erection of two part rings, supported in a cradle, and two full rings which were then cast into the permanent headwall. At approximately 17m below ground level the tunnel levels out by means of 8.25m i.d. tapered rings and continues for a further twenty one 8.25m i.d. rings, which forms part of the lower concourse and houses the lower machine chamber. The tunnel then steps down to 6.5m i.d. and again to 3.85m i.d.. The remainder of the lower concourse will be within the smaller diameter tunnels.

Three 3.85m i.d. cross passages connect the lower concourse tunnels to the Northern Line and pass above the southbound station tunnel by way of new overbridges.

The northern most cross passage connects into the 3.85m i.d. lower concourse tunnel via a short section of 5.75m i.d. tunnel forming a "T" at the end of the concourse. Directly opposite this, and opposite the middle of the three cross passages are stub tunnels (2 rings of 3.85m i.d. tunnel) to provide future access to the proposed JLE. (See Fig.2).

Each structural opening between the concourse and the cross passages was formed by excavating a sizeable chamber in which to erect a structural steelwork frame which was then cast in concrete.

With the exception of a number of replacement segments to the Northern Line at each overbridge, all permanent linings are manufactured from spheroidal graphite iron (SGI), similar to those used during the reconstruction of Angel station. The grey cast iron linings employed for the pilot tunnel were supplied by LUL from old stock. After recovery they were cleaned and returned.

The ground through which the tunnels were driven consist of made ground, approximately 2.5m thick, overlaying 2m of alluvial clay and approximately 4m of Thames Gravel. Beneath the Gravel is the upper part of the London Clay which extends down to a depth of approximately 32m and over lies the Woolwich and Reading Bed Clays. (See Fig.2). Ground level is at approximately +3.0m OD and the surface water table is approximately 0.0.m OD. The high water table was of particular concern and influenced the choice of diaphragm walls for the escalators upper machine chamber with the cement bentonite wall. It was also realised that ground treatment of the Thames Gravel would be required prior to tunnelling to keep the tunnel face dry. Therefore, prior to LUL's main contractor starting work an advance contract was let to Keller Colcrete to chemically grout the Thames Gravel, as the incline escalator shaft passes through this strata. To enable the tunnelling to proceed, Keller Colcrete installed a grid of 200 TAM's (tube-á-manchettes) to a depth of 10m into the ground by rotary drilling. TAM's are grouting pipes with sleeves, so designed that the grout can be placed into the ground at a predetermined depth. The silicate grout gave the Gravel a cohesive strength and reduced their permeability, thus eliminating the risk of any substantial water ingress into the tunnel during construction, and reducing potential for settlement.

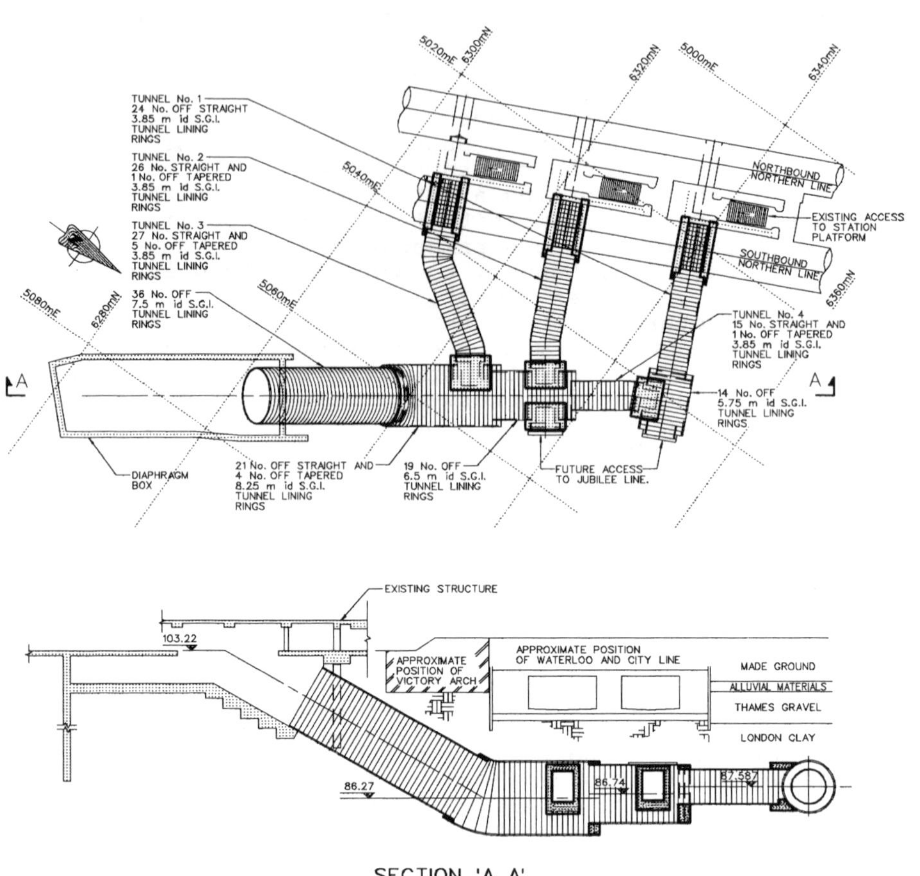

Fig.2 Layout plan of diaphragm box and tunnelling works

In the course of TAM installation an obstruction was encountered at approximately 7.5m below ground level. Keller Colcrete attempted to continue drilling through the obstruction but without success and were stopped by LUL. After extensive checking of record drawings and consultation with other parties, all possibilities excepting a wartime bomb had been discounted and British Transport Police were informed.

The Metropolitan Police bomb squad and British Transport Police visited the site and confirmed that further evidence would be required before the Army could be called in. Arrangements were made for radar scans at ground level above the obstruction to be carried out and the findings were that the object appeared to be 0.45m in diameter, 1.2m in length, unsymmetrical, cylindrical in shape with the possibility of something resembling fins at the back. There also appeared to be disturbed ground behind the object.

On this information the army were called to site and following a limited amount of grouting around the area in question to increase the ground stability and minimise water ingress, a vertical shaft was excavated towards the object by the Army, with the materials plant and supervision provided by LUL. When the shaft was within 2m of the obstruction the Army took control completely.

Excavation continued 24 hours a day and at approximately 7.6m depth a layer of flint was found. Following further excavation to approximately 7.7m depth, a metal object was uncovered. Once exposed this was found to be a metal spike 775mm X 35mm diameter. The spike was removed and it was confirmed by use of a metal detector that no further metal objects were in this area.

4 Design Considerations

An examination of the history of the site in the context of the proposed tunnel and its local surroundings revealed three major problems of concern. These are namely the W&CL tunnels, the Victory Arch and Elizabeth House. In addition old foundations from buildings previously occupying the site were identified. All these restrictions were considered since they would initially influence the design and construction philosophy adopted, and secondly be contemplated regarding the long term effects on existing structures caused by the tunnelling process.

The site was formally occupied in part by offices of the then London & South Western Railway (L&SWR) and the General Post Office. The walls of the buildings continued below basement level and came to rest on concrete foundations bearing on the Gravel below. These buildings were modified and their foundations underpinned during the construction of the W&CL.

The section of the W&CL located immediately above the new tunnels is of cut and cover tunnel construction. The brick walls were built in longitudinal trenches, then the roof erected with iron girders and brick infill. The next stage was to remove the dumpling between the outer walls and the central wall and complete the twin tunnel structure by forming a brick invert. The water table is at present approximately three quarters the way up the outer walls. Although the brickwork is sound, leaks and dampness could be observed in the outer walls of the tunnels. Concern was expressed as to the suitability of the brick structure to be able to withstand the settlement caused by tunnelling so close beneath.

The Victory Arch was built between the World Wars on part of the old offices site. Substantial modifications took place to the old sub-structures during the construction of the arch. Behind the Victory Arch ornamental facade, entrance halls, staircases and offices can be found. The internal structure of the arch itself being quite hollow. Consideration to the sensitivity of Victory Arch with regard to settlement was made owing to the close proximity of the tunnelling operation. Compensation grouting and careful monitoring of the structure during tunnelling provided adequate protective measures.

In 1941 the former L&SWR offices suffered damage from an air attack, the offices and the W&CL tunnels below suffered considerable destruction which fortunately were repaired under emergency works. Finally the foundations of the old L&SWR offices were removed to facilitate the construction of the new office block complex, Elizabeth House, in 1966.

A further difficulty to overcome was the effect that Elizabeth House would have on the new tunnels or vice versa. The North side of Elizabeth House being a 15 storey tower block, with part of its foundation specially designed as a cantilever to relieve the W&CL tunnel of the buildings load. The 3.85m i.d. and 6.5m i.d. passageways pass below this section of Elizabeth House, with the overbridges situated below the 7 storey portion of the building fronting on to York Road.

The loading cases considered, other than the normal ground loading conditions were as follows:

Loading from the Victory Arch.
Loading from W&CL tunnels.
Loading from Elizabeth House.

It can be seen from the previous description of the site that the conditions are far removed from a green field situation. The stress levels within the ground were high due to the loads from the surrounding structures. It was envisaged that the tunnelling would be carried out by employing hand methods because of the short drives and limitations in constructing the inclined tunnel to house the escalators.

SGI castings were considered for the tunnel segment, since they had proved a great success on the reconstruction of London Underground's Angel station. Their low weight/strength ratio was an advantage in hand working and LUL owned the patterns for manufacturing segments.

A geotechnical study was carried out by LUL's consultant the Geotechnical Consulting Group (GCG) to establish the nature of the loading to which the tunnels would be subjected. GCG also advised LUL on the ground treatment requirements of the project and provided the ground treatment design package.

It was established by LUL during the design stage that the SGI linings would cater for the unusual loading conditions prevailing in the ground. The structural steel opening sets provided sufficient support and minimised distortion to the segments after the opening was created in the new tunnels. The opening sets each spanned six 600mm wide rings and balanced the hoop forces generated by the soil and lining interactions by means of a steel frame support structure, consisting of a lintel and cill beam, with jamb plates acting as the compression members.

The proposed overbridges, however, did present a problem. A restraint imposed by LUL was that the Northern Line station must remain open for traffic during normal operating hours. The Northern Line southbound station tunnel is 6.5m i.d. and constructed from grey cast iron segments grade 6 ton/in^2. The LUL restraint on occupying the tunnel meant that no propping of the existing iron could be contemplated within the station tunnel itself. Considering that at least 4 segments in the crown of the tunnel over a width of 7 rings had to be removed to facilitate the construction of each overbridge, a support system had to be devised.

A practical solution to the problem was achieved by introducing a combination of strongback beams and concrete arches. The strongback beams were directly connected to the back of the segments, positioned just below the overbridge floor level, thus enabling the upper segments to be cut and removed to form the openings to receive the bridge. The concrete arches were formed over the top of the station tunnel on both sides of the footbridge openings. The arches provided the necessary headroom and roof support structure for the footbridge construction. The composite action of the arches and strongbacks provided resistance to out of balance forces induced by segment removal that facilitated the openings for the footbridges construction. (See Fig.3).

An assessment of ground and structural loading revealed that the existing tunnel lining was subjected to high hoop forces. The strongback beam, fabricated as a structural steel channel section 750mm in depth, balanced the hoop loads and prevented deformation of the cut rings during removal of the top plates. The end reactions and moments from the strongback beams were directly transmitted to and shared between the first two rings adjacent to the openings.

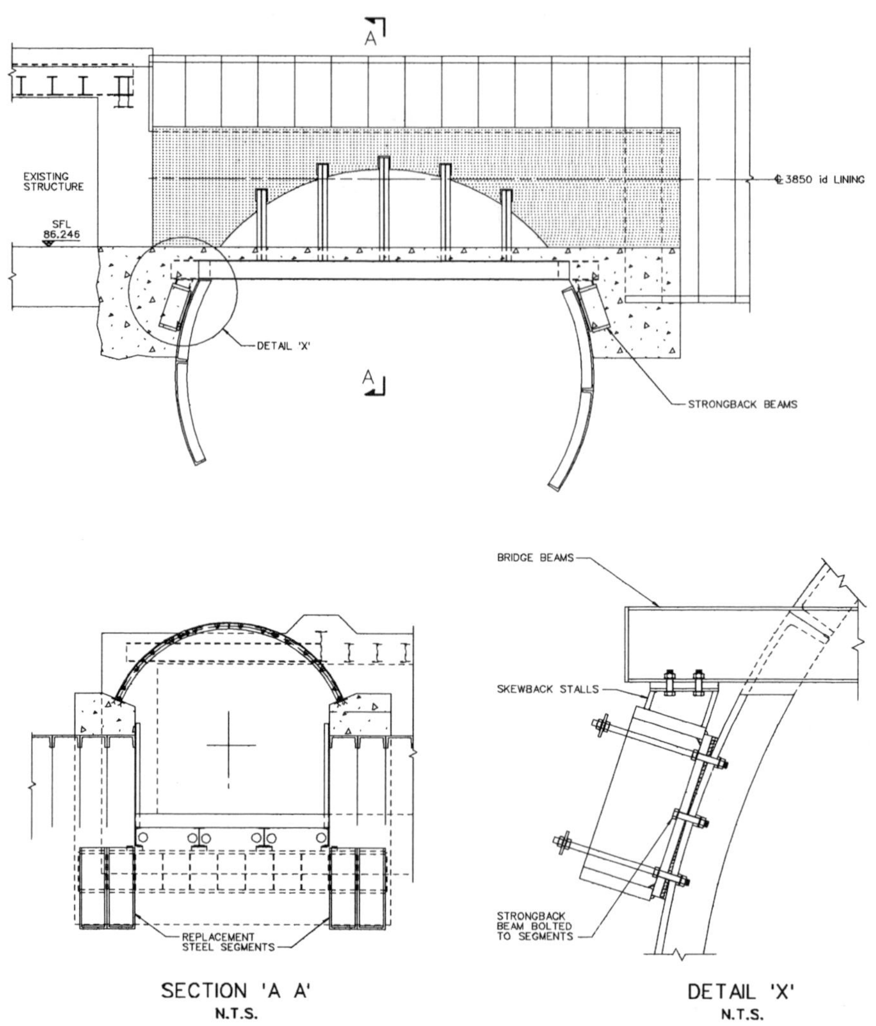

Fig.3 Details of Overbridges

A design check on the existing grey cast iron station lining established that the segments were not strong enough to withstand the combination of moments and end reactions imposed by the strongback connections.

Consequently two cast iron segments were replaced with special segments, fabricated from structural steel, designed to cater for the end loading from the strongback beams. To aid construction the strongbacks were fabricated in three sections and spliced together on site. This enabled the sections to be manoeuvred into position with less difficulty.

It was of prime importance in order to avoid tunnel distortion and minimise ground settlement that the strongbacks were positioned and encased in concrete prior to any dismantling of the top segments creating the openings. This requirement was imposed on the contractor by LUL. It is noteworthy that in the past at LUL's Piccadilly Station a problem occurred when overbridges were constructed without the aid of strongback beams and cracks appeared in the existing linings necessitating their replacement.

The footbridges consist of steel beams supporting a concrete floor. The bridge roof is composed of 3.85m i.d. SGI segments formed into an arch. The roof in turn was supported on concrete arch sidewalls previously mentioned. The strongback beams were also utilised as supports to the footbridge structural members, with the loading from the bridge beams transmitted through steel skewback stalls connected to the strongbacks upper flanges.

The final formidable task was the logistical requirements to construct the new tunnels and ticket hall extension from a small working site with a difficult access within the busy WIT development working site.

5 Organisation of Construction

Sir Robert McAlpine Limited were awarded the main contract for both the ticket hall and tunnelling works. The overall value of the contract was approximately £12.5M, with the tunnelling works accounting for £4.5M, including approximately £0.7M for the geotechnical works. The advance contract for ground treatment was approximately £0.3M.

LUL were responsible for the design and supervision of construction. Work was carried out in a 2 year contract which commenced on the 26th May 1992. The tunnelling works commenced on the 2nd of November 1992 and were completed on the 25th of June 1993. This was followed by a 3 month continued monitoring period of the neighbouring structures and the new tunnels themselves. Completion of the remaining works including refurbishment of the existing ticket hall (phase 2), installation of three new escalators, services and all Architectural works is programmed for July 1994.

The sequence of tunnelling works was as follows:

a) Construction of 3.85m i.d. pilot tunnel in old grey iron linings at inclination of 30 degrees, plus full length of lower concourse. Tunnel No. 4 constructed in new 3.85m i.d. SGI linings.
b) Construction of 3.85m i.d. tunnel No. 1, including the section forming "T" junction with tunnel No. 4.
c) Enlarge incline tunnel to 7.5m i.d.. (See Fig.4).
d) Enlarge lower concourse tunnel to 8.25m i.d. and 6.5m i.d. respectively.
e) Construction of 3.85m i.d. tunnel No. 2.
f) Construction of 3.85m i.d. tunnel No. 3.
g) Enlargement to 5.75m i.d. tunnel at "T" junction between tunnel No. 4 and tunnel No. 1.
h) Construction of Jubilee Line Extension stub tunnels.
i) Construction of overbridges, 1, 2 and 3.
j) Construction of structural openings.
k) Concrete work to all tunnels.

Fig.4 Enlargement of 3.85m i.d. incline pilot tunnel to 7.5m i.d.

Extensive liaison was necessary with other parties including BR, and their Construction Manager Bovis working on the WIT, and also with BR W&CL.

A small working area adjacent to the new terminal building in the "post office yard" was agreed with BR, however the size of compound allowed and restrictions imposed by BR for security reasons made the work involved in removal of spoil and delivery of materials very complex. Close liaison with the relevant outside parties was necessary and weekly interface meetings were held which allowed the works to progress relatively smoothly.

Access became even more restricted in May 1993 when BR replaced the footbridge from the Main Line station to reinstate the link to the Shell building in York Road, which passed directly above LUL's working area. At this stage tunnelling was almost complete and only minor problems were encountered.

6 Compensation Grouting

The object of compensation grouting was to minimize settlement of the Victory Arch and W&CL tunnels, both very sensitive structures, during and immediately subsequent to tunnelling activities, by compensating for ground deformation induced by controlled tunnelling excavation. With normal tunnelling carried out with good practice and care, the maximum anticipated ground settlements at the level of the underside of the Victory Arch and W&CL tunnel foundations were expected to be about 50mm. Without ground treatment, angular distortion of the structures associated with these settlements would have been unacceptable. Compensation grouting was undertaken by the Contractor to reduce angular distortions of the structures to acceptably small limits which they could tolerate without significant distress.

Compensation grouting beneath the structures was carried out in concert with the onset of settlement induced by tunnelling. Accordingly, a system of instrumentation and survey points for monitoring the movements of the structures and underlying ground was specified and this was installed by the Contractor. The structures monitored were the Victory Arch, the W&CL tunnels and Elizabeth House. Regular monitoring of these by the Contractor was required.

Monitoring devices were also installed in the ground above the crown of the proposed tunnels prior to their construction. These devices were installed by means of an inclined borehole from the ground surface and a horizontal borehole from within the lower part of the inclined pilot tunnel, and gave valuable information on the ground behaviour directly above the tunnel during excavation and also during compensation grouting.

For further details of the ground treatment and monitoring systems, reference should be made to the Paper to this conference entitled "Compensation Grouting to limit settlements during tunnelling at Waterloo" by Dr. R.J.Mair et al.

7 Ground Support

On account of the short drives involved, and the various different sizes of tunnels and tapered sections, the contractor elected to undertake all the work by hand rather than employ tunnelling shields. The specification required that the tunnel excavation ahead was restricted to a maximum of 600mm (one ring width). Temporary support was provided by face and crown boarding.

The tunnelling specification also required that the contractor provided method statements in advance of the works, showing his proposals for excavation, temporary support and erection of linings, together with calculations, to allow assessment by LUL Engineers assisted by their consultants.

Other requirements of the specification included the following:

a) Each ring was to be grouted immediately following erection and before excavation for the next ring was commenced.
b) Close poled steel "banker" bars were required to support the top half of the excavation in order to keep ground movement to a minimum.
c) The face was to be fully timbered and grouted should the tunnelling be interrupted at any time.
d) Any tunnel of 5.75m i.d. or greater was to be constructed by first forming a pilot tunnel.
e) Break-ups for enlargement of the pilot tunnel were to be constructed by first forming a timbered chamber in which two rings were fully built and grouted.
f) Openings were to be formed using steel prop frames and ties to support the linings and/or ground during removal of the segments.

In general, timbering and face supports followed traditional methods, as indicated in BS 6164:1990 "Safety in tunnelling in the Construction Industry".

The standard method of support was by the use of close poled banker bars as headboards, 175mm wide and approximately 1m long to the crown, spanning over the ring to be built and extending laterally to support the overhead ground. 225mm x 75mm face boards up to a maximum of 2m long supported the leading edge of the headboards and the exposed face of the excavation. Each line of faceboards were supported on wedged footblocks and restrained by walers at mid span, or by 305mm x 150mm timber walers punched off a 203mm x 152mm steel back kicker to provide continuous face support whilst the ring was built. Further frames similar to this were used below the top frame, and on the largest size tunnels a total of five frames were necessary.

8 Construction of Openings

A timber box heading was constructed from the first ring (Ring No.2) of the cross passage tunnel, and advanced towards the crown of the larger diameter concourse tunnel, with a two plate wide opening formed within the junction as the excavation proceeded.

The heading was then extended half way over ring No. 2 and when completed and fully grouted, 3 No. 203mm x 203mm crown bars were erected, with the concourse end of each crown bar positioned directly above the line of temporary props erected within the large diameter tunnel. (See Fig.5).

The sides of the timber heading were then widened and further crown bars placed at 500mm centres. This method continued until the required width of chamber was achieved, and the remaining headboards were then fully grouted.

The faceboards legs were supported by two 203mm x 203mm steel walers at the position of ring No. 2 of the cross passage, and excavation then proceeded downwards with the sides and face supported by 225mm x 75mm timber boards.

Ring No. 1 of the cross passage could then be built on to ring No. 2 and reinforcement placed to the sides, crown and invert. A "ringwall" around rings 1 and 2 could then be constructed up to top of crown bar level and allowed to gain strength. This then ensured adequate support to both ends of the crown bars and also reduced their span. It should be noted that all timber used in construction of the heading and to remain in position was tanalised.

The cill beam of the opening frame would then be positioned in the invert of the opening. Templates of the opening flanges were then fabricated and the jamb plates drilled accordingly. Following erection of the jamb plates and lintel, the opening soffit was wedged with steel wedges which were then welded to the lintel. The opening was then shuttered and concreted.

Fig.5 Lower councourse tunnel showing temporary props in preparation for opening construction

9 Construction of Overbridges

When excavation of the cross passages had reached the position of the Northern Line station tunnel, the penultimate full ring was built and grouted. The top for the final full ring was then taken over as for normal construction but only excavated to just below shoulder level (ie. approximately 300mm above the crown of the station tunnel level).

The top plates of the last full ring were rolled and built. The key and ordinary segment were also built. Skewback plates with wedges already welded on were then bolted on and a 203mm x 133mm steel I beams installed under the leading edge of the wedge. The bench was then boarded out with 200mm x 25mm x 1200mm long timber boards placed longitudinally and the underside grouted. The top rings were also grouted. This procedure was repeated until this, and four further part rings were constructed.

The part rings were then supported by 203mm x 133mm steel I beams acting as needles, and bolted to full and part rings by "U" frames and packed to support the crown. This allowed excavation for the first section of side walls and the first strongback to proceed. The excavation allowed the side walls under the first 2 overbridge rings to be cast and was supported by 225mm x 75mm faceboards with a steel 152mm x 152mm waler and struts of the same section punched off the last full ring. To allow construction of the chamber for erection of the strongback, two frames were required to support the excavation, and support the ground below the cross-passage.

During engineering hours occupations of the track (approximately 01:00 to 04:30) 4 No. existing cast iron segments were removed at each overbridge, 2 No. either side of the opening to be made, and replaced with new steel segments. During this work it was essential that work in other areas of the tunnels could progress on days, and it was therefore necessary to totally separate the overbridge works from the remainder of the site for safety reasons and also so not to delay the works. This was achieved by erecting a fully sealed steel hoarding at the end of each cross-passage, with a lockable steel door which was removed each night when work on the track could commence.

Removal of one existing segment from each side of the bridge was carried out after removal of the bolts, and the new steel plate was then positioned and the packs and shims installed. The procedure was repeated for the remaining two existing segments.

A template of the strongback was then used to correctly mark out the required position and drill the required holes in new and existing segments, again with all drilling carried out at night. The three sections of the strongback could then be erected at night and fully bolted in place. The strongback was then encased in concrete and then the section of side walls cast. The remainder of the last full ring was then built.

Once the side walls gained strength, tunnelling recommenced allowing a further 4 part rings to be built and grouted. Support was provided by the needles being advanced as work proceeded. Excavation downwards to the Northern Line tunnel was also carried out allowing construction of the next section of concrete side walls. (Fig.6).

Fig.6 Internal view of overbridge construction prior to removal of Northern Line segments

This procedure continued allowing a further five part rings to be constructed, the last being fitted into a chase cut into the concrete head wall of the existing Northern Line overbridge. Excavation down the side of the existing station tunnel was carried out as previously described and the second strongback erected. Following encasement of the strongbacks and completion of the side walls, removal of the crown plates from the existing rings commenced, followed by construction of the bridge deck beams, permanent shuttering and concrete deck. (Fig.7).

Fig.7 View of overbridge construction from Northern Line platform

Permanent connection to the station was then made by breaking through the concrete wall after erection of a 1 hour fire rated hoarding on the operational station side to maintain separation.

10 Emergency Procedures

A Tunnelling Safety Plan was prepared by LUL prior to commencement of the tunnelling, covering all safety requirements and emergency procedures.

A field telephone was in operation between each face and the winch operator in the diaphragm box. Safety gates at the top and bottom of the incline shaft together with a traffic light system provided protection from possible run-away bogies. Air quality was constantly monitored during tunnelling and an adequate supply of self rescue sets provided for labour, staff and visitors.

All personnel on site had to undertake an induction course, provided by the Contractor which covered safety aspects of all parts of the project but in particular the tunnelling works, including emergency procedures and training in the use of the self rescue sets.

A "tally" system was operated for everyone entering the new tunnels so that in the event of an emergency all personnel could be accounted for. The emergency services were invited to site prior to the tunnelling works commencing, and also visited site during construction to allow familiarisation.

If the extensive monitoring equipment indicated problems with settlement or ground loss, the situation at the face would be checked and if necessary the procedure for that particular emergency would then be followed which included notifying BR's control room at Waterloo Station called "SOLARI", who would then follow their own procedures with respect to the Waterloo and City Line and other BR structures. The relevant Emergency Procedures were successfully carried out during a minor fire within the new tunnels, and no injuries or disruption to either LUL or BR services were sustained.

Should a problem have occurred during construction of the overbridges, within the LUL station, the Emergency Procedure called for the LUL control room staff to notify the Line Controller so that the service would not commence in the morning, or if the problem occurred during the day the service would immediately be suspended.

11 Acknowledgements

Thanks are due to LUL, the Client and Engineering Authority for giving permission to write the paper. Photographs were supplied by LUL's Photographic Unit.

Consultants providing advice and design on various aspects of the project were the Geotechnical Consulting Group and Charles Haswell and Partners.

Sir Robert McAlpine was the main contractor for all works with compensation grouting carried out by Bachy Ltd and monitoring instrumentation to neighbouring structures installed and operated by the Building Research Establishment.

Water in TBM drives, Delivery Tunnel South, Lesotho Highlands Water Project

G. A. Featherstone
Ninham Shand, Inc., Randburg, South Africa, and Lesotho Highlands Tunnel Partnership, Maseru, Lesotho

Abstract
This paper deals with a variety of related aspects concerning the dealing with groundwater in excavations driven by tunnel boring machine for the Delivery Tunnel South, a section of the waterway tunnel for the Lesotho Highlands Water Project. The paper commences with the design/tender predictions for water it was envisaged may be encountered, the tender methods specified for dealing with that water and the contractual aspects/sharing of risk/payment mechanisms set out in the tender documents for dealing with water. Following on details are provided of information tenderers submitted regarding their estimates of the time required to carry out cover drilling and grouting and the effect on overall TBM advance rates. The paper then concludes with details of actual water encountered, the manner in which the tender specifications were amended in dealing with that water, the actual effect on TBM production as well as the actual construction methods used.
Keywords: Cover grouting, groundwater, Lesotho Highlands Water Project, TBM excavation.

1 Introduction

Phase 1A of the Lesotho Highlands Water Project (LHWP) under construction (1991-1996) includes the 185 m high concrete arch, Katse Dam, the 45 km Transfer Tunnel, the 72 MW 'Muela Hydropower Plant and the 37 km Delivery Tunnel. The Delivery Tunnel receives water from an intake in the tailpond of the 'Muela Hydropower Plant and discharges it into the Ash River in South Africa and thence into the Vaal River System. Contract LHDA 126 for the Delivery Tunnel South is for the construction of a 15,1 km section of the Delivery Tunnel and is located wholly inside the Kingdom of Lesotho commencing from the 'Muela Intake and running to the Caledon River, being the boundary between the Kingdom of Lesotho and the Republic of South Africa. The nominal design flow for the Delivery Tunnel is 32.6 m^3/sec.

The longitudinal section and horizontal alignment of the Delivery Tunnel South is shown in Figure 1.

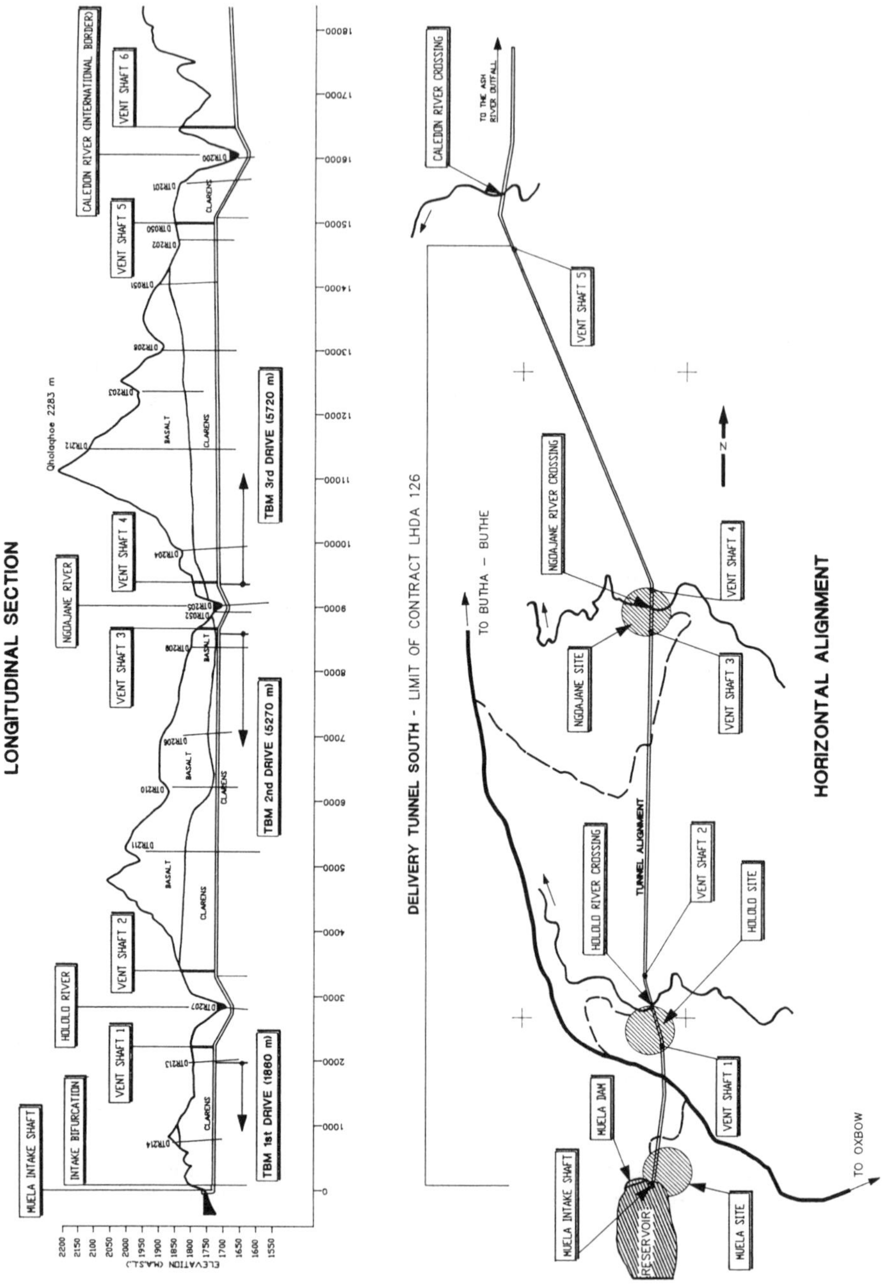

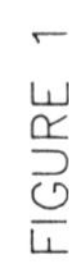
LONGITUDINAL SECTION
MUELA INTAKE SHAFT
INTAKE BIFURCATION
VENT SHAFT 1
HOLOLO RIVER
VENT SHAFT 2
VENT SHAFT 3
NGOAJANE RIVER
VENT SHAFT 4
Qholaqhoe 2283 m
VENT SHAFT 5
CALEDON RIVER (INTERNATIONAL BORDER)
VENT SHAFT 6
TBM 1st DRIVE (1860 m)
TBM 2nd DRIVE (5270 m)
TBM 3rd DRIVE (5720 m)
BASALT
CLARENS
ELEVATION (M.A.S.L)
DELIVERY TUNNEL SOUTH - LIMIT OF CONTRACT LHDA 126
TO BUTHA - BUTHE
TO OXBOW
RESERVOIR
MUELA SITE
MUELA DAM
HOLOLO RIVER CROSSING
HOLOLO SITE
TUNNEL ALIGNMENT
NGOAJANE SITE
NGOAJANE RIVER CROSSING
CALEDON RIVER CROSSING
TO THE ASH RIVER OUTFALL
N
HORIZONTAL ALIGNMENT

FIGURE 1

The tunnel length, excavated at a nominal diameter of 5.1 m, is split into discrete drives by the inverted river siphons constructed by drill and blast under the Ngoajane and Hololo Rivers. The three sections of tunnel so formed by the inverted siphons were all excavated by the same refurbished Robbins hard rock TBM and broad details of each drive are given in Table 1.

Table 1. Details of TBM drives

Drive	TBM drive	Length (m)	Working days (No)	Average daily advance (m/day)	Peak daily advance (m)
1	Hololo to Muela	1,860	62	30.0	55.0
2	Ngoajane to Hololo	5,270	136	38.8	68.9
3	Ngoajane to Vent. Shaft 5	5,720	122	46.9	82.9

Drive No.1 commenced on 19 February 1992 and drive No.3 was completed on 23 August 1993. The advance rates achieved were affected by and include for the time required to grout water bearing features ahead of the advancing TBM face.

The Employer for the Contract is the Lesotho Highlands Development Authority, the Consultant the Lesotho Highlands Tunnel Partnership and the Contractor the Lesotho Highlands Project Contractors.

2 Design evaluations

2.1 Introduction

As part of the Engineer's evaluation included in the Data for Tenderers which was issued as part of the tender documents a hydrogeological evaluation was carried out.

Data gathered during exploration borehole drilling and packer testing was collated and reviewed in the light of the regional geology to provide an evaluation of the regional hydrogeology in the vicinity of the proposed tunnel line. This information was then used to predict inflow problems associated with geological horizons which had been targeted as being potentially problematic.

2.2 Exploratory drilling

Data gathered during the exploratory drilling phase from the 20 boreholes drilled along the tunnel line included.

- Drilling water losses encountered in fracture zones within the Clarens Sandstone, (the rock type through which the majority of tunnelling was to be carried out) as well as significant losses in a fissure encountered 25 metres above the basalt/sandstone contact. In the second of the three TBM drives the basalt/sandstone contact was expected to dip down to within a few metres of the tunnel crown.
- Packer testing was carried out in all but two of the boreholes, these two experiencing complete water loss during drilling. Tests were carried out over a 40 m vertical interval centred on the proposed tunnel line in 6.5 m sections.

2.3 Regional hydrogeology

With regard to the regional hydrogeology it was considered that water may be transmitted through the rock mass via two distinct mechanisms:

- intergranular flow, where water is transmitted through the rock mass itself (primary flow)
- fracture flow, where water travels through an interconnected network of fractures (secondary flow).

Secondary flow was expected to be the dominant flow mechanism and specific features influencing potential inflow are:

- Bedding: although Clarens Sandstone is generally considered as massive there are occasional widely spaced bedding planes along which flow will tend to concentrate.
- Dykes: dyke margins are likely to consist of an interconnected network of cooling joints which may act as preferential conduits for water flow. This effect is likely to be pronounced near surface, where weathering has served to accentuate joints and thus, enhance permeability. However at depth, dyke margins may be relatively "tight" and may yield little water.

  Inflows have been reported [1] as large as 51 000 l/min in the Orange-Fish tunnel. However, it is likely that inflows associated with dykes along the Delivery Tunnel will be in the order of 6 l/sec in areas where there is limited overburden. Inflows will decrease with increasing overburden, to values of less than 0.15 l/sec. Takes of up to 56,5 lugeons (equivalent permeability 7,0 x 10^{-5} m/sec) were experienced during packer testing of one borehole illustrating the effect of a dyke intrusion on the local permeability.
- Major joint zones: calculations indicated that a series of say, ten water bearing joints near surface could yield a total flow of up to 25 l/sec. However, joint zones will become progressively less "open" with increasing depth, and it is likely that most joint zones encountered along the tunnel line will yield less than 5 l/sec.
- River crossings: river crossings may be problematic in terms of inflows, as the tunnel excavation will be close to surface in a zone of weathered material. As river courses in this area appear to be structurally controlled the vertical permeability in these areas is likely to be significantly enhanced by the presence of faults, shear zones or dykes. The magnitude and persistence of inflows at river crossings will be directly proportional to the degree of direct recharge from the water course at the surface, and although no specific information was available for these sites, inflows of up to 5 l/sec for a 10 m section could have been expected.
- Clarens Sandstone/basalt contact zone: the contact zone was identified as potentially problematic in terms of inflow as it has enhanced secondary permeability, due to a pronounced cooling joint network associated with the contact. Significant water loss was experienced in one borehole at this level and extremely high water takes were experienced during packer testing of this zone in another. Although evidence suggests that the contact zone is less permeable than equivalent strata in other regions, flows of 15 l/sec could be intersected. These zones of high inflow could be identified by probe drilling and the nature of the fractures should make them suitable for grouting.

Potential inflow into these specific features are summarised in Table 2

Table 2. Inflows expected in association with structural features

Feature type	Expected inflow (l/sec)
Dykes (overburden less than 100 m)	3,0 to 6,0 (wide >8,0 m)
	0,5 to 1,5 (thin <8,0 m)
Dykes (overburden more than 100 m)	0,5 to 1,5 (wide >8,0 m)
	0,1 to 0,5 (thin <8,0 m)
Faults	5,0
Joint zones (near surface)	25,0 per 100 m tunnel
Joint zones (at depth)	5,0
River crossings	3,0 per 10 m tunnel
Basalt/sandstone contact	1,0 to 3,0 per 100 m tunnel
Clarens Sandstone	0,1 to 0,5 per 100 m tunnel
Clarens Transition	0,0 to 0,1 per 100 m tunnel

2.4 Hydrogeological characteristics of tunnel route

2.4.1. Section, chainage 0 m to 5,650 m
The first TBM drive falls into this section and the primary hydrogeological characteristics which relate to this drive are:

- direct evidence of high permeability in the low cover area adjacent to 'Muela Intake as evidenced by water takes of up to 56,5 lugeons being recorded in association with a dolerite dyke. Significant inflows up to 6 l/sec could be experienced from a single feature
- previous experience of the Clarens Sandstone suggests that well defined partings within the sandstone may yield considerable quantities of water. However the frequency of such partings may be low and background inflow of up to 0,5 l/sec per 100 m of tunnel may occur
- dykes crossing the tunnel are likely to only generate small shortlived flows up to maximum of 6 l/sec.

2.4.2. Section, chainage 5,650 m to 9,000 m
The second TBM drive falls partly into the section noted under 2.4.1 above but primarily into this section and the main hydrogeological characteristics which relate to this drive are:

- this section of the tunnel was to be constructed through the uppermost levels of the Clarens Sandstone well within 50 m of the basalt/sandstone contact and experience of this contact zone gained elsewhere indicates that the contact could have a network of interconnected water bearing features and yields of up to 13,9 l/sec have been recorded in other areas. However drilling and packer tests indicate that flow would be lower say up to 3 l/sec per 100 m of tunnel

- dykes crossing the line are likely to yield inflows of up to 6 l/sec in the section Chg 8,2000 m to 9,000 m and somewhat less down chainage.

2.4.3. Section, chainage 9,000 m to 15,000 m

The third TBM drive falls fully into this section and the main hydrogeological characteristics which relate to this drive are:

- this section was to be constructed through the lower levels of the Clarens Sandstone and the background inflow will be similar to the first section, that is up to 0,5 l/sec per 100 m of tunnel
- seven known dykes cross the tunnel line and each may expect to yield a maximum inflow of up to 5 l/sec.

2.4.4. Tunnel inflow

Based on the above anticipated background and specific features, inflows into the tunnel may be calculated and the potential accumulated inflow determined. The results are presented in Figures 2.1 and 2.2.

A piezometric surface between the 'Muela Intake and the Hololo River is expected to range from 0 to 50 m above tunnel elevation, the surface increasing in elevation southwards from the Hololo River. Between the Hololo and Ngoajane Rivers the piezometric surface is expected to range from 15 to 100 m above tunnel level whilst between the Ngoajane River and Ventilation Shaft 5, it is expected to range from 15 to 200 m above tunnel level.

The various geological horizons will exhibit degrees of confinement in aquifer terminology and are likely to be subject to sub-artesian conditions. This will assist in limiting flow in the short term, but water pressures approximately equivalent to the height of piezometric surface above the tunnel will exist within the undisturbed sidewalls of the tunnel.

3 Tender specifications

3.1 General

Based on the Engineer's evaluations of the possible water inflow as depicted in Figures 2.1 and 2.2 decisions were made on how

- dealing with water would technically be specified in the Tender Specifications
- risk would be shared between the contracting parties should groundwater be encountered.

3.2 Technical specifications

3.2.1. Introduction

The technical specifications were set out to attempt to ensure that by the judicious use of cover grouting ahead of the advancing face that the total inflow of groundwater accumulating down each tunnel drive was at all times kept below 50 l/sec. As such the philosophy at the tender stage was to grout most features to keep the total inflow

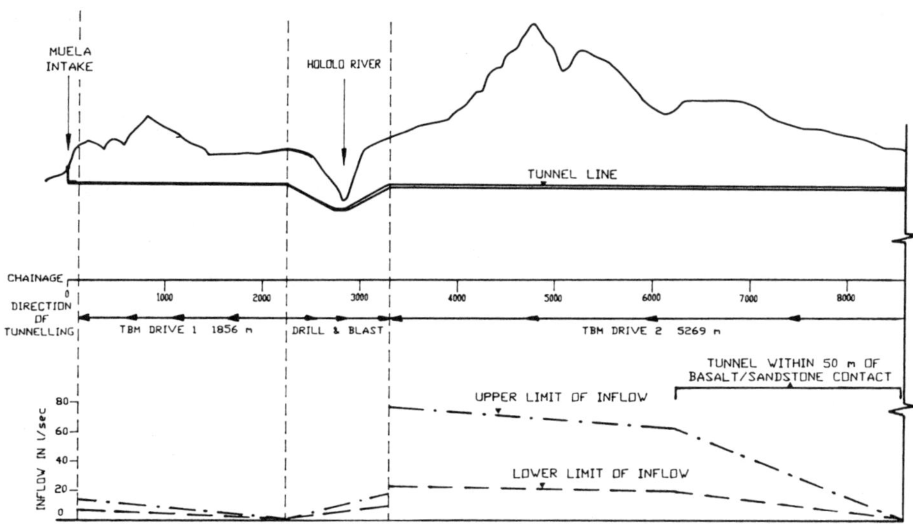

FIGURE 2.1 ACCUMULATIVE WATER INFLOWS - TBM DRIVES 1 AND 2

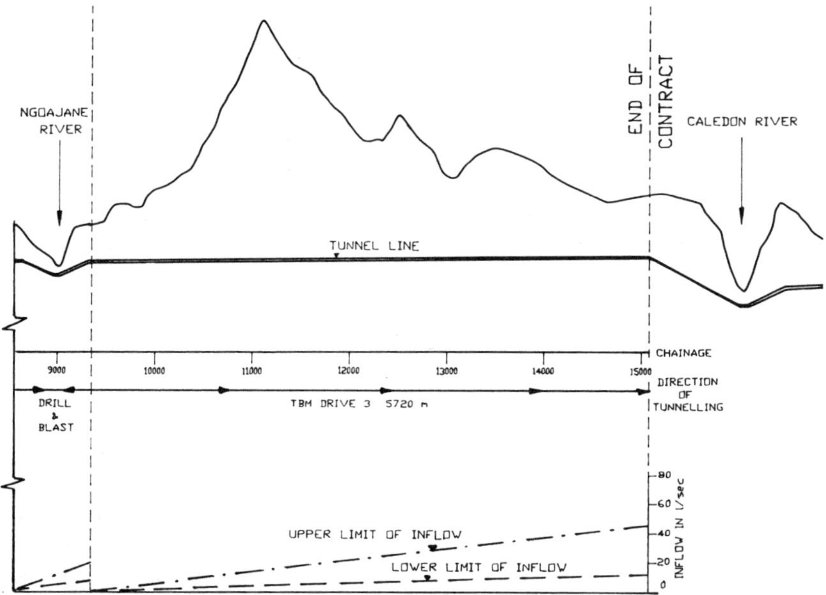

FIGURE 2.2 ACCUMULATIVE WATER INFLOWS - TBM DRIVE 3

well below that amount such that large inflows that may eventuate from forward probe holes could be readily accomodated.

3.2.2. Pumping equipment

The Specifications required that the Contractor provide pumping equipment capable of handling groundwater occurring in each discrete tunnel heading of up to 50 l/sec in total for that heading (in addition to dealing with construction water or water from any other source outside the underground works but which entered it).

The Contractor was required to provide measuring devices to determine the quantity of water introduced into the tunnel for construction purposes and the total amount pumped out.

3.2.3. Probe drilling

The Specifications required that probe holes not less than 48 mm diameter be drilled at least 40 m ahead of the advancing face such that the position of the hole 30 m ahead of the face was no further outside the tunnel periphery than 3,4 m. At all times a minimum length of 10 m of probe hole had to be maintained ahead of the face unless the Contractor elected to reduce the minimum length of overlap to 5m on the provision that he drilled two face probe holes diametrically opposite to each other. In addition the Engineer could instruct that a second probe hole be drilled at any time. One purpose of such a flexibility in probe hole drilling was to provide an opportunity to probe for features running inclined to the line of the tunnel.

3.2.4. Cover drilling and grouting

The Specifications required that cover drilling and grouting be undertaken ahead of the advancing face when the inflow from individual or groups of probe holes drilled ahead of the advancing face exceeded certain limits.

The length of tunnel that it was estimated at tender would require cover grouting follows and the layout of the grouting array and the grouting procedures specified are set out in Figure 3.

- TBM drive 1 50 m out of 1,860 m
- TBM drive 2 1,600 m out of 5,270 m
- TBM drive 3 350 m out of 5,720 m

3.2.5 Vertical alignment of tunnel

All three TBM drives were designed such that they were driven upgrade to allow a free flow of water down the tunnel. The first two drives were at relatively flat grades of 1:1,000 whilst the third was at 1:525. A precast invert segment to be placed immediately behind the TBM on which the back up train would run was shaped to allow for a central drainage channel. The segment is detailed in Figure 4. The construction tolerances on the vertical alignment were tight ($\pm$ 25 mm) to ensure that significant ponding in the channel could not eventuate.

The estimated theoretical flow that the channel could take is tabulated in Figure 4.

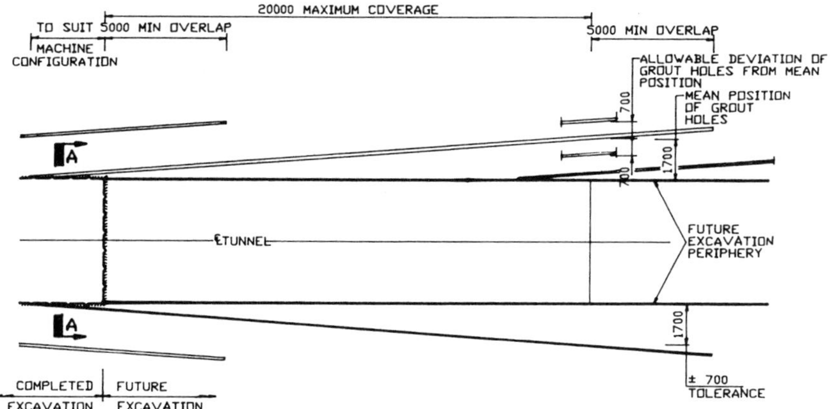

SECTION THROUGH TYPICAL COVER GROUTED LENGTH – BORED TUNNELS

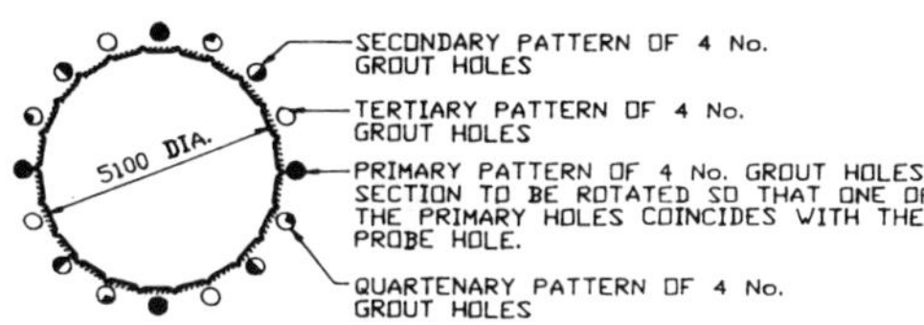

SECTION A–A

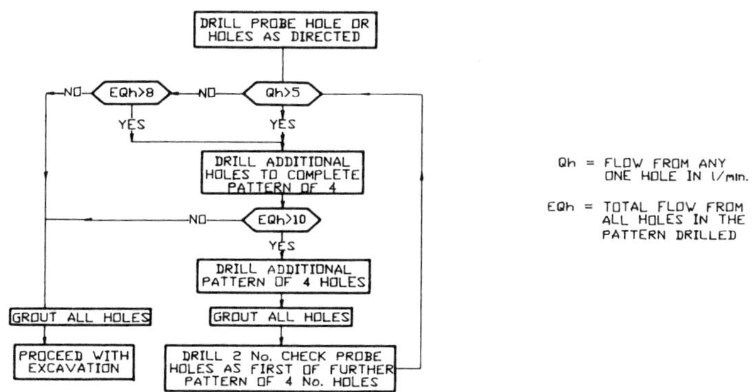

DRILLING, TESTING AND GROUTING SEQUENCE

FIGURE 3

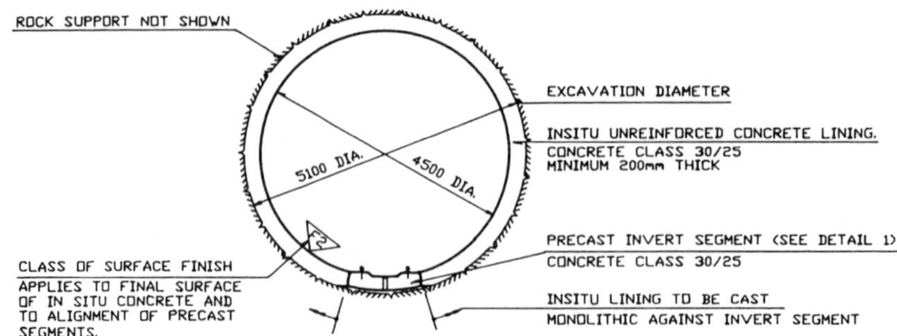

TYPICAL SECTION
UNREINFORCED CONCRETE LINING

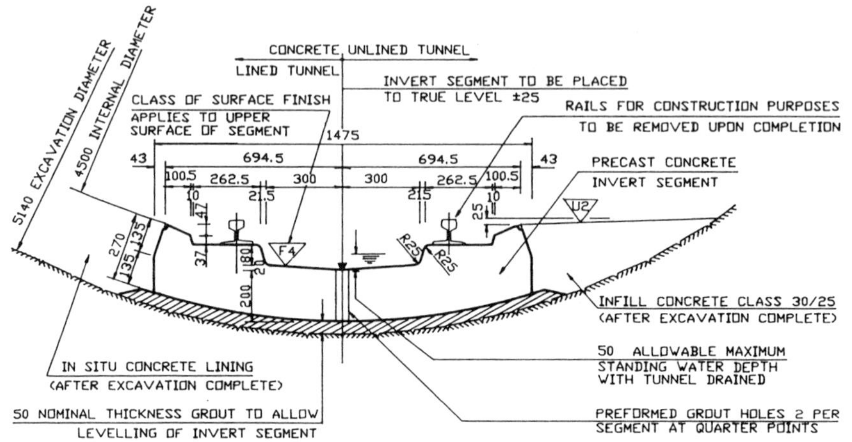

DETAIL 1
PRECAST INVERT SEGMENT

FLOW Q (l/sec)	DEPTH OF FLOW Y (mm) (1)
10	64
20	108
30	127
40	145
50	160
55 - 105	170 (2)
150	214
200	260
250	302
300	343
500	492

THEORETICAL DEPTH VS FLOW RELATIONSHIP
FOR TBM DRIVE - SLOPE 1 : 1000

NOTES:

1. DEPTH OF FLOW FROM BOTTOM OF CHANNEL.
2. FLOW SPILLS INTO SIDE CHANNEL ON EITHER SIDE OF INVERT SEGMENT.

FIGURE 4

3.3 Contractual Arrangements

3.3.1. Pumping of water

Payment for dealing with water would be measured against fixed and time related charges allowed in the Preliminary and General section of the bill of quantities as well as quantity proportional charges for pumping costs. The billed items are set out in Table 3.

Table 3. Bill of quantities, extract for dealing with water

Bill Item	Description	Unit	Quantity
Fixed Charges Provision of Specific Equipment Dewatering Equipment			
A1990	Hololo to Muela	Sum	1.00
A2000	Ngoajane to Hololo	Sum	1.00
A2010	Ngoajane to VS 5	Sum	1.00
Time-related Charges Provision of Specific Equipment Dewatering Equipment			
A4510	Hololo to Muela (Duration..... weeks)	Sum	1.00
A4520	Ngoajane to Hololo (Duration..... weeks)	Sum	1.00
A4530	Ngoajane to VS5 (Duration..... weeks)	Sum	1.00
Tunnel Hololo to Muela, Pumping Water			
B620	> 0 ≤20 l/sec	days	1005
B630	>20 ≤30 l/sec	days	30
B640	>30 ≤40 l/sec	days	30
B650	>40 ≤50 l/sec	days	30
Tunnel Ngoajane to Hololo, Pumping Water			
C900	> 0 ≤20 l/sec	days	200
C910	>20 ≤30 l/sec	days	150
C920	>30 ≤40 l/sec	days	150
C930	>40 ≤50 l/sec	days	610
Tunnel Ngoajane to VS5, Pumping Water			
D790	> 0 ≤20 l/sec	days	340
D800	>20 ≤30 l/sec	days	245
D810	>30 ≤40 l/sec	days	60
D820	>40 ≤50 l/sec	days	30

In the case of time-related charges each tenderer was required to complete the duration for which the facility was to be provided and the duration was to be in accordance with the programme submitted with his tender.

The pertinent aspects of the measurement and payment clauses for pumping of water were as follows:

- measurement for payment would only be for the actual groundwater pumped averaged over a 24 hour period
- measurement would be made separately for each separate heading and if staged pumping was necessary to pump groundwater from that heading such staging would not be considered
- groundwater arising from ancilliary headings or shafts would not be considered separately but the volume cumulated with that arising from the main heading
- no payment would be made for pumping durations in excess of that tendered by the Contractor (plus any extension granted) and payment would only be made while pumping was carried out
- the Contractor would provide, operate, etc. approved measuring devices.

The rates tendered included compensation for operating of pumping equipment and moving it to suit the advancing face as well as for the provision of measurement equipment.

As noted above the pumping installations were required to be able to deal with a total groundwater inflow in each tunnel drive of not exceeding 50 l/sec. In the event that the dewatering installations needed to be increased due to groundwater inflows in excess of that amount payment would be made for such additional installations agreed necessary at cost and disruption to excavation progress while the installations were being increased would be made by extension of the appropriate time-related payments.

3.3.2. Cover drilling and grouting

Cover drilling and grouting equipment was required to be an integral part of the TBM and its ancillary equipment and the tender required that separate fixed and time-related charges were priced for the overall provision of the excavation and associated equipment. In addition quantity proportional items for drilling and grouting were to be priced for each of the separate drives.

Clearly the estimation of the likely volume of grout materials that may be required to seal water bearing fissures ahead of the face was prone to a high degree of inaccuracy due to the nature of the work. However the magnitude of the quantity proportional items was calculated used the following criteria as a base:

- drilling quantities were derived on the basis of cover grouting being carried out in 20 m long sections and that primary grouting would be sufficient for 20% of the time, secondary grouting for 60% and tertiary grouting for the balance.
- anticipated grout volumes were derived on the basis of experience gained elsewhere on pre-grouting and an allowance of 0,2 tonne of cementitious material was allowed per running metre of tunnel.
- substantial water pressure testing was not anticipated and success of the grouting would be measured by residual flow from probe holes.

The billed items are set out in Table 4.

Table 4. Bill of quantities, extract for cover grouting

Bill Item	Description	Unit	Quantity
Fixed Charges			
Provision of Specific Equipment			
Excavation and Associated Equipment			
A1790	Hololo to Muela	Sum	1.00
A1800	Ngoajane to Hololo	Sum	1.00
A1810	Ngoajane to VS 5	Sum	1.00
Time-related Charges			
Provision of Specific Equipment			
A4310	Hololo to Muela (Duration .... weeks)	Sum	1.00
A4320	Ngoajane to Hololo (Duration .... weeks)	Sum	1.00
A4330	Ngoajane to VS5 (Duration .... weeks)	Sum	1.00
Tunnel Hololo to Muela, Cover Grouting			
B890	Set up drilling equipment	Sum	1.00
B900	Set up grouting equipment	Sum	1.00
B910	Drilling of holes	m	750
B920	Re-drilling of holes	m	50
B930	Abbreviated water pressure tests	No	5
B940	Comprehensive water pressure tests	No	5
B950	Grouting of holes	No	30
B960	OPC	tonne	4
B970	RHPC	tonne	4
B980	FA	tonne	2
B900	Sand	tonne	1
Tunnel Ngoajane to Hololo, Cover Grouting			
C1280	Set up drilling equipment	Sum	1.00
C1290	Set up grouting equipment	Sum	1.00
C1300	Drilling of holes	m	20000
C1310	Re-drilling of holes	m	500
C1320	Abbreviated water pressure tests	No	80
C1330	Comprehensive water pressure tests	No	20
C1340	Grouting of holes	No	800
C1350	OPC	tonne	120
C1360	RHPC	tonne	120
C1370	FA	tonne	60
C1380	Sand	tonne	20
Tunnel Ngoajane to VS5, Cover Grouting			
D1120	Set up drilling equipment	Sum	1.00
D1130	Set up grouting equipment	Sum	1.00
D1140	Drilling of holes	m	4,500
D1150	Re-drilling of holes	m	100
D1160	Abbreviated water pressure tests	No	20
D1170	Comprehensive water pressure tests	No	5
D1180	Grouting of holes	No	180
D1190	OPC	tonne	30
D1200	RHPC	tonne	30
D1210	FA	tonne	15
D1220	Sand	tonne	10

A particular feature of the risk sharing principles in the tender documents was that any variations in the amount of cover drilling and grouting from that envisaged at tender would be evaluated based on tender cycles times and savings in time-related charges (TRC) would be shared equally between the parties and additional TRC compensated in full to the Contractor through the appropriate TRC.

This also applied to any variation in the tender rock type/rock support classes and the appropriate specification clause follows:

> "Should the agreed proportions of rock class encountered or support classes installed or cover grouting required in any section of the tunnel differ from the proportions stated or implied in the Contract, the Engineer in consultation with the Contractor shall determine from the advance rates or cycle times in Tender Schedule L for each rock and support class or cover grouting requirement respectively, a revised duration for each item in Section A that is affected.
>
> The revised duration for an item will be applied to the incremental amount (previously calculated) or applicable element thereof and a revised total TRC calculated. Any extra payment thus derived shall be the sole compensation for any increased charges (excepting the quantity-proportional charges) arising from different proportions of the various rock or support classes or different cover grouting requirements.
>
> If, on completion of the work to which any TRC item applies, the total amount derived is less than the tendered sum for that item (i.e. there is a reduction of the Contract Price in respect of that item because of conditions better than those anticipated by the Engineer at tender) then an additional sum equal to half the difference between the amount so derived and the tendered sum will be certified.
>
> If on completion of any Portion of the Works, the overall duration calculated in accordance with this Sub-clause exceeds the tendered duration, the Engineer will in accordance with the Contract grant an extension of time to that Portion of the Works affected, in terms of Clause 44 of the Conditions of Contract".

The tender documents required that the Contractor submit details of the time required to execute cover drilling and grouting and he was required to complete Tender Schedule L following, Table 5.

Table 5. Schedule L, cover grouting cycle time in TBM drives

Activity		Primary array	Secondary array	Tertiary array
No. of holes drilled/grouted		4	10	18
Drill holes	- total time			
	- in line time			
Grout holes	- total time			
	- in line time			
Interference	- in line time			
Total	- in line time			

4 Contractors tender submissions

Tenderers were required to submit as an integral part of their tender the following information for each of the three TBM drives:

- TBM cycle times, which were to be split into the component parts of boring, regripping/steering, mucking, rockbolting, mesh and shotcrete application, maintenance (TBM and back up) and interference
- TBM average weekly advance rates based on the cycle times

This information was to exclude for the effect of cover drilling and grouting and tenderers were thus also required to submit cycle times for primary, secondary and tertiary arrays of cover drilling and grouting.

With regard to the TBM cycle times and advance rates there had to be a separate submission for each of the rock class/rock support combinations that it was anticipated would be encountered along the tunnel length. From this the time required to excavate each tunnel drive could be calculated and to obtain an overall duration for each drive the theoretical time required to cover grout and drill was added to these figures.

Table 6 details the weekly advance rates for each rock class/support class combination (for example TI MI, etc) for each of the seven tenderers. Note that the successful tenderer, coded B 6006 at the evaluation stage, later in agreement with the Engineer modified his cycle times and weekly advance rates to suit the provision of a refurbished machine. It can be seen that the weekly advance rates finally agreed for the refurbished machine do for all rock class/support class combinations lie below the median for all tenderers.

Table 7 details the result of the calculation to determine the theoretical time to excavate each drive (excluding cover drilling and grouting).

Table 8 details the cycle times submitted by each of the tenderers for cover drilling and grouting. With regard to the successful tenderer (B6006) it can be seen that his tender cycle times lie very close to the median of all tenderers representing a balanced bid.

Table 9 then details the calculation of total theoretical time required for cover grouting based on the anticipated lengths required to be grouted and thus the estimated number of grouting arrays required. This time is added to the theoretical time required to excavate the drives as detailed under Table 7 to arrive at a theoretical total programme duration for each drive. The pertinent points arising out of the evaluations in Table 9 are:

- The total durations calculated for the refurbished machine were longer than for any of the tenderers giving a high degree of confidence that the programme would be met.
- The time allowed for cover grouting by the successful tenderer as a percentage of the total tendered drive duration in each of the three drives was:
  - Drive 1 : 2.74%
  - Drive 2 : 17.37%
  - Drive 3 : 4.50%

Table 6. Tendered weekly TBM advance rates

Class	Engineer estimate	Tenderer B1056	Tenderer B5106	Tenderer B2406	Tenderer B4206	Tenderer B3306	Tenderer B1506	Tenderer B6006	Tenderer Average	Refurb. TBM
TI MI	150.4	210.0	181.0	153.2	194.0	185.0	250.0	231.0	200.6	192.0
TI MII	150.4	210.0	145.0	153.2	176.0	179.0	225.0	149.0	176.7	130.0
TI MIII	139.0		125.0	136.0	120.0	149.0	170.0	132.0	138.7	122.0
TI MIV	46.4		75.0	100.7				54.0	76.6	50.0
TI MII+ MV	46.8		88.0	117.4	41.0	50.0	170.0	72.0	89.7	72.0
TI MIII + MV	45.1		80.0	112.1	41.0	47.0	115.0	61.0	76.0	61.0
TII MI	136.6		168.0	138.4	184.0	169.0	225.0	206.0	181.7	192.0
TII MII	136.6	168.0	130.0	138.4	182.0	164.0	200.0	132.0	159.2	130.0
TII MIII	136.6		113.0	121.8	122.0	137.0	170.0	124.0	131.3	122.0
TII MIV	46.3		59.0	85.9	88.0	59.0	115.0	50.0	76.2	50.0
TII MII + MV	46.7		83.0	102.6	42.0	48.0	170.0	66.0	85.3	66.0
TII MIII + MV									0.0	58.0
TIII MI	114.5	126.0	132.0	138.4	173.0	145.0	200.0	165.0	154.2	130.0
TIII MII	114.5		116.0	138.4	173.0	141.0	170.0	124.0	143.7	109.0
TIII MIII	114.5		90.0	121.8	125.0	119.0	140.0	116.0	118.6	103.0
TIII MIV	46.3		57.0	85.9	90.0	55.0	115.0	440.0	140.5	42.0
TIII MII + MV									0.0	58.0
TIII MIII + MV									0.0	55.0
TIV MI	80.3	84.0	105.0	119.2	161.0	121.0	170.0	107.0	123.9	84.0
TIV MII	80.3		105.0	119.2	161.0	118.0	140.0	99.0	123.7	78.0
TIV MIII	80.3			102.1				91.0	96.5	74.0
TIV MIV	45.9		56.0	66.7	92.0	50.0	115.0	41.0	70.1	38.0
TIV MII + MV									0.0	55.0
TIV MIII + MV									0.0	50.0

1. Advance rates in m / week
2. Rock UCS

T1	0 to 50 Mpa
T11	50 to 100 Mpa
T111	100 to 150 Mpa
T1V	Over 150 Mpa

3. Support

M1	Spot bolting
to	to
M1V	Systematic bolting, mesh and shotcrete
MV	1 m bolts for sidewall spalling

Table 7. Theoretical excavation time

Tunnel Section	Refurb T	B1056	B5106	B2406	B4206	B3306	B1506	B6006
Hololo to Muela	15.573	13.684	14.332	14.673	11.220	13.944	9.368	13.557
Ngoajane to Hololo	54.292	28.832	48.415	43.410	38.229	46.468	30.018	44.381
Ngoajane to VS 5	54.690	37.822	48.739	45.535	45.666	50.589	30.464	48.981
Total	124.555	80.338	111.486	103.618	95.115	111.001	69.850	106.919

1. All durations in weeks

Table 8. Tendered cover drilling/grouting advance rates

Tenderer	Engineers Estimate			B5106			B2406			B4206			B1506			B6006			Av. of Tenderers		
Activity	P	S	T	P	S	T	P	S	T	P	S	T	P	S	T	P	S	T	P	S	T
No. of Holes	4	10	18	4	10	18	4	10	18	4	10	18	4	10	18	4	10	18	4	10	18
Drill Holes																					
Time 1	4.60	7.60	15.60	2.17	4.50	7.83	5.00	12.50	22.50	12.00	31.00	55.00	1.50	2.50	4.00	6.66	15.12	26.40	5.47	13.12	23.15
Time 2	4.60	7.60	15.60	0.58	0.58	0.58	2.50	6.25	11.25	12.00	31.00	55.00	1.50	2.50	4.00	3.58	7.81	13.45	4.03	9.63	16.86
Grout Holes																					
Time 1	2.20	4.40	9.20	8.50	19.13	33.00	4.00	10.00	18.00	6.00	16.00	28.00	1.00	1.50	2.50	3.00	6.00	10.00	4.50	10.53	18.30
Time 2	2.20	4.40	9.20	6.58	14.58	25.25	4.00	10.00	18.00	2.00	5.00	8.00	1.00	1.50	2.50	2.00	5.00	9.00	3.12	7.22	12.55
Interference																					
Time 2	0.00	0.00	0.00	0.17	0.17	0.17	8.00	8.00	8.00	1.00	3.00	4.00	0.50	0.50	0.50	2.29	5.49	9.62	2.39	3.43	4.46
Total																					
Time 2	6.80	12.00	24.80	7.33	15.33	26.00	14.50	24.25	37.25	15.00	39.00	67.00	3.00	4.50	7.00	7.87	18.30	32.07	9.54	20.28	33.86

Notes

1. Time 1 is total time (hrs) and Time 2 is in - line time (hrs)
2. P is primary array consisting of 4 holes
   S is secondary array consisting of the primary array of 4 holes plus a further 6 holes
   T is tertiary array consisting of the secondary array of 10 holes plus a further 8 holes
3. Coding B5106, etc. used during evaluation phase
4. Tenderers B1056 and B3306 did not complete the Tender Schedule
5. Tenderer B6006 was the successfull tenderer

Table 9. Total time, excavation and cover grouting

Description	Arrays (no.)	Refurb (hours)	Duration (weeks)	B1056 (hours)	Duration (weeks)	B5106 (hours)	Duration (weeks)	B2406 (hours)	Duration (weeks)	B4206 (hours)	Duration (weeks)	B3306 (hours)	Duration (weeks)	B1506 (hours)	Duration (weeks	B6006 (hours)	Duration (weeks)
9.1 Hololo to 'Muela (tunnel length 1,860 m)																	
Excavation duration			15.573		13.684		14.332		14.673		11.220		13.944		9.368		13.557
Cover Grouting																	
- Primary	1	7.87	0.059	9.54	0.072	7.33	0.055	14.50	0.109	15.00	0.113	9.54	0.072	3.00	0.023	7.87	0.059
- Secondary	1	18.30	0.138	20.28	0.153	15.33	0.115	24.25	0.182	39.00	0.293	20.28	0.153	4.50	0.034	18.30	0.138
- Tertiary	1	32.07	0.241	33.86	0.255	26.00	0.196	37.25	0.280	67.00	0.504	33.86	0.255	7.00	0.053	32.07	0.241
Total			0.438		0.479		0.366		0.572		0.910		0.479		0.109		0.438
Total weeks			16.011		14.163		14.698		15.244		12.130		14.423		9.477		13.995
Grouting as percentage of total time			2.74%		3.38%		2.49%		3.75%		7.50%		3.32%		1.15%		3.13%
9.2 Ngoajane to Hololo (tunnel length 5,270 m)																	
Excavation duration			54.292		28.832		48.415		43.410		38.229		46.468		30.018		44.381
Cover Grouting																	
- Primary	16	7.87	0.947	9.54	1.148	7.33	0.882	14.50	1.745	15.00	1.805	9.54	1.148	3.00	0.361	7.87	0.947
- Secondary	48	18.30	6.606	20.28	7.321	15.33	5.534	24.25	8.755	39.00	14.079	20.28	7.321	4.50	1.625	18.30	6.606
- Tertiary	16	32.07	3.859	33.86	4.075	26.00	3.129	37.25	4.483	67.00	8.063	33.86	4.075	7.00	0.842	32.07	3.859
Total			11.413		12.544		9.545		14.982		23.947		12.544		2.828		11.413
Total weeks			65.705		41.376		57.960		58.392		62.176		59.012		32.846		55.794
Grouting as percentage of total time			17.37%		30.32%		16.47%		25.66%		38.52%		21.26%		8.61%		20.46%
9.3 Ngoajane to Vent. Shaft 5 (tunnel length 5,720 m)																	
Excavation duration			54.690		37.822		48.739		45.535		45.666		50.489		30.464		48.981
Cover Grouting																	
- Primary	4	7.87	0.237	9.54	0.287	7.33	0.221	14.50	0.436	15.00	0.451	9.54	0.287	3.00	0.090	7.87	0.237
- Secondary	10	18.30	1.376	20.28	1.525	15.33	1.153	24.25	1.824	39.00	2.933	20.28	1.525	4.50	0.338	18.30	1.376
- Tertiary	4	32.07	0.965	33.86	1.019	26.00	0.782	37.25	1.121	67.00	2.016	33.86	1.019	7.00	0.211	32.07	0.965
Total			2.578		2.831		2.156		3.381		5.400		2.831		0.639		2.578
Total weeks			57.267		40.653		50.895		48.916		51.066		53.320		31.103		51.559
Grouting as percentage of total time			4.50%		6.96%		4.24%		6.91%		10.57%		5.31%		2.06%		5.00%

5 Construction

5.1 Tunnel boring machine

The tunnel boring machine used on the Works was a refurbished Robbins 186 hard rock TBM. The basic details of the machine were:

• Manufacturer	Robbins	
Type	186-206	
Year of manufacture	1979	
Working principle	Single gripper, 6 cutterhead drive motors	
• Previous contracts	Subway, Buffalo, USA	3 200 m
	Maka Taif, Saudi Arabia	6 420 m
	Milwaukee Sewer, USA	1 707 m
	Montreal Sewer, Canada	4 850 m
	Milwaukee Collector, USA	1 710 m
• Mass of TBM	Approx 270 t	
Length of TBM	15,2 m	
Length of back up	130 m	
• Cutting diameter	5,18 m	
• Electrical load of cutterhead	6 x 250 HP	
Power of advance	7 400 kN	
Power of advance/cutter	200 kN	
Torque of cutterhead	nominal 1 519 kN-m	
Electrical starting kVA	Approx 2 000	
• Cutting tools	432 mm diameter 37 no.	
• Stroke length	1,50 m	
Cutter head speed	7,3 revs/min	
• Gripper system	4 bracing plates, 3 050 mm x 450 mm	
• Directional control	Gripper cylinders during resetting, front support during boring	
• Protection against water	Electric motors totally enclosed, lowest motor 1,35 m above invert	

The machine was extensively refurbished, including a new main bearing, in the USA over the period April to September 1991 and shipped to Lesotho to arrive in November 1991.

The drill used for forward probing as well as the drilling for cover grouting was mounted on a platform at the front of the TBM back up. The drill an Atlas Copco Tunmec R315H was capable of drilling at an angle of as little as 3° to the tunnel perimeter, the only limiting factor being the ability to collar the drill bit. The drill was capable of rotating to cover the full tunnel perimeter. The layout is depicted in Figure 5.

Forward probing was carried out on a daily basis to a depth of about 70-80 m, this being the best possible daily advance of the machine, thus ensuring a forward probe hole ahead of the machine at all times and without the necessity to stop the machine at irregular intervals for forward probing.

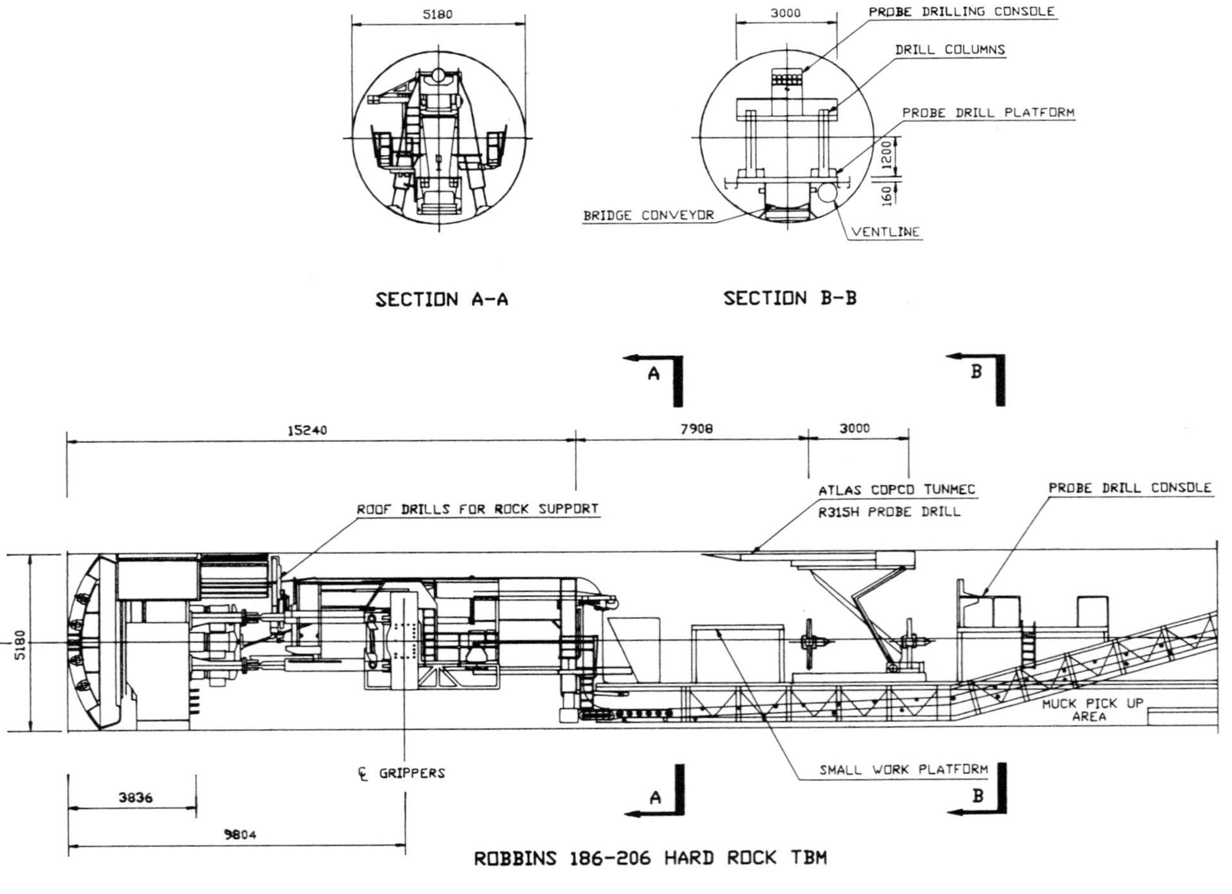

ROBBINS 186-206 HARD ROCK TBM

FIGURE 5

Grouting ahead was carried out with the following equipment which was also mounted on a platform at the front of the TBM back up:

- Keller colcrete grouting module, SD 10 colloidal mixer, 22 kW electric drive, 280 ltr. batch capacity
- agitator, A600 agitation tank, 2,2 kW electric motor
- pump, Mocol 511 grout pump, 7,5 kW electric motor, adjustable flow power
- mechanical packers fixed at the collar of each hole, grout hole diameter 51 mm
- grout mix, w/c ratio 1:1, 3% bentonite, grout hand batched in colloidal mixer.

5.2 TBM drive 1, Hololo to 'Muela Intake

The first TBM drive, the 1,860 m drive south from Hololo to 'Muela Intake commenced on 19 February 1992 and holed out to surface in the Nqoe valley on 25 May 1992. The forward probe holes on limited occasions intersected water of minimal inflow and under little pressure. This water was associated with joints either calcite filled or open (1-5 mm) with no evidence of fill. As the TBM advanced towards the water bearing feature the outflow from the single probe hole decayed at a significant rate indicating little storage within the feature. Whilst the outflow was more than the specified 5 l/min for which grouting was specified it was considered, given the location of the feature in relation to the groundwater model, that little possibility existed for substantially increased inflow to be encountered when the TBM bored through the feature. Records of the rate of decay of inflow from the probe hole were kept. The whole drive was thus excavated without the need to cover grout.

In terms of the Specifications the sharing in time-related charges was calculated as Table 10. It is to be noted that the saving in quantity proportional costs, viz, pumping of water and cover drilling and grouting at bill items accrued fully to the Employer.

5.3 TBM drive 2, Ngoajane to Hololo

5.3.1. General

The second TBM drive, the 5,270 m drive south from Ngoajane to Hololo commenced on 14 July 1992 and was completed on 19 January 1993.

Excavation at the start of the south drive from Ngoajane commenced in much the same manner as for the first drive with occasional inflow to probe holes experiencing rapid decay. The anticipated problems associated with the basalt/sandstone contact did not eventuate. Vertical probeholes drilled by the percussion drill rigs provided for rock support immediately behind the TBM cutter head regularly probed the location of the contact and the condition of the rock to provide the necessary assurance.

However despite the assurance gained from careful investigation of outflow from the forward probe in relation to the groundwater model the Engineer remained concerned that it remained possible that due to lack of diligence it could be possible to bore into a feature/features from which groundwater inflow would exceed, possibly by a large margin, the specified limits within which it was wished to remain. As such the Engineer required that correct and suitable procedures be established which could be clearly understood by all site personnel involved and further required that the aspect of change in risk between the contracting parties be clarified.

Table 10. Calculation of cost saving, Hololo to 'Muela

Class	Refurb. TBM (m/week)	Tender tunnel length (m)	Tender duration (weeks)	Actual tunnel length (m)	Theoretical duration (weeks)
TI MI	192	133	0.691	84	0.438
TI MII	130	106	0.815		0.000
TI MIII	122		0.000		0.000
TI MIV	50		0.000		0.000
TI MII+ MV	72		0.000		0.000
TI MIII + MV	61		0.000		0.000
TII MI	192	834	4.343	721	3.755
TII MII	130		0.000		0.000
TII MIII	122		0.000	3	0.025
TII MIV	50		0.000	13	0.260
TII MII + MV	66		0.000		0.000
TII MIII + MV	58		0.000		0.000
TIII MI	130	438	3.365	960	7.385
TIII MII	109		0.000	5	0.046
TIII MIII	103		0.000		0.000
TIII MIV	42		0.000		0.000
TIII MII + MV	58		0.000		0.000
TIII MIII + MV	55		0.000		0.000
TIV MI	84	191	2.269	24	0.286
TIV MII	78		0.000	25	0.321
TIV MIII	74		0.000	16	0.216
TIV MIV	38	155	4.089	5	0.132
TIV MII + MV	55		0.000		0.000
TIV MIII + MV	50		0.000		0.000
Sub-total		1856	15.573	1856	12.862
Cover grouting			0.439		0.000
Learning curve			1.805		1.805
Temporary suspension			0.150		0.015
Engineer's enlargements			0.361		0.000
Rockfalls			0.181		0.000
		Total Item CLTG8	18.508	Total	14.682
CLTG4 Assemble TBM			4.332		4.332
CLTG5 Commission TBM			1.083		1.083
CLTG6 Introduce to tunnel			0.542		0.542
CLTG7 Initial restricted			1.083		1.083
CLTG9 Remove TBM			4.693		4.693
		Total All Items	30.241	Total	26.415
		LessTheoretical Actual	26.415		
		Saving (weeks)	3.826		
Total value of time related charges		Maloti	5,801,279.51		
Revised time related duration			0.873		= 26.415/30.241
Revised time related charges		Maloti	5,067,269.86		= 5,801,279.51 x 0.87
Shared saving in cost		Maloti	734,009.65		
50 % to either party		Maloti	367,004.82		

5.3.2. Technical procedures

The procedure set up to check whether the groundwater inflow decayed at a suitable rate was:

- Drilling of probehole
- Water pressure test (abbreviated)
  a) Measurement of take at 7 bar pressure
  b) Measurement of water inflow
- Excessive water from probehole

  Should water in excess of 5 l/min from the probe hole be intersected then it was necessary to establish that the decay in total inflow from probe hole(s) lay below the logarithmic discharge curve depicted in Figure 6. Inflow was measured by:
  a) Measurement of quantity of water inflow for a period of 2 hours at 0,5 hour intervals and drill second probe hole
  b) Measurement of water inflow from both probeholes for
     - first 2 hours : every 0,5 hour
     - after 2 hours : every hour
     - after 12 hours : every second hour

  Should the discharge curve resulting from the above measurements lie below the curve and the total inflow had dropped to less than 10% of the total initial inflow from all holes then the TBM would be allowed to continue excavation through the feature provided the total inflow lay below the limits set out in the design assumptions following. In any instance the TBM would be allowed to advance to within 10 m of the water bearing feature whilst monitoring was in progress.

  However should during observance of water inflow there be detected visually either discolouration of the inflow or the prescence of sediments then erosion of the water bearing feature could be assumed and more stringent probing measures would be undertaken particularly related to additional probe holes and to monitoring the rate of decay for a longer period.
- Design assumptions

  The discharge curve does not take cognisance of the amount of water inflow and thus the following criteria would also apply
  a) First event :

     $Q_0 \leq 15$ l/s (initial inflow from one probehole)
     $Q^{(2)}_{15} \leq 1,5$ l/s (from two probeholes after 15 hours)
  b) Second event :

     In the case that a constant waterflow of $\Sigma Q = 25$ l/s exists in the tunnel from the first event, then the following values will be adopted for the second event:
     $Q_0 \leq 10$ l/s (initial inflow from one probehole)
     $Q^{(2)}_{15} \leq 1,0$ l/s (from two probeholes after 15 hours)
  c) Third event :

     In the case that a constant waterflow of $\Sigma Q = 40$ l/s exists in the tunnel, decisions will have to be taken whether cover grouting of the third event or fissure grouting of one or both of the previous events is necessary in order to avoid a total water flow in the tunnel which exceeds 50 l/sec.

TBM DRIVE 3, NGOAJANE TO VENT SHAFT No. 5

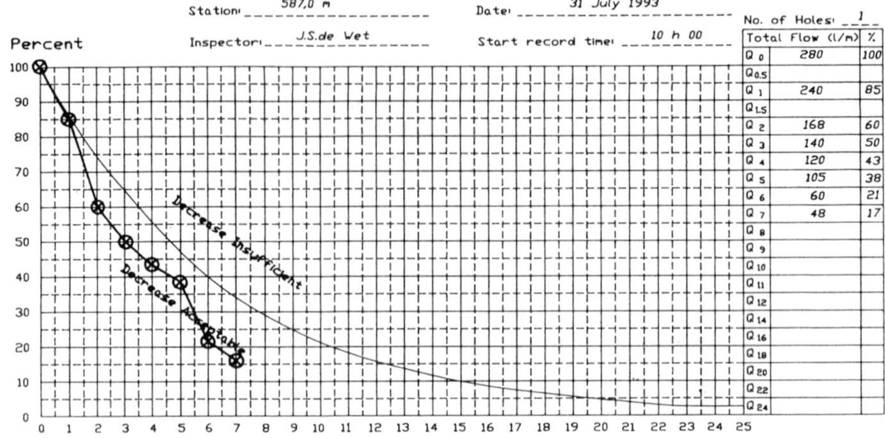

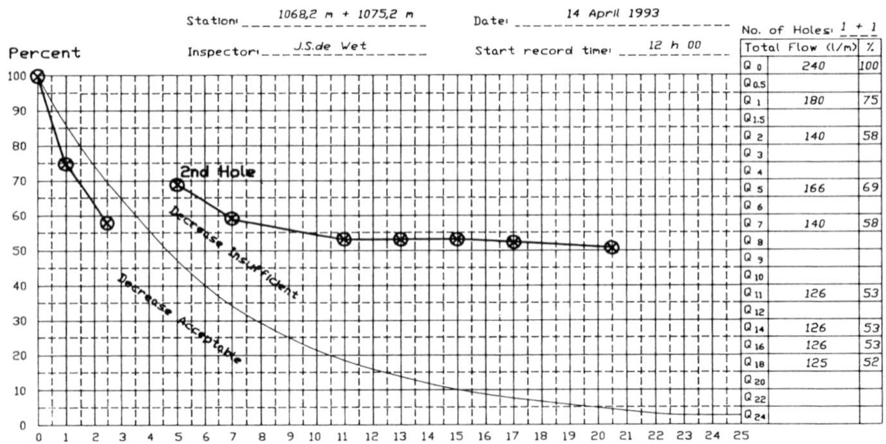

FIGURE 6 : WATER INFLOW, ACCEPTED DECREASE CURVE

A typical set of readings for a specific feature which decayed in a satisfactory manner through which the TBM advanced safely without grouting is shown in the upper chart of Figure 6.

In all instances where the inflow did not fall below the curve cover grouting as specified was carried out.

5.3.3. Contractual considerations

The Engineer required that due to the changed procedure to that specified at Tender that the Contractor carry the risk should total inflow be in excess of 50 l/sec and, inter alia, any delay or associated cost due to excessive groundwater inflow, the provision of additional pumping requirements, time lost or cost incurred in establishing additional pumping requirements would all be to his account. The only exception to this requirement was should inflows in excess of 50 l/sec be encountered from forward probe holes only then the risk and associated cost would lie with the Employer.

The Contractor accepted these conditions since he recognised the significant possibility of saving time by boring through water bearing features without having the necessity to grout.

5.3.4. Grouting of water bearing features

During October/November 1992 whilst excavating the second TBM drive a series of water bearing features were encountered by probing ahead of the face, and this series of features proved to be the worst conditions encountered during the excavation phase and had the most significant effect on progress. Table 11 provides relevant statistics for each of the major features.

Table 11. Summary of water inflow data

Station of fissures (1)	Number of probe holes (2)	Combined flow (l/sec) (3)	Reduced to 70% (l/sec) (4)	Adjusted to full tunnel (l/sec) (5)	Estimated tunnel inflow (l/sec) (6)
3372-75	10	16.3	11.4	114	18
3381-89	8	8.8	6.1	74	18
3407-15	5	20.9	14.6	234	49
3420-25	7	13.9	9.7	136	27
3475-81	9	14.8	10.4	114	29
3601-19	6	7.3	5.1	82	20
3653-74	9	27.2	19.0	209	69
		Totals	76.4	963	230

Table 11 shows the determination of the likely total inflow had grouting not been instructed and is explained as follows:

- Column 1 details the distance from the start point of the drive over which the water bearing feature was encountered

- Column 2 details the number of probe holes from which the cumulative water flow was measured
- Column 3 details the combined flow from the probe holes. In the case of the first feature 16,3 l/sec was an average measured over the first 14 hours which had dropped to 11,4 l/sec (i.e. 70%) after a further 40 hours and was still flowing at this rate some 33 hours later. Clearly this feature was being recharged and would not decay and grouting was required. The subsequent six features encountered were immediately grouted since it was expected that their characteristics were the same as the first one and water inflow would not significantly decay
- Column 4 then uses the experience gained from the first feature to predict that all features would only decay to 70% of the initial inflow
- Column 5 assumes that the borehole inflow provides a representation of the joint pemeability of the feature with the water inflow into the excavation from the entire tunnel circumference calculated pro rata to the combined fissure length around the circumference of all boreholes measured for each feature. The cumulative inflow for all the seven features was then calculated to be 963 l/sec which it was considered may decay in the long term by a further 50% to 480 l/sec.
- The cumulative inflow calculated above was checked by independent calculations and following an estimate of the hydraulic conductivity of the jointed rock estimates of the long term inflow were made using the method presented by Goodman [2]. Column 6 details these estimates. The long term inflow calculated (230 l/sec) is about half that originally derived (480 l/sec) but is well in excess of the specified maximum (50 l/sec). The decision to grout the features was substantiated.

Grouting was carried out not to exclude inflow but to reduce it to reasonable proportions and in the event the combined inflow was reduced to 2-3 l/sec. The quantities of material used for each feature are detailed in Table 12

Table 12. Grouting quantities, Ngoajane to Hololo drive

Station of fissures	Drilling length from face (m)	OPC (kg)	Bentonite (kg)
3372-3375	224,8	11,000	286
3381-3389	666,3	29,325	781
3407-3415	353,2	46,750	1,253
3420-3425	183,2	15,850	466
3475-3481	237,1	3,650	91
3601-3619	194,0	6,500	195
3653-3674	564,5	9,700	273
Total	2,423,1	122,775	3,345

5.4 TBM drive 3, Ngoajane to Ventilation Shaft No.5

5.4.1. General

The third TBM drive, the 5,720 m drive north from Ngoajane to Ventilation Shaft No.5 commenced on 11 March 1993 and holed through into the underground Works of the Delivery Tunnel North Contract for the LHWP on 23 August 1993.

The procedures developed during the first and second TBM drives for dealing with water which had proved successful in the Ngoajane South drive could, following observations of geological and hydrological conditions in the Ngoajane North drive, it was considered, be modified and new criteria developed.

5.4.2. Technical considerations

Initially in the third drive the tunnel was driven though Clarens Sandstone which displayed well developed sedimentary fabrics. Such horizons were not encountered as persistently in the Clarens Sandstone of the two other drives. Further, on exposure, the degree of seepage from the rock mass suggested that these sandstones were relatively more porous and permeable than those generally encountered in the other drives.

It was also seen that further than about 1 km (or 1 month) behind the TBM, seepage was largely confined to a zone about 1-1.5 m above the invert, with the sidewalls and crown being relatively dry. Approaching the TBM, the level below which seepage occurred gradually rose up the sidewalls and into the crown in the TBM back-up area. This suggested that the rock mass permeability was relatively high compared with the completed drives and that the tunnel may be acting as a drain.

The accepted groundwater decay curve developed for the second TBM drive (Figure 6) was based on experience of early incidents of water inflow in the Hololo South and Ngoajane South drives. The drawdown curves showed that the inflows result from limited storage within the fracture cavities with minimum seepage recharge from the adjacent rock mass (upper chart of Figure 6 as a typical example).

However drawdown curves for some of the early water inflow incidents in the Ngoajane North drive show that while there was an immediate rapid decrease in inflow it flattened off between about 50-60% of the initial inflow value (lower chart of Figure 6 as a typical example). However over the next 5 days the inflow was observed to gradually decrease from about 50% to 30-35% of the initial value. It was interpreted that the steep initial decay of flow indicated limited fracture storage around the tunnel and that the decay starts to flatten off due to seepage recharge of the fracture from the permeable rock horizons which it intersects. The inflow continues to gradually reduce as the hydraulic gradient towards the fracture is progressively lowered until equilibruim conditions are achieved.

The following conclusions were thus drawn :

- the sandstone in the north drive is more permeable than in the south drive;
- the tunnel and fractures that it intersects act as a drainage network to the surrounding rock mass;
- seepage from the rock mass and from fractures will continue to decrease once intersected by the tunnel (tentatively, for about 1 km or 1 month behind the face).

It was therefore concluded that no single inflow curve could be used to adequately define the acceptability of an inflow when making the decision to drive through a water bearing feature.

It was considered that provided the fracture storage had been demonstrated to be limited (by having a rapidly decaying initial inflow from probe holes) and the seepage recharge component (represented by the near horizontal portion of the inflow curve) is sufficiently low, then the TBM can advance through a feature without cover grouting.

Arising from these observations new procedures were developed and agreed between the Contractor and Engineer and these are depicted in Figure 7.

The advantage arising from the revised procedure was primarily that there was more flexibility to allow for excavation to continue without grouting and greater cognisance of the actual groundwater conditions was made.

5.4.3. Contractual considerations

The Engineer considered that these revised procedures further increased the risk of total inflow down the tunnel being greater than 50 l/sec. By adhering to the curve shown in Figure 7 it was possible to allow up to 30 l/sec residual flow down the tunnel prior to any grouting being ordered whereas the tender philosophy was to keep the residual flow to manageable amounts as deemed appropriate for the balance of the drive outstanding. As such the Engineer required that:

- In the event that the Contractor excavated the TBM through an identified water bearing feature in accordance with the revised procedures without grouting and total residual flows down the tunnel were realised in excess of 50 l/sec then the Contractor would be responsible for any additional pumping requirements, any delays or costs of any nature associated with an excessive groundwater inflow and the like. This stipulation had previously been accepted for the second TBM drive.
- In the event that probe holes yield groundwater up to 50 l/sec the Engineer believed that the revised procedure increased the risk of total flows in excess of 50 l/sec being realised down the tunnel. For example should a base flow of 30 l/sec be existing in the tunnel and probe holes intersecting a water bearing feature produce 50 l/sec then a total instantaneous flow (before grouting could be carried out) of 80 l/sec would be realised. In this scenario as envisaged at tender cover grouting would have already limited the base flow to say 0 - 10 l/sec and on intersecting the same water bearing feature the total would have been only 50 - 60 l/sec. As such the Engineer believed it to be equitable in the changed circumstances that the Contractor accept the consequences of additional water introduced into the tunnel by probeholes, intersecting a feature ahead of the face, of up to 50 l/sec. It could be argued that this stipulation should have been in force during the second TBM drive.
- In the event that forward probeholes are dry and a water bearing feature is excavated through by the TBM the Engineer recognised that it remained possible for forward probe holes to miss oblique (to the tunnel line) water bearing features. In these circumstances if such a feature is driven though by the TBM and total groundwater inflow in excess of 80 l/sec are intersected then any additional costs

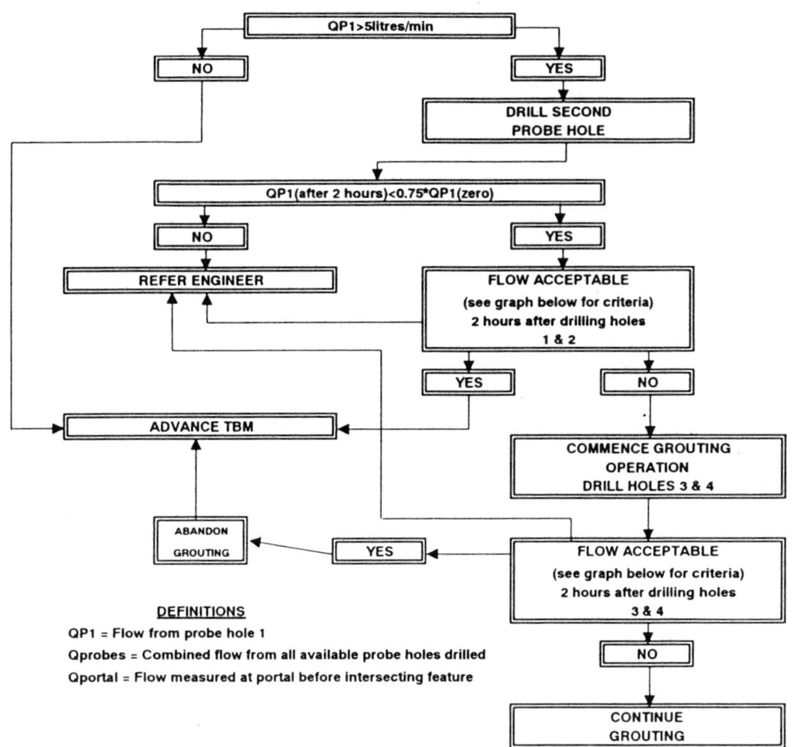
REVISED PROCEDURE FOR DEALING WATER INFLOWS INTO TBM DRIVES
QP1>5litres/min
NO
YES
DRILL SECOND PROBE HOLE
QP1(after 2 hours)<0.75*QP1(zero)
NO
YES
REFER ENGINEER
FLOW ACCEPTABLE (see graph below for criteria) 2 hours after drilling holes 1 & 2
YES
NO
ADVANCE TBM
COMMENCE GROUTING OPERATION DRILL HOLES 3 & 4
ABANDON GROUTING
YES
FLOW ACCEPTABLE (see graph below for criteria) 2 hours after drilling holes 3 & 4
DEFINITIONS
QP1 = Flow from probe hole 1
Qprobes = Combined flow from all available probe holes drilled
Qportal = Flow measured at portal before intersecting feature
NO
CONTINUE GROUTING

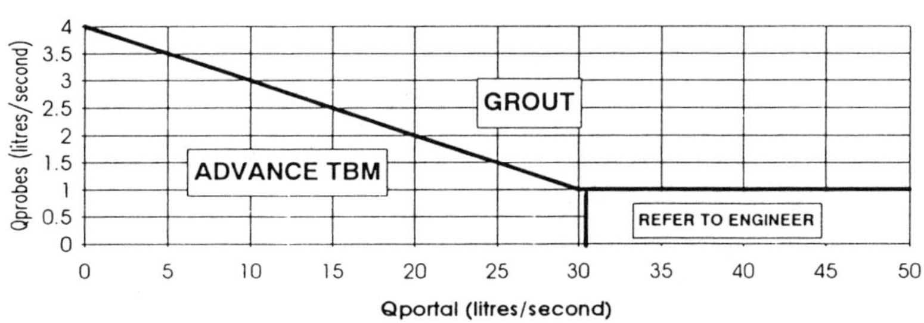
WATER INFLOW INTO TBM DRIVES : CRITERIA FOR ADVANCING TBM
GROUT
ADVANCE TBM
REFER TO ENGINEER
Qprobes (litres/second)
Qportal (litres/second)
0
0.5
1
1.5
2
2.5
3
3.5
4
0
5
10
15
20
25
30
35
40
45
50

Figure 7

associated with dealing with water would be reimbursed to the Contractor in accordance with the Specifications (50 l/sec in Specifications now increased to 80 l/sec.). Again it could be argued that this stipulation was relevant to the second TBM drive also.

The Contractor however was unable to accept the Engineer's contentions and regretably the revised procedures could not be agreed prior to excavation being completed.

5.4.4. Grouting of water bearing features

During June 1993 whilst excavating the third TBM drive from station 3,012 to 3,912 grouting was carried out to reduce the groundwater inflow to reasonable proportions at 5 specific features.

The quantities of material used for the five features are detailed in Table 13.

Table 13. Grouting quantities, Ngoajane to Vent. Shaft 5

Feature	Drilling length	OPC (kg)	Bentonite (kg)
1	300,2	17,125	514
2	368,4	12,600	378
3	414,3	9,600	288
4	179,7	2,900	87
5	260,6	24,500	735
Total	1,523,2	66,725	2,002

6 Progress

Progress on the TBM drives was exceptional and what was reported to be the previous "All African" tunnelling record at 979 metres in one month was bettered on three separate occasions by the TBM, viz:

- August 1992 : 1,018.3 m
- September 1992 : 1,203.4 m
- April 1993 : 1,324.4 m

This was particularly noteworthy in that the back up train accommodated only one muck train at a time and switching was carried out at a California switch up to some 500 m behind the back up.

Clearly progress was affected by cover drilling and grouting and this is summarised in Table 14 and depicted in the pie charts included as Figure 8.

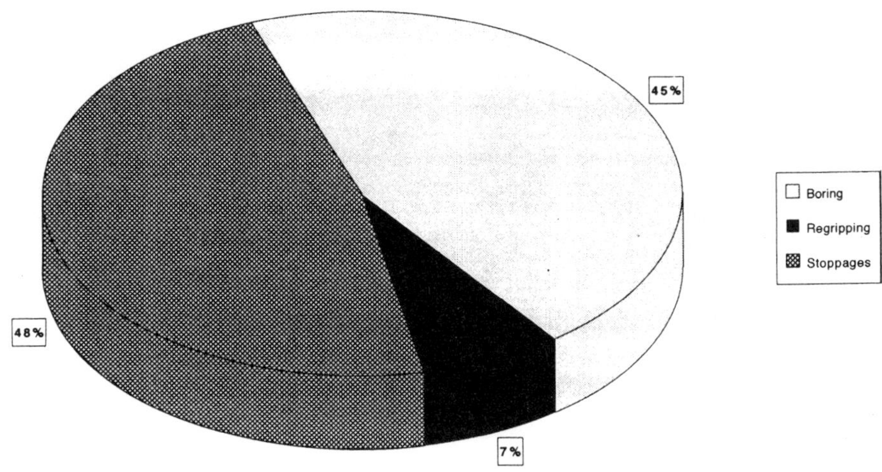

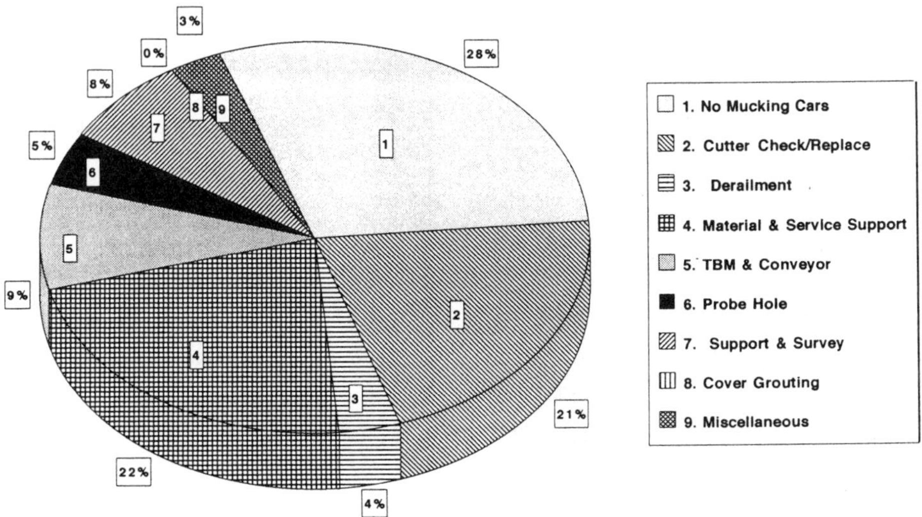

Figure 8.1 Hololo to 'Muela

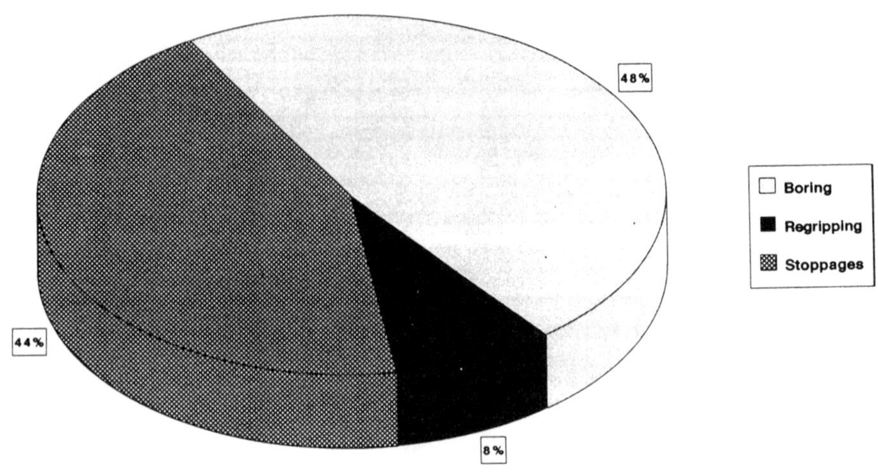

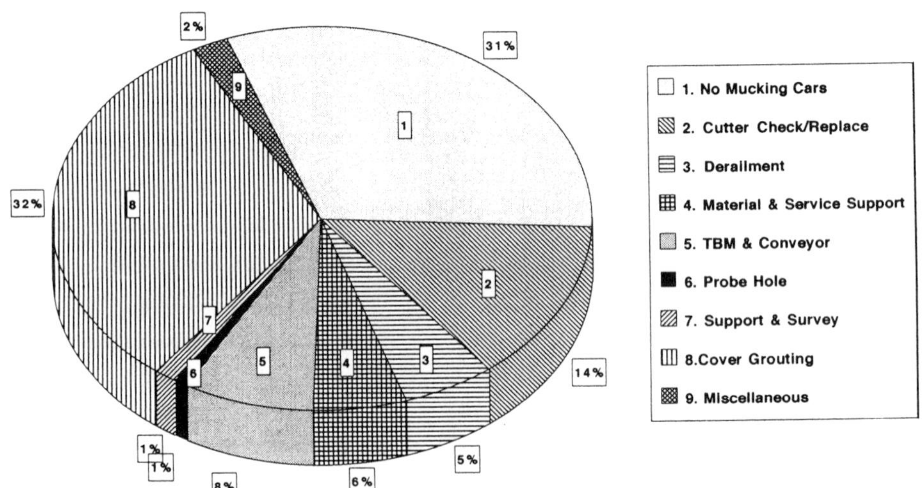

Figure 8.2 Ngoajane to Hololo

TBM Time Utilisation - Ngoajane to Vent Shaft 5 Drive

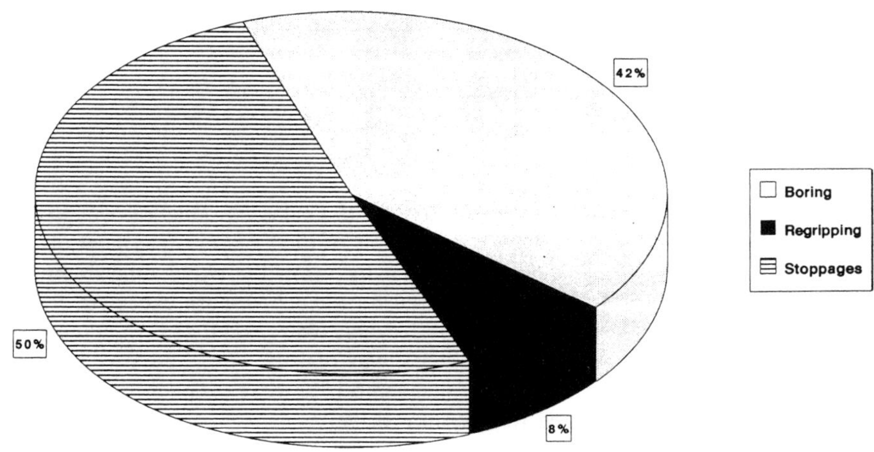

TBM Stoppage Analysis - Ngoajane to Vent Shaft 5 Drive

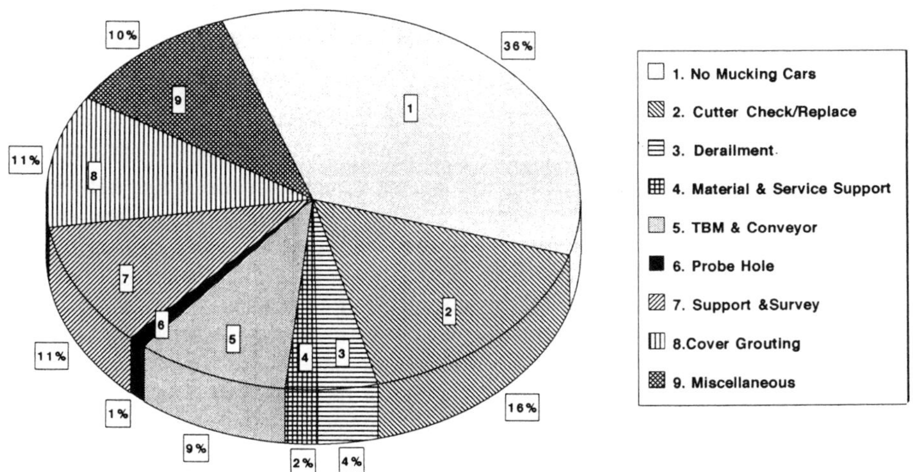

Figure 8.3 Ngoajane to Ventilation Shaft 5

Table 14. Cover grouting as % of total TBM utilisation

TBM drive	Cover grouting as % of stoppage	Stoppages as % of utilisation	Cover grouting as % of utilisation
Hololo to 'Muela	0	48	0
Ngoajane to Hololo	32	44	14
Ngoajane to V.S.5	11	50	5

In the case of the second TBM drive a total of 19 days production was lost due to cover drilling and grouting. However, in terms of the Contract, payment of time for grouting (time-related charges) is calculated by using the cycle times submitted at tender to derive the theoretical time required for the actual work executed. Using the Contractors cycle times a duration of 22 days was calculated showing that the Contractor's tender allowance for the work was realistic (i.e. 22 days in lieu of the actual 19 days indicating that he performed some 15% more efficiently than tendered). Similar results were calculated for the third drive.

7 Conclusions

Arising from the various aspects of dealing with water in the three TBM drives of the Delivery Tunnel South the following is concluded:

1. The tender documents adequately detailed the risks associated with encountering groundwater in each drive and they quantified for tendering purposes the anticipated inflows that could be encountered.

   Actual inflows encountered were somewhat higher than predicted but the documents clearly provided sufficient and proper payment mechanisms for evaluating the work required.

   The documents, further, clearly detailed the apportionment of risk between the contracting parties.
2. A spirit of co-operation existed between the Contractor and the Engineer to safely execute the works yet at the same time modify the tender specifications both technically and contractually to the advantage of both parties. At all times the Employer and the Project Insurers (for Employer provided Contract Works Policy) were kept fully appraised of the measures being undertaken.
3. The Contractor had cost advantages inherent in being able to reduce the value of work required to deal with water in that savings accrued were shared in an equal proportion between the contracting parties.
4. A balanced tender was submitted by the Contractor regarding dealing with water as evidenced by comparison with other tenderers and by the tender time for cover grouting correlating very closely with the actual time required. As such the payment mechanisms in the contract fairly compensated him for the work required.
5. Whilst the total length of each tunnel requiring cover grouting was reduced due to the measures taken there remained a close correlation between the actual percentage

of total time required for cover grouting and that calculated at tender based on the Contractors cycle times, viz.

Drive	Tender (%)	Actual (%)
Hololo to 'Muela	2,74	0
Ngoajane to Hololo	17,37	14
Ngoajane to V.S.5	4,50	5

This correlation came about, in part, in that not only did the Contractor improve his productivity by about 15% on his tendered cycle time for drilling and grouting but also his tendered advance rates for each rock class/support class combination were bettered in practice.

8 Acknowledgements

The author gratefully acknowledges the permission of the Employer, the Lesotho Highlands Development Authority to publish this paper.

In addition the author acknowledges the significant input provided to the paper by work carried out during the excavation phase of the Works by the respective site teams of the Contractor, Lesotho Highlands Project Contractors (Joint Venture of Spie Batignolles, Balfour Beatty, Campenon Bernard, LTA and Ed Zublin) and the Consultant, Lesotho Highlands Tunnel Partnership (Partnership of Lahmeyer International, Mott MacDonald Ltd., Ninham Shand Inc., Van Niekerk Kleyn and Edwards, Steffen Robertson and Kirsten, Keeve Steyn Inc.).

9 References

1. Lesotho Highlands Water Project, (1989). Contract No. LHDA 126, Delivery Tunnel South, Tender Documents, Volumes 1 to 5.
2. Goodman, R., Moye, D., Schalkwyk, A. and Javandel, I (1965) "Ground water inflow during tunnel driving" Engng Geol. 2, 39-56.

Session 7: Tunnel-boring Machines

CEN safety standards for tunnelling machines and air locks

Manfred Bandmann
H. Ramisch
Tiefbau-Berufsgenossenschaft, Munich, Germany

Abstract
CEN standards are European Rules of the Art, which machine manufacturers have to observe. The members of CEN/CENELEC are the national standards organisations of the individual countries. European standards take precedence over national standards. Machine safety standards (Type C-standards) are drawn up in so-called Technical Committees (TCs). The safety standards for tunnelling machines are prepared in Working Group 4 of the CEN/TC 151 Technical Committee "Construction equipment and building material machines". The present speaker is chairman of this Technical Committee, which is broken down into sub-committees, the so-called Working Groups. The countries United Kingdom, Sweden and Germany were in charge of formulating the four standards for tunnelling machines being presented here. These standards deal with specific hazards and appropriate safety measures, provided that the tunnelling machines are used in accordance with their intended purpose and as recommended by the manufacturer.

1 Introduction

A single European market for some 350 million people came into being on 31st December 1992. In addition to the 12 states of the European Community it also includes the 6 EFTA countries and forms a region without internal borders in which the free movement of goods, persons, services and capital is guaranteed. The basis for the abolition of all trade barriers and an end to distortion of competition was created by the Treaties of Rome on 25.03.1957 and the Treaty on the European Economic Community which was signed at the same time. The Single European Act of 01.07.1987 and the Maastricht Treaties of 07.02.1992 confirmed this common will and set deadlines for the implementation of these objectives.

2 European Regulations

Dismantling of economic trade barriers also includes the elimination of technical trade barriers, e.g. national safety regulations and standards which influence the design of machines, e.g. DIN standards or construction and equipment specifications contained in accident prevention regulations.

New criteria have been applied for the performance of this task:

- It is to be assumed that the member states will pursue essentially identical objectives in the field of safety.

- Harmonization of legal provisions is to be restricted to defining basic safety requirements within the framework of directives.

In future the technical specifications for products in Europe will consist accordingly of EC directives and European standards (ES) which will replace national product standards. The European standards organizations CEN and CENELEC will be assigned the task of drawing up these standards which are necessary for completing or defining in specific terms basic safety requirements in the Machinery Directive - 89/392/EEC amended by 91/386. Specific safety requirements are laid down in special standards for particular groups of machinery, e.g for tunnelling machines and air locks. The Machinery Directive quoted above has applied since 01.01.1993 for the construction and equipment of tunnelling machines. Standard prEN 815 applies to the safety of "unshielded tunnel boring machines and rodless shaft boring machines for rock", while the specific safety standards for the other tunnelling machines are ready in draft form and are currently at the enquiry stage.

Safety standards are divided into three categories and defined as follows:

A standards (Basic safety standards): Contain basic terms, design principles and general aspects applicable for all machines, equipment and installations.
B standards (Group safety standards): Deal with one aspect of safety or one type of safety equipment which can be used for a large number of machines, equipment and installations.
B 1 standards: Refer to specific safety aspects (e.g. safety distances, surface temperatures, noise).
B 2 standards: Refer to safety equipment (e.g. two-hand controls, interlocks, contact mats, isolating protection equipment).
C standards (Machine safety standards): Contain detailed safety requirements for a certain type of machine or group of machines.

The draft standards for tunnelling machines and air locks belong to the group of type C standards. CEN standards are EC "harmonized standards" which are published as standards in the EC gazette. They are European technical standards which manufacturers of machinery are obliged to observe if only because of their responsibility in terms of product liability. Membership of the CEN/CENELEC is made up of the national standards organizations of the individual countries.

The following principles apply to European standardization work:

- European standards have priority over national standards.
- Decisions on standardization applications are taken at the Technical Bureau (BT).
- The decision on the beginning and end of standstill obligations in national standardization is made when a standard project is taken up.
- The final vote on EN standards is taken by means of a weighted voting procedure whereby each member country receives a certain number of votes according to its population and its industrial standing, e.g. Germany has the highest number 10 and Iceland the minimum number 1.
- National interests in the drafting of an European standard are represented by each country's standards organization, e.g. in the case of Germany by DIN.

The machine safety standards (type C standards) must be so formulated that the following safety philosophy is fulfilled in the design of the machine:

- Defining limits of application of the machine.
- Identifying hazards and assessing risk.
- Eliminating hazards or restricting risks as far as possible.
- Incorporating protection devices against residual risks.
- Informing users about residual risks.
- Taking every additional precautionary measure into account.

This work is performed by the Technical Committees (TCs). According to the CEN General Technical Report 1992, 299 TCs have been set up to date.

The standards for construction machines and building material machines are elaborated in Technical Committee CEN/TC 151 which is divided into 14 sub-committees known as Working Groups (WGs). Germany has agreed to be responsible for the secretariat and chairmanship of TC 151, and the function of convener of working group WG4 whose task is the standardization of tunnelling machines.

3 European Standards for Tunnelling Machines

The safety requirements for tunnelling machines are drawn up by the WG4 study group as a C-type standard. The purpose of this is to harmonize the large number of national safety regulations which reflect the state of safety engineering in this field in Europe and to embody them in a new European Standard in agreement with the member states. These regulations are product safety standards which contain detailed safety requirements for a particular machine or group of machines.

The safety standards for tunnelling machines and air locks presented here deal with characteristic hazards specific to this type of machinery and corresponding protective measures, subject to the proviso that the machines are used in accordance with the regulations, following the recommendations of the manufacturer.

Not included are hazards which exist with all mechanical, electrical, hydraulic, pneumatic and other components of machines and which are dealt with in superordinate standards. This includes in particular the general design principles of EN 292 which regulate the safety of machines and which are to be observed in the design of tunnelling machines.

How precisely must the state of the art for the safety of machines be defined? The utmost accuracy is demanded of definitions so that the legal situation with regard to supervisory bodies is certain and for reasons of product liability after accidents. On the other hand these could be interpreted as a restriction of the constructor's design freedom and hence as an obstacle to technical progress. Technical, economic and political factors must be finely balanced with the object of creating a uniformly high safety standard and at the same time eliminating technical obstacles to trade within the European economic region.

Tunnelling machines have been divided into three groups: road headers and impact rippers, shielded machines and tunnel boring machines. A standard has also been drafted for air locks. Denmark, Germany, France, Great Britain, Sweden and Switzerland participated in drafting the standards.

The standards presented here are largely harmonized and apply to construction, equipment, marking and testing of tunnelling machines and air locks. Additional regulations apply for their use in coal mining.

The most important effects of these European standards on the manufacturer and safety of tunnelling machines and air locks are illustrated using a few striking examples.

3.1 Structure and Contents of the Standards
The structure of the safety standards for tunnelling machines must comply with the rules for the drafting and layout of European standards (PNE regulations). The elements which together form a standard are divided into the following two groups:

a) Normative elements
contain the main stipulations of a standard in the form of requirements, recommendations, permits or alternatives.

b) Informative elements
1) Introductory elements.
Identify the standard, present its contents, explain background, development and connection with other standards.
2) Supplementary elements.
Provide additional information for applying and interpreting the standard.

The safety standards presented are divided into the following sections:

Foreword
References to:
- Responsible committee "TC 151".
- Responsible subcommittee "Tunnelling machines".
- Date of publication.
- Other organizations which have helped with the drafting of the standard.

Introduction
References to:
- Technical contents.
- Reasons for drafting.
- Commentaries.
Contains no requirements.

Field of application
Definition of:
- Subject and aspects dealt with.
- Limits of applicability.
Contains no requirements

Normative References
List of documents (usually standards) to which normative reference is made in the text in such a way that they are indispensable for the application of the standard.

Definitions
Explanation and definition of specific terms.

List of hazards
List of all hazards which have been identified as serious hazards specific to tunnelling machines in the risk assessment procedure and for which measures to eliminate or reduce the risk are required.

Safety requirements and/or measures (Section 5 of the standard)
This section contains requirements and/or measures relating to hazards during use of tunnelling machines for which special stipulations are required.

Requirements must if possible be specified in the form of protection objectives. It may however be necessary to define technical protective measures in order to ensure that measures are adequate.

Providing these safety standards contain no stipulations to the contrary, tunnelling machines must comply with the applicable regulations of the type A or B standards, in particular EN 292 "Safety of machines, equipment and installations" including Annex A of Part 2. They must also be constructed in compliance with the statutory safety requirements.

Confirmation of the safety requirement and/or measures (test procedure)
Instructions on how the machine's compliance with the stipulated requirements and/or measures can be verified (test procedure) are contained in the safety standards for tunnelling machines, classified according to the particular requirements. The standards must therefore contain all information relating to the procedure for determining characteristics or for verifying compliance with stipulated requirements and for ensuring that the test results can be reproduced.

The verification procedures must be suitable for the type of requirements and/or measures. Verification can be provided by means of investigations, corrections, tests, calculations or other procedures.

Discussion procedures may be suitable for special types of requirements and/or measures, e.g. stability, position of the centre of gravity, design of pressure bulkheads, and if, for example, only destructive test methods are possible.

User information - operating instructions
The machine manufacturer must supply operating instructions with the tunnelling machine containing information for safe and correct operation, repairs and maintenance.

The standards presented here concentrate only on the safety aspects in the operating instructions and deal with the following manuals:

- Operating manual.
- Repair and maintenance manual.
- Spare parts catalogue.

The text must be clear, easily understood, appropriate and complete. All information relating to personal safety must be printed in a type which clearly distinguishes it from the rest of the text. The manuals must be supplied in at least one of the official national languages of the first country of use and in other languages as well on request. The requirements for the contents of the operating manual are reproduced below using the example of a road header (RH):

The operating manual must inform the operator how the RH is to be used. It must contain all instructions which are important for the safe use/operation of the RH.

The operating manual must contain the following information:

- The same information as on the machine data plate.
- An exact description of the area of application of the RH giving precise instructions about the geological conditions for which the RH is designed.
- Specification of hazards and what measures are to be taken in order to use/operate the machine.

- A statement that the operators must have practical instruction in the operation of RH with particular emphasis on the afore mentioned safety precautions including fire-fighting.
- Complete instructions on the operation of the RH.
- A general overview of the RH in the form of drawings, photos and plans of electrical, hydraulic and pneumatic control circuits. All illustrations shall be adequately large and clear.
- A list of names of the main components including function, place of installation and relationship to the RH as a whole.
- Special warnings indicating danger spots and residual risks on the RH in the working area of persons who cannot be protected adequately or at all by protective equipment, such as moving parts, crushing and shearing places, rotating parts (e.g. cutter head), falling objects, high voltage.
  see: EN 292-1, clause 5.5
  EN 292-2, clause 5.1
- Information about known risks in the case of inadequacies in the protection system.
- Details of the steepest inclination for which the RH is designed.
- A warning that during maintenance work on the cutter or impact ripper head of a RH no work or tests on electrical or hydraulic equipment may be carried out which influence the control systems of the cutter or impact ripper head.
- An instruction to the effect that during maintenance and repair work on the cutter or impact ripper head, communication must be assured between personnel at the head of the RH and the machine operator and must be maintained while work is performed on the head of the RH.
- A specification as to how and where the EMERGENCY OFF devices are installed and how they function.
- A description of the control elements and their functions.
- A description of the symbols used.
- An instruction that operating persons must wear ear protection in the event that the average noise level in the working area over 8 hours exceeds 85 dB(A).
- Information about the location, use, maintenance and inspection of fire extinguishers.
- Instruction that e.g. conveyor belts may only be operated if the protective devices are installed.
- Information about size and type of the systems for ventilation, breathing air monitoring and dust prevention.
- An instruction that after a shut-down as a result of excessively high methane concentration, the RH may only be switched on again after the cause of shutdown has been eliminated and this may only be performed by an authorized person.
- Description of additional measures which are necessary in the event of occurrence of methane and other explosive and toxic gas concentrations (e.g. use of gas warning devices).

The contents of the operating instructions must comply with clause 5.5.1 of EN 292-2.

Marking
Every safety standard must contain stipulations with regard to the minimum marking. Machine data plates on tunnelling machines must contain the following information:

- Name and address of the manufacturer.
- EC symbol including year of manufacture.
- Type designation.
- Series number.
- Engine rating.
- Weight in kg with the most used equipment.
- Performance information.
- Dimension of the explosion protection if applicable.

The markings of a tunnelling machine also include warning signs and symbols. These must point out danger spots which are not due to the design and which cannot be guarded with protective equipment and also residual risks by means of suitable warning signs, e.g. indicating:

- Moving parts.
- Crushing danger points.
- High voltage.
- Noise.
- Laser radiation.

Symbols which are easy to understand (pictograms) must be used in preference to written warnings.

Annex A
All three safety standards for tunnelling machines contain informative references in Annex A to other standards which were important for their drafting and to which no reference or no normative reference was made in the text of the standard.

Annex B
All three safety standards for tunnelling machines contain additional requirements for use in explosive atmospheres and in coal mining in Annex B. This section is currently being prepared by the Technical Committee CEN/TC 196 - Mining machinery.

This completes the overview of the general sections of the safety standards for tunnelling machines which are to a large extent harmonized and largely identical.

Safety requirements (Section 5) of European standards for tunnelling machines and air locks are reproduced below in as far as they are unusual and deviate from existing national regulations. (Text of the standard in italics). It should be noted that all 4 safety standards for tunnelling machines quoted here are drafts which may differ from the intended standards.

3.2 pr EN 815 "Safety of unshielded tunnel boring machines and rodless shaft boring machines for rock"

This draft standard has already gone through the public enquiry procedure. At present the objections are being discussed and a final draft worded before the standard is submitted to the CEN member states for the formal vote.

The formal vote, as already explained in Point 2 "European Standardization" is taken according to a weighted voting procedure. If the standard is accepted it is subsequently published by CEN and then adopted in national standards by the CEN member states within 6 months.

The following abbreviations are used in the following text for the tunnelling machines dealt with in prEN 815:

TBM... Tunnel boring machine without shield.
SBM... Rodless shaft boring machine.
BM... Unshielded tunnel boring machine or unshielded reaming tunnel boring machine, or rodless shaft boring machine.

5 Safety requirements and/or measures
5.1.6 Handling of heavy loads
5.1.6.1 Handling of machine parts
Where the weight, size or shape of parts of a BM prevents them from being moved by hand the parts shall be either:

- *fitted with attachments for lifting gear or so designed that it is possible to fit attachments for lifting gear,*
- *or so shaped that standard lifting gear can easily be attached.*

Suitable equipment shall be provided for transportation and lifting of such parts which require regular replacement, e.g. cutter discs.

5.1.6.2 Handling of elements for rock support
When a rock support system requires the lifting of elements weighing more than 50 kg, an erecting device shall be provided.

Erecting devices may be either stationary, with cable winches and levers for moving the elements or rotary for positioning the elements, directly at any position around the circumference of the tunnel. Winches and drive motors shall in all cases be fitted with mechanical brakes which are released during operation....

5.1.7 Gripping and regripping
It shall be ensured that the cutter head drive can be started and thrust force applied, only when the minimum gripping pressure required to prevent rotation or sliding backwards of the BM has been reached.

5.1.8 Access to and egress from operating positions and servicing points
5.1.8.2 Walkways
Walkways within the BM and between the BM and the tunnel wall must have a minimum cross-sectional area of at least 0.5 m^2 within which there shall be a rectangular clear body opening with a vertical dimension of at least 700 mm and a horizontal dimension of at least 450 mm. Walkways should be at least 300 mm wide. Handrails and toeboards shall be provided wherever walkway surfaces are less than 300 mm wide....

5.1.8.3 Access openings
Access openings e.g. to bulkheads and cutter head, shall have a clear opening with a minimum sectional dimension of 400 mm and a minimum area of 0.2 m^2. Whenever feasible the dimensions should be increased.....

5.1.9 Protection against falling objects
5.1.9.1 Falling objects
The control points of the BM shall be provided with a cab or protective roof giving adequate protection for the operator. See clause 5.2.4.

This does not apply if a control point is located in an underground area with an adequately supported roof or the BM is specified to operate in stable rock.

When driving inclined shafts, the front wall of the control cabin must be reinforced to protect the operator from material rolling or falling down.....

5.2 Control points
5.2.1 General
Operators shall not be exposed to harmful dust, gases, vapours or noise. The main operator's control point shall be fitted with a cab or the operator shall be protected by other means.

5.2.4 Cab
If the BM is fitted with a control cab, this shall be designed, constructed and/or equipped to ensure that the operator enjoys safe working conditions (e.g. ventilation, visibility, noise reduction, vibration reduction, protection against falling objects, heating or cooling) and allows rapid evacuation. All materials used for the cab shall, where possible, be fire-resistant and not create toxic fumes in the event of fire.

5.4.3 Control systems
Control systems of BM's shall be so designed that:...

The cutting operation of a TBM can only be started if the dust removal system is in operation. For RSBM's however it is sufficient if the cutter discs are sprayed with water.....

On a TBM, an auxiliary control point connected to the connection box directly at the cutter head shall be provided. This auxiliary control point shall override all other control functions controlling the cutter head and shall only allow the positioning of the cutter head in creep or jog operations. The auxiliary control point must be connected by means of a key-operated switch.
See clause 5.16.3.

When on a TBM the creep or jog operation is actuated from the auxiliary control point, a warning signal shall sound for as long as the creep operation is maintained. See clause 5.4.5.2.

5.4.5 Starting and stopping
5.4.2 Warning system
An audible start-up warning system shall be fitted to fast moving equipment such as open belt conveyors. The start-up signal for conveyor belts must be electrically interlocked to give a minimum preset warning period of 10 seconds before operation starts.

If machine back-up trailers are moved independently of a TBM, an audible or visual warning system shall be provided which has a minimum preset warning period of 10 seconds before the back-up trailer moves.

5.4.6 Emergency stop
Emergency stops on BM's shall be installed where they are suitable for reducing hazards, particularly at the main operator's control point and at additional control points for controlling separate units.

An emergency stop shall be mounted at the entrance to the cutter head which shall stop any movement of the cutter head. Emergency stop devices shall conform to EN 418....

5.5 Towing connection
Every towing connection between BM and back-up trailer shall be designed and constructed to withstand the full tractive force allowing for the following safety factors:

Chain connections	*4 times max. load stress*
Rods and other elements	*2 times max. stretching stress*

All towing connections must be protected against overload.....

5.7 Ventilation, control of dust and gas
5.7.1 General
BM's shall be so designed that appropriate ventilation equipment, air monitoring devices, dust suppression and dust collection systems can be incorporated.

5.7.2 Dust suppression
The cutting face and all other dust emission points shall, if necessary, be equipped with water sprinkling facilities to precipitate dust.

On a TBM, a dust shield or equivalent equipment shall be provided to stop remaining dust from propagating backwards.

On RSBM's and reaming TBM's dust can be removed through the pilot hole by means of the ventilation system.

5.7.3 Atmospheric changes (concentration)
5.7.3.2 Requirements for use in rock formations possibly containing methane
The BM shall be fitted with a permanent methane monitoring system if the possible occurrence of methane is likely....

This monitoring system shall give an optical and acoustic alarm when a preset threshold of 1% is reached or another limit value based on national regulations....

If the methane concentration reaches a preset threshold of 1.5% or another threshold defined in national regulations, all non-explosion-proof electrical and mechanical equipment shall shut down immediately and automatically.....

5.7.4 Operation in an explosive atmosphere
If the concentration in relative volume of methane cannot be constantly reduced to less than 0.3% vol. by means of the methods described in clause 5.7.3.2 and/or continuous duty in explosive atmospheres is foreseen, the BM shall be equipped with explosion-proof electrical equipment. See EN 50014.
All BM's designed to operate in these conditions shall comply with the appropriate regulations for equipment in explosive atmospheres and coal mines. See Annex B.

5.9 Electrical equipment
5.9.1 General
The electrical equipment of BM's shall comply with prEN 60204....

5.9.2 Protective measures
pr EN 60204-1, section 17 shall apply to portable equipment and lighting, with the following additional remark:

The protective measures required to reduce hazards due to difficult environmental conditions in tunnels shall include:

- *residual current protective device with a fault current rating of max. 30 mA,*
- *110 V transformer with grounded centre tap on the secondary side.*

pr EN 60204-1, section 9 shall apply for control circuits with the additional stipulation that manual control devices may only be operated with a maximum control voltage of 50 V. ...

Comments by the author:
The protective measure "110-V transformer with grounded centre tap on the secondary side" permits a touch voltage of 55 V in the event of a fault and thus contradicts EN 60204-1 and IEC 529 to which the text of the standard refers and which restrict the protective measure "Safety extra-low voltage" to max 50 V.

The electrical hazard in using tunnelling machines underground is greater than in the remaining application area of EN-60204-1 and IEC 529 because of restricted space, moisture and high conductivity. This is also apparent from the term "difficult environmental conditions" used in the accompanying text to clause 592 of this standard. For this reason the Tiefbau-Berufsgenossenschaft has objected to this regulation at the TC enquiry stage.

5.9.4 Transformers
Only transformers with air cooling or a PCB-free non-flammable coolant shall be used underground.

5.9.7 Lighting
Illumination in the operator's area shall be at least 100 lux and at least 30 lux in walkways. Where service work is expected to be carried out, sockets for additional lighting shall be provided....

5.9.8 Emergency lighting
Battery-powered emergency lighting which is switched on automatically in the event of failure of the main lighting shall be provided.....

5.12 Fire protection
5.12.1 General
All hydraulic fluids shall contain a strong dye so that leaks are readily visible. Hydraulic tanks shall be fitted with a warning alarm for low level. All hydraulic systems shall be designed so that in the event of rupture of a component, the loss of oil is minimized and early warning is given.

5.13 Storage of safety equipment for personnel
Storage on the back-up equipment shall be provided for personnel safety equipment, e.g. blankets, stretchers, breathing apparatus, etc.

This place shall be clean and protected against dust and moisture. It shall be clearly marked.

3.3 CEN TC 151/WG4 N8 "Tunnelling Machines, Shield Machines, Horizontal Thrust Boring Machines, Tunnel Lining Erection Equipment; Safety Requirements"

This draft standard has gone through the TC enquiry procedure. At present the objections are being discussed and a revised draft drawn up according to the result which will subsequently be subjected to the public enquiry procedure within the CEN member states. The background to the origin of this standard is as described in 3.2.

5. Safety requirements and/or measures.
5.1.5 Cutter head on full face machines......
If the machine is used for work in unstable ground conditions, then proper means of support such as slot-gate closures and/or compressed air shall be provided to allow safe access for inspection and maintenance work on the cutter head (cutting wheel)....

5.1.9.2 Face collapse...
All shield machines where open face excavation can take place shall be provided with mechanical face support systems appropriate to the ground conditions envisaged. These may include face rams, hydraulically operated poling plates and face plates, sand trays or in simple shields, face boards.

3.4 CEN TC 151/WG4 N7 "Road Headers, Continuous Miners and Impact Rippers; Safety Requirements"

This draft standard has gone through the TC enquiry procedure. At present the objections are being discussed and a revised draft drawn up according to the result which will subsequently be subjected to the public enquiry procedure within the CEN member states. The background to the origin of this standard is as described in 3.2.

In the following text,

- Road headers
- Continuous miners and
- Impact rippers

are designated with the abbreviation RH.

5 Safety requirements and/or measures
5.1.5 Handling heavy loads
5.1.5.3 Handling lining material
RH's employed in handling lining material shall be fitted with suitable load handling means and where necessary with working platforms.
See EN 292-2, clause 3.11

5.1.7 Ladders, access ways and platforms
Ladders, access ways and platforms on RH's shall be safely designed, adequately dimensioned and safe to walk on. Ladders and stairways with a height exceeding 60 cm shall be safeguarded by handrails on both sides. Elevated access ways and working platforms shall be safeguarded by side protection. Platforms shall have non-slip surface and are to be protected at the sides by toeboards.
See EN 22867, sections 6 and 7.

5.1.8 Falling objects
The control point shall be protected against spoil rolling down the cutter arm or boom. The control point shall be capable of being equipped with a sufficiently rugged roof to protect against rock fall and falling objects. Verification: by calculation. Load estimates for adequate deformation protection shall be based on those in EN 23449, clause 5.4 where applicable.

5.2 Control point
5.2.3 Cab
The control point shall permit the addition of a dust-tight cab provided with a suitable means of fresh air supply. Safety glazing shall be used for the windows of the cab. The cab shall be equippable with a heating/cooling system.

Comment:
"Permit the addition of" means that the manufacturer takes the required equipment into consideration in the design of the machine and may supply it as additional equipment if required.

5.4 Control devices and systems
5.4.2 Control devices
For lifting, slewing and travel control hold-to-run control devices shall be used. These control devices shall be safeguarded against unintentional operation (e.g. by shift gearing, indexing, two-hands control device)

5.4.3 Control systems
RH control systems shall be configured so that:

- *RHs with dust suppression systems can only be operated when the dust suppression system is in operation at the same time,...*
- *The various work operations can only be started and stopped in the prescribed sequence in order to avoid dangerous operating conditions (e.g. dust suppression system - conveyor system - cutter head)...*

5.4.5.2 Warning devices
RH's shall be equipped with an optical warning device (e.g. flashing warning lights) which switch on automatically for reverse travel.....

5.5 Dust and gas protection
5.5.1 Dust protection
RH's shall permit the installation of effective water-jetting equipment on the cutter head or impact ripper and where necessary of effective dust extraction devices.

Comment: "Permit the installation" means that the manufacturer takes the required equipment into consideration in the design of the machine and can supply it as additional equipment if required.

5.5.2 Exhaust gas control
Combustion engine-powered RHs shall only be equipped with diesel engines, the exhaust gas composition of which shall be suitable for application underground. Suitable in this respect are e.g. engines with two-stage combustion or with exhaust gas filters....

5.6 Noise control
RH's shall be designed and built to minimize the hazards of noise emission on the basis of RH design in accordance with state-of-the-art requirements in noise control and taking into account the relative presence of such means applied.
Noise can be reduced e.g. by the following means:

- *Use of noise-abated equipment,*
- *Reducing vibration*
- *Encapsulation.*

Note: During RH operation, process noise is substantially determined by the overall RH noise emission and is subject to continuous variation....

5.10 Fire protection
5.10.2 Fire extinguishers
RH's shall be equipped with fire extinguishing systems or portable fire extinguishers containing a suitable agent with a mass of at least 6 kg. The fire extinguishing equipment shall be suitable for the type and size of the RH involved. Only such fire extinguishers that are suitable for the purpose may be used underground. Fire extinguishers shall comply with the requirements of EN 3. The extinguishing agent shall be suitable for all classes of fire. Halon gas is not permissible for this purpose.

3.5 CEN TC 151/WG4 N22 "Air Locks; Safety Requirements"
This draft standard has gone through the TC enquiry procedure. At present the objections are being discussed and a revised draft drawn up according to the result which will subsequently be subjected to the public enquiry procedure within the CEN member states. The background to the origin of this standard is as described in 3.2.

5 Safety requirements and/or measures
5.1.1 Calculation, leak-tightness and performance test
Air locks and pressure bulkheads shall comply with EN 286, Part 1, as far as applicable. The reference pressure for design calculations shall be 1.1 times the maximum working pressure. Air locks and pressure bulkheads shall also be capable of withstanding a hydraulic pressure test at 1.5 times the maximum working pressure. A leak-tightness and performance test procedure shall be fully described in the instruction handbook.

Where bulkheads cannot be tested, the reference pressure for the design calculation shall be 1.5 times the maximum working pressure.

5.2 Personnel locks
5.2.1 Number of chambers

Personnel locks shall consist of at least two interconnected chambers.

5.2.2 Dimensions

Personnel locks must meet the following minimum requirements:

	Main chamber	*Second chamber*
Headroom:	*1.5 m*	*1.5 m*
Length:	*see below*	*1.0 m*
Number of persons:	*3*	*2*
Volume per person:	*1.0 m^3*	*1.0 m^3*

The system of air locks shall be large enough and so designed that before decompressing a casualty it shall be possible for him to be placed on a stretcher (at least 1.8 m long) under pressure whilst permitting an attendant to be pressurized and enter the space where the casualty is located. Thereafter it shall be possible to decompress the casualty on the stretcher together with the attendant. If the air lock is designed for a working pressure above 1 bar it shall be possible to keep the stretcher in a horizontal position during decompression.....

5.2.6 Minimum equipment of chambers
Each chamber in the personnel locks shall have the following minimum equipment.

5.2.6.1 Instrumentation

- *Pressure regulator controllable from outside and inside.*
- *Pressure indicator, outside on the control panel and inside the personnel lock chamber for indicating the internal pressure. Reading accuracy to 0.05 bar.*
- *Chamber pressure recorder installed outside the lock, indication accuracy 0.05 bar. Circular chart recorder must rotate at a speed of at least 1 revolution per 4 hours, or at least 120 mm/h for strip recorders. For circular charts, the zero bar pressure scale shall be at the periphery.*
- *Inside thermometer, measuring range + 50°c to 0°c.*
- *Clock, inside and outside, with second display, independent of the mains supply.....*

5.2.7 Other requirements

- *Interior illumination will be at a minimum of 100 lux, measured at head height above the seat.*
- *Emergency lighting shall be provided which is switched on automatically in the event of failure of the main lighting. Light intensity 15 lux for one hour.*
- *Voice communication (intercom system) see also clause 5.1.4.*
- *Sound suppressors for fresh and waste air shall be fitted to ensure that the maximum A sound level in the chambers caused by the operation of the air lock shall not exceed 80 dB (A) measured at head height.*
- *Heating/cooling system to maintain temperature in the chambers between 18°c and 30°c except during pressure changes.*
- *Precautions shall be taken for prevention of the risk of fire and skin burns.*
- *Efficient water fire extinguisher system which is tested at the maximum air lock pressure.*

5.5 Pressure bulkheads
5.5.1 Sealing
Pressure bulkheads shall provide an effective seal between spaces with different pressures.

4 Conclusion

Of the standards for tunnelling machines presented here, prEN 815 "Tunnelling machines..." is expected to be published before the end of 1994 and the other standards in 1995.

During the work of drafting these standards for tunnelling machines it has been found that all European TC member states attach great importance to safety. This is an important step towards realizing the single European market.

Tunnelling machines which are constructed in accordance with stringent harmonized European standards play a part in making work places underground safer and hence making the profession of tunneller more attractive again.

Tunnels below waste disposal tips

Uwe Beckmann
BRP Consult Beckmann, Ruppert + Partner GmbH, Braunschweig, Germany

Abstract
The difficulty in finding new waste disposal sites results in the construction of few, huge and centralised disposal facilities. In order to achieve a save and inspectable drainage system tunnels become part of the drainage system.
Waste disposals, tunnels, drainage system

1 Introduction

The ever increasing volume of waste is a tremendous problem for any civilized society. Therefore, the demand for great volumes of landfill becomes more and more important. On the other hand it is becoming increasingly difficult to find suitable disposal sites.

Everyone produces waste but no one wants to have a landfill site, especially not in one's own back yard.

The regulations for the selection of new sites in Germany have become very restrictive. So called "killer-criteria"; **e.g. the distance to housing areas has to be greater than 300 m, no sites in nature and water-protection areas, etc.,** combined with high geological and hydro-geological requirements; e.g. **a geological barrier of 3 m thickness with a permeability of less than k = 1 x 10^{-7} m/s and a minimum distance of 1 m between the base of the mineral sealing layer and the groundwater level,** result in the fact that very few available areas remain on the map, where, theoretically, waste disposal sites could be located. The consequence is that only a few huge central landfill sites are available, for which the necessary drainage can only be undertaken effectively with the help of tunnels below the landfill. As a result, an increasing number of such tunnels are being constructed for this purpose.

2 Principles

A geological barrier, a sealing system and an effective drainage system are the most essential elements of the total safety system of a waste disposal site needed to avoid environmental pollution.

The requirements for the drainage systems in the German regulations for municipal waste are as follows:

- 1 % inclination of the sealing surface in a longitudinal direction, 3 % in a transversal direction;
- 300 mm thick drainage layer (gravel 16 - 32 mm);
- drainage pipes every 30 m;
- lengths of the drainage pipes less than 300 m, to enable camera inspection on the full length

If the basal ground level of the waste disposal site has no natural inclination, a considerable filling of the center is necessary, and should be of geological barrier quality (i.e. $k = 1 \times 10^{-7}$ m/s). This results in a substantial increase in the construction cost for the landfill.

To avoid these large volumes of fill either a tunnel or shafts are required to provide shorter distances for the flow of the leachate water. However, shaft construction is expensive and difficult to stabilize in landfill depths greater than 50 m and soon requires expensive rehabilitation to guarantee their continued functioning. The German regulations do not recommend that shafts are located in the center of waste mountains. The only practical alternative solution is, therefore, a tunnel that allows a free outflow of seepage water.

Tunnels are advantageous because they generally have no major stabilization problems and can be built completely before the waste is deposited. Also, access to a tunnel is easier than to a shaft. Tunnels can be used for long term control purposes and allow easy control and even rehabilitation of all drainage pipes that terminate in the tunnel. However, some specific construction considerations have to be followed to enable safe tunnel construction to be undertaken in difficult conditions. These are illustrated in the following examples.

3 Tunnelling below a municipal waste disposal tip extension

The municipal waste disposal site "Lachengraben" in the rural district of Waldshut (Baden-Württemberg, Germany) has been in operation since 1970 [1]. The site is located in a valley, filling it in a direction from the bottom to top with an extension of 2 Million m^3 planned recently in the upper part of the valley. The experience gained from the existing drainage system showed that reliable long term dewatering of the new field could not be guaranteed and for this reason a tunnel drainage system was adopted.

The first part of this system comprised a 2.5 m diameter, 190 m long pipe jacked access tunnel to the new field in natural ground (gypsum marl) already covered with 20 m waste. Although a direct contact with leachate water was not expected, for long term safety reasons a concrete with high chemical resistance (BII), a water-cement ratio of less than 0.50 and a water penetration rate (DIN 1045) of less then 30 mm was selected. In addition, owing to the sulphate content of the groundwater in the surrounding gypsum marl a cement with high sulphate resistance was used.

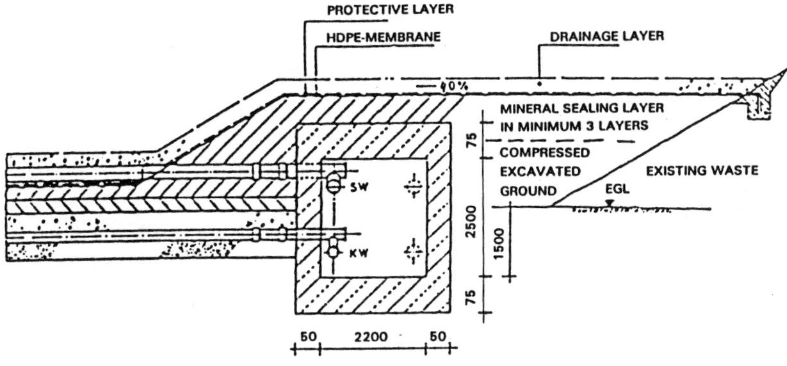

Fig. 1: Typical cross section of the control tunnel below the waste disposal site "Lachengraben"

The access tunnel ends in a distribution chamber from which the control tunnels of the new disposal field branch off in different directions. These have a rectangular cross-section (3.2 m width x 4.0 m height) and wall thicknesses of 0.50 m (sidewalls) and 0.75 m (roof and floor). These tunnels are integrated into the overall sealing system, i.e. they are positioned within the mineral sealing layers and below the HDPE-sealing. This means, that every drainage pipe entering the tunnel must cross and penetrate the HDPE-sealing. These locations require special care in construction, however, they remain points of possible failure.

In this specific valley situation there also exists a second, outer piped drainage system that collects water flowing down the valley sides and bottom. These pipes also lead directly into the tunnel. The tunnel itself was constructed in open cut in 10 m long segments with in-situ-concrete. Special attention had to be paid was given to the construction of the segment joints because considerable settlements were expected owing to the high loadings imposed by the deposited above waste.

In the center of the walls chlorified caoutchouc tapes were used together with bituminous weak fibre plates at the segments joint cross sections. In addition, a compression sealing tape, again made from chlorified caoutchouc, was placed on the outer side of the joint and a permanent flexible plastic material on the inner side.

The tunnels were equipped with electricity, ventilation and a gas concentration control (H_2S, H_2, CH_4, CO_2). The total cost of the scheme was 4.5 Million DM for the entire 400 m of tunnels plus 1.0 Million DM for the technical equipment.

4 Trapezoidal in-situ concrete tunnel

In another waste disposal site in Wiesbaden [2] and [3] a similar 710 m long in-situ concrete lined tunnel was constructed in 10 m long segments. However, this cross section was trapezoidal with a semi-circular roof. The invert thickness was 1.6 m, the wall thickness 0.8 m and concrete quality was similar to that described above (here B 45) with 60 kg/m^3 steel reinforcement. The inside dimensions of the tunnel are 4.0 m in width and 3.6 m in height. One tunnel segment was produced per day using steel shuttering. To reduce the heat development in the invert during concrete curing the amount of cement (HOZ 45 L) was reduced from 385 kg/m^3 to 360 kg/m^3 and the difference was replaced by fuel ash. Temperature control was maintained by permanent measurements and by closing the tunnel cross section and covering the fresh concrete with an isolating membrane. This tunnel is used not only for control and collection of leachate water but also for collecting gas.

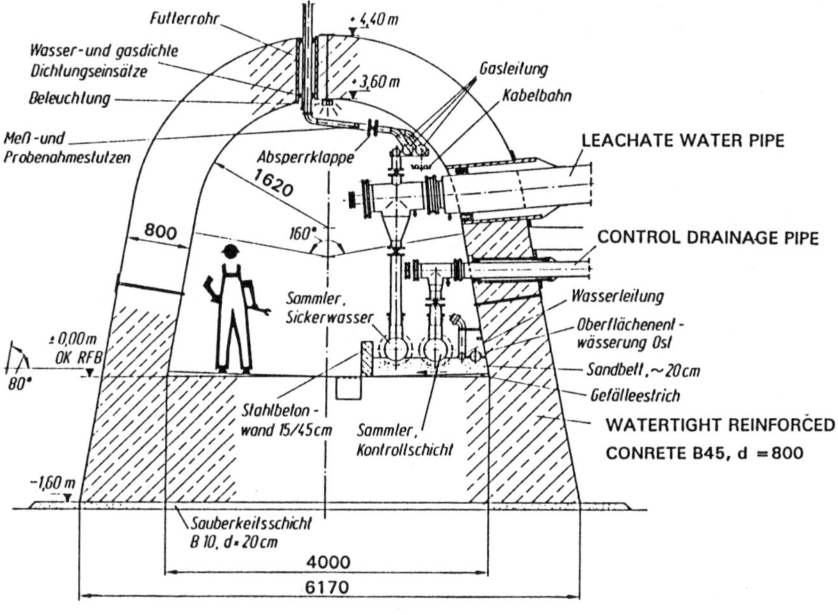

Fig. 2: Control tunnel below the waste disposal site in Wiesbaden

5 Tunnelling below an existing landfill

Again this landfill site (Aßlar - approximately 50 km north of Frankfurt) is situated in a valley, filled to an approximate depth of 60 m with waste [5]. The key elements of the dewatering system are three shafts, which in the process of dumping the waste were considerably damaged to such an extent that they almost ceased to function, in that it was difficult to install pumps to extract leachate water. Rehabilitation of these shafts would have meant the excavation of 60 m deep open pits with no guarantee of a return to a fully functioning system. It was decided, therefore, to construct a tunnel from an entry point at the lowest elevation of the valley and ascending with an inclination of 1.5 % over the first 650 m and 10 % over the remainder (total length 843 m). At the upper end of the tunnel a 38.5 m deep, 1.2 m diameter shaft was bored and installed with a 0.8 m steel pipe.

This shaft provides natural ventilation to the tunnel and at the same time serves as an emergency escape shaft.

The most significant tunnelling problems were that no ground investigation was possible owing to the 60 m deep waste fill and that no damage to the basic sealing system was allowed. A rough geological appraisal based on available generalised maping resulted in an optimization of the tunnel line through a predicted assemblage of a number of variaties of schists and a volcanic diabase. Consequently, six different rock mass and support classes were envisaged for the tunnel excavation, which has an inner cross section of 7 m^2 (2.6 m height and 2.8 m width) and a natural overburden depth of about 2.5 times the diameter. The tunnel was driven by drill and blast methods and special attention was given to the avoidance of too much disturbance to the surrounding rock resulting in excessive settlement, which could have endangered the basic sealing system of the disposal site. The entire tunnelling operation utilised the New Austrian Tunneling Method (NATM) with shotcrete thicknesses of 50 to 250 mm serving as a final lining. An average advance rate of 52 m per month were achieved working 3 shifts, 24 hours per day, 7 days per week and the entire construction took place between September 1991 and July 1993.

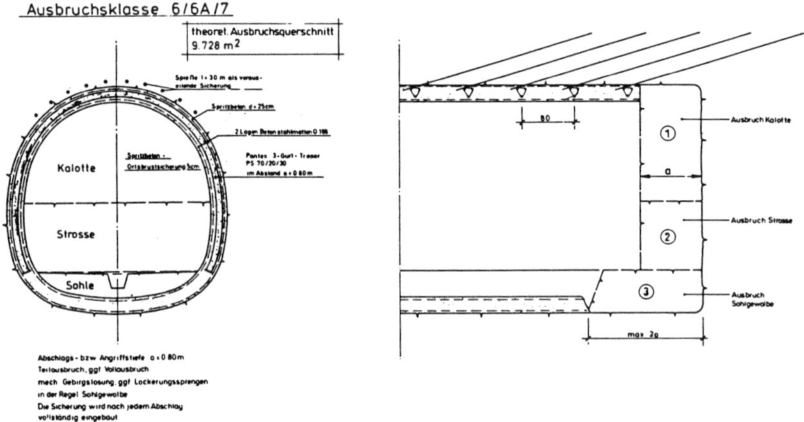

Fig. 3: Tunnel excavation in unfavorable ground

This included the construction of the connections to the three shafts; two of which have short cross cuts of 11 m and 15 m followed by two inclined shafts of 5.5 m and 6.6 m in length ending at the bottom of the shafts. The third shaft was undermined by a 23 m cross cut and the connection between was made by drilling. All three connections at the shafts bottoms were constructed in such a way that no mechanical means are needed to assist leachate flow. The leachate water has a generously dimensioned simple hydraulic access, via a gravel pack thence by free flow in a HDPE-pipeline inside the tunnel. The tunnel also takes in groundwater in separate inlets and along a separate pipeline.

The total cost for the tunnel construction was 9.2 Million DM.

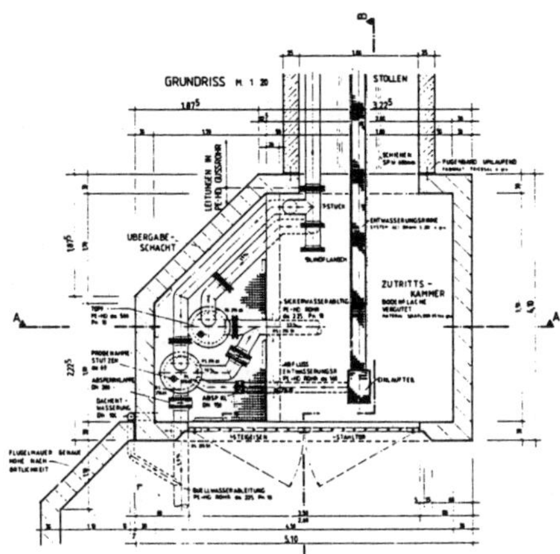

Fig. 4: Access chamber to tunnel

6 Tunnelling in a hazardous waste dipsosal site

Owing to the natural shape of this 250 m x 700 m hazardous waste disposal site in Mainhausen (around 40 km east of Frankfurt) the only economical solution available for collecting the leachate water was to concentrate on one side of the site only, in order to guarantee the required 1 m minimum separation from the groundwater and a minimum of filling at the base whilst reducing the loss of valuable space available for depositing waste. The requirement to control the 26 transverse leachate water pipes involved construction of 26 shafts located either within the disposal area and punched through the top sealing layer or outside the disposal area, which would result in 26 crossings of the side sealing. It was decided therefore to align a tunnel along the inner side of the disposal area, connect the leachate water pipes and to cross the side sealing only at two points located at either end of the tunnel.

Two alternative materials were evaluated for use at this site: conventional concrete tube rings covered with HDPE and a complete HDPE-tunnel. After careful consideration, particularly with respect to the expected settlements in the hazardous waste, the concrete solution was considered too high a risk because of its stiffness and low joint flexibility, which would compromise the long term stability and safety of the tunnel.

HDPE was chosen as the lining material for the 2.3 m diameter circular tunnel with a wall thickness of 100 mm. The following tubes 7 to 8 m long, were welded together on site and great care was taken with the flexible connection of the transversal pipes to enable them to follow any movement without damage. The design enables the inner leachate water pipeline to be opened at any connection point to allow control, camera inspection, flushing and sampling. In normal operation the system is closed that no oxygen can penetrate the transversal pipes and lead to clogging of the system. Ventilation and gas warning devices are integral parts of the tunnel which is also provided with several "doors" to allow inspection and sampling of the hazardous waste.

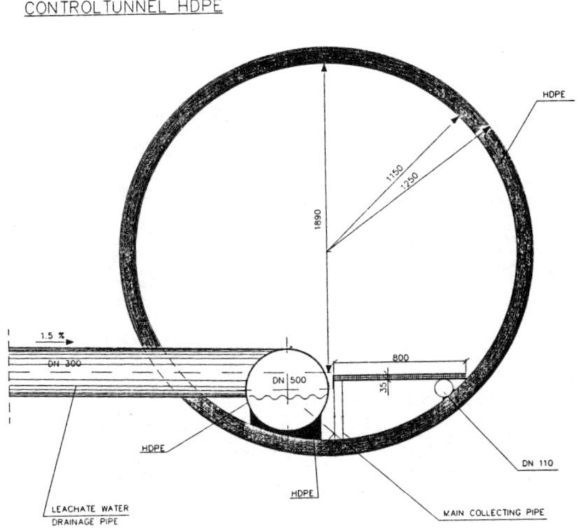

Fig. 5: Control tunnel Mainhausen

7 Tunnelling below concrete hazardous waste disposal shafts

As discussed in the introduction to this paper it is becoming more and more difficult to find sites for waste disposal, especially hazardous waste disposal. In order to create the maximum available volume for disposal of waste on such sites that are avialable whilst maintaining the maximum of safety and economical operation, concrete shafts have been developed to serve as underground storage silos [6]. To date only one such shaft has been constructed at Bornum - around 80 km east of Hannover. This 46 m diameter and 36 m deep shaft holds a total volume of waste of 60.000 m^3 which can easily be kept in dry storage by covering with a roof.

An inner concentric shaft within the core of the main shafts serves to control the stability of the waste and acts as a collector for pumping out leachate water. Its small diameter allows it to be easily integrated into the overall structure and has the added function of a roof support. Peripheral control can be achieved from two or more short tunnels aligned radially like a wheel below the shaft. If any groundwater accumulates in the surrounding geological barrier it can be collected in these tunnels. Although there is not likely to be any leachate water attack under normal operation the tunnels should be provided with an outer HDPE-lining. Access to the tunnels is available via the central shaft.

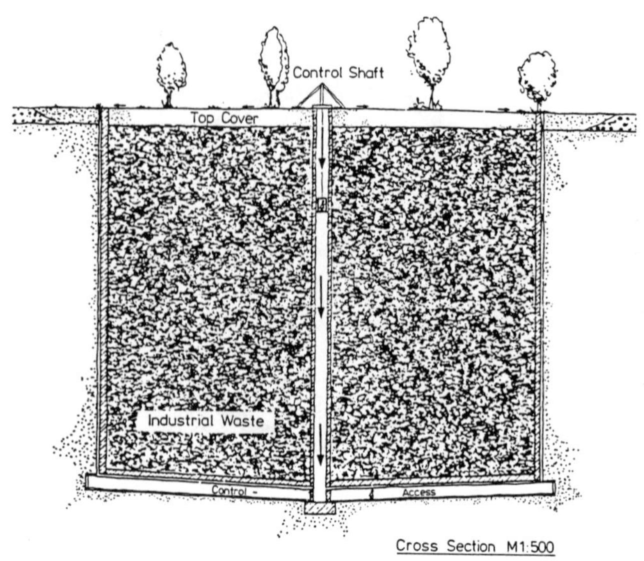

Fig. 6: "In-ground"-Storage TERRA ROTONDE

If total control is required it is possible to construct a complete cellar floor underneath the shaft with columns supporting the shafts sole plate. It should be borne in mind that concrete alone is not a sufficient sealing element for the shafts but may serve as additional sealing, the main sealing system being a combined one, consisting of clay (1.5 m, thick $k \leq 5 \times 10^{-10}$ m/s) and a 2.5 mm thick HDPE-liner. In a recent project for a "field" of six huge shafts (60 m diameter, 45 m depth) in Hoheneggelsen - approx. 40 km southeast of Hannover - which is currently in the process of public authorization, this has not been considered and a diaphragm sealing wall around the whole field has been incorporated instead. Although the construction of the shaft is planned in thick clay, the adversaries of the project take this as one of their numerous arguments against its implimentation, claiming that it does not represent the state of the art.

8 Summary

The difficultly in finding new waste disposal sites results in the construction of few, huge and centralised disposal facilities. As the requirements for modern and long term effective drainage systems, especially those having an optimum inclination of 3 % (longitudinal) resp. 1 % (transverse) are difficult to fulfil in an economical manner, tunnels are becoming increasingly seen as a means of saving valuable storage volume and guaranteeing longterm safety.

These tunnels can be constructed from precast or cast-in-situ concrete, covered with an HDPE-liner if necessary, or even in shotcrete if the tunnel is located at sufficient depth below the waste. For hazardous waste HDPE-tunnels are recommended. All constructions must have some flexibility, since significant settlements can be expected. Also, the connections to the various water pipes leading into the tunnel need to be flexible.

A minimum height of 2.0 m is recommended to allow access for control of the tunnel, maintenance of the inner pipelines and every single connected water pipe including camera access. The basic technical equipment installed in the tunnel must be a ventilation system and a gas control and warning system.

9 References

1. Henseleit, Hansjörg (1992) Kontrollstollen in der Deponietechnik, **Abfallwirtschaftsjournal 4 (1992)**, Nr. 3, 260-263
2. Krüger, Thomas (1992) Zuerst der Tunnel, dann der Berg, **Beton 9/92**, 520-521.
3. N.N. (1992) Kontrolltunnel für Mülldeponie, **Beton- und Stahlbetonbau 87 (1992)**, H. 1, 17.
4. N.N. (1992) Kontrolltunnel behält Müllberg fest im Griff, **Der Baukasten**, 43 (1992), Nr. 6, 3-4.
5. Sommer & Partner GmbH (1993) Entsorgungsstollen Aßlar, **Tunnelbautechnischer und Ingenieurgeologischer Abschlußbericht**, im Auftrage der Hessischen Industriemüll GmbH, Wiesbaden 1993.
6. Beckmann, U./Schiffer, W (1986) "In-ground" storage TERRA ROTONDE for industrial waste, **Proc. Intern. Symposium on Environmental Geotechnology**, Allentown (USA), April 1986.

Machine tunnelling in the northwest of England

B. J. Cole
H. J. Lees
North West Water Engineering, Ltd., Southport, Merseyside, England

The success of any tunnelling contract must rely upon a unique combination of technical and management skills.

A project process must encompass effective and efficient design, a contract strategy that encourages cost effective solutions and innovative ideas, and construction management that delivers the specified product at the least cost.

Estuary and coastal water projects in the north west of England will require ten tunnelling contracts to have been completed between 1988 and 1996. This incorporates the building of 44 km of tunnel ranging in size from 1.8 m dia. to 3.0 m dia. and the use of at least four different manufacturers' machines. To date two contracts have been completed, two are very near to completion, three are currently in progress leaving a further three contracts to be let.

Considerable changes have occurred in the technical requirements of the tunnels and the contractual and construction management development in the drive for economy and efficiency. (See table 1).

Geology and machines

The area encompassed by this paper extends from the Mersey estuary at Liverpool to the Scottish borders and includes the two resorts of Southport and Blackpool (Fylde coast), see map.

Geology changes dramatically across this area ranging from high strength sandstone to Aeolian sands. Types of ground encountered to date include:

1. High strength to fully weathered Sherwood sandstone
2. Weak Mercia mudstone
3. Granite boulders and cobbles
4. Glacial sands and gravels
5. Alluvial sand
6. Aeolian sand
7. Silt
8. Soft to firm boulder clay

Needless to say, mixed face tunnelling has been the order of the day on the majority of contracts, and machine choice has played an important part in contract considerations. A further element in the choice of machine has been the type of tunnel primary lining design, since some are gasketed segmental tunnels and others are pipejacked tunnels.

TABLE 1

PROJECT	TUNNEL INT.DIA. (m)	OVERALL LENGTH (km)	TUNNEL PRIMARY LINING	GASKETS	SECONDARY LINING	TYPE OF CONTRACT
Liverpool MEPAS Stage 1	2.44	2.3	6 part + key trapezoidal tapered ring 1.0 m long	MDPE gasket	150 thick concrete	ICE 5th Edition
Sefton MEPAS Stage 1	1.55	2.2	2.5 m long pipejack pipes	MDPE ring plus secondary MDPE foam gasket	150 thick concrete	Modified ICE 5th Edition with target
Liverpool MEPAS Stage 2	2.44	6.2	6 part trapezoidal tapered ring .75 m long	Hydrophilic rubber	150 thick concrete	Tender target I.Chem.E. Green book
Southport Coastal Waters Interceptor Tunnel	2.82	5.3	One pass 6 part trapezoidal 1.0 m long	Hydrophilic rubber	Low flow channel only	Negotiated target I.Chem.E. Green book (2 Stage tender)
Liverpool MEPAS Stage 3B	2.10	2.5	2.5 m long pipejack pipes	MDPE ring	Nil	Tendered target I.Chem.E. Green book
Fylde Coastal Waters-Tunnels Contracts 1 & 2	2.85	11.7	6 part trapezoidal tapered ring 1.0 m long	Specialised MDPE Gasket	Nil	Tendered targets I.Chem.E. Green book

NWW CATCHMENT AREA

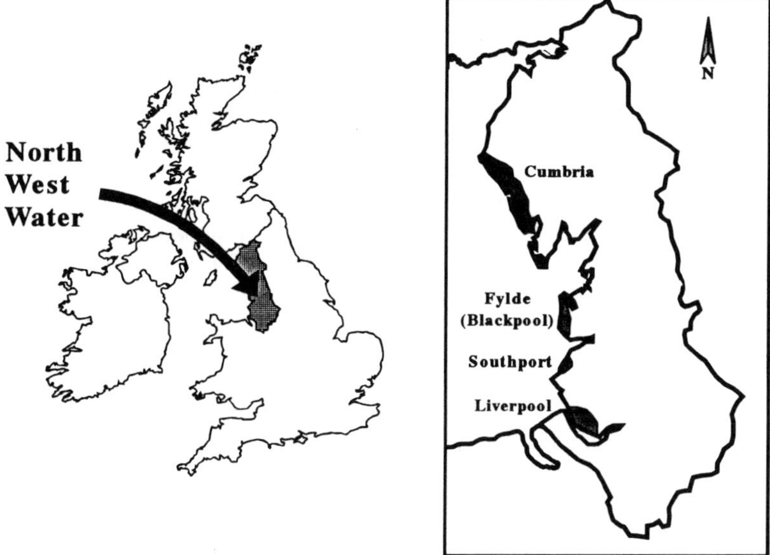

Six machines have or are being used which cover five different types of machine. Compressed air has been used with three of these machines but with pressures not exceeding 1 bar.

A common factor throughout the projects is a high water table, generally being either at or within three metres of ground level, except in one predominantly rock drive where water levels were below the tunnel invert.

Contracts

The first two contracts were let by Liverpool City Council and Sefton Metropolitan Borough Council acting as agents for North West Water (NWW). All subsequent contracts have been let directly by NWW.

Liverpool Mersey Estuary Pollution Alleviation Scheme (MEPAS) Stage 1 was the first contract and was let as a traditional ICE 5th Edition Contract. Considerable difficulties were experienced with this contract which resulted in a high level of claims and late completion. Target contracting was being used within NWW at this time and it was considered that tunnelling contracts would benefit from a new approach.

Sefton MEPAS Stage 1 had been tendered as an ICE 5th Edition Contract but the letting of the contract was considerably delayed by negotiations for entry into the Liverpool docks where a major part of the scheme had to be constructed. An opportunity therefore arose to re-examine the contract in the light of the experiences of the Liverpool MEPAS Stage 1 contract which was still being constructed. It was decided that a new approach was needed, one which enabled the client and contractor to develop a partnership with a joint objective to complete the contract within time and financial limits. Negotiations were entered into with the lowest tenderer that culminated in an ICE 5th Edition Contract modified by the introduction of a Target Bonus system and procurement approvals.

Subsequent tunnelling contracts: two for MEPAS, one for Southport and two for Fylde have all been let using the Institution of Chemical Engineers (I.Chem.E) Green Book Conditions of Contract with NWW modifications for payment and Target.

Various modifications and tendering procedures have been used as the philosophy of Green Book target contracting was pursued and benefits accrued.

Coupled with this change in form of contract was a progressive change in attitude on site between Client and Contractor, this being manifested in the simple arrangement of shared office accommodation. This gave easy access to each other, shortened lines of communication and discussion, thus developing mutual trust, co-operation and a team spirit.

Each Green Book contract has allowed the client to have a considerable technical input into the development of a contract strategy. This results in a shared objective to focus upon methods that give the best chance of achieving the desired results for a known affordable cost. Tunnelling machine procurement readily lends itself to this approach.

Construction

Green Book contracts require different attitudes to site management from the Contractor and Client to those that have traditionally existed. There has to be greater trust and sharing of information, more liaison and discussion and, above all, a greater emphasis on shared risk and joint decisions.

As already mentioned, it is now the practice on these sites to have a joint building with a common entrance for all site staff. Access throughout the building is not restricted so that people are given the facility to make personal contact with others quickly. Common courtesy and privacy have to be observed by both parties. This in itself brings great benefits in as much that many problems or decisions are dealt with quickly at a face to face level which reduces the likelihood of conflict and exchange of protracted paperwork.

The client naturally has an interest in the cost of the works and therefore a fundamental element of the management arrangements is shared cost information and the control of those costs, together with forecasts of future costs. Compatible computer hard and software is an obvious requisite to establishing the offices, but this also extends back into the contractor's head office systems, since much of the site procurement is effected from head office. Decisions on purchase, hiring and sub-contracts are all referred to the client's site representative for approval. A successful contract requires economic solutions to all the problems and it is not uncommon for both client and contractor's site staff to sit down together to discuss the best solutions and most cost effective way of achieving the goal. Having agreed a way forward then the procurement of equipment etc. can be agreed and progressed.

Target movement for additional works or for an agreed variation to anticipated conditions are available to the contractor. Costs are met by the client at the time of the event.

The contractor's cost control and forecasting is checked but not duplicated thus reducing the client's site management costs. This is in effect a small audit each month. However, should errors etc. exist beyond that which could be reasonably expected, then further more extensive checks can be undertaken.

With the advent of BS 5750 (ISO 9001) certificated contractors, the requirement of client quality control is diminishing and staff subsequently has been reduced. Sites will and are being managed by smaller, specialised teams dealing very closely and commonly in partnership with the contractor. Safety on sites is a typical example where the interests of both contractor and client coincide and joint efforts are being made to reduce accident rates. To date this has been very effective.

Tunnelling machines and primary linings

As stated previously, six machines have or are being used. These are listed in Table 2, together with the type of ground that the machine has to handle.

Liverpool MEPAS Stage 1 was the first contract let (under ICE 5th Edition). The machine chosen by the contractor was an Fcb manufactured Kawasaki full face slurry machine. After driving only 700 m this machine was taken out having failed to negotiate a curve and gradient. The tunnel lining was a six part, with key, trapezoidal bolted tapered ring with an MDPE gasket. These units were 1 m long and gave the engineers a number of problems with geometry and build. Modifications were made during the contract. Overall the trapezoidal segments proved to be a considerable advance on standard segments. A section of the tunnels was sub-contracted to Denys of Belgium who proposed the use of their pressurised roadheader with a pipejacked tunnel. This was accepted by the client and that part of the contract was completed using this machine, initially with compressed air in the head, but reverting to slurry for greater stability of the face on a number of occasions. The remaining part of the scheme was completed using drill and blast techniques with the original trapezoidal rings.

Sefton MEPAS Stage 1 was the second contract let. Machine choice followed extensive discussions between the contractor and client and a fundamental shift in machine type and primary lining was agreed. The machine was a Lovat mixed head EP door machine (i.e. conveyor handling of material rather than screw handling) with a pipejack rather than the originally specified segmental tunnel. Ground conditions varied from a full face of sandstone to a full face of boulder clay, passing through an

TABLE 2

CONTRACT	T.B.M.	GROUND CONDITIONS	DEPTH (TO INVERT)
Liverpool MEPAS Stage 1	Fcb Kawasaki Full face slurry machine	a) Full face sandstone, weak to strong; b) Invert sandstone, upper face sands & gravels occasional cobbles & boulders; c) As b) but boulder clay in upper face; d) Water table approx. 2½ - 3 m b.g.l.	13 to 15 m
	Denys pressurised head roadheader Compressed Air or Slurry	As above	13 to 15 m
Sefton MEPAS Stage 1	Lovat M 92 PJ Mixed face machine + specialised clay head. Operating with Earth Pressure Balance Doors with flood doors.	a) Full face sandstone, weak to strong with weathered mudstone; b) Sands & gravels, silts & clay in mixed faces. Occasional cobbles & boulders; c) Full face boulder clay; d) Water table approx. 2 m b.g.l.	9 to 12 m
Liverpool MEPAS Stage 2	Dosco/Decon roadheader With extending hood and breast plate attachments	a) Full face sandstone, weak to strong; b) Short section alluvial sands and silts; c) Water table varies from 3 m b.g.l. to invert level	13 to 15 m
Southport Coastal Waters Interceptor Tunnel	Lovat M 131 SE Operating in earth pressure balance door mode or EPB screw mode. Flood doors fitted.	a) Alluvian and Aeolian sands; b) Boulder clay (cobbles but no boulders) c) Weathered mudstone (very weak) d) Water table generally at g.l.	9 to 13 m
Liverpool MEPAS Stage 3B	Dosco/Decon roadheader With extending hood face rams and fully flameproof	a) Full face sandstone, weak to strong; b) Sandstone invert, alluvial sands upper face; c) Water table 8 m head at invert.	14 to 33 m (Generally 19 to 23)
Blackpool Fylde Coastal Waters Contracts 1&2	Lovat ME 131 SE Full EPB mode	a) Full face sand) Many sloping interfaces b) Full face boulder clay) c) Occasional cobbles and boulders; d) Water table 10 to 20 m above invert.	15 to 30 m

area of sands and gravels. In the event the compromise machine head design was unable to efficiently mine the full face of clay and a new purpose designed clay head was installed at a pre-planned shaft. The changeover took a total of four days and progress rapidly increased for a successful outcome to the project.

Liverpool MEPAS Stage 2 was a continuation of Stage 1 but was predominantly sandstone and therefore a roadheader machine manufactured by Dosco and Decon was utilised. Due to a high number of curves the tunnel was constructed in segmental trapezoidal tunnel rings. Each ring was 750 mm long, consisted of six segments (with no key) and incorporated a hydrophilic gasket. Lytag was injected in the annulus outside the ring at the back of the machine to stabilise the tunnel before subsequent grouting, thus avoiding grout near to the machine. This system proved very effective and was able to be used with low pressure compressed air in one section where soft ground was anticipated. Progress rates of 10 m per shift were achieved.

Southport Coastal Waters Interceptor Tunnel was the next contract to be let. The ground conditions were entirely different from those in the Liverpool area as wet sands were the predominant medium with some clay and a little weathered mudstone. Machine choice centred upon settlement criteria and anticipated progress rates. Due to the long drives and curves a segmental tunnel was required. A Lovat M-131 SE was purchased from the Cairo Wastewater Consortium, modified and refurbished in Canada and then shipped to Southport. This machine can work in EPB mode with an auger screw or with conveyor and EPB doors, the latter being used with low pressure compressed air. The tunnel was constructed in one-pass, six part, trapezoidal segmental tunnel rings 1 m long, incorporating a hydrophilic gasket with no secondary lining. Curved bolts were used on seat joints with dowels at circle joints. This combination of machine and segments has worked very effectively to give high rates of production in difficult ground conditions. Maximum progress of 32 m in a 10 hour shift was achieved.

Liverpool MEPAS 3B is the latest MEPAS scheme where rock drives are the predominant feature. However, nearby old refuse tips have contaminated the ground water giving a high probability of methane gas being encountered in the drives. This tunnel is a smaller diameter than Stage 2 but the roadheader was re-used in a new machine manufactured by Decon. As all drives could be made straight the tunnel is being pipejacked with the new machine having high rate ventilation and flameproof equipment as standard requirements for the drives. One section of the drives is in soft ground which is being stabilised with low pressure compressed air. The machine drive has changed from a pipejack to one pass segments for this compressed air section mid-way between shafts.

The Fylde Coastal Waters Contracts Nos. 1 and 2 are likely to use an EPB mode machine capable of dealing with a number of interfaces between clay and sands. Cobbles and boulders are expected in both the clays and sands. Drive lengths are long and curved and therefore one-pass, segmental tunnel rings will be used. The gasket system has not yet been finalised but a six part, trapezoidal ring is most likely to be adopted. Progress rates need to be high to achieve programme completion dates.

Three further contracts are required to complete the major coastal tunnel projects, these being one each for MEPAS, Fylde and Cumbria. Design and documentation are currently being undertaken. It is likely that two of these tunnels will be driven using EPB machines and the other using a roadheader. Pipejacking is being considered for the MEPAS and Fylde tunnels as these are straight and about 2 m dia. The Cumbria coast tunnel is 3.0 m dia.

Summary

North West Water has gained considerable experience in soft ground and rock tunnelling along its coastal towns.

A number of tunnelling machines have been used and tunnel construction techniques have also varied. Together with project and construction management advances there have been considerable improvements to tunnelling with increased rates of production, cost control, quality delivery and programme completion. Above all, a new and vastly improved working relationship between the client and the contractor has been developed to gain the fullest benefits from the combined knowledge, the sharing of risk and the drive to improve performance and efficiency.

EPB-shield tunnelling for the Taipei Metro, contract 201A

Walter Dietz
Ed. Züblin AG, Stuttgart, Germany

Abstract
The EPB-shield method has proven itself very well in the specific underground conditions of Taipei. In spite of the restricted and complicated starting conditions and the short tunnel lengths, driving rates which are remarkable for EPB-shield drives could be achieved after certain initial difficulties. The wheeled transport proved to be extremely flexible and efficient. Also, although the segmental lining has a thickness of only 25 cm, the completed tunnels are water-tight and of high quality.
Keywords: EPB-Shield Tunnelling, Taipei Metro.

1 Taipei Metro Project

The growing population and the industrialization of Taiwan led to a heavy increase of traffic over the last twenty years. The City of Taipei expects a significant improvement of its traffic infrastructure by constructing a mass rapid transit system.

Of the initial 84.7 km network, some 41 km are underground, 21 km in bored tunnels and about 20 km in cut-and-cover sections, with all underground stations constructed as open-cut boxes. The remaining lengths of the five lines are either at grade or elevated. It is planned to construct this remarkable project in three phases over ten years under the management of the Department of Rapid Transit Systems (DORTS). The total investment cost is expected to be 18 Billion USD. (Fig. 1)

Meanwhile, the construction of the stations and running tunnels is in full swing. In view of the particular ground conditions and a water table, which in most areas is found close to the ground surface and cannot be lowered, it was decided to use the shield tunnelling method for the excavation of the running tunnels. In order to minimize the overall construction time, all works are basically executed in parallel. This has lead to the unusual situation that 27 shield machines are operated in Taipei: 25 EPB-shields and 2 slurry shields. The machines are supplied by manufacturers from Japan, Canada and Germany [1] [2].

After international tendering, one of the first Contracts, Contract 201, was awarded to the Joint Venture Ed. Züblin AG, Stuttgart, Samsung Construction Co. Ltd., Seoul, and International Engineering and Construction Corp., Taipei. The start of the works, however, was significantly delayed, because the site working area could not be handed over. Finally, in January 1991, the size of the contract was reduced to the construction of the two running tunnels and a service shaft and work could begin.

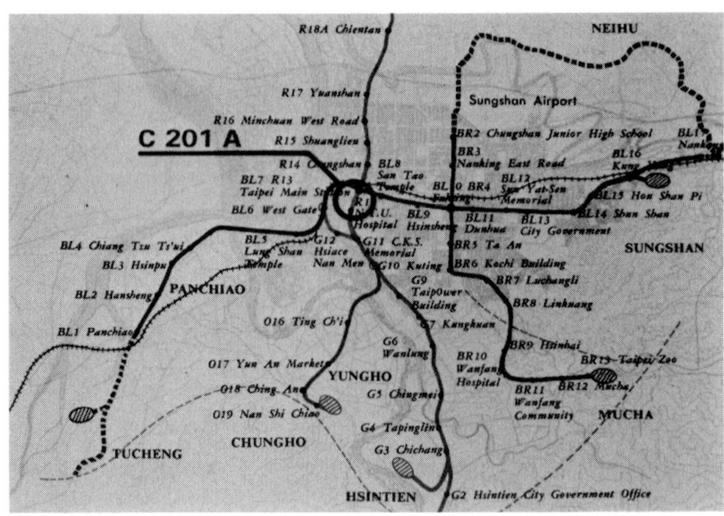

Fig. 1

2 Contract 201A

The site lies close to the main railway station in the town centre, and the tunnel alignment runs underneath a main traffic artery flanked with buildings up to 19 floors high.

The 12 x 44 m large rectangular service shaft is more than 20 m deep and protected by 1.2 m thick diaphragm walls, 35 m deep. They are braced by a ground slab and two intermediate concrete slabs.

The two tunnels each have an internal diameter of 5.42 m and a length of 370 m. They are built to a horizontal curve with a min. radius of 450 m and lie between 15.5 and 25 m below ground in the "Sungshan Formation". (Fig. 2)

At tunnel level, the subsoil consists of medium dense, fine sands with a silt content of approx. 20 %, as well as silty clay, which has a light to medium plasticity. The investigation showed that at least 30 % of the fines have sizes smaller that 0.074 mm. With a ground water level of minus 2 - 4 m the soil is saturated, and the sands and silts have a tendency to run.

3 Shield driving technique and shield machine

The tender documents allowed the choice between a Slurry and a EPB-shield machine. The high content of fines and the presence of ground water present favourable conditions for the use of an EPB-shield. On the other hand, the use of a Slurry machine, would require significant separation. Therefore, preference was given to an EPB-shield.

After detailed technical and economical studies, we decided in favour of a machine built by Herrenknecht GmbH, Germany.

The machine comprises the following (Fig. 3):

The steel shield skin has an external diameter of 6.09 m and a length of 5.47 m, including the tail. The cutter-head has a thickness of 45 cm and is largely closed.

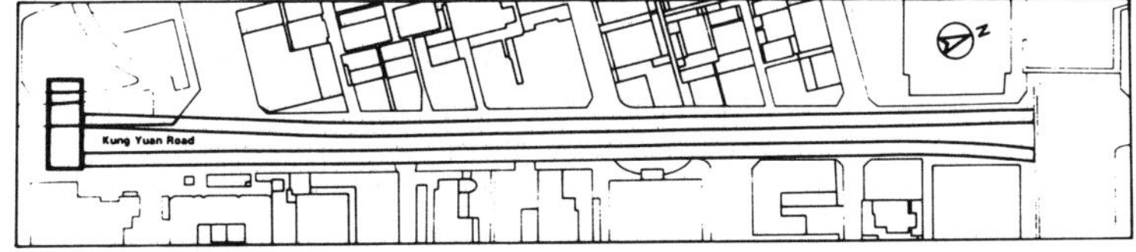

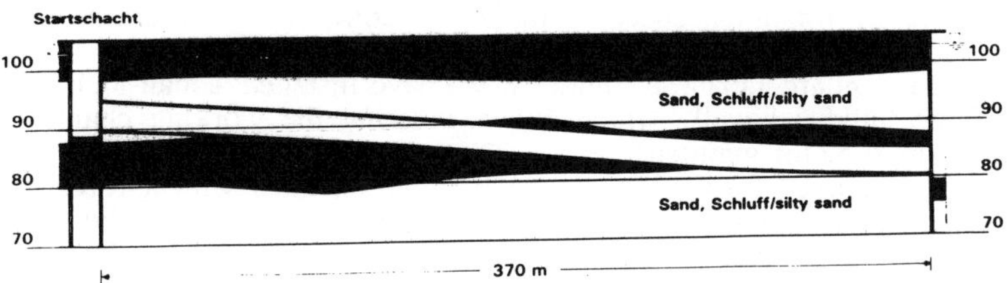

Fig. 2

The openings cover an area of approx. 45 %. They have round and flat picks along their edges. The cutter-head is turned electro-hydraulically by a 640 kW motor with a torque of 5,000 kNm. The turning direction can be changed and the speed varied between 0-3 rpm. Hydraulically operated contour cutters can be used as steering aid.

Paddles on the inside of the cutter-head mix the excavated ground into a homogenious mass. By adding water or bentonite into the working chamber, the ground can be conditioned to the right plasticity. The pressure inside the chamber is monitored by seven sensors on the bulkhead wall of the working chamber.

Arranged on the bottom of the bulkhead wall is the opening for the screw conveyor with a diameter of 70 cm. As no obstructions were expected in this rather

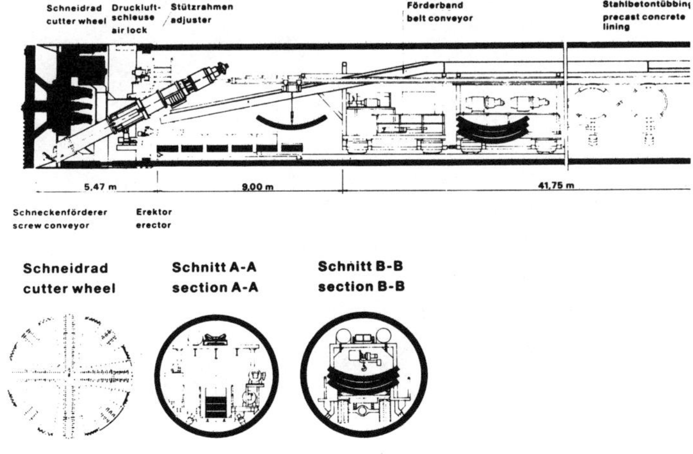

Fig. 3

homogenious ground, we chose a screw with central shaft. It can be moved forward up to 70 cm into the working chamber and also retracted to allow the opening to be closed. It rotates at 0-6 rpm. Water or bentonite can be added in the screw for conditioning the transported material.

It is essential to have correct control of the excavation and jacking processes. This is achieved as follows:

The rotation of the cutter-head and of the screw are adjusted in such a way that the pressure of the ground in the working chamber remains largely unchanged and balances the external pressure. A fall in pressure in the working chamber is met by a reduction of the rotational speed of the screw conveyor, an increase is met by raising the speed.

Contrary to standard Japanese solutions, we have installed a man-lock in the top section of the bulkhead wall, which allows entering the working chamber under compressed air, e.g. for maintenance.

The shield is moved forward by 15 pairs of jacks, each with a force of 2.665 KN, which gives a total thrust of 40,000 KN. Their maximum stroke is 1.45 m. They can be operated alone or in groups, all as required for steering the shield.

When a stroke is excavated, an erector, which is designed for a load of 25 KN and which can be moved 1.50 m in longitudinal direction, picks up the precast segments for installation from the invert where they are positioned on a segment feeder. A ring adjuster helps to achieve precise positioning of the segments.

The end of the shield is sealed with a triple wire brush, into which grease of high viscosity is continuously forced.

The shield machine is followed by a back-up system with the belt conveyor on top, onto which the screw conveyor deposits the excavated material. All the supporting equipment is carried on four gantry sections, including the annulus grouting equipment and the central control station. The back-up system is pulled forward by the shield machine. Together they have a length of 56 m and a weight of 390 tonnes (Fig. 4).

Fig. 4

4 Transport of material

With shield drives it is usual that the segments and the excavated ground is transported on rail when the soil is not pumped. In view of the length of only 370 m per drive of Contract 201A, and with the comparatively complicated starting operation in the restricted space in the workshaft, we opted after detailed considerations for wheeled transport. Lorries with flat platforms transport the segments under the back-up gantries for off loading. Loaded with 7.5 cu.m containers, they take the excavated ground directly from the belt and drive back to the shaft, where a crane lifts the full container to the surface for discharging (Fig. 5).

5 Segmental lining

The tunnel is lined with 25 cm thick precast reinforced concrete block segments, without secondary internal lining. The external diameter is 5.92 m. We chose a ring width of 1.0 m because of the short tunnel length and because we had decided not to use an articulated shield. The ring consists of 5 segments of approx. 3.5 m length and one tapered key-segment, which is erected against the driving direction.

Rings have both parallel and slightly tapered flanges (2 cm) for driving curves and for better steering control. All segments are equipped with Neoprene rubber seals against a water pressure of 2.5 bar. They are fixed together longitudinally and radially by curved M 24 bolts.

The segments have grout holes, which allows the annulus to be continuously filled with pressurized grout while the shield is advancing. The grout pump can be adjusted in such a way that very low settlement values can be obtained.

The Specification required that the tender documents, the segments received a coating of tarepoxy on their outer face and that galvanized steel reinforcement was used. They were produced in a stationary plant in Taipei.

Fig. 5

6 Surveying and steering

A laser theodolite was used in the tunnel and two target plates were installed on the shield. The target data for the shield drive and for the installation of the segments was provided by a tailor-made computer programme. Therefore, and also because the driving lengths were short, no other sophisticated guidance system was needed, which furthermore would have required external service with the danger of delays.

7 The execution of the works

When the site was available, the works began with the construction of the diaphragm walls of the workshaft. Diversion of traffic required that the shaft works were executed in three phases, using the "top-down" method.

The shaft works were followed by soil fracturing to protect buildings with foundations in unfavourable positions. A trial area was also installed in order to be able to determine whether compensation grouting would be necessary to prevent excessive settlement of roads, services and utilities.

The break-out for the shield drive was stabilized and sealed against the water pressure by jet-grouting cubes approx. 8 x 8 m wide, 4 m long for each of the drives.

The diaphragm wall was broken out using high pressure water jets and hydraulic rock splitters. Into this opening a steel ring was concreted. On the inside of this ring a circular Neoprene gasket was fixed, which would serve to seal the gap between the opening and the shield and later the segmental ring.

Meanwhile, the shield was being manufactured over a period of about 12 months and transported to the site by ship and lorry. In late May 1992 shield assembly could commence.

After some pre-assembly on the surface, the shield was lowered into the work-shaft by two mobile cranes and placed onto a steel cradle. The next steps involved the assembly of the cutterhead, the screw conveyor, the segment erector and the tail skin with the wire brush seal. Then the shield machine could be brought upto the break-out with its jacks placed against a steel abutment (Fig. 6).

The first tunnel drive started at the end of July 1992.

Space in the workshaft was extremely limited. The back-up train could only be installed section by section whilst the first 50 m of the tunnel was being excavated. Only then the shield and its equipment were complete. This process required very precise pre-planning and works preparation (Fig. 7).

The above difficulties led to a slow start of the tunnel drive. In addition, when excavating the block of jet-grouted ground, the injected material reacted again in contact with water, so that the cutter head was completely blocked. Profiting now from the installation of the man-lock, the difficulties could be overcome with tole rable effort. A delay of four weeks had however occurred, so that by the end of October only the installation of the shield, back-up system and the first 50 m of the tunnel was complete.

From this moment on, the drive went to plan. The performance, achieved in operating two shifts, was continuously increased up to a maximum of 16 m per working day.

The tunnel drive ended blind against the diaphragm wall of the next station, from which at that time there was no access. Over approx. 300 m an average advance rate of 9.7 m had been achieved (Fig. 8).

Fig. 6

The systems had worked almost perfectly. The measured pressure values corresponded with the theoretical external pressures, which allowed an optimum of pressure control and ground removal. In some sections, bentonite was added for conditioning. During the first 12 m of the tunnel drive up to 20 % of the excavated ground volume was added, reducing to only 4-6 %, and after 30 m none.

As the pre-cast concrete segments, which were delivered from the stationary factory were precise and of high quality, they could be installed with very low tolerances. The completed rings were water-tight and only very few segments were chipped.

The ring adjuster was particularly helpful during grouting, giving better stability against pressure differences.

As a consequence of the continuous pressure control during driving and the careful grout injection to the annulus, the surface settlement reached a maximum of

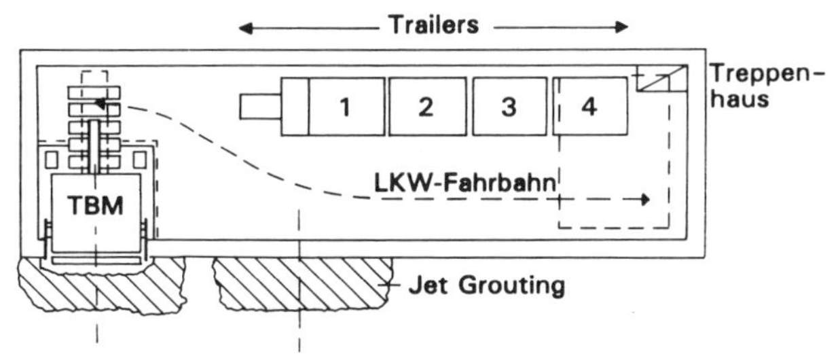

Fig. 7

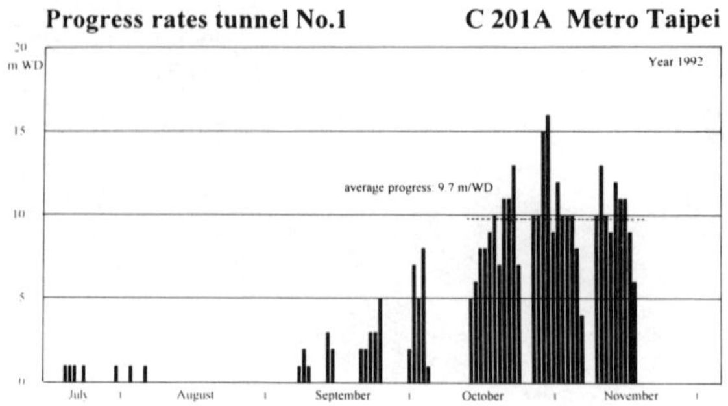

Fig. 8

only 18 mm, so that the envisaged compensation grouting proved not to be necessary.

The simple but reliable survey system allowed respectable advance rates to be achieved whilst staying within the specified tolerances of +/- 50 mm.

Particularly good experience was had with the rubber wheeled transport for segment supply and muck removal. This system proved to be flexible and no flat track was required.

The shield had to be dismantled at the end of the first tunnel drive under the protection of jet-grouting with the skin remaining in the ground. The jet-grouted block, however, proved not to be water-tight. Again profiting from the man-lock, remedial works could be carried out under compressed air and the defect was rapidly dealt with.

A new shield skin had been delivered and the shield machine was dismantled

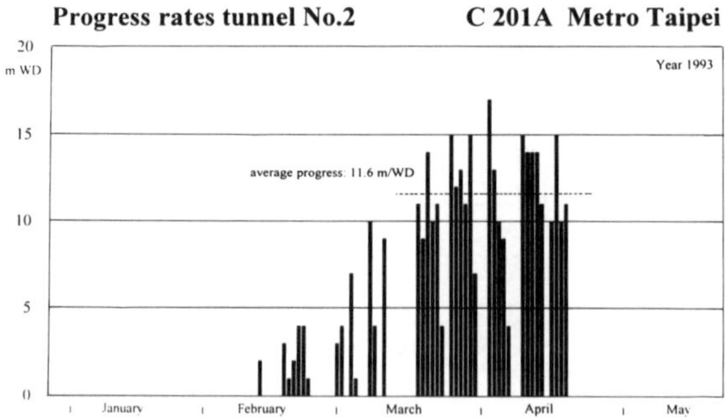

Fig. 9

and re-assembled section after section. These works were completed in late January 1993 and the second shield drive was started with the same procedure as for the first drive.

The second tunnel drive went to plan. After having completed the assembly after 50 m of drive, the maximum performance increased to 17 m and reached an average of 11.6 m over the last 300 m. We consider these to be very respectable results (Fig. 9).

In addition, the settlement of both drives was a maximum of 25 mm.

Again the tunnel drive finished in a blind end, necessitating dismantling in the tunnel.

The end sections of the tunnels and the tunnel invert were lined with in-situ concrete, and the works were completed in August 1993, inside the contract time. Considering the difficult construction conditions in Taipei, this cannot be taken for granted, but has to be regarded a proof of the right choice and operation of the shield machine and all other equipment, of a careful and efficient works preparation and, above all, as a result of the combined efforts of a good team on site.

8 References

1. Wallis, Shani (1992) Tunnelling in Taiwan. **World Tunnelling**, Aug., 300-309.

2. Wallis, Shani (1993) U-Bahnbau in Taipei/Taipei Metro. **Tunnel** 6, 318-329.

TBM drive management by use of computerized systems

D. F. Fawcett
Babtie Shaw and Morton, Maidstone, England
R. Remington
Thames Water Utilities, Ltd., Reading, England
P. J. Chapman
TML (Balfour Beatty International), Thornton Heath, England

Abstract
A TBM drive has the capacity to generate very large amount of recorded data. In the past, either so much data has been collected that analysis becomes impractical, or vital data has been omitted from records. The advent of computer data base handling of the data combined with graphical presentation, has enabled data to be stored and analyzed and then presented in a format usable to busy managers. This paper describes the development and use of computerised monitoring systems on two major projects. Both of the systems described are based on the same philosophy, but were customised to be appropriate management tools for the specific circumstance. A pilot study in 1988 used spreadsheets for data handling. However, these were predicted to be too small in data terms to handle the anticipated projects and both of the commercially used systems handle data using data bases and then present management information using linked graphics packages. The theme throughout the development of these systems has been the requirement to provide a simple user-friendly tool that is of practical use at various management levels. Data collection on site has been designed to be no more onerous than normal tunnel reporting.

The computer side of the system for Transmanche-Link (TML) had to be designed to run as part of the overall project management system that used Artemis 9000 running on an IBM mainframe. The paper goes on to give examples of strategic planning that were facilitated by the availability of reliable records.

Experiences gained in the development and use of the system at TML will be presented and the benefits and disadvantages of the system discussed. For the Thames Water Utilities London Water Ring Main (LWRM) a modified system designed for use on a PC was developed. This constraint arose from the practicalities of using the system on smaller sites and demonstrated that the system could be

transparent to the hardware and software environments in which it was operated. This system was successfully used on all of the Phase II tunnelling works and provided high quality real time management information. Thames Water's experiences are presented.
Keywords: Computerised Systems, Database, Graphical Presentation, Project Management, Shift Reporting, TBMs.

1 Introduction

In 1988 Babtie Shaw & Morton were appointed by TML to develop, and subsequently install, a computer based system for the monitoring and project management of the tunnel boring machines on the UK side of the Channel Tunnel. Initially, a pilot study was carried out using commercially available software for PC computers whereby data from shift reports was processed using a spreadsheet and presented graphically. This pilot study proved the viability of such a system and TML then commissioned the design and installation of a full working system. The full system was specified and a computer program written that could be run in ARTEMIS on TML's main frame, within and compatible to, their existing on site computer project management systems. The computer program was installed and training given to site personnel in both the data collection, computer input and report production. The system was then used continuously with minor modifications for the full duration of all the TBM driving on the UK side of the Channel Tunnel.

Following on from this, Thames Water invited Babtie Shaw & Morton to write a brief report on the adaptation of this system for use on the London Water Ring Main. This report indicated that for the ring main, a similar system, but one which collected significantly different data sets, would be most appropriate for the use Thames Water wished to make of it. The system for Thames Water was designed for use on PC and specified and written based on commercially available database and graphics packages. Once again, the system was installed, staff training given on site and some training provided in the use of reports. This system continued in use throughout the tunnel driving on the London Water Ring Main. Both of these systems have shortcomings and are by no means perfect. However, the systems are significantly better than any systems previously used in the UK (and the authors are unaware of better systems used elsewhere). The fact that both systems remained constantly in use throughout the projects, demonstrates their success in achieving their specified requirements. The object of the systems was to provide readily available, user-friendly, management information that was of real assistance rather than a burden to the site staff.

This paper describes some of the relevant and interesting points in the development of, installation and operation of these two different, but similar, computer monitoring systems.

2 Historical background

Expertise and experience of TBM performance monitoring had been developed on a number of major tunnelling contracts in the UK, South Africa and Cairo. On these projects data relating to TBM performance was collected and processed manually. The result of the need for laborious manual processing was a tendency to data being collected that was never used. Alternatively, potentially useful data was not collected because of the known problems of retrieval and processing. On the projects discussed in this paper the entire process of computation and graphic presentation has been computerised for TML on a Hewlett Packard mainframe, using ARTEMIS software. Following on from this, the Thames system was developed for installation and use on PCs in a site office.

The systems operated previously have been used variously for contractual, management and research purposes. The bias to one of the above headings for any system purpose has been dictated by the system instigator, depending on whether they have been Client, Engineer or Contractor. However, the actual systems have not varied, just the specific use made of them on a particular contract. The two computer based systems developed, follow closely the principles of previous systems, but use the enormous data handling and graphics power of modern computer systems to create highly efficient and effective management tools.

Modern TBMs are highly complex interactive systems where numerous activities are carried out simultaneously. However, experience has shown that by the use of broad headings and the application of some basic rules gained from experience, a highly effective yet simple reporting and analysis system can be created. The latest systems are a result of development having been carried out continuously over the 4 major projects where TBMs of various types have been used. The basic principles involved have changed little during this development. However, recording, analysis and reporting procedures have been refined considerably.

3 System requirements

3.1 General requirements

Both of the systems for TML and Thames have large areas in common. The complexity of modern tunnelling methods leads to a situation where opinions on the performance of various parts of the system vary according to the bias of the observer. This lack of uniform opinion on historical fact leads directly to contractual dispute and/or incorrect management decisions.

To overcome the above problems detailed records can be kept of the work in progress. However, there are three significant difficulties to be overcome if this approach is taken:-

1. Information recording can become a major task causing extreme resistance to it on site. Poor or incomplete data collection can lead to spurious outputs.

2. Analysis of the data collected into usable report format has historically been a very expensive operation, taking up considerable Engineer time.
3. Attempts to limit the data collected to satisfy 1 and 2 above have led to a situation where retrospectively it has been realised that the data collected has not been sufficient to clearly define those aspects of the tunnelling cycle that are eventually required.

Systems have therefore been required that provide the users with a comprehensive, flexible, multi-use management tool. These systems have been specified to be simple to use, cheap to operate and effective for the specified objectives. The systems have provided a level of progress reporting and activity, equipment and systems analysis not previously available on tunnelling contracts.

3.2 Common system philosophy

In a tunnel driven by a modern TBM, many activities are carried out simultaneously. Experience with numerous TBM drives has shown that in almost all cases one activity becomes the critical activity at any one time. The whole reporting depends on the recording and analysis of the single critical activity at any time during tunnel drivage. Experience has shown that the 'critical activity' concept provides excellent reporting and performance analysis results and facilitates data collection. The system has the added advantage that when looking at performance and analysis reports as a contractual or management tool, the critical items are the activities recorded, analyzed and presented. As performance improves so the critical items will change and this will be apparent from the reports.

Tunnel drives are complex production lines, dependent on numerous in-line activities within the 'TBM' area as well as related activities outside this area. The reporting system developed for TML was limited to reporting and analysing data that could be collected at the TBM, i.e. anything that happened forward of the tail end of the machine and its back-up equipment. For Thames Water the system was adapted and extended to include all in-line activities. This becomes practical on these less complex works where it is normal for the personnel on the machine to be aware of the activities or delays occurring elsewhere in the production line. At the same time the TBM data collected was much reduced.

The systems devised have three distinct parts:-

- Data Collection
- Data Analysis
- Reporting

3.4 Data collection

The data required for reporting has been obtained from a single report for each shift in each drive, in a pre-described coded format (see Figure 1).

The boxes at the head of each report sheet provide space to record basic information.

DATE:	THAMES WATER LONDON RING MAIN	WK NO.	REPORT NO.
SHIFT:	TUNNEL DRIVE LWRM 1 T.B.M. SHIFT ACTIVITY REPORT	SHEET NO. OF.	

REPORT CHECKED	DATA ENTERED IN COMPUTER	RINGS BUILT: FROM: TO:

THRUST	TORQUE:	HEAD R.P.M	ADVANCE RATE

TIME			CODE	RING NO.	ACTIVITY
FROM	TO	DURATION (MINS)			
TOTAL DURATION			FACE CHAINAGE AT END OF SHIFT:		

FIGURE 1

The actual data was recorded by providing a continuous record of single critical activities from the official start of the shift to the official end of the shift.

When recording data, durations were only noted to the nearest 5 minutes. In practice it has been found that durations of less than 10 minutes do not normally have any effect on the reports produced. However, frequently repeated short duration procedures (such as TBM re-gripping) can become distorted if not recorded accurately. In this case the actual duration of re-gripping to the nearest minute was recorded.

The data entered into the computer from the reports was limited to the information from the basic data 'boxes' at the top and bottom of the reports and the two columns headed FROM and CODE. From this entry the computer re-created the complete report. The information in the other columns has been required for completeness and to enable errors to be detected and corrected.

3.5 Data headings TML system

With any reporting and analysis system it is necessary to allocate activities to certain headings. The number of headings affects the standard of the report and the useful information that it can provide. To deal with this, code matrices were devised that are specific to each type of TBM (see Figure 2). The matrices are only reference tables that are used to identify the code for an activity. The reference table is set out with vertically listed main category headings and horizontally listed sub-category headings. However, the combination of sub-category headings and main category headings does not always give logical codes. Where codes are illogical, the relevant square is invalidated (black dot).

The quality of the reports is determined by the accuracy of the durations recorded and the quality of the choice of code.

The use of the Matrix Code tables has the following advantages:-

1. About 150 usable activity codes are generated from only 30 headings and 8 downtime causes.
2. Interpretation of reports, where different originators have used different descriptions for similar activities, is avoided.
3. Shift reports can be entered into the computer by non-technical staff.
4. Computer data entry requires the minimum of key strokes.

Code tables are available to the originators of reports so that no intermediate report interpretation is required.

For the LWRM a different set of code tables were devised that were appropriate for the user requirements. On the LWRM the system had to be able to deal with various types of TBM and tunnel lining systems.

To overcome this problem two code tables have been devised, one for each type of TBM. Again, tables are purely reference tables that are used only to identify the code for an activity.

DEC. 89 | REFERENCE TABLE 1 | ISSUE 7

MARINE SERVICE TUNNEL ACTIVITY CODES

GROUP	CODE	MAIN CATEGORY / SUB CATEGORY	NORMAL OPERATION 1	ELECTRICAL BREAKDOWN 2	MECHANICAL BREAKDOWN 3	HYDRAULIC BREAKDOWN 4	LUBRICATION FAILURE 5	BAD GROUND 6	SUPPLY ERROR 7	ALIGNMENT 8	MISCELLANEOUS SUB CATEGORIES
PRODUCTION	Aa	BUILDING RINGS – CONCRETE		●	●	●	●				9 = Bad Build
	Ab	BUILDING RINGS – OPENING		●	●	●	●				
	Ac	BUILDING RINGS – H.D.C.I.		●	●	●	●				
	B	RE-GRIP MACHINE		●	●	●	●		●		
PLANNED DELAYS	C	CUTTING OUTSIDE BUILD	●	●	●	●	●		●		(Only code 'C' if not any other category below)
	D	PROBING		●	●	●	●	●	●	●	9 = Set up 10 = Grouting probes
	E	PLANNED MAINTENANCE		●	●	●	●	●	●	●	
	Fa	CUTTER CHANGES		●	●	●	●	●	●	●	
	Fb	FACE INSPECTION		●	●	●	●	●	●	●	
	Fc	CABLE EXTENSION		●	●	●	●	●	●	●	10 = Ventilation cassette
T.B.M. DOWNTIME (10 = Human error)	Ha	TBM BREAKDOWN HEAD ROTATION	●						●	●	
	Hb	TBM BREAKDOWN THRUST / GRIP	●						●	●	
	Hc	TBM BREAKDOWN FINGERS / SHIELD	●						●	●	
	I	TBM CONVEYOR	●						●	●	9 = Belt
	Ja	TOP ERECTOR	●						●	●	
	Jb	BOTTOM ERECTOR SEGMENT CRANE	●						●	●	
	Jc	SEGMENT LOADER / MAGAZINE	●					●	●	●	
	K	ZED / SURVEY EQUIPMENT	●		●	●	●		●		
	La	PLANT INSTALLATION & COMMISSIONING									
	Lb	PLANT MODIFICATION		●	●	●	●	●		●	
	Lc	TBM MISCELLANEOUS INCL. PROBING EQUIPMENT	●								
OTHER DELAYS		(COLUMN 1 = DELAYS NON BREAKDOWN)									
	N	PUMP FAILURE / FLOODING	●			●	●		●	●	
	O	ENGINEER / SURVEY		●	●	●	●				
	P	CLEAN BUILD AREA		●	●	●	●		●		
	Qa	WAITING FOR SKIPS		●	●	●	●	●		●	9 = TBM SYSTEM DELAYS 10 = RAIL SYSTEM DELAYS
	Qb	WAITING FOR SEGMENTS		●	●	●	●	●		●	
	Qc	WAITING FOR OTHER MATERIALS		●	●	●	●	●		●	
	R	CAVITY GROUTING					●				
	S	SERVICE FAILURE	●					●	●	●	
	T	CREW ABSENT		●	●	●	●	●	●	●	
	V	MISCELLANEOUS OTHER									
	W	SITE CLOSED		●	●	●	●	●	●	●	
	Z	MISSING RECORD		●	●	●	●	●	●	●	

● = INVALID CODE

FIGURE 2

DATE: TBM SHIFT RECORD WEEK NO.
SHIFT: REPORT NO.
SHEET NO. OF.

REPORT CHECKED	DATA ENTERED BY:	RINGS BUILT: FROM NO: TO NO:

ENGINEERING DATA TUNNELLING METHOD: EARTH PRESSURE BALANCE
MODE: SCREW: CONVEYOR: LINING METHOD: BOLTED

RING NO.	OXY. %	WATER INFLOW 1-5	LINE mm	LEVEL mm	EARTH PRESSURE		ROTATION PRESSURE		SHOVE RAM PRESS	INSTANT ADVANCE RATE	PLUMB	SQ.
					HEAD	SCREW	HEAD	SCREW				

TOTAL GROUT TAKE FOR SHIFT BAGS OF GROUT

TIME			ACTIVITY		LAST	COMMENTS
FROM	TO	DURATION	CODE	CAUSE	RING NO.	
TOTAL						

FIGURE 3

CODE TABLE 1
EARTH PRESSURE BALANCE MACHINE

PRODUCTION	A
PLANNED DELAYS	
PROBING	B1
SERVICES EXTENSION	B2
MOVE CROSSING	B3
PLANNED BREAKS	B4
MACHINE DELAYS	
SHOVE RAMS	C1
FRONT SHIELD RAMS	C2
CONVEYOR	C3
ERECTOR	C4
SEGMENT CONVEYOR	C5
LASER/STEERING	C6
HEAD INSPECTION	C7
MISCELLANEOUS	C8
OTHER DELAYS	
DERAILMENT	D
SHAFT CRANE	E
GROUTING	G
SEGMENT SUPPLY (SURFACE)	H
CREW	K
LOCOS	L
NON WORKING SHIFT	N
PIT BOTTOM	P
MISSING RECORD	R
SERVICES FAILURE	S
VENTILATION	V
MISCELLANEOUS	X

FIGURE 4

Whilst production is progressing smoothly, Code A was used to include **all** the various production activities which occur. **Only** when a delay occurred would another code from either Planned Delays, Machine Delays or Other Delays, be used.

The computer was programmed not to recognise or accept non-existent or illogical codes. Codes for miscellaneous items were provided when it is not possible to allocate the activity to any existing code.

Figures 3 and 4 show typical code tables and shift reports for a ring main tunnel.

3.6 Data processing

Historically TBM reporting and analysis has been carried out manually. This has involved site staff in tedious calculations and presentation limitations and has been seriously restricted by the number of headings and codes that could be handled manually.

To extend and improve analysis and overcome the above problems, computer programs have been written to analyze data and generate reports. This has enabled a system to be designed with the following features:-

1. Menu driven system for use by non-computer trained personnel.
2. Basic simple data entry using a designed entry screen.
3. The facility to provide a concise presentation and very rapid calculation of the information required in comparison with a manual output.
4. The ability to transmit reports to any terminal/plotter linked to the system.
5. Large data storage for future specific analysis as requirements arise.
6. High quality presentation of graphical reports.

Computer programs for TML were written using ARTEMIS for use on a Hewlett Packard mainframe computer. Figure 5 shows a typical menu facility.

The system for LWRM was written for the IBM 386 PC using Graftalk and Clipper as the base programs. A fully menu driven approach similar to the TML system was supplied.

3.7 Weekly Reports produced

The purpose of the weekly report has been to present a detailed picture of the week's progress and downtime causes on one sheet of A4 paper.

Typical weekly reports for a TML drive and a LWRM drive are shown in Figures 6 and 8.

A typical TML report comprised 3 main elements:-

1. A pie chart of the four main groups of activities.
2. A table of the hours in the week allocated to each main code heading (the headings down the left-hand side of the relevant matrix code table). This table also includes:-

Average build time per ring

MAIN MENU	SUB-MENUS		
1 - Input Data	1 - Input Data	1.1	Tunnel (North)
		1.2	Tunnel (South)
2 - Amend Data	2 - Amend Data	2.1	Tunnel (North)
		2.2	Tunnel (South)
3 - Validate Data	3 - Validate Data		
4 - Print Data	4 - Print Data	4.1	Tunnel (North)
		4.2	Tunnel (South)
5 - Weekly Reports	5 - Weekly Reports	5.1	Plot Draft Reports
		5.2	Approve Reports
6 - Monthly Report	6 - Monthly Reports	6.1	Plot Draft Reports
		6.2	Approve Reports
7 - Other Reports	7 - Other Reports	7.1	Tunnel (North)
		7.2	Tunnel (South)
8 - Additional Plots	8 - Additional Plots	8.1	Tunnel (North)
		8.2	Tunnel (South)

FIGURE 5

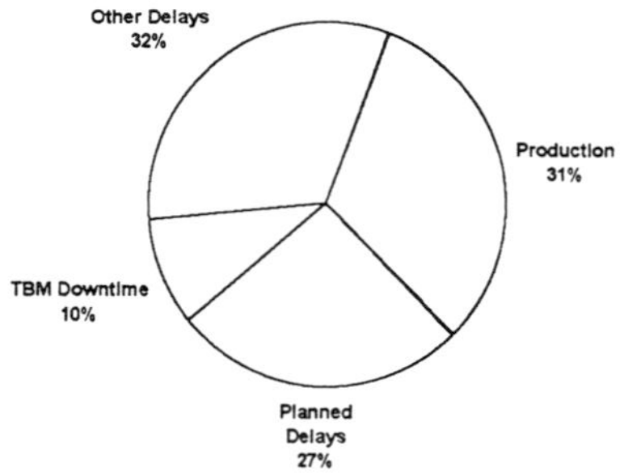

Other Delays 32%

Production 31%

TBM Downtime 10%

Planned Delays 27%

DRIVE LENGTH TO DATE 3366 m

87 RINGS BUILT

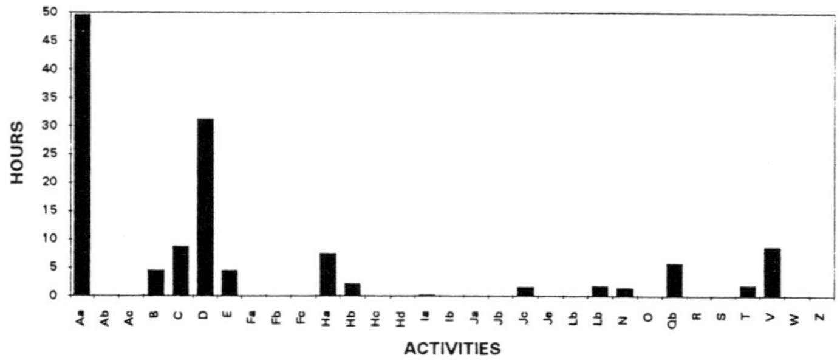

HOURS

ACTIVITIES

Activity	Code	Hours to Date
PRODUCTION		
Building Rings - Concrete	Aa	49.57
- Opening	Ab	
- HDCI	Ac	
Re-Grip Machine	B	4.55
SUB-TOTAL		54.12
AVERAGE BUILD PER CONCRETE RING		0.57
AVE. TIME ONE RING PROGRESS		1.93
PLANNED DELAYS		
Cutting Outside Build	C	8.77
Probing	D	31.17
Planned Maintenance	E	4.5
Cutter Changes	Fa	
Face Inspection	Fb	
Cable Extension	Fc	
SUB-TOTAL		44.44
MACHINE UTILIZATION		
UTILIZATION TOTAL TIME		50.7%
UTILIZATION AVAILABLE TIME		60.6%
TBM DOWNTIME		
TBM Breakdown Head Rotation	Ha	7.5
TBM Breakdown Thrust Grip	Hb	2.17
TBM Breakdown Fingers/Shield	Hc	
TBM Conveyor	I	0.33
Top Erector	Ja	1.62
Bottom Erector Segment Crane	Jb	1.83
Segment Loader/Magazine	Jc	0.67
ZED/Survey Equipment	K	
TBM Miscellaneous	L	1.82
SUB-TOTAL		15.94
MACHINE AVAILABILITY		83.7%
OTHER DELAYS		
Pump Failure/Flooding	N	1.48
Engineer/Survey	O	
Clean build area	P	27.43
Waiting for skips	Qa	7.33
Waiting for segments	Qb	5.9
Waiting for other Mat	Qc	0.67
Cavity Grouting	R	
Service failure	S	
Crew Absent	T	1.92
Miscellaneous Other	V	8.78
Site Closed	W	
Missing Record	Z	
SUB-TOTAL		53.51
RECORDS TOTAL		168.01
TOTAL SHIFT DURATION		168.01

TRANSMANCHE LINK - WEEKLY TUNNEL REPORT

MARINE SERVICE TUNNEL

Chart No : W4 : File : QS2W

Week No : 254

Week ending : 20-NOV-1988

FIGURE 6

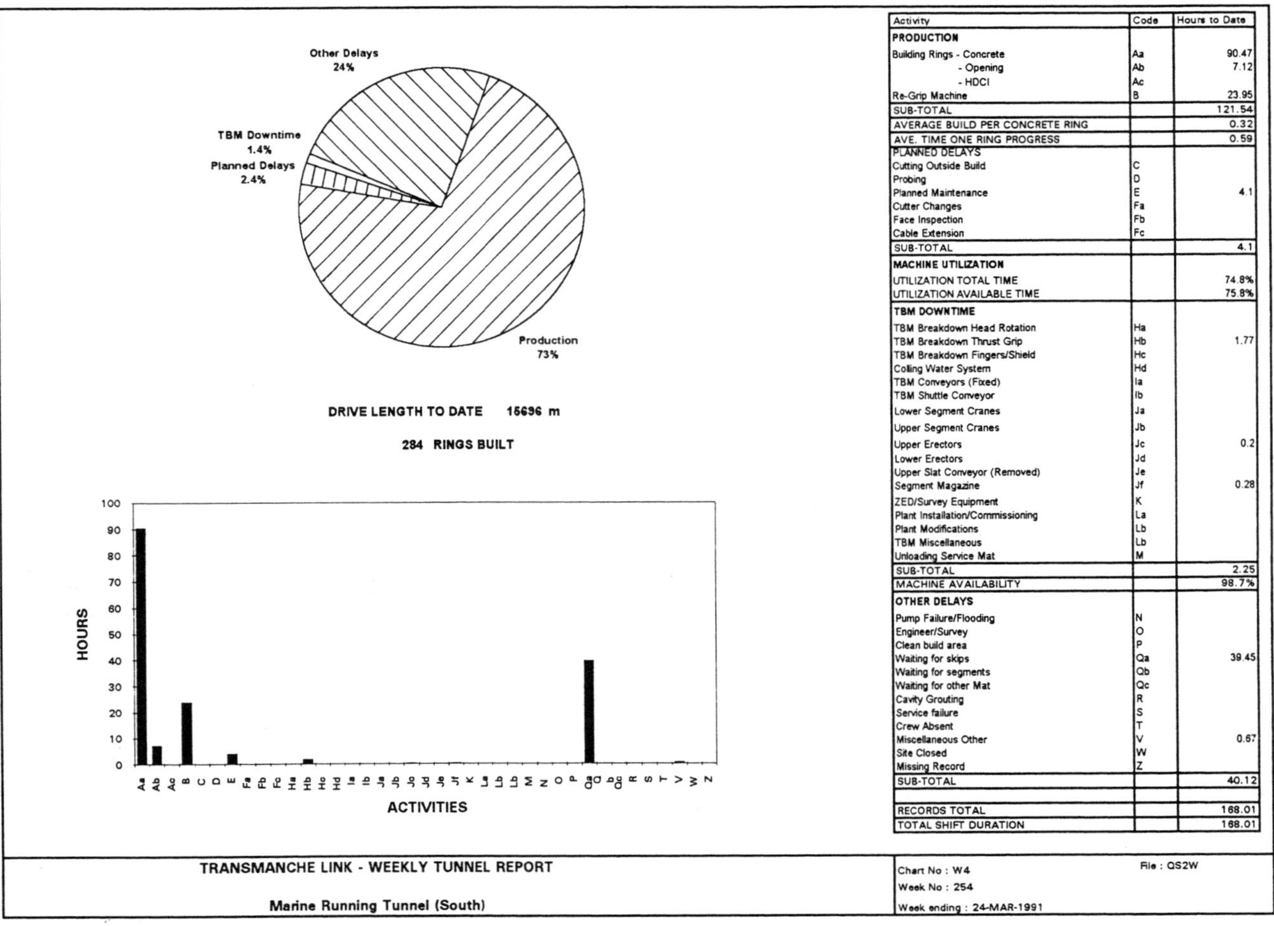

Activity	Code	Hours to Date
PRODUCTION		
Building Rings - Concrete	Aa	90.47
- Opening	Ab	7.12
- HDCI	Ac	
Re-Grip Machine	B	23.95
SUB-TOTAL		121.54
AVERAGE BUILD PER CONCRETE RING		0.32
AVE. TIME ONE RING PROGRESS		0.59
PLANNED DELAYS		
Cutting Outside Build	C	
Probing	D	
Planned Maintenance	E	4.1
Cutter Changes	Fa	
Face Inspection	Fb	
Cable Extension	Fc	
SUB-TOTAL		4.1
MACHINE UTILIZATION		
UTILIZATION TOTAL TIME		74.8%
UTILIZATION AVAILABLE TIME		75.8%
TBM DOWNTIME		
TBM Breakdown Head Rotation	Ha	
TBM Breakdown Thrust Grip	Hb	1.77
TBM Breakdown Fingers/Shield	Hc	
Colling Water System	Hd	
TBM Conveyors (Fixed)	Ia	
TBM Shuttle Conveyor	Ib	
Lower Segment Cranes	Ja	
Upper Segment Cranes	Jb	
Upper Erectors	Jc	0.2
Lower Erectors	Jd	
Upper Slat Conveyor (Removed)	Je	
Segment Magazine	Jf	0.28
ZED/Survey Equipment	K	
Plant Installation/Commissioning	La	
Plant Modifications	Lb	
TBM Miscellaneous	Lb	
Unloading Service Mat	M	
SUB-TOTAL		2.25
MACHINE AVAILABILITY		98.7%
OTHER DELAYS		
Pump Failure/Flooding	N	
Engineer/Survey	O	
Clean build area	P	
Waiting for skips	Qa	39.45
Waiting for segments	Qb	
Waiting for other Mat	Qc	
Cavity Grouting	R	
Service failure	S	
Crew Absent	T	
Miscellaneous Other	V	0.67
Site Closed	W	
Missing Record	Z	
SUB-TOTAL		40.12
RECORDS TOTAL		168.01
TOTAL SHIFT DURATION		168.01

FIGURE 7

SUMMARY OF ACTIVITY HOURS FOR WEEK 16

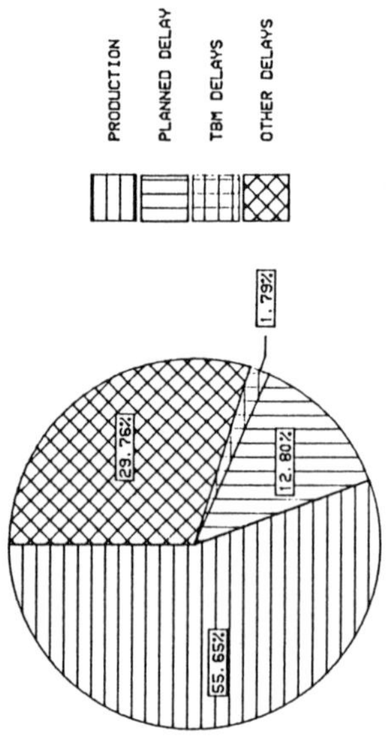

PRODUCTION	55.65%
PLANNED DELAY	12.80%
TBM DELAYS	1.79%
OTHER DELAYS	29.76%

Activity	Code	Hrs Per Week
Production	A	93.50
Planned Delays		
Probing	B1	0.00
Service Extension	B2	0.00
Move Crossing	B3	0.00
Meal Breaks	B4	5.00
Planned Maintenance	B5	16.00
Miscellaneous	B6	0.50
SUB-TOTAL		21.50
TBM Delays		
Shove Rams	C1	0.00
Conveyor	C2	0.00
Erector	C3	1.00
Segment Conveyor	C4	0.00
Laser / Steering	C5	0.00
Excavator	C6	0.00
Miscellaneous	C7	2.00
SUB-TOTAL		3.00
Other Delays		
Derailment	D	2.00
Shaft Crane	E	0.00
Segment Supply	H	0.00
Crew	K	0.00
Locos	L	0.00
Mucking In Shaft	M	0.00
Non - Working Shift	N	48.00
Pit Bottom	P	0.00
Missing Record	R	0.00
Services Failure	S	0.00
Ventilation	V	0.00
Miscellaneous	X	0.00
SUB-TOTAL		50.00
RECORDS-TOTAL		168.00

PROGRESS DATA

TOTAL DRIVE LENGTH	5313
DRIVE LENGTH TO DATE	4188
DRIVE 78% COMPLETE	
RINGS TO DATE	4188
RINGS THIS WEEK	418
RINGS PER SHIFT THIS WEEK	41
RINGS PER SHIFT TO DATE	32

ENGINEERING DATA

	MAX	MIN	UNITS
AIR PRESSURE	16	0	PSI
OXYGEN PERCENTAGE	21.00	4.00	%
WATER INFLOW	16	1	
KEY PRESSURE	2500	0	PSI
LINE	35	-45	MM
LEVEL	30	-30	MM
AVERAGE GROUT TAKE PER RING		0	BAGS

KEY TYPE 0

STONE BANDS

SAND LENSES

WEEKLY DRIVE REPORT

LWRM STAGE V : BARNES - BARROWHILL
Holland Park Ave to Barrowhill

Plotted 30/06/92
Week No 16
Date 28/06/92
Contract Week No 67

FIGURE 8

Average time for 1 ring progress
Machine availability
Machine utilisation of total time
Machine utilisation of available time

3. A bar chart that gives a visual representation of the hours in the week allocated to each of the main code headings.

3.8 Performance analysis reports

Other longer term reporting was specific to either TML or LWRM, but in both cases was designed to show trends for specific activities or delays. For TML a standard 'monthly' report can be requested at any time and will always produce a report up to the end of the last week to have an approved weekly report for that drive.

The standard reports have included 13 pages of plots with up to 8 bar charts on each page. The reports show:-

1. Cumulative data for the drive to date.
2. Stacked bar charts of selected downtime causes over the last 25 weeks.
3. The trend of the ring build and utilisation times and percentages over the last 50 weeks. (Fig. 9)
4. The trend of each main activity heading over the last 50 weeks - these are grouped into Production, Production Delays, TBM Downtime and Other Delays. (Fig. 10)
5. The trend of overall downtime causes over the last 50 weeks for the sub-category headings shown on the matrix code chart.

For the LWRM the other reports were designed to provide:-

TBM performance trends and records
Contractor performance records and trends

These graphs can be plotted for one week, four weeks, twelve weeks and fifty weeks.

Activity hours option which produced a graph of any activity or multiple activities for the 'last' fifty weeks.

An Engineering Data Option was also available that produced graphs of engineering data for a 500 ring section of tunnel.

4 Specification and development of computer programs

The computer programs for both of these systems were specified in detail after consultation with the respective users and then the contract placed with a software house for the writing of the programs in the relevant language for use on the specified

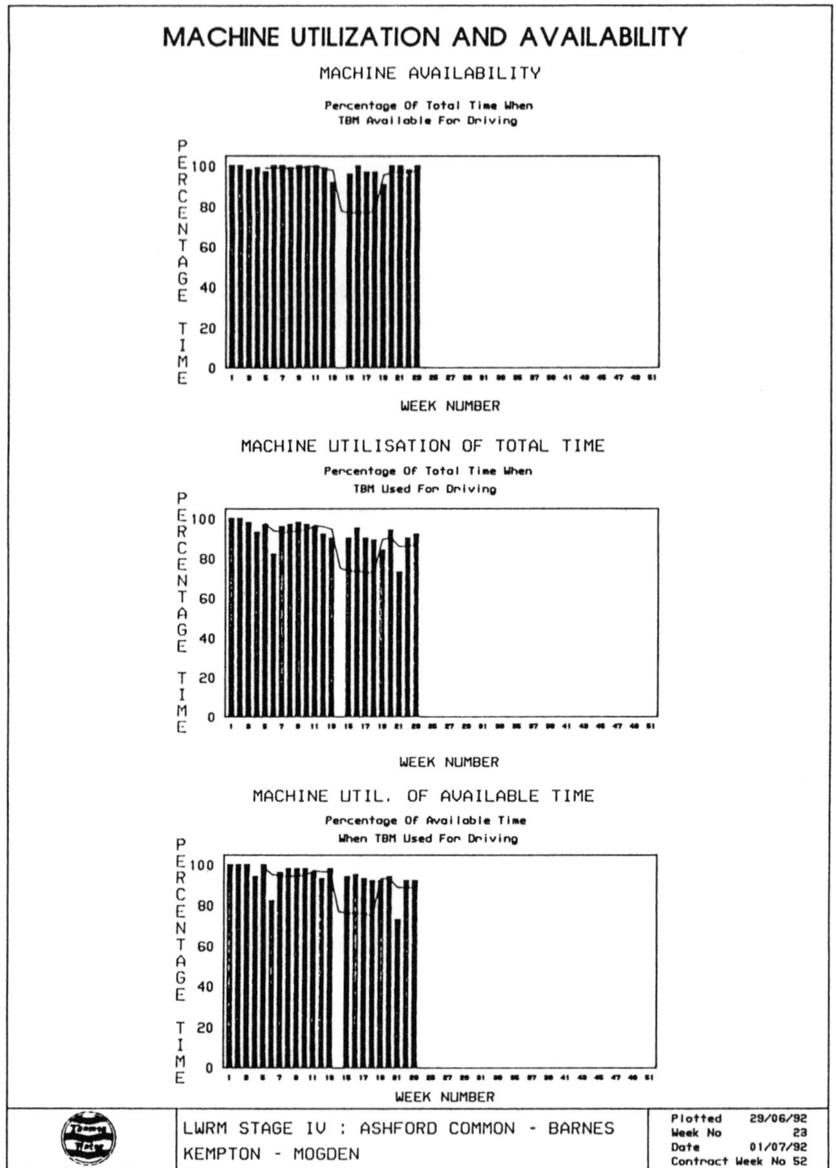
MACHINE UTILIZATION AND AVAILABILITY
MACHINE AVAILABILITY
Percentage Of Total Time When
TBM Available For Driving
PERCENTAGE TIME
WEEK NUMBER
MACHINE UTILISATION OF TOTAL TIME
Percentage Of Total Time When
TBM Used For Driving
PERCENTAGE TIME
WEEK NUMBER
MACHINE UTIL. OF AVAILABLE TIME
Percentage Of Available Time
When TBM Used For Driving
PERCENTAGE TIME
WEEK NUMBER
LWRM STAGE IV : ASHFORD COMMON - BARNES
KEMPTON - MOGDEN
Plotted 29/06/92
Week No 23
Date 01/07/92
Contract Week No 52

FIGURE 9

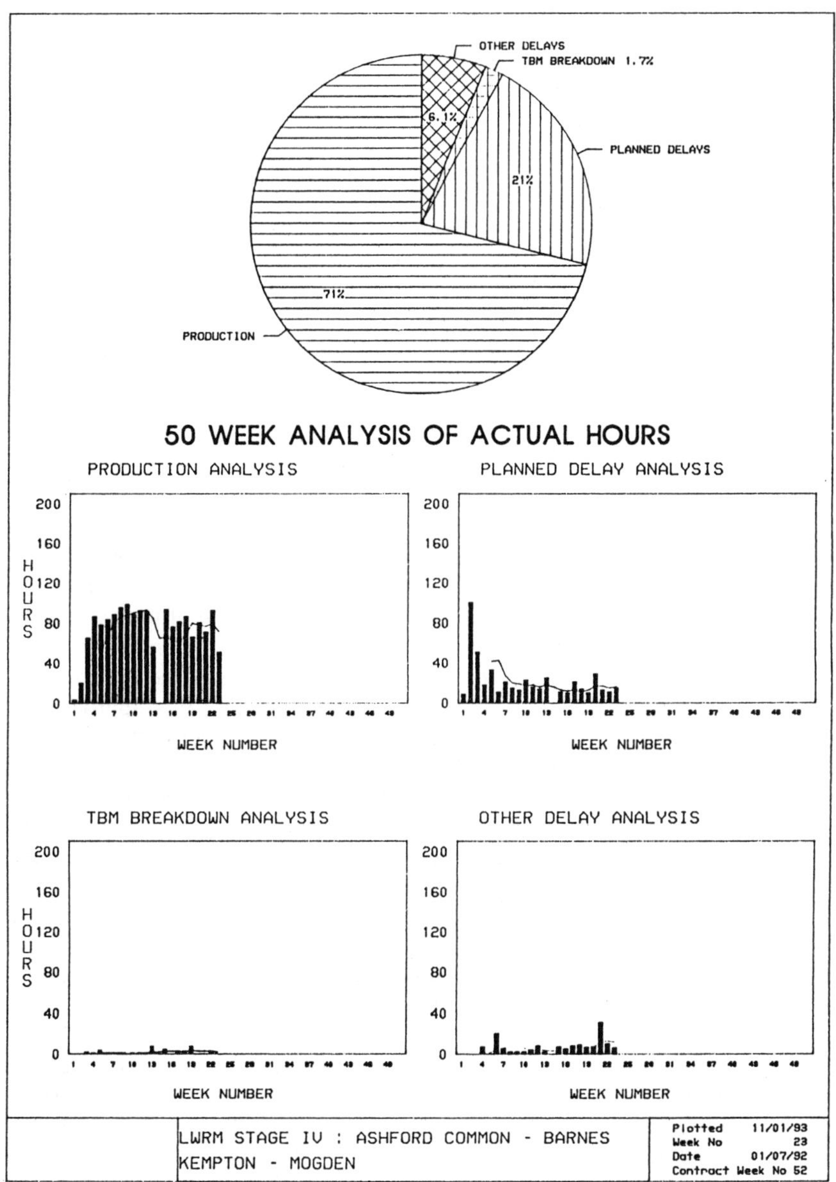
OTHER DELAYS
TBM BREAKDOWN 1.7%
6.1%
PLANNED DELAYS
21%
71%
PRODUCTION
50 WEEK ANALYSIS OF ACTUAL HOURS
PRODUCTION ANALYSIS
PLANNED DELAY ANALYSIS
HOURS
WEEK NUMBER
TBM BREAKDOWN ANALYSIS
OTHER DELAY ANALYSIS
LWRM STAGE IV : ASHFORD COMMON - BARNES
KEMPTON - MOGDEN
Plotted 11/01/93
Week No 23
Date 01/07/92
Contract Week No 52

FIGURE 10

hardware. Both of the programs were written by M.O.M. International Ltd. in about a 4 week period each, and both initial programs largely fulfilled the brief required of them.

Development of the program written in ARTEMIS for use by TML was taken over by TML's own in-house programmers once the system was up and running on site. During the life of the TML system several modifications were made from time to time as circumstances changed on site. The system used by Thames Water has had very little modification with Thames adapting their input and output requirements in order to keep computer program modifications down to a minimum.

5 Practical problems of system installation

The largest hurdle that had to be overcome to enable these systems to work well on site was the scepticism of site staff that any such system could be of any practical value to them. Although a system such as this provides a line manager with a very effective and efficient reporting system, it also provides others with the same information and denies the line manager the historical right that he has had to disseminate information in the way and using the timing at his discretion. Having overcome these problems, either by persuasion or direction from senior management, then installation became fairly straightforward. Due to the scale of the work at TML, Babtie retained an ongoing input into the work for several years after its initial installation. This input was achieved in the form of auditing of the information and proper use of the system at three monthly intervals. Using this approach, TML management were able to gain confidence along with Eurotunnel in the results being produced by the system and after about 18 months, once the staff and management regarded the system as part of the furniture, this procedure became unnecessary and was discontinued. One of the main problems encountered during these initial periods is that no reporting system that can be envisaged can be all things to all men and compromises both in the data input and reporting output are continually required. Many hours were spent with both management, middle management, and staff at reporting levels, negotiating the compromises that were necessary in order to make the system practical and effective. Without this time input, certainly on the TML system, it is believed that the success achieved would not have occurred.

For the Thames systems, this time input was unnecessary due to the very much smaller number of staff involved and the more concise nature of the work being reported on. Problems at Thames ranged from the compatibility of the software with other software being used on the machine, to the problems of handling varying working week lengths in a system of this nature. One of the options that was considered for the Thames system was to enable the system to be tailored from an on-screen menu to suit any tunnel drive. Having looked into this option in some detail it became apparent that although it was entirely possible for the number of drives that Thames envisaged, it would not in fact be an economic proposition. Other problems that occur with systems of this nature relate to the sheer size of data storage and handling required for these large databases and graphics generation. This particular problem did cause inefficiencies at TML in the early days, and was in fact resolved by the introduction of additional computing power. With any computing system where large quantities of data are required to be stored for long periods, then effective and efficient data management archiving, and efficient handling by the computer of these stores, is an essential part of the system.

6 Benefits and shortcomings of TML system

The major benefit of the TML system was that it isolated and reported data on several selected critical functions of the long and complex TBMs, separating hydraulic and electrical functions.

It was established at an early stage that, whilst marginal, building the rings would generally take longer than excavation, and therefore this would be taken as the main critical factor against which other activities would be measured. However, slower excavation through hard ground might be expected which could not be considered as a delay to TBM performance, and was therefore allowed for in planned delays.

It was agreed that the system would be set up to assist operational management and not as a tool for contractual or commercial use. Despite the facility of the system to handle data and the sophisticated methods of data presentation, the process was still vulnerable to the 'rubbish in, rubbish out' syndrome. Education and training was, therefore, given to the engineering staff to ensure that they understood the value of the system, and vigilance was required to maintain a 'standard' approach to the recording and coding of events.

The reports were utilised by the TML construction management in different ways over three distinct periods of each of the TBM drives.

- During the early mobilisation and learning curves, with the particularly difficult conditions initially in the marine TBM drives.
- Refining, streamlining and accelerating each of the TBM drives up to their optimum rates of progress.
- The continuous periods of high production rates over the latter parts of all drives.

During the initial stages of each of the drives TML management were clearly aware of any major areas of difficulty that needed to be addressed, and therefore knowledge of detailed availability factors of individual elements of each TBM drive was of little benefit.

In addition the impact of any major area of difficulty or major element of stoppage could impact not only on the accuracy of reporting the individual progress elements, but also the efficiency at which these elements had been carried out.

However, the reports could clearly quantify the major problem areas and their impact, which could be beneficial where cost or programme benefit studies for modifications or systems adjustments had to be addressed.

As an example, the weekly reports for the Marine Service Tunnel for week ending 20th November 1988, (Fig. 6), clearly quantifies the impact of the known but major difficulties the drive was in at that time. Whilst probing was a necessary operation and therefore entered as a planned delay against the critical activity of building rings, over 30 hours per week was clearly unacceptable, as was then a further planned stoppage for maintenance. The probing system, consisting of two Craelius drilling rigs, was provided at the front of the TBM and was designed to allow two 75 m long holes to be drilled in one 8 hour shift, but considerable practical difficulties were initially found with concurrently drilling the forward probes, resulting in the large delays noted. A series of modifications and tests were made resulting in the acceptance of a single forward probe up to 240m in length which was then generally drilled within, and concurrently with a single 8 hour maintenance shift. As the verticality of such an extended hole was very difficult to predict, a method of ascertaining the level (to within $\pm$ 1m) by micropalaeontogical analysis of drill flushings was developed by the site geotechnical team, and rapid, reliable and informative forward probing was then maintained throughout the drive.

The delays of over 27 hours that week to clean the build area clearly quantified one of the major impacts of the poor ground conditions, with the need to bar down and secure, and then clear up overbreak that had fallen into the segment erection area prior to each build. Similarly, regripping was carried out slowly in an attempt to minimise loosening material in the crown.

Provision of a trailing hood with stainless steel fingers largely overcame this problem. Cutting outside build was a planned delay in the event that hard ground conditions could lead to cutting taking longer than building, which very seldom occurred. The entry of over 8 hours in Fig. 6 was as a result of the need to control overbreak and loose blocks coming into the head, or along the conveyors.

At this time waiting for skips was largely as a result of the need to change muck trains in mid cut, due to the volume of overbreak being encountered.

The main benefits of having clearly identified downtime analysis for individual components of each TBM, came into use after overcoming the initial mobilisation and learning curves, and assisted in structuring maintenance programmes and resource requirements.

In particular the available detailed analysis of the individual performance of the components of the Marine Running Tunnel TBMs greatly assisted in planning the strategy for the modifications to the two TBMs which resulted in the remarkable and consistent rates of progress they achieved for the final eight months of both drives.

Once all drives were achieving optimum rates of progress, the limitations of only having analyses from the front end of the TBMs became exposed, as clearly identified on the weekly report for the Marine Running Tunnel South for week ending 24th March 1991 (Fig. 7) when the remarkable record of 428 m advance in one week was achieved.

Of necessity, the reporting Engineer was always located within the build area of the TBM, extremely busy, and despite good communications, not only along the TBM but throughout the logistics system, with the length and complexity of the TBMs the accuracy of reporting causes of delays outside of the build area was difficult to achieve.

There were good reporting systems in place within the logistics operation, and therefore to avoid attempts of further analysis at the TBM, all apparent external delays were rolled up into "waiting for skips".

With the high rates of progress being achieved in both segment erection and cutting, train turn around times under the TBM, whilst at optimum efficiency, were impacting on and therefore included in this delay time.

Adjustment of the reporting system to introduce a fixed element for train turn round times into planned delays, was not considered beneficial by this stage of the project.

Delegating the reporting to the Shift Boss or TBM operator, who had continuous communication to all areas of the backup and logistics, were not pursued as they were in receipt of production bonus, which was based upon "corrected" versions of the shift production and delay reports.

The monthly report trend graphs were utilised in a number of ways:-

- Forward projection of the trends of individual drives as an aid for programming purposes
- Comparison of the trend graphs for the five concurrent TBM drives were utilised as an aid in establishing and balancing the level of logistical support required in each drive, depending on its criticality.
- Comparison of the various elements of the trend graphs for two similar TBM drives, (e.g. the two Marine Running Tunnels) to ensure the second of the drives adopted the positive elements and avoided the pitfalls of the first.
- Presentation of the other drives trend graphs to the Pit bosses and crews of a particular TBM, acted as a target for them in the friendly, but highly motivated, competition between the various crews on each of the TBMs, which was one of the major elements leading to the success of the UK TBM drives.

7 The Thames system

7.1 LWRM phase I

In 1988 a major problem occurred during tunnelling on Phase I of the London Water Ring Main project when a tunnelling machine was inundated with water under high pressure from the underlying Thanet Sands with 1.2 km of tunnel still to be excavated. To complete the tunnel an Earth Pressure Balance Tunnelling machine was purchased by Thames under a revised cost reimbursable contract. This was the first time that Thames Water had purchased a mechanised tunnelling system the performance of which was critical to the successful completion of Phase I of the project and one where as client/owner of the machine any plant delays during tunnelling would be a cost to Thames Water. In order to monitor performance of the system it was felt necessary to introduce a monitoring system which would provide reliable management data to maximise overall efficiency of the tunnelling operation and minimise costs. Babtie Shaw & Morton were approached in respect of their knowledge of the TML system.

Subsequently the TBM monitoring system was installed on site at Tooting Bec Common and operated throughout the drive. It provided key weekly and monthly management data and highlighted those parts of the tunnelling system which were critical to maintaining progress which in many cases was not always the tunnelling machine but associated activities of train transport and cranage.

7.2 LWRM phase II

Following the successful completion of the Tooting Bec to Radbourne Road tunnel drive, which completed tunnelling on Phase I of the LWRM, planning began on Phase II. Thames Water decided to purchase three tunnelling machines but were intending to tunnel at much higher rates than previously achieved where analysis of the total tunnelling operation was crucial to maintain progress, and again the monitoring system was utilised.

In Phase II several drives were running in parallel so the monitoring system was established at one co-ordinating site where the shift records were sent at the completion of each shift and performance reports sent to each site and back to Thames Water's Engineering office on a weekly basis. The information was viewed by Thames site staff and those of the contractor to analyze and identify potential weaknesses in the total tunnelling operation with a view to minimise downtime and removing potential delays in the system.

7.3 System data collection and output

Data for the system was gathered by Tunnel Inspectors or Shift Engineers who complete proforma sheets for each shift. Data is gathered to cover three specific areas (Fig. 3):-

Progress Data
Engineering Data
Time allocation

Time allocation is used in detail to analyze the delays, a system of sub codes breaks the delay elements down to identify where and how much time is lost i.e. for TBM delays, shove rams, conveyor, erector arm etc. and for other delays, shaft crane, derailment, ventilation, services failure etc.

Data from the drives is input daily and all data processed weekly when a "weekly update" programme is run which prepares the report data and sets the system ready to accept new data. Data can be input on a ring by ring basis as was done on the

Tooting Bec drive where progress peaked at 13 rings per shift. However, progress was such on Phase II that ring data was normally collected and input into the system for every fifth ring. Progress on Phase II peaked at 56 rings per shift.

The system is tailored to provide graphical output for all key reports being the easiest to assimilate quickly. Reports are either pie or bar charts and produced for 1, 4, 6, 12 or 50 week periods. As well as progress and time activity analysis, key data on TBM availability and utilisation is available.

8 Benefits and short comings of the Thames system

When the system was first utilised there was some scepticism as to the benefit that could be achieved. However, the systematic production of valid performance data has clearly identified areas where time savings could be made by minor improvements to the system and also provided reliable trend data upon which to base key decisions.

The system imposes the collection of key data which is often lost on projects and which serves as a substantive record of what actually took place, this consistency of data provided reliable trend patterns against which key management decisions were made.

Both on Phases I and II performance trend data indicated that the tunnelling systems would outstrip tunnel segment supply and that the TBMs would be standing waiting for tunnel rings if manufacture of rings was not increased. In the case of Phase I this meant an increase in the number of segment moulds and in the case of Phase II meant the running of an additional segment manufacturing carousel system for six months. Without these significant additional costs would have been incurred.

An analysis of the planned delays indicated that considerable time could be saved if an improved cable extension system was incorporated on to the TBM, this again was carried out and savings of up to one third of a shift in time were made.

The system has now been used on eleven tunnel drives totalling in excess of 36 km and has provided, reliable, easily interpreted management data clearly identifying the progress of tunnelling operation and critical delay factors.

The system has proved flexible and can be adapted to reflect the differing requirements of the various tunnelling systems available from fully mechanised TBMs of what ever size to partial systems such as Backactors and even hand drives. It can be used to forecast progress rates although this facility has not as yet been utilised.

As present the system is used only to store and report data to enable managers to monitor what has gone on, in some areas data input is cumbersome and inflexible partly to protect the integrity of the data, however the wealth of data now available has made it easier for Thames Water to understand the tunnelling processes and consider all aspects when considering future projects, to minimise machine downtime, increase efficiency and productivity, and reduce costs.

9 Conclusions

Both of the systems described in this paper have proved to be successful cost effective management tools. The key to this success has been the rational and practical approach of designing a system to meet a need as opposed to installing an inflexible tool. The resulting "user friendly" face of the system, combined with a commitment from the users, has enabled the enormous initial resistance to be overcome.

Initial scepticism from both management and particularly site staff has at times seemed to be almost insuperable. Only the commitment of TML senior management and many hours of discussion with site personnel enabled this to be overcome in the early days. Whereas similar problems did occur at LWRM the previous track record from TML was a great benefit.

For the future the systems can be refined to take account of experiences but as with TBMs, each will, to a certain extent, be a prototype as it will only succeed if tailored to meet specific user requirements.

Looking to the future the next obvious step would be to collect much of the data automatically on the TBMs and feed this directly into the computer programme. This would appear to have the advantage of improving accuracy and eliminating the two manual setups of data collection and computer input. The problem still to be overcome to achieve this is that the TBM will need to know not only what it is doing (easy to achieve) but why it is not doing what it should be doing (currently seen as impossible or impractical to achieve).

Given the need, the authors believe that this goal can be partially achieved in a practical manner. There are, however, no current plans for development. Possibly the Jubilee Line or CrossRail will provide the impetus to move to the next step.

References

Chapman, P.J., Wrightham, V.H., Ince G and Lewis R.P. The Channel Tunnel : Machine Driven Tunnels, **Paper No. 9964 ICE Proceedings**, December 1992

Session 8: Research

Long-term measurements of loads on tunnel linings in overconsolidated clay

D. A. Barratt
Transport Research Laboratory, Crowthorne, England
M. P. O'Reilly
Consultant, Camberley, England (formerly, Transport Research Laboratory, Crowthorne, England)
John Temporal
Transport Research Laboratory, Crowthorne, England

This paper reports the results of long-term monitoring of the loads developing in instrumented rings of the segmental concrete linings in the northbound tunnel of the Jubilee Line of the London Underground beneath Regents Park and in the Oxford Trunk Outfall Sewer. The depth to tunnel axis at these locations was about 20m and 15.5m and the surrounding ground was London and Oxford Clay respectively.

At Regents Park, the results indicate that the vertical load on the lining had reached the equivalent of 60 per cent of the overburden pressure after 19½ years: over the same period, the horizontal loading had reached about 40 per cent of the overburden. At Oxford, the loads on the lining had built-up, albeit somewhat erratically, to between 35 and 50 per cent of the overburden, by the time measurements were terminated after 7½ years. Finally, the measurements presented in the paper are compared with other published data on lining loads in overconsolidated clays.

1 Introduction

Although the short and long term ground loads acting on tunnel linings are of considerable interest to engineers, surprisingly few measurements have been made of their magnitude. This paper reports the results of long-term monitoring of the loads developing in rings of segmental concrete lining in each of two tunnels. The first ring was in the northbound tunnel of the Jubilee Line of the London Underground beneath Regents Park and was erected in January 1974, while the second ring was at the Oxford Trunk Outfall Sewer and was constructed in February 1981. The depth to tunnel axis at these locations was about 20m and 15.5m and the surrounding ground was London and Oxford Clay respectively.

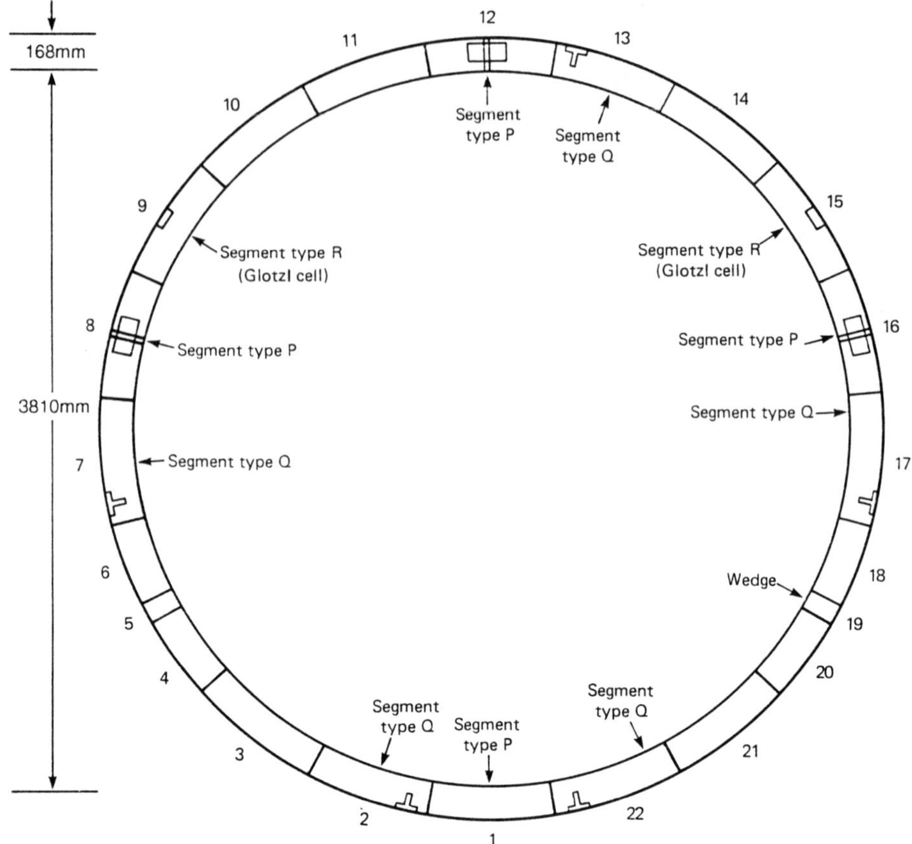

Fig 1 Instrumented ring of segments at Regents Park

2 The instrumented rings

2.1 Regents Park

The instrumented ring installed in the northbound tunnel of the Jubilee Line between Baker Street and Bond Street Stations is shown in Fig 1. All the 600mm wide segments were cast in standard moulds and the instrumentation was fixed to the shuttering prior to casting. Thus for practical purposes, a normal precast concrete ring of lining was being placed and it was possible to make measurements soon after erection with minimal interference to the subsequent tunnelling operations. The instrumentation consisted of load-cells (segment type P) to measure total loads in the ring; earth-pressure cells (segment type Q) and Glotzl cells (segment type R) to measure ground pressures at points around the ring; and embedded vibrating wire gauges to measure strains within some of the concrete segments. In the event, the two Glotzl cells were

Fig 2 Segment type P from Regents Park

damaged at an early stage while interpretation of the strain measurements within the concrete segments proved difficult and tenuous: the data are not presented here, but the reliability and accuracy of different load measuring systems are discussed later in the paper.

Measurements of ground settlements and deformations caused by tunnelling were also made at Regents Park about the time of the installation of the instrumented lining and the complete studies have been described in [1].

2.1.1 Load-cell segments

Segments numbered 1, 8, 12 and 16 in Fig 1 were cast with a single load cell placed at their centre between the mating faces of the steelwork as illustrated in Fig 2. The load in the segment was transmitted by the load cell which was set in pockets in the steelwork assembly. The cell was of the single vibrating wire type and one end of the cylindrical casing was domed to ensure that the load was transmitted axially through it, even when a slight rotation of one half of the segment relative to the other occurred. Steel cover plates on each side of the load cell assembly held the two halves of the segment together during transport and also protected it during casting and erection in the tunnel. While the ring was still supported on the erector arms, the inside cover plate was removed but the plate against the ground was left floating. Erection was then completed by thrusting the wedges into the ring. Just before the ring was erected the load cell in segment 1 was found to be defective and the segment was replaced, leaving only three load cells in the ring. The leads from the cells were placed in a circumferential instrumentation groove, which was cast into the segments right round the ring, and led to a steel terminal box cast alongside the groove. The effect of each load cell was to introduce a further hinge point at the centre of a segment.

2.1.2 Earth pressure measurements

Five prototype earth pressure cells were cast into the back of segments 2, 7, 13, 17 and 22 as shown in Fig 1. These used a vibrating wire gauge to determine the change in deflection of a 140mm diameter circular diaphragm in contact with the ground [1]. Calibration of these cells was carried out prior to their embedment in the segments: the deflection of the centre of the diaphragm was 0.0024mm under the maximum loading of 500Kg, which if uniformly distributed would correspond to a pressure of approximately 320kN/m^2. After erection of the instrumented ring, the cell in segment 17 initially appeared to be defective but subsequently behaved in much the same fashion as the other cells: the earth pressure cell in segment 2 failed during installation.

2.2 Oxford Trunk Outfall Sewer

The instrumented wedge-block ring (Ring No 3636) shown in Fig 3 was installed in February 1981. In this case a coring rig at TRL was used to drill holes in a standard 685mm wide production precast concrete ring to take the four pairs of load cells. The use of pairs of load cells improved the stability of the ring of lining during its erection. The four earth pressure cells installed around the ring were cast, using epoxy mortar, into holes cored in the lining shortly after erection. The earth pressure cells used were of the same type as those at Regents Park.

Again the use of a production ring with the instrumentation inserted into it enabled lining erection to proceed virtually as normal and measurements commenced soon afterwards. Details of the planning and construction of the Oxford Trunk Outfall Sewer have been given in [2] and [3].

3 Results

The records of load-cell and earth pressure measurements at Regents Park are shown in Figs 4 and 5. As the loading on the tunnel lining tends to increase with time but at a decreasing rate, a logarithmic time scale has been used on the horizontal axis. A second vertical axis normalising the loads and pressures with respect to those which would have been imposed by the full overburden is also provided. For simplicity, this

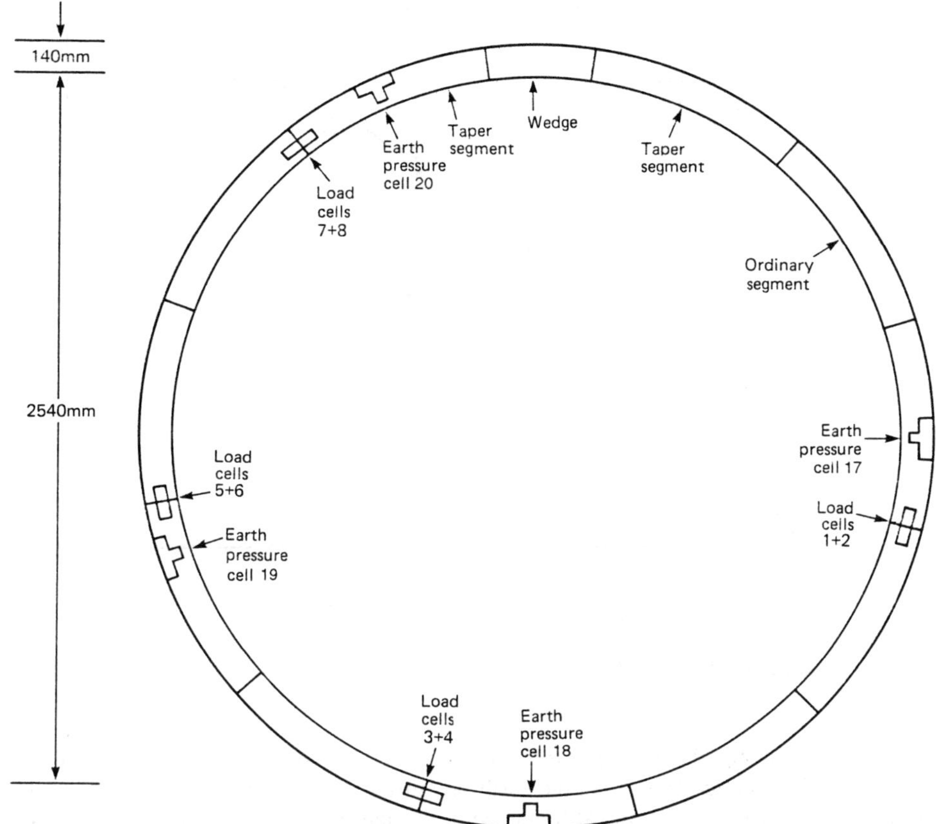

Fig 3 Instrumented ring of segments at Oxford

has been taken as the pressure that would act in the free field at a depth corresponding to the axis of the tunnel.

Measurements have now been taken at Regents Park over a period of 19½ years (up to September 1993) and will continue as long as the instrumentation functions: in September 1993, only two load cells were still giving reliable data. Measurements on the two load cells at the tunnel axis (Fig 4) show that the vertical loads on the lining have now built up to about 60 per cent of overburden while the crown cell indicates horizontal loading of about 40 per cent overburden. The use of a logarithmic time scale gives the impression that the loads in the lining are still increasing significantly. However, closer examination of Fig 4 shows that the average vertical load has increased by only three per cent in the last 15½ years.

The results of monitoring on the prototype earth pressure cells are presented in Fig 5 and show considerably higher loads, with indicated pressures up to the full

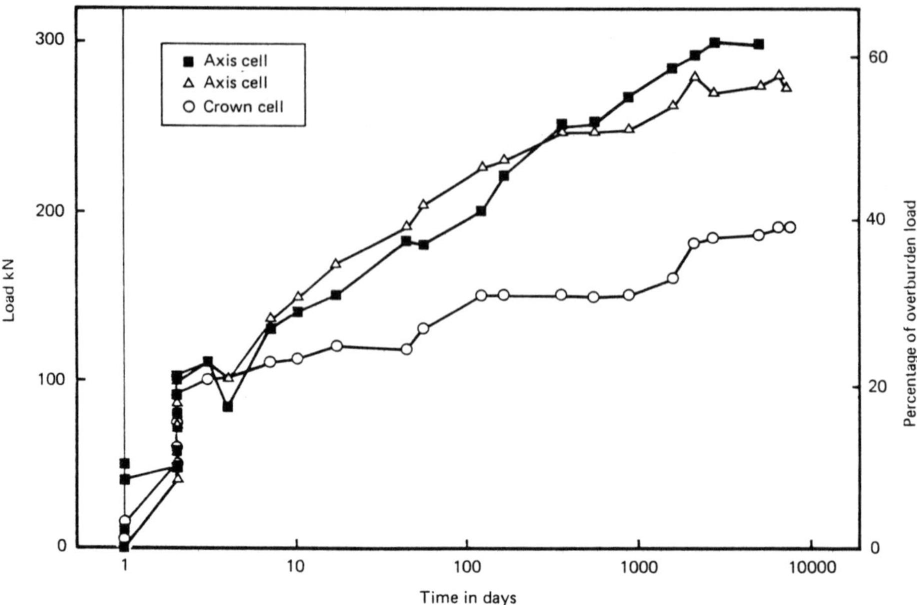

Fig 4 Load cells: Regents Park

overburden and slightly in excess of the top of the calibration range for the instruments. However, the main concern with the earth pressure cell data is the variation with time. These show a continuing upward trend, with average increases of about three per cent a year, compared to the load cell data, which are following expected trends towards asymptotic values.

The measurements of loads and pressures taken at Oxford are shown in Figs 6 and 7 for the 7½ years from February 1981 to November 1988. No readings could be obtained in June 1991 and this probably reflects the more arduous conditions to which instrumentation is subjected in an operational sewer: this is illustrated by the fact that the invert load cell was the first to fail only a month after installation. In Fig 6, the load is the sum of the loads on both the cells at a particular location and, therefore, relates directly to the hoop compressive thrust in the lining. By the end of the measurements, loads on the lining had built-up somewhat erratically to between 35 and 50 per cent of the overburden. The results from the earth pressure cells were again disappointing with negative pressures, up to about 15 per cent of overburden, being recorded at an early stage: final indicated pressures were in the range from 5 to just over 30 per cent of overburden (Fig 7). Other interpretations are possible: for example, if the differences between the highest and lowest recorded pressures for each cell are taken as indications of true earth pressures, the pressures would then range from 15 to 40 per cent of the overburden.

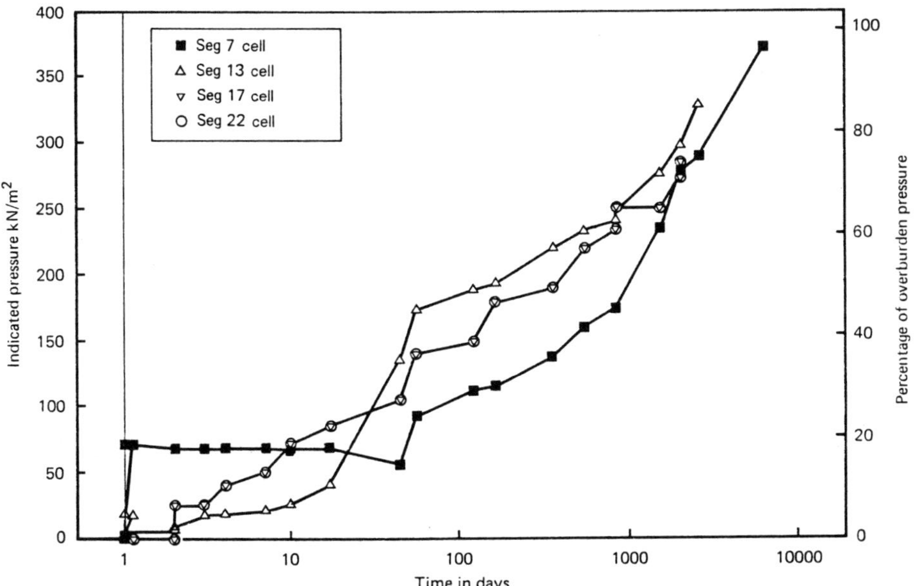

Fig 5 Soil pressure cells: Regents Park

4 Discussion

Studies of loads on different types of tunnel linings have been made in a range of geological conditions. Some of this work, together with analytical methods for predicting and back analysing lining loads, is reviewed in [4]. Further consideration of these, other than for overconsolidated clays, is beyond the scope of this paper.

4.1 Methods of measuring loads

Although some types of strain gauges are complex, expensive and potentially unreliable, in general they represent a cheap and relatively simple method of measuring load effects in tunnel linings. However, for monitoring new construction, rather than 'back measurements' on existing linings, the gauges need to be incorporated in such a way that they do not interfere with erection. This was achieved on the segments at Regents Park by embedding the gauges in the concrete. When interpreting the data, consideration needs to be given to a number of effects.

The concrete will exhibit setting strains, associated with moisture and thermal effects: concrete additives can have a marked effect on these swelling and shrinkage properties. In addition, creep will occur in the concrete over long periods. The strains associated with these effects may well be of the same order as the strains associated with external loading, particularly in relatively shallow tunnels. Some allowance for

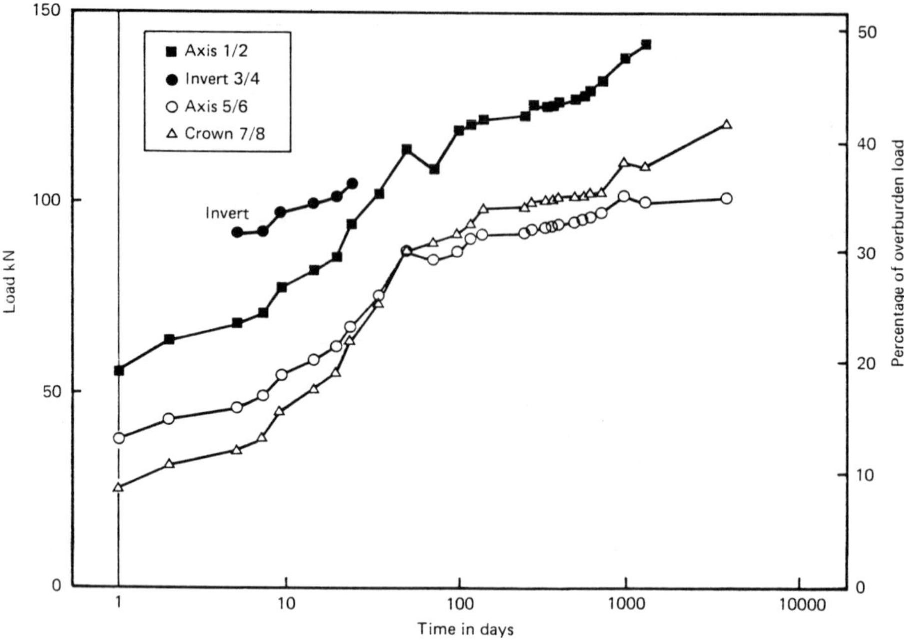

Fig 6 Load cells: Oxford sewer

these effects can be made by incorporating reference gauges in the lining with slots at each end, so that they are not affected by external loading. Reference blocks, of the same mix, can also be strain gauged and kept at the same location in the tunnel. Even after extraction of the strains associated with external loading from the total strains recorded, a modulus of elasticity needs to be assumed for the concrete. This will vary throughout the monitoring period and is unlikely to be independent of the stress and strain history of the concrete. For these reasons, strain gauges cannot be recommended for long term monitoring of tunnel lining loads. A detailed account of the problems of interpreting strain gauge results for an in situ concrete lining is given in [5].

Earth pressure cells are rather more difficult to install in the lining segments than strain gauges, but are unlikely to be as affected by strains in the concrete. There are, however, a number of problems.

The outer surface of the tunnel lining is cylindrical, whilst the active face of the pressure cell is flat. Thus, depending on how the cell is mounted relative to the surface of the lining, either the centre of the diaphragm will be recessed, or the edge will stand proud. In both cases, the contact with the ground will not be uniform and this is likely to affect the calibration. In the former case, the cell is likely to under read, but the situation is less clear in the latter case. To some extent, the situation is similar to the positive and negative projections used in the design of buried pipes. Thus, it is likely that the calibration in the ground will differ from that obtained in the laboratory. This

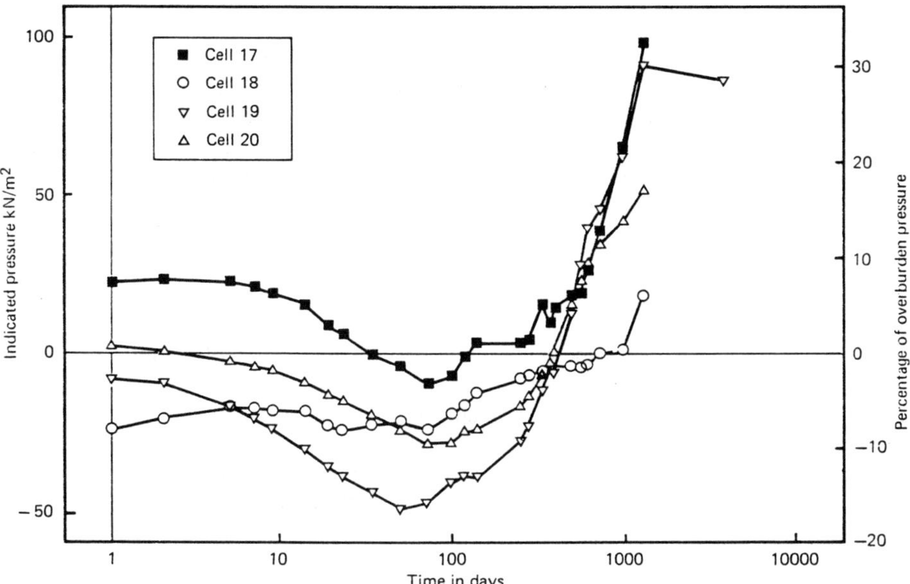

Fig 7 Soil pressure cells: Oxford sewer

would seem to be the most likely explanation of the differences between the earth pressure cell and load cell results reported here. However, the authors are not aware of any investigation of the relations between pressure cell diameter, curvature of the lining segment and the projection of the cell relative to the lining.

Earth pressure cells can, under the right conditions, give useful indications of the stresses obtaining in the ground. One method, which has been used to improve their performance in fills, is that of in situ calibration. In this case, reference cells are buried in the fill to measure the overburden: a calibration can then be performed against the self weight loading of the fill [6]. Such a calibration would not be possible in the case of loads on a tunnel lining, since the loading is influenced by the construction of the tunnel. Although no data were obtained from the Glotzl cells at Regents Park, considerations similar to those for earth pressure cells are likely to apply.

Direct measurements of load appear to provide the most reliable method for tunnel linings. Load cells are, however, difficult to install, requiring specially adapted linings to be cast. An additional problem is that the overall stiffness of the ring containing load cells will be slightly different from a standard ring. By careful design of the load cells used at both Regents Park and Oxford, the overall stiffnesses of the instrumented rings were less than 10 per cent lower than for standard rings. The effect of these differences in stiffness is likely to be small and outweighed by the advantages of direct, rather than inferred, measurements of load, together with the fact that the overall survivability of the load cells was better than any of the other forms of instrument used at either Regents Park or Oxford.

Although direct measurement using load cells is likely to provide the best information on the variation of load with time, 'back measurement' using strain gauges is the only technique which will give a true indication of load. This involves strain gauging a section of lining, removing it and then reloading it in a testing machine to reproduce the strain pattern measured in situ. The disadvantage of this method is that it gives information on load at a unique time, as well as necessitating repair of the lining.

4.2 Instrument durability

As mentioned in the previous section, load cells proved to be more durable than any of the other instruments used for these measurements. This is probably due to their being assembled as sealed units under factory conditions, thus ensuring good encapsulation. However, the cables are prone to damage during installation and under aggressive conditions, such as in the invert at Oxford, the cells can fail due to corrosion attack.

In theory, earth pressure cells should be capable of surviving for similar periods to load cells, but the evidence from both sites is that they were somewhat less durable, primarily because of corrosion. This may be attributable to inferior detailing of the waterproofing.

Strain gauges embedded in concrete in the laboratory can be quite durable, but encapsulation of surface mounted gauges to afford long-term survivability is difficult. Both encapsulation and embedment are even more difficult under field conditions and should be avoided if possible.

One final point that should be borne in mind is that most instrumentation cables are not impermeable to water. Some sheathing materials are less permeable than others and care in the selection of cables is therefore required.

4.3 Results

A large number of measurements was made by the Metropolitan Water Board of the loadings on tunnel linings during their development of the wedge-block lining system for deep tunnels in London Clay [7, 8, 9]. Measurements have also been made on an expanded concrete lining on the Victoria Line of the London Underground, again in London Clay: these were reported in [10] and [11]. The loadings developing on all these linings with time are summarised in Table 1 and compared with those recorded by the load cells at Regents Park. The results from these latter measurements fall well within the range of values found previously. This, of course, is to be expected with the results from Regents Park where the tunnel is in London Clay: the results from Oxford suggest (Fig 6) that the loadings developing on tunnel linings in the Oxford Clay are similar to those in London Clay. The highest loadings recorded in Table 1 have occurred in Donseg Rings in London Clay and this is probably a reflection of the high ram forces which can be needed to insert the closing segment in such rings [8].

The use of logarithmic time scales in representing the data has been discussed earlier. Although the data in Fig 4 give the impression that the loads in the lining at Regents Park are still increasing significantly, reference to Table 1 suggests that the

Table 1 Loads on expanded segmental concrete tunnel linings in London Clay

Location	Type of lining	Internal diameter m	Depth to axis m	Range of hoop loads as per cent overburden	Duration of measurements	Source
Ashford Common	Donseg	2.54	27-28	41-90	9 and 12 months	[9]
Thames-Lee	Donseg	2.71	18.5-54	35-67 31-66 48-72 89	8 and 12 months 4 years 9.5 years 18 years	[9]
Thames-Lee	Wedge-block	2.73	30	35-72 31-52 55 64	12 to 14 months 4 years 9.5 years 18 years	[9]
Victoria Line (Experimental Sections)	Expanded	3.81	26	65	21 months	[10]
Victoria Line	Expanded	3.81	17-33	42-62	1 year	[11]
Regents Park	Expanded	3.81	20	53 57 60	12 months 4 years 19.5 years	This paper

vertical load is stabilising at about 60 per cent of the overburden, the increase being only three per cent in the last 15½ years. Similarly, the horizontal load is stabilising at about 40 per cent of the overburden. It may well be that this is the long term condition and stems from the ground deformations that occurred during construction: see for example [12]. The observation that the horizontal load is lower than the vertical load is supported by observations of lining deformation, which show a shortening of the vertical axis with time: see for example [1] and [10]. Comparison of the results from Oxford suggests that they followed a similar, but more erratic, trend to those from Regents Park.

Observations on circular tunnels in London Clay supported with cast iron segmental linings [13] have shown that the circumferential thrusts in such linings some 50 years or more after construction correspond to the full overburden pressure acting hydrostatically. There is evidence [7, 13, 14] that these loadings can develop quite rapidly and more quickly than in expanded concrete linings. This behaviour probably results from the back grouting of cast iron and segmental concrete linings, which results in a more intimate and immediate contact with the surrounding ground.

The measurement of loads in tunnel linings is a straightforward application of the Observational Method [15], although in overconsolidated clays the times taken to reach equilibrium can be quite long. A feature of the Observational Method is that the outcome can be arrived at without necessarily understanding the mechanisms involved. However, the analysis and assessment of the observations may require assumptions of likely behaviour to be made.

At this stage, all the evidence suggests that it is highly unlikely that the ring loads on single tunnels with conventional segmental linings of either cast iron or concrete ever exceed those corresponding to an all-round pressure equal to the overburden. Indeed, in many instances, the loads are likely to be considerably less. To complete the picture, some measurements of the stresses in expanded concrete linings, which have been constructed for over 20 years, would be valuable. The linings in the experimental section of the Victoria Line are an obvious candidate.

5 Acknowledgements

The Transport Research Laboratory wishes to thank the London Transport Executive (now London Underground Limited), Oxford City Council and Thames Water for their invaluable co-operation in the collection of this data. The work described in this paper forms part of the programme of the Transport Research Laboratory and the paper is published by permission of the Chief Executive.

6 References

1. Barratt D A and Tyler R G (1976). Measurements of ground movement and lining behaviour on the London Underground at Regents Park. **TRRL Laboratory Report,** LR 684, 53p.
2. Watts I L, Northfield J B and Palfrey L F (1982). Wedge-block tunnel for Oxford's main sewer. In **Tunnelling '82** (ed M J Jones), Institution of Mining and Metallurgy, London, pp 291-301.
3. Temporal J and Lawrence G J L (1985). Tunnelling machine performance in the Oxford Clay. In **Tunnelling '85** (ed M J Jones), Institution of Mining and Metallurgy, London, pp 317-327.
4. Whittaker B N and Frith R C (1990). **Tunnelling: design, stability and construction.** Institution of Mining and Metallurgy, London, 460p.
5. Tedd P and Masterton G G T (1983). Behaviour of the concrete lining in the Kielder Water Tunnels. In **Eurotunnel '83,** Basle, Switzerland, pp 47-56.
6. Snowdon R A, Darley P and Barratt D A (1986). An anchored earth wall on the Otley bypass: construction and early performance. **TRRL Research Report,** RR62, 16p.
7. Cooling L F and Ward W H (1953). Measurements of load and strain in earth supporting structures. **Proc 3rd Int Conf Soil Mech and Found Engrg,** Zurich, Vol 2, pp 162-166.
8. Tattersall F , Wakeling T R M and Ward W H (1955). Investigations into the design of pressure tunnels in London Clay. **Proc Instn Civ Engrs,** Vol 4, No 4, pp 400-455.
9. Cooley P (1982). Wedge-block tunnels in water supply. **Journal of Water Engineers and Scientists,** Vol 36, Feb, pp 9-26.

10. Ward W H and Thomas H S H (1965). The development of earth loading and deformation in tunnel linings in London Clay. **Proc 6th Int Conf Soil Mech and Found Engrg,** Montreal, Vol 2, pp 432-436.
11. Smith Osborne R (1970). Discussion on the Victoria Line. **Proc Instn Civ Engrs,** Supplement xiii, Paper 7270S, pp 319-322.
12. Orr T L L (1976). **The behaviour of lined and unlined model tunnels in stiff clay.** PhD Thesis, University of Cambridge, 188p.
13. Ward W H and Chaplin T K (1957). Existing stresses in several old London Underground tunnels. **Proc 4th Int Conf Soil Mech and Found Engrg,** London, Vol 2, pp 256-259.
14. Skempton A W (1943). Discussion to Groves G L. 'Tunnel linings, with special reference to a new form of reinforced concrete lining'. **J Instn Civ Engrs,** Vol 20, pp 29-64.
15. Peck R B (1969). Advantages and limitations of the observational method in applied soil mechanics. **Geotechnique,** Vol 19, No 2, pp 171-187.

Non-destructive investigation of tunnels

Jon Baston-Pitt
G B Geotechnics, Ltd., Bottisham, Cambridgeshire, England

Abstract

This paper explores the use of non-destructive testing (NDT) techniques, primarily Impulse Radar, as a means of investigating tunnels.

The principles of the technology are discussed and then described in more detail using the assessment of tunnel condition and the location of blind access shafts as two specific examples of its application.

Looking to the future, the first combined use in the UK of thermography and Impulse Radar is described and details are given of the preliminary findings.

Case studies are used as illustration.

1. INTRODUCTION

"The only alternative is to bang the inside of the tunnel with a hammer and listen to the response". Such was the comment from a British Rail engineer on the use of non-destructive testing as a technique for investigating their nationwide network of Victorian tunnels.

Built during the heyday of the railway age, many of the original plans for these tunnels are out of date, inaccurate or have been lost over the passage of time.

The volume of traffic which continues to pass through the tunnels, combined with the need to adhere to strict timetables (particularly since the advent in the UK of the Citizen's Charter) makes it increasingly important for any investigation of these tunnels, as part of continuing remedial works or prior to electrification, to be carried out as quickly and efficiently as possible. Possession times are at a premium and need to be utilised in the best possible way.

Non-destructive testing techniques, such as Impulse Radar and thermography are providing British Rail and other owners of tunnels with a means of not only investigating old tunnels, but as a quality control mechanism to confirm the compliance to specification for new construction work. In fact, more than one hundred tunnels have been investigated using Impulse Radar.

Quite simply, to begin to investigate a tunnel without having a clear picture or map of its overall construction and any specific problem areas is akin to tipping out two jigsaw puzzles onto the table, mixing up the pieces and then burning the boxes.

Who, from choice, would relish the challenge of completing two complete jigsaw puzzles without any clue as to the size, shape or composition of the pictures?

There is absolutely no question that the choice of the correct remedial or maintenance strategy for a tunnel will depend upon the quality of the initial information relating to the structure. It is at this first stage in the investigation that non-destructive techniques dominate as the tool of choice. In contrast, destructive methods are notoriously more disruptive and can at best only provide random samples.

NDT methods will provide a map of the complete structure and will thus highlight those areas worthy of more intense investigation,either using traditional destructive means or other non-destructive techniques.

Those areas which can be reported on following a non-destructive survey include:

* location of blind access shafts
* thickness of brick/stone lining and overbreak
* integrity of lining
* integrity of fill material
* other points of weakness (residual timers)
* integrity of country rock

2. THE PRINCIPLES OF IMPULSE RADAR

Impulse Radar is a completely non-destructive testing technique which is based on the interpretation of electrical and electromagnetic signals to locate and map features within man-made structures.

Surveying by this technique is achieved by drawing the transducing radio antennae over the surface to be investigated at a slow speed; pulses of energy are transmitted into the material which are reflected from any internal surface or structural change. The returning signals from each vertical scan build, as the transducer is moved, into a continuous radio profile of the interior of the material, which can be translated into detail of thickness, compaction and inclusions within the subsurface.

A plane of energy, rather than a ray or beam is transmitted. This plane is normally set parallel to the direction of the survey, such that all information streams refer to material directly beneath the survey line.

Selected centre frequencies are used to provide data on different features and depths; the deeper the feature the lower the frequency that is required to identify it. The recovered signals can be recorded on graphic scanning systems, either to allow on-site interpretation of the data in real time, or as data storage for subsequent off-site analysis.

3. TUNNEL CONDITION AND IMPULSE RADAR

3.1 Penshurst Tunnel Case Study - investigating tunnel condition

Built in the mid nineteenth century, Penshurst Tunnel has brickside walls and arch with 5 no. rings of brick lining visible at the portals. The tunnel is twin tracked and is some 75 metres in length. The low cover and low strength of the country rock indicated that the ground would have been considered heavy during construction.

Analysis of the collected radar data indicated that the lining was probably constructed of rings of brickwork laid in three brick bonds. Ring separation had occurred at the boundary of the bonds, and in several instances the mortar work was voided in these areas. Whilst the first bond appeared to be well formed with few internal mortar defects or separation, the second and third bonds were less well formed with separation between brick rings and poor joints between bricks.

The survey identified three vertical features some 250–300mm indimension. It was deemed unlikely that local geology could contain such large open fissures or fault planes and it was concluded that the features were most likely to be man-made. The data indicated that the features could be steel stanchions set to retain some difficult part of the surrounding ground at the time of construction.

4. BLIND ACCESS SHAFTS AND RADAR

4.1 Sources of Information

The location of such well-known features of tunnels as blind access (or construction) shafts, illustrates the value of a technique such as Impulse Radar but also serves to show the importance of interpreting the resultant data within the context of the correct historical period and the known construction standards of the time.

Industrial archaeology and historical records can often provide a valuable source of engineering information, together with original contractors' drawings and any press reports.

Blind access shaft location from radar profile data relies on the accumulation and assessment of a series of factors indicative of the presence of a shaft more than the absolute identification of a single specific form. This is mainly due to the unpredictability of the present form and condition of the shaft,the degree of backfill, and the subsequent distress caused to the tunnel lining by movement of the shaft lining.

For example, information can be obtained from the design details for access shaft construction contained within F.W. Simms,"Practical Tunnelling", combined with drawings from engineers and the library of data collected from other tunnels.

From the various streams of information, a cross section of thetunnel lining at a possible shaft position can be drawn and compared with the apparent lining section in adjacent parts of the tunnel: a level of confidence in the existence of a shaft at any specified location can then be developed.

4.2 Access shaft design and construction

Access shafts are points of potential weakness in the structure of a tunnel, not only as they form an interruption to the arch,but also as they provide a route for surface water to penetrate to the lining. At the surface they may also form a point oflocal instability in the ground for any new development or use of the land.

Access shafts are typically nine to ten feet in diameter, and were originally constructed as a means of access to the line of the proposed tunnel. From the construction shafts small pilot headings were excavated in both directions to provide either ventilation or drainage for the works, whilst the additional faces could be worked by the construction teams, thus speeding up the tunnelling process. They were constructed using the repetitive technique of excavation, timbering and bricking, usually in twelve foot lengths.

The only visual signs, nowadays, are occasional spoil heaps on the surface of the ground or markings that can be detected with aerial photography. Otherwise, the shafts remain hidden as they were bricked over and filled in after use.

Timbering was an important part of tunnel construction. As the tunnel was excavated, timbers, usually fourteen feet in length were used to provide a temporary support for the roof before construction of the final brickwork arch was begun. Although the general policy was to remove all timbers for use in the next section, it was on occasions physically impossible to take them out. These trapped timbers or the voids left after they haverotted away can be located by Impulse Radar, and help to provide information on the thickness and structural condition of the tunnel lining. The entire circumference can be examined, from brickwork in the crown to the invert below the ballast. Along the length of the tunnels, radar data provides information on the condition of the construction behind areas of brickwork, as well as measuring the thickness of brick or arches.

4.3 Survey techniques

Access for the Impulse Radar survey is achieved using a conventional British Rail scissors lift to reach the roof of the tunnel. An initial survey provides continuous data on the entire length of the tunnel, looking at the various structural elements from the lining itself to the overbreak in the country rock and identifying areas of specific interest. A second survey is then undertaken to look at these areas in more detail. Information gained here includes the condition of the mortar and brickwork and the type and compaction of any backfilling and whether any timber remains from the original construction.

4.4 Gonerby Tunnel Case Study - locating blind access shafts

Gonerby Tunnel is a typical example of the brick-lined tunnels built in the mid nineteenth century to open the express routes around the United Kingdom, constructed partly by inventive engineering and partly by the determination, courage and craftsmanship of the labourers who excavated it.

The original survey, using Impulse Radar to locate access shafts was undertaken in 1986.

The results of the survey have since been verified by engineers from British Rail. A line was surveyed over the hill through which the tunnel runs, and stakes were driven in at the positions where the radar survey had indicated the position of the access shafts. Mechanical excavators then removed the soil at these positions. In each case a shaft lining was found within a metre of the position indicated by the radar survey.

5. IMPULSE RADAR AND THERMOGRAPHY

Working in partnership with British Rail, GB Geotechnics has undertaken the first survey in this country combining the use ofImpulse Radar with another non-destructive testing technique,known as thermal imaging or thermography.

Gresley Tunnel in Staffordshire was selected as the site for this survey. Situated near Burton on Trent, it was constructed with a brickwork arch and brickwork sidewalls. It is 570m in length and was originally driven through Permo Triassic sandstone, with up to 20m of cover. There are two tracks with no overhead electrification. It was known that there were two ventilation shafts, but there was no information about the presence of blind access shafts.

The overall objective of this investigation was to extend the understanding of British Rail and GB Geotechnics about the way in which thermal imaging might be used in tunnels and the relationship between thermal images, Impulse Radar, historical data and visual observation can help in examining the condition of a lining.

The more specific objectives of the survey were to identify the possible location of construction shafts within the tunnel and to establish the general nature and thickness of the lining,including areas of inbuilt timber, voids or discontinuities.

5.1 Radar Survey

The main system used was Impulse Radar operating at selected centre frequencies with recovered signals recorded on a graphic scanning recorder, allowing both on site interpretation of data as well as a hard copy for off site interpretation.

Longitudinal profiles were obtained from six parallel survey lines at 0.5 centres extending out from, but not including the crown, to a distance of 1.5m. Complementary data was collected using the 900MHz and 1GHz frequencies at different depth settings.

5.2 Thermography Survey

The thermal imaging system was operated over a narrow band with data recorded both on continuous video and digital frame capture, enabling areas of particular interest to be recorded in detail for subsequent processing.

The thermal imaging system was transported through the tunnel on a raked Permanent Way trolley pushed by the scissors lift carrying the radar equipment. A 20 deg field lens was used, covered with PVC film, to protect it from falling soot and water. Image sensitivity was 0.1 deg C, with the sensors operating in the 8-12nm band.

5.3 Comparing the results

A number of findings can be reported on from a comparison of the two sets of data:

- there appears to be a relationship between moisture in the lining and relative cold areas

- it is also clear that there is a relationship between the thermal lows and the radar plots of both "defective mortar" and lack of compaction to the rear of the lining.

- there is a close correspondence between the locations identified as being access shafts in the radar data and areas oflow thermal emission.

5.4 Preliminary conclusions

Based on the findings of just one tunnel survey, it is clearly not possible at this stage to draw any firm conclusions about the practicality of using the two non-destructive testing techniques to complement each other. In the past, various organisations including British Rail have experimented with the use of thermography, attracted primarily by its apparent speed of operation. However, the results from these surveys have been difficult or confusing to interpret and have failed, generally, to provide meaningful construction details.

However, from the survey in Gresley Tunnel, it appears that thermography may in fact have a role to play in providing a fast overall survey technique, especially bearing in mind the short possession times for tunnels. Having identified those areas from the thermal imaging process which show changes in temperature, these can then be surveyed in more detail using Impulse Radar. Further studies are taking place and it is anticipated that an update on the use of these techniques will be available for the delegates at Tunnelling 94.

6. CONCLUSION

This paper has described the use of Impulse Radar as a non-destructive investigative technique for tunnels. In summary, the technique, which is well-proven and accepted particularly within British Rail, can be used to provide engineers with an overall map of the tunnel structure, as well with specific condition details. More recently, the technique is being used in conjunction with other NDT methods, such as thermography, to provide an even more detailed picture for those organisations whose role is to ensure the efficient and cost-effective maintenance of the country's tunnel infrastructure.

Session 9: Case Histories

Design, construction and performance of a 700-m 'cover and cut' tunnel in soft clay, Norway

F. J. Fredriksen
Norwegian Public Roads Administration, Directorate of Public Roads, Road Research Laboratory, Oslo, Norway
Roald Aabøe
Norwegian Public Roads Administration, Directorate of Public Roads, Road Research Laboratory, Oslo, Norway
G. J. Hasle
Norwegian Public Roads Administration, County Roads Office of Vestfold, Tønsberg, Norway

A 700 metres long "cover and cut" tunnel has been constructed on highway 19 in Horten, a town 100 km south of Oslo. The tunnel is constructed in medium to soft marine clay using friction sheet piles. The clay is very sensitive and partly quick in the northern half of the tunnel. Steel sheet piles are serving as permanent tunnel supports. The sheet piles have no protection against corrosion and therefore have added thickness for long term load conditions.

The groundwater table is typically 1 metre below the top of the sheet piles. Due to nearby buildings it is essential to maintain this level. Nevertheless, a drainage system is installed below the base slab. The sheet pile interlocks are made watertight for the whole sheet pile length. Due to the long drainage path seepage of ground water was expected to be very small.

An extensive control programme has been executed, and in addition to describing technical solution and construction work performance, results from site control and instrumentation are presented.

Long term instrumentation is also provided in an attempt to establish a basis for an IDV analysis (Instrumentation, Documentation and Verification) [1].
Keywords: Construction, design, performance, steel sheet piles, soft clay, tunnel.

1 Introduction

A 700 metres long "cover and cut" tunnel for a two lane road has been constructed on Highway 19 through Horten. Horten, a small town situated about 100 km south of Oslo on the western side of the Oslofjord has for several years had an increasing environmental problem due to the heavy traffic through the main part of the town. Horten has one of the busiest ferry connections in Norway where traffic crosses the Oslofjord between Moss and Horten.

The planning work started at the beginning of the seventies and initially the project was planned as an open cut with a short culvert of about 50 metres for crossing traffic.

Through different stages, both due to neighbour campaigns, but also due to increased emphasis on the roadside environment, the tunnel was increased to a length of 700 metres.

The tunnel passes through a residential area with 50 houses within a distance varying from 6-35 metres. An existing local railway line with low traffic had already opened a corridor passing through this residential area. Only small additional areas had therefore to be expropriated.

Fig. 1 Overview

The tunnel which so far is the longest of this kind in Norway has been constructed using the "top-down" technique with most of the work being conducted under the top slab. A cost saving method using a drained base slab has been chosen even though the tunnel is lying in clay with a ground water table close to the ground surface.

Soil conditions vary from a medium to a soft marine clay, partly quick in the northern part of the tunnel. The clay is somewhat overconsolidated. No permeable layers have been found in the clay.

In addition to describing the technical solution and the construction work performance, technical quality control and instrumentation are also included. An extensive control program has been performed and all the results are documented. The program comprises geometric control of the sheet piles, deformations of the sheet pile wall during the different excavation phases and settlement of the tunnel and the surrounding ground surface as well as the houses in the nearest area. Instrumentation has been performed on the sheet pile wall at two different locations. Earth pressure cells, strain gauges and tubes for inclinometer measurements were installed in advance of the pile driving. Finally, pore pressure gauges are installed at several locations with varying depth and distance from the sheet pile walls. The steel sheet pile wall has no other protection against corrosion than added thickness.

For documentation of long term behaviour the construction will be monitored for several years.

2 Soil conditions. Geotechnical parametres

The ground surface is relatively flat, dipping slightly towards east. Borings indicate more than 20 -30 metres of soil except for the northern part of the tunnel where rock is found at a depth of 13 metres. Ground water table is located 1-2 metres below the ground surface.

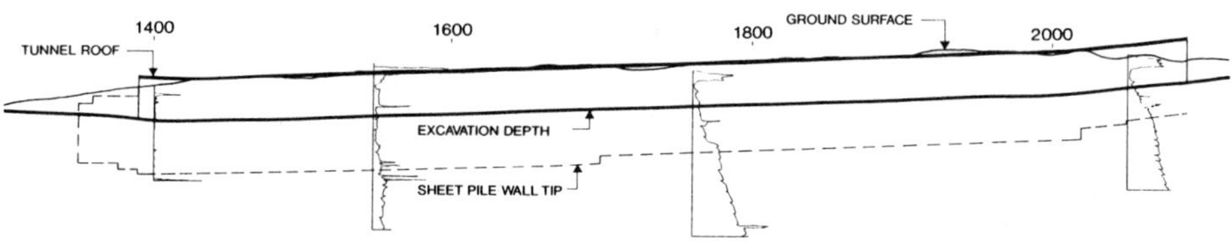

Fig. 2 Longitudinal profile

A large number of borings including rotary pressure soundings, 54 mm undisturbed samples, vane tests and dilatometer measurements have been performed at different stages for this project.

The ground condition in the area consists of an overconsolidated medium to soft marine clay and moraine clay underneath a 1-2 metre layer of nonhomogenious material. Oedometer tests show that the difference between the preconsolidation pressure p_c' and the effective overburden pressure p_o' is almost constant; $p_c' - p_o' = 150 - 200$ kPa. This has been confirmed by dilatometer measurements. The soft marine clay is relatively homogenious with no layers of permeable soil. During excavation in a parallel trench in advance of the construction work, boulders up to 50 cm size were frequently observed in the clay.

In the northern part of the tunnel the clay is getting more sensitive including sections with quick clay.

Even though the clay is changing towards a more sensitive behaviour in the northern part there is no change in strength parameters from triaxial tests. Results from 35 representative triaxial tests (isotropically consolidated, undrained - both active and passive) imply an amazing similarity between the stress paths at the same depth for the whole area. The only difference is that the clay in the northern part of the tunnel shows a more contractant behaviour of the failure part of the stress paths .

A design line for undrained shear strength has been chosen based on the triaxial tests at 2% deformation, dilatometer measurements and also verified by oedometer tests (using the empirical relationship for Norwegian clay $s_u = 0.25\ p_c'$). Therefore routine analyses of undisturbed samples showing 30-40% lower undrained shear strength have more or less been ignored in the design.

Representative soil parametres are given in Fig. 3, Fig. 4 and in Table 1.

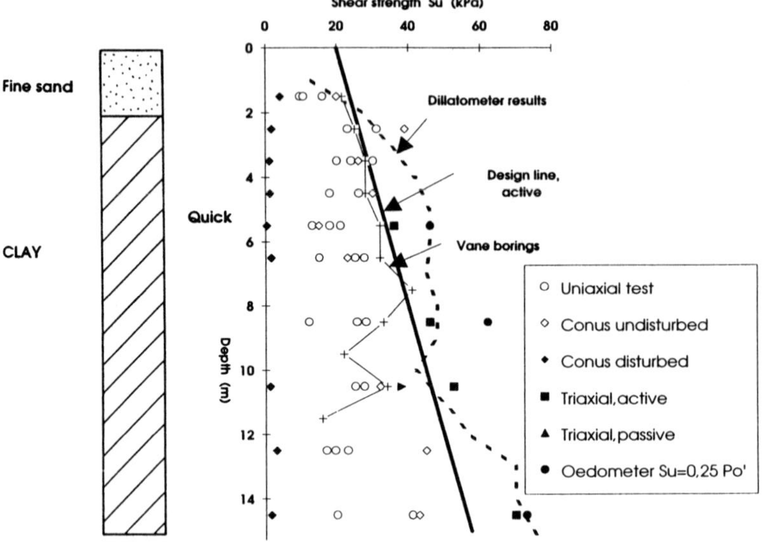

Fig. 3 Representative soil parametres, undrained shear strength.

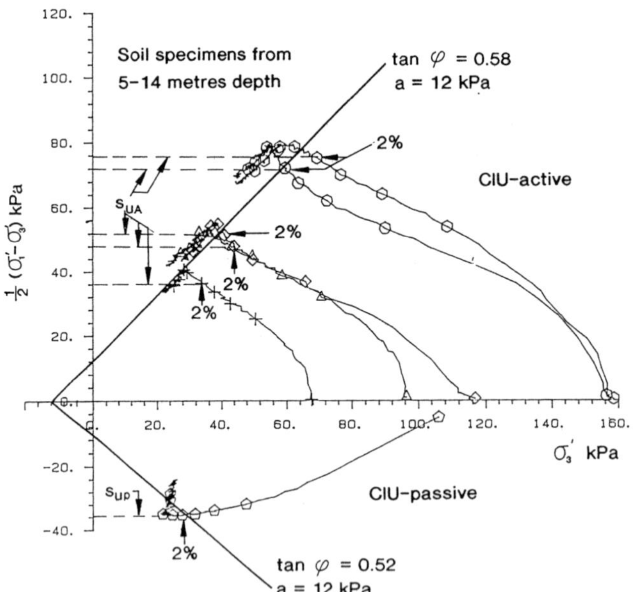

Fig. 4 Representative soil parametres, effective stress parametres from triaxial tests.

Table 1. Summary of geotechnical parametres

Depth (m)	γ (kN/m^3)	a (kPa)	tan ϕ_A	tan ϕ_p	K_0'	M (kPa)	OCR
0-3	20	0	0,7	0,7	1,0	6000	4
3-5	20	8	0,58	0,52	1,0	6000	3
5-8	20	12	0,58	0,52	0,88	6000	2,5
8 →	20	12	0,58	0,52	0,75	6000	2

γ = unit weight of soil
a = attraction = c_A / tan ϕ_A = c_p / tan ϕ_p
c_A = cohesion, from active triaxial test
c_p = cohesion, from passive triaxial test
tan ϕ_A = friction from active triaxial test
tan ϕ_p = friction from passive triaxial test
K_0' = coefficient at rest
M = oedometer modulus in the preconsolidation region
OCR = overconsolidation ratio

Values for undrained shear strength are given by equations 1 and 2 for the active and passive side of the sheet pile wall respectively

$$s_{uA} = 20 + 2{,}5 \cdot z \quad \text{(kPa)} \qquad (1)$$
$$s_{up} = 0{,}75 \cdot s_{uA} \quad \text{(kPa)} \qquad (2)$$
z = Depth below ground surface (m)

3 Design of sheet pile walls

To determine the necessary length of the sheet piles the design is based on undrained shear strength parameters of the soil. The design line for undrained shear strength used on the active side of the sheet pile wall is given in Fig. 3. Undrained shear strength applied on the passive side is 75 % of the active values. Equal design lines are used for the whole tunnel. The undrained shear strength is devided by a material factor (γ_m). Due to varying sensitivity of the clay along the tunnel, the material factors applied differ in value. For the northern part (profile 1340 to 1700) a value of γ_m = 1.35 is applied, while for profile 1700 to 2090 a value of γ_m = 1.25 is applied.

At the design stage, a maximum excavation depth of 6 metres was assumed measured from the top of the sheet pile wall. Sectional performance of excavation and casting of the base slab, after the roof slab is established, is accounted for by increasing and decreasing the limit values for the passive and the active earth pressures respectively. A section length of 15 metres was assumed.

The level of the ground surface relative to the roof slab varies along the tunnel. The vertical loads from the railway also vary as the distance to the railway varies. For most of the tunnel a distance of 2 metres to the railway sleepers is valid. The critical part of the tunnel requires a sheet pile wall of 13 metres length with a moment capacity of about 450 kNm/m (short time load condition, i.e. before the base slab section is established).

According to recommendations for sheet pile walls given by the Norwegian Public Roads Administration [2], a profile type with section modulus of 2000 cm^3/m (steel quality St Sp S, yield stress $f_y = 355$ N/mm^2, equivalent to British Steel BS. 4360:1986 Grade 50A) is required. Utilized steel stress should not be more than 2/3 of the yield stress. If the sheet pile interlocks are situated close to the neutral axis for the sectional area of the sheet pile in question, the section modulus has to be increased by 20%. Hence, for a Larssen profile, a section modulus of 2400 cm^3/m is required.

For long term behaviour, the capacity of the sheet pile wall is controlled by applying effective strength soil parameters (tan ϕ, a; friction, attraction). Tan ϕ is devided by a material factor (γ_m) of 1.4. The soil parametres are listed in Table 1. The sheet pile wall is also analysed for a situation where the at rest stress state is assumed to be re-established.

Two accidents during excavations indicate that the chosen soil parametres are of the right order of magnitude. In the cut in the northern tunnel approach it was necessary to use sheet piles in order to establish a retaining structure.

The sheet piles should have been temporarily supported at the top before excavation to full depth which was 3 metres below the sheet pile top at the maximum. Roads for construction traffic were built behind the sheet pile walls on both sides. The level of these roads were approximately 1.5 metres higher than the sheet pile top.

By mistake the excavation was performed without temporarily support. The mistake was discovered the same day and backfilling was performed immediately. In the mean time the top of the sheet pile walls had moved horizontally as far as 20 cm at the top.

The second accident happened inside the tunnel during excavation work from 2.3 metres to 5.5 metres below the sheet pile top. The soil slope should be given an inclination of 2:1 in the longitudinal direction. At a time the slope was almost vertical and a small slide developed.

The two mentioned accidents did not create any problems for the final construction.

4 Technical solution and construction work performance

The tunnel has been constructed using the "top-down" technique with most of the work being conducted under the roof slab. The first stage was ground clearance including readjustment of 900 metres existing railway. For permanent tunnel supports, two rows of up to 13 metres long steel friction sheet piles were driven through the clay deposits 10 metres apart.

Fig. 5 Sheet piling

A Larssen 606 RD (rolled down) sheet pile wall with a section modulus of 2400 cm^3/m was chosen for the entire construction. The sheet piles were delivered as "triple piles", i.e. 3 and 3 piles were mounted and the interlocks were clamped together.

For driving, a vibratory hammer (type ICE 416) equipped with two clamps was applied. This hammer was extremely efficient for these soil conditions. As an example it could be mentioned that during an 11 hour shift a total of 1680 m^2 sheet pile wall was installed (78 triple piles). Several boulders of 30-50 cm size and even bigger were located in connection with trench works. These did not create any problems for the sheet piling. The triple piles proved to be very stable during driving. The stones were either pushed aside or pushed down, without leading to any deviation of inclination for the sheet pile.

Soil material was then excavated down to 2.3-2.6 metres below the top of the sheet piles and a roof slab was cast in sections of 24 metres. The formwork was built like a carriage which could be rolled along H-beams attached to the upper part of the sheet pile walls. After the carriage was moved to the right position, it was lifted up by hydraulic jacks. Two carriages were built and the construction works were carried out from both ends.

Once the roof slab was cast excavations were continued in sections down to 5.5 metres below the sheet pile top and immediately filled up with road base material. A bottom slab was cast in sections of 12 metres serving the dual purpose of road pavement and lateral support for the sheet piles.

An unconventional solution has been chosen for the road structure. Instead of constructing a watertight concrete floor, a drainage system was installed under the base slab, and the sheet pile interlocks were made watertight by welding in the exposed zone and sealed with a bitumen composit in the interlocks below the bottom slab.

Due to the long drainage path which is forced through the clay below the tip of the sheet piles, seepage of ground water was expected to be very small. Based on a coefficient of permeability of $10^{\div 7}$ cm/sec, calculations show that the leakage will be less than 10 litre/hour for the whole tunnel. Thus, normal rain conditions will keep the ground water table stable.

The calculated leakage has been verified during operation by measuring the amount of water which flows through the drainage pipe and into the last downstream catch pit.

However, a drained solution for the concrete floor causes a steady state seepage process of ground water to be set up. This process gives reduced pore pressures compared to the initial state. The long drainage path leads to small reductions in the pore pressure outside the sheet pile walls. Hence, the settlements of the ground surface will also be relatively small.

A cross section of the tunnel is shown in Fig. 6.

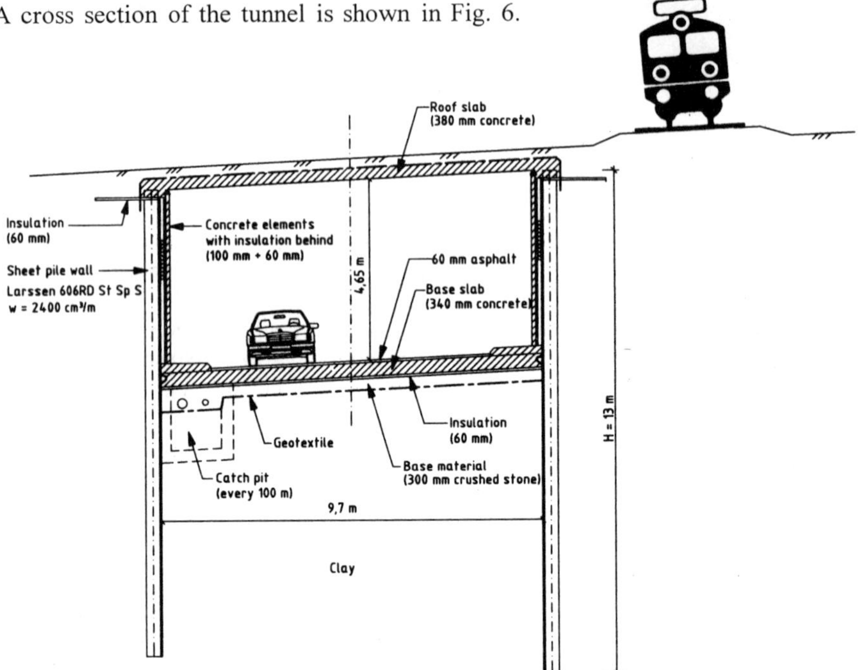

Fig. 6 Cross section of the tunnel

In order to transfer vertical loads from the roof slab (spesific gravity of the concrete and the soil, as well as payload acting at the roof slab) to the base slab, steel brachets were welded to the sheet piles in contact with the base slab. This will also secure the free height inside the tunnel.

The steel sheet pile wall has no other protection against corrosion than added thickness. Hence a reduced moment of inertia had to be used in long term calculations. The Norwegian Geotechnical Institute (NGI) has estimated the corrosion rate to be small due to the soft marine clay. Experience from other projects shows very limited rate of corrosion. The element of uncertainty is the upper 1-2 metre layer with non-homogeneous soil material. A design corrosion rate of 0.05 mm per year has been chosen corresponding to 100 years construction life time. The rate of corrosion will be controlled systematically. For most of the tunnel the remaining sectional area after corrosion is needed for the long term behaviour. There is therefore little to gain by choosing varying sheet pile profiles along the tunnel. Thus, the same sheet pile profile was chosen for the entire tunnel. If the rate of corrosion should increase or the life time of the tunnel prolonged, there is space enough for constructing permanent concrete walls inside the tunnel. The sheet piles are also prepared for cathodic protection.

Using the "cover and cut" method has given better possibilities for taking advantage of optimalisations of the soil mechanic design of the tunnel. Another advantage is the possibility of working late hours without violating the site noise restrictions. The area was very sensitive because of the large number of residents close to the tunnel. The chosen method and the efforts to avoid noise have led to almost no complaints from the neighbours.

5 Geotechnical control

5.1 Technical quality control

During the sheet-piling, routine controls were performed for approximately every 10 metres. The driving was stopped when the top of the sheet pile was projecting about 4 metres above the terrain. The inclination was then measured using a level and a three metre straightedge. The inclination was measured along, as well as perpendicular to, the tunnel direction.

After driving, the top of the wall was surveyed both in relation to height and horizontal projection.

The tolerance requirements were as follows:

- Vertical deviation, both directions: ± 1%
- Height of the sheet pile top: ± 5 cm
- Deviation from horizontal position: 0/+ 10 cm

Despite the stringent requirements, the contractor had generally no problems in fullfilling these. Apart from the last 40-50 metres of the sheet pile walls, there were few deviations which exceeded the tolerance limits. The contractor had to work under pressure the last couple of days in order to finish the work within the time limit. This was probably the reason why several and considerable deviations were detected for this part of the tunnel.

5.2 Settlements and deformations

The total settlements of the ground surface may be divided into two different contributions; settlements due to consolidation and settlements due to shear deformations. The zone of influence for the shear deformations reaches approximately

15 metres from the sheet pile wall, while the influence zone of the consolidation settlements is estimated to be 30 metres. This means that at a distance of 15 metres from the tunnel wall the total settlements of the ground surface will have contributions both from consolidation and shear deformation.

A drained solution for the concrete floor causes a steady state seepage of groundwater to be set up. This process gives reduced pore pressures compared to the initial state. Consolidation settlements in this case are due to the drained solution for the concrete floor.

Settlements of the ground surface due to shear deformations will occur during the construction period, while the consolidation part needs much longer time to develop. It was estimated that 90 % of the consolidation settlements are reached after more than 30 years. 50 % of the consolidation settlements are reached after 6-12 years.

Expected settlements of the ground surface are listed in Table 2.

Table 2. Expected settlements of the ground surface

Distance from the sheet pile wall (m)	Settlements due to reduction of pore pressure (mm)	Settlements due to shear deformations (mm)	Total settlements of the ground surface (mm)
0	60	60	120
5	40	40	80
10	25	20	45
15	20	0	20
20	15	0	15
30	~ 0	0	0

In advance of casting the concrete roof, the soil was excavated down to maximum 2.6 metres below the top of the sheet piles. The ground surface level vary from 0.5 metre under the sheet pile top to 1 metre above the sheet pile top. The walls had no support until the roof section was cast (one week or more).

Measured horizontal deformation of the sheet pile top varies between 10 and 40 mm. Expected ground surface settlements due to shear deformations in Table 2 are based on casting sections of 15 metres and that no backfilling of crushed stone is performed before the whole section area is excavated. The horizontal movement of the sheet pile wall at the floor level was expected to be as much as 50 mm.

However, the Public Roads Administration chose to operate with smaller section length (12 metres). In addition, crushed stone was successively filled back to serve as a foundation platform for the excavator. The casting cycle for the base slab was 4 days.

Therefore, one should expect the horizontal movement to be less than 50 mm. Fig. 7 shows results from inclinometer measurements. As seen, the horizontal movement at the floor level (4.60-5.00 metres from the sheet pile top) is only 12-20 mm. These results are representative for the whole tunnel as they agree with routine measurements which were performed for every section.

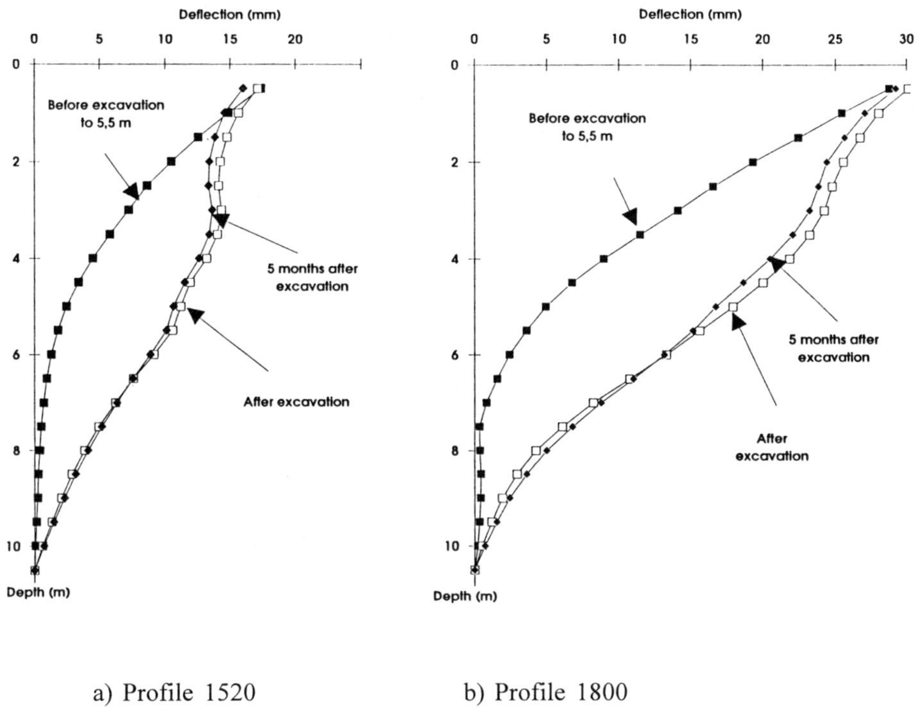

a) Profile 1520 b) Profile 1800

Fig. 7 Results from inclinometer measurements

The reason why the deflection at the top is different for the two diagrams in Fig. 7, is that for Profile 1800 the soil was excavated down to 2.6 metres before the roof slab was cast, while it for Profile 1520 was excavated down to 2.3 metres below the top of the sheet piles.

Bolts for measuring settlements of the ground surface are also installed at these cross sections. Results are shown in Fig. 8. These are also considerably less than the values for shear deformations listed in Table 2.

The total settlements of the ground surface measured so far are plotted in dimensionless form in Fig. 10. Also shown on this figure are the zones suggested by Peck [3] based on field data from deep excavations in several types of material.

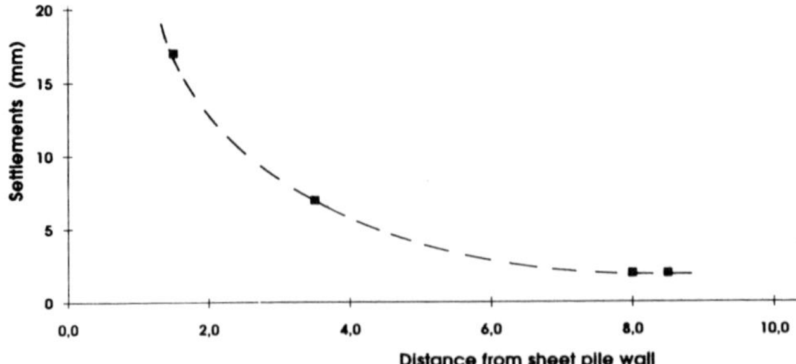

Fig. 8 Ground surface settlements

Terrain settlements versus time are given in Fig. 9. The settlements have stabilized after 1-3 months.

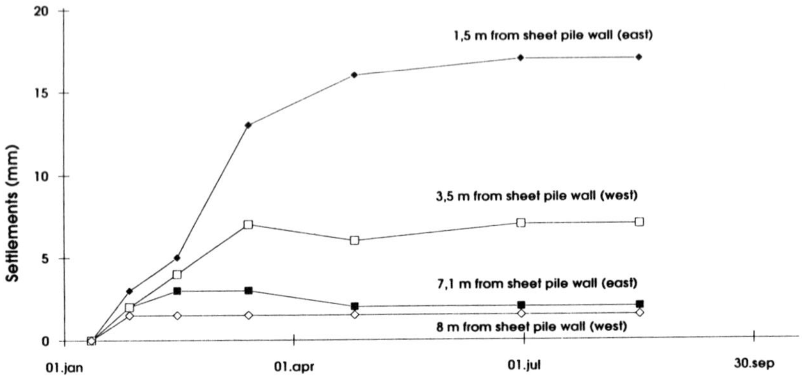

Fig. 9 Ground surface settlements versus time

The settlement of houses within a distance of 6-35 metres from the sheet pile walls are followed up by levelling. Levelling is also performed regularly on the tunnel roof. So far, neither any of the houses nor the tunnel, have been exposed to settlements.

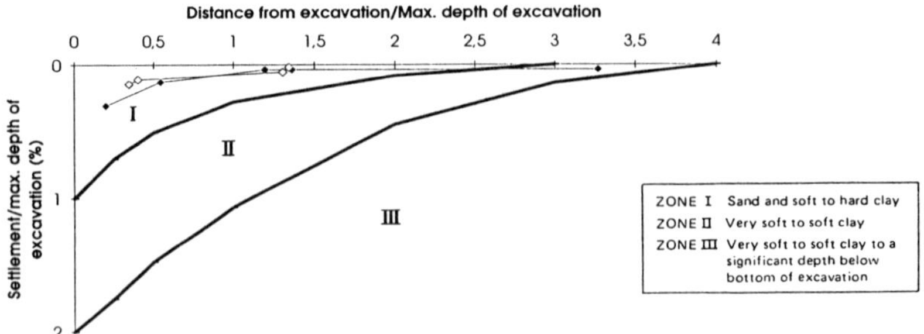

Fig. 10 Comparison of ground surface settlements with values given by Peck

5.3 Pore pressure measurements

A total amount of 34 pore pressure gauges are installed for monitoring the pore pressure development before, during and after the construction period. Both hydraulic and electric gauges are in use. Two and two gauges are installed at different depths at the same location.

Fig. 11 shows the pore pressure development for profile 1800, 5 metres from the sheet pile wall. At the same cross-section monitoring with inclinometer, earth pressure cells and strain gauges is being performed.

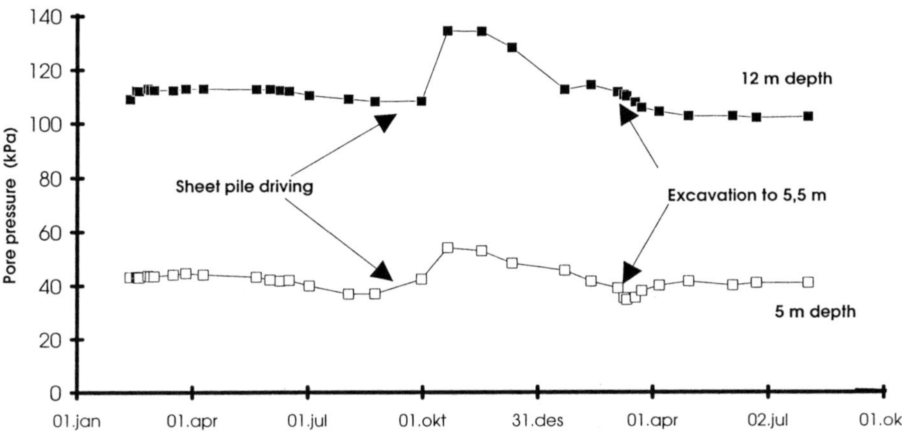

Fig. 11 Profile 1800, pore pressure development 5 metres from the sheet pile wall

The sheet piles were driven at this location during the first half of October 1992. As seen on Fig. 11 the sheet pile driving caused an excess pore pressure of 15-30 kPa. The excess pore pressure reduces to normal after 3-4 months. Excess pore pressure is also recorded at locations further away from the wall. 15 metres away, the excess pore pressure was in the range of 10-15 kPa, and even 30 metres away from the sheet pile wall a slightly increase in the pore pressure was observed (~ 5 kPa).

Full excavation at this profile was performed between 5th and 11th March 1993. In this period quite a number of readings were taken in order to observe any suction. At five metres depth a slight drop in the pore pressure was observed. However, the main conclusion has to be that the pore pressure is almost unaffected by the excavation.

The pore pressures have now stabilized. The drop in pore pressure due to the steady state seepage was expected to be 10-15 kPa at 12 metres depth, and about 5 kPa at 5 metres depth for a location 5 metres away from the sheet pile wall.

Based on readings from all the pore pressure gauges it seems that the pore pressures will stabilize at expected levels (October 1993).

5.4 Instrumentation

Identical instrumentation was installed at two different locations, at Profile 1520 and Profile 1800. At Profile 1800, the clay is insensitive while at Profile 1520 it is medium to very sensitive.

9 earth pressure cells are installed at each place as shown in Fig. 12.

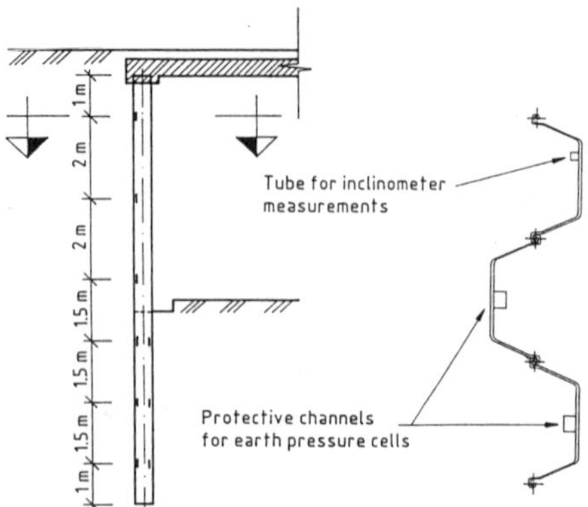

Fig. 12 Distribution of earth pressure cells at the sheet pile

The earth pressure cells are of the Glötzl type. The installation of a cell is shown in Fig. 13. A rectangular hole is cut at the sheet pile surface. A 15 mm thick steel plate is then welded at the back side covering the hole. The plate has a circular hole for running the flexible airtubes through. The cell is then fixed to the steel plate by using epoxy in such a way that it stays in same level as the sheet pile surface. At the back side, the flexible tubes are protected by a steel channel which is welded at intervals to the pile.

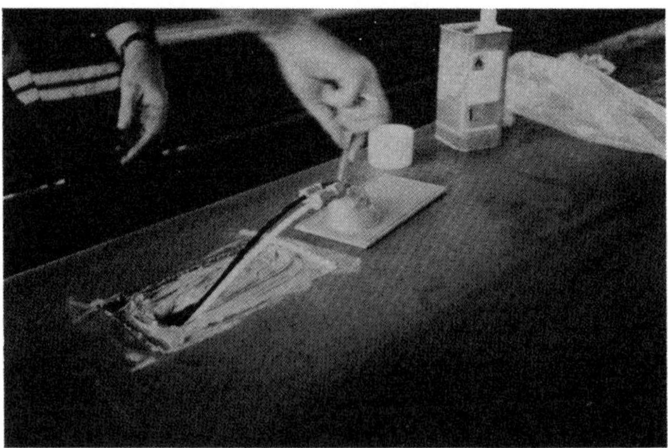

Fig. 13 Mounting of an earth pressure cell

Strain gauges of the "half bridge" type were attached to the sheet pile surface located as shown in Fig. 14.

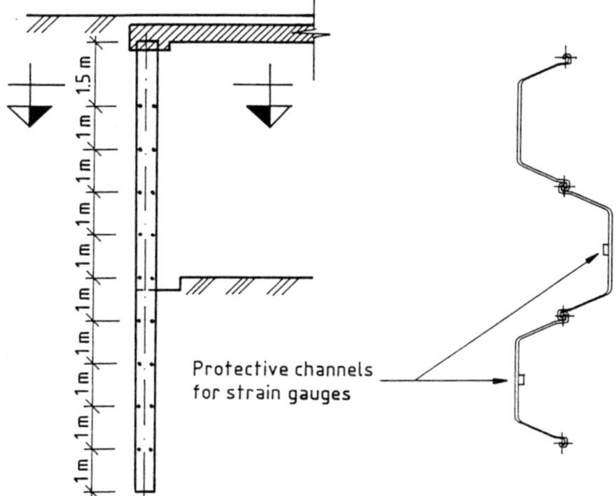

Fig. 14 Distribution of strain gauges

Steel channels are also welded at intervals for protecting the strain gauges. However, the welding was performed in advance of attaching the gauges. Rectangular holes of about 40 x 80 mm were made by each point at the channel, and the gauges were attached through these. Caps were then put on for covering the holes, and finally, the joints were sealed by using a sealing compound (silicone) in order to prevent water getting in.

5.5 Vibrations

5.5.1 Vibrations connected to the construction work

As mentioned earlier the tunnel passes as close as 6 m from some of the houses in the residential area. The following permissible vibration level were prescribed:

- Continuous vibrations measured on the foundation wall should be less than 7 mm/s
- Transient vibrations should not exceed 12 mm/s.

Very few measurements show values exceeding these limits. Even though most of the houses are built with sensitive building materials (eternit, concrete blocks etc.) no damage has been detected except for a few cracks in one house.

5.5.2 Vibrations in permanent situation

An estimate of future vibrations due to traffic in the tunnel has been performed by the Norwegian Geotechnical Institute (NGI) [4]. The intention was to observe the magnitude of vibrations with respect to human annoyance and/or complaints about interference when the tunnel was finished.

Based on a horizontal distance of 12 metres from the house wall to the centre of the tunnel, vertical vibrations have been estimated, see Table 3.

Table 3. Estimated vibrations for a house situated 12 metres from the centre of the tunnel

Category of vehicles	Vibrations at the foundation wall, 12 m from the centre of the tunnel (mm/s)	Vibrations at the middle of the floor, 18 m from the centre of the tunnel (mm/s)
Trailers	0.17	0.13
Trucks	0.11	0.09
Motor-cars	0.02	0.02
Ref.cars, Scania Vabis 10t	0.16	0.12

The estimations are based on vibration measurements from, on and near a road with 23 cm concrete surface with similar soil conditions and soil dynamics properties as determined from ground surface seismic measurements.

Control measurements were performed after the tunnel had been finished. At the middle of the floor vibrations from the reference car were registered to 0.15 mm/s. Vertical vibrations were entirely dominating. At the foundation wall the vibrations were only one third of what was registered at the middle of the floor.

6 Comparison of calculated and measured results

6.1 Computer program used

Besides hand calculations, a computer program SPUNT-A2 [5] has been used designing the sheet pile wall.

SPUNT-A2 was developed by SINTEF (The Foundation of Scientific and Industrial Research at the Norwegian Institute of Technology, Division of Geotechnical Engineering). The program is a simple finite element program for analyzing sheet pile walls. The wall is assumed to be linear elastic, while the soil is modelled in a nonlinear manner. The spring stiffnesses at the nodal points may be computed both from an effective stress soil model and a total stress soil model.

Limit values for active and passive earth pressure for the nodal point in question are assumed to be reached by following hyperbolas in an earth pressure - deformation diagram. Initial stiffness is given for the initial stress state.

The different excavation phases and horizontal support from the concrete roof slab may be accurately simulated.

Results are produced in terms of horizontal displacements and rotations, horizontal shear forces and moments, and finally, earth pressures and degree of mobilization for both sides of the wall.

Due to the finite element formulation, measured deformations may be given as specified displacements to the programme, in order to calculate the corresponding earth pressures and moments for the sheet pile wall.

6.2 Calculation of "as-constructed"-situation

As mentioned in 5.2 it was chosen to operate with excavation sections of 12 metres. Crushed stone was gradually filled back to serve as a foundation platform for the excavator. To account for the gradual backfilling, the excavation depth applied in these calculations is reduced from 5.5 metres below the top of the sheet piles to 5.2 metres.

Results from an execution of SPUNT-A2 are given in Fig. 15. A total stress soil model is assumed. In order to simulate the real situation, the undrained shear strength is not divided by material factors (or safety factors).

The results will be compared with measured data in 6.3.

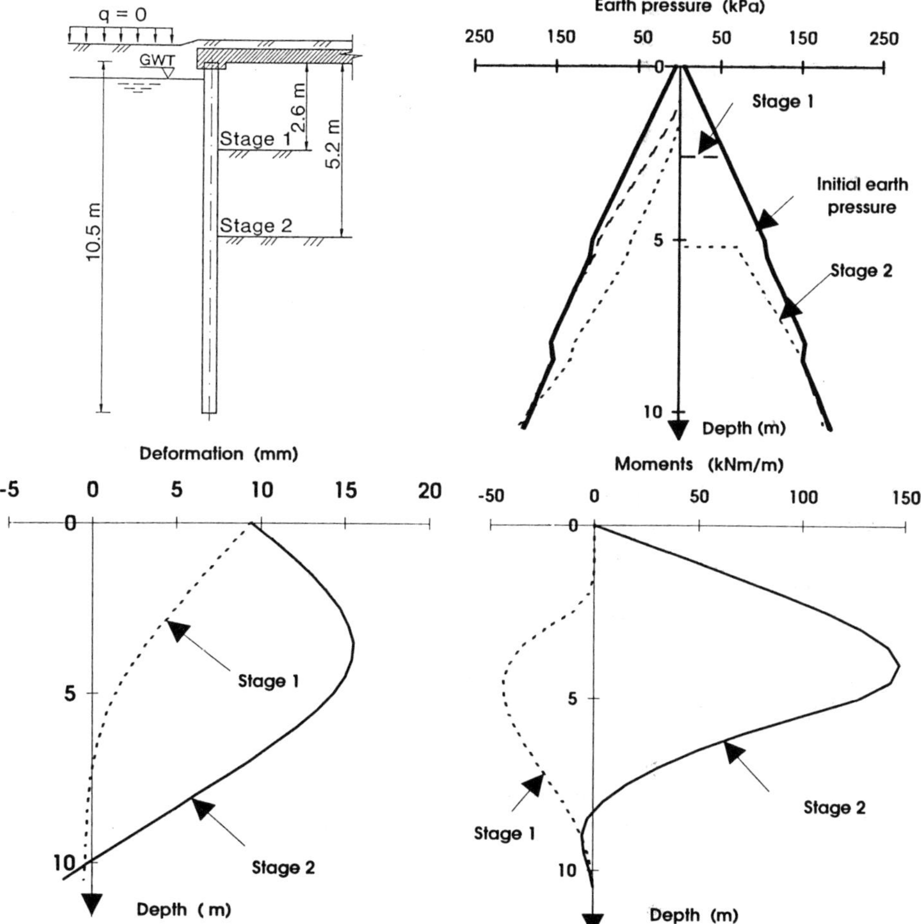

Fig. 15 Results computed by the sheet pile program SPUNT-A2

6.3 Comparison of computed and measured data

6.3.1 Initial earth pressure

Dilatometer measurements were performed in order to evaluate the initial earth pressure for input to SPUNT-A2. Assumed initial earth pressure based on dilatometer measurements is compared with measured values from earth pressure cells in Fig. 16.

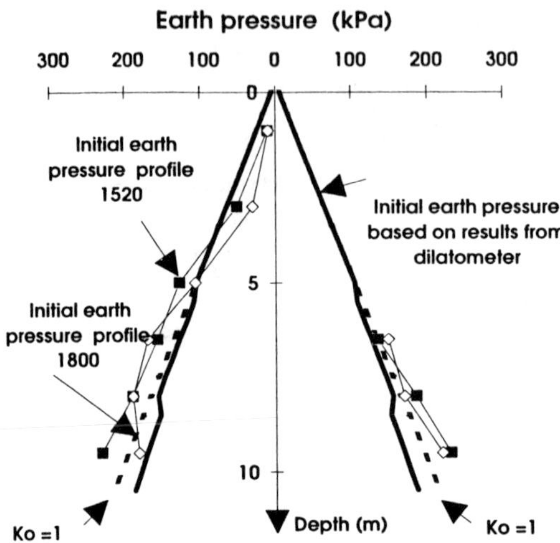

Fig. 16 At rest condition (total stress)

Dilatometer measurements were performed prior to the construction period, while readings of the earth pressure cells in order to determine the at rest condition were performed two and a half months after the sheet pile driving had passed the instrumented section in question. Hence, the excess pore pressure due to driving should at that time be dissipated.

As seen in Fig. 16, the measured initial horizontal earth pressure is even higher than the overburden stress state below 5 metre depth.

6.3.2 Measured earth pressures at different stages

Fig. 17 gives an overview of results from earth pressure cells for the different excavation stages.

In Fig. 18 comparisons are made between readings observed a short time after excavation to 5.5 metres and for five months later, i.e. after the tunnel is entirely excavated. Surprisingly small changes in earth pressures have occurred during that period.

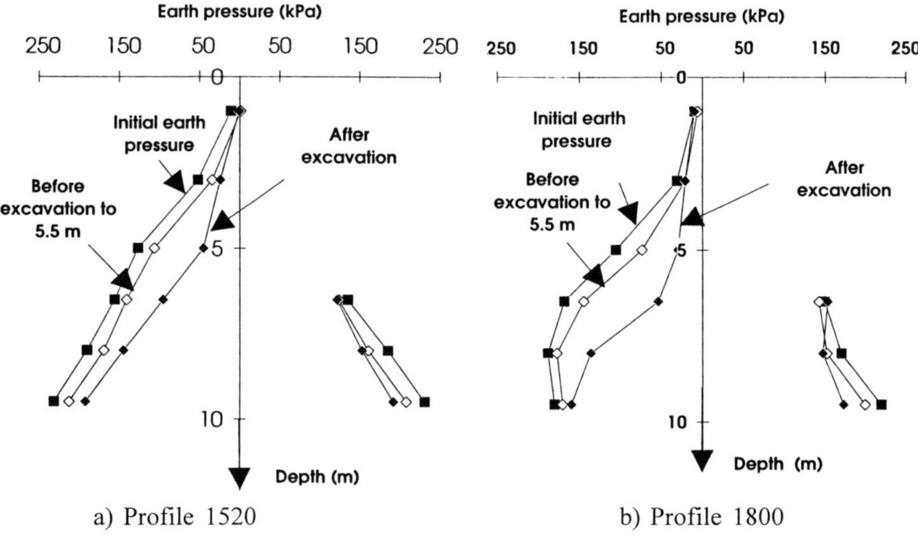

Fig. 17 Results from earth pressure cells for different excavation stages (total stress)

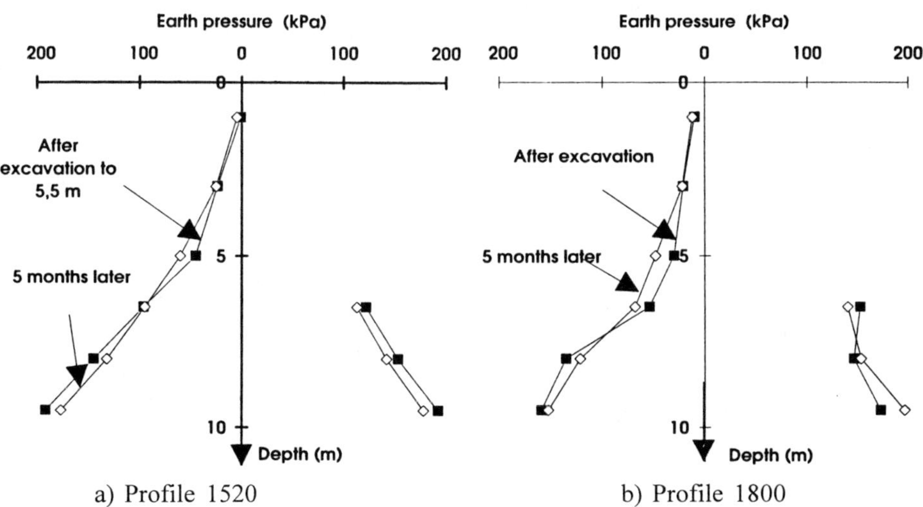

Fig. 18 Results from earth pressure cells short time after excavation to 5.5 metres compared with readings done five months later (total stress)

Earth pressure development versus time after driving is given for 4 selected earth pressure cells in Fig. 19.

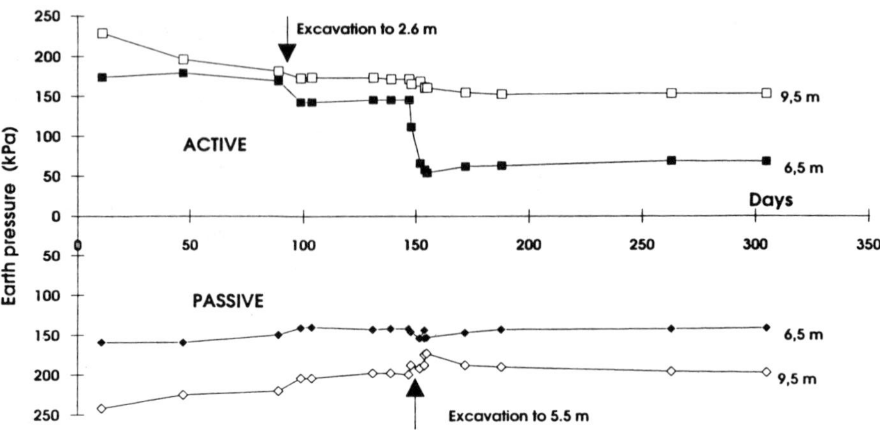

Fig.19 Earth pressure development for 4 earth pressure cells in Profile1800

6.3.3 Computed and measured earth pressures

Comparisons given in Fig. 20 represent readings performed shortly after excavation to 5.5 metres. Results from simulation of excavation by the SPUNT-A2 program are identical to those given for stage 2 in Fig. 15. Earth pressures obtained by giving specified displacements as input to SPUNT-A2 are also included. The specified deformations given are in accordance with the inclinometer measurements in Fig. 7. The limit pressures are reached only if the soil is subjected to full shear mobilization, i.e. degree of mobilization equal to 1.0. Exactly the same limit pressures are applied in hand calculations.

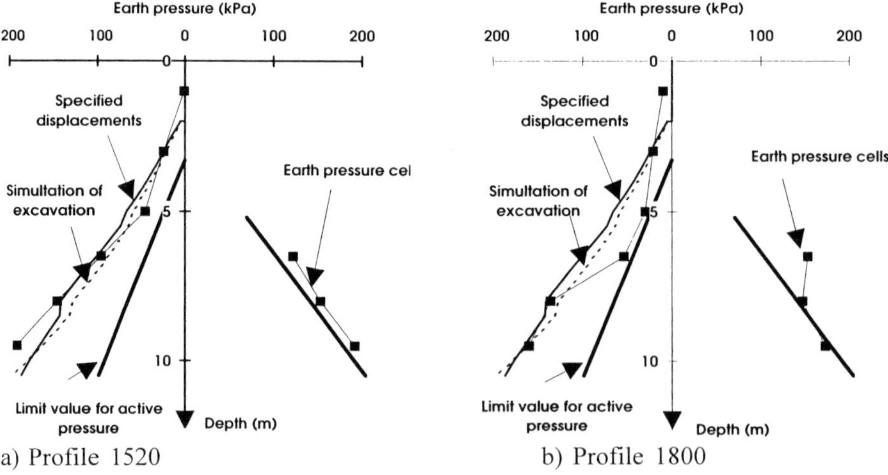

a) Profile 1520 b) Profile 1800

Fig. 20 Comparison of computed and measured earth pressures short time after excavation to 5.5 metres (total stress)

On the passive side results obtained by SPUNT-A2 are not plotted. Here, earth pressures from specified displacements as well as from simulation of excavation, both have reached the limit value for passive earth pressure, except for the lowest 2 metres (although the earth pressure here is close to the limit line).

At Profile 1520 there are very good agreement between measured and calculated earth pressures. At Profile 1800 the cells at 5.0 and 6.5 metre depth on the active side show earth pressures very close to the limit line. On the passive side the cell at 6.5 metre depth shows a much higher value than calculated. Suction and excess pore pressure on the active and passive side respectively, are not set up, as later readings show rather stable values for these cells (see Fig. 19 and Fig. 11).

6.3.4 Bending moments

Even though the strain gauges were installed in a very thorough and conscientious manner, and in accordance with recommendations given by specialists, the attempt to measure strains in the sheet pile steel seems to have failed. So far, we are not sure of what has gone wrong, but one possibility may be that a leakage has occurred during driving.

However, moments may be calculated based on the deformation shape obtained by inclinometer measurements. Moment values are taken from results produced by SPUNT-A2 where specified displacements were given as input. The moment diagrams which are given in Fig. 21 represent the situation shortly after excavation to 5.5 metres as also seen for the corresponding deformation shapes in Fig. 7.

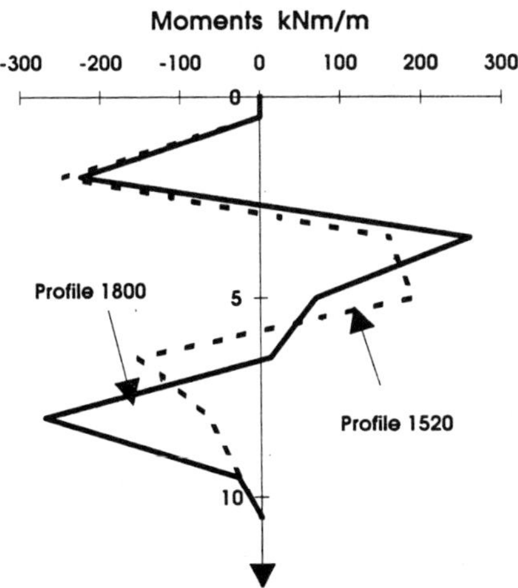

Fig. 21 Moments based on the deformation shape of the sheet piles

Obviously, minimal agreement is found between measured (Fig. 7) and calculated (Fig. 15) deformations (or moments) of the sheet pile wall.

7 Costs and progress. Main contractor

The sheet pile works started up in the middle of September 1992 and were finished two months later. The first roof slab section was cast at the south end of the tunnel late October and from the north 2 months later. A casting cycle of one week was followed at both ends and the roof concrete works were finished on 18th March 1993.

Excavation inside and casting of the base slab sections were started up immediately after the first roof slab section was finished. The last base slab section was cast on 13th May 1993.

The tunnel was opened for traffic on 22nd October 1993.

The County Roads Office of Vestfold was operating as main contractor and did most of the works. Several subcontractors were engaged and the sheet pile works represent the biggest tender. The cost was NOK 14.3 million (excl. VAT) and includes delivering, driving and sealing/welding of the interlocks.

Construction cost for the tunnel is NOK 65 million, i.e. NOK 93.000 per metre tunnel. Included are 40 metres of retaining concrete walls at both ends (at both sides). Readjustment of 900 metres of railway is not included.

8 Concluding remarks

The tunnel has been given a very simple technical solution resulting in low cost per metre. The solution is however somewhat unconventional because the tunnel is situated in medium to soft clay (partly quick) without rock-contact. Also a drained base slab is chosen even though the ground water table is staying close to the ground surface.

Results from instrumentation and technical quality control agree well with the assumptions and calculations performed in the design stage.

For documentation of long term behaviour the construction will be monitored for several years after completion.

9 References

1. ØSTLID, H. (1993). **The Norwegian road to bridge quality.** Nordic Road & Transport Research No. 2 1993, 18-19.
2. Norwegian Public Roads Administration, Directorate of Public Roads (1990). Veiledning 016, **Geoteknikk i vegbygging;** 2. utgave. (in Norwegian).
3. PECK, R. B. (1969). **Deep excavations and tunnelling in soft ground:** state of the art report. Proc. 7th Int. Conf. on Soil Mechanics and Foundation Engineering, Mexico, State of the art vol., 225-290.
4. NGI-report (The Norwegian Geotechnical Institute) 913020-1 (1992). **Rv. 310 Horten-Borre, Miljøtunnel. Vurdering av rystelser fra trafikk** (in Norwegian).
5. SINTEF report STF69 F88007 (1988). **SPUNT-A2. Et program for analysering av spunt- og slissevegger. Dokumentasjon og manual. Versjon A2.** (In Norwegian).

Shotcrete in the construction of Pen-Y-Clip tunnel, North Wales

D. A. Hobson
Howard Humphreys and Partners, Ltd., Leatherhead, England
D. C. Gilchrist
Messrs. Sandberg, London, England

Abstract
This paper details the use of shotcrete during the construction of a road tunnel constructed through a microdiorite headland. It details the use of accelerators with dry process shotcrete with particular reference to the effects on strength gain and variability. Comparisons are made between the use of liquid and powder accelerators.
Keywords: Shotcrete, Tunnels, Accelerators, Additives

1 Introduction

The Pen-Y-Clip Tunnel, together with its approach roads, forms one of the final links in the 85km long A55 North Wales Coast Road which provides a high standard dual carriageway between the UK's national motorway system and North Wales. The 930m long tunnel carries one carriageway of the road through a microdiorite headland.

The scheme was let as a single Contract under ICE 5th Edition to John Laing Construction in 1989. The portal and tunnelling works were sub-contracted to Cementation Mining. The Engineer for the works was Travers Morgan in association with Howard Humphreys & Partners.

The primary tunnel support consists of two main systems, these being; steel arch ribs or, rockbolts with mesh reinforced shotcrete. In addition to being specified for use in the primary support, shotcrete was also required to re-profile the excavation outline to a constant profile as shown in Fig.1. However, apart from a forty metre length, the Contractor opted to use concrete to reprofile the Tunnel excavation outline. This decision reduced the total shotcrete required by approximately 6000 m^3.

For all shotcrete used in the tunnel construction the Contractor opted to use Dry Process Shotcrete, this being supplied both by a local Ready-Mix company and from a site based batching plant.

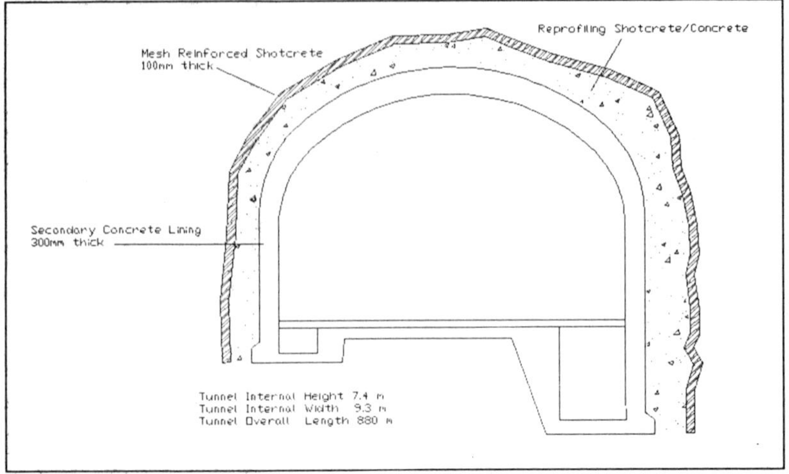

FIGURE 1: Schematic section of tunnel profile

An estimated 15,000 m^3 of shotcrete was supplied to the Tunnel for spraying for both temporary and permanent works applications. Temporary works shotcrete was sprayed to the excavated periphery, to certain of the faces and to secure unstable excavations; the bulk of the temporary works shotcrete incorporating a set accelerator. Permanent works applications comprised primary support to both tunnel profile and portal excavations and as re-profiling to overbreak.

This review details the initial specification requirements, the method of application, the strength achieved and the methods of test used.

2 Specification Details

The Specification did not detail the materials sources to be used or specify the grading curves required. It did specify, however, the minimum 28 day strength required, the maximum and minimum proportions to be used in respect of the mix components and the initial and final set times to be achieved for accelerated shotcrete. It also specified the quality control measures to be adopted and the rates of testing required.

The minimum specified strength requirement for all shotcrete was 30 N/mm^2 to be achieved at 28 days. Additionally, for accelerated set shotcrete, a minimum strength of 4 N/mm^2 was to be achieved within 8 hours and 10 N/mm^2 at 24 hours. A further class of shotcrete was introduced for a number of variations with a minimum strength for accelerated shotcrete of 25 N/mm^2 at 28 days. The Specification also permitted the use of admixtures, subject to the Engineer's approval.

The quality control measures to be adopted were by means of the spraying of Works Test Panels and by in-situ coring. The Contractor was required to establish the parameters of the proposed mix by means of trials and the subsequent performance via core specimens cut from 750 x 750 x100mm panels. For the preconstruction trials the Specification required that three number test panels be sprayed both in the downhand and overhead positions. The purpose of these trials was to assess the strengths achieved in respect of both sidewall and top-heading shotcrete application. For routine test purposes a panel was required to be sprayed for each $10m^3$ of shotcrete applied in the Works. Whilst consideration was given to reducing the frequency of testing once consistent results had been achieved, in practice this never proved possible as a satisfactory level of consistency was never achieved.

In-situ testing took two forms. Routine Windsor Probe Tests were to be carried out, with one set of readings to be taken for each $30m^2$ of shotcrete sprayed. The testing regime actually adopted and the results recorded are discussed in detail in Section 5. Where the core specimens cut from the works panel failed to achieve the required 28 day strength, additional, 100mm dia. core specimens were required to be cut from the in-situ concrete.

3 Application

The Contractor opted to use dry-mix shotcrete for all applications and utilised Aliva 240 and 260 rotating barrel machines for all spraying operations in the Works. The bulk of the spray application was carried out using either 50mm or 75mm delivery hoses and standard nozzles. The mixed Shotcrete was fed to the Aliva machines via a mobile sled unit that consisted of a split discharge hopper feeding two inclined conveyor belts which in turn discharged into two Aliva machines located beneath.

The two types of set accelerator used (Powder and Liquid) were added to the Shotcrete as follows :

i) Powder Accelerator : This was trickle fed over the mixed materials as they were transported along the inclined conveyors to the Aliva units. Dose rates were controlled via a metered screw feeder mounted above the conveyors. The sled and screw feeder were manufactured by Aliva and the general arrangement is shown in Fig.2.

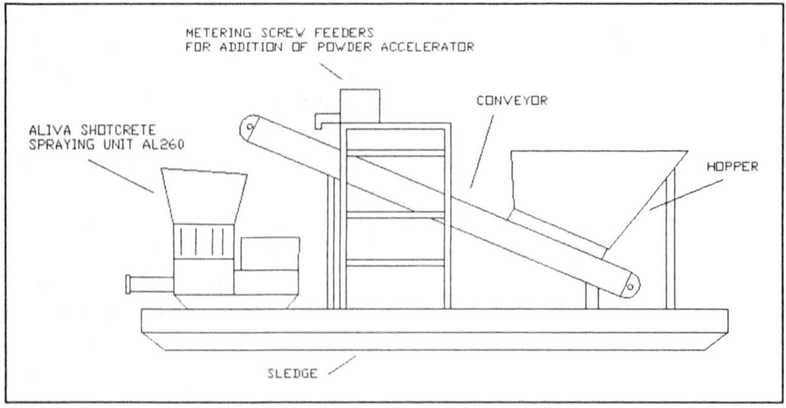

FIGURE 2: Dosing arrangement for powder accelerator

This system was not found to be particularly satisfactory as the trickle feeding of the set accelerator resulted in uneven dosing and as no additional forced mixing was employed distribution of the accelerator through the mix was variable. Probably as a consequence of this poor mixing 28 day core strengths were variable and a considerable number of core specimens did not achieve the specified minimum strength.

ii) Liquid Accelerator : This was fed from a bulk storage tank to the nozzle via an high pressure dosing/mixing unit. In addition to the mixing unit a special mixing head and nozzle were used. The Kapillar mixing head consisted of a number of minute holes (generally 0.45-0.9mm in dia.), through which the water/accelerator mix was pumped at a pressure of up to 100bar. The jet bores were arranged in such a way as to form a consistent pattern of closely meshed jets through which the pneumatically carried Shotcrete had to flow. The turbulence occurring within the mixing zone promoted the take-up of the water and improved the mixing of the cement and water. In addition as the water flow was adjusted by the nozzleman the pressure drops recorded were used as signals for the flow controllers integrated into the mixing/dosing unit. These then maintained a constant dose rate directly proportional to the volume of water being used.

This system was manufactured by Sehurenberg Beton-Spritzmachine and is distributed in the U.K. by Burlington Engineers Ltd. The use of this system resulted in a significant reduction in rebound, permitted a reduction in dose rate to 3% by weight of cement compared to 5% for the powder accelerator, produced an increase in the 28 day core strengths and reduced the variability in the results recorded.

The site batching plant used initially was a Benford 21/14 drum mixer, having a theoretical output of approximately 6m^3 per hour in respect of shotcrete production. Problems were experienced with this mixer due to material build up on the blades, this resulting in premature discharge of the partially mixed shotcrete. This necessitated a reduction in volume from 0.3m^3 to 0.25m^3 per batch and an increase in maintenance to reduce build up on the mixer blades. As a result of these problems and to increase output the Benford plant was replaced by a Camco pan mixer having an output of 12m^3 per hour.

The Specification stipulated that no material other than oven-dried or bagged pre-mix should be used later than one hour after the addition of cement to the aggregate. Trials where subsequently conducted whereby shotcrete test panels were sprayed with material which had been batched two hours previously. Core specimens cut from these test panels achieved satisfactory strengths and so it proved possible to use material that had been batched two hours earlier providing weather conditions were temperate. However, as the drive advanced it was found that the number of core strength failures being recorded increased. These failures were attributed in part to premature hydration of the cement as a result of an increase in the ambient temperature within the Tunnel. As a consequence the one hour limit was reinstated.

Problems were also experienced during the early stages of the project when low ambient temperatures resulted in core strength failures being recorded in respect of the accelerated set shotcrete. Shotcrete by reason of the thin sections generally applied generates very little heat of hydration. The use of set accelerators which increase the rate of hydration exacerbate this phenomena, as a consequence the strength gains achieved are considerably reduced. In view of the nature of the surfaces to which shotcrete is applied and the large surface areas involved, the provision of thermal insulation is largely impracticable. In order to mitigate the effects of low ambient temperatures the Contractor implemented a number of measures which included the provision of heaters within the batch plant enclosure, sheeting and heating of the aggregate storage bins and the provision of an in-line water heater. The implementation of these measure considerably alleviated the effects of low temperatures on the fresh shotcrete.

4 Shotcrete Trials

The Contract Specification stipulated that all mix designs must comply with the following criteria :

Material	Proportion of Mix
Cement	20% Maximum
Fine Aggregate	50% Maximum
Coarse Aggregate	30% Minimum

Table 1: Specified mix proportions

Additionally the cement type was limited to Ordinary Portland Cement complying with BS12 and the aggregates were to comply with the requirements of BS 812. The use of admixtures was also permitted subject to the approval of the Engineer, but within the above constraints design of the Shotcrete mix was the responsibility of the Contractor.

The initial mix design assessments were carried out in June 1990 using a pre-mixed shotcrete supplied by a local Ready Mix company. The trials were generally successful with mean core strengths being in the order of 39.0 N/mm^2. The Contractor also required a mix design which incorporated a set accelerator in order to provide: high early strengths for initial support, to stem water inflows and to give high build-up characteristics in order to maximise shift outputs. Extensive trials were conducted using a number of set accelerators before a satisfactory mix design meeting the Contract strength criteria was achieved.

Where set accelerators were to be used the following additional criteria were required to be met :

- Initial set of cement/admixture paste to be 3 minutes maximum.
- Final set of cement/admixture paste to be 12 minutes maximum.
- 8 hour strength of mix to exceed 4 N/mm^2
- 24 hour strength of mix to exceed 10 N/mm^2.

Considerable problems were experienced initially in determining the initial and final set times of the cement/accelerator paste due to the rapid setting times involved. The setting times were determined in accordance with a modification to the method detailed in BS4550:Part 3 using the Vicat Apparatus. This British Standard relates to the assessment of cement/water pastes which have much longer setting times and different set characteristics. If the BS procedure was followed the true initial set would be destroyed and as a consequence longer unrepresentative times would be recorded. Shorter mixing times during sample preparation and mould filling are essential if true results are to be recorded, as prolonged mixing destroyed the initial set of the accelerator paste. The mixing time for the cement/accelerator/water paste, was reduced from the 240 $\pm$ 5 secs. stipulated in the British Standard, to 15secs., with filling of the mould being completed within a further 15 seconds. With this modification the results recorded for the initial and final set of the paste were found to comply with the Contract criteria. However, the very short mixing times involved present additional difficulties in ensuring that homogeneity of the cement/water/accelerator paste has been achieved. For this reason the use of this modified procedure is not considered completely satisfactory and an alternative method of test does therefore need to be developed.

Two further problems were also encountered in producing a satisfactory mix design. These were; the significant drop in 28 day core strength as a result of the use of set accelerators and the requirement that in order to avoid any possibility of Alkali-Aggregate reactivity, the sodium oxide equivalent value of the mix not exceed three kilograms. This latter requirement resulted from the approved mix design being required to comply with the provisions of the Department of Transport's Specification for Highway Works Series 1700. It was found that a number of set accelerators contained very high proportions of alkalis and even at relatively low dosages their use could result in the maximum permitted level of equivalent sodium oxide in the mix being exceeded.

The problem of strength loss was mitigated by the use of a 10% addition of micro-silica, which also yielded a significant reduction in the volume of rebound produced. The fine pit sand component was also deleted from the mix, it being replaced by an increase in the mass of crushed limestone fines. This change allowed a reduction in the amount of water being needed to wet the mix (i.e. a reduction in the free water/cement ratio)and hence gave an increase in strength. An additional benefit derived from the omission of the pit sand was that it resolved the problem of potential alkali-aggregate reactivity as the remaining components of the mix were classified as non-reactive.

The materials and sources of supply used for the shotcrete are given in Table 2. Fig.3 below shows the grading curve resulting from the blend of coarse aggregate and crushed rock fines used in the approved mix design. The upper and lower grading limits plotted against the design curve are those which are generally accepted as being applicable to dry process shotcrete and were the limits used by the Contractor in selecting his proposed mix proportions as shown in Tables 3,4 and 5.

Table 2: Approved mix designs - materials sources

ORDINARY PORTLAND CEMENT	CASTLE CEMENT, PADESWOOD WORKS
COARSE AGGREGATE	10-6mm CRUSHED MICRODIORITE,exARC LTD. PENMAENMAWR QUARRY
FINE AGGREGATE 1	PIT SAND, ex LLOYDS SAND & GRAVEL LTD;MAES MYNNON, CLWYD
FINE AGGREGATE 2	CRUSHED LIMESTONE FINES, ex ARC LTD;LLANDDULAS QUARRY
MICRO-SILICA	CARBON ENTERPRISES LTD
POWDER ACCELERATOR	SIKA LTD; SIGUNITE
LIQUID ACCELERATOR	SIKA LTD; SIGUNITE WALD

Table 3: Unaccelerated shotcrete mix design

Material	Kg/m^3
O.P.Cement	420
Coarse Aggregate	720
Fine Aggregate 1	480
Fine Aggregate 2	480

Note : From June 1991 on-wards the fine aggregate component consisted of crushed rock fines only.

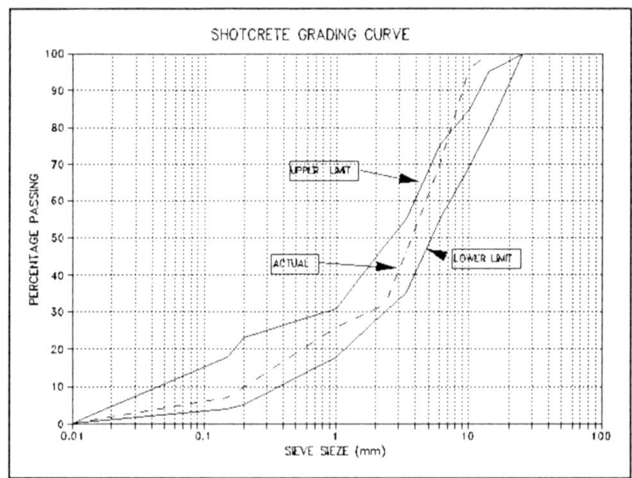

FIGURE 3: Aggregate grading curve for shotcrete

It should be noted that successful trials were conducted omitting the addition of micro-silica. However, as the Shotcrete deliveries were frequently split between areas where both the powder accelerator and liquid accelerator were being used it was necessary to add the micro-silica to all batched shotcrete.

Table 4: Powder accelerated shotcrete mix design

Material	Kg/m^3
O.P.Cement	420
Coarse Aggregate	720
Fine Aggregate 2	960
Sigunite Powder	21
Micro Silica	42

Table 5: Liquid accelerated shotcrete mix designs

Material	Kg/m^3
O.P.Cement	420
Coarse Aggregate	720
Fine Aggregate 2	960
Sigunite Liquid	12.5
Micro Silica	42

5 Works panel strengths

The Specification required that one works test panel be sprayed for each 10m^3 of Shotcrete applied in the Works. Additionally, of the panels produced during each weeks production, it was required that at least one was to be sprayed in the overhead position.

From each panel four core specimens were to be cut for test at 3, 7 and 28 days. Each core specimen was to be 100mm in dia. and was to be tested in accordance with the procedures detailed in BS1881:Pt.120:1983. The compressive strength recorded was subsequently adjusted to produce an equivalent in-situ cube strength, the calculation being as detailed in clause 7.2 of the foregoing British Standard.

The Specification stipulated that the test panel moulds were to be 750mm x 750mm x 100 deep and to be constructed of timber. The Contractor subsequently proposed that he be allowed to use rigid steel moulds 150mm deep and, following a number of successful trials, this proposal was agreed. It was found necessary, however, to place timber sheets in the base of the moulds in order to increase adhesion and to limit damage to the moulds during coring of the panels.

The results recorded for the test panel core specimens are summarised below and the 28 day strengths presented graphically in Figs 4, 5 and 6.

5.1 Unaccelerated shotcrete

Some 473 panels were sprayed with unaccelerated shotcrete between October 1991 and September 1992. From the core specimen results recorded the statistical results shown in Table 6 were obtained.

	3 DAY STRENGTH (N/mm^2)	7 DAY STRENGTH (N/mm^2)	28 DAY STRENGTH (N/mm^2)
Sample Mean	26.5	31.5	37.5
Sample standard deviation	7.0	6.2	6.7
Range of 28 day Results			16.5 - 53.0

Table 6

Figure 4 shows a reasonably consistent rate of gain between 3 and 7 days of 5 N/mm^2 and 6 N/mm^2 between 7 and 28 days. Also from Fig 4 it can be seen that a number of peaks are recorded in the moving averages, these peaks are generally the result of increased control by the Contractor following an earlier period when a downward trend in strengths had been recorded.

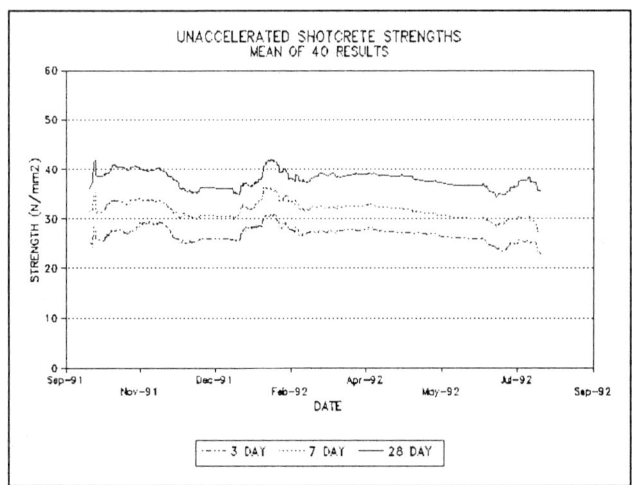

FIGURE 4: Unaccelerated shotcrete strengths

5.2 Powder accelerated shotcrete

A total of 400 number panels were sprayed with powder accelerated set shotcrete between June 1991 and September 1992. A statistical analysis of the core strengths recorded gave the results given in Table 7.

If the extremes of the strength range are discounted it can be seen from Fig.5 that fairly consistent rates of gain of between 5 N/mm^2 and 5.5 N/mm^2 are recorded for strengths between 3 and 7 days and 7 to 28 days respectively.

	3 DAY STRENGTH (N/mm^2)	7 DAY STRENGTH (N/mm^2)	28 DAY STRENGTH (N/mm^2)
Sample Mean	19.0	23.0	28.0
Sample standard deviation	6.7	6.4	7.2
Range of 28 day results			12.0 - 49.5

Table 7

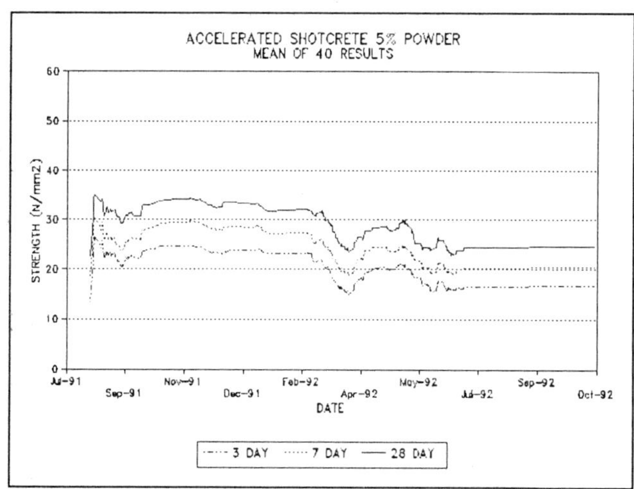

FIGURE 5: Powder accelerated shotcrete strengths

5.3 Liquid accelerated shotcrete

A total of 98 panels were produced using liquid accelerated set shotcrete between March 1992 and August 1992. From the core strength results recorded the statistical data shown in Table 8 was derived.

	3 DAY STRENGTH (N/mm^2)	7 DAY STRENGTH (N/mm^2)	28 DAY STRENGTH (N/mm^2)
Sample mean	20.0	24.5	29.0
Sample standard deviation	6.8	5.1	5.6
Range of 28 day results			18.5 - 40.5

Table 8

It can be seen from Fig.6 that whilst the rate of gain between the different ages of test are broadly similar to those achieved by the powder accelerated core specimens the mean strengths achieved are significantly higher.

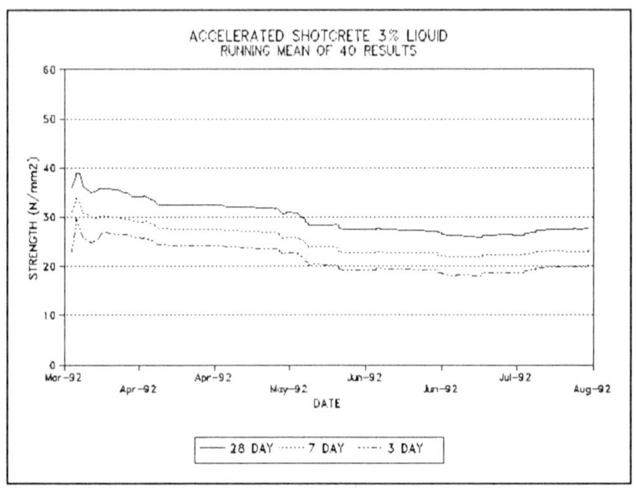

FIGURE 6: Liquid accelerated shotcrete strengths

6 In-situ testing

The Specification instructed that two systems of in-situ testing should be carried out by the Contractor. These systems were ; the drilling of 100mm dia. cores to determine shotcrete thickness and quality of shotcrete in the Works, and the carrying out of Windsor Probe Testing. With regard to this latter method three number readings were required to be taken at positions indicated by the Engineer in order to assess the compressive strength.

It was also required that prior to in-situ Windsor Probe tests being carried out, the probe readings were to be correlated to compressive core strengths taken from the Preconstruction Test Panels. The Windsor Probe apparatus consists of a gun which fires a set of 3 hardened steel alloy probes into the concrete at high speed using an exactly measured explosive charge. The depth of penetration is measured and the compressive strength of the concrete is directly related to the resistance to penetration of the crushed aggregate and cement matrix : this is determined by the distance required to absorb the specific amount of kinetic energy of the probe. The compressive strength of the concrete is empirically related to the penetration which varies with the hardness of the aggregate. This relationship is recognized by determining the Mor's Scale Hardness value of the aggregate and applying a correction factor to the penetration.

A considerable number of comparative tests were carried out by the Contractor, which involved carrying out both Windsor Probe tests and coring of test panels to compare the compressive strength values recorded by each. However, despite this extensive testing no correlation could be established between the results for the probe tests and those for the cores due to the inconsistent results. All results were related to 7 day strengths as it was required that an early indicator of potential 28 day strength be established. It should be noted however, that a number of comparative checks were made at 28 days after spraying, with similar results being recorded.

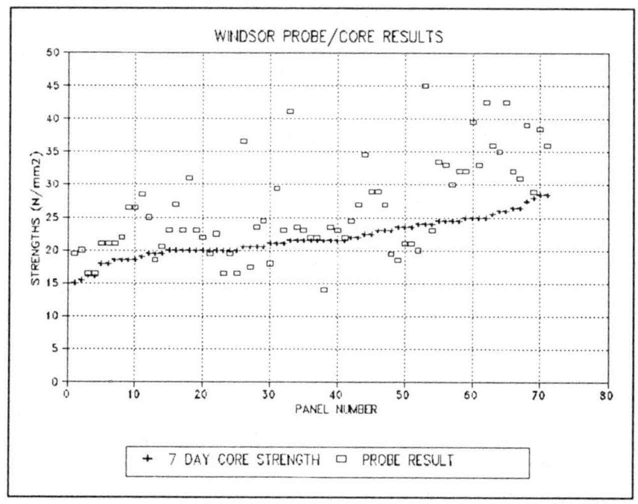

FIGURE 7 Comparison of Windsor Probe and cube results

Fig. 7 shows the results recorded for the core specimens plotted cumulatively and the corresponding results for, the probe tests. As can be seen a considerable disparity exists between the two sets of data. As a consequence of this failure to establish a correlation between probe and core results, the use of the Windsor Probe was discontinued. A possible reason for this disparity in the results recorded could be the variation in Mor's Hardness Number between that of the coarse aggregate and that of the matrix. From tests conducted by the Contractor a mean Mor's Hardness Number of 6 was adopted in determining the appropriate correction factor to be used. However, the limestone fines which formed the matrix with the cement paste had an estimated hardness number of 3, whilst for the microdioritic coarse aggregate it was 6.5 -7. As a result of this variation in hardness number it is probable that in many instances the true Mor's Hardness Number was different from the adopted mean value.

In-situ coring was carried out in all areas where Works Panel core strengths had been found to fail to comply and/or where it was required to determine the depth or integrity of the in-situ shotcrete. It was found in the majority of cases that a good correlation existed between the results recorded for the panel cores and those for the in-situ core specimens.

7 Discussion of results

Unaccelerated Shotcrete - Generally the results show that the rates of gain between 3, 7, and 28 day core strengths and the 28 day equivalent in-situ cube strength remained consistent throughout the contract period. The frequency graph, Fig.8, shows the results to follow a normal distribution with an approximate mean strength of 40 N/mm^2. Reference to the cumulative frequency graph, Fig.9, shows that approximately 9% of all the results recorded failed to achieve the required 28 day strength.

Powder Accelerated Set Shotcrete - Whilst generally a consistent rate of gain between 3, 7, and 28 day core strengths was recorded, the range of strengths recorded was considerable. It will be noted from Fig 8 that the results are skewed towards the lower strength values, this being a function of the loss in strength which occurred due to the use of the set accelerator. Fig. 9 shows that approximately 65% of all results failed to achieve the originally specified minimum strength of 30 N/mm^2. Even when compared to the revised strength targets of 25 N/mm^2 and 20 N/mm^2 the failure rates remain high at 45% and 12% respectively.

Liquid Accelerated Shotcrete - The results recorded show that whilst the rates of gain in strength are generally similar to those recorded for the powder accelerated set shotcrete the mean strength is significantly higher. These results clearly illustrate the benefits gained due to the greater control and lower dosage rated associated with the use of a liquid accelerator. It should be noted however, that the use of both the Kapillar nozzle and high pressure pump were also in part, responsible for the improved results recorded.

Fig.9 shows that the failure rates for target strengths of 30 N/mm^2 and 25 N/mm^2 remained high at 58% and 30% respectively. For a target strength of 20 N/mm^2 the failure rate was less than 5%.

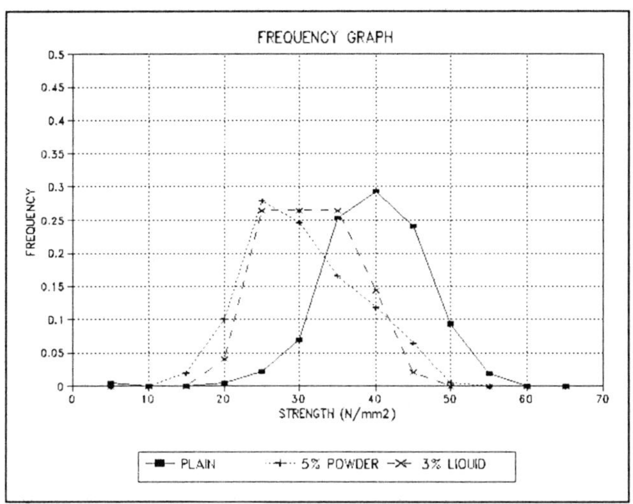

FIGURE 8 Frequency Graph

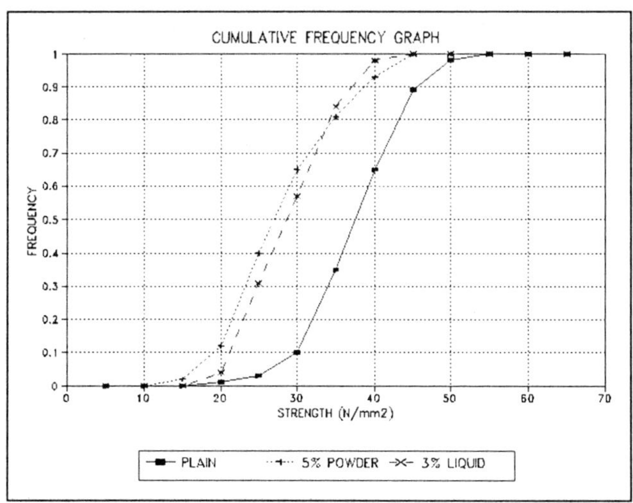

FIGURE 9: Cumulative frequency

It should be noted that in all cases where the shotcrete failed to achieve the specified strengths it was either removed and replaced with conforming shotcrete or a thicker layer of shotcrete was applied to compensate for the lower strength thus preserving the integrity of the works.

8 Conclusions

The experiences gained on the Pen-Y-Clip contract highlighted a number of areas of production, application and control where improvements could have been made. These can be detailed as follows :

8.1 Mix Design

Careful consideration needs to be given to the mix proportions selected to ensure that optimum results are achieved. In particular it would appear that considerable benefits can gained in respect of both the strength achieved and in the reduction of rebound through the use of micro-silica. The use of this material in shotcrete mix designs clearly needs further investigation. The potential for alkali-silica reaction also needs to be considered particularly where set accelerators are to be used, as these frequently contain very high levels of alkalis.

8.2 Set Accelerators

Whilst the use of set accelerators will be unavoidable in many situations extreme care should be taken in their selection and the method of dosing to be used. In particular the use of powder accelerators should be treated with extreme caution. It is recommended that their use should be restricted to, temporary works applications, to stem water inflows and to provide immediate support in excavations. Their use during the winter months in particular should be treated with extreme caution as the rapid strength loss resulting from their use is considerably exacerbated at low temperatures. With both powder and liquid accelerators great care needs to be exercised to ensure their correct use and to ensure that only the minimum dosages necessary are used.

8.3 Production and Application

Prior to any work commencing details need to be established regarding the methods to be adopted in the production and application of the shotcrete. In general it will only be necessary to follow good concrete practice in respect of the production methods adopted. However, it should be borne in mind that the time available for use will be considerably less than is the case for normal concrete, particularly in underground situations where elevated temperatures may occur. Care also needs to be taken in the selection of the application equipment to be used and the methods of spraying to be adopted. It is also recommended that both plant operatives and nozzlemen be carefully assessed to ensure that they are aware of the good practices to be followed and that these are applied throughout the construction of the Works.

8.4 Control

Prior to the commencement of shotcreting the Contractor should be requested to provide a written method statement detailing the quality control measures to be adopted. On larger contracts this may involve the provision of an on-site laboratory together with separate quality control staff. On smaller contracts this quality control function could be carried out by a supplier or local independent laboratory. Contractors should also ensure that all personnel involved in the production, application and control of shotcrete, receive adequate technical training. On larger contracts consideration should also be given to establishing a detailed quality plan either as part of an existing quality scheme or as a separate quality function.

9 Acknowledgements

The authors wish to thank the Welsh Office for permission to publish this paper and their colleagues on the Pen-y-Clip site for their assistance in its preparation.

Face support for a large Mix-Shield in heterogeneous ground conditions

Sandor Jancsecz
Wayss & Freytag AG, Frankfurt am Main, Germany
Walter Steiner
Balzari & Schudel AG, Bern, Switzerland

Abstract
Control of face support is a major issue in slurry shield tunnelling. Besides the regular support with slurry also compressed air-support of the face during maintenance of the tools in the working chamber is necessary. The required support pressure is designed with a limit state model to balance water and earth pressures at the face. Experience with face support and associated support mechanism are described. Special procedures and slurry mixes were developed to ensure a safe support of the face without uncontrolled loss of slurry or air.
Keywords : Bentonite, compressed air-support, design of face support,polymers, sand content, slurry shield, support pressure.

1. Introduction

The Grauholz Tunnel is a double-track high-speed railway tunnel located north of Bern, Switzerland (Figure 1.).

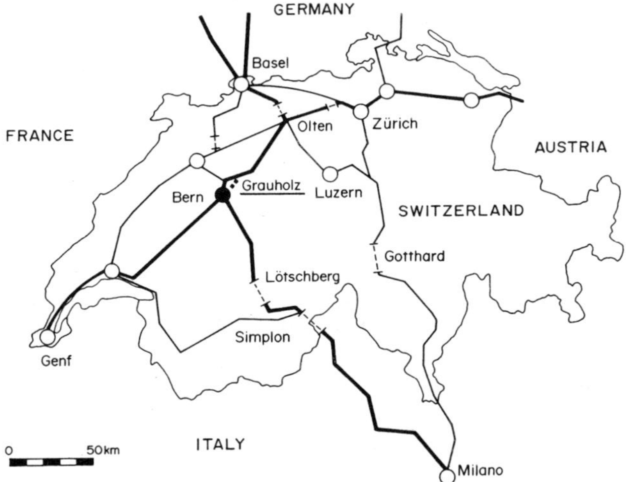

Figure 1. Location Map of Grauholz Line in Switzerland

It serves to alleviate the congested situation at Zollikofen Station (Figure 2.) and forms the first section of a new high-speed railway line to Olten. More details on the design of the Grauholz line have been published elsewhere [1], [2].

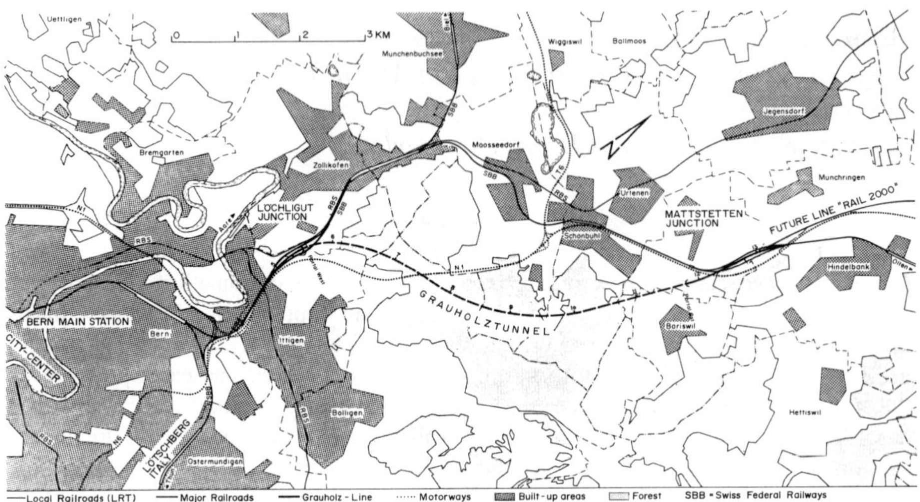

Figure 2. Alignment of Grauholz Tunnel

An alternative proposal by a contractor joint venture had been selected proposing a unilateral tunnel advance with a Mix-Shield, being able to excavate through soil, rock and mixed-face conditions.

The contract was let in 1988. Tunnelling with the Mix-Shield started in November 1989 and the 5.4 km long mined section of the 6.3 km long Grauholz Tunnel was holed through on May 6, 1993.

The changing ground conditions, with a wide range of soil types under various hydrogeological and stress conditions, as well as rock and mixed-face sections, led to substantial experiences.

2. Ground conditions

Ground conditions are briefly summarized here. More details have been presented earlier [2], [3]. The Grauholz Tunnel is located where two major alpine glaciers conflued. The eastern branch of the Rhone glacier flowed in from the west and from the south the Aare glacier flowed in. Glacial deposits in general are complex. Matters were further aggravated by the interaction of the two glaciers. Quarternary deposits are underlain by tertiary bedrock. A geologic-hydrogeologic-geotechnical longitudinal section is presented on Figure 3., together with the types of tunnelling applied and the general characteristics of the ground types through which the tunnel passed.

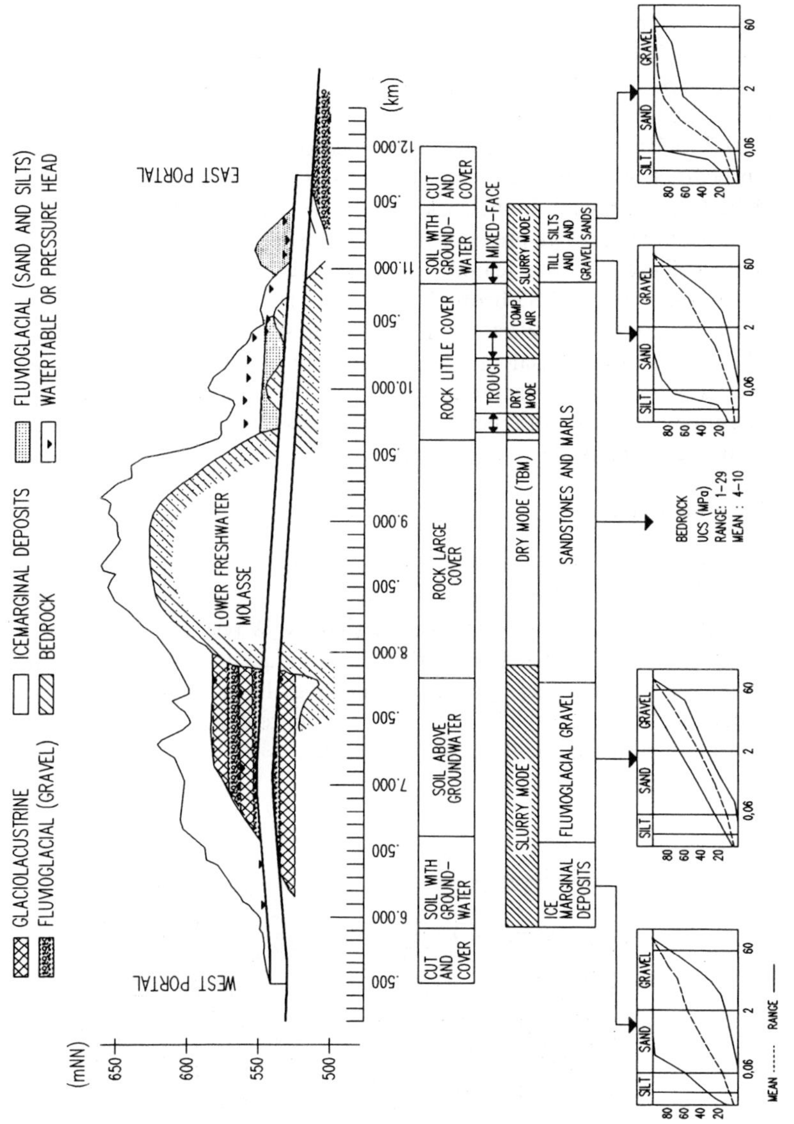

Figure 3. Geologic - geotechnical conditions

Tunnelling started from the east portal. The first section (km 11.3 - 11.0) contained irregularly alternating glacial silts and sands. The content of fines less than < 0.06 mm was on the average 60 %. Most of the silts were of low plasticity, the plasticity index being less than 4 %. Some layers of clayey silt with a maximum plasticity index of 11 % were also present. Then from km 11.0 - 10.7 glacial tills with plasticity indices from 5 to 12 % alternating with clean gravel were encountered. The variation of permeabilities between the different layers were substantial. Water pressures were a few meters above the crown.
From km 10.7 to 10.6 the tunnel entered into bedrock with a mixed-face. From km 10.6 to 9.6 bedrock with small cover (a few meters) was encountered. Between km 10.32 and 10.17 a depression in the rock surface caused mixed-face conditions, with sand in the crown and a water pressure of 25 m above the crown. Between km 9.75 and 9.7 another depression was feared, the rock cover was there, but of insufficient thickness. From km 9.6 to 7.85 bedrock cover was larger. Low strength sandstones and shales were encountered. West of km 7.85 again soil was encountered. From km 7.85 to km 6.5 fluvioglacial gravel without groundwater was encountered. Groundwater was perched in glaciolacustrine clay. From km 6.5 to the portal at km 5.92 icemarginal deposits with perched watertables at various levels had to be crossed up to a few meters above the crown.

3. The Mix-Shield system

With the assumed instability of the face in glacial deposits with and without ground water, and relatively good face stability in the bedrock (molasse), a convertible mix-shield machine - based on a patented design by Wayss & Freytag AG - was selected by the client. This machine worked in the glacial soil deposits as a slurry shield with pressurized face and hydraulical mucking. In the molasse it worked as a TBM with dry mucking. The shield is described in more detail in [2]. As Figure 4. shows the Mix-Shield has a double chamber system.

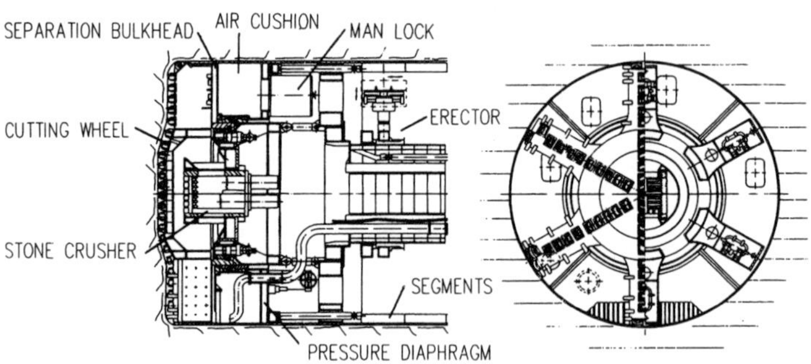

Figure 4. The Mix - Shield system (D = 11,6 m)

The pressure chamber for controlling the slurry pressure by an air cushion is separated by a diaphragm bulkhead from the working chamber of the cutting wheel. To reduce the number of entering into the working chamber there was a stone-crusher installed in the centre, for breaking boulders up to 1 m^3 in size.

3.1 Stages of face support

Beside the described adaptability there is an other advantage of this type of shield, namely; maintenance on the cutting wheel can be undertaken with partial or full drawdown of the slurry under compressed air. Figure 5. shows the typical supporting modes and the distribution of support pressures on the tunnelling face. However, the support with compressed air - either half or full - was a rather exceptional case.

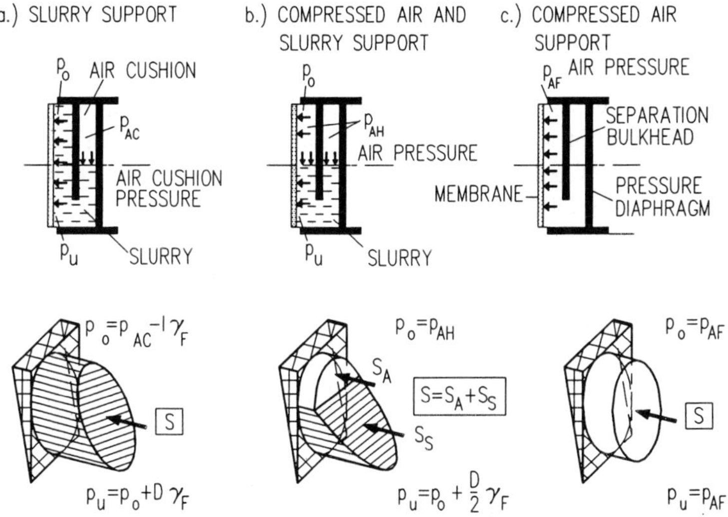

Figure 5. Supporting modes and distribution of face supporting pressures

When driving in slurry mode the face is supported by the slurry pressure. In this case the slurry pressure distribution is very similar to the distribution of earth and water pressures. The pressure - transmitting medium between liquid and soil is a membrane or cake, filtrated from the bentonite slurry onto the tunnelling face.
The two most important questions regarding face support in slurry mode are; how high the support pressure should be (earth statical aspect) and how is it possible to reach a safe and quick membrane developing (fluid mechanical aspect).
An important element of the Mix-Shield system is the automatic fluid pressure regulation with the help of an air cushion as shown in Figure 5. This makes it possible to provide a constant support pressure on the face under continuous and parallel charging and discharging of the slurry into and out of the working chamber. With the face partly or fully supported by compressed air, various problems

can arise: The support pressure is less ideal (constant over face). Compressed air influences the ground in different manners : soil is more permeable to air, large losses of air may occur, finally resulting in a blow out. In some soils drying out may occur under compressed air support and the face may thus become unstable after a certain period of time (stand-up time).

4. Theoretical approach of the face stability in non cohesive soils

To find out the right support pressure for a tunnelling face in granular soils, it is first of all necessary to consider the forces on a possible and probable failure model. Similar to other solutions in soil mechanics, application of limit equilibrium analysis can deliver a limit earth pressure acting on tunnelling a face. Selection of an appropriate factor of safety assisted by deformations or settlement control allows determination of support presssure which is not too high or not too low. If the support pressure is too low, face collapse or extreme settlements occur. If it is too high excessive slurry or air penetration with surface upheaval may occur together with loss in effective support, particulary with air. This may result in face collapse.
The first calculation of the support pressure for a Hydroshield (an earlier designed slurry shield from Wayss & Freytag AG) based on a simple three dimensional model, as shown in Figure 6.,

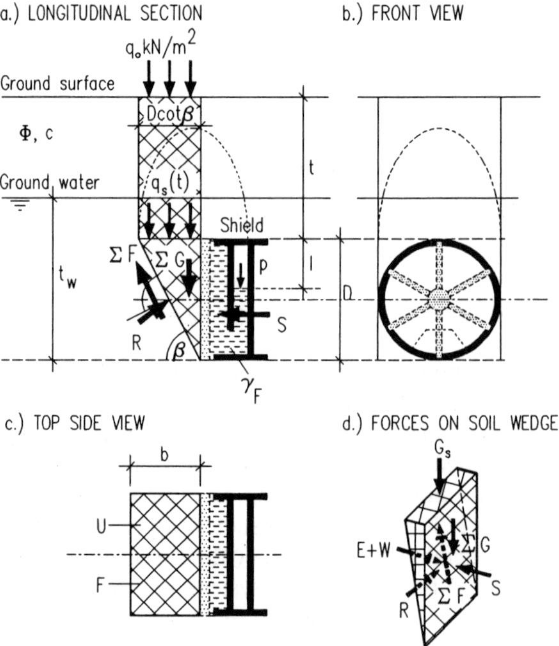

Figure 6. Three dimensional limit equilibrium model

was carried out approx. six years ago [4]. After six years experience in using this failure model on different projects [5] it appears well proven , that there is no particular need for more complicated or sophisticated calculation methods. Due to many uncertainities like : soil parameters, boundary conditions, constitutive law etc. it is more than adequate to use the simplified three dimensional model from Figure 6. It is more important to collect further practical experiences and carry out different measurements to be able to adapt or apply the theoretical results to them. [6]
The three dimensional failure model consists of two parts ; a soil wedge (lower part) and a soil silo (upper part) above them. For the first, the vertical pressure resulting from silo and acting on the horizontal surface of the soil wedge will be determined separately from the lower part. According to Terzaghi's solution the vertical pressure is given by [7]

$$q_S(t) = \frac{\frac{F}{U}\gamma - c}{\lambda\tan\Phi}\left\{1 - \exp\left(-t\frac{U}{F}\lambda\tan\Phi\right)\right\} \qquad (1)$$

Where γ is the unit weight, Φ is the angle of internal friction, c is the cohesion, λ is the earth pressure coefficient, U and F are the circumference and area of the horizontal plane from the soil wedge. As the expression shows, the vertical overburden pressure acting on the top plane of the soil wedge is also a function of the slip angle. This angle is the linking member between upper and lower part of the failure model. The conditions of equilibrium for the lower part are (see also [8])

$$\sum F + R\sin\Phi + E\cos\beta - \sum G\sin\beta = 0 \qquad (2)$$

$$R\cos\Phi + E\sin\beta + \sum G\cos\beta = 0 \qquad (3)$$

and with it the earth pressure is given by

$$E = \frac{\sum G(\sin\beta - \cos\beta\tan\Phi) - \sum F}{\tan\Phi\sin\beta - \cos\beta} \qquad (4)$$

The maximum of the earth pressure is to derivate by fulfilling next differential condition :

$$\frac{dE}{d\beta} = 0 \qquad (5)$$

The solution for $\beta = f\left(\frac{t}{D}; \Phi\right)$ after solving the differential equation is obtained through a series of iterations. Figure 7. shows the variation of the slip angle ß over

Φ and $\frac{t}{D}$. It is important to recognize that the effect of the overburden pressure on ß - if $\frac{t}{D} \geq 2$ - is not that great. With knowledge of ß it is possible to define a three-dimensional earth pressure coefficient :

$$K_{A3} = \frac{\sin\beta\cos\beta - \cos^2\beta\tan\Phi - \frac{K\alpha}{1.5}\cos\beta\tan\Phi}{\cos\beta\sin\beta + \tan\Phi\sin^2\beta} \qquad (6)$$

Where $K \approx \frac{1 - \sin\Phi + \tan^2\left(45 - \frac{\Phi}{2}\right)}{2}$ and $\alpha = \frac{1 + 3\frac{t}{D}}{1 + 2\frac{t}{D}}$. Figure 8. clearly indicates the predominant influence of the internal friction on the earth pressure. The diagrams of Figures 7. and 8. can be applied for estimating support pressure in different cases. For practical use and parametric studies a computer program was developed by [9] using a Windows / Mathcad spread sheet.

Other solutions of the support pressure for cohesionless soil have been published by [9]. Their theoretical solutions for statically admissible stress field (lower bound) and for kinematically possible collapse mechanism (upper bound) were found to be in close agreement with experimental results. Therefore this theoretical method seemed to be applicable, as a calculation for comparison, to the three-dimensional problem from the tunneling face especially - in fluvioglacial gravel above the ground water level - in western part of the Grauholz - Line. The necessary magnitude of uniform normal support pressure on the exposed periphery and face were estimated after (7) and (8) :

$$\sigma_{TL} = \frac{\gamma D}{K_{P2}^2 - 1} K_{P2} \qquad (7)$$

$$\sigma_{TU} = \frac{\gamma D}{4\cos\Phi}\left(\frac{1}{\tan\Phi} + \Phi - \frac{\pi}{2}\right) \qquad (8)$$

provided that $\frac{t}{D} \geq \frac{1}{\sin\Phi} - 1$. Where D is the diameter of shield, γ is the unit weight of soil, Φ is the angle of internal friction and $K_{P2} = \tan^2\left(45 + \frac{\Phi}{2}\right)$ is. Substituting D = 11.6 m and γ = 22 KN/m^3 into the above equations the lower

$\frac{t}{D}$	β				
	$\Phi=20°$	$\Phi=25°$	$\Phi=30°$	$\Phi=35°$	$\Phi=40°$
0	60,340	62,611	64,802	66.900	68,918
1	61,890	64,161	66,286	68,283	70,177
2	62,197	64,464	66,580	68,566	70,423
3	62,322	64,592	66,706	68,683	70,527
$45+\frac{\Phi}{2}$	55,0	57,5	60,0	62,5	65,0
$K=\frac{K_0+K_2}{2}$	0,574	0,4916	0,4167	0,349	0,287

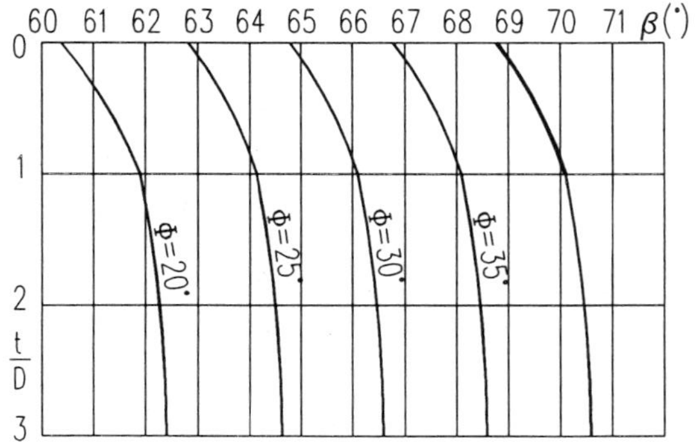

Figure 7. Variation of the slip angle β

$\frac{t}{D}$	K_{A3}				
	$\Phi=20°$	$\Phi=25°$	$\Phi=30°$	$\Phi=35°$	$\Phi=40°$
0	0,386	0,310	0,248	0,199	0,159
1	0,354	0,279	0,222	0,177	0,141
2	0,348	0,273	0,217	0,173	0,138
3	0,345	0,271	0,214	0,171	0,136
$K_{A2}=\tan^2(45-\frac{\Phi}{2})$	0,49	0,41	0,33	0,27	0,22
$K=\frac{K_0+K_{A2}}{2}$	0,574	0,4916	0,4167	0,349	0,287

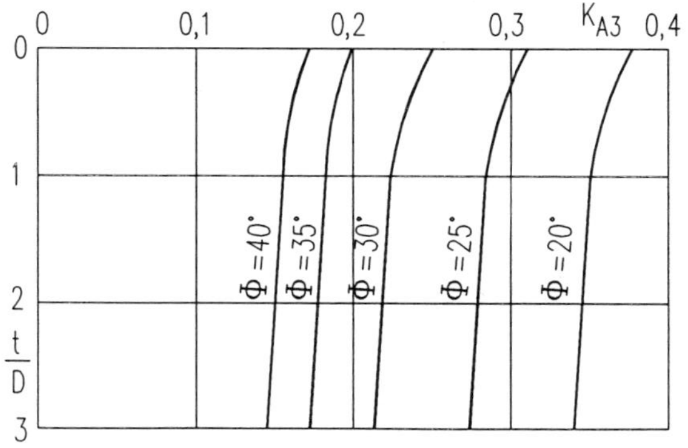

Figure 8. Three dimensional earth pressure coefficient

and upper limit of support pressure were calculated and plotted versus the angle of internal friction Φ as shown in Figure 9.
Calculations based on simplified three - dimensional model delivered very similar results at Φ = 35 °- 40 °.During tunnelling the stability of the face was verified by reducing the air support pressure in the soils of the western tunnel section. For safe and stable tunnelling a support pressure was judged necessary, corresponding to the lower bound limit for gravel with a friction angle of Φ = 38 °. During the tests, with reduction of air pressure, it was noticed that instabilities developed with support pressures in the range of 15 - 25 KPa. This corresponds to the upper bound solution for a friction angle from 38 ° to 44 °. Since such friction angles are possibly close to the origin, the agreement between the analytical solution after [9] and experience was cosidered as good.

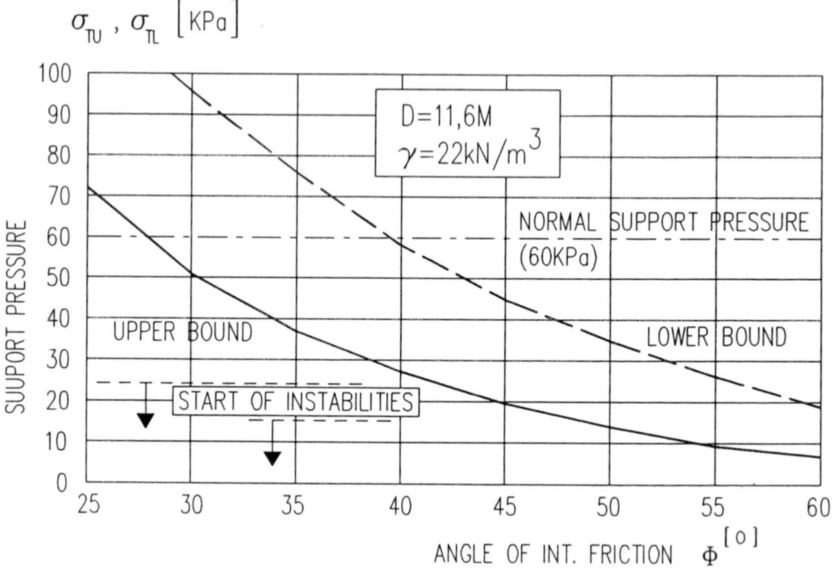

Figure 9. Support pressures after Atkinson & Potts

5. Support pressures along the tunnel alignment

The required support pressures along the tunnel axis were calculated with the help of the above discussed methods. Figure 10. and Figure 11. shows the air cushion and air pressure for to the centre lowered (half drawdown) slurry in the eastern and western part of the tunnel. In slurry mode one has to recognize, that the support pressure in the working chamber is of course not equal to the air cushion pressure. The fluid pressure in the crown-point can be derived as follows $p_o = p_{AC} - l\gamma$ where p_{AC} is the air cushion pressure, γ is the unit weight of slurry and l is the distance between the crown-point and slurry level behind the bulkhead (see Figure

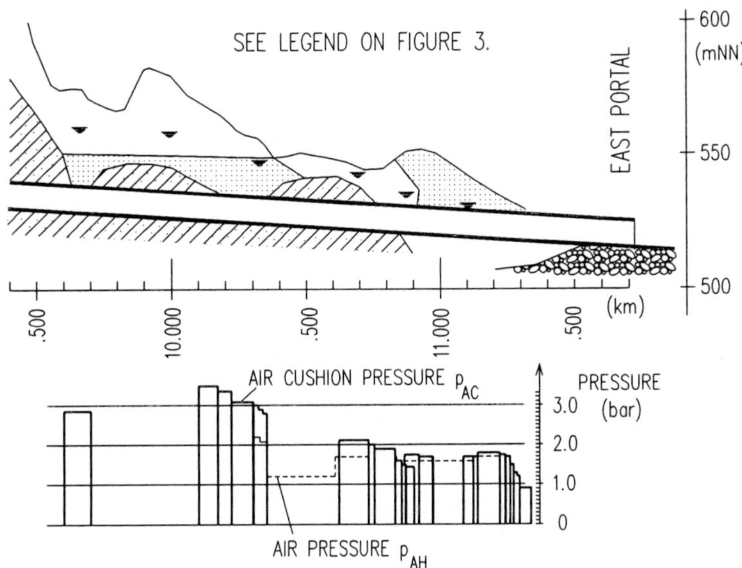

Figure 10. Support pressures in eastern part of tunnel

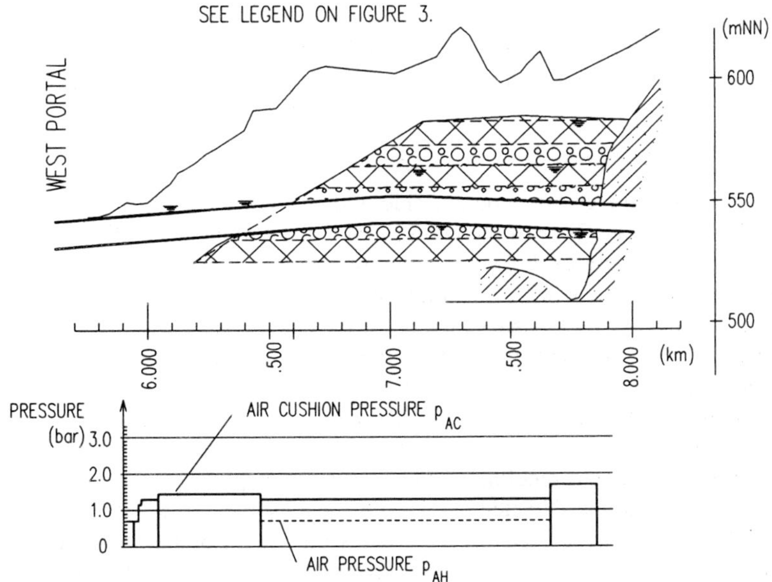

Figure 11. Support pressures in western part of tunnel

5.). During tunnelling in granular soils it has been proven that a factor of safety - against collapse according to the simplified model - not higher than 1.2 to 1.3 was necessary to ensure a stable face.

5.1 Experiences with face support

In bedrock with sufficient rock cover no face support was necessary and the Mix-Shield was operated as an ordinary TBM. In soil and mixed-face conditions the Mix-Shield was operated as a slurry shield. During slurry shield tunnelling no substantial problems with face support developed. However, one has to be aware that with little overburden the slurry pressure may cause heave. This limiting condition has to be checked.

Problems developed essentially only with compressed air support, which had to be applied for the purpose of maintenance of the tools. When pervious gravels were encountered loss of air support was experienced twice under different conditions. In both cases a collapse of the face occurred. The cutting wheel had to be freed either from a shaft or by grouting from inside the shield [3]. Both operations were tedious and time-consuming.

After the first incident protective roofs were frozen in maintenance stations, located every 50 meter . This reduced loss of compressed air and prevented the development of running ground.

After the second incident a special slurry was developed which formed an essentially air-tight membrane on the surface of the ground. Also controlled entry procedures were then prescribed. First the loss of supporting slurry was monitored after increasing the face pressure by 0.5 bar (50 KPa). The slurry at the face was then lowered in stages and at each stage the loss of compressed air monitored. With these procedures further incidents, with uncontrolled loss of face support pressure could be avoided.

5.2 Settlements

The tunnel passed under free country-side and several roads, a petroleum pipeline and a major motorway. Two of the roads were close to the portals. Since instabilities could not be completely excluded with the overburden varying between 6.5 to 8 m, the traffic was diverted when tunnelling directly underneath the road. No incidents developed. With the first road crossing, less than 100 meters after the start of tunnelling, and 6.5 m overburden above the crown, a central settlement of 25 mm was monitored. This translates to a loss of ground of 0.4 %.

When passing under the motorway with 60 meters overburden the survey at the ground surface did not detect any settlement. Also when crossing 20 meters underneath the petroleum pipeline, which had been emptied prior to passage of the tunnel, no settlement was detected. The same was observed when passing underneath the road close to the westportal. In these cases this might be partly attributed to the local ground conditions with fine-grained layers of silt to clayey silt which deformed less during tunnelling and allowed a complete grouting of the tail void. Interestingly in zones where collapses to the surface had occured, no settlements were detected at some distance (15 - 20 m) from the sink-holes and no settlements indicated imminent collapse.

6. Penetration of slurry into the soil at the tunnelling face

A soil-bentonite-water slurry doesn't penetrate into the pores of the soil if the grain size distribution of solids in the slurry is coarser than the pore-size (= constriction size) distribution of the soil . Constriction size (D_C) is defined as the diameter of the largest sphere that will pass through a particular constriction (see Figure 12.).

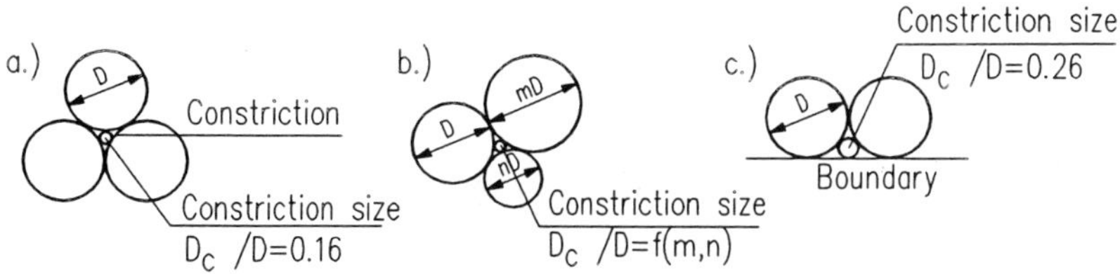

Figure 12. Constriction size in dense assemblages of spheres after Kenney et al.

The distribution of constriction size is dependent mainly on the grain size distribution and probabilistic methods can be used to estimate the distribution for a given gradation, e.g. from [11]. Figure 13. shows an example for the constriction size distribution in

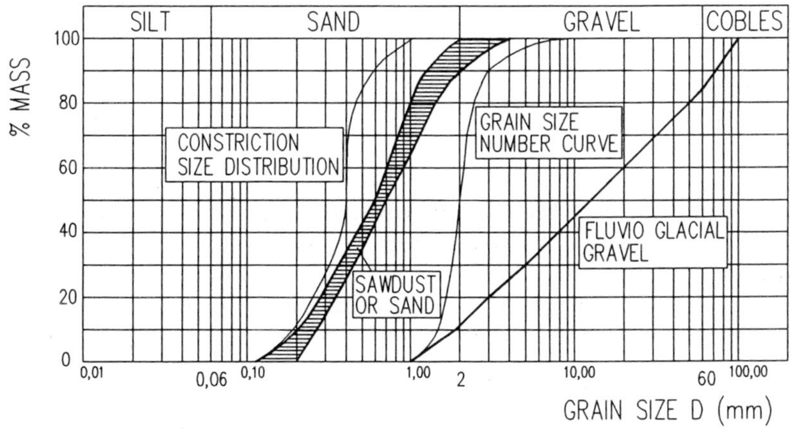

Figure 13. Particle size and constriction size distributions

comparison with the mass distribution of gravel and sawdust or fines. Granular materials are composed of a range of particle sizes and, therefore, contain a range of constriction sizes. The filtering capability of a soil at the face is dependent on the minimum size of constrictions along flow paths. [12]
According to [13] the penetration length (Figure 14.)

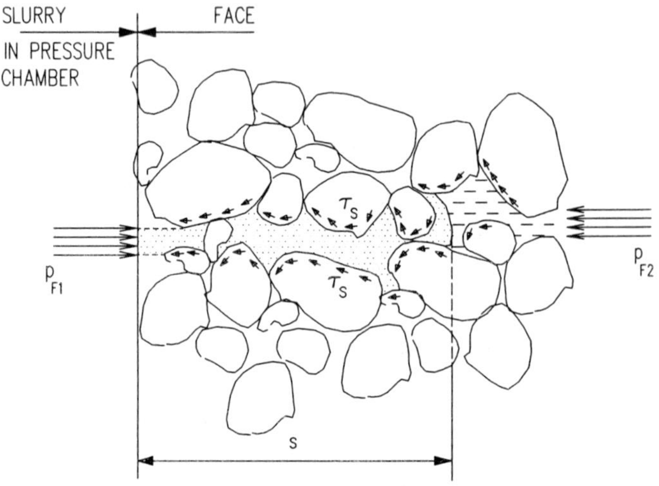

Figure 14. Penetration length in state of stagnation

of a bentonite slurry in a pore channel of soil is given by

$$s = \frac{\Delta p d_{10}}{3.5\tau_S} \qquad (9)$$

Where Δp is the difference between slurry pressure and water pressure, d_{10} is the effective particle size and τ_S is the shear resistance of the slurry. Figure 15. illustrates an example for the penetration in soils with different d_{10} and at constant pressure difference and at two different shear resistance. Experiences have proved, that the slurry intrusion will be unacceptably high when d_{10} > 2 - 3 mm, even if the stability criteria of a single particle or particle group after [14] $\tau_S \geq \frac{d_{10}(1-n)(\gamma_s - \gamma_S)}{\tan \Phi}$ is the fulfilled. Where n is the porosity, γ_s is the specific weight of soil particle and γ_S is the unit weight of slurry. In some similar cases the tunnelling face collapsed because the stabilizing effect of the penetrated slurry (state of stagnation) outside the possible failure area is practically equal to zero. On the other hand in strong heterogeneous soil the above mentioned equilibrium criterion is less or not valid because forces between and torques on single particles are not considered in it.

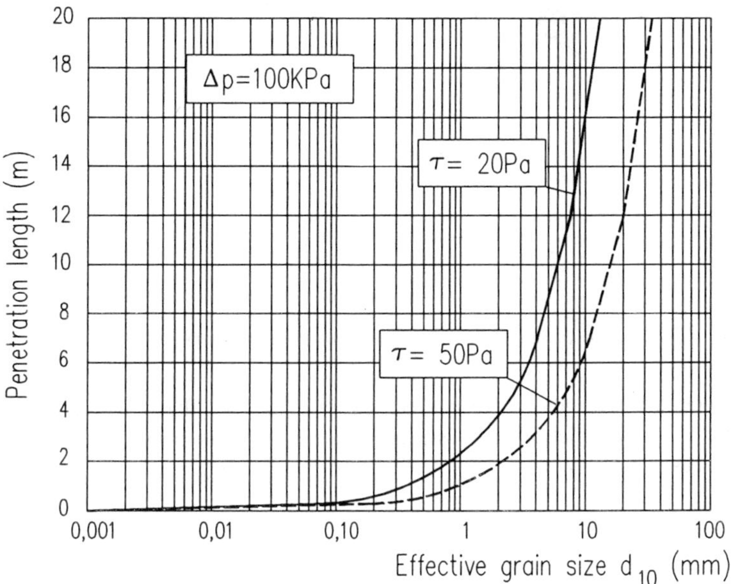

Figure 15. Penetration lengths in different soils

Over years collected field experiences, series of laboratory and in situ tests have formed the philosophy at Wayss & Freytag AG for a quick developing membrane on the face without large intrusion of slurry into the pores of any kind of soils, fullfilling criterias for pumpability ($\tau_S < 40 - 50$ Pa), separability and last but not least the cost efficiency. In soils with higher sand content there is no problem to reach a quick cake developing with a pure bentonite slurry.
In the western section of the Grauholz tunnel (km. 7.6 - 6.5) the sand content in the slurry circuit was artifically kept high (density of slurry 10.8 to 11.2 KN/m^3) by taking out of operation half the capacity of the multicyclones and recirculating it to the face. The slurry had a shear strength of only 10 - 20 Pa.
In addition to bentonite (50 kg / m^3) also polymer was mixed to the slurry (0.4 kg / m^3). The polymer with long strings forms a " net " over the pores causing a rapid clogging almost without respect on any filter law. Figure 16. tries to visualize this special situation. The soil-bentonite slurry would penetrate relatively deep into the pores if there were no polymer strings in the slurry mixture.

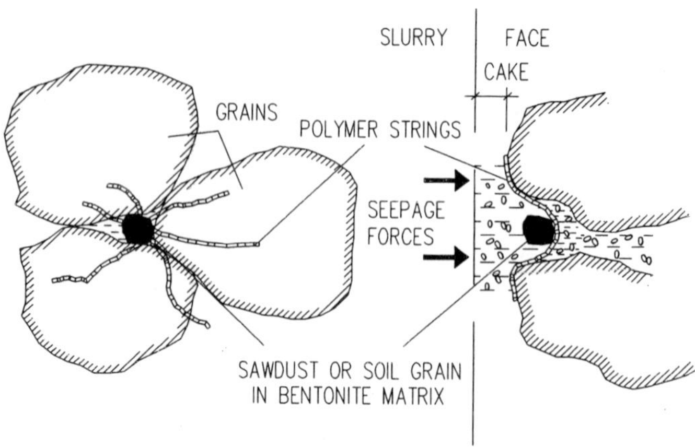

Figure 16. Membrane (cake) developing with special slurry

Quintessence of the above discussed experiences is that, when developing a membrane without penetration of slurry into the pores is possible, then the role of the shear resistance is secondary. In Figure 17. an attempt has been made to demonstrate that a

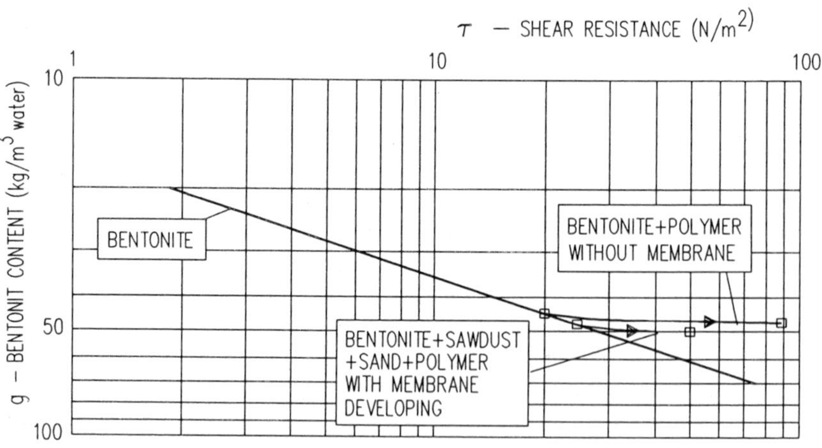

Figure 17. Effect of additives on membrane developing

very high shear resistance alone (up to 90 N/m^2) is not enough for membrane developing in a very coarse soil like the fluvioglacial gravel. Results taken from [15].

The theory of the small slurry penetration into the pores leads to a relatively thick and resistant slurry mixture. Even if this could be described as a tendency to earth pressure or slime balancing method, an important difference has to be emphasised : to adjust an appropriate support pressure in a slurry shield is more exact, safer and technically easier. It is not so dependent on the workmanship of the crew as in a earth pressure balanced shield. Also the applied pressure acts directly at the interface of the tunnel. A slurry shield acts in terms of effective stresses which are applied to the face.

It is necessary to limit and relate the above postulated thesis of the quick cake developing for a " high speed " face advance in a medium coarse sandy gravel. In such a case, also depending on the pressure difference and bentonite content, there is no time for a deep slurry penetration into the soil even if the mixture doesn't fulfill the filter law or contains additives. Consequently no significant pressure loss will be obtained at the face. This " overriding " method is only practicable under really favourable conditions.

Another major issue is in very coarse soils support of the face - partly or fully - with compressed air. Not only great air losses but the danger of a blow out and / or face instability can occur.

In the fluvioglacial gravel, above the ground water level a special slurry : formed from a mixture from fines + sawdust + polymer + bentonite and water was used to obtain an air-tight or less air-permeable membrane. The role of the sawdust is similar to the fines from soil (clogging) with one difference, that a cake which contains sawdust particles is more compressible at the beginning of the filtration and therefore the cake will be developed safer and will be denser than with sand only. Tests with a slurry containing the same amount of sand by volume did not give an impervious membrane. The sawdust of the same gradation gave an impervious membrane. [3], [14]

Repairing works in the working chamber were systematically carried out on Mondays. After the last shift on every Friday the above described special slurry was pumped into the shield chamber and pressurized over the normal pressure needed (up to 1,7 - 2,0 bar air cushion pressure). This was continued over the weekend. Slurry losses between 15 - 30 m^3 were recorded. In fact these were liquid or water losses until a very dense non permeable cake was achieved. The result was a very stable air supported tunnelling face in the coarse gravel. In this connection it is important to remember the height of the air supported face from 11.6 m. Figure 18. shows the air consumption measured in a Monday-shift on 23rd. Nov. 1992, first between 08 -13^{30} hours with a very low max. air loss (30 m^3 / min) for the full-face compressed air support. After a short pressurizing, between 13^{30} - 16 hours, again with special suspension, the working chamber was entered with to the center lowered suspension i.e. half-face compressed air support. In this case a higher air consumption tendency was obtained but still within a very acceptable range. From

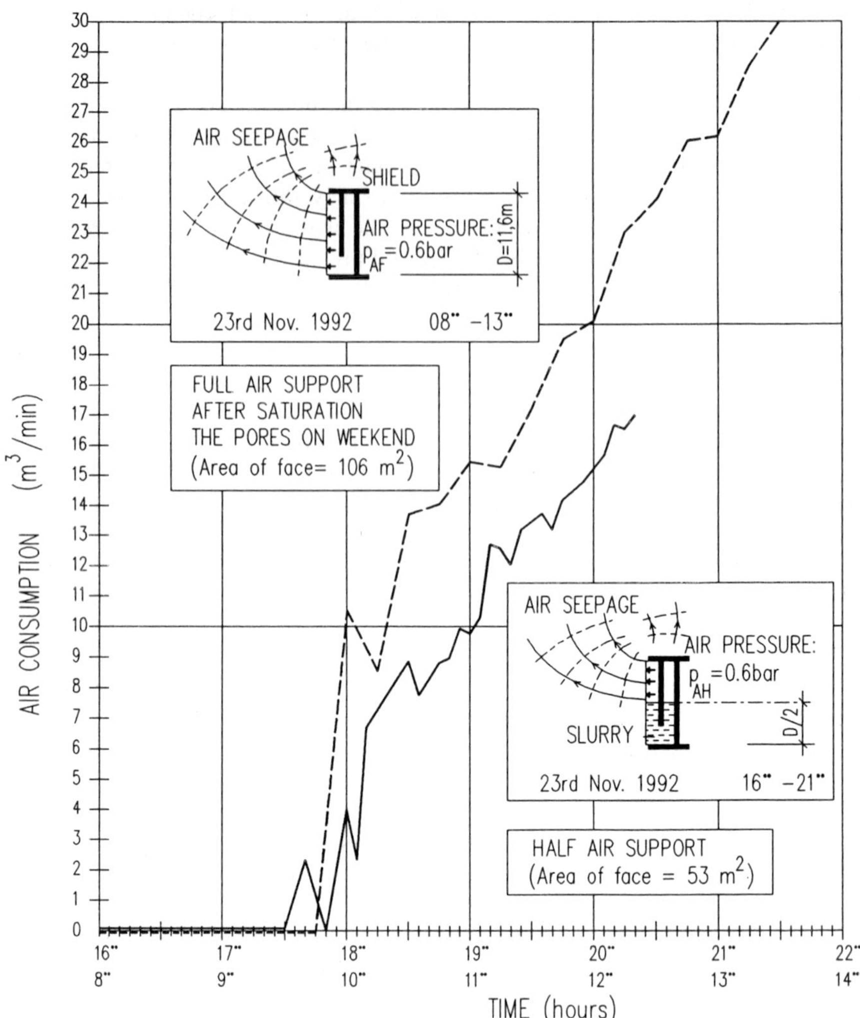

Figure 18. Air consumption during face support with compressed air

Figure 18. it is also interesting to recognize the relatively long " no air loss " periods; in both cases approx. 1.5 hour, directly after entering the working chamber. This phenomenon could be explained with a consolidation process within the cake (two phase or saturated condition) followed by an successive emptying of pore channels from water (three phase condition).

7. Conclusions

The ground conditions at the Grauholz Tunnel were quite demanding due to the heterogeneous ground conditions. The great variety of ground conditions on the same project allowed insight and direct comparison in the functioning of the Mix-Shield in particular when operating as a slurry shield.
The double chamber slurry shield greatly facilitates maintenance. With the slurry shield face support can be actively and positively controlled. Slurry pressure must balance water and earthpressure. Design procedures for determining support pressures were established and verified. In order to prevent loss of air pressure special slurry mixes and entry procedurs were developed.
The Mix-Shield, with the lessons learned at the Grauholz Tunnel, has developed to a method for tackling complex ground conditions, mixed-face and rock conditions. However, the importance of well-trained experienced personnel with the necessary understanding of geotechnical factors should not be overlooked.

Acknowledgements

The authors wish to thank for the kindly permission of the Swiss Federal Railroads to publish this paper. Very special thank to the staff of the joint venture, especially to the engineers Mr. Aebersold, Mr. Sträßer and Mr. Bönsch for their cooperation and useful discussions.

References

1. Steiner, W. (1988) : **The Grauholz high-speed railway tunnel in difficult ground.**
   Proceedings of Int. Conf. on Tunnels and Water, Madrid. Balkema, Rotterdam, Netherlands, Vol. 3, pp. 1381 -1385
2. Steiner, W. and Becker, C. (1991) : **Grauholz tunnel in Switzerland : large mixed-face slurry shield.**
   Proceedings of RETC, Seattle, WA., pp. 329 - 347
3. Steiner, W. (1993) : **Experience with an 11.6 m diameter Mix - Shield : The importance of the ground machine interface.**
   Proceedings of RETC, Boston, MA., pp. 759 - 779
4. Jancsecz, S. (1987) : **Three dimensional calculation of required support pressures for the Hydroshield tunnelling in Berlin U-8, Lot D 79.** (in German) Report submitted to the Senate of City Berlin, FRG. 27th. April. 1987

5. Jancsecz, S. (1988 - 1992) : **Support pressure calculations for Mix-Shields** (all in German)
Duisburg (D = 6.52 m), City Rail TA6, Lot 22N. 1988 -1989. Reportss ubmitted to the City Duisburg.
Berne, Grauholz Tunnel (D = 11.6 m).1990 - 1993. Reports submitted to Swiss Federal Railway.
Cologne (D = 6.52 m), City Rail, Lot M1. 1992.Reports submitted to the City Cologne.
Essen (D = 8.33 m), City Rail, Lot 34. 1992. Reports submitted to the city Essen.
Düsseldorf (D = 5.55 m), Main Sewer, Lot 1. 1992. Reports submitted to the City Düsseldorf
6. Jancsecz, S. (1991) : **Definition geotechnischer Parameter für den Einsatz von Schildvortriebsmaschinen mit suspensionsgestützter Ortsbrust.** Tunnelbau - Neue Chancen aus europäischen Impulsen. Vorträge der STUVA - Tagung in Düsseldorf, F. + P. 34
7. Kezdi, A. (1962) : **Erddrucktheorien.**
Springer - Verlag, Berlin / Göttingen / Heidelberg
8. Prater, E. G. (1973) : **Die Gewölbewirkung der Schlitzwände.** Der Bauingenieur, 48, pp. 125 - 131
9. Jancsecz, S. (1993) : **Windows / Mathcad* based computer program for support pressure calculation.** Not published.
*Windows is a trade mark from Microsoft Corp., USA
Mathcad is a trade mark from MathSoft. Inc. Cambridge, MA, USA
10. Atkinson, J. H. and Potts, D. M. (1977) : **Stability of shallow circular tunnel in cohesionless soil.** Geotechnique 27, No. 2, pp. 203 -215
11. Ziems, J. (1968) : **Beitrag zur Kontakterosion nichtbindiger Erdstoffe.** Dissertation, TU Dresden
12. Kenney, T. C., Chahal, R., Chiu, E., Ofoegbu, G. I., Omange, G. N., Ume, C. A. (1985) : **Controlling constriction sizes of granular filters.**
Canadian Geotechnical Journal, Vol. 22, pp. 32 - 43
13. DIN 4126 (1986) **Ortbeton - Schlitzwände. Konstruktion und Ausführung**
14. Müller- Kirchenbauer, H. (1972) : **Stability of slurry trenches.**
Proc. of 5th Europ. Conf. SMFE Madrid, Vol. I.,pp. 543-553
15. Puntke, W. and Wichmann, W.-R. (1993) : **Laboratory tests on differently composed suspensions for Grauholztunnel. Lot West.** (in German)
Report (No. 599) submitted to Swiss Federal Railway

Session 10: Design

Tunnel ring design, development, testing and manufacture for the London Water Ring Main

J. P. Farrow
Thames Water Utilities, Ltd., Reading, England
P. M. Claye
Thames Water Utilities, Ltd., Reading, England
R. B. Warren
Taylor Woodrow Construction, Ltd., London, England
M. A. Ginary
Taylor Woodrow Construction, Ltd., London, England
R. Matthews
Taylor Woodrow Construction, Ltd., London, England

The London Water Ring Main project includes 80 km of tunnels, linking the water treatment works within the Thames and Lee valleys. To improve tunnel construction time, Thames Water Utilities Ltd (TWUL) developed a new type of expanded wedgeblock concrete lining, for the second phase of the project, with new tunnel boring machines. Taylor Woodrow Construction (TWC) undertook in conjunction with TWUL to design, develop and load-test four types of segmental rings, subsequently followed by manufacture of 275,000 segments.

Design calculations were prepared using 'standard' empirical formulae, then optimised for various parameters. Detailed finite element analysis work was carried out in parallel, investigating visco-elastic as well as elastic behaviour of the clay, including creep related long-term behaviour, and also considering time-related movements of the ground due to the excavation and construction process.

Each ring type was subjected to geometry/build tests, and load tests: on individual segments and complete rings, using specially designed test rigs, incorporating heavy prestressed concrete to provide the necessary reaction for applied loadings. The complete-ring testing rig is capable of up to 8,800 tonnes over a wide range of loading configurations and is believed to be unique in Europe and the USA.
Keywords: Water supply, tunnel, concrete, segment, design, finite element analysis, testing, London Water Ring Main.

1 Background to the London Water Ring Main Project

Thames Water Utilities (TWUL) supplies an average of 2100 million litres of treated water every day to Customers in London. Until recently, most of this had to be pumped at high pressures over considerable distances from the treatment works in the Thames and Lee Valleys, through a series of ageing cast iron and steel trunk mains; an energy-intensive system, prone to leakage and periodic major bursts. With consumption continuing to rise, the decision was taken in 1985 to adopt a new, revolutionary strategy: to construct a tunnel ring main deep beneath the Capital [1]. This would allow water to flow under gravity to a series of new underground pumping stations, to be constructed at strategic locations, to lift the water direct into local distribution zones (Fig 1).

The Ring Main strategy therefore avoided the need to excavate many miles of new large-diameter pipe trenches through congested London streets, and the major disruption which would have been caused. The tunnel is 80 km long, at an average

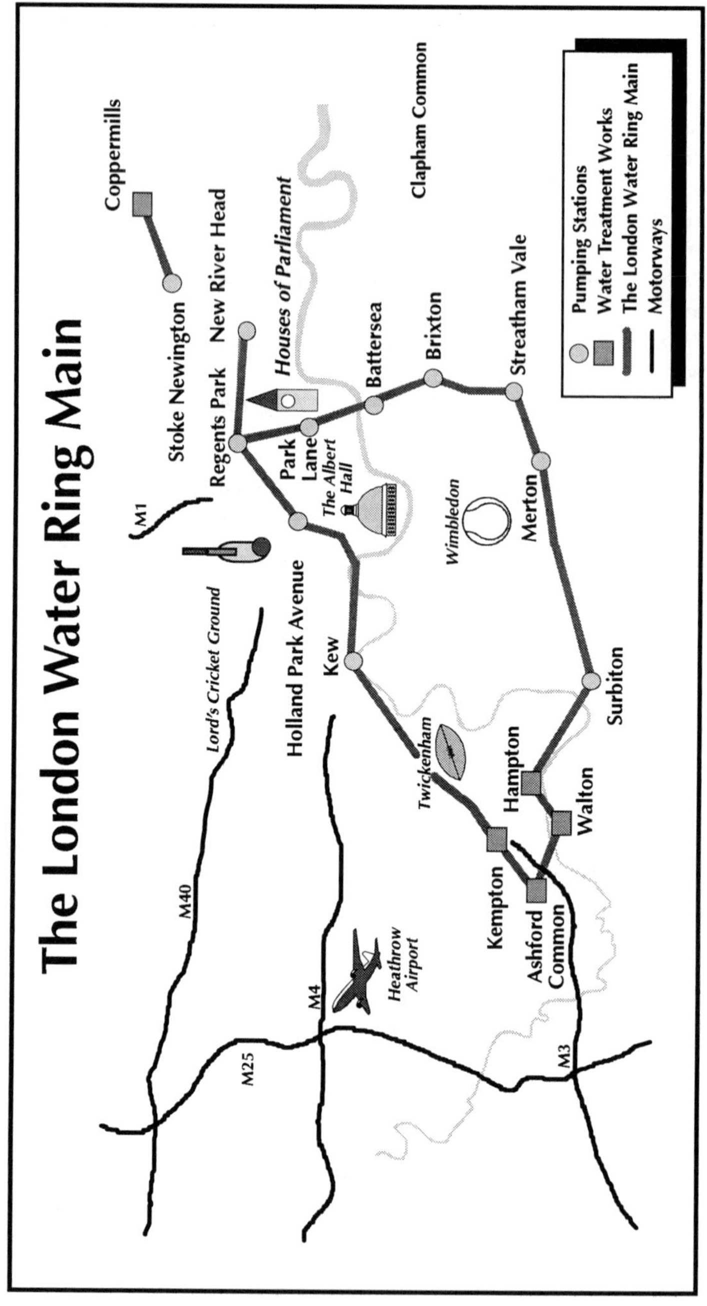

Fig. 1 London Water Ring Main

depth of 45 metres, and passes beneath most existing transportation, electricity and telecommunications tunnels. The Ring Main supplies over 1300 million litres of treated water every day, over half of Londons' needs, allowing the old trunk mains to operate at lower pressures, thereby prolonging their useful life.

2 Programme and construction phases

2.1 Phase 1

Design commenced in 1985, with the target that the whole £250 M London Water Ring Main project should be complete and fully operational in 1996. The project was split into two phases, initially concentrating on south/ central/ north-east London (Phase 1), with the aim of bringing these tunnels and pumping stations into use as soon as possible. Mechanical, electrical, instrumentation and control equipment installations would follow closely behind civil engineering construction.

The geology of the Phase 1 tunnel route essentially comprised made ground, overlying London Clay, Woolwich and Reading Beds, Thanet Sand and Chalk (Fig 2).

Some 60% of Phase 1 was constructed using the "wedgeblock" system in London Clay, one of the most efficient tunnelling systems ever developed. It comprised an expanded concrete segmental lining, incorporating eleven segments plus a tapered key (Fig 3), originally developed by Thames Waters' predecessors in the 1950's. Tunnels in strata other than London Clay were mainly constructed using bolted lining systems, followed by full secondary reinforced concrete linings. The exception was a 1.5 km length from Coppermills (Walthamstow) towards Stoke Newington, constructed by pipe-jacking.

Construction commenced in 1986 and the final Phase 1 pumping station, at New River Head (Islington) was commissioned in 1993. The original wedgeblock segments had evolved as unreinforced concrete 720mm x 685mm x 140mm thick, essentially sized to suit mainly manual handling/ erection systems. Several machines, mainly open faced tunnel shields, usually incorporating back-acter excavators were used, achieving average production rates of approximately 115 metres per week in London Clay. However, two serious problems, due to ground conditions, occured at Pimlico and at Tooting Bec. Compressed air and freezing techniques were necessary and ultimately, a new Earth Pressure Balance Machine (EPBM) had to be purchased, to complete a 1.5km length of tunnel beneath Tooting Bec Common.

2.2 Phase 2

Towards the end of 1990, before embarking on Phase 2, the entire system of tunnelling and methods of approach to construction were reviewed, with the following objectives:

- to increase safety.
- to minimise risks.
- to increase potential productivity and thereby reduce overall costs, and ensure completion to programme.

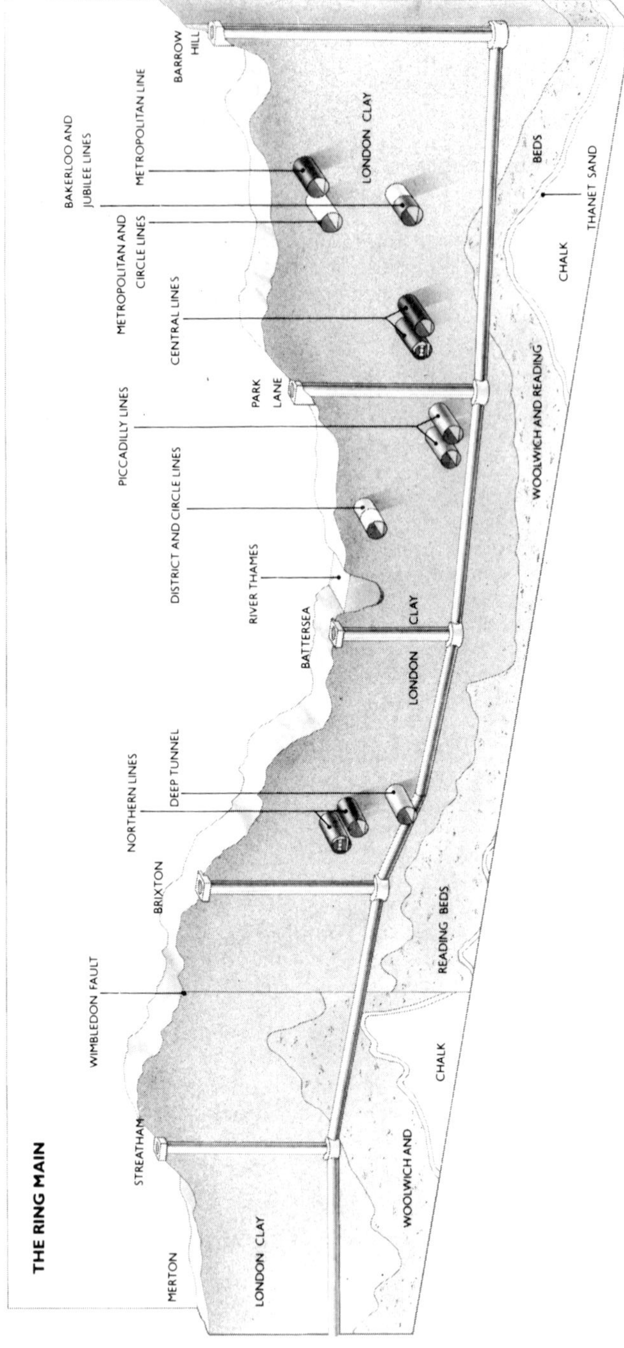

Fig. 2 The Ring Main through south and central London

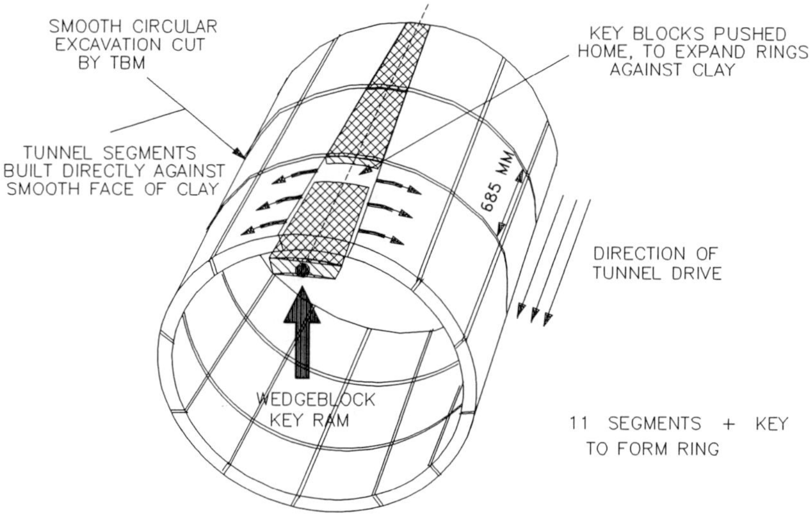

Fig. 3 100 inch wedgeblock tunnel system

The main points of TWUL's new strategy for achieving these objectives were:

- the purchase of three new dual-mode tunnel boring machines (TBM's).
- the design and manufacture of new, larger tunnel segments.
- the adoption of a new, more appropriate form of contract for construction.
- review/re-design of the entire system of construction and handling from site entrance to tunnel face.

All of the above were necessary in order to achieve the Phase 2 objectives. The new TBMs and comprehensive handling/erection systems are described elsewhere [2]. The new contract strategy adopted the I.Chem.E 'Green Book' form [3] and incorporated TWUL's Target Cost mechanism. The TBM's, tunnel segments and proposed methods of working were developed by close teamwork, involving:

- TWUL Engineering.
- Three main tunnelling contractors {a}.
- Tunnel Boring Machine manufacturer {b}.
- Taywood Engineering Ltd, for detail design and testing of tunnel segments.

Design commenced in December 1990; the Phase 2 programme required tunnelling to start in August 1991 and a total of 275,000 segments of various types were needed. Ground investigation revealed London Clay for the whole route of the tunnel around west/north London. Wedgeblock type segments would therefore mainly be used, but alternative bolted segments were required in case unexpected ground conditions were to be encountered.

The old Phase 1 wedgeblock segments were small with many joints, which minimised bending moments and stresses in the unreinforced concrete sections. The new design required the basic simplicity of the wedgeblock system to be retained with simple joints and use of unreinforced concrete for economy, but with segments as large as possible to minimise handling and to speed-up erection times. To take advantage of the most up-to-date technology, TWUL appointed a consultancy with the capability to design and comprehensively test a complete new range of interchangeable tunnel rings. The wedgeblock rings were desired to be 1000m long, with the minimum possible number of segments per ring. The alternative bolted rings were required to be geometrically similar, but suitable for erection within the tail of the TBM.

Two basic sizes of tunnel rings were required: 2.5m and 3.0m nominal internal diameter, to suit the hydraulic requirements for Phase 2. The exact dimensions were designed in conjunction with the TBMs, to the following specified methodology:

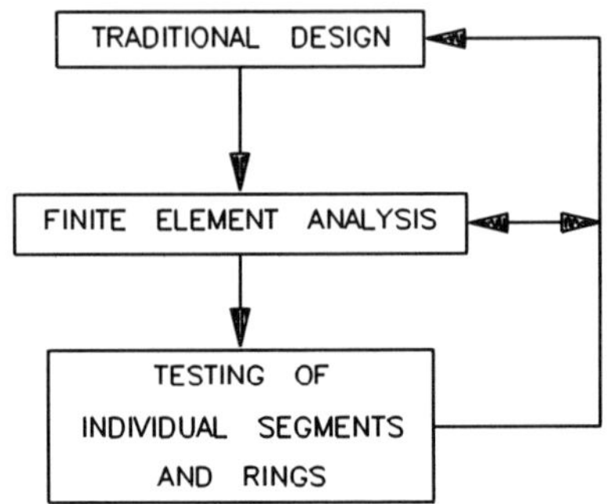

Fig. 4 Design, development and testing flowchart

The Design and Testing Contract was awarded to Taywood Engineering Ltd in January 1991 and subsequently a contract to manufacture 275,000 segments was won by Taylor Woodrow Civil Engineering Ltd, in March 1991. Their proposal included the purchase of a major part of the facility which had been used for manufacture of segments for the Channel Tunnel, on the Isle of Grain in Kent. This allowed the large number of newly designed segments to be manufactured in a short time-scale. The plant was moved to Southall, Middlesex and became operational in June 1991.

The new tunnelling system achieved a long term average production rate of 240 metres per week, over 33km of tunnelling. In the most productive week, 501 metres of tunnel was constructed during ten normal working shifts.

3 Design of segmental wedgeblock linings

3.1 Design approach

The development of the lining was governed primarily by the requirement to match the newly ordered tunnel boring machines and to improve the installation process leading to overall economies in programme and costs. The tunnel was to be driven mainly through overconsolidated London Clay and the lining was required to be of a single pass wedgeblock system with all materials suitable for carrying treated water. The principal requirements for the lining design were therefore:-

- installation behind a TBM in "open mode".
- expanded directly against the excavated ground.
- least number of segments.
- segments to be of unreinforced concrete.
- length of segments to be the maximum compatible with the TBM design.

At approaches to pumping and access shafts or in blocky ground, the expanded linings would be replaced by bolted linings. If water or bad ground was encountered the TBM would be converted to "closed mode" and a smoothbore lining incorporating a bolted system with an insitu concrete secondary lining would be employed.

3.2 Ring Parameters

In order to minimise the construction period, it was determined the ring would be built using the minimum number of segments. However, this requirement had to be balanced against a number of other major issues as follows:-

- reduced number of segments in a ring leads to reduced articulation and higher bending moments induced under operational loadings.
- handling and storage capacity of the precasting facilities.
- transportation of the segments.
- capacity of the erector equipment on the TBM.

The last three considerations above are influenced by the length of the ring adopted and the thickness of the segments. The ring length must be as long as possible to achieve a high speed of tunnelling. It is, however, limited by the extent of the bored clay which can be left exposed as well as the extent of the ram travel of the TBM thrusting jacks. The lining thickness is governed by structural considerations.

Tunnels of two alternative diameters were required.

- 2590 mm dia. - 16 Kms tunnel

- 2910 mm dia. - 17 Kms tunnel

For both sizes of tunnel a ring length of 1 metre was adopted. Both tunnels are formed using seven segments and a wedge shaped tapered key (Figure 5). The odd number of segments are installed symmetrically about the vertical axis with the wedgeblock located at the crown of the tunnel. The tapered wedgeblock is slightly shorter than the normal ring length and is installed between the upper two segments, each tapered on one face.

The wedge is jacked into position using the jacking ram on the TBM. This allows the lining to be expanded to the circumference of the bore.

3.3 Design Optimisation

The determination of lining thickness is dependent upon a number of parameters including the configuration of the tunnel ring, the applied loading and material properties. The major variables considered were:-

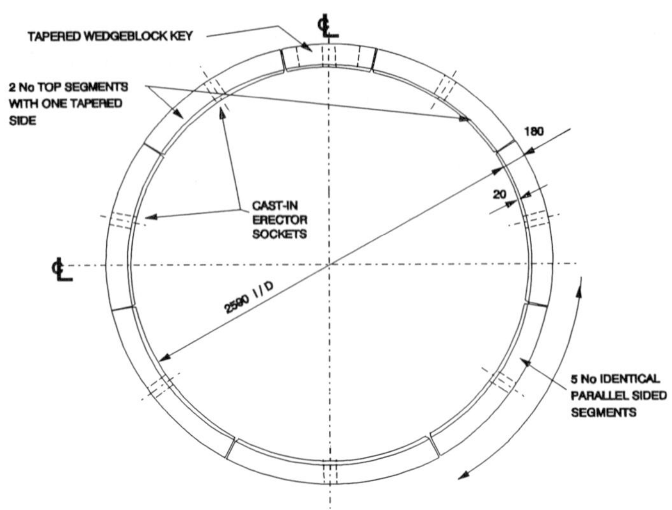

Fig. 5 Segmental wedge block tunnel lining

- number of segments in the ring.
- shape of the radial joint faces of the segments.
- strength and properties of the concrete.
- properties of the surrounding ground.
- depth of the tunnel and the applied loading.

Optimisation studies investigated these variables in order to determine the thickness to be adopted for each of the tunnel diameters. This information was needed early for the design of the new T.B.M.'s.

These studies were based upon the use of the classical empirical methods as published by A M Muir-Wood [4] and D J Curtis [5].

3.4 Number of Segments

The original tunnel for the Phase 1 sections of the Ring Main contained 11 segments. In order to achieve the primary aim of improving the construction programme the selection was based on the use of 5 or 7 segments in a ring together with a key. Considerations of design, handling and erection led to the final selection of 7 segments as the optimum.

3.5 Concrete

The selection of the concrete for the lining was primarily governed by the following considerations:-

- durability for a 150 year life.
- high strength to minimise section size and for load transfer at joints.
- suitable tensile capacity to allow use without reinforcement.
- dense medium for low permeability

The outcome of these studies was that a designed mix of grade 60 concrete containing granite aggregate was selected.

3.6 Geometry of Joints

In order to achieve economies in the precasting process it was determined at an early stage that the radial joints would be principally flat faced. However, to avoid local crushing of the corners, chamfers were essential. A smaller face dimension would aid the articulation of the joint but coincidentally the stress concentrations would be increased. A balance had to be sought.

Optimisation studies including some preliminary finite element analysis subsequently substantiated by joint tests led to the adoption generally of a 30mm chamfer (Figure 6). The chamfer on the intrados was shaped to form a caulking/ pointing groove.

A typical circumferential joint is also shown in Figure 6.

3.7 Ground Parameters

The tunnels under consideration were for construction mainly in overconsolidated London Clay. Geotechnical data was available from ground investigations along the route and this was supplemented by generic published information. The major parameters affecting the design of a tunnel are the elastic modulus (E) and the lateral pressure coefficient (Ko).

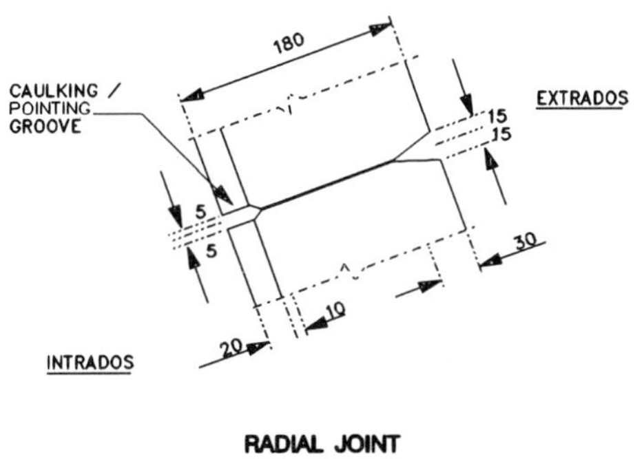

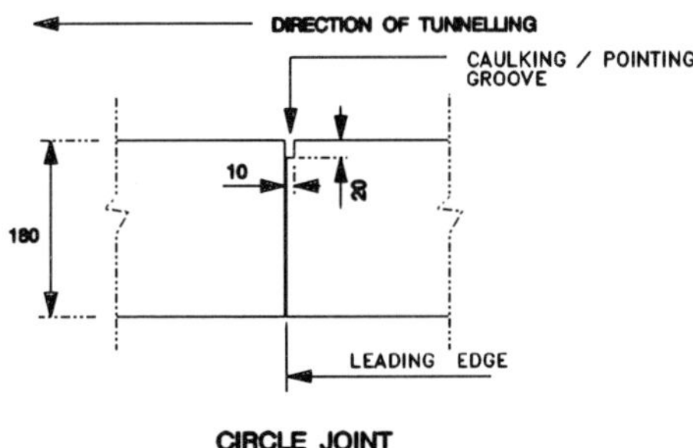

Fig. 6 Segmental wedgeblock joint details

The elastic modulus of clay is related to its undrained shear strength (Cu) and there exist empirical formulae relating these two soil parameters. Furthermore, the undrained shear strength of London Clay is known to vary with depth and there is considerable published data on this variation. A combination of these empirical rules was adopted to establish sufficient data to optimise the relationship of the lining thickness with the installation depth of the tunnels.

A further important parameter of the soil affecting the design is the Ko factor which gives the relationship of the insitu horizontal pressure in the ground to the imposed vertical overburden pressure. Ko for London Clay is known to reduce with depth, with a value at ground level of about 3.0 reducing to 1.0 at significant depths. A value of 1.75 was adopted for the design of the Ring Main for the operating conditions. A conservative condition with Ko of 2.0 was also investigated. However, this was applied in combination with appropriate reduced safety factors.

3.8 Optimum Sizes

Optimisation investigations taking into consideration the variable parameters led to the adoption of the following thicknesses for the unreinforced sections:

Tunnel Internal Diameter (mm)	Thickness (mm)
2590	180
2910	200

These optimised diameters were finalised and the sizes used for all further TBM designs.

The investigations also led to the conclusion that the unreinforced precast segments could be used to a depth of 55m. Beyond this depth the linings needed to be reinforced or manufactured with a higher grade concrete.

3.9 Finite Element Analysis

The objective of the finite element computer analysis was to allow investigations to be concentrated on:-

- confirming the optimised design.
- refining the most critical parts of the designed lining in accordance with the findings of the analytical work.
- confirming laboratory test results.

Two computer programs were utilised to achieve the above objectives. The program used for the ground/lining interaction analysis was ADINA developed by ADINA R & D of Boston, USA. The model for the analysis was developed using a mesh of the ground and the tunnel with the boundaries of the area sufficiently distant from the zone of interest.

A typical mesh is shown in Figure 7 for a plane strain analysis.

A variety of analyses were carried out, initially to develop the model and then to investigate the following:-

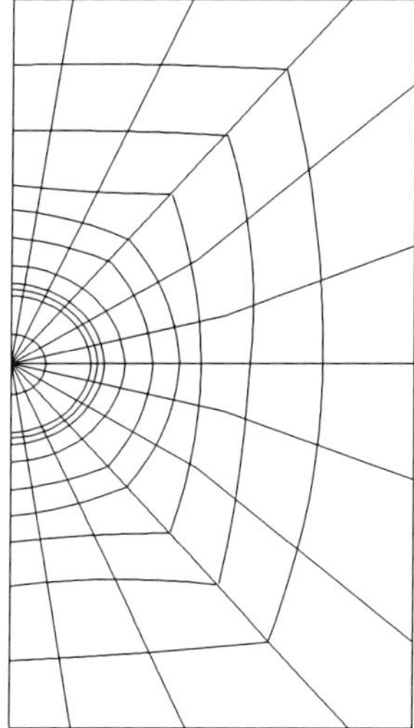

Fig. 7 Typical plane strain mesh

- sensitivity of the interaction of the ground and lining under varying lining stiffnesses, representing degrees of articulation of the segments.
- effect of non-linearity of lining material.
- the effects of the construction process.
- long term creep/visco-elastic behaviour of the clay.

The construction process requires representation of the lining installed at a certain distance behind the cutting edge of the TBM. Under these circumstances the stresses in the ground are affected in a three dimensional manner and a plain strain model would not be representative. Hence, an alternative axisymmetric model was developed to represent this complex three dimensional behaviour. The analysis using this model hinged upon the "birth and death" options available in ADINA. The results of these finite element analyses confirmed the design method adopted.

The PAFEC finite element package was used to investigate induced stresses in the segments due to loading applied by the rams during forward movement of the TBM. A typical ram layout is shown in Figure 8. Based on the results of this analysis recommendations for sizing of ram shoes were put forward.

An analysis was also carried out to investigate the wedgeblock key during its installation. Boundary conditions were introduced in this analysis to represent restraint caused by adjacent segments.

4.0 Design of bolted linings

4.1 Design Specification

Fully reinforced bolted wedgeblock linings were required for use in the following situations:

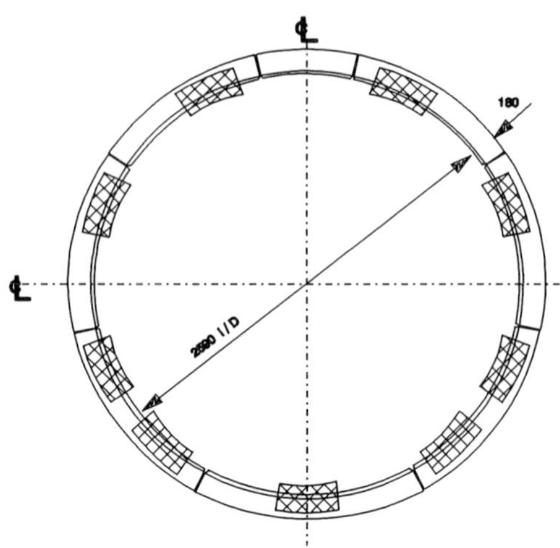

Fig. 8 Shove ram locations

- adjacent to pumping shafts as junction pieces between the shaft and the expanded lining, where the ground was disturbed due to shaft construction.
- through blocky clay where the effectiveness and security of normal expanded wedgeblock linings could not be achieved.

In addition, fully reinforced bolted smoothbore linings were designed, to be used compositely with an inner insitu reinforced secondary concrete lining, where the tunnels passed through poor or water-logged ground. This occured mainly through the Woolwich and Reading Beds and the Thanet Sands. The interspace between the excavated face and the lining extrados would be pressure grouted. The internal secondary lining was designed to the water retaining code, to withstand the full ring tension developed by the internal water pressure.

4.2 Ring Geometry

Both bolted wedgeblock and smoothbore rings consist of seven segments with a key as shown in Figure 9. The segments and the key are bolted together at the radial joints and adjacent rings bolted at the circumferencial joints. The rings are designed to be self- supporting under external loading and the bolts mainly provide stability during construction prior to grouting. Hence, the design of the linings was based on the same principles as the expanded lining using segments of similar thicknesses to the expanded type.

The bolted smoothbore rings varied from the wedgeblock and bolted wedgeblock in having a narrower parallel sided key, to facilitate building inside the tail of the TBM, in poor ground.

Bolted joints, however, produce some restraint and tend to set up bursting stresses. Hence, reinforcement was provided in these segments to reduce the possibility of cracking.

All joints were provided with shaped grooves to receive hydrophilic joint sealer.

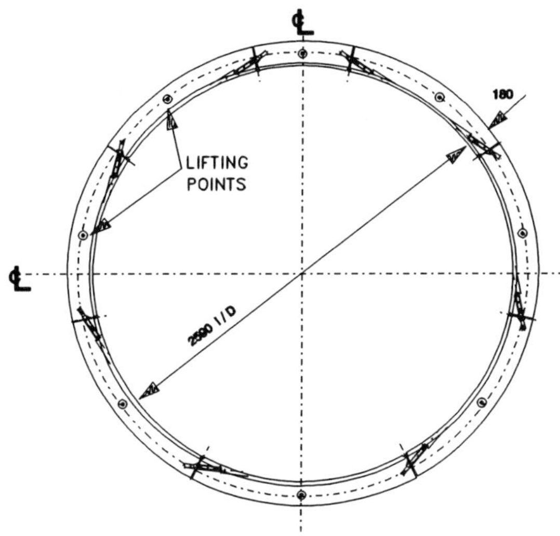

Fig. 9 Bolted tunnel linings

Provision was also made in each segment for grouting behind the extrados.

The bolting system for both radial and circumferential joints was developed with inclined bolts to avoid the need for flanged connections. A typical bolting assembly is shown in Figure 10.

5.0 Design of high strength wedgeblock segments

The original design of the wedgeblock lining had shown the unreinforced segments could be used only to a depth of 55m and that reinforcement would be required beyond this depth. The London Water Ring Main tunnels were required to a maximum depth of 70m.

Considerations of long term durability of tunnel segments and of the significant cost of providing reinforcement inevitably led to the conclusion that wherever possible unreinforced segments should be adopted. Some further development work was undertaken to examine the feasibility of using higher strength mix designs for producing unreinforced segments for use at greater depths.

A design was developed which could be introduced with minimal or no disruption to the existing precast production line. It was established that the unreinforced segments could be used to a depth of at least 70m using concrete with a characteristic strength of 85N/mm2.

A number of trial mixes were prepared at the Taywood Engineering laboratories and a successful micro-silica mix was developed to meet the requirements.

Approximately 1.4km of tunnel between 55m and 70m deep was constructed using these segments.

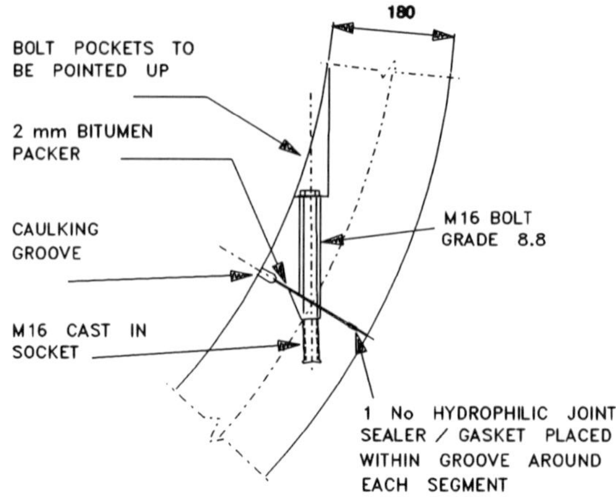

Fig. 10 Radial joint detail for bolted segments

6.0 Testing of segmental linings

6.1 General

A programme of testing on both reinforced and unreinforced rings was under-taken to support the design development. The following tests were conducted for each ring type considered:-

a) tests to check the geometry/build
b) load tests on individual segments
c) load tests on complete rings

Prior to commencement of the test programme, it was necessary to develop and construct the unique facilities required to undertake the tests. The very tight tunnelling programme required that the testing be completed within 15 weeks from award of contract. The scale of the testing was such that the development of the test reaction frames formed a major part of the testing programme.

The reaction frames for the load tests in b) and c) incorporated the use of heavy pre-stressed concrete structures to provide the necessary reaction for the loading applied during testing. Hydraulic jacks were used to simulate both uniform as well as differential loading conditions. A total of 6 jacks on two levels were provided for loading individual segments and 44 jacks on two levels for complete rings, as shown in Figures 11 and 12.

Deflections of segments during the tests and the loading applied through the jacks were monitored through a computerised data logging system.

6.2 Geometry/Build Tests

The objective of this test was to check the geometry and build of a complete ring. The segment circularity, segment thickness and radiality of mating faces were checked using a purpose built test frame which provided a level base, external circumferential restraint and a means of driving in the wedge segment.

The tests demonstrated that the segments used for the test programme fully met the specified tolerances.

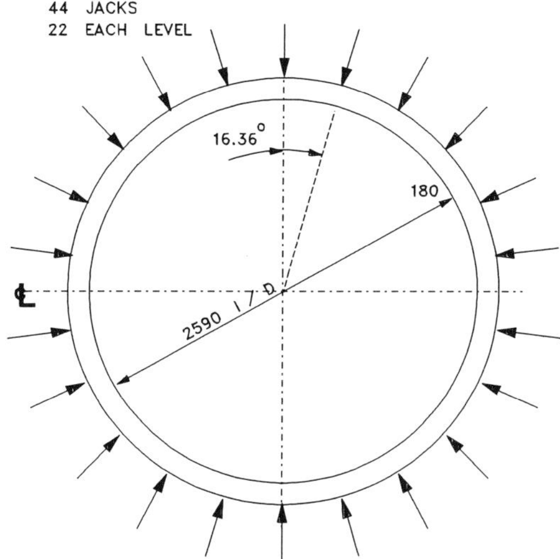

Fig. 11 Jack locatations - tests on complete rings

6.3 Load Tests on Individual Segments

The objective of this test was to examine the response of individual segments to loading up to 1.2 x the design load for a range of Ko values to failure or until a load factor of 2.0 (at a Ko of 1.75) was reached.

The load regime equivalent to the imposed loading on the tunnel installed in the ground yielded four unique segment loadings. Three segments were tested (for each lining type) subject to the three most severe segment loadings. A fourth segment was tested to simulate a "lipping" condition that can occur for misaligned segments in a ring and subject to the most severe segment loading.

These tests were undertaken in the test frame shown in Figures 13 & 14. The load was applied using 6 hydraulic rams installed on two levels. In an attempt to model the segment behaviour underground, they were supported on a plane surface at one end and a roller provided at the other end to simulate rotation under load.

In all cases, the segments withstood a load factor of 2.0 without any significant indication of distress.

6.4 Load Tests on Complete Rings

These tests were undertaken in the test frame shown in Figures 15 & 16. Complete rings were subjected to a load regime equivalent to the design loading on the tunnel in the ground.

The equivalent radial loads applied to the segments at the jack positions were based on the bending moments and constant axial thrust in the ring as obtained from the design process. These could not fully include the variable component of thrust which is due to the radial as well as tangential effects of the ground/lining interaction. However, by application of varying radial loads in the jacks around the ring circumference, some of the variable thrust effect was included in the test. This limitation led to conservative results.

The application of the varying radial loads was controlled by means of a

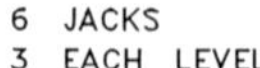

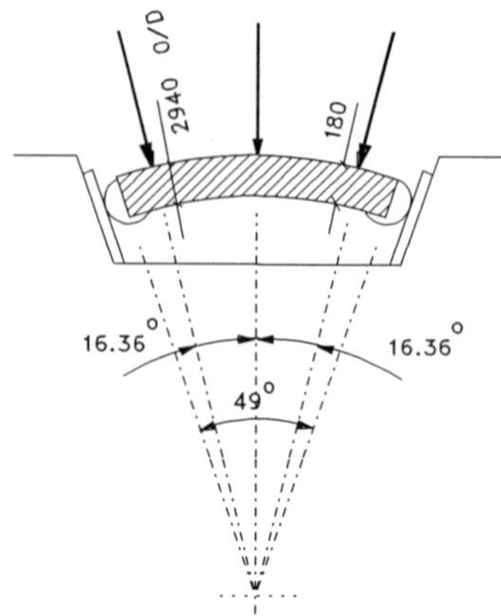

Fig. 12 Jack locations - tests on individual segments

micro-computer linked to a data logger with the load applied incrementally on opposite pairs of jacks until the design loading pattern had been reached. The loading was then increased proportionately until failure occurred.

Segment Test Results - Factors of Safety

Ko FACTOR	DESIGN	FE ANALYSIS	TEST
1.75	2.1	1.73 to 2.26	1.8 to 2.0
2.0	1.26	1.33 to 1.80	1.55 to 1.7

All rings developed circumferential cracking on the intrados at the level of the jacks. It was considered that this was a result of the discrete nature of the jack loads and would not occur under ground loads.

All unreinforced rings failed due to excessive longitudinal tensile strains. It was possible that this failure mode was encouraged by the initial horizontal cracking already mentioned. Failure could only occur in this way because the single ring test specimens were not restrained longitudinally as the real tunnel lining would be. In this case, the test results represent a lower bound on the factors of safety.

- In all cases, some sliding occurred at the joints.
- The reinforced rings tended to fail at the joints.
- The test results for all rings exhibited deformations within satisfactory limits.

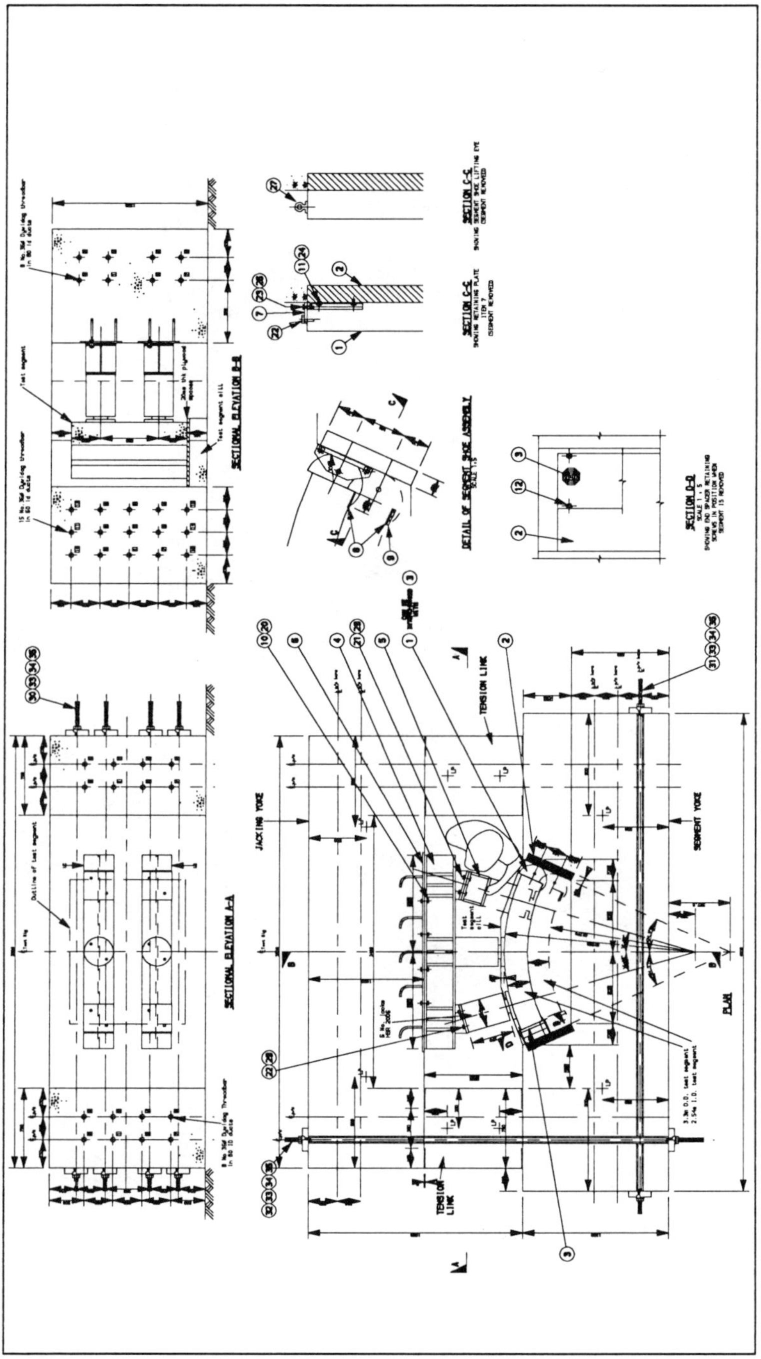

Fig. 13 General arrangement - individual segment test rig

Fig. 14 Individual segment under test

Fig. 15 Complete ring under test

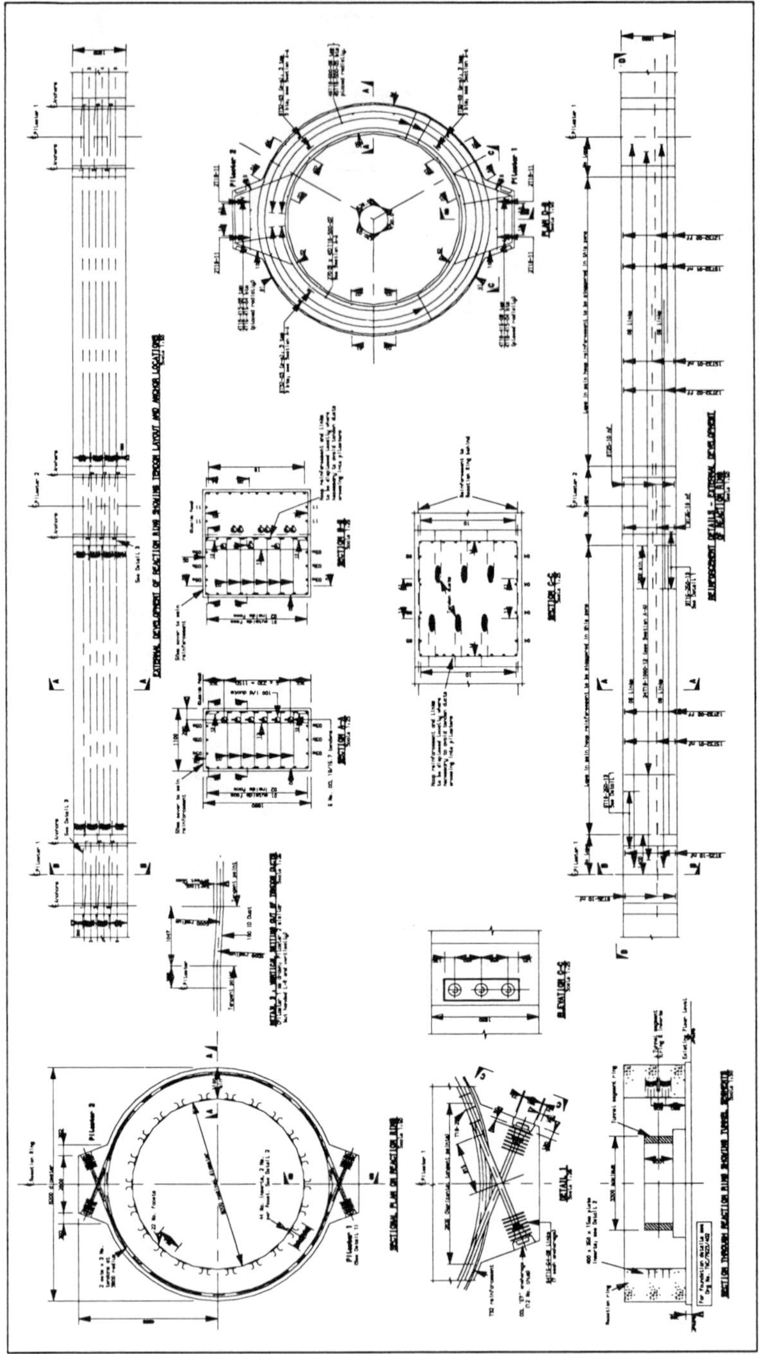

Fig. 16 General arrangement - complete ring test rig

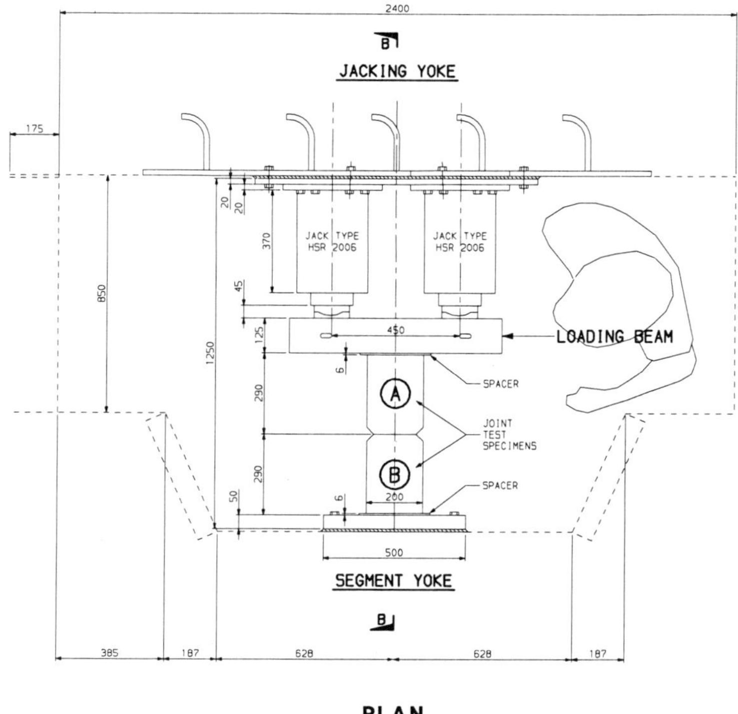

PLAN

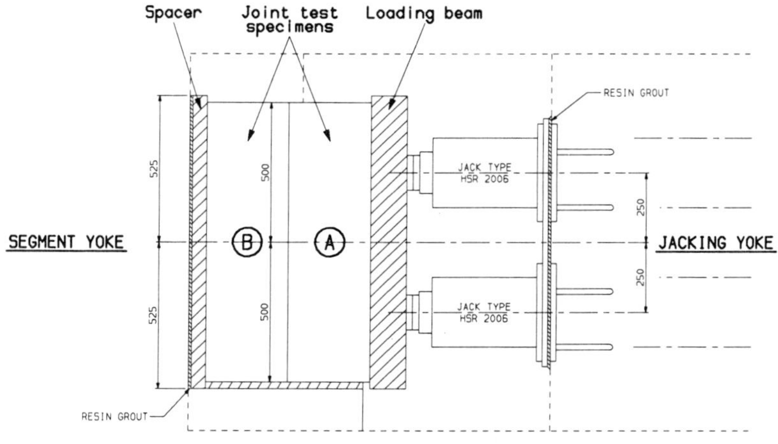

SECTIONAL ELEVATION B-B

Fig. 17 Individual segment test frame

6.5 Tests on Plane Joints to Segmental Linings

It was identified in the design analysis that a more economical design could be achieved by increasing the size of the chamfers at the radial joints. This would make the lining more flexible and lead to a reduction in maximum bending moments in the lining.

However, increasing the chamfer size also reduces the bearing area, although the line of action of the thrust is forced closer to the section centroid. It was felt prudent that the effect of the increase in chamfer size on the bearing capacity of the joint should be investigated further using finite element analysis and demonstrated practically through a test programme.

A set of short, straight, unreinforced (to give a lower bound on bearing capacity) segments were cast especially for this purpose.

They were subjected to a known load and eccentricity in a modified form of the Individual Segment Test Frame described previously (see Figure 17). Segments having a 50mm chamfer were also included, in addition to those having a 10 and 30mm chamfer, to gain as much information as possible from the tests.

Known loads with known eccentricities were applied through a loading beam using four hydraulic jacks on 2 levels. Strains and deflections of the segments and the test frame were monitored during loading. The joints were loaded until failure at their design eccentricity.

The test results were correlated to the design concrete strength based on results from cubes cast with the test specimens.

Results showed that;

1. The final factors of Safety for the joint types were:

   10mm Chamfer - 2.2
   30mm Chamfer - 1.8
   50mm Chamfer - 1.4 (not a practicable design)

   A factor of Safety above 1.5 was considered to be satisfactory.

2. The average ultimate bearing stress deduced from the test results was exactly 0.9fcu which is in line with the recommendations of BS 8110. [6].

3. The test results demonstrated that the larger chamfer joints were more flexible generally as deduced from the design calculations.

4. The ultimate test results showed that the bearing capacity was found to reduce with increasing chamfer size. However, a certain chamfer size is needed to limit the bending moment generated in the segment and to avoid spalling the edges of the joints.

It was therefore concluded that the optimum chamfer size to satisfy the above requirements was 30 mm.

7.0 Segment Manufacture

7.1 Precast Site

A total of 275,000 segments were manufactured for 33km of London Water Ring Main tunnels at Taylor Woodrow Civil Engineering Limited's precast facilities at

Southall, Middlesex and Thamesport, Isle of Grain. The Southall facility was established to produce up to 500m of segmental tunnel lining per week and the Thamesport production line ran for 6 months in order to meet peak demand which exceeded 1000m per week.

Working on the principle of high volume, product standardisation and long production runs, the semi-automated production lines could cast and cure segments in 5 hours, each line producing up to 500 segments per 12 hour shift.

Two independent carousel lines were set up at the Southall factory and one at Thamesport. Each line involved a series of work stations and a steam curing tunnel. Each station was set up to carry out one of several functions as follows:-

- mould stripping
- segmental removal
- mould cleaning
- reinforcement placing (bolted rings only)
- mould reassembly
- concreting
- concrete finishing

7.2 Concrete Production

Concrete was batched on site and transported to each carousel via a conveyor belt. The plant at Southall had a total production capacity of 50m3 per hour.

The 60N/mm2 mix design for the concrete had to take into consideration the delicate balance of obtaining early age strength with temperature tolerance because of the rapid curing process in the high speed production line. The mix included 30% cement replacement by P.F.A. to reduce the heat of hydration and to give controlled exothermic reaction. This also provides sulphate resistance allowing O.P. Cement to be used. All aggregate used was Scottish Glensanda granite including crushed granite for the fines content.

Superplasticiser was included in the mix to produce a workable concrete whilst maintaining a low water cement ratio of 0.34. The concrete mix was sensitive to change and with the high volume production run, a constant quality control was instituted including records from every batch.

As mentioned earlier, for 1.4km the tunnels ran at depths exceeding 55m and a higher strength concrete was used to eliminate the need for reinforcement. This higher strength mix included micro silica to provide strengths exceeding 85N/mm2 at 28 days.

7.3 Segment Production

Upright steel moulds were used for casting the segments as opposed to the method of horizontal casting. The main reason for adopting this method was to ensure that the segment extrados surface together with the other critical surfaces were formed to the precise profile and dimensions.

After the mould had been filled and finished, it was moved to the curing tunnel, where it was subjected to a steam curing process for 4.5 to 6 hours to achieve a minimum compressive strength of 15N/mm2 before it was lifted from the mould.

Each segment had a sheradised lifting anchor cast into the circle joint on its leading edge to enable the unit to be handled from the precast factory to the storage yard and then to site. Segments were delivered to site in bundled rings ready for despatch straight to the tunnel face. The plastic socket matched to suit the handling and assembly equipment for the TBM was also cast into the intrados of each segment at its centre of gravity.

Segments were tested on a random basis, with one segment out of every 100 measured. The overall rejection rate for the segments was less than 0.2%.

8.0 Performance in construction

The decision to invest in this programme of design, development and testing for Phase 2 of the London Water Ring Main has proved once again the value of investment. However, the benefits could not have been achieved without the simulataneous decision to design the matching tunnel boring machine described in the construction paper [2]. It was essential to review all aspects of the process including design, manufacture, handling, transportation and erection of the segments.

The outcomes from the programme in respect of the segmental wedgeblock lining system were:-

- Unreinforced segments of high grade concrete.
- Improved durability.
- Segment length increased to 1000mm.
- Number of segments per ring reduced.
- Reduced number of segments handled.

The number of segments per kilometre of tunnel was reduced from 17,800 in Phase 1 to 8,000 in Phase 2. This in turn considerably reduced the length of joints in the tunnel with longitudinal joints reduced by 33% and circumferential joints by the same amount. This reduced the number of possible leakage paths from the tunnel.

However the overall objective was to develop a system which in conjunction with the new TBM's was capable of dramatically increasing the rate of tunnelling. The evidence that the objective was achieved has been proven by the number of world records for tunnelling which were achieved on Phase 2 of the London Water Ring Main.

9.0 Acknowledgements

a) Summary of major civil contracts all let under I Chem E Green Book Conditions:

AMEC Civil Engineering Ltd
LWRM Stage 4, Ashford Common to Barnes, 17Km.

J Murphy & Sons
LWRM Stage 5, Barnes to Barrow Hill, 8Km.

Costain Civil Engineering Ltd
Raw Water Tunnels, Moor Lane to Ashford Common, 8Km.

b) Tunnel Boring Machine Manufacturer:
Lovat Tunnelling Equipment Inc.

c) Steel Mould Manufacturer:
Steel Moulds Ltd.

10.0 References

1. Dickens W.J. and Bensted I.H. London Water Ring Main. **Proceedings of the Institution of Civil Engineers,** Part I, 1988, 84, June, 445-474.

2. Remington R Development and performance of tunnel boring machines on phase 2 of the London Water Ring Main. **Tunnelling '94.**

3. **The Institution of Chemical Engineers.** Model Form of Conditions of Contract for process plants. Suitable for reimbursable contracts in the United Kingdom. July 1976.

4. Muir Wood A.M.(1975). The Circular Tunnel in Elastic Ground. **Geotechnique 25,** No.1, 115-127.

5. Curtis D.J. (1976). Discussion of: The circular tunnel in elastic ground, by A.M. Muir Wood. **Geotechnique 26,** No.1, 231-237.

6. **BS 8110: Part 1 and 2: 1985,** Structural use of concrete.

Medway tunnel detailed design

H. Kruizenga
Tarmac HBM JV, Chatham, Kent, England
C. R. Weeks
Tunnel Division, Mott MacDonald, Ltd., Croydon, England

Abstract
The Tarmac HBM Joint Venture is currently constructing an immersed tube tunnel for Kent County Council across the River Medway adjacent to the Chatham Maritime development in North Kent. The dual two lane tunnel is the central part of the Medway Towns Northern Relief Road, which will provide a much needed alternative route to the congested A2 between Rochester and Strood and will improve access to commercial and industrial developments in the area.

The paper gives an overview of the project and describes the detailed design of the immersed tunnel elements, waterproofing techniques, closure of the casting basin and final jointing. The approach structures, aquifer protection and M&E services are briefly reviewed together with temporary works including marine works, casting basin and cofferdam construction. Also discussed are the various interfaces between temporary and permanent works designs, which are particular to immersed tunnel construction.
Keywords: Casting basin, design and construct contract, dilatation joints, immersed tunnel, longitudinal jointing, Medway Tunnel, ventilation, watertight concrete.

1 Introduction

The Medway Tunnel contract forms part of the Medway Towns Northern Relief Road and includes approximately 1,500 metres of dual carriageway highway of which 720 metres is in tunnel.

The contract between Kent County Council and the Tarmac/HBM Joint Venture is for design and construction with Mott MacDonald as the named Designer for all the permanent works, both Civil and Mechanical & Electrical. The Contract has no Clause 12 and requires the Contractor to bear the risks associated with adverse physical conditions and any artificial obstructions that may be encountered.

Mott MacDonald also assisted the Tarmac HBM Joint Venture in its winning alternative tender design. A comprehensive tender design was developed from the bid outline design and included innovative approaches to the tunnel ventilation system and the casting basin. The latest concrete technology was used to develop a system whereby watertightness of the tunnel would be achieved without the use of an external membrane.

While the tube section will be founded on chalk, the approaches pass through alluvial clay and sandy gravel requiring various construction techniques including tunnels, troughs and slabs supported from diaphragm walls or piles. Significant restraints on design and construction are caused by the existence of a fresh water aquifer and special measures such as a river bed sealing membrane will be employed to prevent long term contamination of this water supply. Other environmental issues involve dredging and preservation of buildings of historical value adjacent to the site.

2 Project Description - The Project

The Medway Tunnel crosses the River Medway adjacent to Whitewall Creek and Tower Hill on the west bank and Chatham Maritime and the Historic Dockyard on the east bank. See Figure 1.

Fig.1 Location Plan

Local constraints dictated the form the crossing has taken. A bridge would have had to be at a high level to accommodate the passage of the ships to the port of Rochester, and would have rendered road connections on the east bank impractical. A bored tunnel would also have caused road connection difficulties and an opening bridge would have been too disruptive to traffic on such a major by-pass route. Hence an immersed tunnel form of construction was chosen.

The portal to portal length of the tunnel is 720 metres, of which 370 metres are immersed tunnel and the remainder are the cut and cover approach tunnels. At tender two options, for a short or long east approach tunnel were required for the bid. The client chose the long tunnel option some 140 metres longer which allows more usable land area above the tunnel as part of the Chatham Maritime development zone. See Figure 2.

The road is dual 2-lane and the horizontal alignment is virtually straight for the whole length of the project, which extends from the new roundabout on Anthony's Way to the new split level interchange at Pembroke Road. This interchange is in a completely separate contract. The vertical alignment at the river is dictated by the required navigable depth of water in the river and is also set such that no part of the tunnel or its protective layer of rock comes above the existing river bed level.

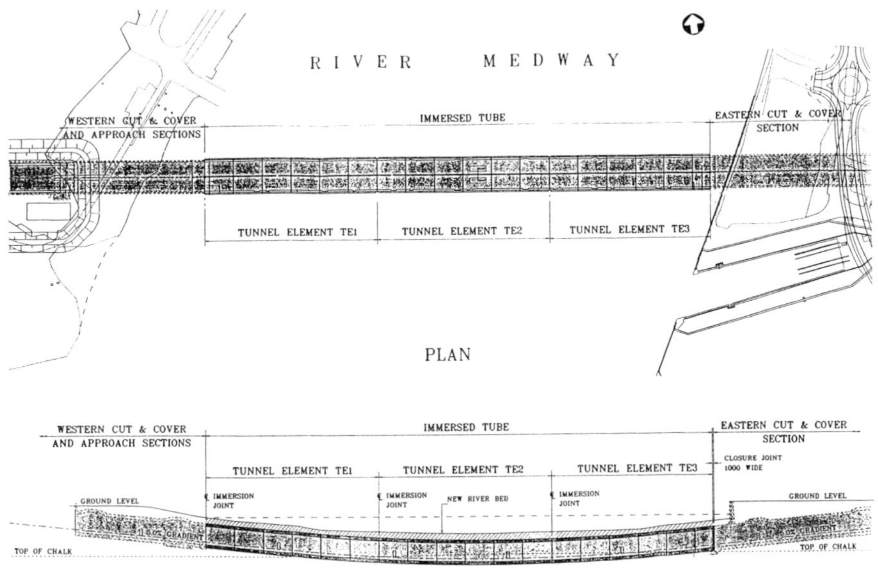

Fig. 2 Plan and Longitudinal Section

The immersed section of the tunnel project is constructed of discrete elements of tunnel structure, each in the region of 120 metres in length. These are prefabricated in the dry within a purpose built casting basin which is partially on the line of the eastern approach. The casting basin entrance is protected by cellular cofferdams. See Figure 3.

The tunnel will be provided with forced ventilation, artificial lighting, and LV power supplies with emergency facilities, safety systems and pumped drainage together with remote display and monitoring control systems. A main service building will be provided adjacent to the west portal and a secondary service building at the east portal. The tunnel also accommodates a potable water main for Southern Water.

The client is Kent County Council, and their Engineer is Travers Morgan Ltd in association with Rendel Palmer & Tritton. TM/RPT progressed the project to the outline design and also prepared the contract documents for the design and construct contract.

Tarmac/HBM JV are the Contractor, with Mott MacDonald as their named Designer for the permanent works. The contractor carried out the temporary works designs and coordinated the numerous interfaces between these and the permanent work. Tarmac Construction Engineering Services carried out the design of the land temporary works. The marine engineering is the responsibility of HBW-Engineering in Holland.

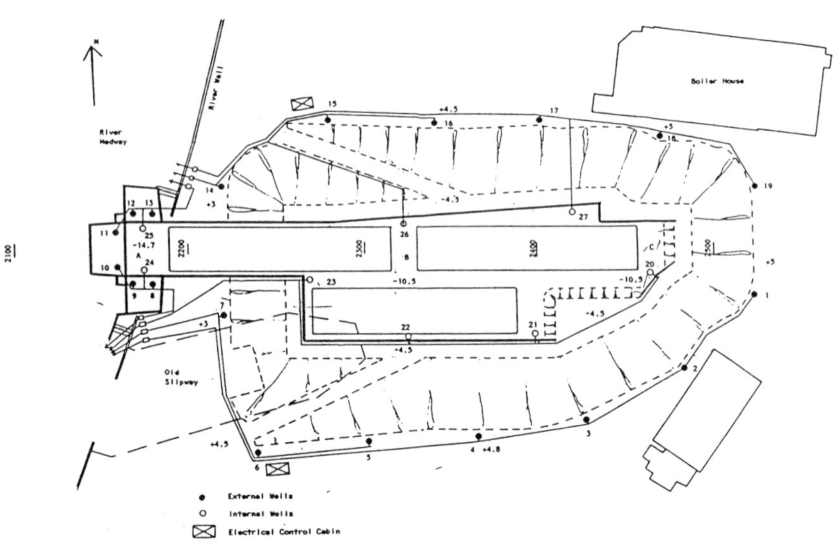

Fig.3 Casting Basin Layout

The tender design was well developed to significant detail. The proposal included a casting basin for the tunnel elements which was located partially on the line of the eastern approach. The tender proposals included a design fully conforming with the Employers Requirements and an additional design which proposed a number of cost saving options. These included forced ventilation from the ends of the tunnel in place of numerous fans throughout its length, and the deletion of the waterproof membrane around the immersed tunnel elements. The Client accepted the former proposal from the outset of the contract and has subsequently taken up the latter after in-depth consideration.

The contract documents call for the Contractor to take full responsibility for the design, and to this end a system of self certification is employed. However, the design is still subject to the approval of the Engineer. Before detailed design can be progressed, technical approval in principle has to be sought from the Engineer. On completion of the detailed design this is then submitted to the Engineer for approval, together with the appropriate design and independent check certificates, which are then countersigned by the Engineer.

During the construction Mott MacDonald have a team of site staff, who ensure that the Works are being constructed in accordance with the requirements of the approved design. On completion of sections of the work, Mott MacDonald issue construction certificates to this effect. Throughout this phase the Designers site staff work in close liaison with the Contractor's own supervisory staff and quality control teams. TM/RPT have a Resident Engineer and a team of engineers resident on site, monitoring the construction phase on behalf of the Client.

3 Geology

The Chatham District lies within the southern portion of the London basin. The proposed tunnel approaches generally comprise 2 to 5 metres thick of made ground, overlying alluvial soft clays and silts, sandy gravels and then the upper chalk. The soil profile within the river channel appears similar to that of the tunnel approaches with the exception that the made ground is absent. See Figure 4.

The superficial geology is dominated by the presence of a buried channel of the Medway, which has incised into the chalk, and has been partly infilled with alluvial deposits. The buried channel is steep sided and trends northeast - southwest with a relatively flat base at about -13.5 metres OD, being locally over-deepened on the west bank of the present Medway.

The upper chalk is typically white chalk with flint bands at 0.5 to 1.5 metres vertical spacing; within each flint band the flints may be typically 20mm to 150mm across, spaced at 0.2 metre to 1.0 metre horizontally. There is evidence of hard band of chalk at depth.

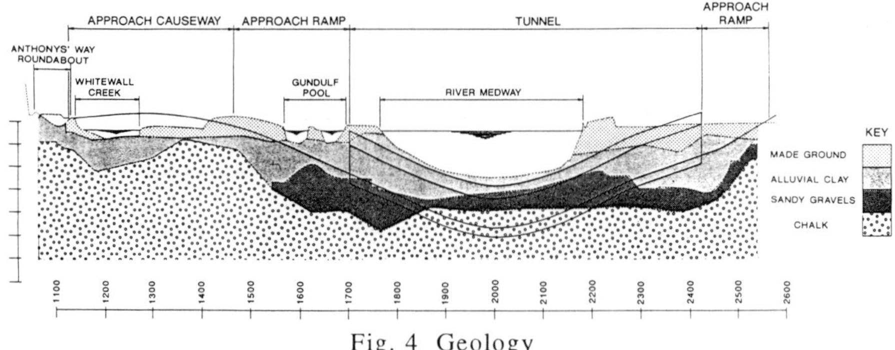

Fig. 4 Geology

The recent deposits comprise, generally, a medium dense alluvium of sandy gravels and clayey silts. The thickness of the alluvium appears to vary due to the presence of minor channelling in the chalk.

The alluvial clay is of intermediate to high plasticity, and clayey peat within the alluvium is of very to extremely high plasticity. The moisture contents are found frequently close to or above the liquid limits. Undrained shear strength generally ranges between 10kPa and 30kPa.

The nature of the made ground is very variable, with materials ranging from loose to medium dense sandy gravels, brick rubble and wood fragments to organic clay in places.

Readings taken from piezometers installed on the west side of the river with tips sealed in the sandy gravels and the upper part of the chalk strata indicate that the groundwater levels generally lie at +2.0 metres and -1.5 metres OD. The water levels fluctuate in phase with the tide, but with a muted response giving values up to 1.5 metre lower and 2.0 metres higher than the river water levels at high and low tide respectively. These differences reduce with distance from the river. The long term groundwater level in the made ground appears to be approximately at + 1.0 metre OD.

4 Approach Structures

Both the Western and Eastern Approaches and Structures are substantial undertakings and incorporate a range of construction techniques. However, as the paper is primarily directed at the design and construction of the immersed tunnel only a brief outline of the approaches is presented.

On the Western side, cut and fill operations include the hydraulic placement of over 30,000 tonnes of marine dredged sand across Whitewall Creek. The placement of the fill is a staged operation, taking approximately 18 months for surcharging and primary consolidation to occur.

Where the carriageway enters cutting, piled raft construction is employed, the piles being utilized to achieve a stable pavement. As the cutting deepens the function of the piles changes to act in tension to resist uplift forces.

For the deeper open cut section, and the cut and cover tunnels on the West Bank, diaphragm walls have been constructed. The final 50m of cut and cover tunnel will be constructed within the temporary cofferdam and a bottom-up construction, of similar cross section to the immersed tunnel units, will be formed.

The Eastern Approach structures will be formed within the area of the casting basin, and can only be constructed when the immersed tunnel units have been completed, sunk into position and the casting basin has been drained.

Normal bottom-up construction is employed, but heavier than normal structures are required to avoid uplift where the tunnel section widens to accommodate the Pembroke Road slip roads.

5 Immersed Tunnel Structure

5.1 Cross section

The cross-section of the tunnel is a straightforward twin cell reinforced concrete structure. See Figure 5. Excluding the ballast concrete and carriageway surfacing the internal height and width of each cell is 6.8 metres, and 10.45 metres respectively. These dimensions accommodate two lanes of traffic with a vertical clearance of 5.35 metres and verges of 1.25 and 1.3 metres with a vertical clearance of 5.1 metres. The ballast concrete

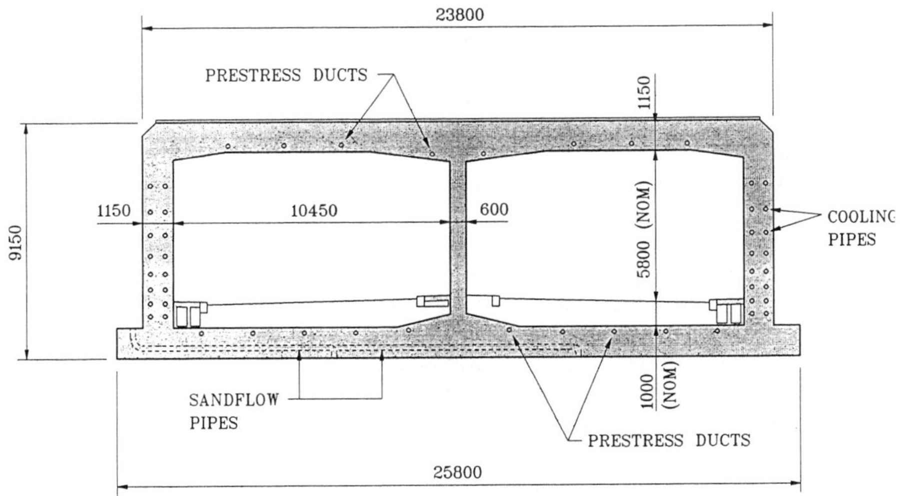

Fig. 5 Typical Cross Section

and roadway is of nominal thickness 1.0 metre and has a crossfall of 1:40. Clearance above the traffic envelope is 450mm at the centre line of the carriageway, which is sufficient to accommodate lighting and signs only. Forced ventilation is achieved by axial flow fans mounted in raised sections of the tunnel roof in the cut and cover approach tunnels. The main advantage of this fan configuration is to minimise the height of the immersed tunnel section. With less void the amount of concrete needed to provide the appropriate factor of safety against buoyancy is significantly reduced, the dredged trench is less deep, the approach tunnels are shallower and the approach gradients shorter, all of which assist the economy of construction.

Design of the cross section was by a two-dimensional linear frame analysis. Limit state design principles in accordance with BS 5400 were used. In addition to strengthening to accept loads from marine operations, the tunnel is designed of units of highway loading, vehicle collision and external loading from river vessels.

5.2 Buoyancy

Buoyancy considerations are a major factor in the design of the cross-section of the immersed tunnel. The actual weight of each 126 metre long tunnel element is in excess of 24,000 tonnes. During early stages of detailed design importance was given to determining the aggregate source for the concrete to allow an accurate assessment of the likely concrete weight to be made. From the structural viewpoint the member thicknesses could be reduced, but they have been chosen to provide the structure with almost neutral buoyancy in the temporary float out condition. The weight of tunnel section is 1958 kN/m, whereas, if fully submerged, the weight of displaced water is approximately 2254 kN/m.

The muted response to the tidal cycle of the groundwater levels within the chalk results in an allowance having to be made for an additional uplift under the base of the tunnel at a low tide, equivalent to 1 metre head of water. In the permanent condition a factor of safety of 1.1 against buoyancy is required. This is achieved by the addition of ballast concrete internally and by taking into account the weight of backfill material on the extended toes of the base slab together with the rock protection across the whole of the tunnel.

In addition to providing weight the ballast concrete also accommodates numerous cable ducts, the carriageway drainage system and the watermain. The majority of the mid-tunnel drainage sump in Tunnel Element 2 is also within the ballast concrete thickness although it has been necessary to provide some recesses in the structural concrete locally for the pumps themselves. This avoids the major structural and construction difficulties which would have been associated with a change to the external cross-section of the tunnel at this point.

5.3 Trench and backfill

Within the trench in the river bed the tunnel elements are founded on bedding sand, locked in with granular backfill, and covered by the trench membrane and rockfill protection. Finally, the amenity backfill is added to restore the river bed profile. See Figure 6. The trench membrane is made of an impermeable material, and its function is to avoid the long term contamination of the groundwater in the chalk and gravels by preventing infiltration of the river water through the granular backfill around the tunnel.

5.4 Immersion joints

Longitudinally the immersed tunnel is made up of three elements, two are 126 metres in length and the third is 118 metres. They are connected in the river by special immersion joints. See Figure 7.

The joint is effected by means of a special rubber gasket mounted on the leading end of each element which then bears against a flat plate on the secondary end of the previous element. The rubber gasket has a soft tip to provide an initial seal and a harder body to accommodate the full hydrostatic forces. This type of gasket is known generally as the 'Gina' profile. The 'Gina' mounting plate and the corresponding bearing plate each have to be constructed to a high degree of accuracy. This is achieved by initially casting I-Beam sections into the concrete forming the end sections of the elements. The mounting and bearing plates are then welded to the flanges of the beams under strict survey control to a tolerance on plane of ± 5mm. The resulting voids between the plates and the webs of the I-Beams are carefully grouted. Exposed steelwork is protected using a coal tar epoxy paint system.

The immersion joint between adjacent tunnel elements incorporates, in addition to the Gina profile, a second watertight seal known as the Omega profile. After the first seal has been made the Omega profile is clamped across the joint on the inside, protected with formwork and the joint is then

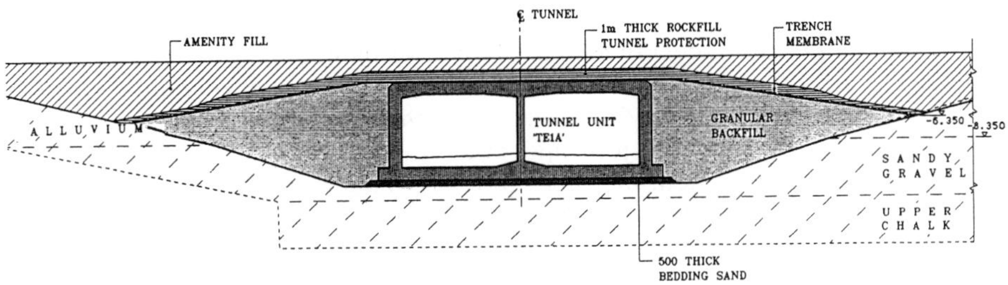

Fig.6 Cross Section in Trench

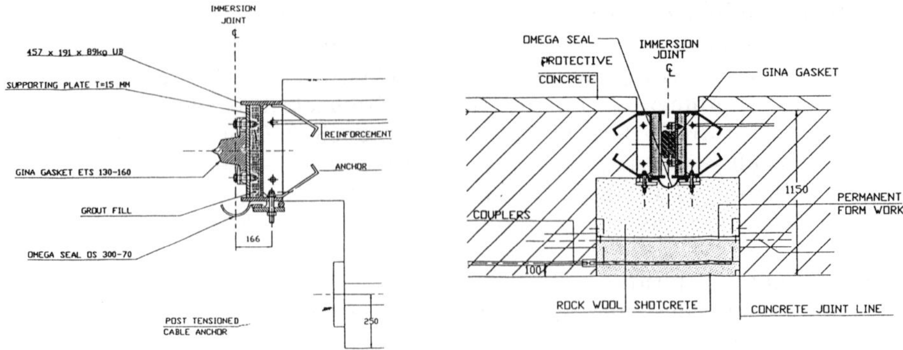

Fig.7 Immersion Joint Detail

concreted in the base and walls, and shotcreted across the roof. Incorporated in the base slab at the immersion joint are a number of dowels to provide shear resistance across the joint.

5.5 Dilatation joints

Although straight in plan the elements are curved in profile. Each element is made up of 6 sections, cast individually, of 21 metre length. Between each of these sections is a specially formed shear joint which will allow a small degree of articulation. This is known as a dilatation joint. See Figure 8. It is waterproofed by means of a specially developed flexible water stop which has steel plates bonded along both sides, with foam strips attached to the steel plates. Injection tubes at regular intervals allow the waterstop to be resin injected past the foam strips after the concrete has cured, thus sealing any possible water paths.

The longitudinal reinforcement is not continuous across the dilatation joints and hence in the temporary condition it is necessary to longitudinally post-tension the elements for the float out, sinking and founding phases of the construction. The post tensioning cables are later cut at the dilatation joints to allow the element to articulate as necessary.

The longitudinal design also used a two-dimensional linear frame analysis. Differential founding conditions were simulated by varying spring stiffness. By progressively modifying parameters the analysis was used to reflect the staged nature of the jointing and founding of the tunnel elements. Engineering judgement based on practical experience is an important factor in creating and interpreting the analysis model.

5.6 Watertightness

The immersed tunnel and the cut and cover tunnels of similar construction do not have an external waterproof membrane. Watertightness will rely on

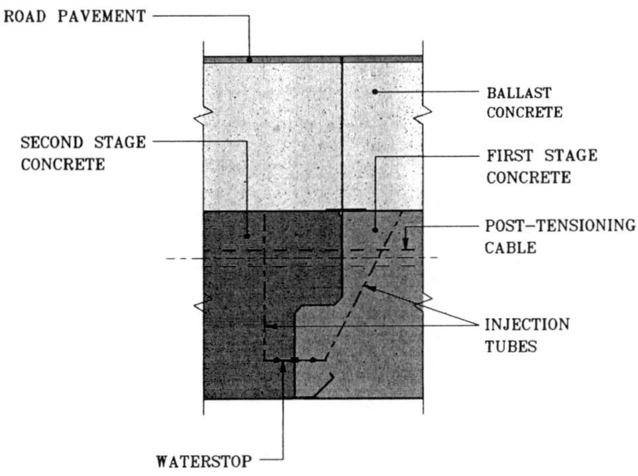

Fig. 8 Dilatation Joint Detail

a fully integrated design which includes the concrete mix, reinforcement, dilatation joints and cooling of the newly placed concrete in zones of high thermal restraint.

There are a number of different reasons why concrete may crack during the construction period. During hardening cracks can occur by plastic settlement, shrinkage and constructional movements. After hardening cracks may be caused by physical, chemical, thermal and structural conditions. All these items have to be taken into account in the detailed design. Specified temporary measures and construction methods are part of the integrated detailed structural and concrete curing designs.

6 Curing Concrete by an Internal Cooling System

On the Medway Tunnel Project, an internal cooling system is used to minimise the early age thermal cracking. See Figure 9. The technique being employed has been used successfully in many tunnels constructed in Holland. The cross section is poured in three separate stages; the bottom slab, followed by the central wall, and finally the roof slab and external walls together. This construction sequence will cause an external restraint between the bottom slab and the outer walls during hardening of the walls creating the tendency for vertical cracks in the walls, which can extend right through the section. The design specification requires maximum tensile stresses not exceeding 50% of the tensile strength of the immature concrete. In the early stages after pouring, the strength development is approximately linear with rate of hydration. Cooling the concrete in the walls, reduces the peak

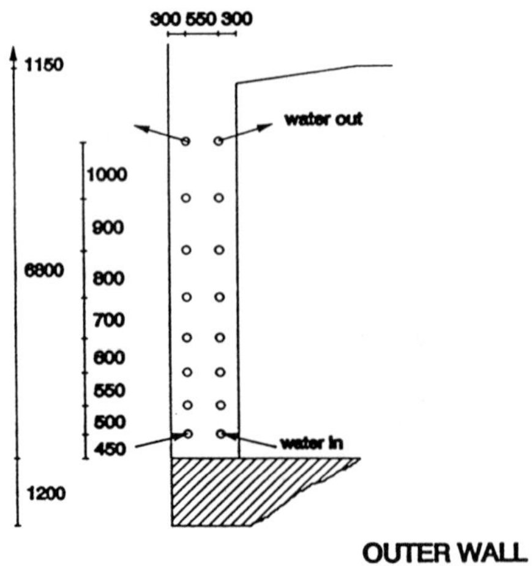

Fig. 9 Cooling Pipes: Section Through Wall

thermal differential between the walls and the base slab, resulting in a reduction of the tensile stresses.

Detailed calculations have to be carried out to design the required cooling capacity. Based on data available from previous projects, notably in Holland, a specific computer program has been developed to design the optimum cooling capacity and the layout of the cooling pipes. Consideration has to be given to:-

- dimensions of the concrete structure
- type of formwork
- sub-grade reaction modulus
- insulation
- fresh concrete temperature in Summer and Winter
- ambient atmospheric conditions of temperature and wind

A detailed temperature monitoring system is built into the system which is connected to data loggers provided with a special software package.

The internal cooling system consists of two rows of 10 cooling pipes cast in the concrete and a 100 kw chiller set up adjacent to the pour. From the start of the concrete pour, chilled water at 6°C will be pumped through the system. The pipes are set closer together near the construction joint and have a wider spacing at the top of the wall. The bottom slab will remain at approximately the ambient temperature.

Adiabatic Test Results - Concrete Mix C40 140 kg/m³ OPC; 270 kg/m² GGBFS; 1760 kg Sand/Gravel; w/c 0.39				
	1 Day	**2 Days**	**3 Days**	**7 Days**
Compressive Strength (MPa)	6.6	14.0	18.3	32.3
Modulus of Elasticity (MPa)	16,300	25,700	29,000	33,500
Tensile Splitting Strength (MPa)	0.6	6.5	1.8	3.1
Max Temperature Gradient		43.3°C		

Table 1 Adiabatic Test Results

The cooling is aimed at ensuring that the temperature of the fresh concrete at the base of the walls does not exceed the base slab temperature by more than 15°C. Without cooling this temperature difference would be in the region of 40°C. A further requirement is to ensure that the temperature of the uncooled roof zone does not differ from the base slab by more than 37°C. See Figure 10. The consequence of the two requirements is that the temperature difference between the uncooled roof and the cooled wall will be approximately 25°C.

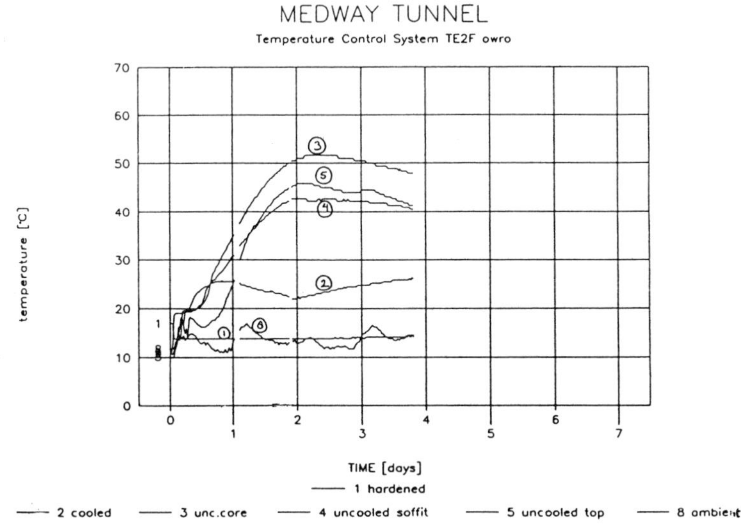

Fig. 10 Graph of Cooling Results

7 Closure Joint

An interesting aspect of the design has been the treatment of the final closure joint between the third immersed tunnel element and the east approach tunnel. See Figure 11. When sunk into its final location the trailing end of the third element remains within the narrow neck of the casting basin, and is founded on a stepped concrete slab cast as an integral part of the in- situ approach tunnel base slab. The element is then wedged against the step and the gap between the element and the cofferdam is closed by means of temporary side and top panels braced against the slab and the top of the element respectively.

The casting basin may now be dewatered and the remaining construction of the east approach tunnel carried out, leaving initially a short gap at the end of the third element.

The river wall is constructed over the approach tunnel and the backfill placed. Thus with the majority of the permanent loading place, the short gap is finally closed with in situ concrete, the remaining backfill completed and the temporary closure panels are removed.

8 Temporary Works - Cofferdam

The temporary works are heavily influenced by the specific complex geotechnical and special foundation requirements of the site.

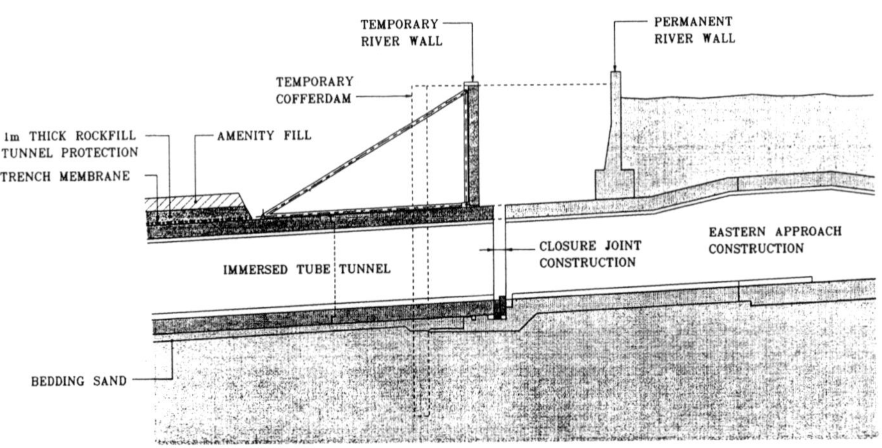

Fig. 11 Closure Joint Principle

The temporary retaining structures at the river bank have to be designed for the following functions:-

- A temporary river wall protecting the on-shore activities from high floods.
- Protection of the construction of the cut and cover sections to which the immersed tunnel elements are to be connected.
- Protection of the river banks during the dredging and the immersion activities.

These temporary structures have been designed as rectangular tied cellular cofferdams of "Combi-Wall" construction. A set of 3 cofferdams on each side of the river permits the construction of the tunnel in a very restricted location.

The piling consists of tubular piles of 1400mm diameter at 2600mm centres with a double Larssen infill between piles. On both sides of each tubular pile special clutches are fully welded over the required length to achieve a watertight construction between the tube and the sheetpiles.

In the river, additional sheet piles are installed parallel to the river banks to support the slope of the dredged trench near the shore line.

9 Casting Basin and Dewatering

Two of the three immersed tunnel elements are built on the centre line of the tunnel and the third element is built off line. See Figure 3.

The chosen excavation depth is -10.5 metres OD. With a draught of 9.0 metres the elements can be floated out during the high tides. Between the river and the wide part of the casting basin a 30m wide float out channel has been constructed by using the same type of combi-wall as for the cofferdams. This float out channel is propped at high level only which allows the elements to pass underneath the props.

During construction of the elements the dewatering requirement is to drawdown to 0.5 metre below the formation level to maintain dry working conditions. Internal and external deepwell systems were used.

10 Dredging and Backfilling River Trench

A specific requirement of immersed tunnels is the need to dredge a trench through the river bed.

The depth of the trench has to take into account an appropriate sand bedding layer underneath the tunnel, the height of the tunnel and the required cover to protect the tunnel. The maximum excavation depth is approximately 13 metres, including dredging tolerances. At MHWS the river depth is approximately 9 metres. The equipment will therefore have a dredging depth of at least 22 metres.

The total quantity to be dredged is 350,000m^3: silt clay 200,000m^3, sandy gravels 100,000m^3 and chalk 50,000m^3.

Tolerances for the dredging will be ± 300mm vertically with a horizontal tolerance of ± 1000mm. The bottom of the trench has a minimum width of 31 metres. The excavation width at river bed level will be in the order of 150 metres.

Immediately prior to immersion of the tunnel elements the bottom surface will be cleared of silt to avoid contamination of the sand bedding.

After the immersion and sandflow operations the trench is backfilled around the tunnel using sand resourced from the Thames Estuary. Each side of the tunnel has to be backfilled at a similar rate to avoid differential loading.

When backfilling is complete, the protective membrane for sealing the river bed is placed using a specially equipped pontoon. The membrane is kept in place by a thin layer of ballast material, above which a final layer of rock with nominal stone sizes of 150 - 200mm will be placed to protect the membrane and the tunnel. Finally the river bed will be reinstated with layers of amenity fill. See Figure 6.

11 Immersion Operation

11.1 General

The immersion activities fall into the following main groups of operations:-

- preparations in the casting basin
- preparations in the river
- float up in the casting basin
- marine fitting out
- transport, immersion and sandflow operations

The marine engineering associated with all these activities has a significant impact on the detailed design of the tunnel elements. All fixings and temporary construction phases have to be taken into account.

11.2 Preparations in the casting basin

Each tunnel element is designed as separate sections, connected together in the temporary condition by a post tension system. The post tension cables are built into the base and roof slabs and are stressed and grouted before flooding of the casting basin. During all the immersion activities the tunnel element acts as a monolithic structure.

Each element has a temporary 'nose' support structure mounted on the central wall at the primary end and a corresponding 'chin' at the secondary end. See Figure 12. Also there are two jacking rams on the inside of the element within recesses in each outer wall at the secondary end. These temporary support arrangements support the tunnel element for the initial period after immersion. The central wall of the west bank cut and cover tunnel also requires a 'chin'. The maximum service load on the primary support after ballasting is 6500 kN and the maximum service load on each secondary support is 6000 kN.

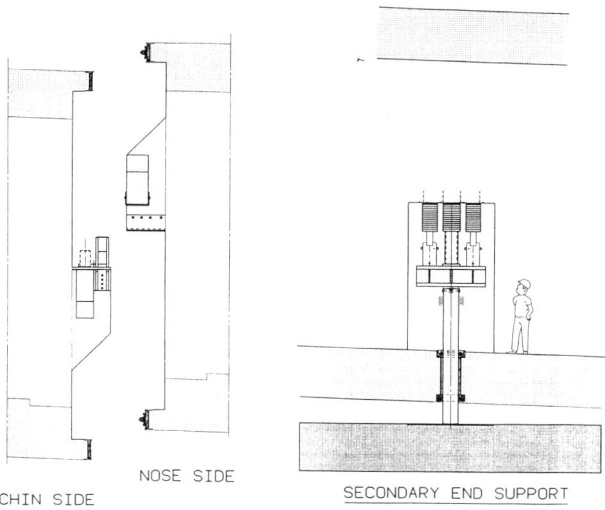

Fig. 12 Primary / Secondary and Supports

Each end of each element is closed by means of a temporary bulkhead. The bulkheads must be watertight and take the full water pressures during and after immersion. The bulkheads are 250mm thick reinforced concrete supported by steel beams.

On top of the roof slab anchors are cast in for the attachment of bollards, fairleads, hoisting structures, access shaft and survey tower. The bollards have a capacity of 500 and 1000 kN. Each lifting point has been designed for 1500 kN vertical and 300 kN horizontal loads.

Inside the element is a system of ballast tanks and pumps. Water is introduced into the ballast tanks and the amount is adjusted in order to keep the tunnel elements under control in all temporary conditions. The maximum volume of the ballast tanks is 6000m^3 and their distribution within the tunnel is significant for the loading design.

11.3 Preparations in the river

After dredging the trench, the temporary mooring and anchor points for the marine equipment are installed in the river and on shore.

On the bottom of the dredged trench, two sets of concrete foundation plates, 6 x 5 metres in size, are installed. These take the loads from the jacking rams at the secondary ends of the elements.

A system of sandflow pipes is cast in the base slabs of the elements, which will be used for the sand bedding underneath the element after placing.

11.4 Float up in the Casting Basin

Prior to flooding the casting basin in a controlled manner, the ballast tanks will have been filled with water to prevent the elements from floating.

Float up starts with deballasting of the tanks inside the tunnel element as the tide drops. The float up has to be achieved with the rising tide.

11.5 Fitting out and transportation

When the required buoyancy of the element has been achieved, the main winch lines on the East and the West banks are connected and made operational. The short access shaft will be removed.

The element will be winched through the narrow channel and just into the river, where the final outfitting for the sinking operation will be done. See Figure 13.

The transport of the element in the river to the right position is achieved at high tide. The immersion of the element requires ballasting to a submerged weight of 3000 kN. The final horizontal and vertical movement of the element is done in small steps, fully controlled by the survey team, and jointing with the Gina profile is carried out in the conventional manner.

The water in the ballast tanks can now be replaced step by step by permanent ballast concrete and the temporary post tensioning will be cut at each dilatation joint thus allowing each 21 metre long section to settle on its own by articulation of the joint.

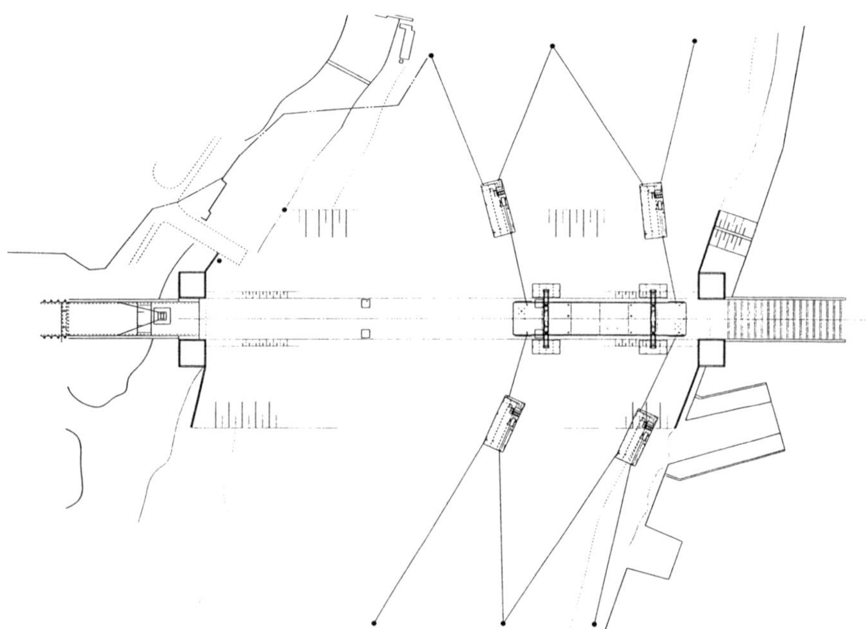

Fig. 13 General Arrangement of Float Out

12 M&E Installations

12.1 General

The Contract also requires the design and installation of full Mechanical and Electrical services. These are controlled at the main Services Building which is located south of the tunnel portal at the West Bank. A secondary services building is located adjacent to the portal on the East Bank. Both buildings are unmanned.

12.2 Power Supply and Distribution System

Two independent high voltage electricity supplies are provided to the West Service Building. For practical and economic reasons these supplies are extended to the East end of the tunnel by running two cables through the tunnel. Low voltage supplies are provided by transformers at each end of the tunnel.

The electrical loads, are connected equally to two supplies. This reduces the effect on the tunnel of an electrical power supply failure. In addition to this, essential loads such as emergency lighting, communications etc are connected to an uninterruptable power supply system. These loads and the sump pump systems are also backed-up by a diesel generator.

12.3 Tunnel Lighting

The tunnel allows for the safe use of vehicles entering from a relatively high ambient light level. This is achieved by having high lighting levels at the entrance and exit from each bore. The lighting is arranged in four zones of diminishing intensity from the portal, followed by an interior low level zone and a high intensity zone near the exit portal. High intensity lighting is achieved using high pressure sodium luminaires.

12.4 Tunnel Ventilation

The ventilation system has two primary objectives. Under normal operating conditions it is required to maintain an acceptable environment for tunnel users; and under emergency conditions, ventilation is required to move fire smoke away from users who are trapped by an incident. Taking into account these objectives and the geometry of the tunnel, a longitudinal ventilation system has been selected using jet fans. Under normal free flow and static traffic conditions, these operate to enhance the vehicle induced air flow in the tunnel. The system is controlled by carbon monoxide and visibility sensors which cause the fans to switch on when pre-determined levels are reached.

If a fire occurs during normal one way traffic operations the fans will be used to blow smoke in the direction of normal traffic flow. If contra flow is in operation the fans will blow the smoke to the nearest portal. The system has been designed for a 50 MW fire with a 20% redundancy in the one way traffic condition. In order to reduce the height of the immersed tube section of tunnel, larger than average diameter fans have been selected which are arranged in three banks of four per bore, positioned within aerodynamic niches in the cut and cover sections.

12.5 Emergency Services

A fire main through both bores of the tunnel supplies Fire Hydrants throughout the approaches and the tunnel. Emergency doors are situated at every 100m in the dividing wall. Emergency panels containing fire hose, fire extinguishers and emergency telephones are positioned every 50 metres in the tunnel nearside walls.

The Services Buildings have a fire alarm system, fire extinguishers and selected rooms are protected by inert gas flooding.

Each sump is protected by an inert gas foam discharge system.

12.6 Drainage system

This system comprises three sumps each with three duty submersible pumps and one standby pump. These discharge surface water into the River Medway. The mid-river sump discharges at the East Bank. The portal sumps discharge to their respective river banks. Any significant spillage of hazardous materials in liquid form will be collected either via floating skimmers or directly by specialist tanker vehicles.

12.7 Communication and control systems

12.7.1 Telephone systems

An emergency telephone system similar to the Motorway Emergency Telephone System is provided in the tunnel, with telephones located approximately every 50 metres connected to Rochester Police Station. In addition a tunnel operators telephone system is provided together with smoke control telephones at each of the four tunnel portals.

12.7.2 Radio relay system

A radiating cable system in the tunnel caters for Emergency and maintenance services. The system will also cater for future public broadcast services.

12.7.3 CCTV surveillance

The tunnel bores and approaches are provided with a CCTV system transmitting to Rochester Police Station and Kent County Council (KCC), HQ, Maidstone.

12.7.4 Police control

A new control room is situated at Rochester Police Station and accepts information on plant, emergency and traffic control. The control room is equipped with a VDU to monitor and control plant and traffic. Two TV monitors with controls are provided for tunnel surveillance via CCTV. A

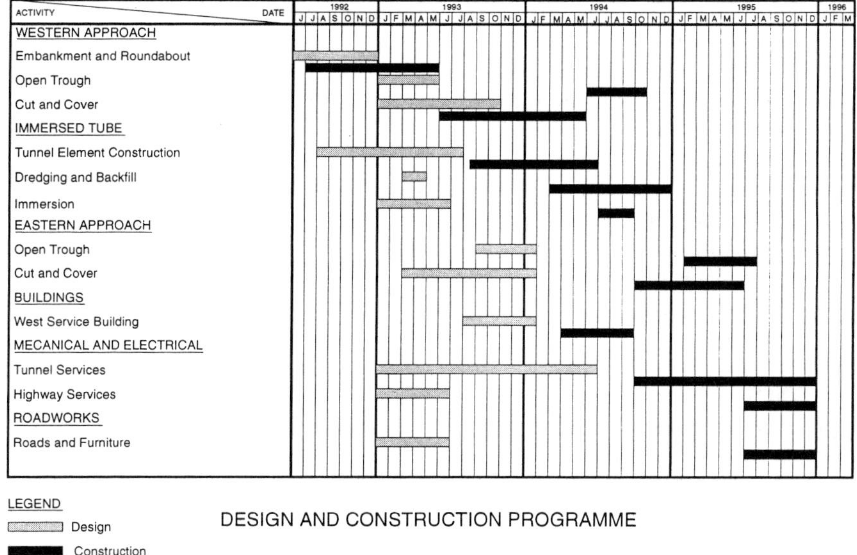

Fig. 14 Construction Programme

control panel is also included to deal with the emergency telephone systems.

12.7.5 Environmental control and plant monitoring system
A comprehensive Environmental Control and Plant Monitoring System is provided for the following:
- Air pollution monitoring
- Tunnel ventilation monitoring and control
- Portal luminance monitoring
- Tunnel lighting monitoring and control
- Service building plant monitoring
- Systems link monitoring
- Traffic control

12.7.6 Traffic control
The principal means of traffic control is achieved by the use of variable message signs (VMS) situated in both tunnels and approach roads, similar to those used by the Department of Transport on motorways and trunk roads.

13 Construction programme

Construction of the Medway Tunnel Project commenced in July 1992. See Figure 14.

On the west bank early construction activities were concentrated on Anthony's Way roundabout and the filling of the Whitewall Creek area. The roundabout was handed over to the client in September 1993. For the western approach, priority was given to the diaphragm walled and cut and cover sections which needed to be completed prior to the connection to the immersed tunnel.

On the eastern side, the early construction concentrated on the formation of the casting basin. This commenced in July 1992 and the first concrete pour for the immersed tunnel elements themselves was that for the base slab of tunnel element TE2F in mid-August 1993. Tunnel element TE2 was completed just before Christmas 1993, a 4-month construction.

The immersion operation of the tunnel units is scheduled between July and October 1994.

The completion date for the eastern approach works is August 1995.

The final construction items are installation of M&E services, roadworks furniture and fittings. The Medway Tunnel is scheduled to be opened to traffic at the end of 1995.

Developments in precast concrete tunnel linings in the United Kingdom

T. R. Winterton
C. V. Buchan Ltd., Northwich, Cheshire, England

Abstract
Precast concrete segmental linings have been in use for many years in both tunnel and shaft construction. Since their introduction over 50 years ago, they have become the standard method of lining a tunnel in soft ground conditions in this country.

The last review of tunnel linings practise in the U.K. was covered in TRRL Supplementary Report 335, 15 years ago. Since that time there have been a number of new developments in the linings available and the methods of construction.

This paper reviews the developments in precast linings and methods of construction in the U.K. from the unique position of a tunnel lining manufacturer.

The topics covered include:-

Smoothbore linings using proprietary fixings and sealing systems.

Fully tapered rings with trapezoidal shaped segments and gaskets which allow the building of tunnels, on curve and straight, through water bearing ground without the use of packings.

Theoretical and practical tolerances for linings and tunnels, with particular reference to sealing gaskets.

Methods of joining segments and rings, including curved bolts, bolt plates, proprietary fixings and the use of plastic connections.

The long term durability of segmental linings, with particular reference to mix design, cover to reinforcement and cast-in items.
Keywords: Precast concrete tunnel linings, smoothbore, tolerances, tapered rings, sealing gaskets, segment joints, segment durability.

1978 Review

At the time of the last review [1], standard bolted linings were the predominant means of lining a tunnel in the U.K. There were also a number of proprietary linings providing a smoothbore single pass lining. Brief details of these lining types are given below:-

Standard bolted linings

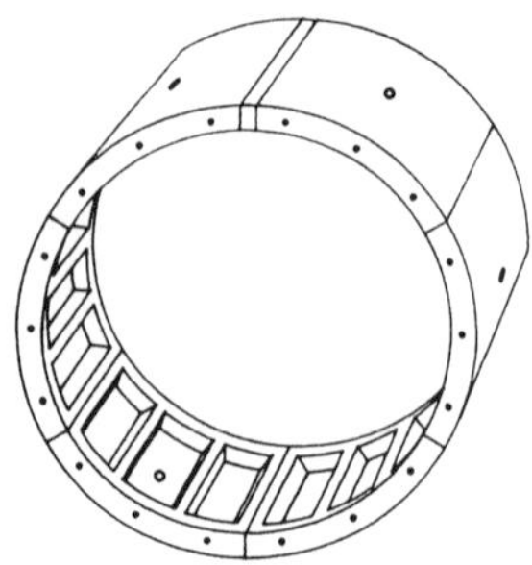

Fig. 1

A standard range of bolted concrete tunnel linings was introduced by the tunnelling contractor and precast concrete manufacturer Kinnear Moodie Ltd. after 1945. They have become an integral part of traditional underground construction since that time and there are over forty different sizes available from several manufacturers, ranging in diameter from 1.52m to 35.5m; the larger sizes are used in shaft applications.

A typical tunnel ring (Fig. 1) consists of four ordinary segments, two slightly shorter top segments and a small parallel sided solid key. Each ring is usually 610mm wide. The segments are bolted to each other circumferentially to form a ring, two bolts connecting each segment. The key is provided to allow the placing of the last segment in a ring and is the last piece to be inserted. The rings are annulus grouted immediately after erection. These types of rings have been used in all types of ground conditions normally encountered in the U.K. and the simple method of construction allows them to be built with or without a tunnelling shield if the ground conditions allow.

Expanded concrete flexible linings

Expanded linings were developed to allow fast erection in self supporting and homogenous clays where they could be expanded directly against the ground. A tunnel shield is required to ensure that a true circle is excavated.

Expanded linings can be considered to have fallen into two broad groups, depending on tunnel usage: Water transmission or rail transport.

A 2.54m I.D. (100") wedgeblock lining was developed by the Metropolitan Water Board in 1955 for the transmission of water. This type of lining became the standard for small diameter tunnels in firm to stiff clays and is available from several manufacturers.

The standard 2.54m I.D. (100") Wedgeblock lining is a 140mm thick un-reinforced lining which consists of 11 segments and a wedge-shaped key.

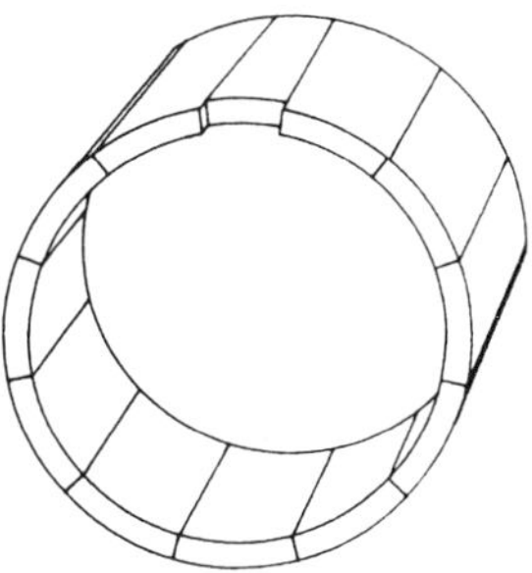

Fig. 2.

(Fig.2). Longitudinal and circumferential joints are plane except for the helical key joints. Longitudinal joint surfaces are coated with a high build bitumen/cement to relieve high contact stresses.

The linings are expanded directly against the excavated bore by a wedge key, 50mm shorter than the 685mm long rings. The gaps at the keys are filled with dry pack concrete after ring erection.

The smoothbore segments are cast vertically in concrete moulds in order to maintain a high degree of accuracy on the longitudinal joints and extrados. Keys of different sizes are available to cope with varying ground behaviour and a caulking groove is provided on the trailing edge of the rings.

This design of lining is available in sizes of 2.05m, 2.54m and 2.91m I.D. and has been used at depths of up to 60m in London clay.

3.81m I.D. expanded concrete linings were developed for the London Transport Executive (L.T.E.) and were first used on the Victoria Line in 1961. Different linings have been developed with either concave/convex or convex/convex joints. The ring designs varied between 11 and 20 segments and were 0.6m wide. Rings were expanded by insertion of jacks at the knee joints on earlier designs but these were later superseded by the use of wedge keys.

A 10.29m I.D. expanded lining was developed for the Heathrow Cargo Tunnel in 1966 - 68. This ring consisted of 27 segments, 0.6m wide with convex/convex radial joints. The rings were expanded by jacks at axis level and the jacking space filled with drypack concrete.

Smoothbore grouted linings

These linings are generally unbolted and require a former or other form of temporary support to hold the ring shape until grouting is carried out and is sufficiently strong to support the lining. They have been used for a range of ground conditions from soft silts to hard rock but are not generally used in water bearing ground.

A number of proprietary designs have been developed suitable for use with or

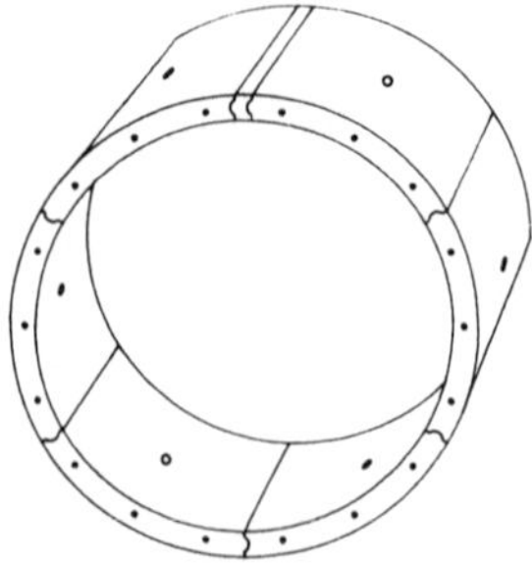

Fig. 3.

without a shield depending on ground conditions.

Various joint shapes have been developed to facilitate segment to segment location (Fig. 3) and allow some degree of articulation and some have dowels on the circumferential joint to ease erection.

A range of diameters, between 1.0m and 3.0m, I.D. is available from the manufacturers.

Expanded grouted lining

This is a smoothbore lining which is expanded on to the ground by a wedge shaped key and then grouted. Pads projecting from the rear of the lining (Fig. 4) allow the ring to be expanded against the ground for initial support and then subsequently grouted to fill the annulus and overbreak.

This type of lining was used in the service tunnel of the Stage 2 contract for the Channel Tunnel in the abandoned scheme in the 1970's. It was subsequently used in the final Channel Tunnel scheme and is described more fully later in this paper. This lining is appropriate in hard ground rock conditions where an initially self-

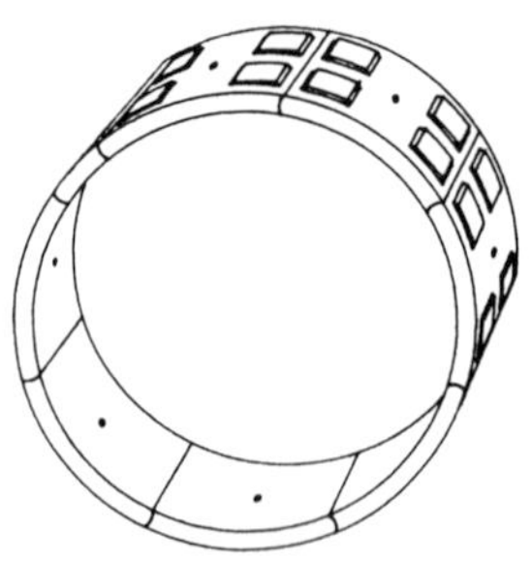

Fig. 4.

supporting bore can be excavated by a T.B.M., to provide an excavated surface good enough to allow a ring to be expanded against the ground. The grout provides a partial seal against water ingress.

Lining developments since 1978

Since the time of the last review there have been a number of developments in precast linings. Developments in standard bolted linings and smoothbore grouted linings have often been complimentary and these are therefore described together in roughly chronological order. Expanded linings and expanded grouted linings are discussed separately.

Mechanically jointed smoothbore

In the late 1970's the smoothbore grouted ring which had been used in many ground conditions was improved and started to be used more widely.

The "Onepass" (Fig. 5) ring was developed from the Kinnear Moodie Universal lining [2] and incorporated a mechanical steel connection on the longitudinal joint and a dowel connection on the circle joints.

"Bucline" [3] tunnel rings are a similar variant with a hidden bolted connection at the longitudinal joint and a combination of dowels and hidden bolts on the circle joints.

These rings are partly self-supporting and this has extended the range of ground they can be used in. They can be built with or without a shield in most types of ground from soft silts to hard rock. Grouting up to axis is carried out immediately to provide side support and the joint connections allow the ring to keep its shape and prevent segments moving relative to one another.

These rings provide a smooth lining which requires a minimum of cleaning out and needs only caulking/pointing of the joints and small connection recesses.

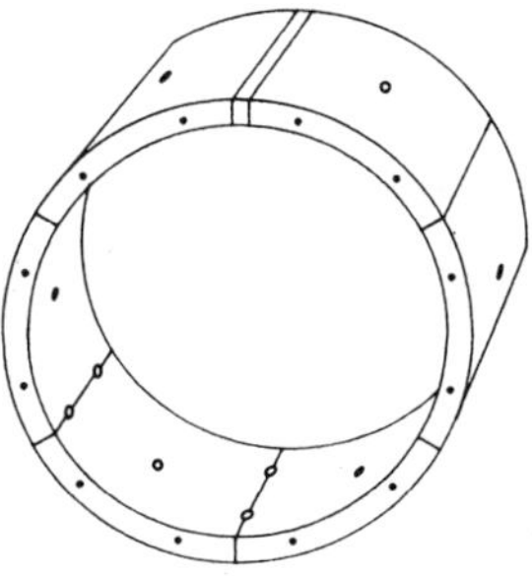

Fig. 5.

Tapered bolted rings

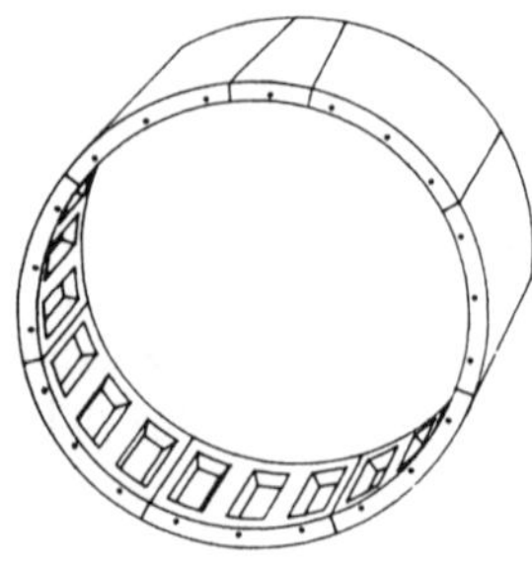

Fig. 6

The next stage in the development of linings occurred with the increasing use of Tunnel Boring Machines and the introduction of tapered rings and gaskets in Germany and the U.S.A.

In 1985 a fully tapered bolted lining with Neoprene rubber compression gaskets was used on Stage 1, Contract 2 [5] of the Don Valley Interceptor Sewer in Sheffield. The 4.11m I.D. lining was a development of a standard bolted lining. This 750mm long ring consisted of seven segments and a tapered key installed in a similar manner to a wedgeblock key.

A single type of tapered ring (Fig. 6 and 7) was used with a 30mm taper across the axis; the ring was built in a single position and the segment erector allowed the ring to be rotated through 360° when built. Tunnel alignment was corrected by rotating the ring to vary the taper position.

In 1987 a fully bolted 2.74m I.D. lining with Ethylene Propylene Diene Monomer gaskets was used on the M.E.P.A.S. Stage I scheme in Liverpool. The ring consisted of five segments and a wedge shaped key. The rings had a 15mm taper across the axis of the 1.0m long ring and a left and right hand ring were used to allow the key to be built in the top half of the ring at all times. Straight drives were built by alternating the taper with left and right hand rings; 255 and 415m radius curves could be negotiated by building different ring sequences. (Fig. 8). Tunnel alignment could be corrected by rotation of the rings to vary the taper position. To achieve a correction of 11.5mm "Lookup" (bottom edge of ring 11.5mm advanced relative to the top) a combination of rings type I and type II was used with the key and positions 2 and 6 respectively.

The EPDM rubber compression gaskets were designed to accommodate joint gaps of up to 12mm with an external water pressure of 2 bar. Particularly aggressive trade effluent necessitated the use of an un-reinforced insitu concrete

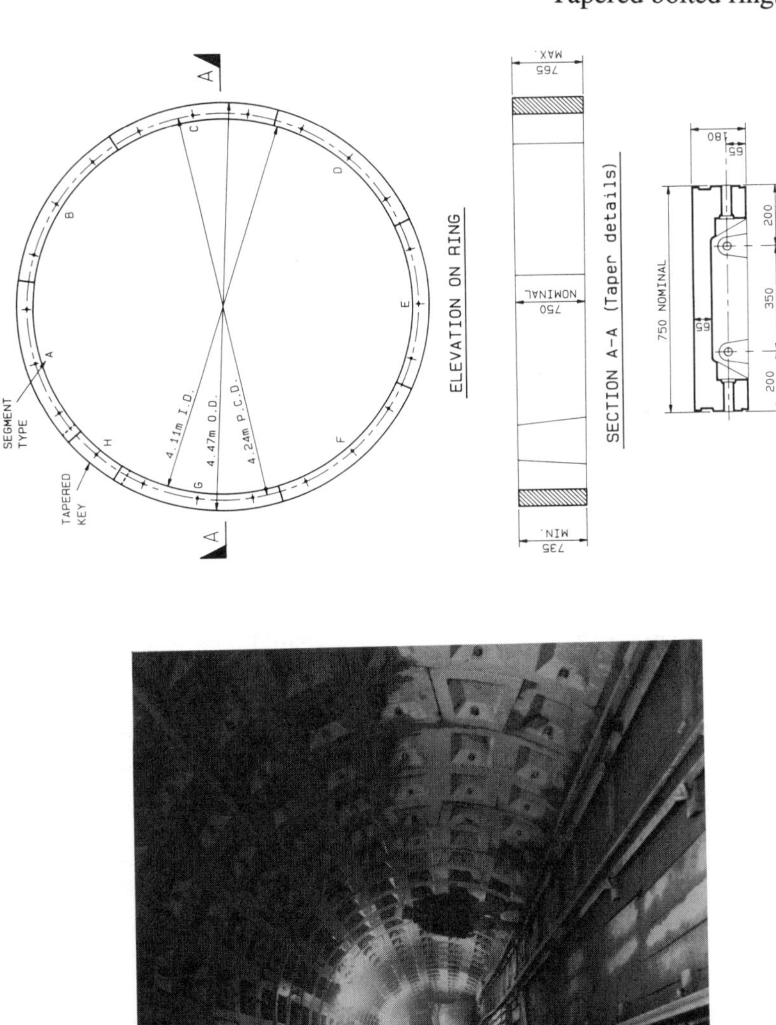

Fig. 7 - 4.11m I.D. Lining - Don Valley Interceptor Sewer, Stage 1, Contract 2

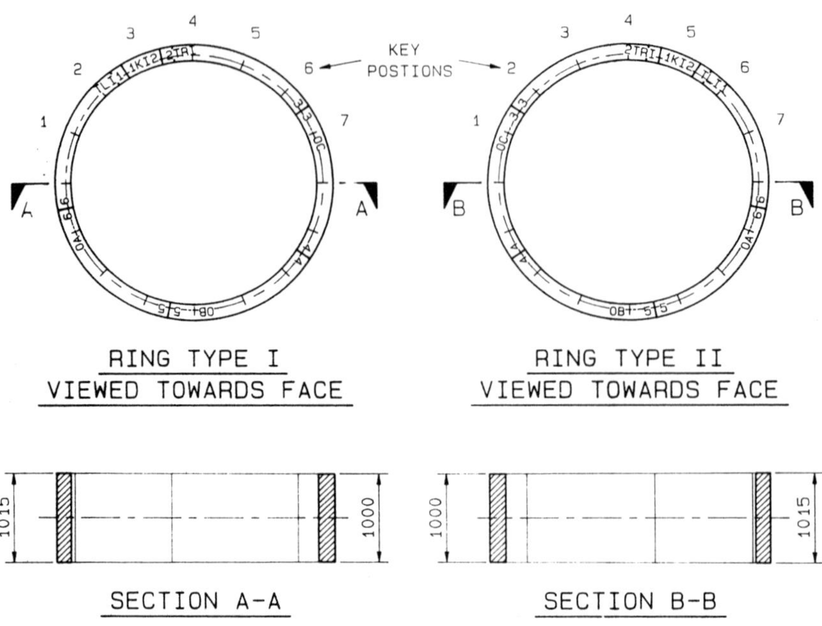

STEERING INFORMATION FOR TAPERS								
ADJUSTMENT (mm)	LOOK UP		OVERHANG		RIGHT LEAD		LEFT LEAD	
	RING TYPE	KEY POS.	RING TYPE	KEY POS.	RING TYPE	KEY POS.	RING TYPE	KEY POS.
11.5mm	I + II	2 6	I II	4 4	-	-	-	-
15.0mm	-	-	I	7	II	5	I	3
21.2mm	I + II	1 7	I + II	5 3	II + II	3 7	I + I	1 5
27.7mm	I + II	6 2	-	-	II + II	4 6	I + I	2 4
30.0mm	-	-	I + I	7 7	II + II	5 5	I + I	3 3

Require 4 ring sequence to give = 415m radius (ie 3xII@5+1xI@3)
Require 4 ring sequence to give = 255m radius (ie II@3+II@7+II@4+II@6)

Fig. 8 - Ring Building Sequences - 2.74m I.D. Lining, M.E.P.A.S., Stage 1 lining, 180mm thick.

A number of similar linings with compression gaskets were used about this time and were successful in significantly reducing pumping costs during construction and subsequent sealing costs.

Examples where this type of ring was used included Don Valley in Sheffield on Contracts 2 and 3 [6] with 1.83m, 1.91m and 4.11m I.D. linings; M.E.P.A.S. Stage 1, Contract 2 in Liverpool, mentioned above; Oldham Interceptor Sewer, a 1.80m I.D. lining, and at the Royal Docks, Phase 9 in London with a 1.80m I.D. lining used with Bucfil infill pellets.

Not all of these types of lining were tapered; a number were parallel sided rings with gaskets whose performance allowed the use of timber packings to correct for alignment in the traditional manner.

Bolted smoothbore with gaskets and curved bolts

In 1988 a 5.64m I.D. smoothbore lining was used to line the rail freight tunnel link to Standsted Airport [7]. This ring used EPDM sealing gaskets tested to 3 bar with 6mm joint gaps and curved bolts on all the joints. This was the first use of curved bolts in the U.K. (Fig. 9 and 10).

This lining tapered 8mm across the horizontal axis and the 1.0m long rings consisted of seven segments and a wedge shaped key. The design incorporated a large invert segment (2.2 tonnes) with a flat invert and corbels for track support. The key joints incorporated a shear ledge to prevent the key dropping and a reformer ring was used to hold the ring shape after erection until the grout had hardened sufficiently to support the ring.

A similar smoothbore design with curved steel bolts on all joints and a diameter of 3.70m I.D. was used on the Bradford Beck flood alleviation scheme in 1991. This 200mm thick parallel sided lining consisted of 5 segments and a wedge-shaped key.

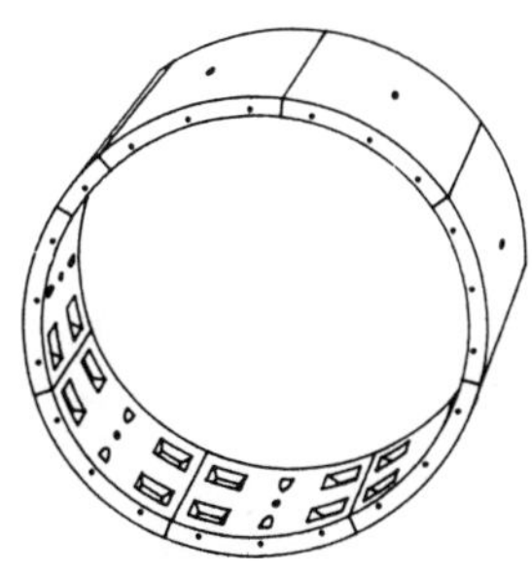

Fig. 9.

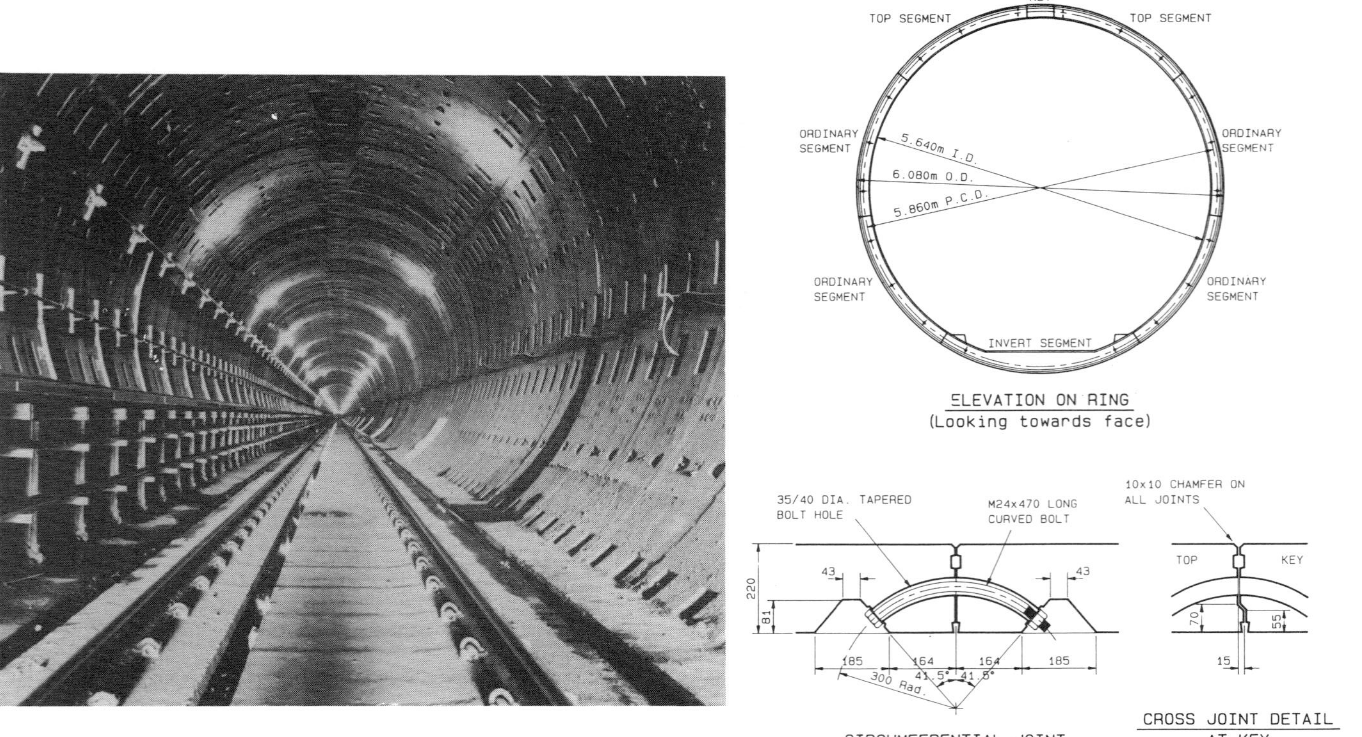

Fig. 10 - 5.64m I.D. Lining - Stansted Airport

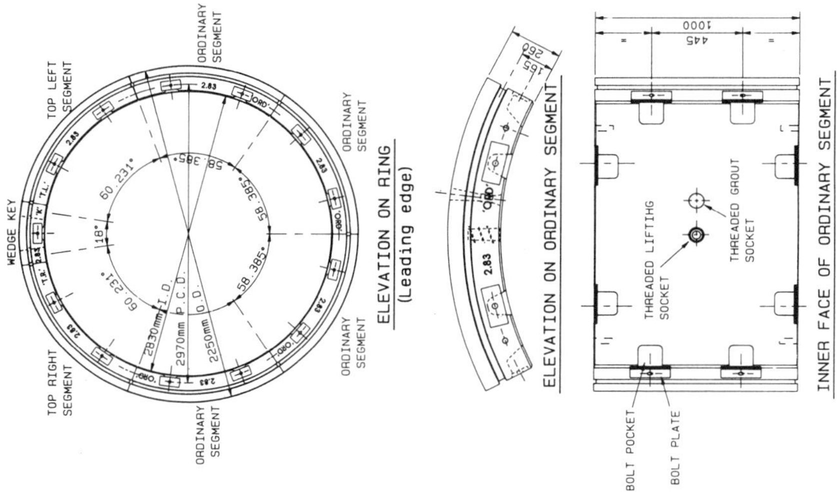

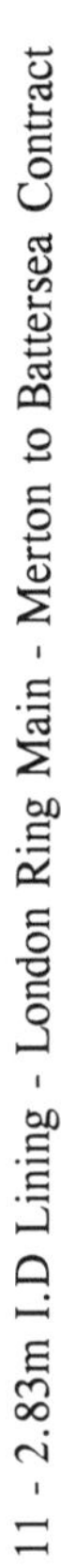
Fig. 11 - 2.83m I.D Lining - London Ring Main - Merton to Battersea Contract

EPDM compression gaskets were used and tested to resist 3.5 bar external water pressure with joint gaps of up to 10mm. A Grade 50 concrete was used in these reinforced linings which were used at depths of up to 17m in water bearing and faulted coal bearing measures.

Bolted smoothbore with gaskets and bolt plates

A smoothbore bolted variant of this type of ring, using steel bolt plates, was designed for the Green Lane Interceptor Sewer on the Wirral in 1990. This 2.44m I.D. smoothbore ring consisted of five segments and a wedge shaped key. The 1.0m wide segments each weighed 660 kg and were picked up through a steel lifting/grout socket at the centroid of each segment. Segments and rings were bolted together using sherardized bolts and cast-in steel bolt plates. The parallel sided rings were driven through competent Sherwood sandstones and the gaskets were designed to accommodate 2 bar external water pressure with joint gaps of up to 10mm.

In 1990 a smoothbore bolted lining with an EPDM gasket was used to restart a flooded tunnel drive on the London Ring Main contract at Tooting Bec Common. The existing 2.80m I.D. tunnel and shield were submerged beneath 40m of water and the tunnel drive was rescued by dropping a new 6.0m diameter shaft and ground freezing. A Lovat EPBS machine was installed and a smoothbore lining designed to cope with the new conditions.

The new 2.83m I.D. lining consisted of five segments and a wedge shaped key. (Fig. 11). The lining was 265mm thick with large EPDM compression gaskets and was fully bolted using 15mm thick sherardized bolt plates and M24 bolts. The lining was designed for depths of up to 50m in the over-consolidated London clay and the 40m head of water. The lining was parallel sided and packing of up to 6mm was allowed for alignment correction. The 1.0 tonne segments were lifted centrally with a 50mm coarse threaded steel screw attached to the erector; the screw located in a threaded socket formed in the concrete.

Six segment trapezoidal bolted linings

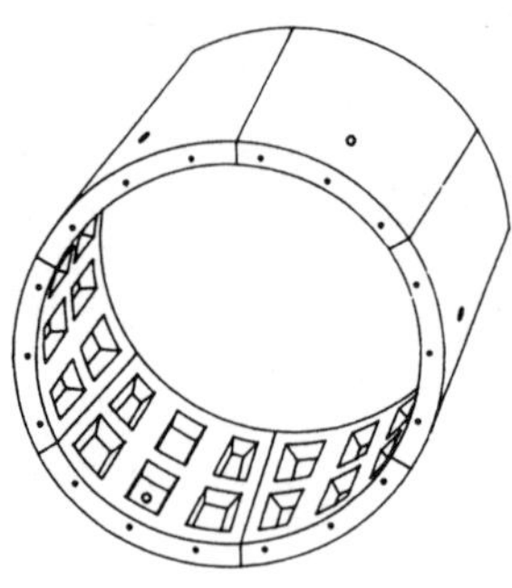

Fig. 12.

In 1989 a 2.9m I.D. trapezoidal bolted lining was designed for Stage IV of the Don Valley Interceptor Sewer in Sheffield. The 180mm thick, 1.0m long rings were a development of the standard panelled bolted linings and incorporated EPDM rubber compression sealing gaskets in preformed grooves. The tunnel was excavated using drill and blast techniques in fissured and faulted coal bearing measures at depths of up to 25 metres. (Fig. 12 and 14).

The lining was erected behind a basic shield using a simple erector which lifted the segments by the bolt holes. The rings were built with segments always in the same orientation, a full segment in the invert, and had the advantage that cruciform joints (where the corners of four segments meet) were avoided. This had always proved an area prone to water ingress in previous designs with gaskets.

A similar ring of 2.74m I.D., using hydrophilic sealing strip, was used on Contract 3 of the M.E.P.A.S. scheme in Liverpool in 1992.

Six segment trapezoidal smoothbore linings

In 1992 several smoothbore trapezoidal linings were used incorporating hydrophilic sealing strip in preformed grooves.

A 3.35m I.D. tapered lining with curved bolts was used for the construction of a Bulk Handling Conveyor Tunnel under the River Avon in Bristol. These linings had a 10mm taper across one axis and were 180mm thick with M16 bolts curved to a radius of 300mm. A 500m vertical curve was constructed under the river and horizontal and vertical alignment adjustment was achieved by rotation of the rings. (Fig. 15).

Similar designs of ring are under construction at the time of writing; a 3.35m I.D. smoothbore lining providing the water cooling tunnels for Barking Reach Gas Fired Power Station at Dagenham. These linings incorporate a small rubber compression gasket to cater for external water pressures up to 40m with joint gaps of 5mm.

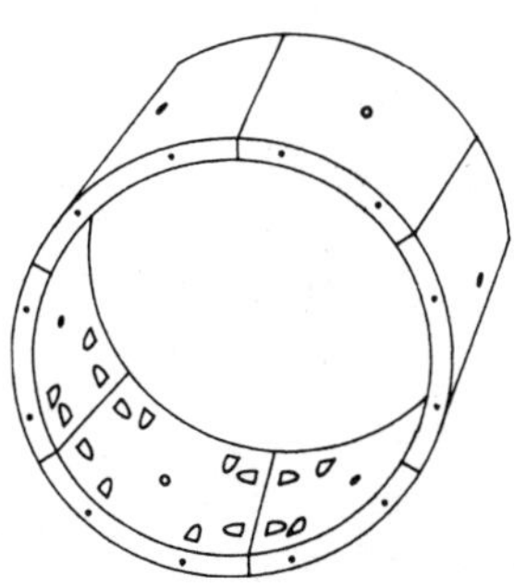

Fig. 13.

Fig. 14 - 2.9m I.D. Trapezoidal Lining - Stage IV of the Don Valley Interceptor Sewer

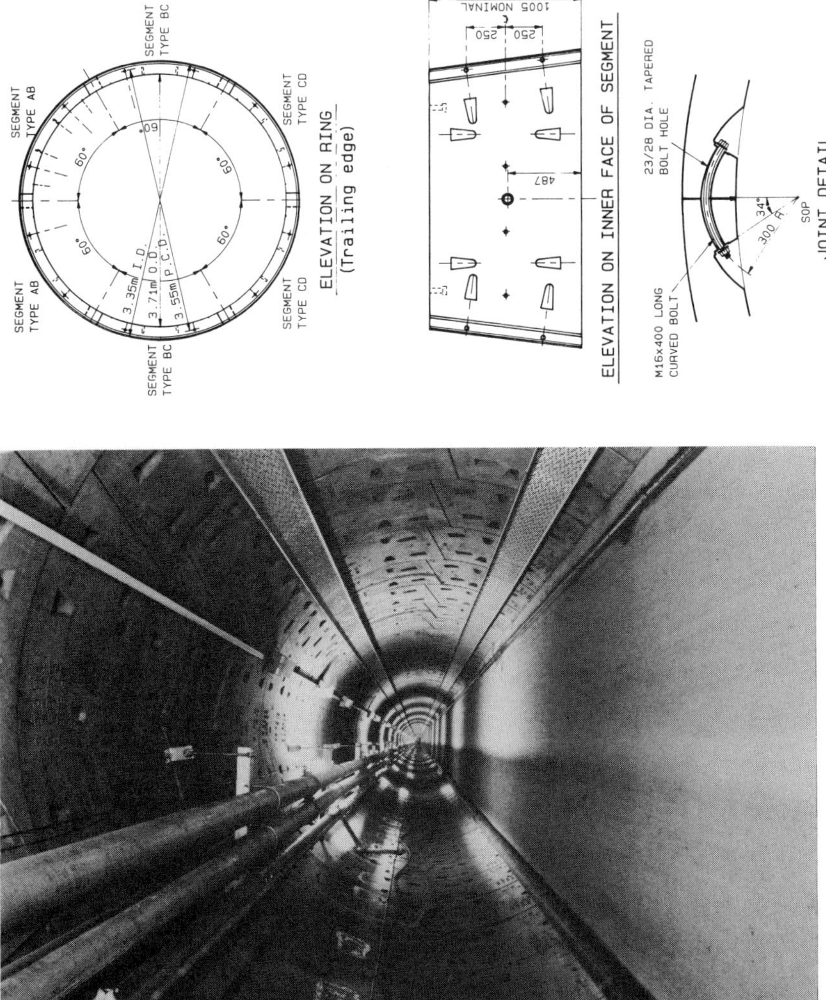

Fig. 15 - 3.35m I.D. Smoothbore, Bulk Handling Conveyor Tunnel, Bristol

A similar 2.44m I.D. x 180mm thick smoothbore lining is being used for the 4 km long main storage tunnel on the Cleethorpes Sewerage Scheme; the rubber compression gaskets are designed to cater for external water pressures of 2 bar with joint gaps of 6mm.

The erection of all these linings is similar behind Lovat TBM's. Segments are supplied to the erector in a pre-determined sequence, dependant upon the alignment required. The rings consist of A, B and C type segments and are erected in a set sequence to simplify the ring build. The ring orientation is altered by supplying segments in a different order to the rotating arm erector which picks up segments using a specially developed coarse thread steel screw and ball attachment.

Straight driving is achieved using a three ring sequence with the rings rotated at 120° to each other, as shown in Fig. 16.

With this type of ring and method of building, the last segment erected is always in the top half of the bore. The trapezoidal joints help the ring keep its shape prior to grouting, which follows closely behind the TBM, and have been built to a tight tolerance.

Similar designs of ring using hydrophilic rubber seals have been built on the London Ring Main and currently on the Southport Interceptor Sewer. These rings have a locking steel dowel connection in place of curved bolts on the circumferential joints.

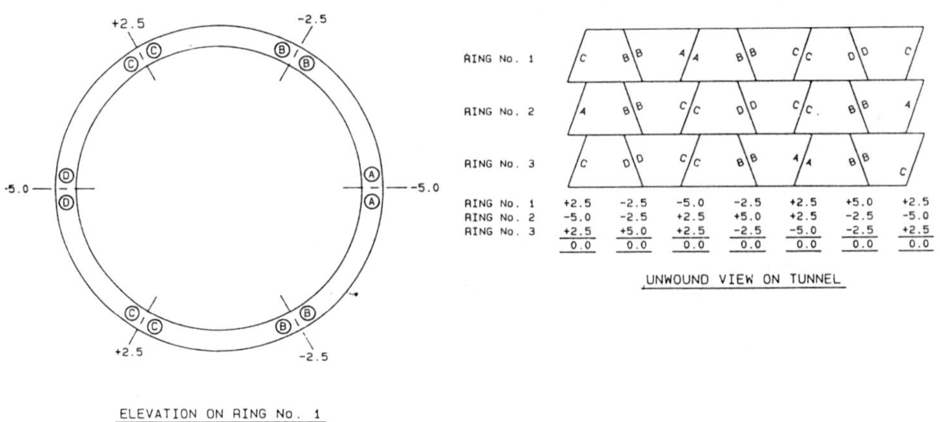

Fig. 16 - Straight driving sequence for six tapered rings

Expanded linings

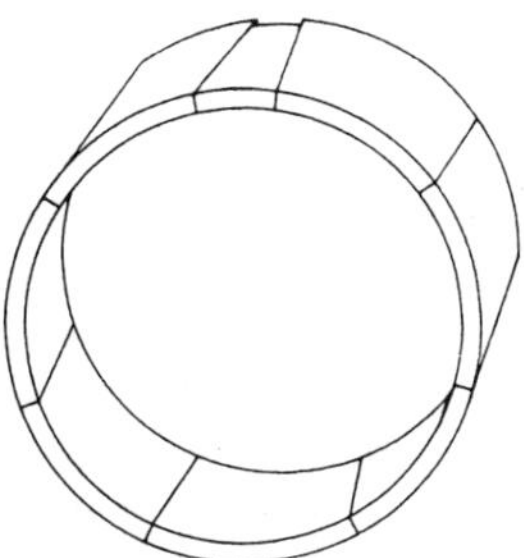

Fig. 17.

In 1983 a 3.81m I.D. expanded lining was used on the extension to the Picadilly Line for London Regional Transport at Heathrow Airport. The 750mm long rings consisted of sixteen segments plus two wedge keys, 550mm long, driven into place at the knee joints by hydraulic rams on the tail shield to expand the lining. A Grade 50 concrete was used in this 152.5mm thick un-reinforced lining. Most of the tunnel was driven in stiff London clay at depths of up to 17m. The longitudinal and circumferential joints were plane, except for the helical joints at the keys, with no mechanical connection.

In 1991 new 2.59m and 2.91m I.D. (Fig. 17) wedgeblock linings were designed for Thames Water for the remaining contracts on the London Ring Main. The Thames Water rings were wider (1000mm) and had fewer segments per ring (7 + key) than the original 685mm long, 100" wedgeblock with 12 segments per ring. The longitudinal and circumferential joints were plane, except for the helical joints at the key. A Grade 60 concrete was used in these 180mm thick segments. The rings were designed to be used in conjunction with the three Lovat tunnelling machines provided by Thames Water for faster building rates. The segments were unreinforced at normal depths in London clay (up to 50m), reinforcement was added to extend their use at depths of up to 70m.

The significant development at this time was the development of a bolted version of this ring, compatible with the erection system of the wedgeblock. This extended the range of ground conditions that could be dealt with and allowed the expanded lining to be used in drives where poor ground was anticipated.

Expanded grouted linings - Channel Tunnel

The linings designed for the U.K. side of the Channel Tunnel [8] in 1989 were similar to those used in the aborted construction in the 1970's. The tunnels for the Channel Tunnel were located in the relatively soft and impermeable Lower Chalk Marl at depths of up to 160m. An articulated smoothbore lining built behind an open faced TBM was used. The lining was expanded directly against the excavated ground as this allowed the fastest construction cycle.

The 7.6m I.D. main marine running tunnel linings consisted of 8 segments and a wedge-shaped key. All segments had 20mm high concrete pads on the extrados which came into contact with the ground when the lining was expanded. The 20mm annulus and any additional overbreak were grouted immediately behind the TBM, the grouting fulfilling the additional function of reducing water ingress into the tunnel. The 1.5m long rings were expanded by a 1.3m long wedge-shaped key, the 200mm wide gaps at the key were filled in as part of the grouting procedure. Longitudinal joints were curved to allow articulation and reinforced against bursting stresses. Both convex/convex and convex/concave joint shapes were used in the U.K. linings.

The linings had no sealing system but diverted water through channels on the intrados of the joints to a drainage system in the invert. Overbreak in the upper half of the bore necessitated packing out behind the lining with cement bags or treated timber as is often the case with expanded linings in these conditions.

Of significant interest was the modification of the TBMs by the addition of trailing steel fingers on the rear of the shield. (Fig. 18).

The fingers slotted between the pads on the extrados of the ring and spanned between the TBM and the previous ring. The key was inserted and held in position by the erector and only rammed home when the rear of the TBM had been advanced for the next ring build. Curved aluminium former rings were also used to help minimise lipping between segments until the ring was grouted.

The service tunnels were of similar design to the running tunnels but consisted of 6 segments and a wedge key. Grade 60 concrete was used and lining thickness varied between 270 to 410mm and 270 to 540mm for the service and running tunnels respectively.

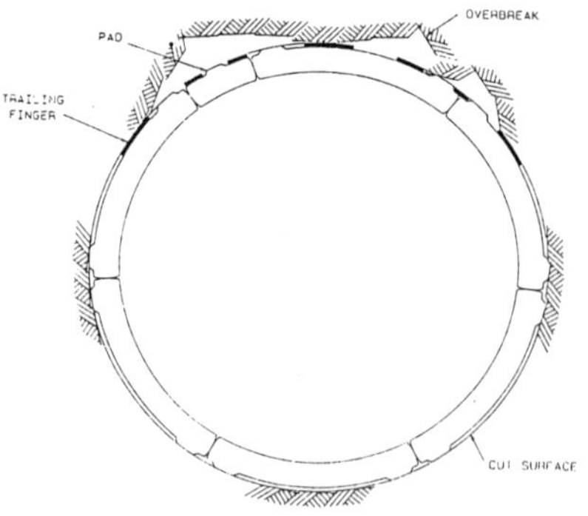

Fig. 18.

Watertightness

The traditional method of achieving watertightness in shafts and tunnels had been the combined use of back grouting the annulus between the lining and ground and a caulking groove provided on the inner face of all segments.

In the early 1980's the principal caulking material, an asbestos/cement cord, PC4 produced by Expandite, was withdrawn from manufacture because of the health risks associated with asbestos. No immediate replacement was found and alternative methods of sealing linings were investigated.

Hydrophilic rubber

This material was first imported from Japan and, as its name suggests, absorbs and swells on contact with water.

The hydrophilic neoprene rubber strip is supplied as a continuous strip which is cut off to length to suit segment size. Initially it was glued to two edges of standard segments and when adjoining segments were bolted together it expanded to provide a seal. Whilst providing a reasonable seal, it impeded the build of the rings and was soon placed in precast grooves around all four sides of every segment. It can be used on standard bolted rings and on smoothbore grouted linings with mechanical fixings.

This is still a popular way of sealing linings and most standard linings available from manufacturers now provide grooves for its use if required. Groove sizes vary in depth from 1.5 to 3mm as standard, depending upon the grade required. The groove size and the material type are selected according to the pressure to be sealed against and the gap to be sealed. A typical joint detail is shown in Fig. 19. It has the disadvantage that the full sealing potential is not instantaneous and requires the

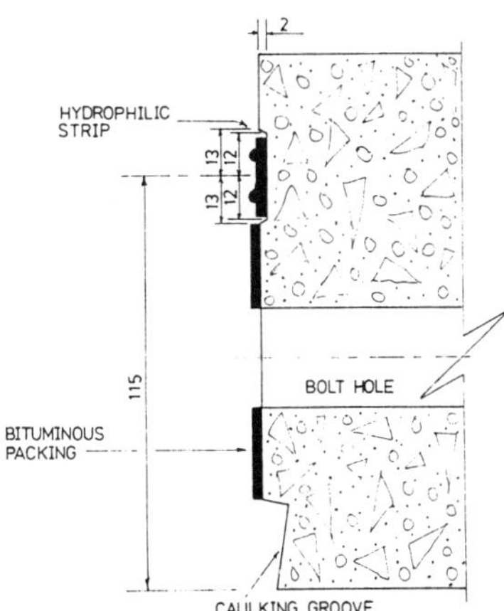

Fig. 19.

continuous presence of water to effect a seal; in a situation with fluctuating water levels the seal may be intermittent. The swelling characteristics of the rubber are reduced in saline water. The long term sealing performance has not been established.

The use of timber packing for line and level adjustment of the tunnel alignment generally means that an instant seal is not possible.

Rubber compression gaskets

At about the same period, rubber compression gaskets were introduced with bolted segmental linings. These gaskets are usually made from EPDM (Ethylene Propylene Diene Monomer) or Neoprene rubber and are manufactured as a complete rectangular frame to fit into preformed gasket grooves on all four sides of every segment. Gasket corners are joined by vulcanising. The gaskets are glued into the grooves to a snug fit and may be fitted at the factory or on site. (Fig. 20).

They form a permanent seal designed to cope with the construction tolerances normally associated with tunnels.

Jointing at the corners of segments is critical and requires special attention. Water ingress is most likely to occur at a cruxiform joint (where the corners of 4 segments meet) and this should be avoided where possible. Damage to the concrete around the gasket groove will also impair the efficiency of the seal, as will dirt between the seals.

The pressure and gap sealing properties vary with the gasket manufacturer and can be tailored to suit individual specifications. The gap and pressure sealing capacities will seal at pressures of up to 5 bar water pressure with gaps of up to 10mm. Performance can be varied by altering the groove depths; gasket size and shape. These gaskets protrude significantly from the edges of the segment and hence the final segment to be inserted must be wedge-shaped to allow the segment to be positioned.

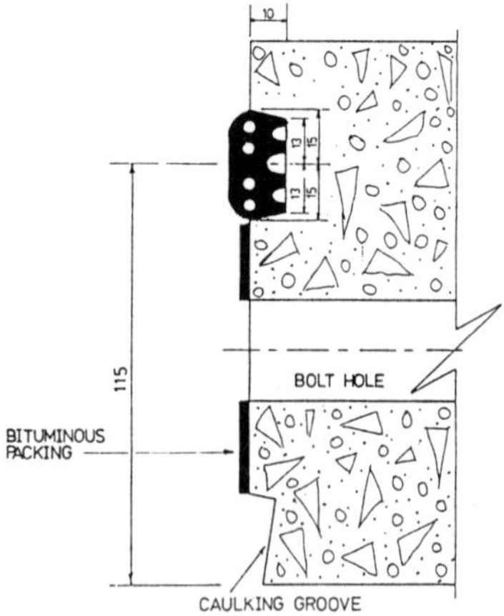

Fig. 20.

Methods of joining segments and rings

There are four main types of joint used in precast linings; plane, concave/convex, convex/convex and tongue and groove. The last three do not usually have any physical connection and the details of these joints have been described elsewhere [1].

Plane joints are the predominant joint type used in the U.K. in standard linings and a great variety of fixings have been used to connect these joints.

Traditional flanged joint

Traditional bolted linings are fully bolted, segment to segment and ring to ring. Standard thread bolts pass through reinforced concrete flanges; grommets are used to seal the bolt holes and washers to prevent punching failure. A 3mm thick bituminous packing separates longitudinal joints and all inside edges have a caulking groove. The bituminous packing reduces high contact pressures and the gel grummets seal the bolt holes against water ingress. The joint is easily adapted to accommodate sealing gaskets; a typical detail is shown in Fig. 21.

This provides a strong, simple joint suitable in all ground and water conditions and is easy to build.

Standard linings utilize bolt sizes varying between M20 to M30, depending on the lining size and flange thicknesses. Typical flange thicknesses vary between 65 and 120mm with bolt pull out strengths of 2.5 tonnes to more than 5 tonnes. Shear strengths are similar.

When a gasket is used it is usual to provide a recess on the segment behind the gasket to prevent spalling due to ram loading or handling damage.

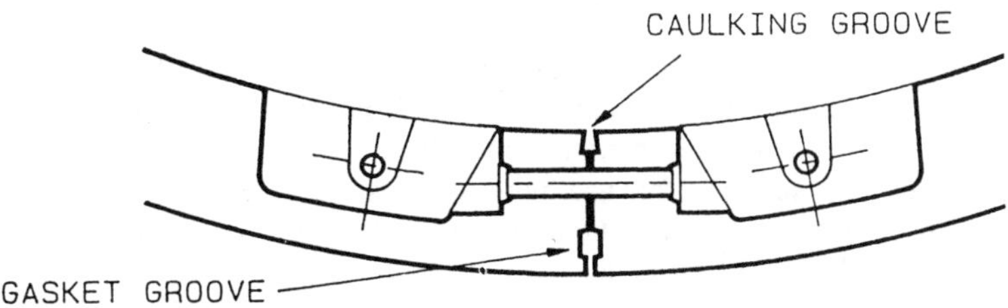

Fig. 21 - Bolted Flange Joint

Bolt plates

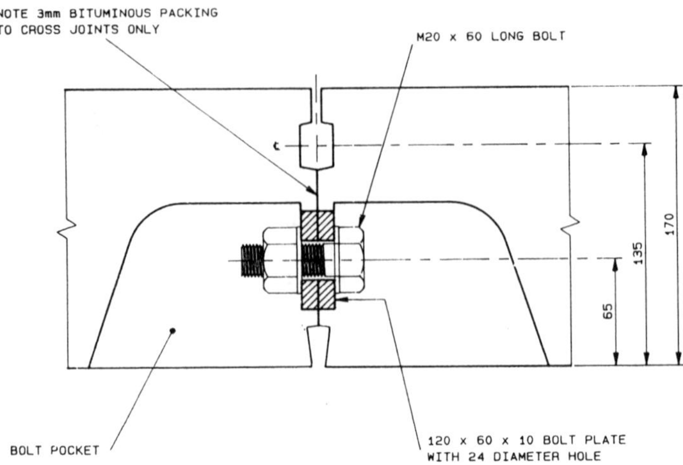

Fig. 22 - Bolt Plate Joint.

Originally developed by the Japanese, bolt plates have been used to provide a fully bolted connection between smoothbore segments and rings. (Fig. 22).

Steel bolt plates are cast into the segments in similar positions to the bolt holes on traditional linings. The bolt plates are fully anchored into the segments by steel anchor bars.

The bolt plates and anchors which are designed to suit particular bolt sizes are sherardized or galvanized.

The joint is easily tailored to meet most circumstances and can be used with sealing gaskets. Typical bolt plate thicknesses vary from 8mm to 16mm and have design strengths varying from 1 to 5 tonnes in bolt pull out and shear.

The bolt plates are suitable for use with gaskets but a minimum segment thickness is required to accommodate the bolt plate and the gasket and still provide satisfactory cover to the steel bolt plate.

The bolt pocket requires an infill pellet or insitu mortar to provide a smooth finish and to provide cover to the bolt plate and bolt.

Timber dowels

First use of these dowels in the U.K. was at the Carsington Aquaduct Tunnel in 1980.

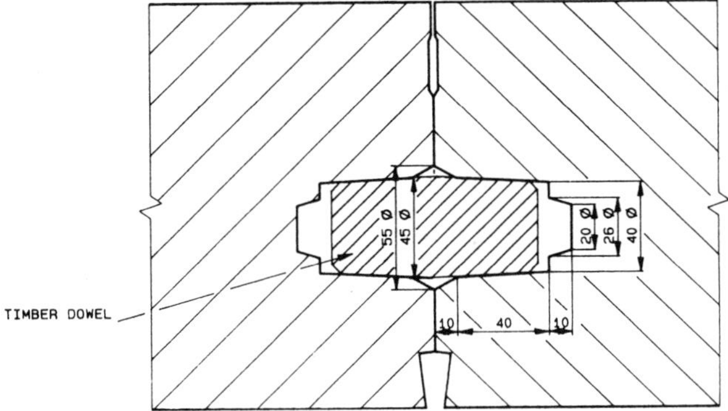

Fig. 23 - Timber Dowel Joint

In 1984 a 6.0m I.D. smoothbore lining was used to reline the 180 year old [8] Blisworth Canal tunnel near Northampton.

The 225mm thick segmental lining with integral concrete fenders at axis consisted of twelve segments. Longitudinal joints were connected through 16mm galvanized steel bolt plates and 20mm bolts.

Circumferential joints were connected using double conical timber dowels to give speed and accuracy in erection. The timber dowels were a nominal 50mm diameter and were manufactured from white oak. The dowels were supplied at a controlled moisture content and sealed in polythene bags until immediately prior to use. The dowels were designed to have a shear strength of 2 tonnes and a pullout strength of 1.5 tonnes.

Different sizes have been used on a number of other contracts since then, most notably on 2.8m I.D. smoothbore rings on several contracts on the London Ring Main for the Thames Water Authority. (Fig. 23).

Curved bolts

Curved bolts had been used on several contracts overseas, on the Hong Kong and Singapore Metros, but were not used until 1988 in the U.K. at Stansted Airport [7].

M24 x 470mm long galvanized bolts bent to a 300mm radius were used in curved holes that tapered from 35 to 40mm to allow the withdrawal of hole formers after casting. Even with this tolerance the dimensions of the curved bolts had to be carefully controlled to ensure bolt entry.

The bolting action of the joint depends upon the stiffness of the bolt; this is usually designed to be flexible enough to allow the line of bolting action to pass through the pitch circle diameter rather than on a straight line between the nut and bolt head.

The curve of the bolt reduces the bolting force by the ratio of the sine of the angle between the bolt head and the plane of the joint.

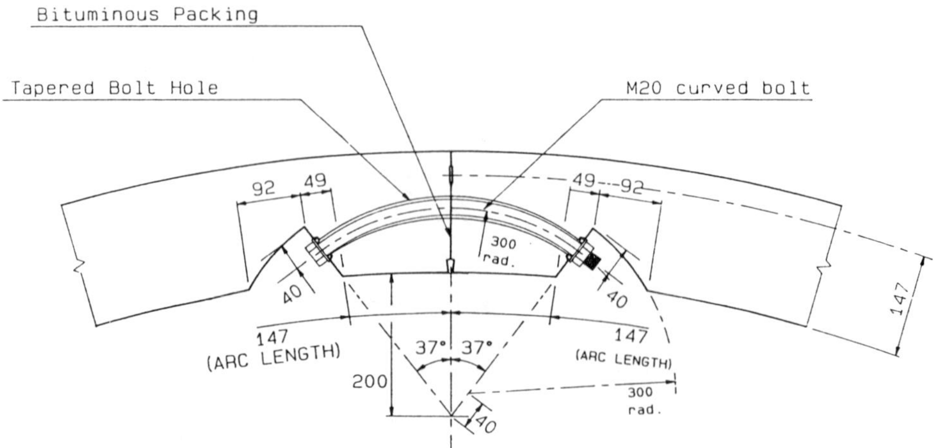

Fig.24 - Curved Bolt Joint

The joints are readily used with gaskets and give a strong bolted connection with a smoothbore ring but the geometrical restraints of this joint design usually result in a lining thickness greater than that required by ground loading.

Bolt curvatures of between 250 - 300mm satisfy the conflicting requirements of a small bolt recess, satisfactory bolt action and concrete joint strength. Rings with gaskets have been designed with curved bolts of M20 and M16 diameter in smaller diameter rings with a thickness of 180mm. (Fig. 24).

The joints are reinforced in a similar manner to a flanged joint.

Proprietary joint fixings

Several lining manufacturers have developed joint fixings to provide a mechanical connection between joints and provide a lining which requires minimal finishing works to achieve a smooth lining with a fast erection cycle.

The joint details are suitable for use with hydrophilic sealing gaskets.

Longitudinal joints

Several methods are available on standard ranges of tunnel linings produced by the manufacturers.

C. V. Buchan produce a range of "Bucline" [3] tunnel linings with a hidden bolted connection in which a fast thread mushroom headed bolt locates in a steel slot. The bolts are inserted in one side of the joint prior to erection and the bolt heads positioned into recesses in the opposite side; the segment is pushed sideways by the shield rams to engage the steel slots; the bolts are finally tightened through a gap in the caulking groove using an open ended spanner to tighten a hexagonal nut integral with the bolt.

Charcon Tunnels produce a "Onepass" [4] range of linings which connect segments by an expanding fastener between interlocking steel reinforcing loops cast into the ends of the segments. The fastener is inserted parallel to the joint and driven home by a power tool.

Circumferential joints

C. V. Buchan produce a hidden bolt detail on the circumferential joint similar to those on the longitudinal joints and are currently introducing a "Buclock" plastic dowel with a locking action. This has the advantage of giving a self-aligning action, a securely locked mechanical fixing and retaining the fast erection cycle of the timber dowel from which it was developed.

Non-ferrous dowels are located on one side of the joint and retained by a serrated plastic rod cast in the segment. The joint is made by pushing the segments into position using the shield rams. Segments are pushed individually against the previously built ring.

Charcon Tunnels [4] provide a "fastlock" dowel system. Dowels are placed unlocked into sockets in the previous ring and the whole ring pushed back after erection to close the gap. The ring under construction is built 150mm away from the previous ring. The dowels are positively locked by gripper washers in each socket.

Double conical timber dowels and plain steel locating dowels have also been used as previously described.

Segment tolerances

Manufacturing tolerances should be realistic and appropriate to the lining design, usage and erection requirements. It should be borne in mind that tight tolerances add significantly to production costs and that the finished tunnel is always built to a tolerance several orders of magnitude larger than those specified for segments.

Tolerances for lining types differ, but they should always meet the following criteria:-

1. Buildability: The rings must be capable of being built into a tunnel or shaft to the degree of accuracy required in normal construction.
2. Structural Design: The tolerances must be consistent with the requirements of the structural design.
3. Sealing Systems: Tolerances should be appropriate to the sealing systems employed.
4. Practical: Tolerances should be achievable using practical and economic manufacturing techniques.
5. Quality Control: It must be demonstrated that the stated tolerances are achieved by regular checks during manufacture.

There is no British Standard or Code of Practice specifically related to the design and manufacture of tunnel linings. Design and materials are normally controlled in accordance with the general principles of B.S.8110 [9]. and the Water Industry Specification [10]. These are given below for segments and joints:-

Circumferential length	$\pm$ 1.5mm
Radius of Curvature	$\pm$ 3mm
Thickness	$\pm$ 3mm on back face
Width	$\pm$ 1.5mm
Lips at Joints	$\pm$ 5mm

These are generally satisfactory for the standard bolted linings for which they were written.

Tunnelling consultancies produce their own lining specifications and tolerances for individual ring designs on major contracts.

A tighter set of tolerances are appropriate for segments with gaskets (generally with tapered rings) in particular on the following dimensions:-

1. Segment width
2. Segment thickness
3. Groove depth
4. Groove mismatch at corners

The following specification meets the above criteria and has been generally achieved in production over the last few years:-

Tolerances for tapered rings with compression gaskets:-

Rings - type tests

Two test rings are erected on a flat and level base; the top ring to be rotated 180° relative to the bottom ring.

The following dimensions shall be checked:-

1. Internal diameter	$\pm$ 5mm
2. Height of two rings	+ 3mm - 1mm
3. Lip between adjacent segments on internal diameter	$<$ 3mm
4. Gap between longitudinal segment joints	1mm feeler gauge not passing

Segments

1. Segment thickness	+ 2mm - 2mm
2. Segment width	+ 1.5mm - 0.5mm

3. Flatness of sides	1mm feeler gauge not passing beneath a 1.0m long straight edge
4. Width of groove	± 1mm
5. Depth of groove	+ 1.0mm - 0.5mm
6. Mismatch of groove at corners	< 1mm

Type tests are carried out at moulding stage and are satisfactory for checking diameter, buildability, warp, flatness/squareness of ends and taper.

Groove depth, width, mismatch at corners and flatness of sides, as well as segment thickness and width, must be checked on segment production; usually at the sampling rate of 1:20 segments.

It is important to remember these tolerances when choosing a gasket; this is often neglected. Rubber gaskets are produced by continuous extrusion, they are normally extruded to within a variation of ± 0.5mm. If this tolerance is added to the groove tolerance then it can be seen that a theoretical variation in protruding height of 2.5mm can occur when a gasket is in position. When both sides of the joint are considered there can be a theoretical variation of 5mm in the worst case. This should be taken into account when testing is carried out.

Durability

Precast concrete linings were first used in the U.K. in 1903 but it was not until after 1945 that they came into general use. Most of the tunnels constructed at this time were for the transmission of water or sewage. These were secondary lined with bricks, insitu concrete or precast concrete infill pellets. There is little data available on the performance of these tunnels.

In 1982 a CIRIA Report [11] was published on the hydraulic roughness of precast concrete segmental linings; the fifteen tunnels surveyed for the Report consisted of the four most common smoothbore lining types: Wedgeblock, Knuckle-jointed Smoothbore, Smoothbore Bolted and Standard Reinforced Bolted with Infill Panels. Most of the tunnels were in the region of 10 years old and were all in good condition.

As previously mentioned, the majority of segments in the U.K. are designed to comply with the Water Industry Specification [10] and B.S.8110 [9]. The Mix Design Specification is superior to that required for similar products such as that required in B.S.5911; [12] the British Standard concerning precast concrete pipes, and has the following properties:-

High cement content	(> 400 kg/m^3)
Low water/cement ratio	(< 0.45)
Well compacted and dense concrete	
Low porosity and absorbtion	

The majority of the earlier precast linings were manufactured using Ordinary Portland Cement but Sulphate Resisting Portland Cement (S.R.P.C.) has been used extensively since the 1970's.

A typical current Mix Design for Standard Segmental Linings is given below:-

S.R.P.C.	410 kg/m^3
Coarse Aggregate 14mm single sized Quartzite	1180 kg/m^3
Fine Aggregate Grade 'M' washed Quartzite Grit Sand	700 kg/m^3
Water/Cement Ratio	0.42

Chemical resistance

Major concerns regarding the durability of buried concrete are its resistance to attack by chemicals in the environment; specifically sulphates and low pH values in the ground water and surrounding soil and industrial effluent in sewage tunnels. Clear guidance on this subject is given in BRE Digest 363; [13] standard precast concrete linings produced by the authors company all comply with Class 4 in the Digest.

Most standard linings are annulus grouted, thus providing a high pH coating to protect the linings. If a particularly aggressive environment is anticipated, perhaps in situations with a highly mobile ground water with a low pH value (< 3.5 pH), segments have been spray coated at the factory with a coal tar pitch epoxy coating in two layers. Several sewers in the West Midlands have used this protection where a tunnel passed through Victorian industrial waste tips.

Cover to reinforcement

Corrosion of reinforcement is another major factor affecting the long term durability of linings. Cover to reinforcement is important and should be chosen to give the optimum balance between durability and the effectiveness of the reinforcement.

Small diameter (< 4.0m) standard bolted linings have a **minimum** cover to reinforcement of 15mm; this corresponds to a "moderate exposure" in the B.S.8110 classification with a Grade 50 concrete. this is considered appropriate for concrete subject to condensation or in contact with non-aggressive soils. Smoothbore and larger standard bolted linings have a minimum cover of 20mm; this is appropriate for "severe exposure". Occasionally a minimum cover of 40mm is required in contract specifications. This is usually as a result of adherence to the Code of Practice [14] for water retaining structures. The author considers that this level of cover is excessive in sewage and water transmission tunnels, where the concrete lining is usually in compression; such a high level of cover seriously reduces the effectiveness of the reinforcement. In very aggressive environments it is better to take other measures to increase durability, as described elsewhere in this paper.

Permeability
Given the constraints of the physical geometry of tunnel linings, it is more effective to concentrate on the permeability of the concrete to gas and water and thus reduce the effects of carbonation and the ingress of sulphates and chlorides in solution.

This was the approach adopted for the Channel Tunnel; a concrete of very low permeability and diffusibility was specified for the linings which have a design life of 120 years. A high strength Grade 60 concrete was specified which used a 70/30 blend of OPC/PFA, a water/cement ratio of 0.35 and had a cement content of 430 kg. The cover to reinforcement was specified as a nominal 35mm (30mm minimum).

Permeabilities in the order of 1×10^{-13} m/s [8] were achieved in this concrete.

This approach has started to spread throughout the industry and a number of major contracts since then have specified similar mix designs.

The specification for the concrete linings on the proposed Jubilee Line Contracts, with a required design life of 400 years, stipulates that the coefficient of permeability to water should be a maximum of 1×10^{-13} m/s for cores conditioned at 75% RH at 20°C and tested at 35 days; samples should also have an oxygen diffusion coefficient value not greater than 5×10^{-8} m/s.

Research [15] carried out to determine the mix designs required to meet these requirements indicates that this can be achieved with closely controlled production methods with good curing and compaction. The most cost effective approach, and the mix design most suited was found to be very similar to that used on the Channel Tunnel; this required the addition of PFA to achieve a high fines content and a super plasticizer to achieve a low water/cement ratio.

Cast-in items
Cast-in steel connections and fittings in segments have been in use for over 30 years and the author is not aware of any reported corrosion problems in this country. The connections are galvanized or sheradized and are positioned so that they always have a cover to the intrados or extrados of the lining greater than that specified for the reinforcement.

The connections are not necessary for the structural performance of the standard linings and are always isolated from the steel reinforcement. Connection recesses are filled in after construction with a high strength low shrinkage mortar or caulking material.

However, cast steel fixings are considered to be a potential weakness by some engineers and non-ferrous alternatives are being sought.

Threaded steel grout sockets have been in use for over 15 years and can be strengthened to allow lifting via segment erector. These have recently been replaced by coarse threaded plastic grout/lifting sockets. A typical detail is shown in Fig. 25.

A coarse thread is moulded into a plastic socket which is cast in at the centroid of the segment. The thread has 3 starting points for ease of insertion of the steel lifting pin by which the segment is lifted by the segment erector.

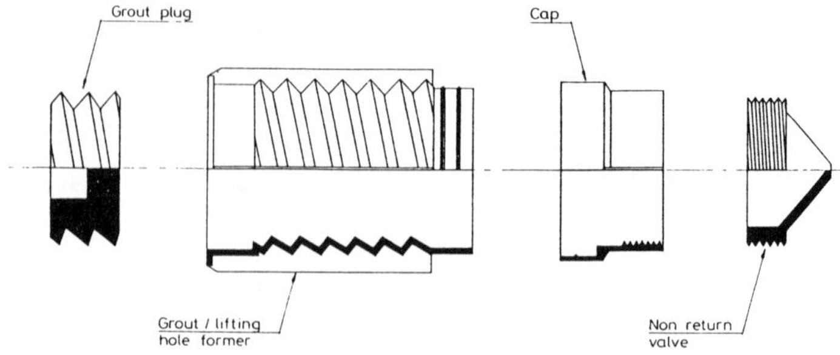

Fig. 25 - Plastic Grout/Lifting Socket

The design is based on the optimum balance of ease of fixing, cleaning and pull out strength.

The design shown has a pull out failure strength in excess of 5 tonnes in a Grade 50 concrete.

A simple four petal split cone plastic non-return valve can be easily fitted to the rear of the socket to prevent grout loss. It is most appropriate when the lining is used with a TBM with tail seals on the shield, particularly when a bentonite plug is maintained directly behind the shield. The values have been tested to 3 bar external water pressure.

A plastic threaded grout plug is also available.

Discussion of developments

In pursuit of faster rates of progress, and reduced handling and sealing costs, segments have become wider and longer and the number of connections in rings has been reduced to a minimum of two per joint. The introduction of heavier segments has coincided with the move away from manual construction with the increasing use of TBM's and segment erectors; this has been further encouraged by the greater awareness of health and safety at work and restrictions on manual lifting.

The introduction of sealing gaskets has encouraged greater awareness of accuracy of construction and has lead directly to many developments; tapered rings developed to allow the control of alignment and the negotiation of curves without packing; wedge shaped keys developed to allow the placing of the final segment without fouling the gasket; six segment trapezoidal rings developed to allow a single ring type to be used with a final segment closure above axis and a herring bone pattern of segment joints to eliminate the coincidence of four gasket corners.

The ring developments previously described are shown together in Fig. 26. The

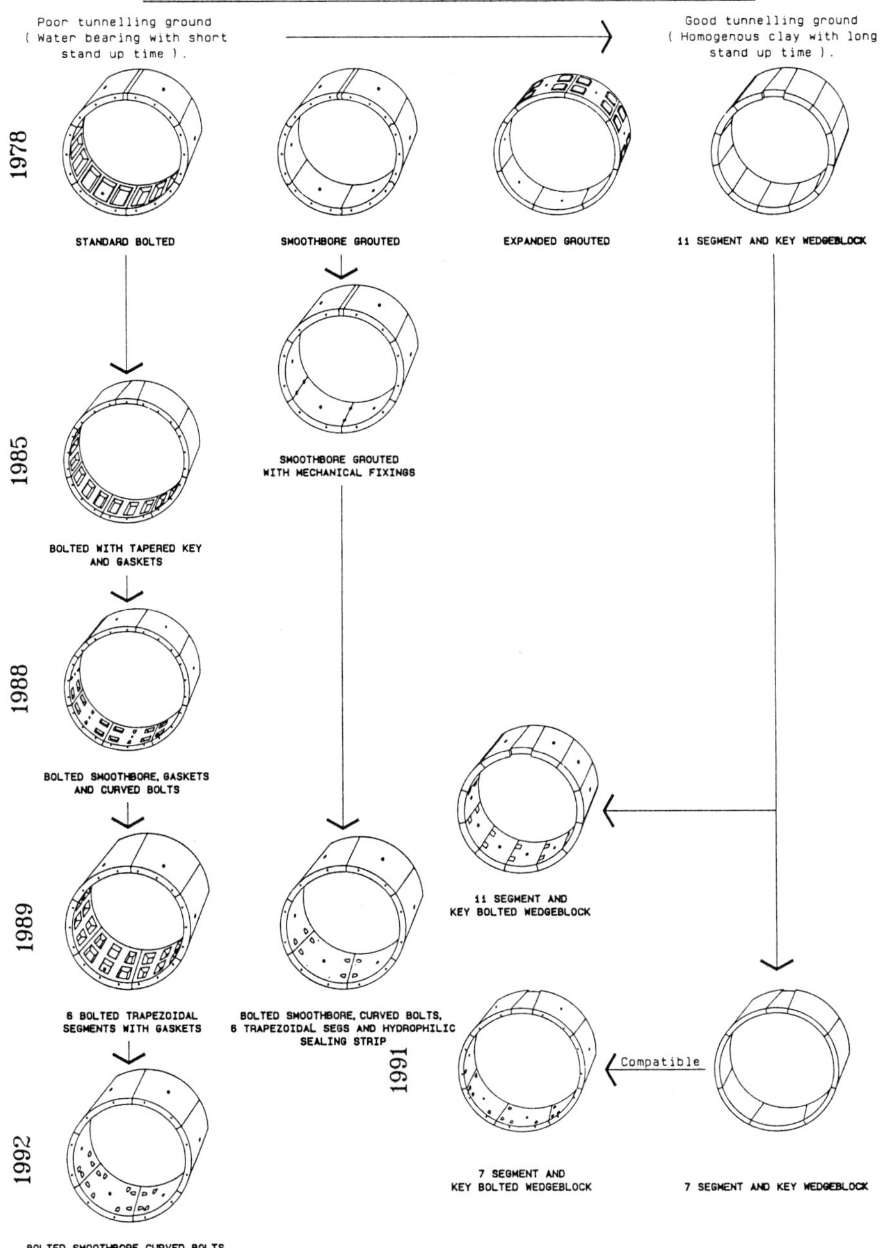
Development of Precast Linings - Since 1978.
Poor tunnelling ground
(Water bearing with short
stand up time).
Good tunnelling ground
(Homogenous clay with long
stand up time).
1978
STANDARD BOLTED
SMOOTHBORE GROUTED
EXPANDED GROUTED
11 SEGMENT AND KEY WEDGEBLOCK
SMOOTHBORE GROUTED
WITH MECHANICAL FIXINGS
1985
BOLTED WITH TAPERED KEY
AND GASKETS
1988
BOLTED SMOOTHBORE, GASKETS
AND CURVED BOLTS
11 SEGMENT AND
KEY BOLTED WEDGEBLOCK
1989
6 BOLTED TRAPEZOIDAL
SEGMENTS WITH GASKETS
BOLTED SMOOTHBORE, CURVED BOLTS,
6 TRAPEZOIDAL SEGS AND HYDROPHILIC
SEALING STRIP
1991
Compatible
1992
7 SEGMENT AND
KEY BOLTED WEDGEBLOCK
7 SEGMENT AND KEY WEDGEBLOCK
BOLTED SMOOTHBORE, CURVED BOLTS,
6 TRAPEZOIDAL SEGS AND GASKETS

Fig. 26.

diagram represents the development of ring types suitable for different ground conditions; from left to right ground conditions progress from poor to good, the development of linings is shown chronologically from top to bottom.

The convergence in the development of standard bolted linings with smoothbore grouted linings becomes apparent in the fully bolted tapered ring with curved bolts, trapezoidal segments and sealing gaskets.

A similar development has occurred with expanded linings. Variable ground conditions and the search for fast progress rates on the London Ring Main contracts led directly to the use of a bolted wedgeblock lining and the development of an expanded lining with fewer segments. Wedgeblock and bolted linings have been combined into a compatible two ring lining system which can be built behind the same shield and use the same erection system.

Future developments

We can expect a continuing trend towards off site preparation, with linings supplied to site with gaskets and packings factory fitted or cast-in. The linings will be supplied in rings and perhaps shrink wrapped in polythene to protect the gaskets and aid curing of the segments.

The search for enhanced durability will direct the design of concrete mixes towards greater strength and impermeability with increasing use of super plasticizers and blended cements. With this will come the demand for a reduction or elimination of steel reinforcement and steel cast-in items, perhaps to be replaced by high strength plastic with good concrete bonding properties.

The ultimate in soft ground ring design would be a totally smoothbore lining requiring no caulking or finishing and incorporating a small sealing gasket. It would be manufactured from a high strength, very low permeability, low cost concrete and have no cast-in items. Built behind a TBM, it would be expanded against the ground as it left the tail and then grouted to fill the overbreak.

This is not feasible at the moment but it is now possible to envisage a system of compatible smoothbore linings suitable for all ground conditions. Such a lining system might have three or four interchangeable lining types with the same internal diameter; an expanded lining for good clays; an expanded grouted lining where overbreak becomes significant; an expanded, bolted and grouted lining where additional support is required and finally a fully bolted and grouted lining for bad ground conditions.

All the linings would have 6 - 8 wedge-shaped segments; the rings would be tapered to allow alignment correction and would all have provision for sealing gaskets. The rings would be built behind a TBM with several types of detachable tail skin but using the same erector and segment train.

This is an ideal to work towards and would only apply to linings for large contracts where mould costs could be justified. The majority of small and medium sized contracts will continue to require economic and versatile designs, tailored to meet the specific ground conditions and machinery available. The range of standard linings and special designs produced by manufacturers reflects these needs and are continuing to develop and improve in line with current trends.

References

1. **A Review of Tunnel Lining Practice in the United Kingdom**. T.R.R.L. Supplementary Report 335. 1978.
2. Doughty. M. **Sewer Reconstruction at Westham uses One-Pass Concrete Lining.** Tunnels and Tunnelling, September 1980. P. 35 - 37.
3. C. V. Buchan Ltd. - Tunnel and shaft brochures.
4. Charcon Tunnels Ltd. - Tunnel and shaft brochures.
5. Martin. D. **Neoprene Gaskets Seal Lining on Sheffield Sewers**. Tunnels and Tunnelling, March 1985.
6. Traherne. D. **Contract 3 Moves Forward to Successful Conclusion**. Tunnels and Tunnelling, April 1988.
7. Wallis. S. **Tunnel Taxies Under Airport Runway**. Tunnels and Tunnelling, June 1989.
8. Curtis et al. **The Channel Tunnel: Design, fabrication and erection of precast concrete linings**.5th International Symposium "Tunnelling 91". P. 161 -172.
9. British Standards Institution. **Structural Use of Concrete**. B.S.8110: Part 1: 1985.
10. Water Authorities Specification: Civil Engineering Specification for the Water Industry (3rd edition). Publ. Water Research Centre.
11. Pitt. J.D. and Ackers. P. **Hydraulic Roughness of Segmentally Lined Tunnels**. C.I.R.I.A. Report 96, 1982.
12. British Standards Institution. **Precast concrete pipes, fittings and ancillary products**. B.S.5911.
13. Building Research Establishment. **Sulphate and acid resistance of concrete in the ground**. BRE Digest 363. 1991.
14. British Standards Institution. **Design of Concrete Structures for Retaining Aqueous Liquids**, B.S.8007: 1987.
15. C. V. Buchan Ltd. **Internal Report on Permeability Testing of Concrete. 1993**.
16. British Standards Institution. **Guide to Planning and Construction of Sewers in Tunnels**, B.S.8005, Part 3, 1989.

Session 11: Case Histories

Channel tunnel: French undersea crossover design and construction

Y. Leblais
Simecsol, Paris, France
L. Leblond
Department of Engineering, TML French Site, Calais, France

Abstract
This Paper covers the whole of the reasons which led to the design of the French crossover, including various types of constraints : construction programme, boring schedule and boring methods of the running tunnels, geometrical, geotechnical and structural. After describing the geotechnical and structural aspects, the design methods and the construction techniques, it presents the outstanding facts of the construction from the point of view of a back-analysis of the design of the structure. Eventually, the knowledge gained from this realisation is emphasized. Finally, it makes some comparisons between the French and the U.K. crossovers.
Keywords : Chalk, Channel Tunnel, Construction methods, Design, French crossover, Geotechnical, Structural, Tunnel lining.

1 Function and basic design

1.1 Function

The Channel Tunnel is 50.4 km long and has four crossovers. These crossovers are special structures that allow trains to pass from one running tunnel to the other. Trains can still run in both directions using only one running tunnel while the other is used for maintenance of tunnel structures and fixed equipment.

The first crossover is located outside the Beussingue portal on the French side, two are undersea, the last one is underground at Hollywell on the UK side. The undersea crossovers are located at approximately one third of the underground length from each portal, for operating reasons. The UK undersea crossover is 7.5 km from the UK coast. The French undersea crossover is 12.5 km from the French coast (Fig. 1).

1.2 Geological setting

The geological plan and section of the Channel Tunnel is shown in Figure 1. Undersea, immediately beneath the superficial seabed deposits, the Chalk is divided in two layers, the Grey Chalk overlying the Chalk Marl. The Chalk Marl is underlain by 2 or 3 m of Glauconitic Marl beneath which lies the Gault Clay.

All the available data induced the tunnels and caverns to be excavated in the Chalk Marl. This layer is a dark grey bluish clayed carbonate mudstone deposited thinly and regularly, in a sequence of sedimentary cycles with some sponge beds near the base. The Chalk Marl is considered to have a low permeability, essentially bound to the discontinuities. From the earlier studies, it was considered a very good tunnelling material.

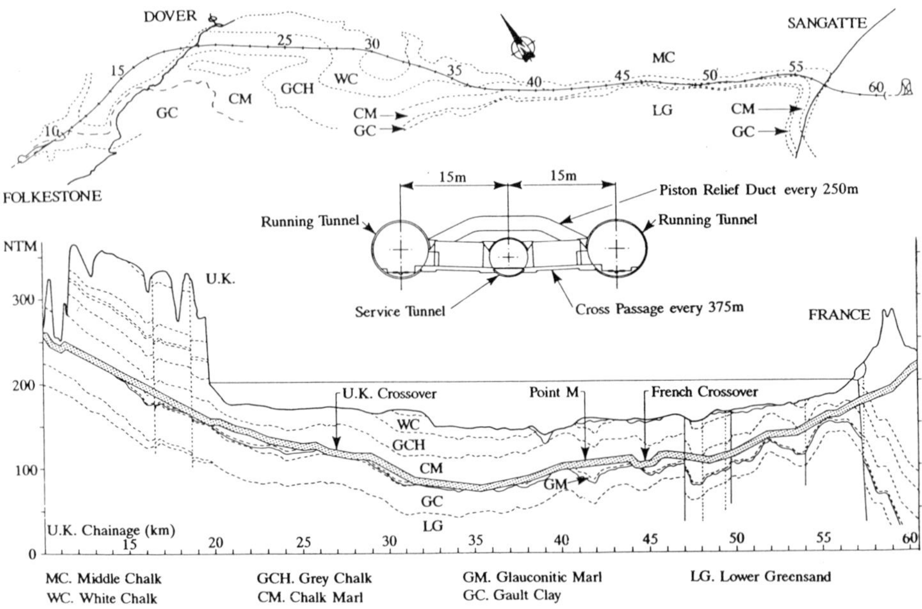

Fig. 1. Channel Tunnel. Geological plan and section

1.3 Basic design

It was initially envisaged to have the crossover function realised by two adits linking both running tunnels, but this scheme was abandoned (except at Hollywell) and a large cavern scheme adopted for the following reasons :

- this scheme was too difficult to place in the available sound and impervious chalk ;
- it needed four connection structures between the tunnels and the adit, each of which as big and as difficult to build as the single cavern itself ;
- these structures were not compatible with programme constraints ;
- the cost of the four structures would have been too high.

The single cavern appeared to be the optimum structure in terms of safety programme and cost for both the French and the UK crossovers. This meant the deviation of the service tunnel from between the two running tunnels to the north of the route [1]. North was chosen because :

- the chalk marl layer is inclined downwards in the north direction at the crossover location ;
- evacuation of material from the crossover to the service tunnel would have the benefit of gravity ;
- construction personnel would not be trapped in case of major water ingress in galleries and could move to a safe place in the service tunnel.

To minimise the cavern width the first step was to define what could be the minimum distance between the two running tunnels. This was derived from the analysis of the chalk pillar that would remain between the tunnels in the vicinity of the crossover.

The final figure adopted for both the French and the UK crossovers is 10.5 m between centreline of running tunnels, i.e. a chalk pillar of about 1.5 m minimum thickness.

The necessary cavern size to achieve the required function is 170 m long, 19 m wide and 8 m high. These impressive dimensions result from the shuttle gauge size and alignment criteria :

- the maximum speed within the crossover is 70 km/h, leading to a minimum horizontal radius of 1.000 m since it is not possible to have any cant in this area ;
- the track crossing, UE 15 type, has an angle of 4 degrees which is the maximum angle resulting from the required radius given the distance between tunnels.

For lining design reasons and considering ground characteristics on the French side, the minimum required height is 12 meters.

As a consequence of this scheme large longitudinal sliding doors had to be designed and installed to isolate both running tunnels during normal operations for safety reasons. Equipment rooms could be installed within the space available above the tracks. Two passenger evacuation cross passages are necessary from the crossover to the service tunnel.

2 Choice of construction methods

Two fundamental criteria governed the choice of the construction methods : SAFETY and PROGRAMME.

2.1 Safety

Priority was given to safety during construction. The following criteria were imposed :

- prevent major water ingress ;
    - perform preventive ground treatments ;
    - install powerful temporary pumps ;
    - have small cross sections for any opening ;
- place the crossover within the best possible ground ;
- use simple and classical boring techniques.

2.2 Programme

The fundamental programme constraint on the French side was that the undersea cross-over was on the critical path of the whole project due to its distance far from the coast at the vicinity of French and UK running tunnel junction. The French crossover was then the biggest and last underground structure to build on the project (Fig. 2).

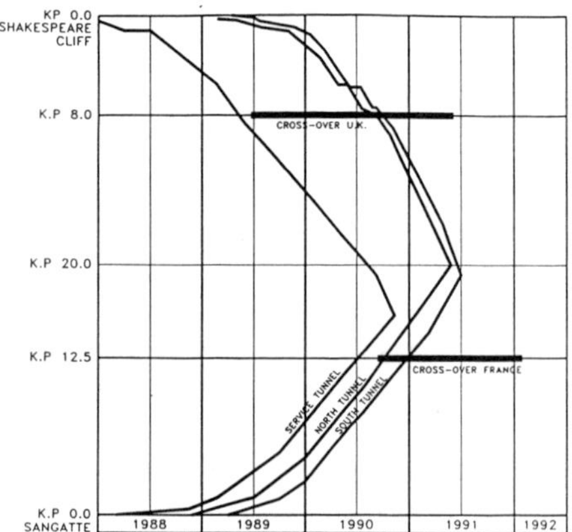

Fig. 2. Undersea tunnels and crossovers
Construction programme actual dates

This resulted in the following criteria in order to minimise construction time :

- allow crossover construction to start as soon as the service tunnel would reach the location of the crossover ;
- allow running tunnel TBMs to pass through the crossover at any time during its construction without building any additional structure prior to TBM arrival or during TBM crossing ;
- avoid any interference between boring of running tunnels and crossover construction ;
- minimise any technical risk that could have resulted from the use of sophisticated construction equipment, then use classical equipment and standard supply ;
- use very flexible construction methods.

2.3 Construction methods

These criteria could not be achieved with construction methods based on large face opening. After technical and economical comparisons of many alternatives the following principles were adopted (Fig. 3a to 3c) :

- the crossover structure would be built from two inclined ramps driven from the service tunnel. It would consist of small galleries surrounding the future cavern ;
- the internal dimensions would be such that the TBMs could pass through at any time ;

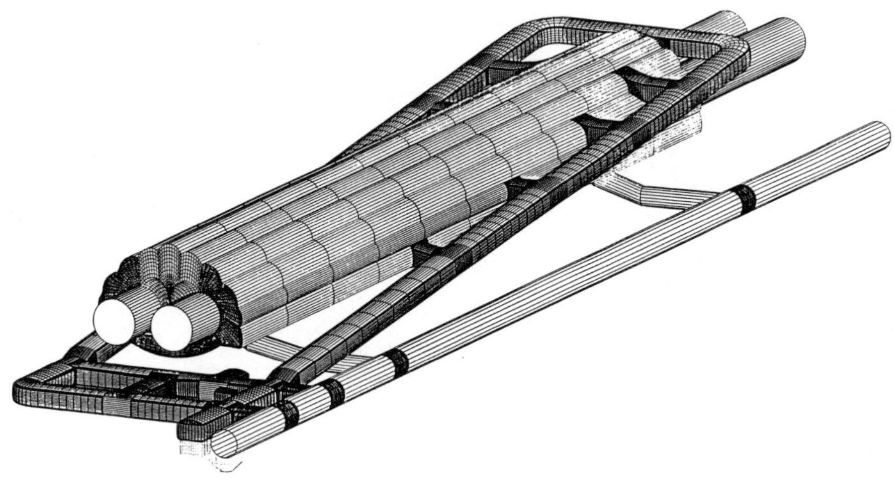

Fig. 3a. French Crossover. Axonometric view

- the galleries would be built either with small mechanical tools or Lynx machines ;
- the galleries would be filled with concrete without any sophisticated shutters. These adjacent galleries would constitute the final lining of the cavern ; since the minimum height of the galleries would be 2 m - i.e. more than the minimum thickness required for structural strength - there would be no need for a high strength concrete or for reinforcement ;
- most of the chalk within the cavern could be removed by the TBMs therefore allowing the use of running tunnels muck carriers rather than those of the service tunnel that could concentrate on transporting the chalk from the galleries.

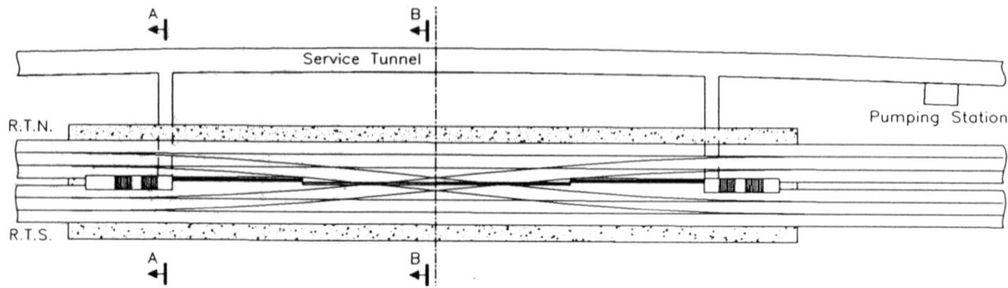

Fig. 3b French Crossover. Plan layout

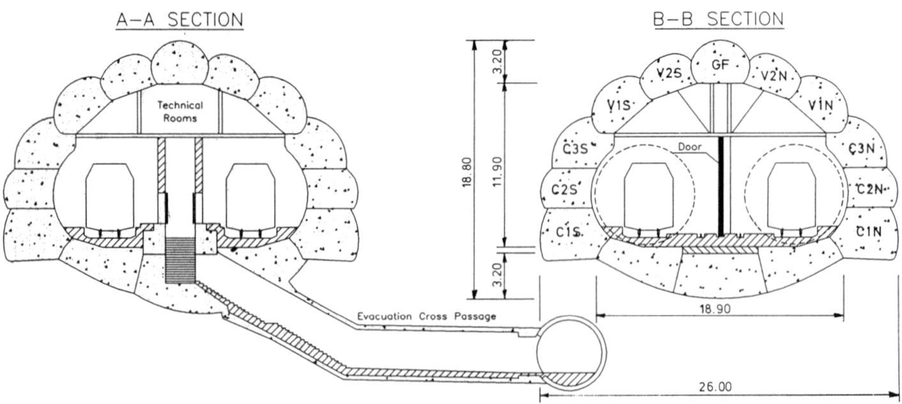

Fig. 3c. French Crossover. Typical cross - sections

3 Geotechnical design

Since numerous published papers deal with the geological design of the Channel Tunnel [2 to 7], this paper emphasize only the geotechnical design of the French Crossover.

3.1 The 1985 geotechnical data

When the geotechnical design of the Channel Tunnel started again in 1986, some geotechnical data were available from the previous site investigations between 1958 and 1985 (Table 1).

It was considered that the viscous behaviour of the chalk could be taken into account through a Kelvin-Voight viscoelastic model with a relationship between a long term strain modulus (E_∞) and a quasi instantaneous or undrained one (E_0), using the creep factor ϕ , $\{E_\infty = E_0 /(1 + \phi)\}$.

Table 1. Channel Tunnel. Geotechnical data in 1985

	W (%)	γ_d (kN/m^3)	UCS (MPa)	E_o lab (MPa)	E_o field (MPa)	Creep factor ϕ
Turonien	10.0	20.7	6.3	2500	-	1 to 2
Cenomanien						
White Chalk	14.0	19.5	8.0	1700	-	< 1
Grey Chalk	11.0	20.3	8.5	1700	-	≅ 1
Chalk Marl	9.2	21.2	6.0 to 9.0	UK : 920	-	
				FR : 2340	FR : 2500	≅ 1
Glauconitic Marl	8.7	21.8	6.0	1600	-	-
Gault Clay	19.0	17.9	3.0 to 4.5	UK : 620	UK : 250	-
				FR : 350	FR : 350	

Since the strength of the chalk was largely higher than the initial stresses at the tunnel depth, it looked quite clear that the chalk behaviour would be an elastic one. Thus the next investigations would have to emphasize on its stiffness rather than on its strength.

3.2 The 1986-1987 geotechnical data

The 1986-1987 geotechnical site investigation on the French side consisted of nine marine boreholes firstly used to control the geophysical survey. They were cored continuously into the Lower Greensand in order to get a good correlation with the strong geophysical reflectors that are the top and bottom of the Gault Clay. Subsequently the cores were used to perform laboratory testing.

In the first stage the potential sample sections to be tested were determined according to the water-content using continuous gamma-densitometer tests. In the second stage the samples were selected in these sections and it was emphasized on the weakest strata to get a more conservative design. It appeared clearly that the test results were very scattered mainly due to the carbonate content.

As the excavation of a tunnel induces a global stress release in the surrounding ground, it seemed adequate to perform laboratory tests with such stress paths. Thus the strain moduli were determined through uniaxial and triaxial tests with different stress paths in compression, extension and constant average stress tests.

The creep behaviour was studied through two tests :

- step by step tests, with successive 20 to 60 min steps loaded each with an increase of a tenth of the failure stress ;
- long term tests with a constant load during 90 days.

The creep factor (ϕ) was determined using two non convergent - conservative - relationships between the strain (ε) and the time (t) :

$$\varepsilon = \varepsilon_o \left(t / t_o \right)^{\beta} \qquad \text{and} \qquad \varepsilon = \varepsilon_o + a \log t / t_o$$

The in situ stress state was introduced through the K_o at rest coefficient. It was determined by means of hydrofracture tests performed in deep boreholes near the Sangatte shaft and oedometer tests. The in situ tests indicated a K coefficient of 0.7 ; the laboratory tests gave a K_o value of 0.5. That value was used both transversely and longitudinally for all the calculations on the French side.

Table 2. French side. Geotechnical data in 1986-1987

	γ_h (kN/m^3)	UCS (MPa)	E_o (MPa)	Creep factor ϕ	Mass Permeability (m/s)	K_o
Grey Chalk						0.5
PK -0.2 to 1.0	22.5	9.0	2000	1.5	10^{-6}	
PK 1.0 to 8.5	22.5	9.0	1500	1.5	5 10^{-6}	
PK 8.5 to 16.3	22.5	9.0	1300	1.5	1 to 5 10^{-6}	
Chalk Marl						0.5
PK -0.2 to 1.0	23.5	9.0	1500	1.5	10^{-7}	
PK 1.0 to 8.5	23.5	9.0	1500	1.5	5 10^{-7}	
PK 8.5 to 16.3	22.5 to 23.0	6.0 to 9.0	900 to 1300	1.5	1 to 2 10^{-7}	
Glautonitic Marl	23.5	5.0	900	1.5	5 10^{-7}	0.5
Gault Clay	21.5 to 22.0	1.8	200	2.1	10^{-9}	0.5

3.3 The 1988 site investigation

In addition to the previous survey, a specific site investigation was performed in 1988 at the location of the French crossover. It consisted of a geophysical survey with two deep boreholes, FX01 and FX02. They were located on a line parallel to the route nearly 40 m north, and symmetrically to the axis of the cavern. The distance between them is 190 m.

These boreholes were continuously cored 5 m into the Lower Greensand to get undisturbed samples in order to perform additional laboratory testing.

3.4 Position of the crossover within the ground

The additional site investigation confirmed that it was possible to set the cavern within the Chalk Marl (Fig. 4) taking into account the dip in the strata and considering the following criteria :

- cover of 25 m minimum above the cavern, larger than its overall height ;
- cover of 5 m to the top of the Chalk Marl ;
- distance of 3 m minimum to the Gault Clay.

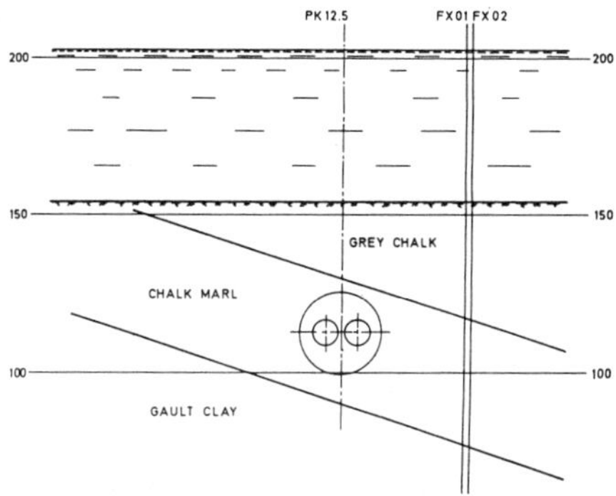

Fig. 4. French Crossover. Geological cross-section

3.5 Geotechnical results

It was confirmed that :

- the Chalk was slightly fractured with a low mass permeability ;
- the Gault Clay was stiff with a very low permeability ;
- there was no strong geological or permeability marker between the Chalk Marl and the Grey Chalk.

3.5.1 Chalk Marl

The R.Q.D. in the Chalk Marl at the level of the cavern was 90-100%, with fractures of 0.7 to 2.5 per metre. Considering the good quality of the rock mass, it was decided not to use any classification system ; the mechanical properties of the Chalk Marl were directly taken from the average values from the laboratory testing.

Regarding the dimensions of the crossover it was necessary to find an adequate geotechnical qualification to take into consideration the cyclic way in which the chalk was deposited and the strong influence of the carbonate content on its mechanical characteristics.

For that reason the carbonate content (% $CaCO_3$) and the longitudinal sound velocity (VL) were determined every 10 to 15 cm all along FX01. A strong correlation between the two parameters (Fig. 5) was confirmed ; the high values of the velocity are noticeable and indicate the good quality of the samples.

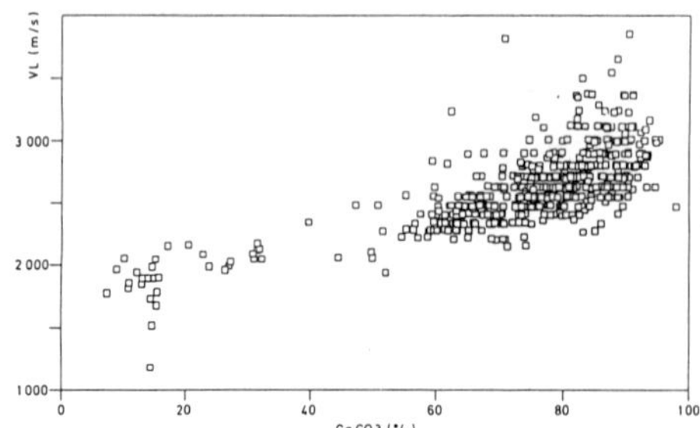

Fig. 5. FX01. Longitudinal Velocity versus Calcium Carbonate Content

Due to a better statistical definition of the carbonate content, that parameter was retained and determined on the same way for FX02. The two logs appeared strongly similar (Fig. 6).

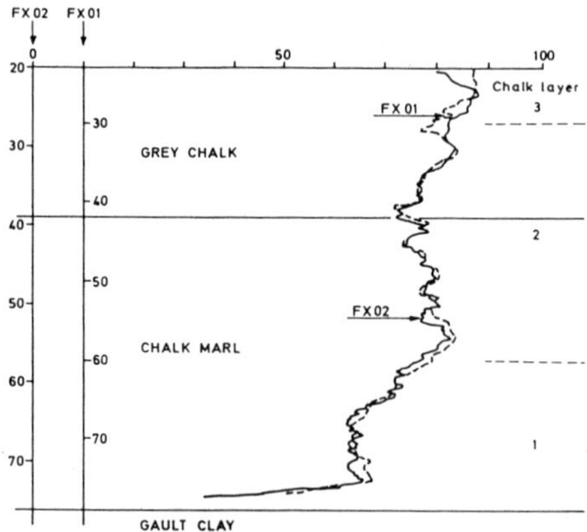

Fig. 6. FX01 - FX02. Logs of Carbonate Content

Three main chalk layers were determined according to the average value of the carbonate content and to its range of variation (Fig. 6) :

- the Layer 1 extends from the Glauconitic Marl to 60 m below the sea level ; the carbonate content presents high variations (60 to 80 %) on small lengths (≅ 10 cm) and increases from the bottom to the top ;
- the Layer 2 extends from 60 to 30 m ; the spacing between the low content cycles is greater and the variations are smaller ;
- the Layer 3 is above ; the log shape is quite similar to that of the Layer 2 but the content increases.

Additionally, the decimetrical chalk cycles (more than 1000) were classified in four elementary families :

Chalk Family A	% $CaCO_3$ ≤ 70 %
Chalk Family B	70 % ≤ % $CaCO_3$ ≤ 80 %
Chalk Family C	80 % ≤ % $CaCO_3$ ≤ 85 %
Chalk Family D	85 % ≤ % $CaCO_3$

The Chalk mass shear strength was estimated from the average UCS values. The Chalk mass stiffness was based on the undrained modulus tangent to the load origin (E_0), as this modulus is always less than that from the first load cycle (E_1) and quite equal to that from the second cycle (E_2) (Fig. 7).

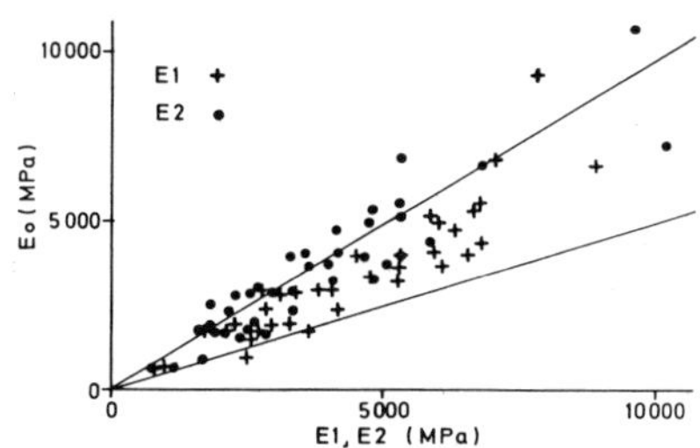

Fig. 7. Chalk Layers. E_0 versus E_1 and E_2

The shear strength profile (Fig. 8a) reflects a strong decrease down from the top of the Layer 1 ; the E_0 stiffness profile (Fig. 8b) is quite different as the decrease is beneath the upper layer.

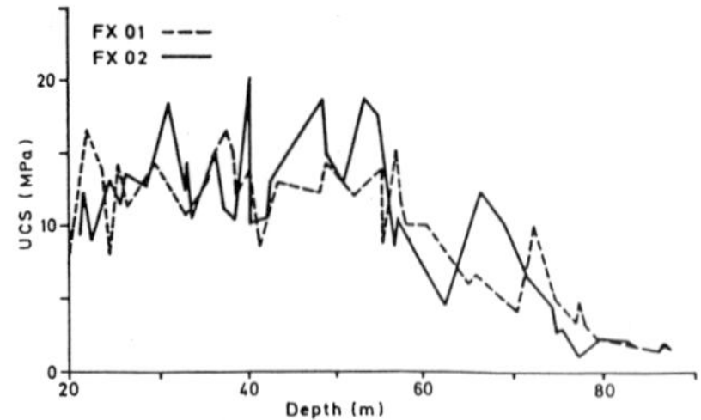

Fig. 8a. Intact UCS profile

Fig. 8b. Intact stiffness profile

The results from laboratory testing were classified according to the families, each of them being qualified by its mechanical properties (Table 3).

Table 3. Chalk Families - Mechanical properties

	Fam. A	Fam. B	Fam. C	Fam. D
UCS (MPa)	7.5	11.9	13.8	12.9
E_0 (MPa)	1426	3068	4511	9660

The proportion of each family was determined in each chalk layer (Table 4).

Table 4. Family percentage per Layer

	Fam. A	Fam. B	Fam. C	Fam. D
Layer 1	68.6	20.9	7.2	3.4
Layer 2	9.0	51.9	28.8	10.3
Layer 3	0.3	9.3	16.3	74.2

Then using an adequate analysis [8], these data were used to qualify the strain anisotropy of the Chalk.

3.5.2 Glauconitic Marl. Upper Gault Clay
Their characteristics were determined in the same way used in the previous investigation. In addition the swelling potential of the Gault Clay was determined through the volumetric swelling observed in both Huder-Amberg oedometer swelling tests [9] and triaxal swelling tests.

3.6 Geotechnical design data

The geotechnical cross-section of the Crossover is shown in Figure 9. Representative geotechnical parameters are listed in Table 5.

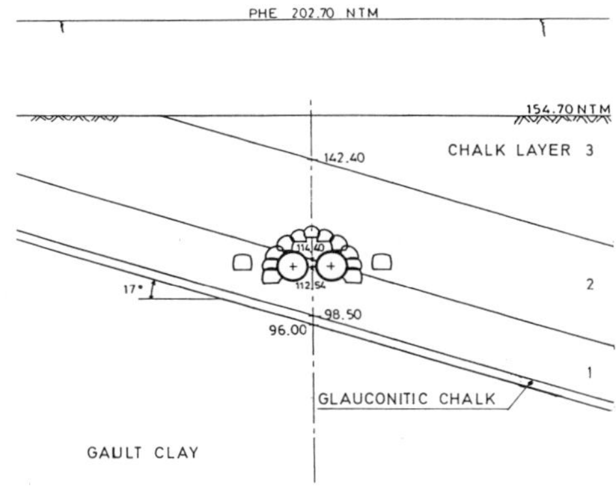

Fig. 9. French Crossover. Geotechnical cross-section

Table 5. French Crossover. Geotechnical design data

	γ_h (kN/m^3)	UCS (MPa)	E_x (MPa)	E_z (MPa)	G_{xz} (MPa)	ν_{xz} ν_{zx}	Creep factor ϕ	K_o
Chalk Layer 3	22.4	10.0	5400	4700	1800	0.25	1.0	0.5
Chalk Layer 2	23.3	10.0	4000	3300	1300	0.25	1.0	0.5
Chalk Layer 1	22.3	6.0	2300	1800	700	0.25	1.0	0.5
Glauconitic Marl	24.5	3.6	800	800	320	0.25	1.0	0.5
Gault Clay	21.5	1.9	600	600	222	0.35	2.1	0.5

An intermediate modulus, equal to the average value between the short term modulus and the long term one, was retained at the end of the construction stages.

The swelling power of the Gault Clay was characterised through the swelling factor $C_g = 0.05$

4 Structural design principles

4.1 Design criteria

One major requirement was to get a watertight structure for maintenance reasons (a drained one would have been acceptable regarding only the low permeability of the ground).

The design was governed by the interaction between the ground and the structures. The following load and material factors were adopted for the ultimate limit state :

• load factors	permanent gravity loads :	1.35
	variable and imposed loads :	1.50
	accidental loads :	1.10
	construction loads :	1.20

- material factors

for normal load combinations	concrete in compression :	1.50
	steel reinforcement :	1.15
for exceptional load combinations	concrete in compression :	1.15
	steel reinforcement :	1.00

According to contractual requirements, the permanent lining had also to be checked under the direct application of the total ground overburden and water pressure calculated at the horizontal axis of the cavern.

4.2 Design data

In addition to the geotechnical ones, the concrete design data were as follows :

- Vault B30 (minimum strength of 30 MPa at 28 days)
- Invert and Internal Structure B35 (minimum strength of 35 MPa at 28 days)

The contact between the tunnel linings and the ground was taken either as perfect adhesion, or as perfect sliding. The contact between the cavern lining and the ground was taken as perfect adhesion. In the case of adhesion, the models allowed a potential detachment between the lining and the ground due to the water pressure.

4.3 Design methods

The design of the cross-section of the cavern both during the construction stages and in the long term stage was based on the convergence-confinement method [10].

The chalk was considered as a linear-elastic medium with a check of the failure criterion. The Gault Clay was taken as linear-elastic up to failure, and perfectly plastic after failure.

Most of the design was verified using the finite-element computer program CESAR (LCPC France) that modelled the slightly fractured ground as a continuum. The models were bi-dimensionnal, in plane strain state, except for the endwalls of the cavern for which a three-dimensional analysis was performed to check the bending effect between the endwall and the cavern lining as well as the connection with the running tunnels.

4.4 Design stages

Three main design stages were considered :

- the design of the pillar between the two running tunnels along the crossover area ;
- the design of the whole lining of the cavern taking the minimum concrete thickness and the design of the endwalls ;
- the design of each individual construction stage interacting with the previous ones.

4.4.1 Chalk pillar between the two running tunnels

It was necessary to minimise the cavern width and to define the minimum thickness of the chalk pillar between the two tunnels. It depends both on the chalk mechanical properties and on the tunnel lining loading.

The stresses induced in the pillar were determined using both analytical approaches [11] to get a rough estimation of the pillar thickness and FEM calculations. The chalk was modelled as a continuum but the pillar was divided in parts to perform calculations either on the stiffness of the chalk, or on the potential damage done to the pillar due to the first tunnel boring [12].

The different models were used to estimate the influence of both the contact between the segmental lining and the surrounding grouted mortar and the stiffness of the mortar itself. Additional calculations were made to determine the potential effect of steel H shape ribs installed in the first tunnel prior to the boring of the second one. The main results of these studies are as follows :

- the chalk behaviour remains elastic even if the main part of the pillar is destroyed. The minor stress is quite horizontal and is a compression that means a low risk of discontinuity opening ; the maximum deviator is always less than the UCS value ;
- the distribution of the normal loads in the pillar, the grouted mortar and the linings is not in direct proportion to their stiffness ;
- the type of contact has no major influence on the pillar stress state ;
- steel ribs would have no influence on the stiffness of the first lining even if they were contiguous. For that reason only two H120 ribs were installed per segmental ring in the first tunnel for safety reason in the event of cracks.

4.4.2 Whole lining of the cavern

The main purpose of these calculations was to check the stability of the vault as long as the invert will be under excavation ; in that case the average chalk deviator beneath the vault feet never exceeds 2.1 MPa with a pick value of 6.1 MPa.

The potential heave which might affect the invert was calculated ; it might not exceed 15 mm due to the creep effect. At last the results confirmed that the water pressure induces the essential load on the cavern lining.

4.4.3 Construction stages

Actually the two running tunnels passed through the crossover area prior to the cavern excavation works. The design had to cope with that situation which allowed the cavern lining to be tangent to the tunnel linings, hence reducing the internal width of the cavern by one meter to 18.90 m (Fig. 3b).

The main purpose of these calculations was to determine the loads in the three chalk pillars during the successive construction stages. Any failure might have induced a strong increase in the arch width (Fig. 10) and potential instability of the area.

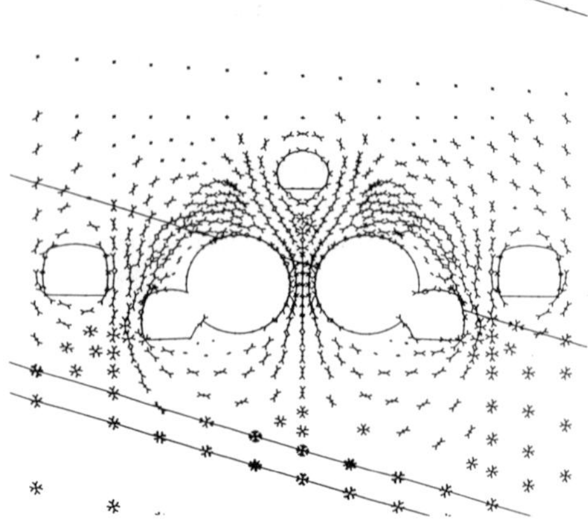

Fig. 10. Typical stress distribution in the pillars

In the first step the model included the linings of the tunnels with the grouted mortar taken as chalk to increase the loads in the pillars. The two ramps were located at C2 level to decrease the width of the lateral pillars. In accordance with the actual working plan, the calculation stages were as follows :

- stage 1 : north running tunnel (NRT) excavation ;
- stage 2 : south running tunnel (SRT) excavation and NRT lining ;
- stage 3 : both north and south ramps excavation and SRT lining ;
- stage 4 : C1N excavation ;
- stage 5 : C1S excavation ;
- stage 6 : GF excavation ;
- stage 7 : C1N concreting and C2N excavation ;
- stage 8 : C1S - GF concreting and C2S excavation ;
- stage 9 : C2N - C2S concreting and C3N - C3S - V2N - V2S excavation ;
- stage 9b : C2N concreting and C3N - C3S - V2N - V2S excavation ;
- stage 10 : C3N - C3S - V2N - V2S concreting and V1N - V1S excavation ;
- stage 11 : V1N - V1S concreting

In the second step some additional calculations were done corresponding to the C1N and C1S excavation with the ramps at the same level. The linings of the running tunnels were not modelled to increase the loads in the ground especially in the lateral pillars. In that case the average deviator does not exceed 4.7 MPa with a pick value of 6.5 MPa to be compared respectively to 4.2 MPa and 5.8 MPa in the above model.

5 Detailed construction methods

5.1 Phasing

The first operation was the drilling from the service tunnel for a general ground treatment. It was designed to achieve a permeability of 10^{-6} m/s.

Construction started from the service tunnel through two standard cast iron opening sets [13, 14]. This method avoided enlargement of the service tunnel that would have taken about 5 months based on the UK experience.

Two horizontal adits 44 m long with a cross section of 20 square metres were excavated under the east end of the crossover . From these adits the excavation of the two lateral ramps started. They were 182 m long with a slope of 10 degrees upwards, to allow easy circulation of men and machines. A transverse horizontal adit linked the two ramps at their top for ventilation and access to the upper gallery. From the two ramps through small access drifts it was possible to excavate the eleven longitudinal galleries . The phasing of excavation is given on Figure 11.

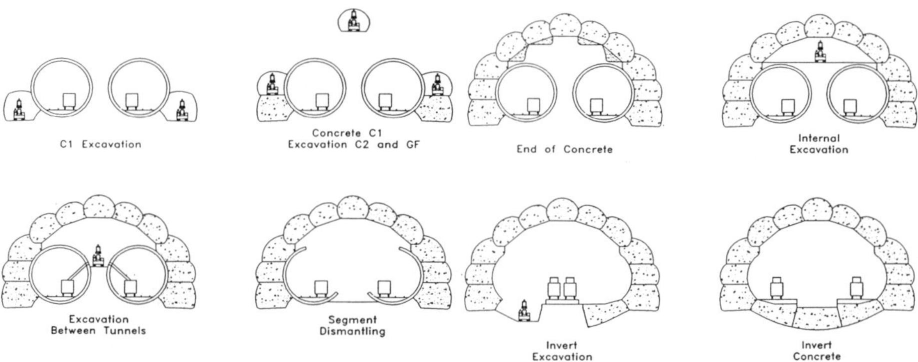

Fig. 11. French Crossover. Main construction phases

This phasing results from the limitation to three of the number of galleries opened at the same time and from the simple reason that it was not possible to open a gallery as long as the gallery beneath had not been filled with concrete.

As the two TBMs had passed through the crossover area well ahead of schedule programme it became possible to use the running tunnel segments as shutters for the bottom galleries and significantly reduce the volume of concrete to be used. As a consequence temporary steel structures had to be installed in the tunnels and in the lower galleries to ensure stability of the segments.

Once all the galleries filled with concrete it was possible to start excavation in the cavern, above the two tunnels, using road headers. The phasing of the invert excavation was such to allow at least for one temporary track to be available in each tunnel for TBM' trains, equipment trains and safety trains.

5.2 Excavation

Excavation basically used four WESTFALIA LYNX B110 road headers (110 kW each) and small hand equipment (Fig. 12). Excavation speed was between 5 to 12 m per day depending on the cross section of each gallery.

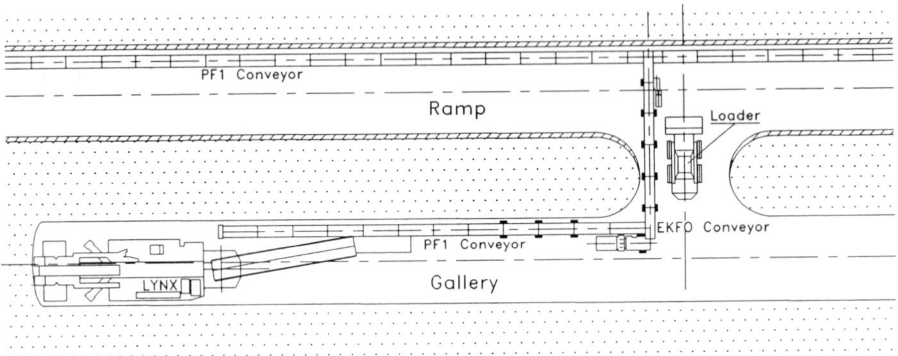

Fig. 12. French Crossover. Gallery excavation

5.3 Bolting and shotcrete

Since there would be no overstress in the central pillar, theoretically there was no need of ground reinforcement in it. Considering some potential weaker chalk parts, it was decided to set up a ground reinforcement (Fig. 13) associated with the steel ribs all along the length where the average deviator in the pillar would be more than 1.5 time the UCS value of the Chalk Layer 1 - 380 rings i.e. 708 m.

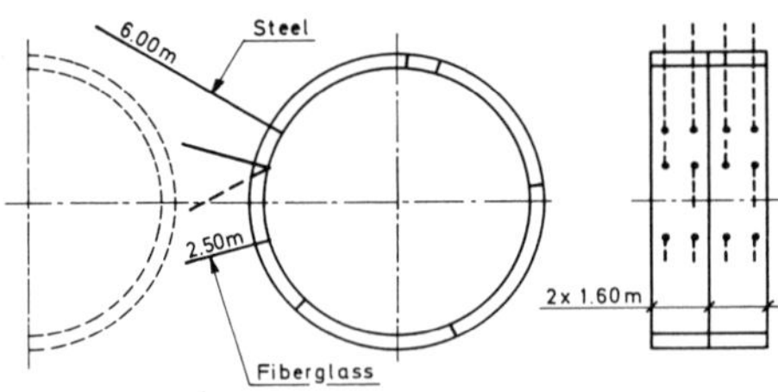

Fig. 13. Central chalk pillar. Reinforcement

Excavation equipment was designed to installed rock bolts while progressing. Different bolt lengths (1.8 m, 2.1 m and 2.4 m) and bolt types (10,000 SWELLEX and 5,000 LENOIR ET MERNIER) were used. Long bolts up to 6 m were used in limited locations between the ramps and the lower galleries. The typical bolting rate was of one bolt per 2.2 m^2 or 1.5 m apart. Depending on rock quality, plastic or steel wire mesh and shotcrete were used.

5.4 Muck transport

Materials were loaded directly by the road headers onto the WESTFALIA PFO and PF1 chain conveyors in the galleries and then loaded on belt conveyors in the construction adits to the service tunnel.

In the service tunnel excavation materials were loaded on muck trains. Two hours were necessary to reach Sangatte shaft where muck was transformed into a quasi liquid material to be pumped to Fond-Pignon dam 1.2 km away.

5.5 Concreting

Concrete was prepared at Beussingue portal, 16 km from the crossover. Due to duration of transport a retarder was added to the concrete to allow concreting operations to take place up to eight hours later. Concrete was brought to the crossover through the running tunnels on six MULHAUSER wagons with a total capacity of 60 m^3. Concrete was remixed at the crossover where an accelerator was added just prior to placing it. The remix stations (SECATOL type) were installed in the running tunnels.

Concrete was pumped from the running tunnels through the ramps to the galleries with PUTZMEISTER KOS 1402 pumps on 125 mm diameter lines. It was possible to pour up to 200 m^3 of concrete in each 12 hour shift.

Concrete blocks were 11 to 16 m long. No vibration was used for concreting. Low heat concrete was used to prevent thermal cracking of lining (Table 6).

Table 6. Concrete characteristics

MATERIAL	QUANTITY
Cement CLK 45 Marine Lumbres)	255 kg / m^3
Fly ashes	100 kg / m^3
Aggregates 3 / 8 mm	230 kg / m^3
Aggregates 4/ 12.5 mm	700 kg / m^3
Sand Boulonnais SH 0004	890 kg / m^3
Water	177 l
W / C ratio	0.50
Retarder CHRYSTARD	0.50 l / m^3
Superplasticiser DURCIPLAST	4.11 l / m^3
Slump	22 cm

5.6 Running tunnel lining removal

The segments were first cut with a road header from the inside of the cavern. Dismantling of the lining was done by hydraulic concrete hammers fixed on mechanical diggers.

5.7 Ventilation

Four fans were installed in the ramps and the top gallery. Three of them, equipped with dedusters, extracted air from the working areas at the vicinity of the road headers through 600 mm diameter air ducts (Ventubes).

Air was blown from the service tunnel to the southern ramp, upwards, and to the service tunnel from the northern ramp, downwards (Fig. 14).

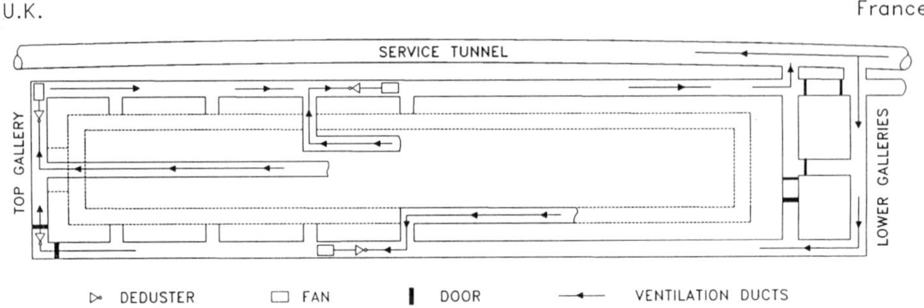

Fig. 14. French crossover. Ventilation scheme

6 Some lessons

6.1 Grouting effects around tunnel lining

The first lesson is that the actual short term loads in the segmental lining resulted mainly from the grouting effects around tunnel lining, as the chalk is rather stiff.

This was checked while excavating the two bottom galleries :

- since the north tunnel crossed the area first, the requirements were to seal strongly its lining to the ground in order to better attract the loads in the lining and mortar during the excavation of the south tunnel ; this was done successfully, but during the excavation of the C1N gallery some major cracks occurred in the segments of the north running tunnel tangent to the gallery ;
- for the south tunnel the requirements were not to load the lining along the cavern and it was quite different for C1S along which only thin fissures occurred.

6.2 Type of contact between the segmental lining and the ground

The second lesson is that the contact between the tunnel lining and the ground is more or less perfect adhesion.

The calculations were done using two types of contact : perfect adhesion and perfect sliding. The crossover gave the opportunity to back analyse this point.

For construction reasons it was necessary to excavate a gallery between the two tunnels, 1 m wide and 2 m high. The calculations indicated that in the case of a perfect sliding contact it was necessary to install a strong temporary lining (H 120 ribs every 80 cm ; 20 cm thick shotcrete) to prevent the chalk to be largely over stressed. For operational reasons there was a long delay between the opening of the gallery and the shotcreting, thus allowing to make a precise geotechnical observation of the chalk of the central pillar. The stability of the unsupported gallery was perfect and no sign of any overstress was seen ; the loads were obviously concentrated in the tunnel linings in accordance with the perfect adhesion contact model.

An instrumentation was installed in the north tunnel eastward from the cavern (ring T2-7653) in order to monitor the effects of the south tunnel boring transmitted through the chalk pillar. The instrumentation consisted in wire gauges in the segments and convergence measurement across the tunnel. The analysis (Fig. 15) shows clearly, at least for the normal loads, that the adhesion contact model is very consistent to the actual loads.

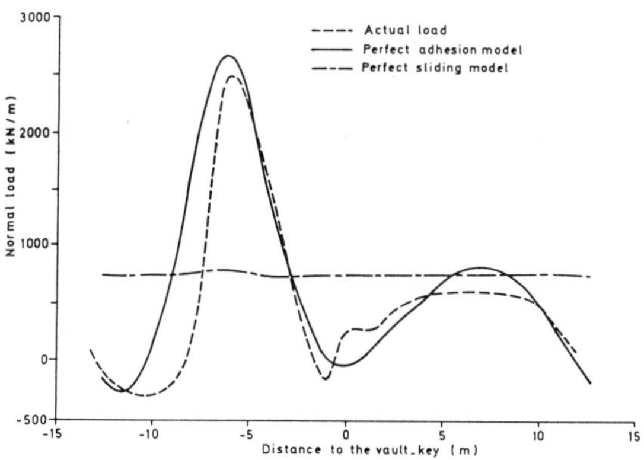

Fig. 15. North tunnel-Ring 7653.
Computerised normal load and actual one

7 Main data

Table 7. Cavern dimensions

Length overall	180.0 m	Length internal	162.0 m
Height overall	19.0 m	Height internal	12.0 m
Width overall	25.0 m	Width internal	18.9 m
Dept undersea (at track level)	100 m	Chalk cover	30.0 m

Table 8. Excavation and concrete volumes

	Cross Section (m^2)	Length (m)	Excavation Volume (m^3)	Rate (m^3/day)	Concrete Volume (m^3)	Rate (m^3/day)
TEMPORARY GALLERIES			10894		10927	
R1	22.0	44.0	968	50	968	111
R2	20.3	44.0	893	50	672	124
R3	20.0	11.0	220	50	237	38
R4	19.1	12.0	229	50	256	44
R5	19.1	12.0	229	50	199	30
N RAMP	19.6	182.0	3567	50	3609	323
S RAMP	19.6	182.0	3567	50	3430	316
TOP ACCESS	18.8	32.0	602	50	602	95
ADITS	16.3	38.0	619	50	954	150
PERMANENT GALLERIES			25072		22854	
C1 NORTH	16.9	168.6	2849	40	2660	93
C2 NORTH	10.1	168.3	1700	80	2030	143
C3 NORTH	13.6	168.4	2290	57	2018	38
V1 NORTH	15.1	168.3	2541	88	2085	130
V2 NORTH	11.9	181.0	2154	78	1771	83
TOP GALLERY	11.5	174.3	2004	63	2010	120
C1 SOUTH	16.9	168.6	2849	69	2657	150
C2 SOUTH	10.1	168.3	1700	86	1749	121
C3 SOUTH	13.6	168.4	2290	79	2018	59
V1 SOUTH	15.1	168.3	2541	65	2085	95
V2 SOUTH	11.9	181.0	2154	64	1771	90
INVERT	60.6	160.0	9690	201	8005	200
INTERNAL			12000	222		

8 Programme

Construction therefore took 16 months (from 15 Sep 90 to 16 Janv 92) (Fig. 16).

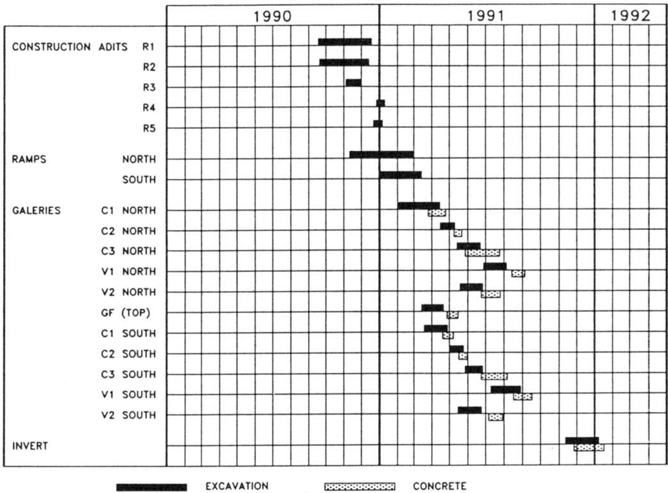

Fig. 16. French crossover
Programme Excavation and concrete

9 Basic comparisons with the UK Crossover

The functional scheme and the general arrangement of the UK crossover [15] are very similar to these of the French crossover. However it differs on the following points :

- programme constraints were different. More time was available to build the UK crossover since it was not on the critical path of the project ;
- it was considered that the NATM method could be used with large face openings (based on ground characteristics, it could be also for the French crossover) ;
- most of the technical rooms are placed outside of the cavern in additional underground structures ;
- overall primary lining of the cavern had to be built prior to the running tunnel TBMs arrival at the crossover to allow TBMs to go through the cavern ;
- TBMs had to go through the cavern prior to the construction of the secondary lining because of reduced internal width of the crossover (18.1 meters) ;
- the internal height is smaller due to reduced width and different ground characteristics ;
- a watertight membrane was placed between the primary and secondary lining as well as drain materials.

Construction of the UK cavern including primary and secondary lining took 23 months from June 1989 to May 1991.

As a conclusion

The French crossover is probably the world largest undersea cavern built to date. Construction programme was the fastest ever achieved for such a large structure.

Acknowledgement

The Authors are grateful to Eurotunnel and TML for their permission to publish this paper.

References

[1] Carmes J. Le crossover France : Conception d'un ouvrage très spécial **Congrès AFTES Lille 1990**

[2] Bour N. Optimisation du tracé du tunnel et des ouvrages spéciaux **Le Tunnel sous la Manche, géologie et geotechnique, ENPC 1989**

[3] Crighton G.S., Leblond L. Tunnel Design **The Channel Tunnel, Conference, ICE and ISF, Paris, 1989**

[4] Larrière A., Leblais Y. Etude géotechnique de la Craie et de l'Argile du Gault **Le Tunnel sous la Manche, géologie et géotechnique, ENPC 1989**

[5] Leblond L. Les ouvrages souterrains du tunnel sous la Manche **Le Tunnel sous la Manche, géologie et géotechnique, ENPC 1989**

[6] Birch G.P. et *al* Geotechnical aspects of the design and construction of the UK crosssover **Tunnelling'91**

[7] Varley P., Radcliffe E. Geology, alignement and survey **The Channel Tunnel Part 1 : Tunnels, ICE 1993**

[8] Wardle L.J., Gerrard C.M. The equivalent anisotropic properties of layered rock and soil masses **Rock Mechanics 1972**

[9] Huder J., Amberg G. Quellung in Mergel, Opalinuston und Anhydrit **Schweizerische Bauzeitung Vol. 88 N° 43 1970**

[10] Panet M. Understanding Deformations in Tunnels **Comprehensive Rock Engineering, Vol.1 Fundamentals, Pergamon Press, 1993**

[11] Kirsch G. Die Theorie der Elastizitat und die Bedürfnisse der Festigkeitslehre. **Veit. Ver. Deut. Inj. 42, 1898**

[12] Samama L., Leblais Y. Tunnel sous la Manche, Crossover côté France, Analyse du rapprochement des tunnels ferroviaires **Géotechnique et Informatique, ENPC 1992**

[13] Delvoie P.E. French undersea crossover chamber. **World Tunneling, December 1991**

[14] Leblond L. et *al* Le croisement ferroviaire **Tunnel sous la Manche, TOS n° 114 nov. dec. 1992**

[15] Fugeman I.C.D., Myers A.G. Major underground structures **The Channel Tunnel Part 1 : Tunnels, ICE 1993**

Design and construction of Metsovon road tunnel, Greece

Yannis Malios
Yannis Malios & Associates, Ltd., Athens, Greece

ABSTRACT

The Metsovon Road Tunnel 3,500 m long with a ventilation side adit 1,900 m long under an overburden of 600 m is placed in North - West Greece and is a part of the East-West new motorway of Greece.

The tunnel passes through schist (sandstone till very soft argillic schist), peridotite (a large variation concerning the cerpendination). Said geological materials are overthrusted and a large fault zone has been met.

During the construction, there occurred large scale deformations and sliding of the footing of the support. This, in some cases, led to a full collapse of the support of the tunnel.

Said problem resulted in the termination of the contract and the continuation of the project with a new contractor and consultant.

The adopted principles of N.A.T.M. enable the contractor to give solutions to the deficiencies and perform the excavation and support with an efficient way.

Analysis by F.E. has been done (two D and three D) and the results have been continuously compared with the site investigations and measurements.

Large scale deformations occurred due to the relatively high overburden, the swelling characteristics of the rock mass and the existence of high water table, but they have been successfully handled.

Due to the relatively long time that took place for the finalisation of the project, it was possible to study the long term behaviour of the rock mass, and also the interaction of the excavation of two tunnels (main tunnel and side adit) that have been driven in an axial distance of 25 m.

Initially the contract for the design and construction of this road tunnel was awarded to an Italian firm (1986) which did not complete the project. During initial construction, large scale deformations occurred as well as sliding of the footing, which led to a full collapse of part of the tunnel. This problem resulted in the termination of the initial contract and the continuation of the project with a new contractor and consultant.

The Project

The Metsovo tunnel is situated in North-Western Greece between the provinces of Trikala and Ioannina, at an altitude of 1.100 m. The construction of the tunnel is imperative in order to deviate from the "KATARA" col which lies on the National road of Ioannina-Trikala. This Project is a part of the future Motor way Igoumenitsa-Volos and of Egnatia Highway.

The project Metsovo Tunnel includes the road tunnel of length approximately 3,500 m with two lanes and clearance profile 65 m², and a ventilation tunnel of length 1,900 m and clearance profile 14 m². The final width of the main tunnel is 9.50 m (Fig. 1). The tunnel runs in a curve at the entrances with a radius of 900 m at the Western Portal (Anilio) and a radius of 700 m at the Eastern Portal (Malakasi) and has a longitudinal gradient of 0.5%. The final width of the horse-shoe shaped ventilation tunnel is 3.70 m (Fig. 2).

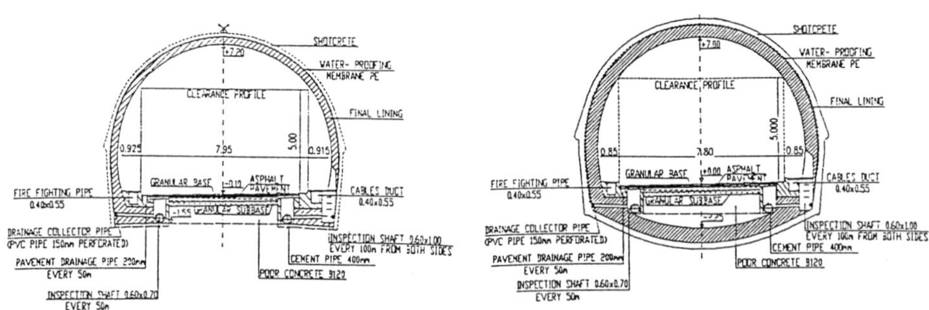

Fig 1 Main tunnel cross sections

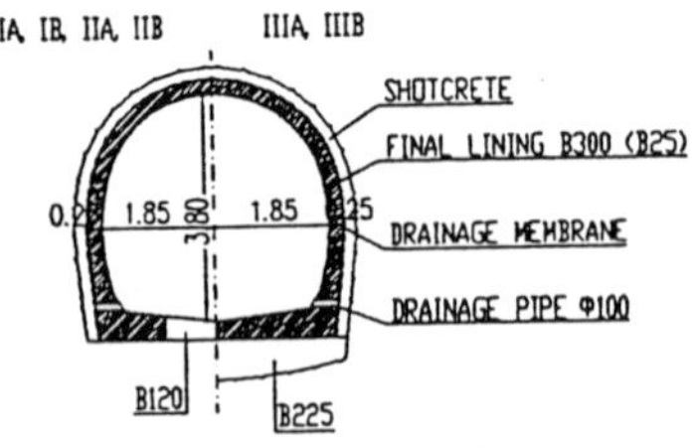

Fig 2 Ventilation tunnel cross section

The Problem

The project began with the reconstruction of the tunnel length between ch. 0+362.00 and ch. 0+425.00, which due to the unstable and loose overlying formation caused large scale deformations and sliding of the footing which led to full collapse of the structure.

The initial contractor had chosen the "side - drifting" method for restoring the damage. This method, though, could not give effective results because it could not activate the adjacent rockmass by applying passive pressure which through redistribution of the stresses would result in an increase of the bearing capacity of the rock mass - temporary support system.

Therefore, for the restoration of the damage, the final contractor decided to follow the New Austrian Tunnelling Method (NATM).

At the same time the damage of the support between ch. 3+146 and ch. 3+222 (Eastern face) was also restored. Within this tunnel length, after excavation of the top heading and part of the bench, large deformations of the support (70-80 cm) were noted with simultaneous fracturing of the shotcrete, distortion of the steel arches and movement of their footings.

The various geological investigations which were carried out in the area of the damage, showed that the tunnel passed through formations with poor to very poor geomechanical characteristics. Due to a lack of anchors the shell of the temporary lining disconnected from the rock mass and deformed uncontrollably.

For the part of the tunnel that had not been as yet driven as well as for the ventilation tunnel, the New Austrian Tunnelling Method (NATM), which allows flexible adaptation of the tunnelling measures to the encountered conditions, was successfully used.

The final design, including complementary soil mechanical investigation and evaluation, was totally prepared in Greece.

Geology

a) Main Tunnel

The formations which compose the geologic subgrade of the area along the tunnel axis are separated into two categories (Fig. 3) :

1. Flysch of the geotectonic zone of Pindos.
2. Serpentine

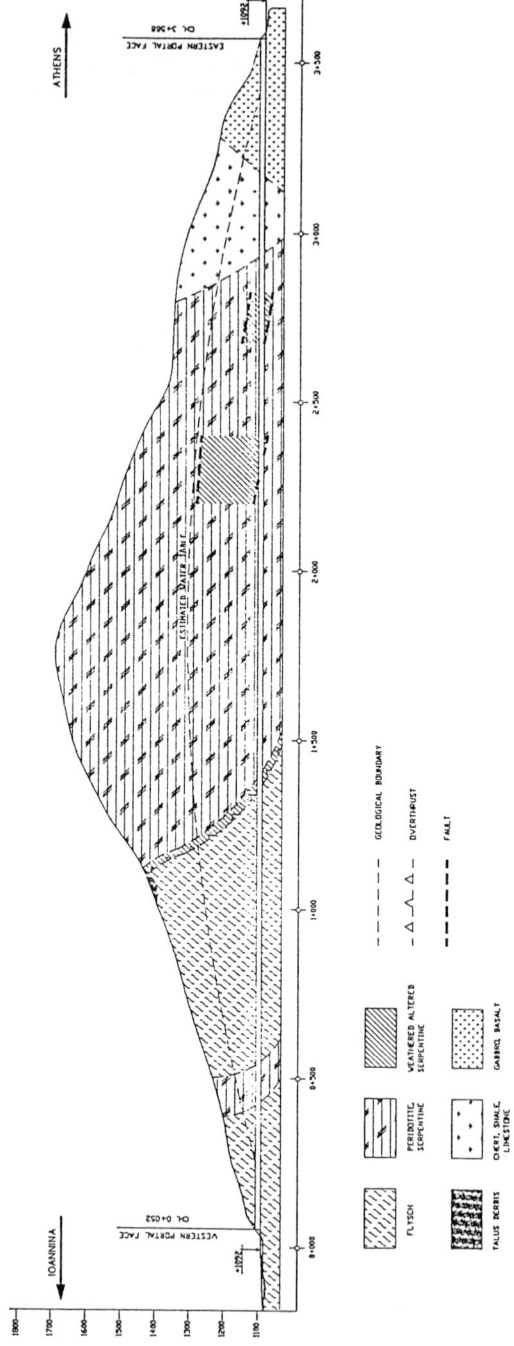

Fig 3 Geological longitudinal alignment

Flysch is a unit of clastic sedimentary rock. It is highly inhomogeneous due to the fact that it does not consist of one rock type but of a conjunction of rock types.

Various rocks were encountered in the rock mass unit of Flysch, such as siltstone, sandstone, argilaceous schist, serpentine lenses, and serpentinized peridotite lenses.

Serpentine rock masses consist of basic and ultrabasic rocks such as gabbro, peridotite, serpentinized peridotites, serpentine as well as a schist hornstone formation which consists of radiolites and schist hornstone.

The most important fact is that the mass of the serpentine has been overthrusted on the Flysch of Pindos because of very strong tectonic movement which belongs to the recent Alpine orogenesis cycle to which the development of the whole mountain chain of Pindos is due.

In general a systematic water flow was noted in the form of moisture or drops which can reach a flow under small pressure in crossings of large rupture surfaces or disintegration of the rock. The water flow had a negative influence on the tunnel drive because it caused a further reduction of the mechanical characteristics of the minerals which fill the discontinuities causing rock subsidence. In these cases anisotropic loads on the temporary lining were noted.

Due to the water flow in the tunnel, geotechnical laboratory tests were carried out in order to verify if there is any creep potential and how much. The results showed that the creep phenomenon was intense in clayey sedimentary rock and specifically in argillaceous schist and siltstone. Therefore, the necessary measures were taken to cope with this phenomenon, because, due to the time lag with which it occurs, influence on the final lining of the tunnel is anticipated.

b) Ventilation Tunnel

The ventilation tunnel was excavated in the Serpentine formation, which consists of basic and ultrabasic rock masses such as peridotites, dunites and gabbro. There were also encountered Serpentinized peridotites, Serpentines and a Schist hornstone formation which consists of radiolites and schist hornstone. Due to the stress field the whole mass of the Serpentine has been displaced and overthrusted over the Flysch of Pindos. This last formation though was not met during the excavation of the ventilation tunnel.

In general, the mechanical characteristics which were encountered during the excavation were poor to very poor with compression phenomena, as well as intensive weathering and tectonism in certain zones. The gabbro was an exception, which presented characteristics of a rock mass with relatively good quality, and for this reason, excavation through this rockmass was achieved with the use of explosives.

The presence of water was encountered in several conditions, from light humidity to water inflow under high pressure, which primarily influenced the serpentines and the serpentinized peridotites because it drastically minimized their geomechanical characteristics.

Tunnel Drive

During tunnel excavation and installation of temporary support there arose some serious problems regarding the stability of the tunnel cross-section. Special measures had to be taken so as to avoid or to restore damages. The areas where the most serious damages took place are cited below.

a) Restoration of damage at Western - Portal

The project started with the restoration of the damage along the tunnel length ch. 0+362 to ch 0+425 on the Western face.

The initial contractor had refilled with earth the above mentioned tunnel length, leaving a 2,5 m passageway, so as to continue works in the next tunnel section, due to large deformations of the preliminary lining. From on site investigation carried out by the final contractor, it was ascertained, wherever possible, that the tunnel in this area had passed through formations with poor geomechanical characteristics and to a great extent serpentinized. Also, the existence of water, even in the form of humidity, had drastically minimized the geomechanical characteristics of the rock. The side-drifting method did not prove to be an effective measure because the time duration which the sections remained unsupported was large and as a result the overlying rock mass loosened over an extensive height.

The final contractor re-excavated this tunnel length using a mechanical shovel. The re-excavation took place in two stages. During the first stage the top heading was excavated and supported with shotcrete of minimum thickness 25 cm, double steel fabric reinforcement T139, steel arches IPB140 and anchors with bearing capacity 400KN and length 6.0m.

The area of the bench was especially strengthened with double anchors and an increase of the shotcrete thickness to 50 cm.

In the second stage the bench and invert was re-excavated and was supported with shotcrete of minimum thickness 30 cm, steel fabric reinforcement T139, and anchors of bearing capacity 400KN and length 6.0 m in the area of the bench.

b) Overthrust

During tunnel excavation damages in the top heading were noted in the area from ch 1+316 to ch. 1+429, which belongs in a broader area of the overthrust zone of peridotites on the Flysch subbase of Pindos. The damages were in the form of significant convergences (from 10 cm to 70 cm), serious deformations, distortion of steel arches and fracturing of the shotcrete in the northern wall of the tunnel. Simultaneously there appeared phenomena of loosening of the rock from the tunnel face, as well as a significant water inflow under pressure.

Initially the contractor strengthened the area of the damage with 4-8 anchors of bearing capacity 400KN. Never-the-less strong phenomena of instability accompanied by continuing deformations and fracturing of the temporary lining appeared and continued to a smaller degree in spite of the additional measures taken for strengthening the temporary support (additional shotcrete, anchors).

Because the convergences that had been measured where very large and the minimum required thickness for the final lining could not be achieved the contractor re-excavated and restored the inner lining of the final support in the above mentioned tunnel length.

c) Damage at Eastern Portal (during Bench excavation - ch. 1+113 - ch 1+189, ch. 1+189, ch. 1+245)

Serious damages appeared between ch. 1+113 and ch. 1+189 during bench-invert excavation (Fig. 4). Extensive deformations where measured at ch. 1+189 and ch. 1+245 for the same reason.

Failure of the temporary support was especially intense in the lower parts of the wall where there appeared serious convergences which were accompanied by vertical displacements. This resulted in fracturing of the shotcrete, rupturing of the steel plates of the anchors with bearing capacity 200KN and slight distortion of the steel plates of the anchors with bearing capacity 400KN.

To cope with the above phenomena the contractor, in agreement with the owner and consultant, after discontinuing bench-invert excavation, reinforced the excavated top-heading with 6 anchors of bearing capacity 400KN and length 8.0 m at each steel arch, 6 bolts of bearing capacity 200KN and length 6.0 m in each panel between the steel arches and one anchor of bearing capacity 400KN of length 8.0 m at the lower ends of each steel arch.

After the immediate adaptation of the additional measures of reinforcement of the bench of the tunnel, the deformations decreased significantly and as a result the contractor proceeded to excavate the invert.

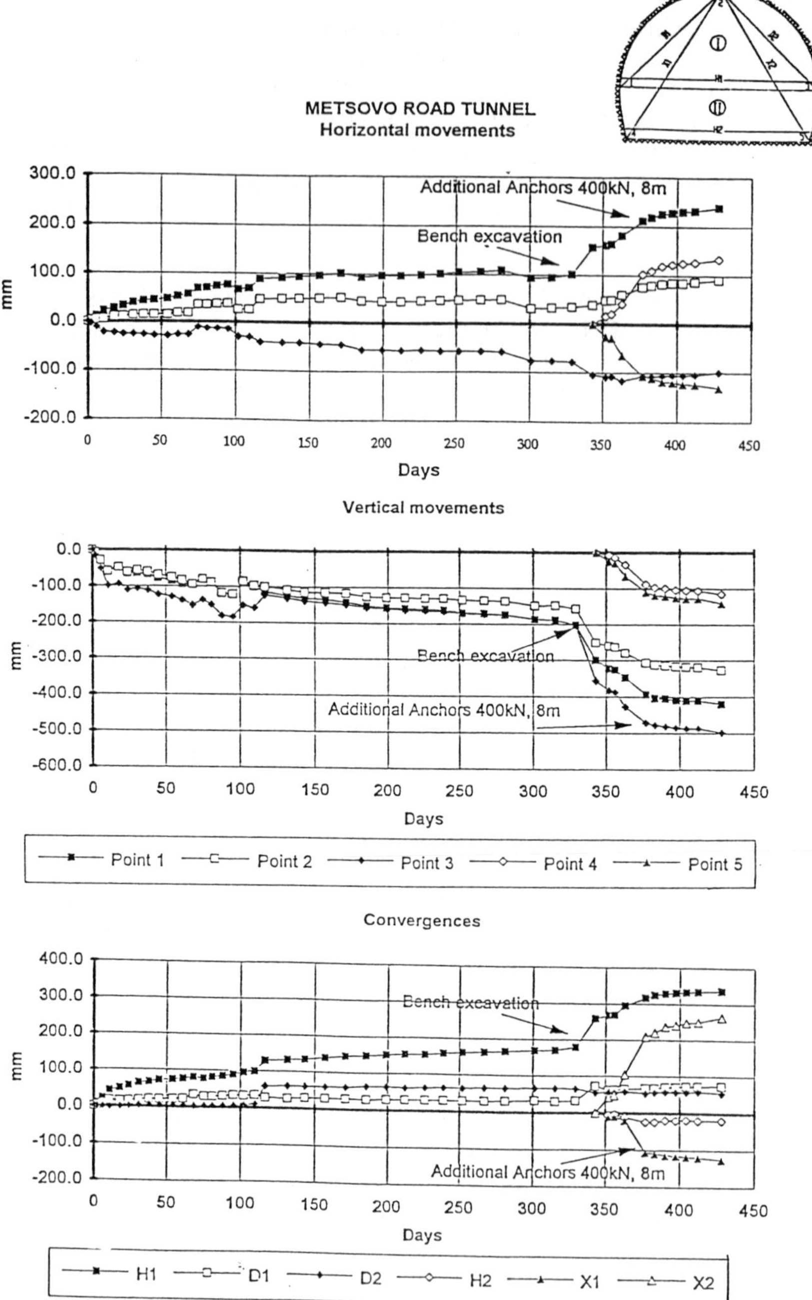

Fig 4 Deformations measured in Station 13 (c.h. 1+158.7)

d) Damage at Eastern face (ch. 3+146 - ch. 3+222)
Finally, serious damages appeared on the already excavated tunnel length between ch. 3+146 and ch. 3+222 at Eastern face . In this section, the initial contractor had excavated and had placed the temporary support of the top heading and had also started the excavation of the central part of the invert.

There appeared serious deformations on the inner lining of the tunnel which where accompanied by fracturing of the shotcrete, distortion of the steel arches, displacement of its footings and in general of its contour. The displacements of the inner lining reached 90 cm and were localised in the schist hornstone formation.

This formation, due to the tectonic phenomena in this area, had crushed creating a weak zone and the prescence of underground water aggravated the situation and corrupted the formation.

On the whole, the damages showed that in this specific area the use of steel anchors, with a length large enough to anchor in the strong rock mass, was necessary. This would increase the bonding between lining-rock. For this reason the contractor, in order to restore the damage, reinforced the existing temporary support with prestressed anchors, and proceeded to excavate and support the above section.

e) Ventilation Tunnel
During advancement of the face of the ventilation tunnel failure of the temporary lining was noted, especially in the mass of the schist hornstone, serpentines and serpentinized peridotites. The failure was in the form of intense deformations of the walls and the crown of the ventilation tunnel followed by fracturing of the temporary lining and distortion of the steel plates of the bolts. These failures were dealt with by placing additional bolts of bearing capacity 200KN and length of 3.00 m as well as grouting.

The most notable failures are presented analytically below.

a1) Section between ch. 0+715 to ch. 0+725
In an intensely compressed serpentine, fractured and intensely weathered in sections, there where encountered quickly increasing deformations of the temporary lining, which, despite the placing of the additional anchors, lead to complete loosening of the rockmass and as result convergence of the inner lining (>1.00 m). The Contractor, to avoid total collapse of the section, used additional support measures in the form of wooden beams. The Contractor then removed the steel arches from the damaged temporary support and after re-excavating placed new steel arches and increased the thickness of the shotcrete (d>30 cm). Finally, new bolts were necessary with a length of 3.00 m and 18 pieces per panel.

a2) Section between ch. 0+869 to ch. 0+871
In this section serpentines were encountered, highly weathered with peridotites and saturated. As a result, during excavation of the face for the placing of the steel arches the tunnel collapsed to a great extent completely. The volume of the material that collapsed was approximately 30 m^3.

To deal with the collapse and driving of the face, the contractor covered the face and the cone of the material that collapsed with shotcrete. Then the contractor placed in the periphery alluvial lances for grout injection along with the use of steel lances and decreased the excavation pull to l = 0.50 m. The top heading was now excavated in three phases.

Design of excavation and support

a) Main Tunnel

The method of construction of the tunnel which the contractor adopted was the New Austrian Tunnelling Method (NATM), as mentioned before.

The excavation of the main tunnel took place in three stages. The first and second stage included the excavation of top heading and bench and the third stage included the excavation of the invert (only in the mouth shaped sections). When the ground conditions were bad or very bad the face of the excavation needed to be inclined with partial or total temporary support with shotcrete in some places.

The design of the support had the aim of maintaining and improving the bearing capacity of the surrounding ground.

According to the expected tunnelling conditions the classification system included six categories of excavations (Fig. 5 - Fig. 10). From these categories the first three were horse-shoe shaped and the last three were mouth-shaped. This classification system describes the excavation and support measures for the full range of the expected tunnelling conditions.

The temporary support of the tunnel, which takes place immediately after the excavation, included the systematic installation of shotcrete armed with steel arches and steel fabric reinforcement SN type fully grouted and prestressed rock bolts. When needed, alluvial lances were installed to support the ground ahead of the face.

Specifically, the temporary support included steel arches of IPB 140 or equivalent, fully grouted rock bolts 25mm and 28mm SN type with bearing capacity of 200KN and 400KN respectively and lengths varying between 4m to 8m. The shotcrete armed with one or two steel fabric reinforcement T139 had thickness varying from 5cm to 30cm.

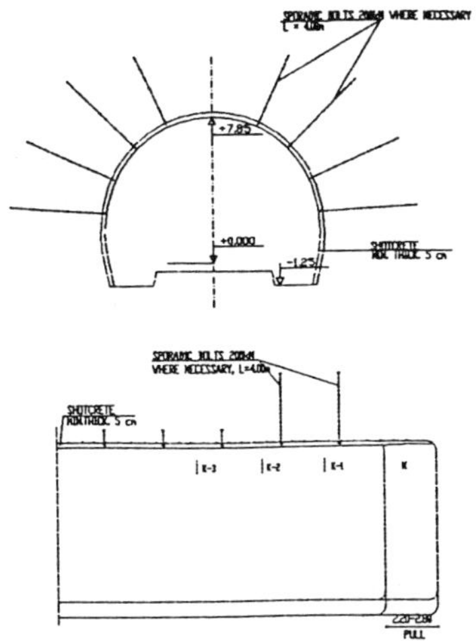

Fig 5 Typical cross section Ia and longitudinal section

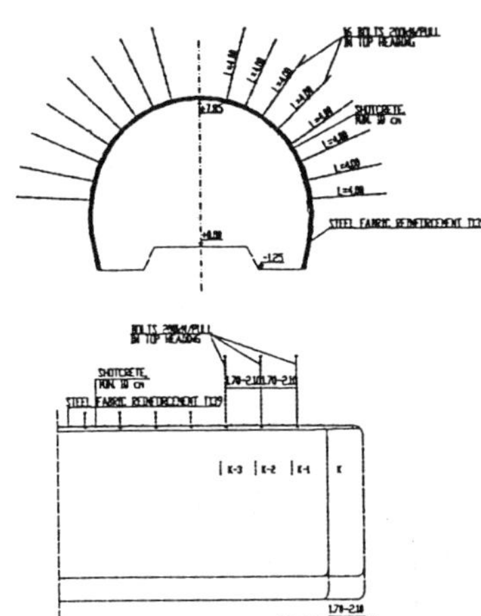

Fig 6 Typical cross section Ib and longitudinal section

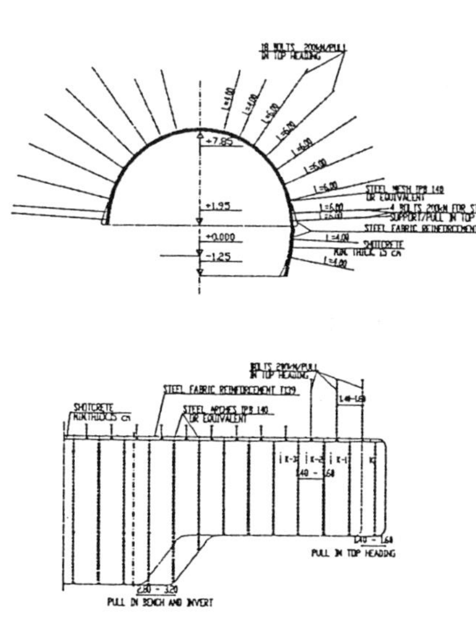

Fig 7 Typical cross section IIa and longitudinal section

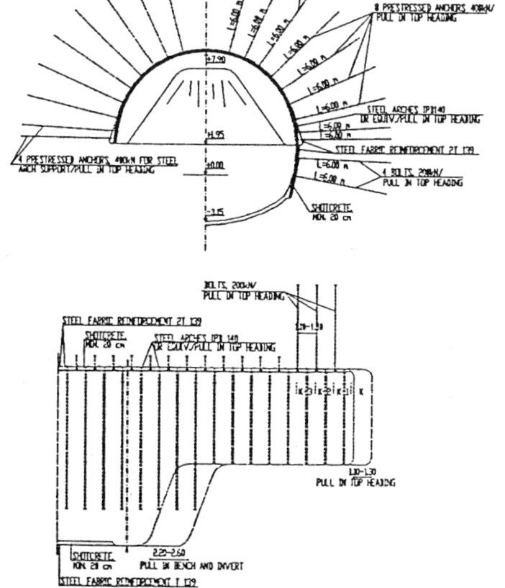

Fig 8 Typical cross section IIb and longitudinal section

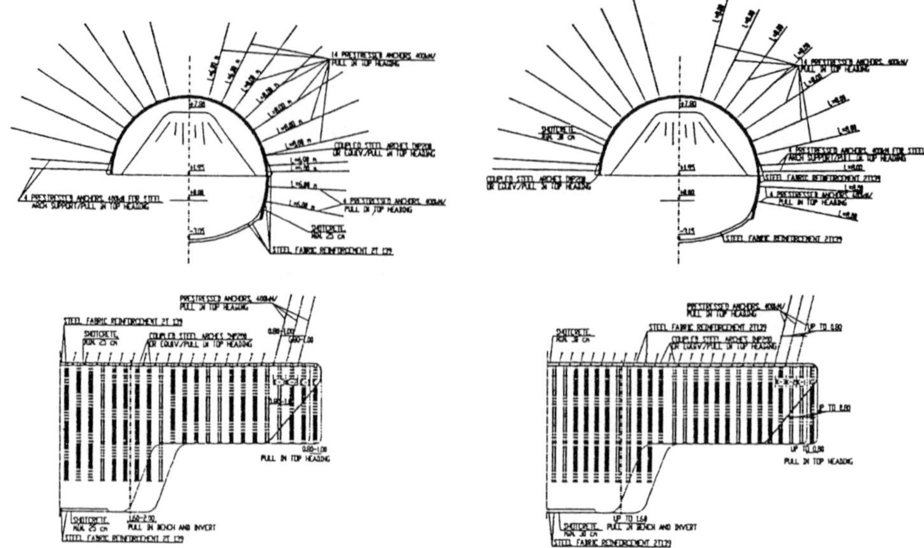

Fig 9 Typical cross section IIIa and longitudinal section

Fig 10 Typical cross section IIIb and longitudinal section

Before the excavation of the main tunnel took place the damages at Western face and Eastern face were repaired as mentioned before.

The restoration of Western face damage started with the filling of the side-drifts with concrete. There followed the strengthening of the temporary support before and after the damaged area as well as the strengthening of the foundation of the invert in order for it to be capable of carrying the load of the repaired work. And finally there took place the re-excavation and the installation of the new temporary support in the damaged part.

On the other hand the restoration of Eastern face damage started with the strengthening of the temporary support with prestressed rock bolts of bearing capacity 400KN and 6m length in order to reactivate the surrounding rock mass. And finally there took place the excavation and the installation of the temporary support of the damaged part of the main tunnel.

b) Ventilation Tunnel

In accordance with the main tunnel, for the excavation of the ventilation tunnel six categories of excavation and support were used, depending on the rock mass quality that was expected to be met. The tunnel cross section was horse-shoe shaped. In the first four categories the bottom was open and in the last two categories the bottom was closed.

The temporary support measures include shotcrete of thickness 5 cm to 25 cm reinforced with double steel fabric reinforcement depending on the category, steel arches IPB120 in the three latter categories and 6 to 10 steel bolts per section with bearing capacity 200KN and length 3.5 m in the four last categories. For the first two categories, because the excavation was in ground with good geomechanical characteristics (gabbro), the use of bolts was not necessary.

Design analysis

As mentioned above, the works were executed with an overburden varying from 20m to 550m and in very changing ground conditions (Fig. 2).

The whole design of the excavation is based on the 2D-Finite Element Method (Fig 11). The third dimension and the time influence for load and support strength developments were introduced following Bundendistel by stage calculations (Fig. 12).

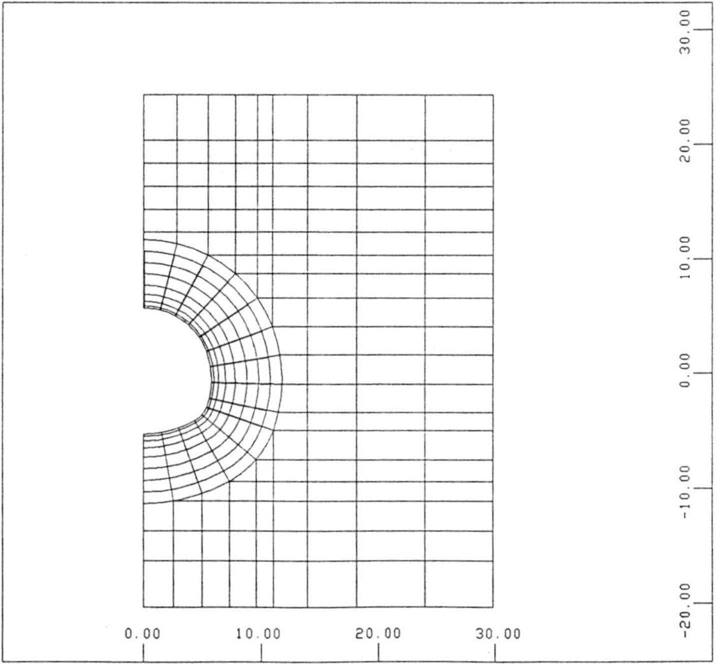

Fig 11 Finite element net of Main Tunnel

CONSTRUCTION STAGES SIMULATED BY FEM

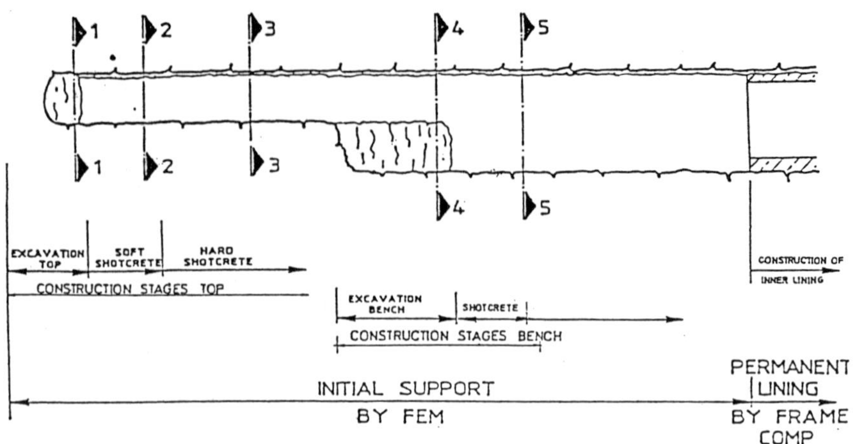

Fig 12 Construction stages simulated by FEM

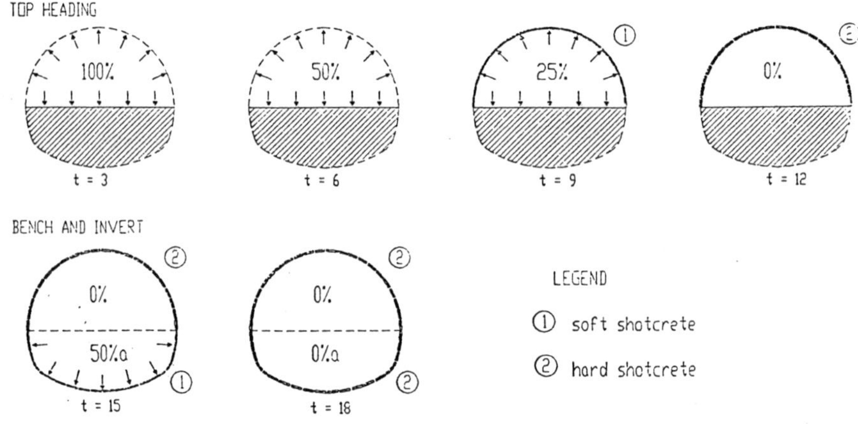

Fig 13 Load distribution during tunnel construction stages

For this purpose, the tunnel section was analysed with a 2D-Finite Element mesh. The elements referring to the excavation were taken away leaving in their place the reactions which were gradually reduced (Fig 13). With this method the loads coming with the excavation process are gradually rearranged in the surrounding ground.

The law which has been used in order to describe the behaviour of the materials was the "Mohr-Coulomb Criterium" elastic - totally plastic.

All the materials referring to the initial support were introduced into the F.E. mesh gradually with a linear elastic - plastic law.

The results of the F.E. analysis have been continuously compared with the Site investigations and measurements in order to correct the initial acceptances and make the F.E. model more realistic.

Final Lining

Assuming the temporary support to be out of action, the final lining was designed to carry the loads alone. It was calculated as a bedded polygonal frame. The reaction of the surrounding ground was taken into consideration with the action of "spring" elements around the frame. Wherever one spring element was in tension it was taken away and the calculation was repeated. Finally, all the remaing springs had only compression loads.

The final lining was installed in two sections, the invert and the arch. Each block had length 7.60m for the main tunnel and 10m for the side tunnel.

The high reinforcement of the final lining took into consideration combinations of many loads. These loads were the design loads of the Contract Documents, the expected water loads, the temperature loads and the drying shrinking loads.

Instead of water impermeable concrete, sheet insulation was used. A separating plastic layer was installed between the temporary and the final lining, in order to gather the water into the drainage pipes, reducing the water pressure.

The section of the final lining which refers to the intersection of the main tunnel with the ventilation tunnel at ch. 1+807.25 proved to be of special interest. The particularity of this section was that the cross section of the ventilation tunnel 6.50 m before its final connection with the main tunnel (the axis of the ventilation tunnel meets the axis of the main tunnel perpendicularly) broadens in a funnel shape (Fig. 14). The scope of this broadening was to decrease the velocity of movement of the wind along this section, so that no depression will be created when entering the main tunnel.

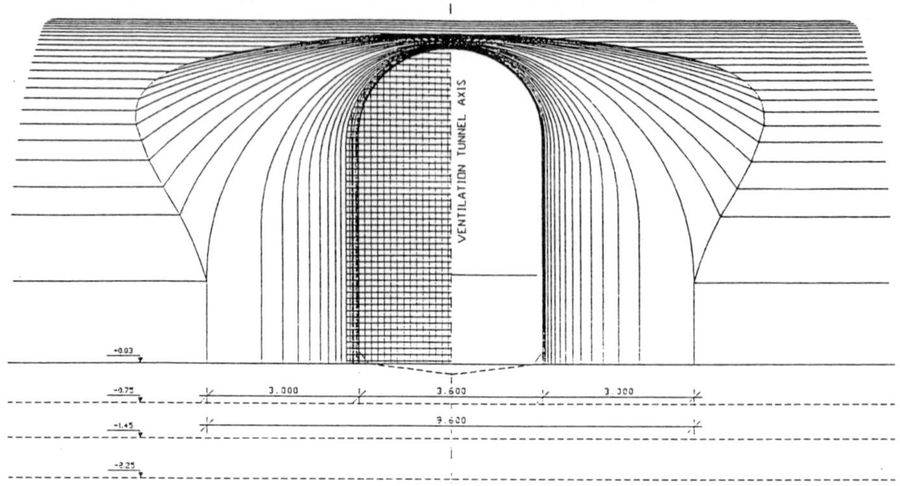

Fig 14 View of ventilation tunnel entrance at intersection with main tunnel

To solve this specific section, a 3-D finite element analysis program was used. The final lining was simulated by three dimensional surface finite elements, elastically supported (Fig. 15). From the results the isostatic lines

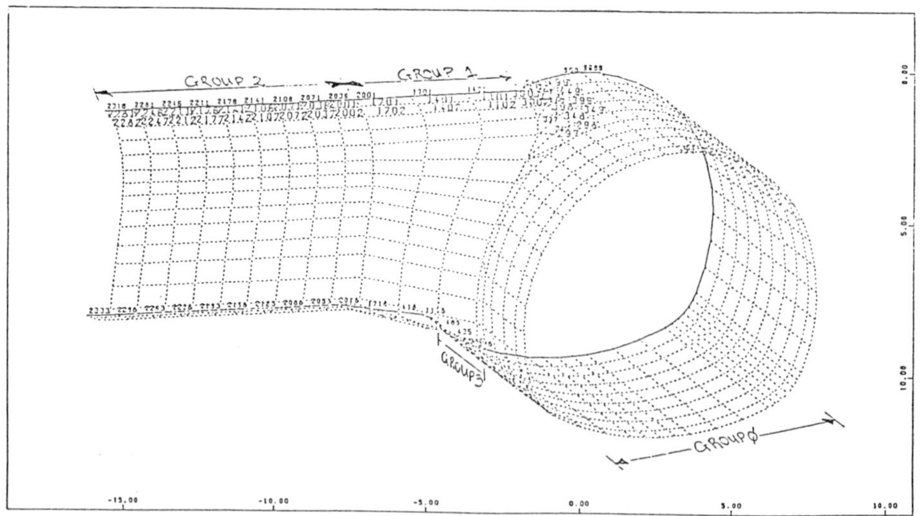

Fig 15 3D Finite element net at intersection of main tunnel with ventilation tunnel

of the loads which fatigue the lining were found, and based on these the lining was dimensioned. The load combinations taken into account are the same as those used for solving the final lining, of the main tunnel as described above.

Geotechnical measurements

During the excavation of the tunnel, cross sections were equipped with multiple extensiometers (in the top heading) and pneumatic piezometers (in the bench). The extensiometers had lengths varying from 3 m to 9 m as figure 16 shows.

At the same time many Convergence Stations (Fig. 17) were installed on the tunnel axis, following the installation of the initial support 1.50÷2.00m of the Tunnel face.

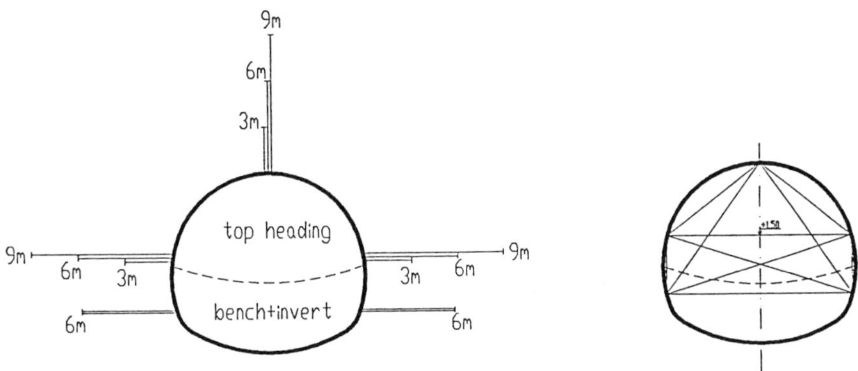

Fig 16 Multiple extensiometers and pneumatic piezometers installed around the tunnel

Fig 17 Convergence measurement cross section

According to the results large scale deformations occurred due to the relatively high overburden, the swelling characteristics of the rock mass and the existence of high water tables, as was mentioned before. In every case all the above problems have been successfully handled by measures taken to strengthen the temporary support, according to the size of the deformations.

SUMMARY

The construction of Metsovo Road Tunnel has lots of benefits, as it will improve traffic conditions along the international transportation axis connecting Western and Northern Europe with Middle East.

Economy was achieved due to the shortening of the route between Ioannina - Trikala by approximately 30 km and communication between Epirus, Thessaly and Macedonia was improved (without danger of being blocked during winter).

The problems which were encountered during excavation were many and difficult, but with correct and timely interventions for each case, they were coped with successfully.

At the same time the experience gained for the owner, the contractor and the consultant was considerable.

Finally, future construction of a second parallel tunnel is foreseen, similar to the one already excavated, and at a distance 30 to 40 m which will serve the second branch of the motorway.

Construction of the 'Passante Ferroviario' link in Milan, Italy, lots 3P, 5P and 6P: excavation by large earth pressure balanced shield with chemical foam injection

J. Y. Peron
Neyrpic Framatome Mécanique, Tunnelling Machine Department, Lyon, France
P. Marcheselli
Passante Scarl, Milan, Italy

Synopsis
A twin tube tunnel of 2 000 meters long and an internal diameter of 6.90 m is being excavated through alluvial ground, with low overburden, using a 8.0 m diameter Earth Pressure Balanced tunneling machine with foam injection and extruded concrete behind segment.

This paper describes the tunnelling work of the first tunnel with three major areas :

- Description of the project and tunnelling conditions
- The use of the latest technology in Earth Pressure Balanced shield
- Progress of the work (difficulties encountered and adjustments made)

1 Description of the project

In Lombardy, the present rail system results from the convergence in Milan of a number of lines from other towns. This situation causes great difficulties arising from the mixing of international, national and local traffic on the same lines which are used for both goods and passenger trains. This diversity leads to overloading, difficulties in managing the installations and negative consequences on the regularity of services.

Another special feature of Milan is the existence of terminal stations, in which trains have to enter and leave the same way on their way to their final destination. This situation leads to significant wastes of time which have repercussions on the organization of the overall system.

Other towns in Italy and in Europe have similar problems. Several cities such as Munich, Oslo, London and Brussels have adopted solutions of which one of the main features is the construction of a direct underground link between several stations, making it possible to partially transform the old "header" (or terminal based) rail network into a "passing" (or direct link) system.

This "passing" solution means that trains coming into a station continue on their way by going through the town via these underground lines, considerably improving rail traffic, with obvious advantages for the town, the urban district and the region.

In Milan, the "passing" solution was accepted by the F.S (National and International Rails Networks) and by the FNM (North Milan regional railways). It has thus been possible to plan a single, integrated "passing" transport system between the national and regional rail networks. An agreement was signed between the Lombardy Region, the city of Milan, the State railways (FS) and the North Milan railways (FNM) to build the new rail infrastructure. This work, known as the Passante Ferroviario di Milano, provides an underground link between the FS stations of PORTA GARIBALDI and PORTA VITTORIA to the FNM and FS networks and to the stations of BOVISA (FNM) and CERTOSA (FS). The new underground line is also connected, at 3 points, to the Milan Metro lines.

a) The route

The urban network from the station of BOVISA (FNM) and CERTOSA (FS) as far as the station of ROGOREDO (FS) and the junction for TREVIGLIO, can be divided into three main sections :

1.	Lancetti - Vittoria (FS)	=	5.80 km
2.	Lancetti - Bovisa (FNM)	=	2 km
	Lancetti - Certosa (FS)	=	3.73 km
3.	Vittoria (FS) - Rogoredo (FS)	=	3.66 km
	Vittoria (FS) - Junction for Treviglio	=	3.10 km

b) Construction methods

The choice of construction methods is conditioned by a whole series of factors, both on the surface and underground (e.g. road and tram networks, social and commercial activities on the surface, the stability of buildings, the type of subsoil).

In general, the "cut and cover" method has been favoured. However, for certain sections, given the surface infrastructures, especially the existing railway lines and the fact that routes are laid down in unconsolidated soil, the use of a tunneling machine has turned out to be the only way of performing the work with a minimum of interference to the surface and with greater safety and excavation efficiency.

It was decided to use a tunnelling machine for the single track, double tube section, referenced lots 3P, 5P and 6P, of the railway line between VILLAPIZZONE - LANCETTI and PORTA GARIBALDI.

Fig. 1 View of the site

Fig. 2

MILANO:
COLLEGAMENTO
FERROVIARIO PASSANTE
SESTO 1° MAGGIO F.S.
SESTO RONDO
SESTO MARELLI
VILLA
PRECOTTO
GORLA
TURRO
ROVERETO
PASTEUR
LORETO
CASCINA
CRESCENZAGO
CIMIANO
UDINE
LAMBRATE F.S.
PIOLA
LIMA
CERTOSA F.S.
BOVISA F.N.M.
VILLAPIZZONE
LANCETTI
MACIACHINI
ZARA
SONDRIO
M. GIOIA
CAIAZZO
CENTRALE F.S.
GARIBALDI
REPUBBLICA
VENEZIA
MOSCOVA
TURATI
MONTENAPOLEONE
PALESTRO
S. BABILA
DATEO
TREVIGLIO
VITTORIA F.S.
OT 8
LOTTO
AMENDOLA FIERA
BUONARROTI
PAGANO
CONCILIAZIONE
LANZA
CAIROLI
CADORNA
CORDUSIO
DUOMO
WAGNER
DE ANGELI
GAMBARA
BANDE NERE
PRIMATICCIO
S. AMBROGIO
S. AGOSTINO
MISSORI
CROCETTA
PORTA GENOVA F.S.
PORTA ROMANA
LODI TIBB
ROMOLO
BRENTA
CORVETTO
S. CRISTOFORO
FAMAGOSTA
PORTO DI MARE
ROGOREDO F.S.
CHIESA ROSSA
MISSAGLIA
S. DONATO
PAVIA
PIACENZA

1.1 Characteristics of the work to be carried out by the tunnelling machine

The two tunnels have a finished internal diameter of 6.9 meters and a length of about 2 600 meters, with :

First Tube :

A section of 2 680 m, with 2 054 m of bored tunnel and two main existing structures to be passed through :

- interconnection structure (APRICA) L1 = 146 m
- LANCETTI station L2 = 480 m

Second tube :

A section of 2550 m, with 1950 m of bored tunnel passing through the two existing structures :

- interconnection structure (APRICA) L1 = 120 m
- LANCETTI station L2 = 480 m

The total length of the tunnels to be excavated by the shield machine is 4 000 m.

Tunnelling has started from the VILLAPIZZONE section, constructed in cut and cover. Because the surface facilities for shield driving were set up there, the second tunnel will have to be driven from the same starting open trench.

The two tunnels pass under roads and railway lines, and occasionally under buildings (silos and apartment blocks). The tunnels are not very deep. The maximum overburden is 15 m at GARIBALDI, with a minimum of 4 m and an average of 8 m for an excavated diameter of 8 m. The ratio of excavated diameter / overburden is less than 2 and over most of the tunnel route, it is close to 1.

The minimum curve radius is 360 m, but the lining characteristics are set up for curves of 300 m radius. The tunneling machine itself can handle curves of 200 m radius (articulated shield body).

1.1.1 Main characteristics of the tunnels

Excavation diameter : 8.03 m

Finished internal diameter : 6.9 m (minimum toleranced diameter = 6,75 m)

Lining :

- by segments thickness = 300 mm, width = 1 200 mm
- by extruded concrete to fill the void between the segments and the ground : thickness = 250 mm

1.1.2 Geology

The tunnels are excavated throughout very open alluvial, sandy and gravelly ground, above the water table. The Milan subsoil is made up of fluvioglacial sandy and gravelly alluvial deposits, with large pieces (rolled stones of very varied geological origins, reaching sizes of up to 100 mm and even 300 mm in some cases).

This recent alluvium features widely variable mixtures alternating between gravel, sand and silt to a quite minor extent. The granulometric distribution on the cutting front is therefore very heterogeneous throughout the length of the route. Sections of very rough sand and gravel with very few binding elements (clays) are encountered.

Friction angle Ø ≈ 35° (32° - 38°)

Density $\gamma \approx 2$ gr/cc

Cohesion Cu ≈ 0

Typical grain size distribution : Size < 0.074 mm = 20% (clay < 5%)

Size > 2 mm = 70%

Size > 10 mm = 20% (up to 100 mm)

This is very open ground (porosity ranges from ~ 33 to 38%) and generally unconsolidated in the upper part, leading to instability of the cutting front and the absolute necessity of supporting the face during excavation.

Abrasivity of the sand is quite high.

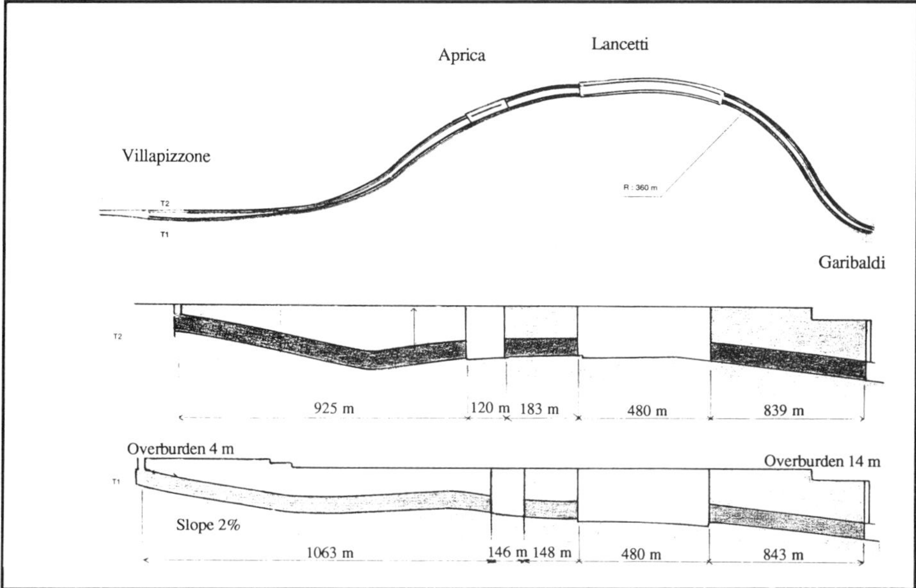

Fig. 3 Longitudinal section of tunnels

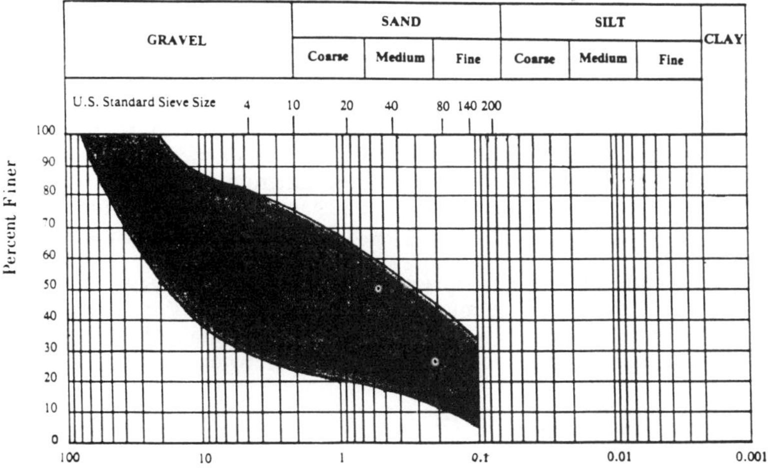

Fig. 4 Grain size distribution curves

1.2 Choice of the excavating method

Given the characteristics of the subsoils and the level at which the tunnels were laid down with a very low overburden, the excavating method selected was the shield driven approach with Earth Pressure Balanced shield and injection of fluidization agents at the front, together with the continuous filling of the void between the segment ring lining and the ground with extruded concrete to limit the risk of surface settlement.

This Earth Pressure Balanced method was chosen in preference to the two other methods studied for this project, i.e:

- Excavation using a mechanical shield machine, controlling the stability of the front by varying the opening ratio of the cutter head. This method was not confirmed because of the difficulty of guaranteeing optimum advance, while controlling the stability of the front under a very low overburden by reducing only the rate of opening of the cutter head.
- The use of a Slurry shield. This method, suitable for excavating in a wide variety of alluvial soils, had the following disadvantages in the context of this project:
  - in very open gravel formation, with no water to impregnate the ground, an excess consumption of mud to form the cake and properly control the stability of the front was likely. In addition, given the low overburden and high porosity of the sand, significant losses of slurry towards the surface could occur, causing settlement and large caving-in due to sudden front instability.
  - Impregnation of the ground by the bentonite mud could, with time, cause movements of the ground after the tunnelling machine has gone by.
  - The high heterogeneity of the ground granulometry, with many areas of gravel and pebbles made it difficult to control the hydraulic mucking, with the risks of instability described above and amplified because of the low overburden and the route in dry soil.
  - It was not possible to retrieve the spoil without a prior, more expensive muck treatment. Retrieving the spoil (≈ 190 000 m3), especially the gravel and pebbles was one of the economic attractions of the project.

2 The earth pressure balanced shield machine

The Earth Pressure Balanced Shield technique was developed in Japan. The first application dates from 1974 and involved boring a collector of 3.35 m in diameter and 1 900 m long. This technique developed quickly and it is today the most widely used in Japan, requested in almost 80% of cases and with excavation diameters of up to 11 m.

In the Earth Pressure Balanced Shield technique, the ground excavated by the cutter head is mixed and accumulated under pressure in a chamber (called the muck chamber) and is then extracted at atmospheric pressure by means of an Archimedes screw (called the screw conveyor).

The principle of earth confinement is based on the use of the excavated material itself which, when mixed and filled into the muck chamber, makes a mass of varying viscosity (depending on the fluidity and plasticity characteristics), maintained under controlled pressure to balance the pressure exerted by the ground, thereby ensuring front stability as the shield progresses.

The pressure exerted on the cutting front, to balance the pressure of the ground and possibly the hydraulic load, varies as a function of the force of the thrust exerted by the shield on the cutting front and also of the rate of muck extraction (screw speed).

As the tunneling machine advances, pressure is maintained on the front by having the screw extraction speed (variation in speed of rotation - rate of opening of the screw discharge gate) and the thrust controlled by pressure detectors mounted on the rear wall (bulk head) of the muck chamber (4 detectors for a 8 m dia. shield). This system of regulating, directly influences control over front stability. This means that the excavated material has to be kept under pressure in the muck chamber and consequently that the volumes of material excavated and discharged have to be kept equal at each instant; this is why the speed of rotation of the screw has to be controlled relative to the constant advance speed of the shield, the pressure in the muck chamber being kept greater by about 0.2 bar than the pressure exerted by the ground (earth pressure + the hydraulic load).

$$P = \text{earth pressure} = \frac{1 - \text{Sin } \varphi}{1 + \text{Sin } \varphi} \times \gamma \times H + 0.2 \text{ (bar)}$$

where			
	φ	=	internal friction angle
	γ	=	density of the soil
	H	=	height of the ground above the axis of the tunnel

For an overburden of 8 m ($H \sim 12$ m) and for the ground encountered ($\varphi \sim 33°$ to $38°$ and $\gamma \sim 2.2$ to 2.4 gr/cc), P, pressure in the muck chamber, should be kept at about ~ 1.3 bar.

This pressure is set within a range taking the passive earth pressure (theorical earth pressure) as an upper limit and the active earth pressure as a lower limit. The pressure should be adjusted during the first length of excavation keeping control of the boring parameters and surface settlement.

Injection of fluidization agents at the front :

The ground excavated, used to transmit the supporting pressure, must have appropriate characteristics of fluidity and plasticity :

- for front support : it must not be too compressible,
- for mixing and removal via the screw for efficient pressure control, it must have a sufficient plastic flowability.

The earth pressure balanced shield, in its simplest form, is well suited to tunnelling work in grounds containing up to 20 - 30% percent of binding elements (silt and clay). These fine elements naturally give to the ground the required plasticity and fluidity.

As the Milan subsoil does not have the right fluidity for optimum earth pressure confinement, foam is injected.

Adding foam to the excavated earth and injecting it directly at the cutting face, considerably improves the fluidity, the plasticity and the pressure transmission characteristics of the supporting mixture. The foam also acts as a lubricant, thereby reducing friction and consequently the power level (torque) consumed when cutting the ground as well as limiting the wear of the cutter head (structure and cutting tools).

The quantity of foam to be injected must be adjusted so as to ensure an adequate support force, while providing a mixture of the proper plastic flowability. Should the volume of foam injected be greater than the volume available within the ground (porosity), the fluidity of the mixture increases but the support force decreases.

The excavated soil with optimum plastic flowability, being stored and pressurized in the muck chamber, has a very good stabilization function for the tunnel face. If the chamber is properly filled with this muck, it is impossible that this compressed material is replaced by anything else and the earth in the tunnel face cannot be exchanged with the muck into the chamber.

It is necessary to analyze the mixture of earth discharged to determine the optimum volume of foam which must be injected into the ground. This analysis can be performed by means of "slump value tests". Foam injection flow rate is controlled using boring parameters (advance rate, working torque and screw removal conditions).

In the case of Milano, excavation is carried out above the level of the water table and therefore through dry soil. If the ground is laden with water, the situation would be more favorable, since excavation in materials which are saturated with water would reduce the consumption of foam.

This fact was verified during excavation of the last 200 m when the tunnel entered into the water table (0,3 bar). Volume of foam injection was then reduced by nearly 30%.

2.1 Main characteristics of the tunnelling machine

2.1.1 Shield

a) Cutter Head

Excavation diameter: 8.03 m (with a fixed overcut of 15mm radius)

Spoke type cutter head made up of 4 arms and 4 half arms with plates to restrict the rate of opening.

The opening ratio was optimized, taking into account the granulometry of the soil with the presence of pebbles of 200 mm, and the required free passage to ensure correct transfer of the material excavated at the working face to the muck chamber and screw conveyor when operating in the Earth Pressure Balanced mode.

The edges of each arm are fitted with cutter bits. Disks cutters are installed on the face of each arm.

Protective tools are also installed on the cutter head, particularly in the peripherical area, to limit wear of the structures.

Number of cutter bits : 66 for each direction of rotation (132 in total, bolted on the arms)

Number of disk cutters : 20 triple disks roller cutters (15 1/2") with 6 at the periphery and 14 on the face

The cutters are drag type bits for loose ground. Their shape and position are adapted so as to facilitate cutting and removal of ground materials. These tools are fitted with tungsten carbon tips to improve abrasion resistance during excavation.

The disk cutters are installed for excavation through gravels and pebbles layers and protect the drag bits from excessive wear.

Two cutter bits are fitted with a wear detection device. Any abnormal wear is detected by a sudden drop in pressure in an hydraulic circuit built into the special cutter bits. This detection is monitored in the shield.

The bolted cutter bits can be dismantled from inside the cutter chamber.

Two over-cutting tools, mounted on hydraulic jacks, are installed for excavation of an "out of profile".

The cutter head support system is of intermediate beam type with large diameter bearing (3 rows of cylindrical rollers) and bearing seals (lip seal type).

The cutter head is hydraulically driven with variable speed in both directions (anti - rolling effect) :

14 reduction gear motors and 10 motor driven pumps
Speed of rotation = 0 to 1 rpm in both directions

Installed power = 110 kW x 10 pump units (1 110 kW)
Working torque = 900 t.m. at 1 rpm (200 bar)
Unlocking torque = 1 350 t.m maximum

The cutter chamber is fitted with fixed and mobile mixing bars. In EPB shield with foam injection, the mixing of foam with the ground material mainly occurs in front of the cutter head before entering into the muck chamber. To improve control of the front stability with earth pressure, mixing bars are installed to ensure an homogeneous mixure into the muck chamber and avoid sticking. To fulfill this important function, the size and positions of the bars are well optimized.

The design of the cutter head has been based on MITSUBISHI's very wide experience in EPB technology for the type of soil encountered in Milan.

b) Shield body

The shield's forward section is articulated in 4 directions, with a leak tight "active mechanism" of articulation.

Articulation of the shield when tunnelling a sharp curve (for Milano R = 300 m) or while compensating tunnel misalignment, prevents excess of overcutting, improves guidance and minimizes deterioration of the excavation face which in turn limits risks of settlement.

Articulating jacks link the shield sections midway with a two layer diaphragm seal system (lip seal packing and grease packed sponge rubber seal).

Body length	=	8 040 mm
Articulating jacks	=	14 x 290 t x 150 mm (articulation angle is 1°)
Thrust	=	5 510 t (290 t x 19 jacks) equivalent to 109 t/m^2
Extension speed	=	7 cm/min (stroke = 1650 mm).

Thrust jacks are selected individually and the thrust is obtained by all the jacks selected.

4 extensometers are installed to measure the stroke and the speed of thrust jacks. This measurement permits the control of foam injection flow rate and the guidance of the shield.

c) Screw conveyor

Type of screw	=	center shaft type screw
Casing diameter	=	800 mm
Screw pitch	=	650 mm
Speed of rotation	=	0 to 17 rpm
Discharging capacity	=	280 m3/h
Drive	=	2 hydraulic motors (Power: 2 x 90 kW = 180 kW)

Access doors together with injections holes are provided to improve, if necessary, muck removal.

d) Erector

Annular type single arm erector, with mechanical gripping in the middle of the segments :

Radial extension	=	950 mm
Longitudinal displacement	=	650 mm

Speed of rotation	=	0 to 2 rpm (hydraulic drive)
Angle of rotation	=	220° (left, right)

Swing movements around the longitudinal and radial axes enable final adjustment of the segments.

Segments gripped by a threaded insert (mushroom) in the center and locked on hydraulically.

The erector is mainly operated with a radio controlled panel.

e) Tail seal and back grouting

The void between the ground and the segments is continuously filled with extruded concrete. Therefore, leak tightness between segment lining and shield is achieved by the sliding frontal shuttering system (called Stop End Shutter ring) with rubber and a flexible metal shell type seal. The SES ring is moved hydraulically.

Concrete is injected at 8 points distributed over the face and through the SES ring.
The sliding of SES ring is controlled by 18 jacks.
Max. back-grouting pressure = 5 bar.

Refer to paragraph 4.2 for more details about the tail seal and back grouting system with the extruded concrete method.

f) Injections

Foam injections: 5 nozzles for injection at the front through the rotating joint and from the cutter head face plates.
2 orifices for injection into the muck chamber

Liquid chemical grout injection for ground treatment from the tunnel : 16 orifices for injection through the outside of the shield body are provided.

g) Measurement of earth pressure

4 pressure detectors are placed on the rear wall (bulk head) of the muck chamber to measure the earth pressure and its gradient :

- 2 on the median line of the bulkhead
- 2 on the upper section of the bulkhead.

The data being indicated by the pressure detectors in the muck chamber are collected and processed.

Reliability of such read out is essential for the tunnel face stability control and overall shield advance. The earth pressure detectors installed are of the type currently used in EPB shield technology.

2 more detectors were added during shield operation.

h) Guidance system

By ZED 260 with an articulation interface unit.

2.1.2 Back-up

A back-up comprising five main gantries which accomodate main shield equipment and three further gantries for supporting the medium voltage cable, primary and secondary ventilation as well as some auxiliary equipment (pipe storage, compressed air unit, ...).

A double railway track (750 mm apart) and two outer rails are installed in the tunnel on metal sleepers placed on the invert of the lining and constructed at the front of the back-up between gantry n° 1 and the segment conveyor.

- The track formed by the two outermost rails is for the running of the back-up gantries.
- The two tracks in the tunnel and inside the back-up are for the service trains to run.
  Those two tracks haul :
  - 2 service trains for mucking and segment supplying
  - 1 mortar train

Fig. 5 Haulage tracks into the tunnel

The service trains are made up (from the front to the rear of the tunnel) of the following waggons:

- 1 double flat waggons to house one segment ring (large segment axes parallel to tunnel axis)
- 10 muck wagons (8.5 m^3 capacity)
- 2 locomotives

The mortar train is stationed on the right track inside the back-up for preparing and supplying the extruded concrete from aggregates stored on the surface. This mortar train is made up of a mixer / pump unit in position in the back-up and a mobile unit with 2 waggons (8 m^3 + 10 m^3)to convey the pre-proportioned aggregates.

Segments are supplied to the erector by means of a conveyor table running on the invert and which can store one segment ring and transfer it to the erector by means of a hydraulically operated fork lift.

The segments are off loaded from the flat waggons, turned and loaded on the conveyor table using a hoist / monorail device. This hoist device is also used for the unloading and the installation of the railway tracks (rails of 6 m long, sleepers of 5 m long).

Muck removal :

A belt conveyor, in a **single** articulated section, transports the muck from the screw conveyor outlet to the rear of gantry n° 5 and then discharged this muck onto a shuttle conveyor which loads it into the waggons, the service train being stationary on the track.

Length of belt conveyor	=	58m (one single through belt)
Capacity	=	600 t/h
Width	=	800 mm

This conveyor can handle curves of 200m radius and the flow rate is compatible with the maximum shield advance.

Outside the tunnel, a tilting type boxcar unloader is used to unload the muck train into a trench. The muck is then taken out of the trench in an intermediate hopper which discharges it onto a continuous belt conveyor transportation system. This surface belt conveyor transports the muck to a group of silos where it is removed by lorry and transported out of Milano.

Ventilation :

The primary ventilation is of blower type (20 m^3/s). The air is supplied through a Ø 1200 mm ventube. The ventube cassette and its hoisting device are installed on the last gantry.

A secondary ventilation with sound proof fan and dust separation is also provided for extraction of air from the front of the tunnelling machine to the rear of the back-up.

2.1.3 Other characteristics

Total length of the tunnelling machine	=	90 m
Total length of the back-up train	=	70 m
Total weight of the tunnelling machine	=	750 t
Total installed power	=	1 600 kW

Fig. 6 Shop assembly of the tunnelling machine

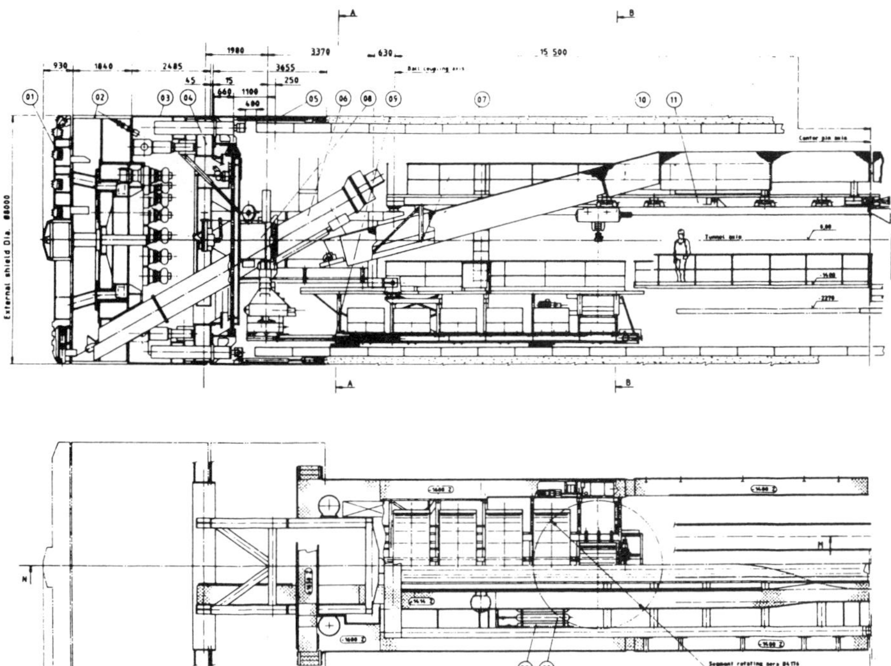

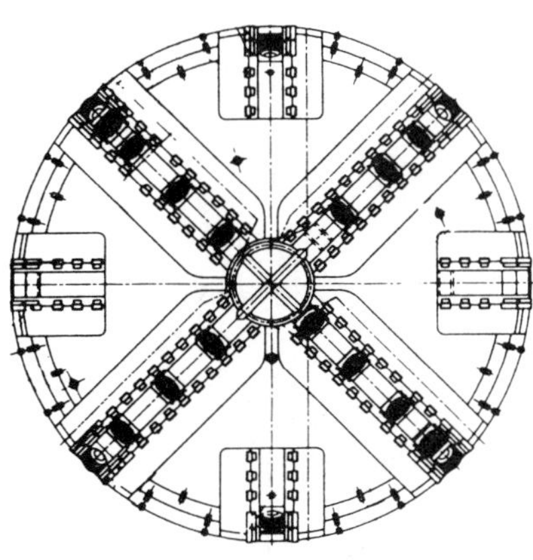

1	Cutter head
2	Front body
3 and 4	Intermediate body
5	Rear body (tail skin)
6	Screw conveyor
7	Belt conveyor
8	Erector
9	Segment conveyor
10	Hoist
11	Connecting beam
12	Control cabin
13 and 14	Greasing
15	Cutter head drive pumping unit
16	Cooling unit
17 to 22	Foam injection unit
23 to 27	Power units
28 and 29	SES ring control
32	Oil tank
33 and 34	Electrical panels
35	Main transformer
36 and 37	Medium voltage cable and protective cell

Fig. 7 Shield general arrangement and cutter head

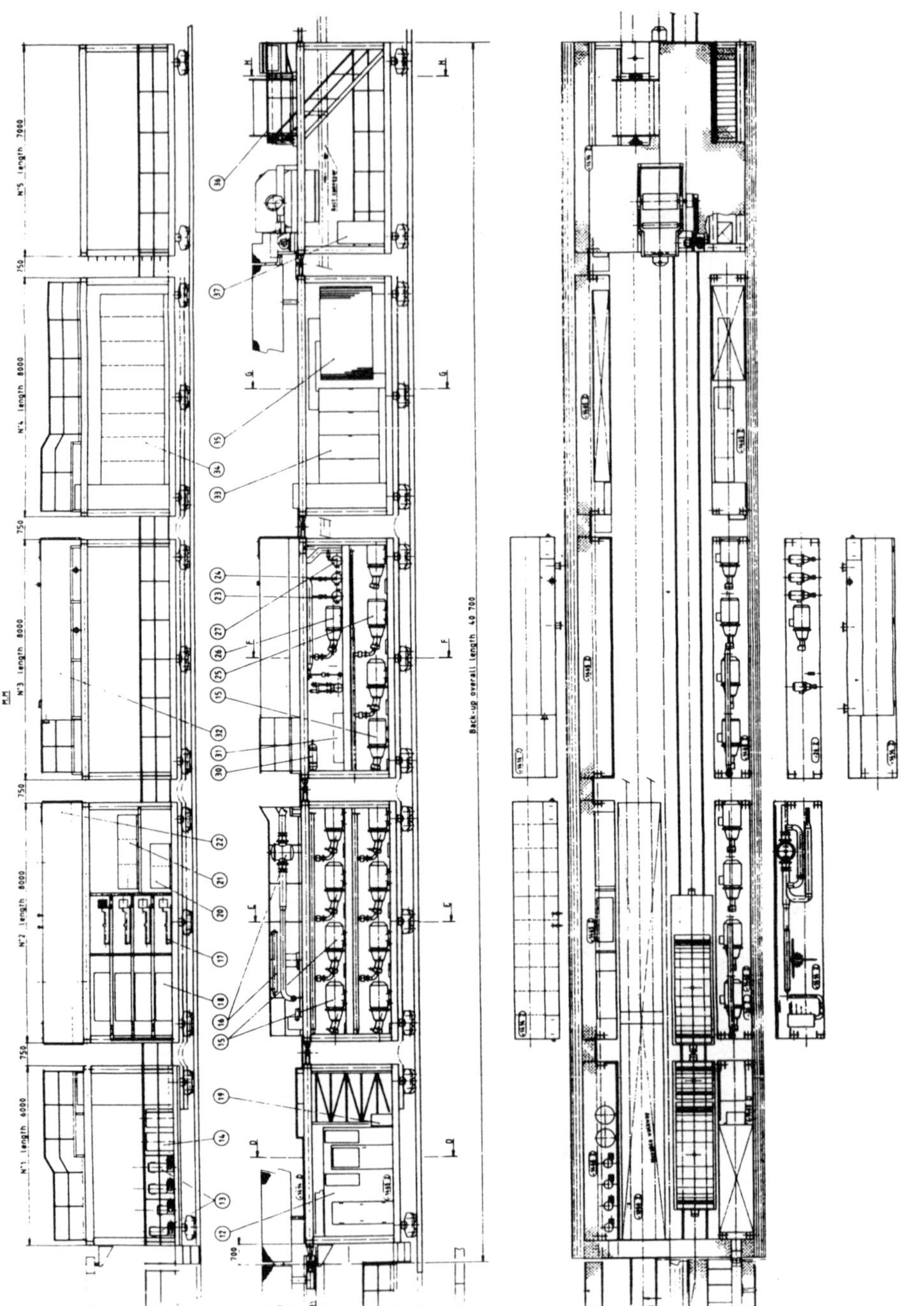

Fig. 8 Back-up arrangement

2.2 Tunnel lining scheme

The double skin tunnel lining is composed of :

- A precast concrete segmented lining of 300 mm thick
- A cast in situ lining of 250 mm thick by extruded concrete filling in the annular space between the ground and the extrados of the segment lining.

2.2.1 Segment lining

Precast reinforced concrete segments

internal diameter	= 6 900 mm
external diameter	= 7 500 mm
thickness	= 300 mm
width	= 1 200 mm
weight of a segment	= 3.6 t
number of elements	= 6 + 1 longitudinal key

"Universal" type segments, designed so as to make it possible to handle curves of 300 m radius and to obtain all straight and curved sections of the tunnels by simply positioning the key and orienting the segments in relation to each other,

The segments are connected in the circumferential joints by means of self-locking CONEX plastic dowels. The segments are of trapezoidal or rhomboidal shapes so as to allow the dowels to be inserted into their housings without segments coming into contact in the longitudinal joints. The segments are centered in relation to each other longitudinally by means of neoprene guide rails.

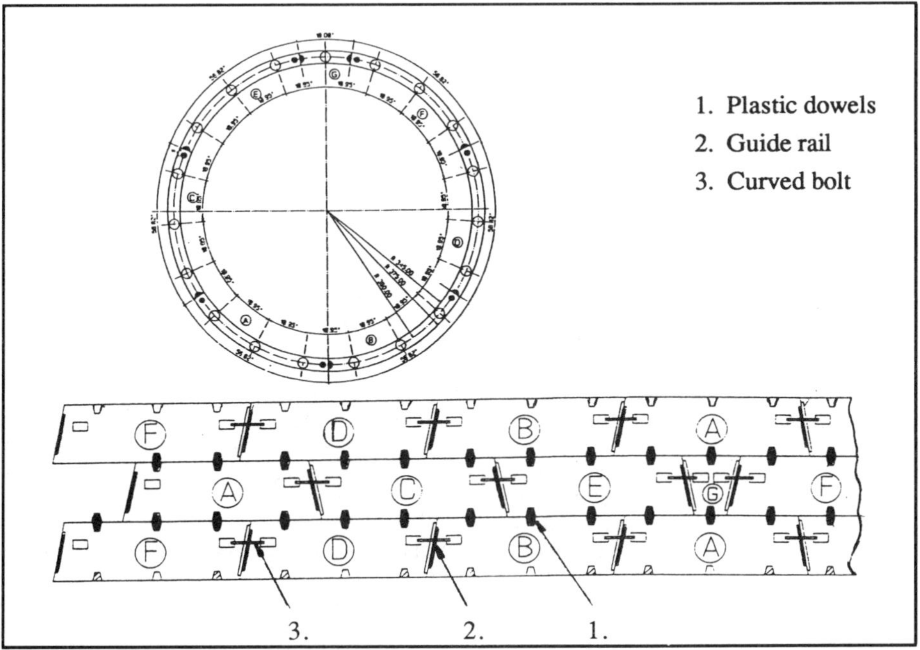

Fig. 9 Segment installation

Three plastic dowels per segment are used. (Fitting of the dowels allow the segment to remain in place without pushing with the thrust jacks).

One radial curved bolt for holding the segments to each other during erection.

A gasket is also used to provide leak tightness when filling in with extruded concrete.

Lifting and gripping point by one insert in the center of each segment.

The CONEX system ensures that the segments are properly centered in relation to each other, reduces inequalities in laying and gives to the tunnel a remarkable finish with a high degree of circularity.

The system has the potential to shorten lining construction. This advantage was not fully utilized in the present project. Safety wise, the system has also proved its advantage and workers are far less exposed to hazards during erection compared to the bolted segment system.

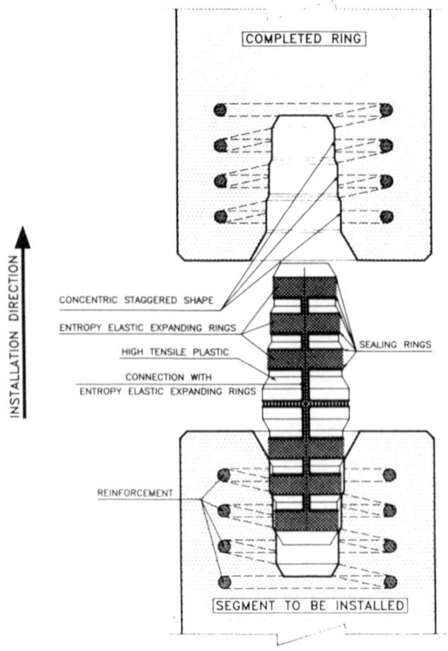

Fig. 10 Dowel arrangement

2.2.2 Extruded concrete

The extruded concrete makes it possible to immediately fill in the annular space created between the diameter of the shield (8 m) and the outside diameter (7,5 m) of the segment lining as the shield moves forward.

HOCHTIEF has developed this new process. The liquid concrete is pumped through injection nozzles (8 in number) located on a sliding ring (Stop End Shutter) in the tail skin of the shield, which forms the front shuttering of the ring space. The void to fill in is limited on the outside by the ground, on the inside by the segment lining and to the front by the front shuttering as it moves forward. The design of the Stop End Shutter, the instrumentation and control systems for movement (18 jacks) make it possible to guarantee constant pressure of the concrete injected into the annular space.

As the extruded concrete is kept under a constant excess pressure, it exerts a radial compression force on the surrounding ground. It is therefore impossible for the ground to decompress or to be dislocated when extruding the concrete. For this reason, the risk of surface settlement are quite minimized as the tunnelling machine moves forward.

The Stop End Shutter activated by hydraulic jacks enables the differences in volume which may occur during extrusion of the concrete, to be compensated for. The hydraulic circuit of the jacks is connected to a compressed nitrogen cushion which expands or contracts to compensate variations in volume as shield advances and maintain a constant pressure in the concrete. If a void is created at the back of the shield with a larger volume than can be injected at the same time, the expansion of the nitrogen cushion allows to maintain a constant pressure on the extruded concrete. When the pressure of the extruded concrete exceeds a preset value, the Stop End Shutter moved by compression of the nitrogen cushion. The nitrogen pressure can be regulated and adjusted to balance the ground pressure. This system provides instantaneous and efficient balancing of volume variations in the support liquid, which in turn ensure support of unstable ground continuously and in good safety.

The extruded concrete is a normal concrete but with improved rheological qualities. A typical mix formula is given hereafter :

Sand (0 / 4) + gravel (4 / 8 and 8 / 13)	:	1900 kg / m^3
Cement	:	250 kg / m^3
Fly ash	:	350 kg / m^3

Admixtures :

- Super fluidifier (with the cement) : 2 %
- Retarding agent (when necessary)

Water

The concrete must have a good liquidity for pumping and for meeting the requirements of controlled extrusion and must keep this liquidity long enough to allow for some stoppages as it is transported along the piping system to prevent it from plugging. It must also be free from permeating phenomena when pressurized (fly ash added).

The concrete is prepared in the tunnel from components stored at surface. A large storage unit made of several silos was installed near the service pit. A special mortar train running on one of the track of the service railway line, conveys the pre-proportioned aggregates and prepare the concrete at a distance of about 60 m from injection joints.

The concrete is pumped to a rotary distributor. Eight flexible pipes connect the distributor to eight injection nozzles on the Stop End Shutter. The distributor supplies each of the eight nozzles in rotation. Injection is controlled by the advance of the shield and stops the shield from advancing if the pressure levels are not attained. Pressure sensors are placed as close as possible to each injection joint. A computer controls the changes in pressure along the circular shuttering in order to determine through which nozzle and in which quantity the concrete must be pumped in.

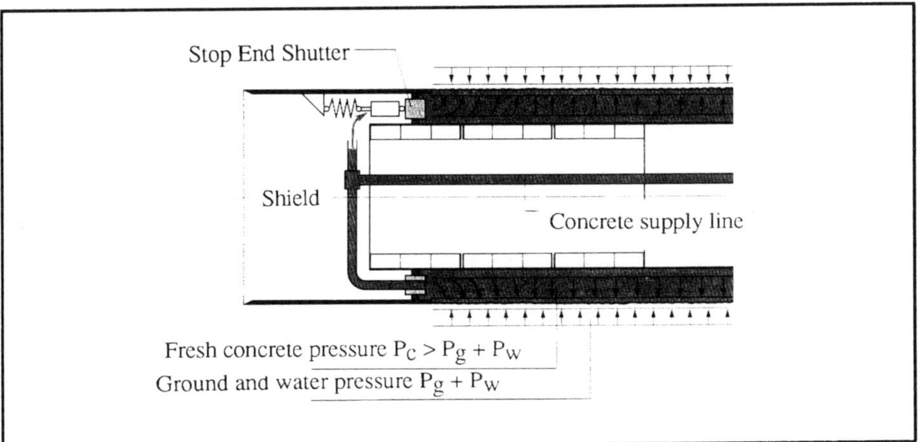

Fig. 11 Extruded concrete process

2.3 Foam injection

The system to inject foam was developed in Japan by OBAYASHI and is now very widely used with the Earth Pressure Balanced technique. A large number of references (over 150 projects) can demonstrate the validity of the method.

This method improves the fluidity and reduces the permeability of excavated materials inside the muck chamber with the use of foam generated by mixing foaming solutions and compressed air.

Economically speaking, the process has a number of definite advantages in comparison with other more conventional injection methods which use clay / bentonite high density slurries :

- By limiting the quantities of liquid products injected and retrieved with the spoil (The foam dissolves naturally and when the air disappears, the liquid based on surface active agents and stabilizers which remains is itself biodegradable).
- By making it easy the disposal of the spoil while respecting the strict environmental regulations in Milano. The addition of defoaming agents has not been envisaged. The spoil is removed as it is and disposed without subsequent treatment.
- By improving the conditions in which the ground is excavated. Lubricating action of the foam increases the flow of excavated material, reduces risks of clogging in the muck chamber, and reduced working torque.
- By increasing the lifetime of parts in contact with the ground (cutting tools) by a reduction in the abrasiveness of the ground when it is impregnated with foam.
- By providing, via the properties of the injected product itself, a better control of the cutter face stability. The cushioning effect of the foam made of micro air bubbles, limits the fluctuation of pressure at the cutter face.
- By improving the conditions in which the product is removed by belt conveyors (less liquid).
- By improving the machine operating conditions, as the products used make it possible to keep the installations in good conditions of cleanliness (a significant advantage over the bentonite-based high density slurry method).

All above advantages resulting in a smoother excavation with improved work environment.

The foam solution used in Milano is prepared from a basic formula consisting of :

- a foaming agent (surface active agent) of the FOAMEX type. (Surface tension at 1% dilution is $\approx 34.5 \times 10^{-3}$ N/m with a viscosity of 42 Cst)
- a cellulose polymer based additive (stabilizers) of the DRILLAM type to increase the stability of the foaming agent.

These two products, similar to products OK1 and OK2 developed by OBAYASHI, are marketed by LAMBERTI CMC.

Usual proportions for 1 m3 of water is 0.7 per cent of stabilizer and 1.5 per cent of foaming agent.

The foam is made from an air-liquid mixture with a foaming ratio (air / liquid solution ratio) which can vary between 5 and 8, depending on the nature of the ground encountered and shield progress.

The foam is created from a unit for controlling the air-liquid mixture and via foam generators to ensure that the foam has the best possible properties when injected.

The volume of foam to be injected can be determined from the void index of the ground and the grain size distribution. In general, a Foam Injection Ratio is defined as the ratio between the volume of foam injected and the volume of excavated soil.

$$\text{FIR} = \frac{\text{Volume of foam to be injected}}{\text{Volume of excavated soil}}$$

For very clayey soil,	FIR≈ 20 %
For sandy - silty soil,	FIR≈ 30 %
For gravel layers,	FIR > 40 % and the required ratio can be calculated from the grain size distribution curves with the given formula :

$$\text{FIR} = \frac{a}{2}\left[(60 - 4 \times X^{0.8}) + (80 - 3.3 \times Y^{0.8}) + (90 - 2.7 \times Z^{0.8})\right]$$

where
x = percent by weight passing 0.074 mm
Y = percent by weight passing 0.42 mm
Z = percent by weight passing 2 mm
a = coefficient, a function of the uniformity ratio of the granulometric Uc.

if $Uc < 4$ $a = 1.6$
$4 \leq Uc \leq 15$ $a = 1.2$
$Uc \geq 15$ $a = 1.0$

The Foam Injection Ratio means the volume of foam under pressure in the muck chamber.

For Milano and for a typical section of ground with sand and gravel, we find FIR = 68 per cent.

Example of chemical foam preparation :

FIR	= 70% (700 L of foam for 1 m^3 excavated)
Foaming ratio	= 7
Pressure in muck chamber	= 0.9 bar

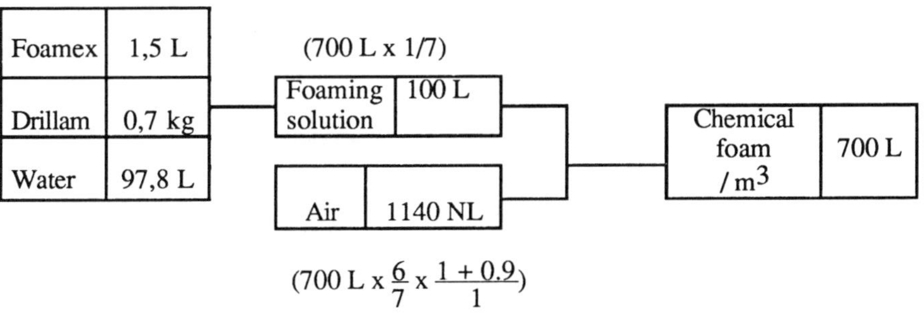

As the ground is very open with few fine particles and outside the water table, the volumes to be injected are calculated from the FIR formula and may vary between 60% and 80%. The values actually observed during excavation are within this range.

2.3.1 The Installations

a) The foam solution is prepared on the surface by a unit which comprises

- a 10 m^3 tank for the foam agent,
- a powder measure (for the DRILLAM),
- a 3 m^3 tank for putting the additive into solution and mixing it,
- a 10m^3 buffer tank (plus a second tank in the back-up),
- 3 transfer pumps,

b) The foam preparation and injection system into the back up comprises

- a control cabinet (with computer and key board to fix the parameters)
- a control panel located in the tunnelling machine control cabin,
- five independent injection lines for the foam solution, each line comprising:
  - a screw type pump,
  - a unit for controlling the air flow and the supply of foam solution,
  - a foam generator, installed as closely as possible to the injection point.
- an air compressor.

The injection system is completely computer assisted and the programme is controlled by the speed of the shield moving forward (extension of the thrust jacks) in order to determine the volume to be injected from 3 parameters :

- the pressure of the earth into the muck chamber (range value)
- the foaming ratio (set value)
- the FIR (set value)

The response in volume of foam to be injected to changes in load on the cutter head is controlled through an integrator which boosts the volume of foam if the cutter head torque exceeds a prefix value.

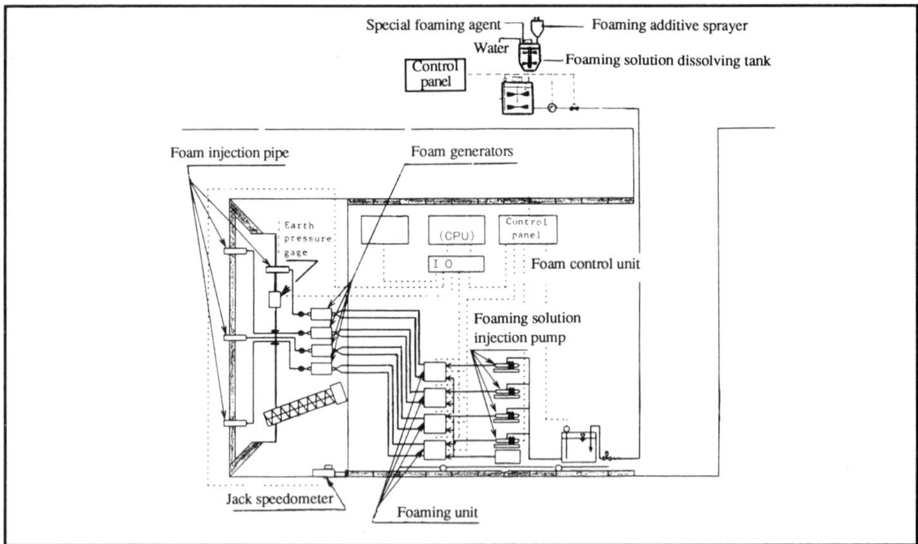

Fig. 12 Foam injection system

3 Progress of the tunnelling work

3.1 Performances

The tunnelling machine began excavation in early September 1992, reaching an average speed of nearly 10 m per day following a start-up period of about one month.

Considering the new technologies used for the first time in the Milano subsoil (Earth Pressure Balanced Shield, foam injection and extruded concrete), with operating teams who were unfamiliar with this type of operation, it must be said that a very satisfactory rate of advance was quickly obtained, in conditions which were very difficult (an overburden of less than 7 m and having to go beneath railway lines).

Total length excavated on 7 November 1993 (breakthrough) : 2 054 m with an average monthly progress (working months only) of almost 160 m (including start-up and transfer through stations).

The attached progress graph shows the total advance of the shield during excavation of the first 2054 m long bored tunnel.

If we discount the transfer periods to cross the stations and the first month of running (starting period), this average monthly rate of excavation is almost 220 m.

The average daily rate of excavation (working in 3 shifts) is 8 rings, or 9.6 m per day. Since September 1993 the average daily rate has been 12 m (10 rings).

a) Best rates of excavation

Start of the excavation	:	3 of September 1992
First 150 m completed by	:	16th of October 1992
Best day (3 shifts)	:	14 rings or nearly 17 m.
Best 5-days week (3 shifts)	:	43 rings or 52 m (18 to 22/01/93)
Best 7-days week (3 shifts)	:	88 rings or 105.7 m
Best monthly rate (3 shifts 26 days)	:	320 m (4 to 31/10/93)

b) The performance between two stations (APRICA - LANCETTI) should also be noted

Length to be excavated = 146 m.

Total duration of the operation, from putting the shield in contact with the front until it was removed in the other station : 22 days comprising :

- 2 days for installing the thrust reaction beam
- 4 days for excavating an amount equivalent to the length of the machine and installing 12 rings of segmental lining.
- 12 effective days of excavation (including 1 day off).
- 4 days for removing the tunnelling machine.

or an average of almost 10 m per day in 3 shifts work.

c) Typical rates of progress (work in 3 shifts)

from 12-01-93 to 15-01-93 (4 days)	=	31	rings (37.2 m)
from 18-01-93 to 22-01-93 (5 days)	=	43	rings (51.6 m)
from 05-04-93 to 14-04-93 (10 days)	=	72	rings (86.4 m)
from 30-08-93 to 12-09-93 (14 days)	=	100	rings (120 m)
from 04-10-93 to 10-10-93 (7 days)	=	67	rings (80.4 m)
from 25-10-93 to 31-10-93 (7 days)	=	88	rings (105.7 m)
from 01-11-93 to 07-11-93 (7 days)	=	81	rings (96.6 m)

d) Organization of the work

Work in 2 shifts (2 x 10 hours) during the start up period.

Move to 3 shifts (5 days out of 7) early October 1992.

Move to 3 shifts (7 days out of 7 - 25 days per month) on 24 May 1993 until the breakthrough of the first tunnel.

e) Operating team

- Personnel in the tunnelling machine : 11 operators per shift
- Personnel for the logistic of the shaft : 7 operators per shift
- Personnel outside the tunnel (staff) : 6

Fig. 13 Site erection and start-up

PASSANTE FERROVIARIO MILANO

Fig. 14 PROGRESS OF TUNNELLING WORK - FIRST TUNNEL

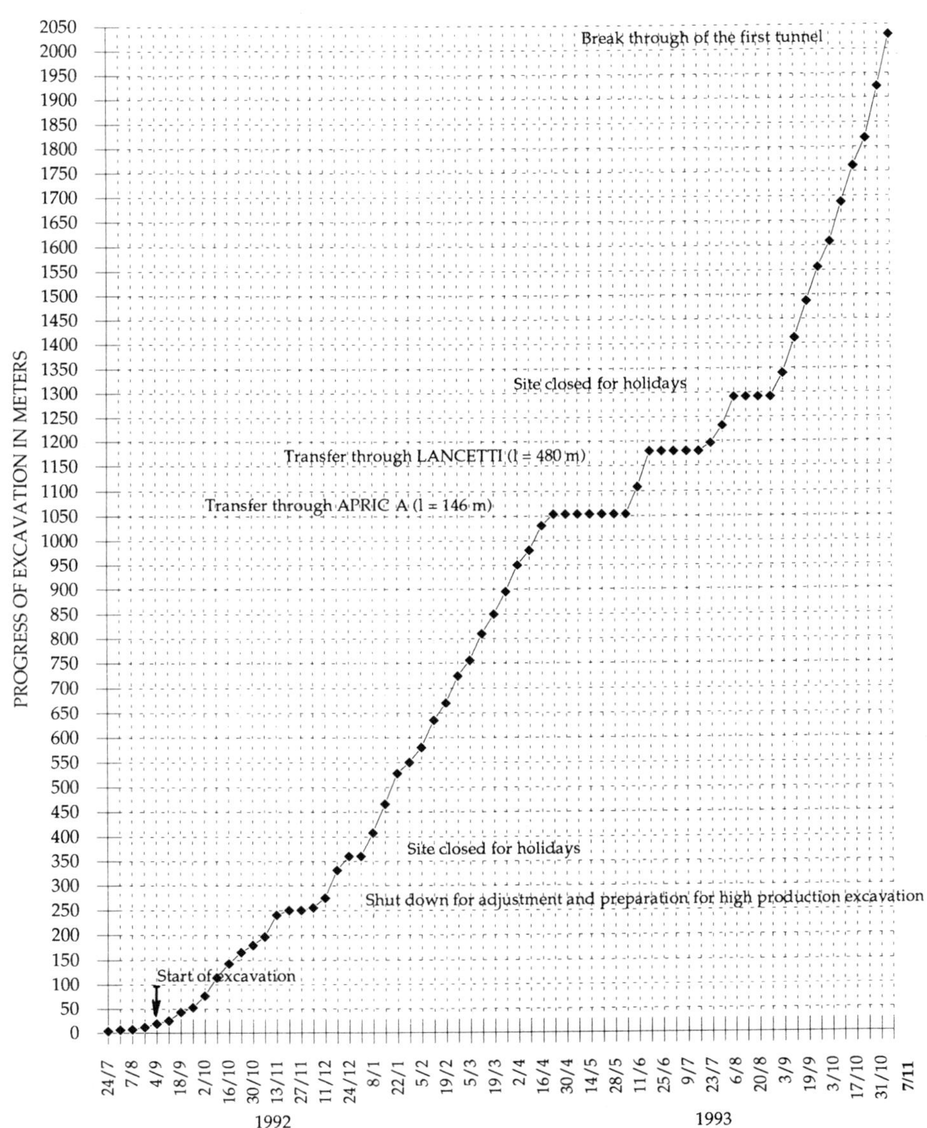

3.2 Excavation parameters

advance	:	2.5 cm to 3.2 cm per minute (limited by concrete injection rate)
cutter head rotation	:	0.6 to 0.7 rpm
thrust	:	1900 t to 2300 t
earth pressure (tunnel axis)	:	0.9 to 1.4 bar
foam injection (% of earth excavated)	:	40-80%

ring no	399 9/1/93	403 19/1/93	433 22/1/93	452 29/1/93	483 8/2/93	497 10/2/93	509 11/2/93	21/9/93
earth pressure (bar)	1.1	1	1.1	0.9	1	1.1	1.1	1.4
advance (cm/min)	3	3	3.3	2.5	1.5	2.9	3	2
thrust (no of jacks)	2200 (14)	1900 (13)	2300 (14)	2400 (14)	3300 (14)	2300 (13)	2300 (13)	2400 (14)
speed of rotation (RPM)	0.6	0.6	0.6	0.6	0.6	0.6	0.6	0.7
torque (t.m)	560	560	600	580	540	580	540	650
screw rotation (rpm)	6	8	9	5	4	7	7	5
foam injection (% of excavated volume)	40	39	60	60	70	70	70	80

Advance cycle: excavation = 40 to 60 mins
segment erection = 30 to 40 mins

Excavation through consolidated sand / gravel layers under surface load (buildings)

Compressive strength	≈	100 bar
Earth Pressure	=	1.6 bar
Thrust	=	3700 t
Working torque	=	700 t.m.

3.3 Main difficulties encountered and adjustments made

At the start, the operating teams had difficulty in optimizing the boring parameters, taking into consideration :

- excavation : optimizing the advance and the pressure in the muck chamber. Setting the right pressure in the muck chamber was definitely the most critical point.
- foam injections : optimizing the volume to be injected in keeping with the required advance (lubricating action) and controlling settlement (plasticity of the mixture). Optimising the foam characteristics.
- the extruded concrete : optimizing the composition of the concrete and the conditions of injection in keeping with the advance of the shield (injection pressure and volume to ensure that the annular void is properly filled).
- the tunnel machine guidance system (ZED)
- laying the segments with the centering dowels.

The start-up conditions were particularly difficult too with a very low overburden (4 m) and a section under railway and road bridges in the very first 150 m.

a) Clogging in the cutter chamber and occasional jamming of the cutter head as a result of

- a too high earth pressure (1.6 to 2 bar) in the chamber.
- too slow shield advance, partly due to problems encountered with the extruded concrete (1 cm/min).
- too high cutter head rotation speed (0.9 to 1 RPM).

The main difficulty was to determine the right earth pressure which would enable the shield to advance with an optimum working torque (less than 30% of the nominal torque) and to control the stability of the front. This pressure had been fixed too high initially.

For optimum control of the boring parameters, the prime objective was also to obtain proper flowability of the mixture in the muck chamber. This optimum fluidity and plasticity depends on the volume of foam injected and the mixing capacity of the cutting chamber.

Too frequent changes in parameters made it impossible to assess their effects.

This clogging occured mainly in the central section of the cutter head and in some extend had the effect of turning the central section into a solid rotating disk of material occupying about 20% of the cutter head area. This results in reducing the available space in which excavated material could enter the chamber and in limiting the advance rate, excavation being carried out by applied force rather than cutting the ground.

Material being compacted into the muck chamber central area, the foam injected in too high quantity compared to the volume available, found its way to the surface through open and dry ground in low overburden condition (4 to 6 m).

Clogging of the cutting chamber brought about a significant degree of friction, leading to overheating of structures and excessive wear of the cutter tools with damaged of roller cutters. Most of the cutting tools were changed after the first 250 m, before starting on the continuous excavation zone under the existing railway lines (800 m of tunnel length before reaching APRICA station, without any further tool replacement).

b) Ground settlement was observed when starting after a lengthy period of shutdown of the tunnelling machine

(a typical instance of this is when restarting excavation after stopping for the weekend). When the tunnelling machine was stopped, soil became tightly packed in the muck chamber with a reduction in the volume occupied by the foam (diffusion into the ground and reduction of the stabilizing power). This phenomenon was accentuated by too great an amount of foam injected during the last meters before the tunnelling machine was stopped, leading to a greater impregnation of the ground and therefore a much greater risk of consolidation with time. The total void in the muck chamber was also resulting of the natural void of ground after swelling and the void due to some lack of volume control.

In order to eliminate this phenomena and to minimize the risks of settlement when starting up, better preparation of the front before stopping is necessary. This preparation involves the following action taken during the last advance cycles with control of the volume of soil excavated and stored in the muck chamber (filling up the soil into the chamber) :

- reduction in the volume of foam injected
- reduction of the advance and increase of the pressure in the muck chamber
- maintaining the thrust

- injection of 2 or 3 m^3 of long life time foam or polymer solution into the muck chamber during the last revolutions of the cutter head.

These actions carried out in parallel make it possible to compact the material in the muck chamber with respect of the volume excavated and to be extracted, and to reduce consolidation phenomena with time, thereby limiting the risks of settlement when excavation is resumed. Injection of polymer solution demonstrates its effectiveness.

A pressure drop from 1.2 bar to 0.7 bar was generally observed, 1.2 bar being the earth pressure value. Without polymer injection, this drop increases further and pressure in the chamber was down to 0.3 bar.

For lengthy periods of shutdown, the support pressure must from time to time be re-established by rotating the cutter head, injecting polymer and possibly re-pressurizing the front by using the thrust jacks.

c) Getting the extruded concrete right as the shield was advancing was a source of difficulty which slowed things down during the first 300 m of excavation

The extruded concrete had to be adapted to the conditions of use under pressure (permeating phenomenon) with the high permeable ground. Some modifications were also made to the concrete supply system (piping and rotary distributor).

The S.E.S. ring jammed due to some concrete ingress in the system. This problem was overcome by adding a grease packed seal.

Unblocking the piping (60 m long, Ø 80 mm) took up a great many man-hours during the start-up period and beyond.

Nevertheless and as an overall performance, we can notice that the extruded concrete lining is of very good quality.

On the segment lining, some breaks occurred at the corners of the segments when these were being installed. In fact, the segments were not designed with bosses to minimize parasite forces related to the laying of the rings and which generally cause cracking. Those local crackings occured maily during the advance in curved section with radius less than 450 m.

d) Obstacles

The shield encountered a non detected old water well made of a casing of 600 mm diameter with a wall thickness of about 10 mm. Shield advance was kept and nearly 400 kg of metal were discharged by the screw conveyor.

Some damages occured on the cutter head which could be repaired after the breakthrough during the transfer of the shield from GARIBALDI station back to the VILLA PIZZONE section.

3.4 Observations

a) Modifications and adjustments

Some modifications were made to the tunnelling machine. The following should be noted :

- two additional pressure sensors were installed for optimum control of the earth pressure, (one in the upper part of the muck chamber and one in the upper skin plate of the shield body)
- installing an extra foam injection line. (5th line)
- adding an extra bar in order to optimize spoil mixing in the muck chamber.
- installing an high pressure water injection line into the muck chamber.
- Protecting the SES ring with an additional seal packed with grease.

The two additional pressure sensors allow a better control of the pressure distribution in the muck chamber and of the filling of this chamber, consequently a better control of the good balance between the volume excavated and the volume to be discharged.

Proper control of the earth pressure was achieved through the monitoring of the sensor installed on the median line of the bulkhead. The earth pressure gradient was continuously verified with the two additional sensors installed. Control of this last value leads to predict higher risks of settlement because a void may occurs in the top portion of the muck chamber.

b) Volume of foam injected
Over the all tunnel length, the volume of foam injected was :

- From PIOPPI to APRICA (FIR ≈ 50 %) + 5 % of water in the muck chamber
- From APRICA to GARIBALDI (FIR ≈ 80 to 85 %) + 5 % of water in the muck chamber
- Last 200 m reaching GARIBALDI and within the water table (FIR ≈ 50 to 60 %)

Fluidification of the material was maintained at a slump level not greater than 10 cm and not less than 5 cm.

Very little spillage in transferring the muck from the screw conveyor to the belt conveyor was observed during the whole excavation of the first tunnel.

Water injection improved the fluidification of the excavated material and reduced the amount of foam to be injected.

c) Consumption of cutting tools
2 sets for 2000 m of excavation with a half set replaced after the first 250 m of excavation.

Nearly 800 m of excavation were done without replacing tools after proper adjustment of boring parameters.

d) Volume of extruded concrete
The actual volume of concrete extruded was about 8.5 m^3 per ring (1.2 m long ring) for a theoretical volume of 7.8 m^3 per ring.

e) General comments
The technical assistance of a joint team made up of staff from the constructor NFM, from MITSUBISHI and from OBAYASHI, who are very experienced in the use of Earth Pressure Balanced shields, enabled all the staff responsible for the operation to acquire understanding and expert knowledge of the system, particularly of the control of the earth pressure.

The measures taken in November 1992 with respect to the tunnelling machine operating instructions after the first 250 m, together with the fact that the operating staff were more familiar with the ground conditions and the equipment, rapidly bore fruit. The monthly rate of progress went from 100 m in October 1992 to almost 200 m as of December 1992.

Once again, it should be noted that the period devoted to starting up and adapting to special problems was relatively short, despite difficulties at the start of excavation which were the results of working with new teams and a machine which was to a certain extent a prototype.

Tunnelling conditions may change day by day and the operating team has to pay much attention to such change from the data gathered during excavation and particularly :

- boring parameters to keep an optimal and a safe rate of advance
- characteristic of fluids injected
- nature of muck discharged by the screw (slump test values, water content, granulometry).

3.5 Surface settlement during shield drives

Measurements were made continuously as the tunnelling machine advanced. Throughout the route, the values measured are very much lower than 10 mm. Singularities have been observed with values of 25 mm on certain occasions.

The followings are examples of settlement measurements made as the tunnelling machine advanced.

It was observed, when tunnelling with an overburden of 10 m, that 25% of the settlement occured at 4 m in front of the shield, 65% at the shield level and 10% after the shield.

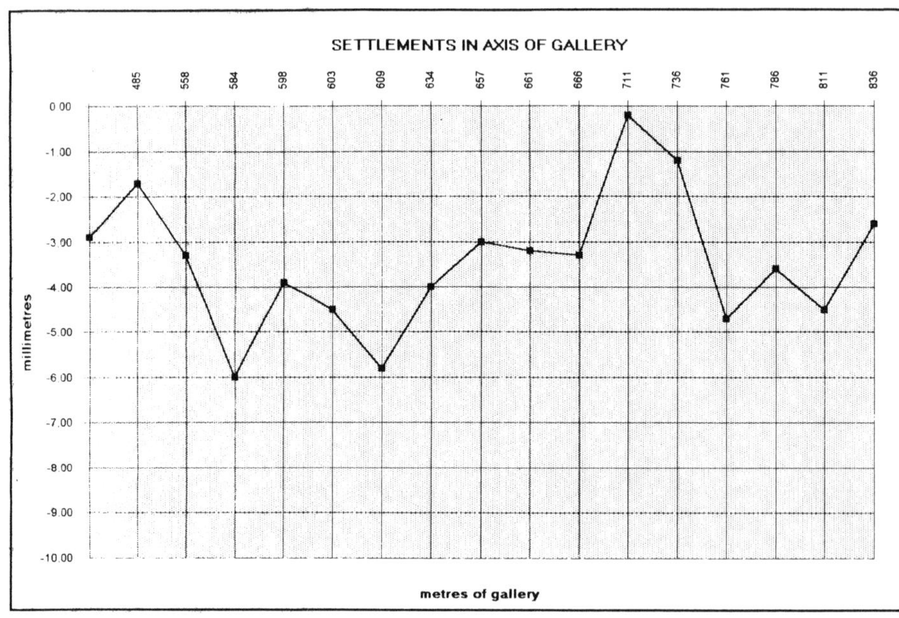

Fig. 15 Settlement during shield drives

Conclusion

The good performances achieved during excavation of the first tunnel, confirm the possibilities of Earth Pressure Balanced shield for tunnelling work with low overburden through unstable ground consisting of sand and gravel layers. These conditions are generally considered to be very difficult for shield tunnelling with the control of surface settlement within 5 mm.

Overall performance of the method was aided by the use of chemical foam injections, proper control of the back grouting with extruded concrete method and smooth installation of the segmental lining with CONEX plastic dowels system.

Needless to say that the success of the tunnelling work is fully depending upon the skill of the operating team to acquire understanding and expert knowledge of the all system which represents in some extend the state of the art of Earth Pressure Balanced shield machines.

The transfer of the shield, back to the start of the second tunnel took 2 months. Excavation of this new tunnel started on January 28th, 1994 and more than 160 m were achieved in 12 days with a maximum best day recorded performance of 15 rings (18 m), keeping the surface settlement at less than 5 mm with a tunnel route passing under a road bridge with an overburden of less than 5 m.

THOSE INVOLVED

Owner :

Lombardy Region - Community of Milano
Italian State Railway (FS)
North Milan Railway (FNM)

Prime Contractor :

M.M. Strutture ed Infrastrutture del Territorio SPA de Milano

Construction of lots 3P, 5P and 6P :

Design :

Italprogetti Srl - Rocksoil Spa - CP engineering Srl - REICO Spa

Construction :

Consortium of :
TORNO (representative) • COGEFARIMPRESIT • LODIGIANI • CMB •
COLLINI • IFG - TETTAMANTI • PROGETTI & COSTRUZIONI

Tunnelling Machine :

Neyrpic Framatome Mecanique under licence of MITSUBISHI Heavy Ind.

Foam Injection: OBAYASHI system

Extruded Concrete: HOCHTIEF system

Pen-Y-Clip road tunnel, North Wales: achieving a dry lining through a pervious rock formation

P. Catling
Cementation Mining, Ltd., Bentley, Doncaster, England
J. Scholey
Howard Humphreys and Partners, Ltd., Leatherhead, England

Abstract
The paper describes aspects of waterproofing and lining of the 930 m long single carriageway Pen–y–Clip road tunnel. Tunnel supports comprise either rockbolts and shotcrete or steel ribs. In consequence, the in situ tunnel lining consists of two distinctly separate details. In the rockbolted section, the lining is a constant 300 mm thick, achieved by profile regulation, and is fully waterproofed with a continuous membrane. The linings to the steel ribbed sections of tunnel are a minimum of 600 mm thick to incorporate the 300 x 300 steel rib supports at 750 mm or 1500 mm centres. Seepage drainage is provided by collector drains around the excavated profile between the steel ribs.

Keywords: Concrete, crack limiting, crack treatment, drainage and waterproofing membrane, rockbolts, shotcrete, steel ribs, tunnel lining, tunnel support.

1 Introduction

1.1 Project description
Pen–y–Clip tunnel, together with its approach roads, forms one of the final links in the 85 km long A55 North Wales Coast Road (Fig.1) which provides a high standard dual carriageway between the UK's national motorway system and North Wales. It also forms part of Euroroute E22, from the Baltic port of Sassnitz, in Germany, to Dublin in Eire.

The 2 km long Pen–y–Clip scheme is a coastal dual carriageway along the line of an existing single carriageway road in a narrow corridor of sidelong ground between existing housing, a mountain and a main line railway. The 930 m long tunnel carries one two lane carriageway of the road through the microdiorite headland. It runs parallel and close to the existing road and railway, both of which run partly in tunnel and on viaduct (see Fig. 2).

The 2 km long scheme was let as a single Contract under the ICE 5th Edition to John Laing Construction in 1989 for the Tender sum of £51.5M. Laing sub–contracted the portal excavation and tunnelling works, which constituted some 25% of the Contract, to Cementation Mining of which Trafalgar House Construction (Tunnelling) is a division. The Engineer for the Works was Travers Morgan in association with Howard Humphreys &

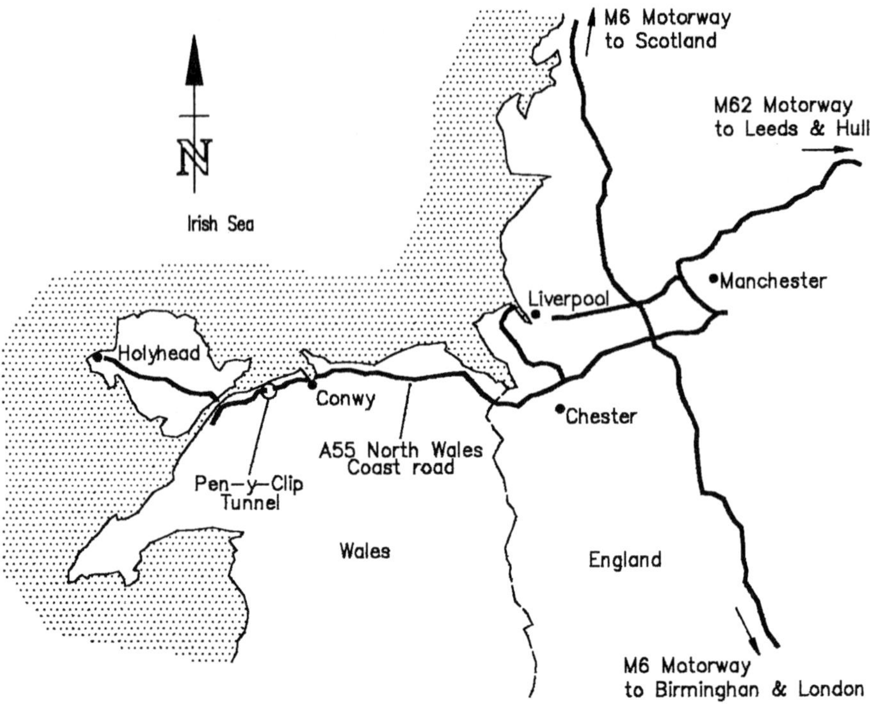

Figure 1. Location of Project

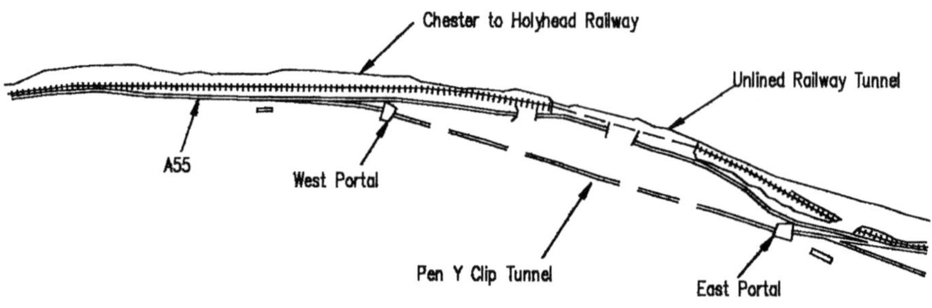

Figure 2. Layout Plan.

Partners Limited, who were responsible for the design of the tunnel and its portals together with mountainside rockfall protection measures. Howard Humphreys staff were also present on site in key supervisory roles. Tunnel excavation and lining took some 2 years of the 4 year Contract period. The tunnel was opened to traffic in October 1993.

1.2 Tunnel profile

The tunnel profile is a "D" section with a flattened crown. The lined tunnel is 9.95 m wide with a maximum height above road level of 7.1 m. The tunnel accommodates a single 7.3 m wide carriageway flanked by two service troughs of approx 0.8 m and 2 m depth respectively. A typical section is shown in Fig 3.

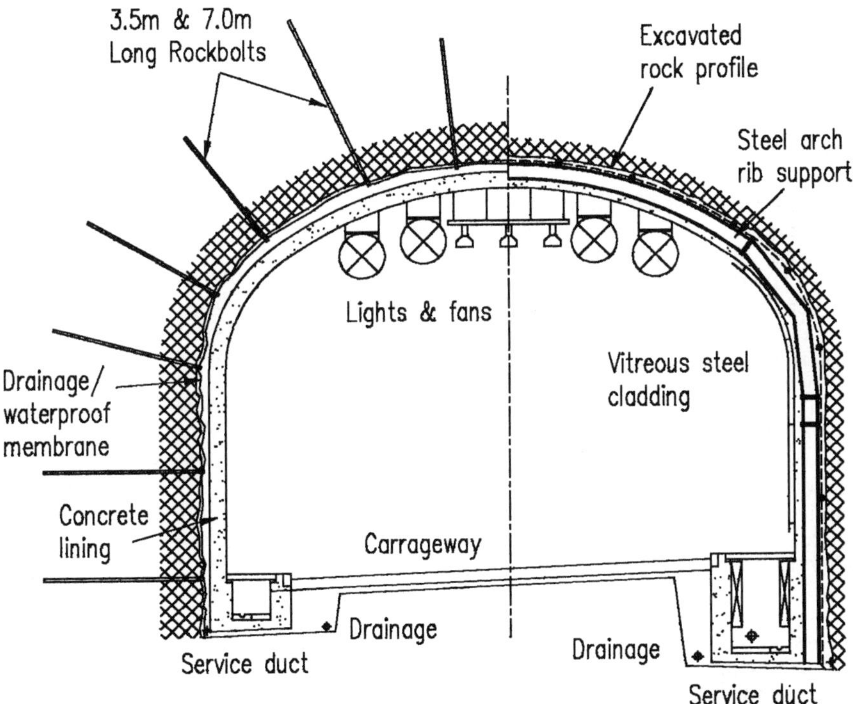

Figure 3. Typical Section Through Tunnel

1.3 Geology

The Pen–y–clip Headland is a microdiorite intrusion which rises steeply from the foreshore. On both east and west sides of the headland the rock is more broken as a consequence of faulting and glacial actions and is overlain by scree, partly infilled with glacial deposits and quarry related debris. The tunnel alignment is constrained by highway geometry, the need for portal areas at the face of the intrusion to be relatively clear of overlying materials and the need to ensure adequate rock cover. (Photograph 1: Aerial view of West Portal showing rockfall protection fences).

The east portal is sited 100 m from the new road junction with the bypass west of Penmaenmawr. The west portal is sited at a rock outcrop close to where the road built by Telford in 1830 joins the existing A55 east of Llanfairfechan. The tunnel is 930 m in length which includes the two 25 m portal rockfall protection shelters. The layout of the project is shown in Fig 3.

The jointing in the igneous microdiorite rock is generally close and the compressive strength of the rock is very high. Excavation by drill and blast was envisaged. The tunnelling conditions ranged from good to extremely poor and structural support was required throughout in the form of shotcrete and rockbolts, with steel rib supports where rock cover was shallow or quality very poor to exceptionally poor. (Photograph 2: West Portal showing skew of portal and minimal rock cover).

2 Tunnel support

2.1 Rock mass classification

Rock support was decided on the basis of geological mapping of each excavated face and assessment by means of both the C.S.I.R. Rock Mass Rating (RMR) and the N.G.I. "Q" system. The latter proved more accurate and sensitive than RMR, particularly in weak rock. Three support classes were detailed in the Contract; Classes 1 and 2 consisted of rockbolts and mesh reinforced shotcrete whilst Class 3 comprised steel rib support.

Rockbolts were installed in ground assessed at having a "Q" value greater than 0.01 ("extremely poor"). Support classes 1 & 2 both comprised pattern bolting and reinforced shotcrete with Class 2 applying where "Q" was less than 0.4 ("very poor"). Class 2 support comprised more closely spaced rockbolts than Class 1, with mesh reinforced shotcrete extending over the complete periphery; in Class 1 support, shotcrete above springing level only was reinforced. Rockbolt arrays comprised a variety of combinations of 3.5 m and 7 m bolts as determined by wedge analysis to suit the encountered joint set combinations.

2.2 Steel rib support

Arch rib support was designated for lengths of tunnel at the East and West Portals and for a known central section shear zone, with a provision for further steel rib support where ground conditions were expected to require Class 3 support. These were defined as having a C.S.I.R. rock mass rating < 20 and N.G.I. Q value < 0.01 described as weak to very weak, highly weathered microdiorite, heavily sheared and possibly yielding.

The steel arch ribs were designed to carry the full overburden pressure at the portals and were erected at 750 mm centres and designed as a steel

Photograph 1 Aerial view of West Portal showing rockfall protection fences

Photograph 2 West Portal showing skew of portal and minimal rock cover

structure to BS 5400. The crown consisted of 305x305x158 kg/m universal beam rolled with a radius of 6 m. This was seated on and bolted to cranked shoulders of the same steel section formed at a radius of 3 m. The width at springing level was therefore approximately 10.5 m. The cranks were erected on continuous wallbeams at springing level. The sequence of erection, following initial shotcrete, was therefore to set the wallbeams to line and level, fix the cranks including tie–bars, and then mount the crown. (Photograph 3: Portal section of tunnel showing steel ribs).

The suitability of the construction plant used for erection was checked for operational capability by undertaking full scale trials within a number of ribs erected in the compound. The JCB 820 and associated bracket used for crown erection was also demonstrated to representatives of the Health and Safety Executive prior to its utilisation since its application was non–standard.

Top heading excavation and support was followed by bench excavation, which normally lagged at least 50 m behind the face. The wallbeams were undercut to install the arch rib legs. A maximum of three arch ribs could be undercut i.e. supported on the continuous wallbeam, and posts had to be installed within two days. Posts were 305x305x137 kg/m universal column and extended to the full depth of the troughs, those on the south side being longer than those on the north.

Rib posts were installed using the JCB 820, but this time with the bucket in place and using chains. The void beneath the post footplate, to cleaned rock, was shuttered and filled with a fast setting grout.

Load transfer from the ground to the arch ribs was by means of hardwood blocking. Once positioned and wedged tightly, shotcrete was applied in around their contact with the rock to limit subsequent displacement following blasting. Blockings were spaced at between 600 mm and 1000 mm centres being most closely spaced on the crown and cranks and least on the posts; blockings were cast in with the final lining.

2.3 Rockbolt support

Lengths of tunnel to be supported by rockbolting were driven full face, with the trough excavation carried out as a secondary operation behind the face. The advance for each round was a maximum of 4.2 m. Rockbolt arrays were installed following mucking out of blasted tunnel spoil and scaling of the face and after application of the initial 50 mm of shotcrete, which incorporated a joint sealing layer of accelerated shotcrete where required as temporary works.

No Permanent Works rockbolts were to be installed closer than 3 m to the tunnel face. Contractually any bolt positioned in this zone was considered to be temporary support only and did not contribute to the permanent support.

During the early stages of full face excavation the Universities of Bradford and of Aberdeen continued their research project, commenced at Penmaenbach Tunnel four years earlier, into the effects of blast vibration on full column resin bonded rockbolts. For this research project, rockbolts were installed as close as 0.7 m to the face at Pen–y–Clip. No significant load loss or debonding was experienced as a consequence of blasting.

In Class 1 rock, rockbolt arrays were drilled out using a Tamrock 305 drill jumbo with two drill carriage booms and a basket. After flushing the holes clean with water they were charged with resin capsules. Rockbolt installation continued later, once one boom had commenced drilling face blast

Photograph 3 Portal section of tunnel showing steel ribs

Photograph 4 Installation of rockbolts in Class 1 rock

holes. One boom was used for spinning the rockbolts through the resin whilst access for stressing was gained from a Normet access basket. In this way rockbolting was overlapped with face drilling and later with meshing and shotcreting by working consecutive operations on north and south sides of the face. (Photograph 4: Installation of rockbolts in Class 1 rock).

Rockbolts were 3.5 m and 7.0 m long, and 25 mm diameter. The 7.0 m long rockbolts were supplied in two approximately equal lengths with an equivalent strength slim–line coupler. This simplified handling and installation in comparison with the single length 7.0 m long bolts installed at Penmaenbach.

Rockbolts, bearing–plates and hemispherical washers were all galvanised. Rockbolts were installed into capsules of medium and slow setting resin to permit post tensioning and provide full column bonding and double corrosion protection in association with the galvanising. Rockbolts were post–tensioned to 100 kn with one in ten bolts proof loaded to 150 kn.

2.4 Reinforced shotcrete

Dry process shotcrete was used for all tunnel works. Details of plant, materials and methods may be found in Ref 1.

Following rockbolt installation the surfaces to be supported were meshed with steel fabric reinforcement to BS4483 comprising 4 mm wires on a 100x100 grid; laps were 200 mm. Square pieces of mesh were installed with the rockbolt bearing plates providing fingers for lapping and tieing the mesh panels. Thereafter the mesh was secured and contoured with mechanical fixings inserted into drilled holes.

Shotcrete was then applied to cover the mesh to a depth of 50 mm providing overall a minimum of 100 mm mesh reinforced shotcrete. Shotcrete for permanent works required a minimum strength of 30 MPa and was generally sprayed without the accelerated set additive used for temporary works shotcrete.

Both powder and liquid accelerators were used on the Contract in different applications and with the addition of microsilica to reduce rebound and recover some of the long term strength lost through the use of set accelerators. Although the 28–day strength of accelerated shotcrete was close to but below the minimum required for permanent work, a conscious decision was taken not to use accelerator unless ground conditions so required. Instead, the shotcrete thickness was built up over several short periods with unaccelerated shotcrete within the shift cycle.

The use of steel fibres was also investigated and a mix was approved for use under the contract. Some use of this mix was used for reprofiling shotcrete but its application as an equivalent to mesh reinforcement was not pursued once the face cycle was established with concurrent activities. It was not thought viable to dedicate time in the heading for the purpose of spraying steel fibre shotcrete nor was it considered viable to undertake this operation as a concurrent activity.

3 Tunnel lining

3.1 Final lining

There is a final in situ 30 mpa concrete lining throughout the tunnel although its detail varies between rockbolted and steel ribbed sections. As full permanent support is provided by either the rockbolts or steel ribs, the lining fulfils a secondary role and is not designed to provide structural support. All lining bays were separated by contraction joints. Reinforced concrete to the service troughs preceded arch lining. The construction joint at the foot of the side walls was so positioned that it would ultimately be hidden by the precast concrete service trough covers.

In rockbolted sections, the lining is 300 mm thick and lightly reinforced to be self supporting. This reinforcement was in the form of single layers of radius curved A393 panel mesh (10 mm wires on a 200 mm square grid). The lining is isolated from the surrounding works by a full profile drainage and waterproofing membrane described in section 3.4.

Reprofiling in shotcrete was specified to ensure the thickness was maintained to tight tolerances, to control cracking in the 10 m maximum length bays. Waterstops were incorporated in the contraction joints between bays and in the horizontal construction joint between the lining bay and the trough pour. The specification also called for a trial 10 m bay to be cast. This was followed by an instructed 60 day period of assessment prior to any other lining operations commencing.

Fixed to the shutter skin were grout pipe inserts which were cast into the lining for the purpose of allowing contact grouting which followed lining by a minimum 30 days. Contact grout holes could not have been drilled because of the drainage and waterproofing membrane. The ends of the inserts were sealed with expanded polystyrene which fitted against the drainage membrane. These could be easily raked out without potential damage to the membrane.

The lining to steel ribbed sections is 600 mm thick, to encase the steel ribs with a minimum concrete cover of 75 mm. The Specification restricted lining pours in steel ribbed sections to 5 m bays. Because of blocking between the ribs and the excavated profile, there was no opportunity to incorporate a continuous waterproofing membrane around this part of the lining. Instead, intermittent geotextile wrapped conduits were installed between the ribs to some effect. This is described in Section 3.6. (Photograph 5: 5m Stelmo shutter awaiting introduction to tunnel).

The lining was placed behind rail mounted shutters fabricated by Stelmo. These shutters were fabricated in 2.5 m units bolted together and each shutter mounted on a "horse". The "horse" had an opening sufficiently large to allow the passage of the loading shovel, muck–away trucks and concrete wagons in use at the face, as lining commenced some 8 months before the end of excavation and ran concurrently without serious disruption to either operation. Progress of excavation and concrete lining is shown in Fig. 4.

Initially there were two shutters, of 10 m and 5 m length respectively. The shutter for the rockbolted sections was fabricated from four 2.5 m units to give a 10 m shutter i.e. the maximum pour length permissible. In the ribbed sections two 2.5 m units were used with a 250 mm extension to provide a shutter length of 5.25 m, to which was added a tailskin fitting inside the previous pour. The pour length was such that the contraction joint fell mid–way between the 750 mm centred ribs. A third 5.25 m shutter was subsequently brought into use in response to an increase in the length of steel ribbed support and thus the number of 5.0m long pours.

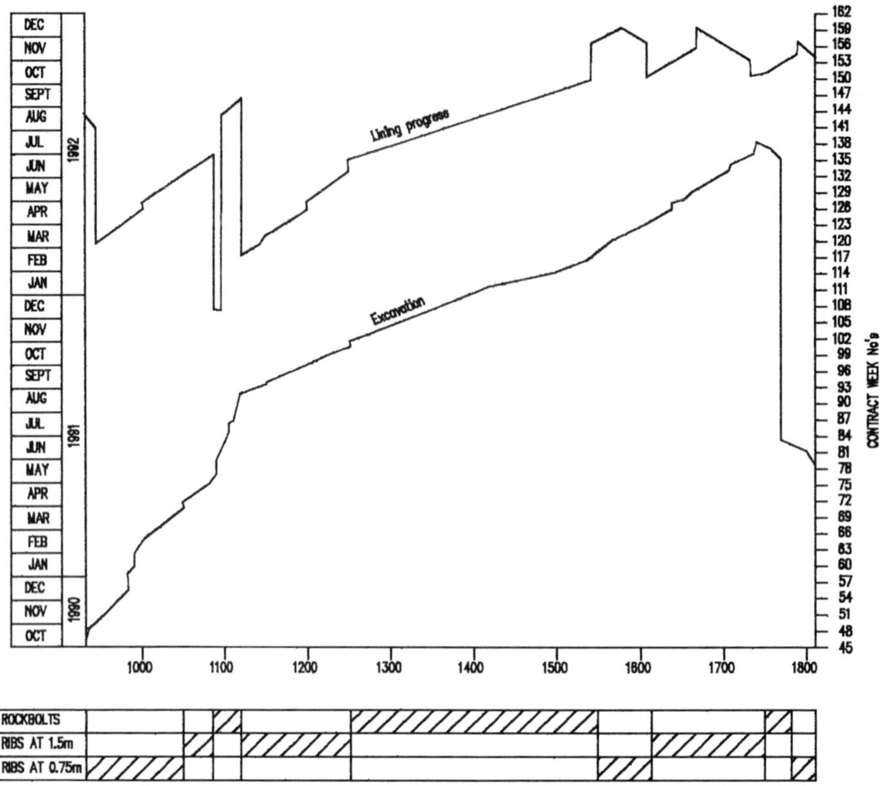

Figure 4. Excavation & lining progress showing support types.

Waterstop was incorporated in the transverse contraction joints between lining pours and in the horizontal construction joint between arch and trough concrete. Subsequently for ease of installation, the contraction joint waterstop in ribbed sections was charged to a combination of "Rearguard" attached to the shotcrete and "Hydro–tite" hydrophyllic compound on the joint face. The horizontal construction joint waterstop throughout the tunnel was 305 wide flat dumbbell (Servicised Servitite 305 CJ).

Concrete was placed using a mobile pump positioned at the rear of the shutter. A 30 Mpa pump mix was used including a superplasticiser to provide the desired workability. Strike time was constrained by the Specification to a minimum of 48 hours from the end of the pour. Curing methods had to be compatible with the final paint system and, in certain parts of the profile, silane impregnated, thus ruling out the sole use of curing membranes. Water curing by spray was practised initially although this was later replaced, due

Photograph 5 5-m Stelmo shutter awaiting introduction to the tunnel

Photograph 6 Transition between ribbed section and primary lining

to the operational difficulties of this method. Subsequently, pours were covered with light-weight polythene below the "paint line" to retain humidity whilst a membrane was sprayed on to the unpainted crown. Shortly before break-through, strip curtains were installed at the portal to prevent the prevailing wind creating a "wind tunnel" to the detriment of the recently cast concrete.

3.2 Reprofiling shotcrete as a regulating layer

The contract required the filling of overbreak in all sections of tunnel to the theoretical excavation profile prior to fixing the drainage and waterproofing membrane. This served two main purposes in that the final lining was of a constant cross section to control cracking and a surface free of sharp edges was produced onto which the membrane could be fixed without risk of puncture.

To accommodate the first requirement, very tight tolerances were placed on the shotcrete which was specified to be within +25 mm and -0 mm of the theoretical profile shown on the drawings. Consideration was given by the contractor to achieving this by, inter alia, spraying to piano wires, spraying to a fixed or rigid former, "cutting" an oversprayed thickness to the required tolerance, spraying a gunite finishing layer and "plastering" or working the final sprayed layer.

Although smooth wall blasting was specified, and controlled blasting provided scope for limiting overbreak, the actual geology of the Clip was such that significant overbreak was encountered. Large volumes of shotcrete were necessary and consideration was given as to whether this element of the work could be undertaken within a rolling window, and to what extent, given the large volumes involved, wet shotcrete could be used.

This was not considered to be an economic solution; given that the necessary plant and skills were required at the face for mesh reinforced shotcrete as part of the tunnel structural support would have meant an amount of duplication.

A visit by the contractor to Europe to view wet shotcrete in use indicated that this would not be a panacea. The continentals on the sites visited had ground conditions which allowed the ground to be supported without the application of shotcrete for some time permitting large volume shotcrete setups. The wet shotcrete processes entailed combinations of additives for which HSE approval was unclear and which might not comply with the requirements of the contract. Large volumes of rebound were witnessed which did not indicate much improvement over the dry shotcrete method.

Consideration was again given to placing shotcrete behind a former. A solution would have had to be found to allow the exclusion of rebound from the works. Investigation of experiences in Japan revealed that only short lengths of trial tunnel had been lined using the technique of placing shot-crete behind a travelling/rotating former; this had been heavily subsidised by their government. A trial on construction site was programmed at that time but again this would not be a commercial venture.

Having looked at wet shotcrete, including a pumpable concrete, and at a rigid travelling former it was natural to look at regulating the profile by placing concrete behind a shutter. This solution employed tried and tested technology, provided the tolerances which were required for the subsequently installed drainage membrane and eliminated the problem of controlling shotcrete dust in the tunnel atmosphere.

This option was costed and reviewed against shotcreting within the programme and was found to be feasible. The proposal was agreed in principle and design of the shutter and the so-called "primary lining" commenced.

3.3 Primary lining

The 10 m long shutter was designed by Stelmo who already had an order for the 10 m and 5 m long final lining shutters, and to the same principles: a rail mounted gantry carrying a crown section and two side wings.

The shutter was struck from the lining by collapsing the two wings inwards and lowering the crown piece on hydraulically operated rams. Whereas the final lining shutters were lapped on the reinforced concrete walls of the North and South troughs the primary lining shutter wings were bolted onto a lower "boot" section. This section extended the wing down to the invert level of the trough and incorporated the longitudinal recess to accommodate the foot of wall drainage from the membrane.

Since the primary lining had to be acceptable as part of the permanent works in replacement of the regulating shotcrete required under the Contract it had to be designed within the standards acceptable to the U.K. Department of Transport. It therefore had to perform in the same way as the shotcrete it was replacing and must not place any loading on the final lining which had been designed as a free standing arch. The designers of the final lining had already established that the minimum thickness for a free standing arch in 30 MPa concrete was 300 mm. (Photograph 6: Transition between ribbed section and primary lining).

As the primary lining was a filling to overbreak, its thickness could reduce to zero and it could not therefore be designed as a simple arch structure. It was not the contractor's intention to over-excavate a minimum 300 mm to accommodate the primary lining and some other analytic approach was therefore needed. The design of the primary lining undertaken for the contractor by Mott Macdonald was based on anchoring it back to the surrounding profile as with the original shotcrete.

A grid of rockbolts in the upper arch at 2 m centres approximating to 3 No bolts per 2 m chainage was installed in advance of the shutter. These were closed down to 1.5 m centres depending on the overbreak experienced. Rockbolts were fully galvanised and resin bonded within the drilled hole; external to the drilled hole in addition to the galvanising, the bolts were wrapped in Denso tape. From these, immediately above the shutter, panels of A393 mesh were installed with 50 mm spacing between panels; the weight of mesh installed varied with over-break (in proportion to the volume of concrete necessary to fill the void). The rockbolts were installed so that the first layer of mesh had 50 mm cover to the surface.

Placing of the primary lining used the same equipment as the secondary/final lining, based on a mobile concrete pump supplied by Pochin, and was fitted into the same routine. Concrete was poured on a six-day week with each of the three shutters being poured twice per week. Whilst the secondary lining was constrained by the Specification to a 48 hour strike time, the primary lining shutter was struck after 36 hours based on the crushing results of early cubes.

After lining, the specified grid of drainage holes was drilled through the regulating layer into the rock beyond. Thereafter the drainage membrane was installed, an operation which benefited from the smooth concrete perimeter in terms of "steering" the membrane sheets and ensuring coincidence/register of the dimples at overlaps.

3.4 Drainage and waterproofing membrane

Between the regulating layer and the final lining, Dorken MS 20 membrane was installed. Dorken MS 20 is a semi-rigid plastic, supplied in rolled sheets, formed into 20mm high dimples such that it resembles an egg-box in structure. Its use in Pen-y-Clip Tunnel is beleievd to be the the first application of the material in the U.K. and also for a tunnel under construction as opposed to relining work, with which it is normally associated.

The membrane serves as a full profile waterproofing and drainage layer and is connected to an invert drainage system running the full length of the tunnel.

Water within the microdiorite intrusion drains through arrays of 40mm diameter drain holes drilled through the reprofiling layer and the microdiorite to the interface between the Dorken membrane and the shotcrete/concrete regulating layer. Holes were drilled on a 2m square grid to penetrate beyond the reprofiling layer into the microdiorite beyond to a depth of 1m.

The high relief of the membrane provides channels down which water can trickle, gravitating to a recess in the regulating layer incorporating a perforated pipe surrounded with no fines concrete. These 100mm diameter pipes (one on each side of the tunnel, below trough level) carry the collected water to cross-cutting pipes feeding into manholes at approximately 50m centres, ultimately feeding the water into the 225mm diameter carrier pipes running under each side of the carriageway. (Photograph 7: Dorken membrane and lining shutter).

3.5 Membrane installation

The membrane was fixed to its shotcrete or concrete substrate by nailing with 37mm long nails on a 1m square grid. The nails were fired through the base of the dimple i.e. that part in direct contact with the infill below. The head of the nail incorporated a steel washer and a neoprene seal to seal the perforation; subsequently the dimple was filled with cold applied bitumen rubber sealant (Expandite Plastiseal).

The individual 2m wide strips of Dorken membrane were overlapped by 300mm. The water-tightness of the membrane was maintained by trapping between the overlapping sheets two adhesive strips of butyl rubber (38mm wide, 1.5mm thick, Expandite Seelastrip PIB500) spaced 150mm apart. Between these two strips the lapped joint was then nailed at 300mm centres. Nails were fired using conventional Hilti tools with cartridges. However in parallel a pneumatic system was used employing nails and tools supplied by Aerosmith; this was selected in the belief that it was a less hazardous tool to use. In the event neither tool was involved in any accident.

In order to eliminate problems during concreting, the crown was instructed to be oversprayed with a thin layer of gunite held in place by a light wire mesh anchored by nailing at 300mm centres.

3.6 Drainage in steel ribbed sections

A full profile drainage and waterproofing membrane could not be provided in the steel ribbed sections because of the timber blocking between the ribs and the profile. The Contract included no drainage provisions in these sections

Photograph 7 Dorken membrane and lining shutter

Photograph 8 Lining shutter in West Portal ribbed section

and detailed staged consolidation grouting to reduce inflows.

Poorer than anticipated ground conditions extended the length of tunnel supported by steel ribs with a consequent impact on excavation progress. In an attempt to mitigate such delays, the designers assessed that the rib spacing could be increased from 0.75m to 1.5m if there were no hydrostatic loading on the supports. This could be ensured if a way were found to provide a drained periphery. During the construction period a proprietary product called Servidrain had become available which fulfilled this need.

This took the form of a 900mm wide semi–rigid dimpled strip wrapped in a continuous geotextile "sock" which was permeable to water but which prevented the passage of fines and grout through to the channels within. Water paths could thus be provided by strips of the material attached to the profile prior to concreting.

Strips were cut from a roll of material and fixed in the spaces between arch ribs. Strips were fixed across the crown and from the shoulder down terminating in a receiving hopper installed in line with the backwall drain running in an excavated recess behind the arch leg. Fixing was by Aerosmith nails i.e. a pneumatically fired nail penetrating the membrane and the shot–crete beneath. The head of the nail carried a 50mm diameter plastic disc to prevent damage to the Servidrain membrane or former.

Joints between the crown and shoulder pieces were simply made by lapping the two pieces together and binding the joint with wide adhesive–backed tape.

Although the geotextile was impermeable to cement particles, the designers were concerned that it might become permeable or rupture if subjected to cement grout under pressure. Contact grouting had been detailed in steel ribbed sections to ensure that all ribs were fully encased for corrosion protection. However if there was some overbreak above the ribs, it was adjudged that satisfactory concrete embedment could be achieved by inspection though the stop end. As the steel rib support was based on a fully blocked steel design, total filling of all voids was not considered essential. (Photograph 8: Lining shutter in West Portal ribbed section).

To take advantage of the space created by overbreak, reprofiling in the ribbed sections of tunnel was limited to achieving a regulated surface not varying by more than 1:10 overall. This was also generally in line with the contractor's philosophy of substituting concrete for shotcrete where this was possible. Prior to membrane fixing, arrays of weep holes were drilled through the shotcrete and into the rock for a depth of 1m. Holes were 40mm diameter and at 2m centres.

Two modifications were made in the method of installing the Servidrain. The first involved the hopper receptacle into which the Servidrain was connected to the 100mm diameter toe drain. The proprietary unit was an expensive article imported from Europe and subject to delays in supply. It also required adaptor and connector pipes to connect to the toe drain, which made the job of installation labour intensive. This hopper was therefore replaced with a more simple and robust design based on a modified length of uPVC pipe used as the toe drain but fitted with lips into which the Servidrain membrane was slotted. It proved possible for these units to be produced locally.

The second modification also entailed the development and local manufacture of a fitting, this time a connector to enable trough construction, with its attendant servidrain and toe drainage, to precede whatever arch

Photograph 9 Finished section of lining

Photograph 10 Section of lining with finishing under way

reprofiling shotcrete proved necessary. A certain lead for trough concrete construction ahead of arch lining was necessary as the trough concrete was required to achieve a minimum strength to carry the arch shutter. The Servidrain connector permitted a joint to be incorportated at carriageway level and permitted the top of the servidrain to be temporarily plugged. The trough could then be decked out at carriageway level to simplify removal of rebound from the localised reprofiling shotcrete operations.

Although the application of Servidrain had been introduced in additional lengths of steel ribbed tunnel to achieve a drained periphery, use of the material was extended to all steel ribbed sections, where ribs were at the closer centres of 750mm. A narrower Servidrain membrane 300mm wide had to be used in these locations. Consolidation grouting was limited to the west portal zone where the most open jointed ground had been experienced. In this portal zone, Servidrain was still installed except in the locations of consolidation grouting arrays which were at approximately 2.5m centres.

3.7 Concrete mix design

Cracks in the final lining were limited by the design of the concrete mix and by the method of construction. The specification required that there should be no cracks in excess of 0.1mm width within 36 hours of striking the formwork.

The principle of mix design was to consider, inter alia, the use of low heat cement, the use of pulverised fuel ash as cement replacement, the use of coarse aggregate with a low coefficient of thermal expansion, the use of admixtures to improve workability and the reduction of cement content and water/cement ratio.

Following laboratory trials to select an appropriate mix, batch trials were conducted with a super–plasticised concrete with 275kg of Ordinary Portland cement, 100kg of PFA and a w/c ratio of 0.41 giving a target slump of 160mm. Finally, as specified, a trial bay was cast to form part of the permanent lining. This bay was cast in an unribbed section of tunnel and in addition to reviewing the concrete mix, placing equipment and methodology, it demonstrated the adequacy of the installation of the drainage membrane and underlying regulating shotcrete.

The trial bay was successfully constructed and was incorporated into the permanent work. Thereafter the remaining final lining was cast and all the bays were accepted as permanent work. (Photograph 9: Finished section of lining).

3.8 Crack limiting

Whilst, as described above, efforts were made to limit cracks in the lining, provision was made within the contract for treatment of cracks associated with ageing and long–term shrinkage. Epoxy–resin was specified for sealing to prevent water ingress. In the central section of the tunnel where the construction was a free–standing mass concrete arch with a drainage and waterproofing membrane behind no crack treatment was required.

Where vertical cracks occurred with age, the integrity of the waterproofing system was unaffected. These cracks occurred in the vertical wall section and were vertical and parallel with the contraction joints and occurred in the centre of some 10m bays suggesting that a 5m maximum bay length might have been more appropriate.

Longitudinal cracking was frequently observed at and above springing in pours to the steel ribbed sections. Such cracking usually became apparent within a week of striking the formwork. These cracks could be related to the bracings between the ribs and in particular to the flanges of the continuous wallbeam. In order to identify the factors influencing the developement of these cracks, a number of trials were directed in successive bays, as follows.

Fabric steel reinforcement was welded to the arch ribs, regulating shotcrete was sprayed to limit the overall concrete thickness to 600mm, bracings between steel ribs were removed and the continuous wallbeam cut away between ribs, the striking time was reduced based on crushing cubes at earlier and more frequent intervals together with temperature–time plots from cast–in thermocouples and hot water curing was investigated.

None of these trials demonstrated any significant benefit and sooner or later the flange of the continuous wallbeam acted as a crack inducer. Finally it was decided to fit crack inducers to the formwork to concentrate cracking at the two wallbeam flange horizons in straight lines to improve regulation of crack treatment.

One further trial involved casting a 10m long bay in the ribbed section on the premiss that if limiting the bay length to 5m did not totally eliminate cracking, there might not be any difference if a larger bay were cast. This trial did produce a vertical crack compared with none in the 5m pours and therefore no change was implemented.

3.9 Crack treatment

Cracks were treated after a minimum of 60 days after casting. Dry cracks greater in width than 0.1mm were treated with an injected epoxy resin. This product was selected following injection trials the results of which were verified by coring.

Where water was present in a crack, this was stemmed from deeper into the concrete by first injecting a urethane which expands considerably in the presence of moisture. Thereafter sealing continued with epoxy resin. (Photograph 10: Section of lining with finishing underway).

3.10 Finishes

The concrete lining was finished to Department of Transport Specification for Highway Works Class F3, that is with a smooth finish of uniform texture and appearance using formwork that imparts no blemishes to the surface. In the central, rockbolted part of the tunnel the walls were painted in an off white colour to a height of some 5.5m above road level, i.e. about 1m above arch springing. The paint system as applied by the Main Contractor was Keim Granital silicate based paint. Preparation for this system was confined to pressure washing of the finished concrete. Light abrasion or grit blasting to remove small blow holes was not required since there was no exothermic reaction associated with curing of the paint which would cause the paint to blister.

The choice of wall colour was restricted by the need to achieve certain reflectance levels for safe operation of the tunnel. A strip at the base of each wall some 0.3m high in the zone subject to discolouration by road dirt from traffic spray was painted in a brown colour using the same paint system.

Painting of the sections nearest the portals, which tended to be mainly steel ribbed was limited to the above strip of brown at the base of each wall. The walls to these sections were considered to be prone to the most frequent and extreme temperature changes and were clad with a tertiary

Photograph 11 Finished section of tunnel with tertiary lining in place

lining, again by the Main Contractor. This comprised a coloured ceramic coated thin steel sheet manufactured in 2m panels with stainless steel cover strips.

Panels were fixed to the concrete lining with drilled fixings which were covered with stainless steel flashings such that the whole could be maintained by cleaning using mechanical rotating brushes. The lower sill of the cladding is formed as a channel which is sealed against the concrete lining. Any future ingress of water behind the cladding will not therefore attract road dirt but collect in the sill channel to be drained into the trough and thence into the tunnel drainage system. (Photograph 11: Finished section of tunnel with tertiary lining in place).

Examination of the flows in the manholes shows that both the full profile waterproofing and drainage layer in the central section and the partial profile drainage installed in the ribbed tunnel sections are collecting water whilst the linings remain water tight.

4 Acknowledgements

The scheme was commissioned by the Welsh Office Highways Directorate whose permission to present this paper is gratefully acknowledged. The views expressed are those of the Authors and not necessarily those of either the Client or their Employers.

5 Reference

1. Hobson, D.A. and Gilchrist, D. (1994) **Shotcrete in the Construction of Pen- y-Clip Tunnel,** Conference Proceedings, Tunnelling '94.

Session 12: Hydropower

Tunnel design in the design and construct context—the Pergau experience, Malaysia

A. D. Murray
Knight Piésold and Partners, Ashford, England
Bruno Marrai
Knight Piésold and Partners, Ashford, England
P. G. Davies
Sumber Bumi Knight Piésold, Sdn. Bhd., Kelantan, Malaysia

The 600 MW Pergau Hydroelectric Project in Northern Malaysia is being constructed as a design and construct contract. 33 km of underground tunnels and shafts and a large underground powerhouse complex were to be constructed in accordance with the principles of NATM. Limited site investigation was enhanced by additional investigations put in hand as construction began. After preliminary designs were prepared a site design team was set up to supervise and direct the excavation phase. Although tunnelling started well investigations indicated low rock stresses at depth which was verified as the tunnels neared the cavern itself. Following this verification a major review of the design of the high pressure system was carried out resulting in significant last minutes changes to tunnel and shaft geometry and lining design. In the design and construct context the contractors construction requirements must be dealt with as well as the evolving geological conditions which requires close cooperation between all parties involved to ensure the smooth evolution of the design.
Keywords: Design and Construct, NATM, Low Rock Stresses, Hydropower.

1 INTRODUCTION

The 600 MW Pergau Hydroelectric Project is located in a remote area of northern Malaysia, in the state of Kelantan near to the border with Thailand. It is being constructed by an Anglo-Malaysian joint venture, Kerjaya Balfour Beatty Cementation Sdn. Bhd. (KBBC),with international funding assistance from the British government. The Pergau Project will feed peak power into the Malaysian national grid, and in an effort to overcome current power shortages caused by an economic boom with a corresponding 15% increase in electrical demand, the Malaysian Electricity Company (TNB) has designated the project as an essential part of its rapid expansion in generating capacity. Generation is planned for early 1996 giving an overall design and construction period of about five years.

A contract was awarded to KBBC in January 1991 for the design and construction of the scheme based on the Feasibility Study prepared earlier by Snowy Mountains Engineering Corporation (SMEC), who were also appointed as Supervising Engineer. Knight Piesold and Partners (KPP), in association with Ewbank Preece Ltd. (EPL), were appointed as Design Engineers to the Contractor, with design liaison services on site being provided by their local associate Sumber Bumi Knight Piesold (SBKP).

The Pergau Hydroelectric Project is a major engineering undertaking, particularly in relation to underground works, encompassing as it does some 33 km of tunnels, shafts and underground caverns of varying size, shape and length. In the time and space allocated to us here it is not possible to do more than give a description of the project layout, outline the design development and give some idea of the changes to the design which occurred during construction for geological, technical and other reasons.

2 PROJECT DESCRIPTION

The layout of the project is shown in Figs. 1(a) and 1(b). Essentially the scheme is in two distinct sections; the first, consisting of the dam, power tunnels and powerhouse, is generally referred to as the power scheme, and the second, consisting of the pumping station and water transfer tunnel, as the aqueduct.

The power tunnel system as originally proposed consists of a low pressure headrace tunnel connected to a high pressure tunnel by a pressure shaft. The internal diameter of the headrace system is 6 m for the concrete lined sections reducing to 4.2 m on the steel lined sections. The net internal static head on the low pressure tunnel under normal operating conditions is approximately 100 m and on the high pressure tunnel approximately 500 m. The overall length of the headrace system is about 2 km and a surge shaft, initially 7 m internal diameter, is required for system stability. The tailrace tunnel is also 6 m internal diameter and concrete lined. It is approximately 2.5 km long and subject to a net operating internal head of about 40 m. A tailrace surge system is provided by a 7 m wide by 8 m high tunnel about 300 m long adjacent to the powerhouse cavern. The design flow in the pressure system for full generation is 141 m^3/s.

The powerhouse is located in an underground cavern some 96 m long by 30 m wide and 39 m high with a separate transformer cavern 80 m long by 15 m wide by 12.5 m high about 10 m away. Access to the cavern complex is via an unlined tunnel 7 m wide by 8 m high approximately 1 km in length. The power cables are taken to the surface in another unlined tunnel 6 m wide by 7 m high which also serves as part of the main ventilation loop. There are a number of other adits in the vicinity of the cavern ranging in size from 5 m to 6.5 m equivalent diameter.

The aqueduct tunnel is approximately 24 km long and its function is to transfer pumped water from the Terang river to the Pergau, and, using deep vertical dropshafts,to pick up additional flow from six other rivers intersected en route.. It is nominally unlined but has a concrete invert to improve the hydraulic characteristics. It is excavated to a horseshoe profile with an equivalent diameter of 3.5 m. The design flow for the aqueduct is 17 m^3/s made up of about 10 m^3/s from the pump station at Terang and 7 m^3/s as gravity feed from the intersected catchments.

3 GEOLOGY

The dominant rock type in the project area is granite, which is part of the Main Range Granite forming the backbone of Peninsular Malaysia, and is probably of Upper Jurassic to Triassic age. To the north of the area, in the vicinity of the outlet portal of the tailrace tunnel, Permian metamorphics occur, comprising biotite schists limestone and quartzose meta-sediments.

Deep weathering occurs throughout the project area and thicknesses of up to 40 m of completely to moderately weathered granite are common. Soft ground tunnelling techniques are, therefore, required at portals and shafts to reach the slightly weathered to fresh granite.

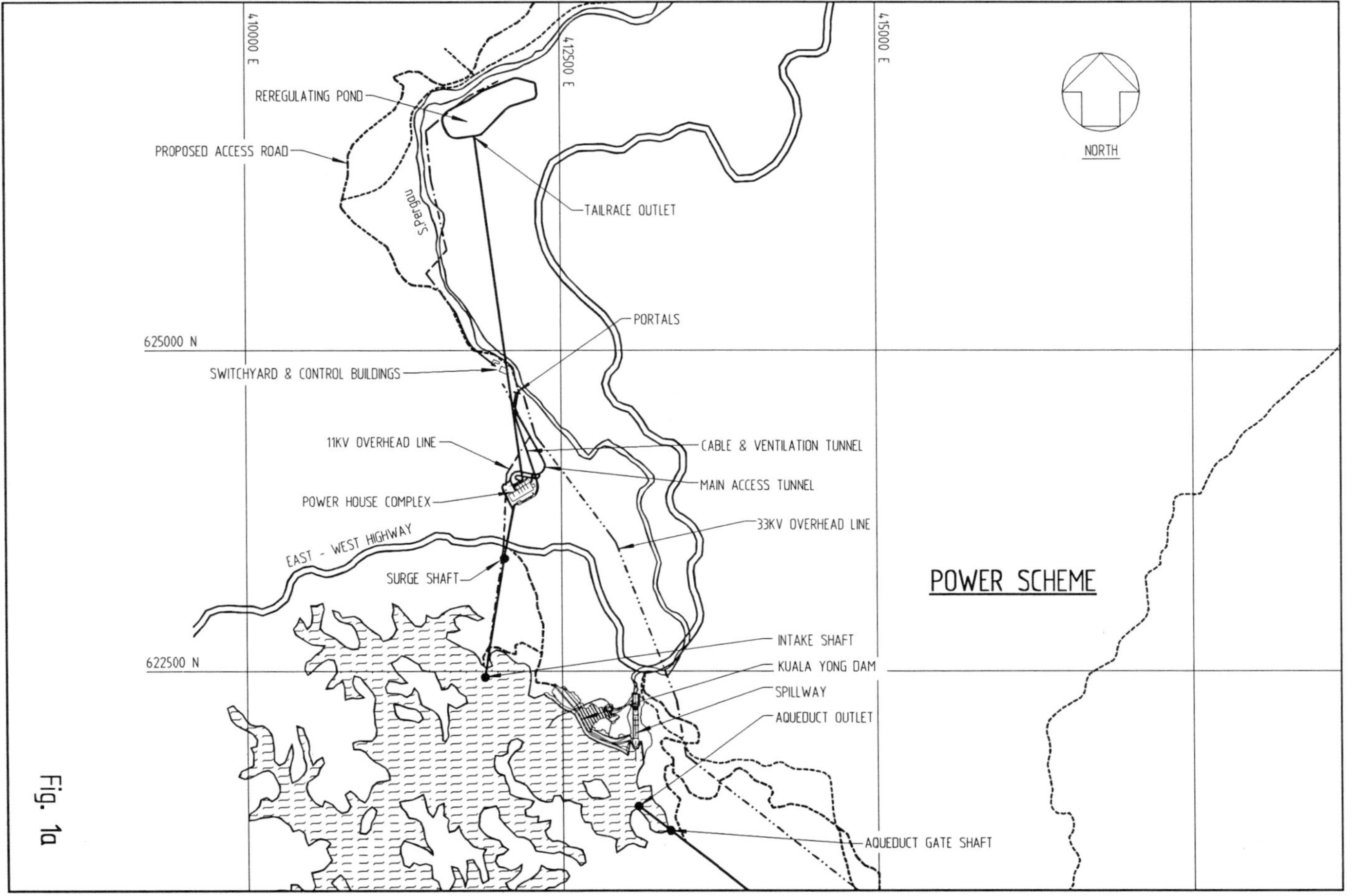
POWER SCHEME
NORTH
410000 E
412500 E
415000 E
625000 N
622500 N
REREGULATING POND
PROPOSED ACCESS ROAD
TAILRACE OUTLET
S.Pergau
PORTALS
SWITCHYARD & CONTROL BUILDINGS
11KV OVERHEAD LINE
CABLE & VENTILATION TUNNEL
MAIN ACCESS TUNNEL
POWER HOUSE COMPLEX
33KV OVERHEAD LINE
EAST - WEST HIGHWAY
SURGE SHAFT
INTAKE SHAFT
KUALA YONG DAM
SPILLWAY
AQUEDUCT OUTLET
AQUEDUCT GATE SHAFT

Fig. 1a

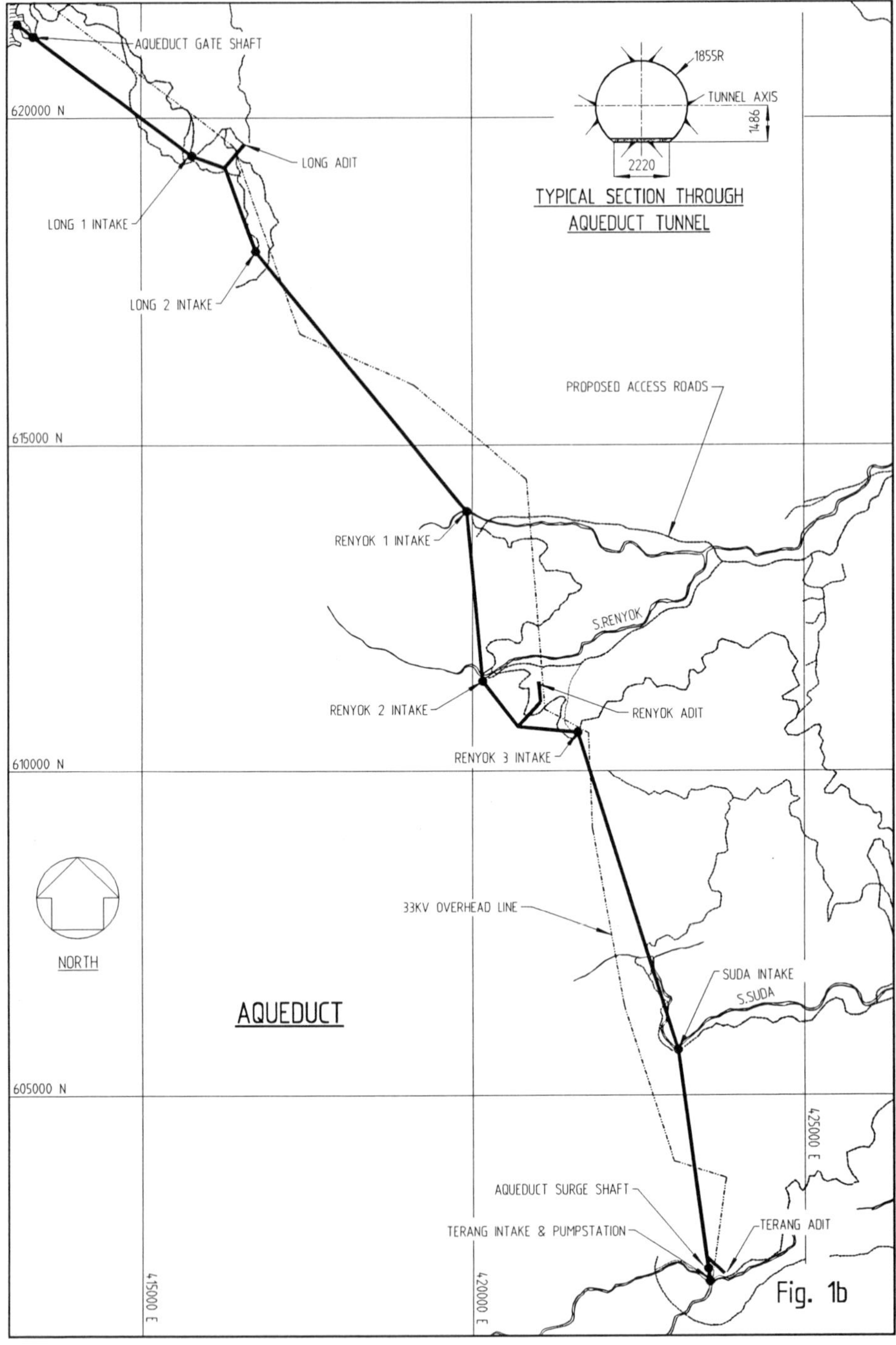
AQUEDUCT GATE SHAFT
620000 N
LONG ADIT
LONG 1 INTAKE
LONG 2 INTAKE
1855R
TUNNEL AXIS
1486
2220
TYPICAL SECTION THROUGH
AQUEDUCT TUNNEL
PROPOSED ACCESS ROADS
615000 N
RENYOK 1 INTAKE
S.RENYOK
RENYOK 2 INTAKE
RENYOK ADIT
RENYOK 3 INTAKE
610000 N
33KV OVERHEAD LINE
NORTH
SUDA INTAKE
S.SUDA
AQUEDUCT
605000 N
425000 E
AQUEDUCT SURGE SHAFT
TERANG INTAKE & PUMPSTATION
TERANG ADIT
415000 E
420000 E

Fig. 1b

The fresh granite is strong to very strong. However, the granites have in places been subjected to late stage chemical alteration by gases and fluids which permeate in the semi-solid rock. This alteration weakens the granite and can lead to complete decomposition of the rock to form clayey materials.

The Feasibility Study identified a well defined lineament system and that several of these lineaments crossed the lines of the proposed tunnels. The lineaments were attributed to zones of intense chemical alteration, faulting and fracturing.

Following the methodical logging of the underground excavations, and in situ testing and field tests during the construction period, it was found that topographic variations, or lineaments, on the surface resulted in unreliable indications of poor ground at depth. In many cases the severity and type of poor or water bearing ground was found to bear no relationship to the overlying topography and was first detected by the forward probing from the excavation face.

The estimation of the rock mass quality for the underground excavations is based on the Norwegian Geotechnical Institute (NGI) method which determines the parameter Q from six factors, principally defined by examination of the geotechnical face logs, drillhole cores and drillhole logs. A typical distribution of the Q values in one of the power tunnels is shown on Fig. 2. The geological records were input into a computerised database which amongst other functions allowed analysis of joint orientations, water inflows and other relevant factors.

4 TUNNEL DESIGN - DEVELOPMENT PHASE

The design of tunnels for hydroelectric projects generally falls into three sections:

1 - the hydraulic design
2 - the stability support requirements for the underground excavation and overlying ground slopes
3 - the design of any permanent tunnel linings.

The final alignment and profile of the tunnels comes from an optimisation of these.

The most common method of implementing a hydroelectric project is to appoint consultants to prepare tender designs and specifications, generally from the basis of a feasibility study, and then to tender both the civil works and the electrical and mechanical plant as separate contracts. The tender design period is generally between 15 and 24 months and during this time additional site investigations are put in hand to improve the specific knowledge of the site before commencing the final designs for the project. In relation to the tunnel system hydraulic designs would be completed and the design parameters for the turbines determined. Tenderers would then bid to meet those parameters.

On the Pergau project, a negotiated design and construct contract, this design development was somewhat different. From the receipt of the Feasibility Report from TNB a period of about three months was allowed for the submission of the bid together with all the financing proposals. Six weeks of this time was allocated to the Design Engineer to review the Feasibility Report and prepare preliminary designs, specifications and bills of quantities for the Contractor to price. The construction programme submitted with the bid required the hydraulic design of the power waterways to be frozen within three months of contract award in order to release the turbine designer. To meet the requirements of the turbine erection programme tunnelling for the access to the powerhouse needed to get underway as soon as possible after the Contractor's mobilisation was completed. Portal locations

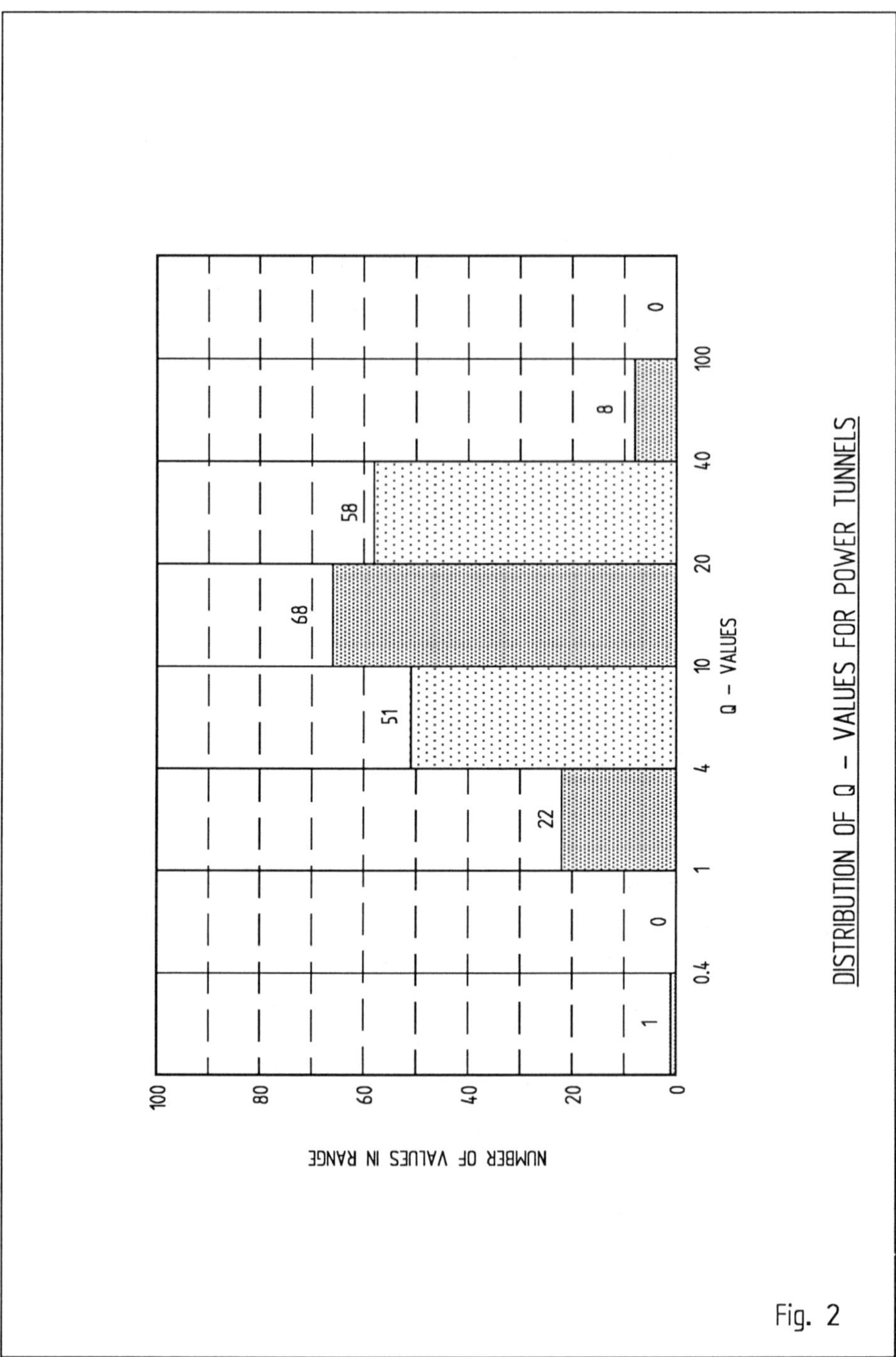
NUMBER OF VALUES IN RANGE
0
20
40
60
80
100
Q - VALUES
0.4
1
4
10
20
40
100
1
0
22
51
68
58
8
0
DISTRIBUTION OF Q - VALUES FOR POWER TUNNELS

Fig. 2

for the aqueduct adits were also urgently required as the overall construction of the aqueduct lay on the critical path for the project.

4.1 Hydraulic Design

The layout of the pressure system in the feasibility study is shown in Fig 3(a) and consisted of a low pressure tunnel, a surge shaft, a sloping pressure shaft and a short high pressure tunnel, all 6 m internal diameter (i.d.) and concrete lined, except for the surge shaft which was 7 m i.d. and a short section of which was steel lined. The headrace tunnel connected to a short steel lined section of tunnel 4.2 m i.d. which in turn connected to the powerhouse through a steel lined four branch manifold. The contract imposed guarantees on the Contractor for turbine performance based on the hydraulic parameters indicated in the feasibility report. In conjunction with this, faster turbine response times had also been imposed during the negotiation period. A hydraulic design was carried out using in-house computer programmes on the original scheme layout to determine the effect of the faster response time and the base parameters for any future changes.

For construction and economic reasons the Contractor wished to construct a vertical rather than a sloping pressure shaft. He also requested that we consider the use of unlined power tunnels as significant programming advantages were possible if these could be adopted. The hydraulics of the system were determined for all these possibilities and confirmed that acceptable solutions could be achieved with changes to tunnel and shaft sizes. However, for other reasons it was decided that the unlined tunnel option was not viable and so the final arrangement adopted a larger diameter vertical pressure and surge shaft while retaining the tunnel diameter at 6 m.

These reviews were carried out within the allotted time spans and the hydraulic parameters confirmed to the turbine designer.This continual review of the hydraulic characteristics of the scheme is one of the particular features of the design and construct contract, enabling the contractor to assess the effects of proposed construction changes and assist in his construction planning.

4.2 Stability and Support Requirements

The alignments of the power tunnels and aqueduct were also governed by the need to maintain adequate rock cover to confine the internal water pressures and inhibit leakage that could destabilize the overlying slopes. The contract required that the tunnel design be carried out in accordance with the principles of the New Austrian Tunnelling Method (NATM) and the application of NGI rules. This generally means that excavation support measures will consist of rock bolts and shotcrete reinforced as necessary by the use of steel mesh and lattice or colliery arches. The intention is to create long term stability in the excavated rock arch enabling it to be self supporting and hence reducing any long term rock loads which could be imposed on permanent linings. This type of design is of necessity site driven as the design engineers must react to the geological conditions being uncovered to optimise the design. Generally initial predictions are made from site investigation results by the head office design team and these predictions are then confirmed or otherwise through methodical inspection and instrumentation monitoring by the site design team and the support measures are adjusted to suit. KPP, through their local associate company SBKP, set up a site design team to deal with the tunnel design requirements, and procedures were developed in conjunction with the Contractor for the inspection of the excavated rock faces, the selection of the rock support classification (RSC) and the issuing of the necessary directives for underground support (DUS).

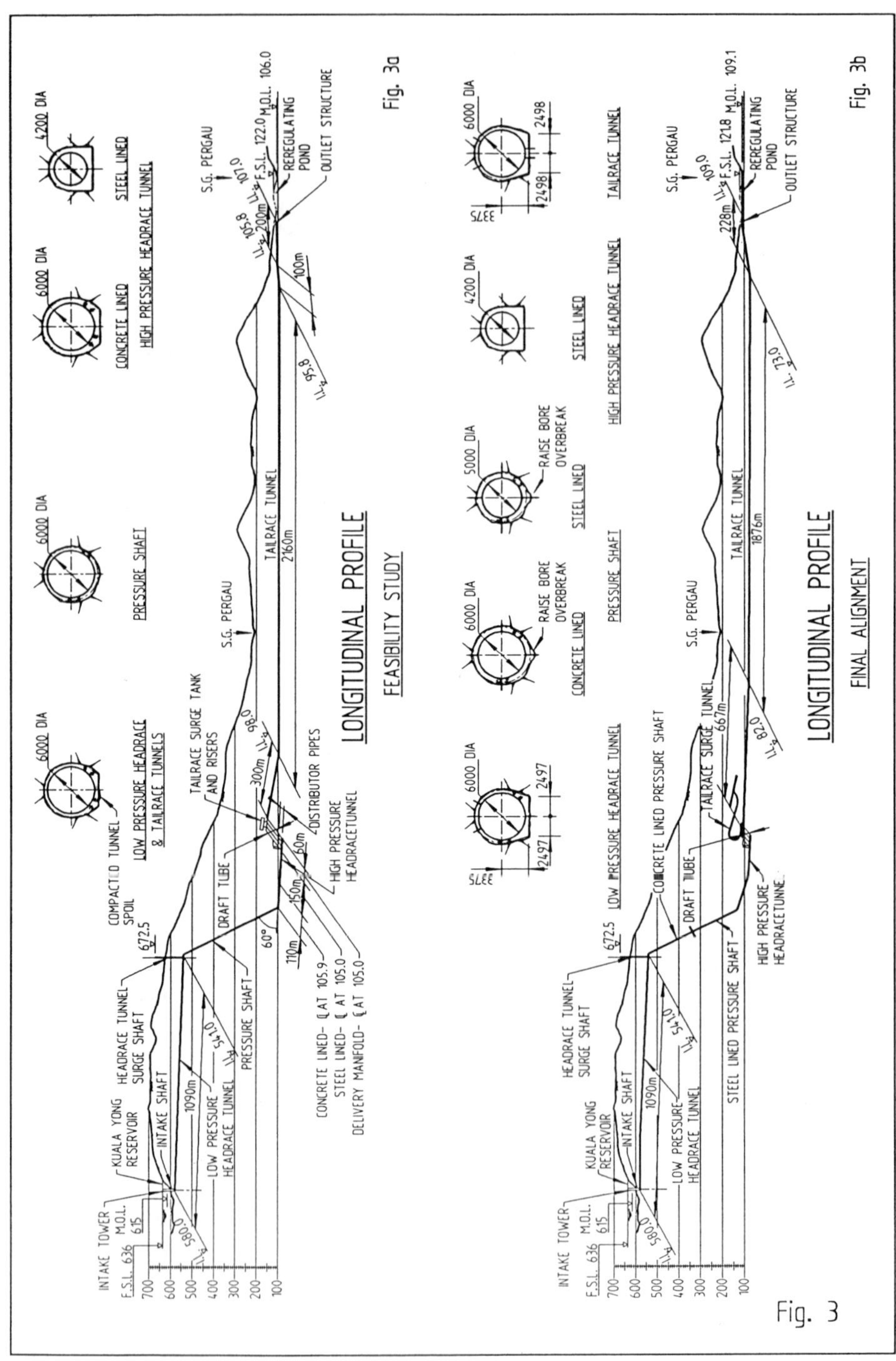

LONGITUDINAL PROFILE
FEASIBILITY STUDY
Fig. 3a
LONGITUDINAL PROFILE
FINAL ALIGNMENT
Fig. 3b
Fig. 3

In agreement with the Contractor it was decided to log in detail all the excavated faces and forward probe holes in each tunnel and shaft. As tunnelling operations were underway on many faces at one time and for up to twenty four hours in the day a large team of geologists and engineers was required to carry out the geological logging, set up and monitor the instrumentation and supervise other site investigation work. However, the variable conditions described later in this paper justified the amount of effort which went into this exercise.

4.3 Tunnel Lining Design

Three types of tunnel lining are used for the Pergau power tunnels - plain concrete, reinforced concrete and welded steel - the choice being dependent on the geological conditions pertaining to any particular location. Where possible the linings are designed to resist the internal pressures generated by the operation of the scheme with the assistance of the rock mass. The linings were not generally required to resist external loads other than water loads as it was assumed that the excavation support installed by the site design team was a permanent support and hence capable of supporting the rock mass in the long term.

For those areas where lining permeability was not a concern and where sufficient rock cover existed, unreinforced concrete linings with weepholes for pressure balancing were proposed. In areas of poorer rock the concrete linings were reinforced and the weepholes eliminated to reduce the lining permeability. Steel linings were adopted when a completely impermeable lining was required, designed both hydraulically and structurally using our own computerised design method based on the recommendations prepared by the Comité Européen de la Chaudronnerie et de la Tolerie (C.E.C.T.) for the design of penstocks.
Initial designs were based on the site investigation data included in the Feasibility Report but in relation to tunnel design this data was limited. Shortly after contract award the Contractor initiated a sub contract to carry out the additional investigations recommended in the Feasibility Report plus other investigations identified by the Design Engineer and the Contractor as essential for the successful execution of the design and the Works. However as these investigations were put in hand so too did the Contractor start his tunnelling works. Preliminary lining thicknesses were determined for planning purposes which would have to be confirmed when the additional geological data became available and as the hydraulic design was firmed up.

5 TUNNEL DESIGN - CONSTRUCTION PHASE

Modern tunnel design is not static; it must react to the geological conditions as they are revealed. As tunnel designers we expected that the designs, as on traditional contracts, would have to be confirmed and adapted as additional information became available but that these revisions would generally be fairly minor. The nature of a design and construct contract is entirely different. Our client was a contractor who felt that notwithstanding his initial construction planning, anything and everything was subject to review as conditions changed and as he firmed up on construction, or possible construction, methods. To illustrate the effects of this we shall address each tunnel section separately.

5.1 Diversion Tunnel

Diversion tunnel layouts are illustrated in Fig 4. In the Feasibility Study the diversion tunnel was located on the right bank of the river. Site examination of the portal areas showed that the downstream portal was set in a near vertical outcrop

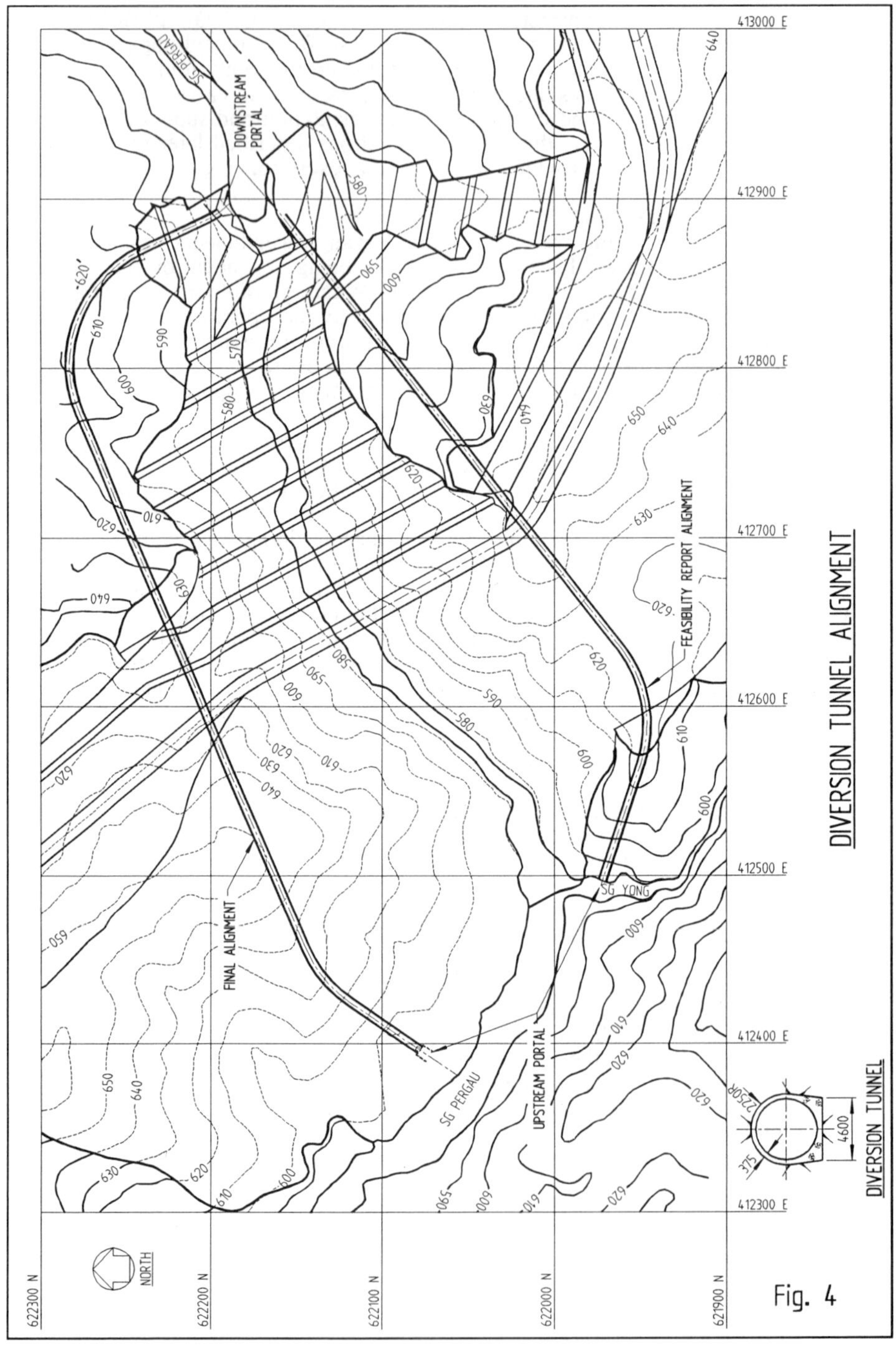
DIVERSION TUNNEL ALIGNMENT
DIVERSION TUNNEL
DOWNSTREAM PORTAL
UPSTREAM PORTAL
FINAL ALIGNMENT
FEASIBILITY REPORT ALIGNMENT
SG PERGAU
SG YONG
NORTH
413000 E
412900 E
412800 E
412700 E
412600 E
412500 E
412400 E
412300 E
622300 N
622200 N
622100 N
622000 N
621900 N
4600
2250R
375

Fig. 4

of sound granite right on the bank of the river. The layout of the scheme requires permanent access to the downstream portal and although this was an ideal geological setting for a tunnel portal, to provide this access would have involved extensive hard rock excavation up the river channel. The upstream portal location was situated adjacent to the confluence of two rivers in an area strewn with large rock boulders and although a portal could have been established extensive earthworks would have been required first. The left bank of the river, however, offered good portalling conditions at the upstream side although the downstream side was not so good. The tunnel length on the left bank was also some 50% longer but the Contractor still felt the left bank offered the best overall position particularly as he had decided to try to accelerate the diversion works in order to gain advantage on the dam construction. Additional site investigation was put in hand to determine final portal locations and the tunnel alignment and designs in the alternative location quickly prepared and approved.

5.2 Low Pressure Headrace Tunnel

The low pressure headrace tunnel (LPHT) is shown in Fig 5. Water enters the tunnel from the power intake through a short vertical shaft. The LPHT ends at the junction with the surge shaft approximately 1 km away. The initial site investigations had gaps in the vicinity of the intake and there was some concern that on the proposed alignment there would be inadequate - or even no - rock cover. Additional drill holes were commissioned which did indeed confirm the low rock cover resulting in the tunnel invert being dropped about 10 m.

Access to the tunnel was initially proposed via an adit portalling near the intake structure and entering the tunnel about one third of the way down its length. In order to commence tunnelling proper as quickly as possible the Contractor required that the tunnel adit length be as short as possible with the maximum acceptable grade set at 1:7. The original adit portal location was no longer suitable and a new location was required. Several alternative sites were examined but all the options required some soft ground tunnelling before a rock face could be established. At this stage the Contractor was not keen on soft ground tunnelling as this required a different set of skills and so we were requested to examine the possibilities of portalling from the other end. This required a review of the entire tunnel alignment and lining design in order to raise the far end of the tunnel as high as possible to cut the adit length down. The feasibility of the proposal was proved but unfortunately the only suitable portal location would have undermined a stretch of the recently constructed main east-west highway across Malaysia and so this idea was abandoned.

During the excavation of the low pressure tunnel it was predicted that about 750 m from the intake portal the tunnel would pass through a geological feature or lineament identified from surface mapping. Drillholes had been put down both during the feasibility study and in the construction phase in an attempt to intersect the lineament but these had only been partly successful. The diversion tunnel had passed through such a lineament during its construction without problems but other tunnels had encountered problems particularly with relation to the ingress of large volumes of water. The condition encountered 700 m into the LPHT was different to any previously encountered as it consisted of a series of relatively closely spaced bands of altered rock intersecting the tunnel in a crisscross pattern. These bands were up to 5 m thick and were infilled with clayey sand so soft that the material could be removed by hand. Extensive support was required to enable tunnelling to proceed and it was obvious that the ground in this section of tunnel could not be made self supporting. The tunnel lining design had to be quickly checked as the lining would now have to carry the long term rock loading and we could no longer

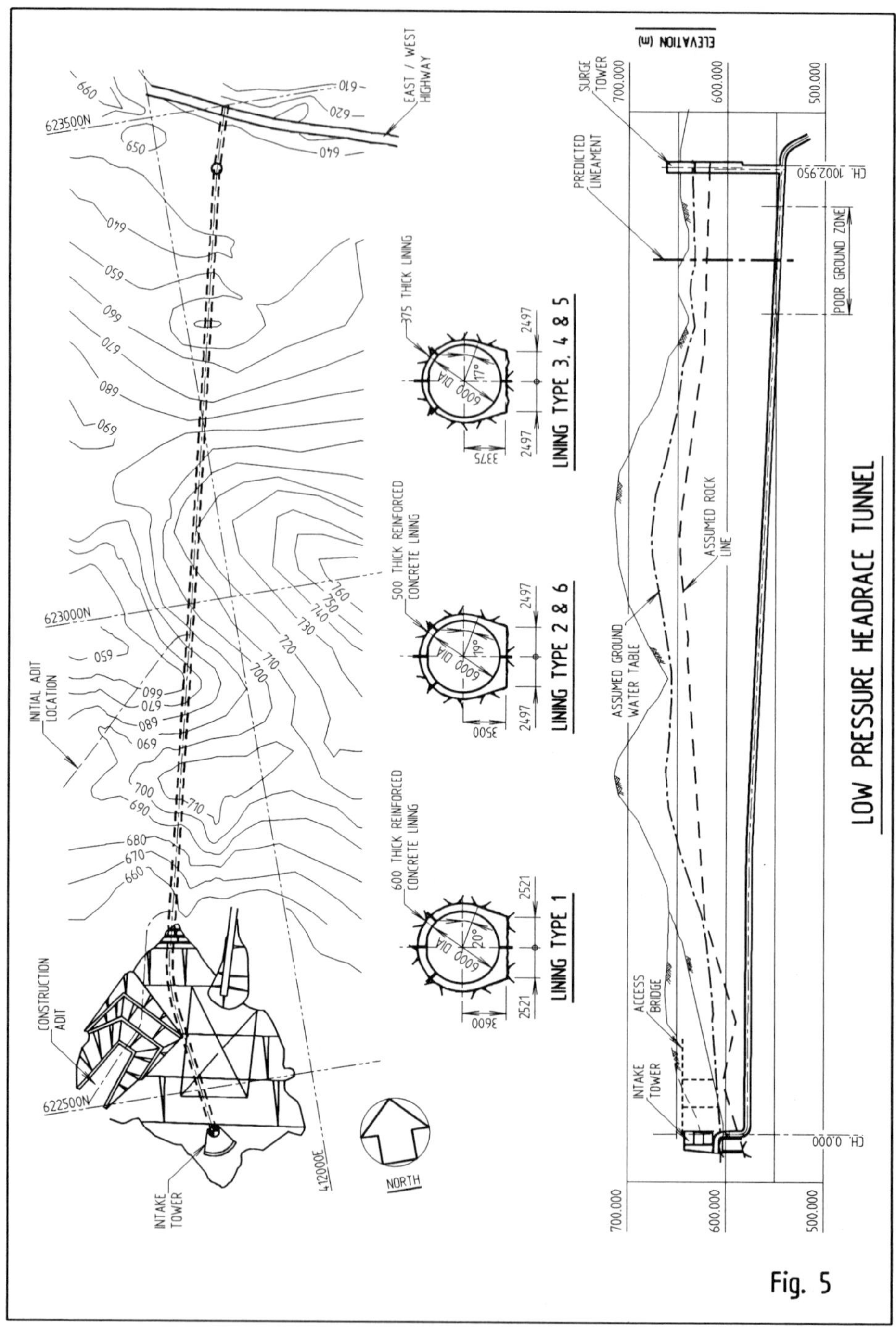
EAST / WEST HIGHWAY
623500N
623000N
622500N
412000E
INITIAL ADIT LOCATION
CONSTRUCTION ADIT
INTAKE TOWER
NORTH
600 THICK REINFORCED CONCRETE LINING
500 THICK REINFORCED CONCRETE LINING
375 THICK LINING
6000 DIA
LINING TYPE 1
LINING TYPE 2 & 6
LINING TYPE 3, 4 & 5
ELEVATION (m)
SURGE TOWER
PREDICTED LINEAMENT
ASSUMED ROCK LINE
ASSUMED GROUND WATER TABLE
ACCESS BRIDGE
INTAKE TOWER
CH. 0.000
CH. 1002.950
POOR GROUND ZONE
LOW PRESSURE HEADRACE TUNNEL

Fig. 5

rely on support from the surrounding rock mass against internal pressure. The presence of a high ground water table in this area meant that a completely impermeable lining was not required and a thicker heavily reinforced concrete section was adopted. The design excavation profile was enlarged and the redesign and additional rock stress testing took place fairly quickly so that the Contractor, held up by the poor ground, was able to continue working with minimal delay.

5.3 Powerhouse Complex and Access Tunnels

The first tunnelling works to get underway on site were for the main access tunnel (MAT) and the cable ventilation tunnel (CVT) leading to the powerhouse caverns. The layout of the MAT and CVT and the cavern complex is shown in Fig 6. This layout is somewhat different from that proposed in the Feasibility Study which implied that the MAT and CVT were the same length and could be straight between the common portal and the cavern complex. A review of survey information for the area which had not been available before submitting the bid indicated that the distance between the portal site and the cavern was too short to adopt a straight line alignment and that some form of curved approach would be necessary. It was decided that the two tunnels did not have the same grade requirements and hence could have different lengths. The MAT grade was 1:10 maximum due to permanent access restrictions giving an overall length of about 1100 m to reach the cavern but the CVT with no such restrictions could have the maximum construction grade of about 1:7 and only needed to be about 800 m long. Other requirements of the Contractor to suit his proposed construction methods for the cavern, to reduce the overall length of any adits or to relocate them were taken into account and resulted after much calculation and deliberation in the final layout shown.

Additional site investigations were put in hand to determine rock strength parameters for use in the design of both the powerhouse cavern and the high pressure system and to optimise the location and orientation of the powerhouse. Borre down-the-hole overcoring stress testing in deep drillholes together with hydraulic fracture testing was carried out under a specialist sub-contract by Swedpower. The results of the overcoring testing suggested for the first time that the rock stresses might be abnormally low and caused some concern. The hydraulic fracture testing tended to confirm these low stresses. At first there was some reluctance to accept these results as all the tunnels under construction were being excavated in sound rock with little difficulty, and the down-the-hole testing had not exactly been a smooth performance inspiring great confidence. The whole testing programme had been beset with problems from the outset with difficult access, monsoon weather, incompatibility of equipment and the losing of packers stuck in the drillholes amongst them. The cores from the drillholes and the drilling records did, however, indicate the presence of a large band of altered and voided granite in the powerhouse location.

As the excavation of the MAT and CVT continued and the results of testing and an underground seismic survey came forward a geological picture of the area was developed. This picture predicted the presence of several major bands of altered and open jointed rock around the cavern location and the decision was taken to move the cavern and change the orientation. This resulted in changes to the MAT and CVT alignments already well under construction. Indications that all might not be well came in the CVT when forward probing hit a large inflow of hot water under considerable pressure. More probing was carried out from the face and this indicated that the first water was only a few metres away from that face . The Contractor immediately decided to construct a concrete bulkhead at the face as a precaution against a face collapse. Additional probing was carried out to the sides

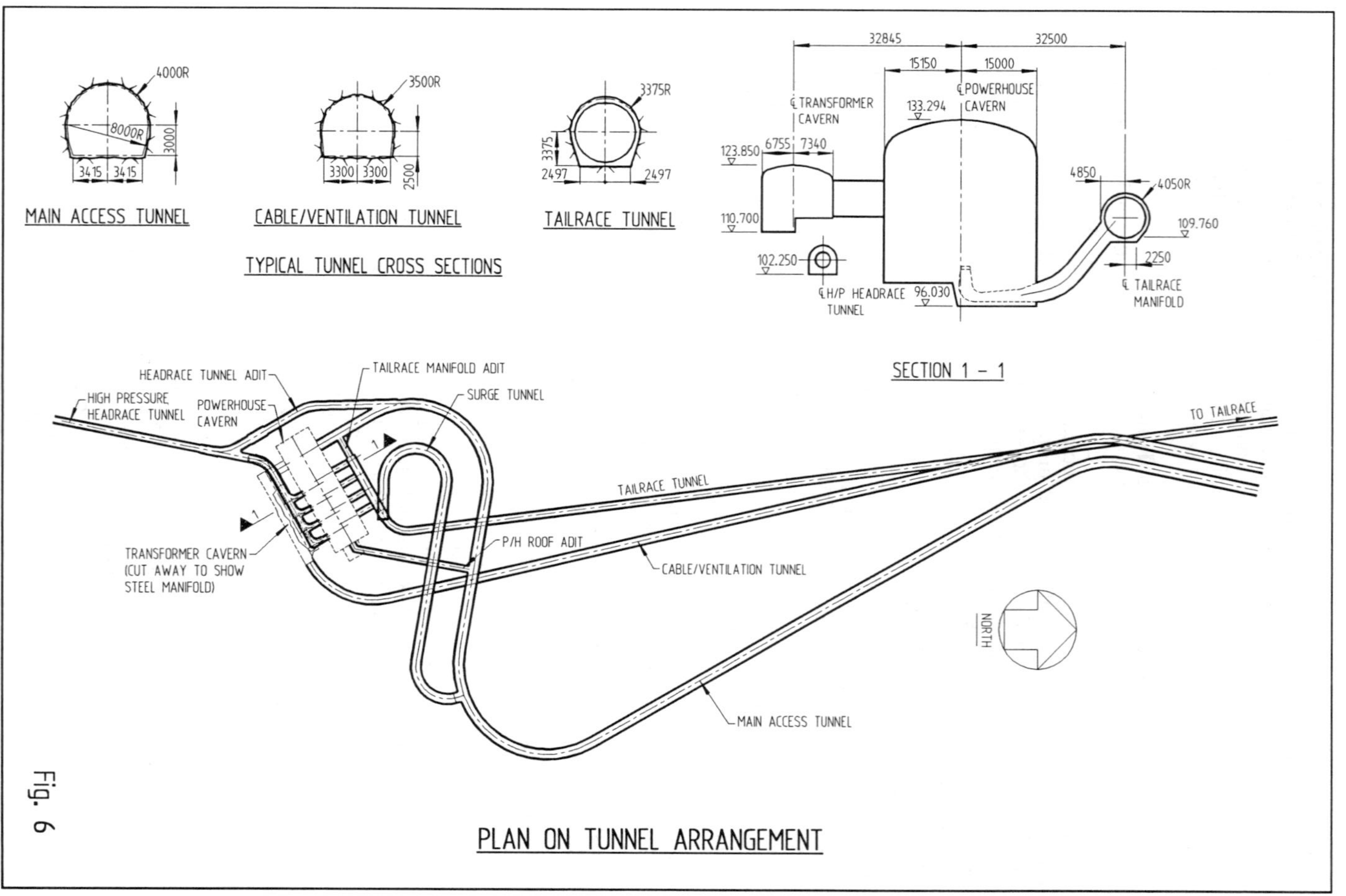
4000R
8000R
3000
3415 3415
MAIN ACCESS TUNNEL
3500R
3300 3300
2500
CABLE/VENTILATION TUNNEL
3375R
3375
2497
2497
TAILRACE TUNNEL
TYPICAL TUNNEL CROSS SECTIONS
32845
32500
15150
15000
℄ TRANSFORMER CAVERN
133.294
℄ POWERHOUSE CAVERN
123.850
6755
7340
110.700
102.250
℄ H/P HEADRACE TUNNEL
96.030
4850
4050R
109.760
2250
℄ TAILRACE MANIFOLD
SECTION 1 – 1
HEADRACE TUNNEL ADIT
HIGH PRESSURE HEADRACE TUNNEL
POWERHOUSE CAVERN
TAILRACE MANIFOLD ADIT
SURGE TUNNEL
TO TAILRACE
TAILRACE TUNNEL
P/H ROOF ADIT
CABLE/VENTILATION TUNNEL
TRANSFORMER CAVERN (CUT AWAY TO SHOW STEEL MANIFOLD)
NORTH
MAIN ACCESS TUNNEL
PLAN ON TUNNEL ARRANGEMENT

Fig. 6

of the tunnel as well to see if it was possible to bypass the feature but high pressure water was encountered in these probes as well. Eventually it was decided to grout up ahead of the excavation to try to strengthen the ground and to reduce the water flow to a low enough value to enable tunnelling to proceed again safely. The site design team duly directed grouting requirements and after a delay of some weeks to drill and grout 40 probeholes tunnelling got underway once more. However, the good progress which had been possible up till then was never to be repeated and progress was seriously hampered in many areas by the need to cement grout the rock. In fact over 1500 tonnes of cement have been used to date (September 1993) to improve and seal the rock around the powerhouse.As both tunnels progressed the geological conditions changed with more and more open jointed zones being encountered, some, but not all, of which carried large flows of hot water at high pressure (up to 30 l/s at pressures of around 10 bars in the 75 mm diameter probeholes and temperatures ranging from 40 to 55 °C). Dewatering tests were carried out close to the powerhouse. These indicated that the hot groundwater originated from an extensive area and that inflows could persist for a virtually unlimited period of time.

One of the most prominent zones was carefully mapped and probed and found to form a large rock wedge behind the downstream edge of the cavern excavation with a mass of around 5000 t. This wedge was stabilised by the progressive installation of 82 no. 225 t anchors as the cavern was excavated.

We will return to this subject shortly but first it is necessary to describe the high pressure system as it was also to be affected.

5.4 High Pressure System

The layout of the original proposal for the high pressure system is shown in Fig. 7. Initially an inclined pressure shaft was proposed leading to a concrete lined high pressure headrace tunnel (HPHT) which in turn connected to a steel lined tunnel and penstock manifold. Although the contract was submitted with an inclined shaft the Contractor had always intended to change to a vertical shaft as he felt this type of construction offered both cost and time advantages and this concept was adopted in the design.

The design of the concrete linings for the pressure shaft and HPHT were based on the precept that the rock mass had sufficient strength to carry the internal water pressures. For this reason thick unreinforced concrete linings were proposed incorporating weepholes to balance internal and external water pressures i.e. the linings were basically to improve the tunnel hydraulics and had to carry only the short term pressure transients imposed from turbine operation.

Close to the powerhouse, as a leakage control measure to prevent high pressure water from the tunnels entering the cavern, a steel lined section of tunnel was proposed with an overall length of about 200 m including the penstock manifold. These steel tunnel liners were to be let as a sub-contract to the main contract and we prepared a number of alternative designs in both mild steel and high tensile steel using both stiffened and unstiffened pipes in order that the Contractor could obtain competitive pricing from local manufacturers who lacked the design capability to design these specialist items, as well as obtaining prices from recognised international penstock suppliers. The sub-contract was duly let to a local manufacturer, TORSCO, based on our design.

As the designs for the pressure system were progressing the Contractor was examining alternative construction methods for the pressure shaft. He was particularly keen on the possible option of constructing the shaft in a single shot using a large diameter raisebore. The hydraulic design had resulted in an overall excavation diameter of around 8.5 m for the pressure shaft which was felt to be too

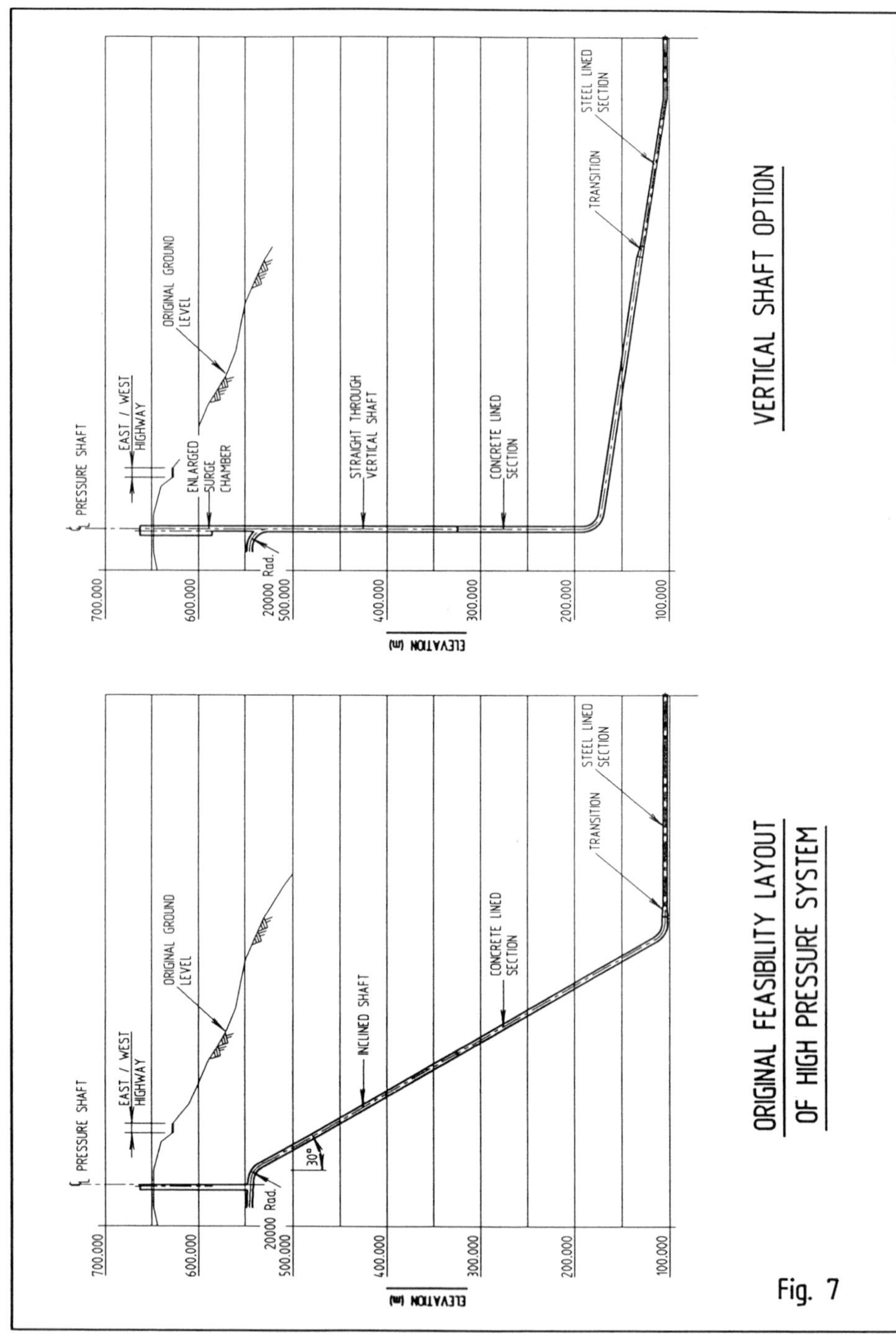
VERTICAL SHAFT OPTION
PRESSURE SHAFT
EAST / WEST HIGHWAY
ENLARGED SURGE CHAMBER
ORIGINAL GROUND LEVEL
STRAIGHT THROUGH VERTICAL SHAFT
CONCRETE LINED SECTION
TRANSITION
STEEL LINED SECTION
20000 Rad.
700.000
600.000
500.000
400.000
300.000
200.000
100.000
ELEVATION (m)
ORIGINAL FEASIBILITY LAYOUT OF HIGH PRESSURE SYSTEM
PRESSURE SHAFT
EAST / WEST HIGHWAY
ORIGINAL GROUND LEVEL
INCLINED SHAFT
30°
CONCRETE LINED SECTION
TRANSITION
STEEL LINED SECTION
20000 Rad.
700.000
600.000
500.000
400.000
300.000
200.000
100.000
ELEVATION (m)

Fig. 7

much for a single raisebore and so we were requested to review the hydraulic design to see if we could reduce this diameter by trading off elsewhere if necessary. The best result achievable only reduced the diameter to about 7.2 m which was still felt to be too large and so this approach was abandoned.

This then was the state of play when the geological conditions in the MAT and CVT were manifesting themselves. As the cavern was approached the presence of these geological conditions began to convince us as designers that the rock stress values might well be as low as the testing had indicated and we began to sound further cautionary notes to the Contractor. Also a 500 m deep hole drilled to investigate the pressure shaft indicated the possibility of low stress conditions similar to those around the cavern. We assessed what the effect of low stresses might be on both the cavern and high pressure system design and came to the conclusion that the high pressure system would be the worst affected. Low rock stresses of the order predicted from the stress testing would mean that the rock was not able to carry the internal water pressures generated in the system and hence an impermeable lining would be required in order to prevent hydrojacking of the surrounding rock mass.

Naturally the Contractor needed to be convinced. We prepared position papers detailing our arguments and conclusions, the major of which was that the steel linings would probably need to be extended to a point about half way up the pressure shaft. The Contractor was not alone in his concern and the Supervising Engineer was not convinced either and required some rock stress testing in the vicinity of the HPHT before he would accept that the rock stresses were indeed so low as to require an extension to the steel linings. Notwithstanding the need for this testing the tunnellers did not want to stop and had actually started to excavate the first sections of the steel lined HPHT at the time the decision was taken to order additional testing. We agreed that the excavation of the HPHT could proceed to the end of the existing steel lined section and that the additional testing would take place in specially constructed niches off this tunnel. If the low rock stresses were confirmed then we would amend the designs as necessary from the end of the existing steel liner.

Additional hydrofracturing and hydrojacking testing was carried out in these niches by the Australian firm of StrataTek in February 1993 and the low rock stresses were duly confirmed. At the end of February it was agreed by all parties that the steel linings would have to be extended and the Design Engineer was charged to come up with a revised design as soon as possible as the Contractor wanted to minimise the delay to his operations and in fact still intended to try to maintain his original completion dates.

In the redesign exercise it was virtually back to the beginning. Both vertical and inclined pressure shafts were considered as each had their own particular construction problems. In this exercise the true advantage of the design and construct contract was illustrated as all parties pulled out all the stops in order to achieve the optimum solution. The hydraulics had to be completely reviewed to assess the effects of extending the steel liners and altering the waterway passages which involved close cooperation between the Design Engineer and the Turbine Designer. At the same time the Contractor's construction needs had an effect on the hydraulics as the only way that the linings could be installed was to lower them down the surge chamber and shaft and hence down the pressure shaft.

The final solution adopted for the high pressure system is shown in Fig 8 and for the entire scheme in Fig 3(b). As can be seen we had reverted to the inclined pressure shaft for economic reasons with the steel liners stopping about half way up. The configuration at the junction of the LPHT, pressure shaft and surge shaft had also been revised to satisfy both construction and turbine design needs. During

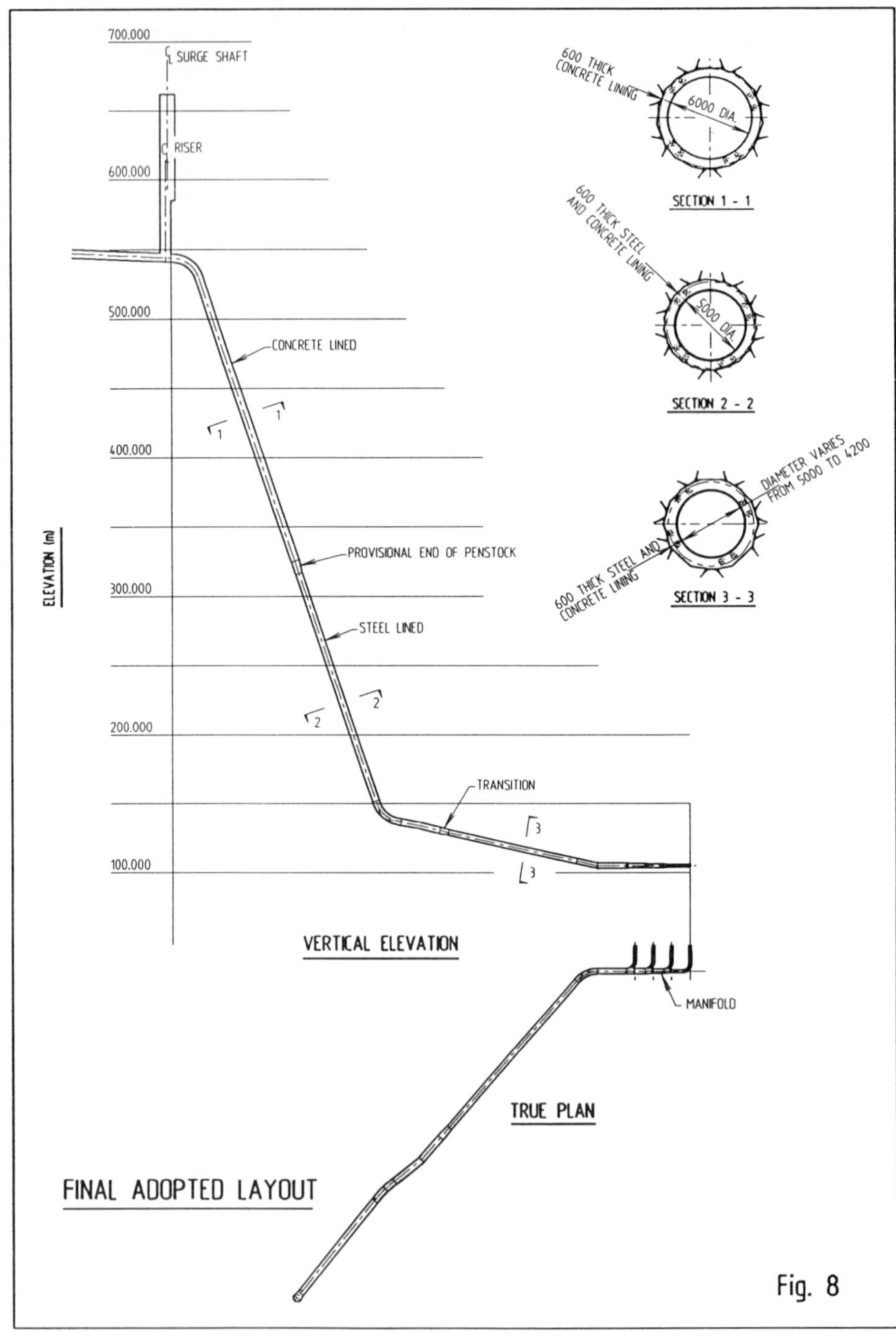
700.000
SURGE SHAFT
RISER
600.000
500.000
CONCRETE LINED
400.000
PROVISIONAL END OF PENSTOCK
300.000
STEEL LINED
200.000
TRANSITION
100.000
ELEVATION (m)
VERTICAL ELEVATION
600 THICK CONCRETE LINING
6000 DIA.
SECTION 1 - 1
600 THICK STEEL AND CONCRETE LINING
5000 DIA.
SECTION 2 - 2
DIAMETER VARIES FROM 5000 TO 4200
600 THICK STEEL AND CONCRETE LINING
SECTION 3 - 3
MANIFOLD
TRUE PLAN
FINAL ADOPTED LAYOUT

Fig. 8

this time stress testing had also been carried out in the LPHT to verify the stress levels in the tunnel and at the top of the pressure shaft in order to confirm the LPHT lining design and predict the level at which the linings could terminate. Unfortunately although it was confirmed that the LPHT design was satisfactory the testing was inconclusive as far as the location of the terminal point in the pressure shaft was concerned and additional testing will have to be carried out in the pressure shaft itself in order to finalise the design. At the time of writing (September 1993) the sloping pressure shaft has been raisebored and the final slashing down is about to proceed. A sub-contract for the extended steel liners has been prepared, tendered and awarded to FILIPPO FOCHI Construction (Malaysia) with technical assistance from Whessoe Projects and manufacture of the liners is imminent. The lining design has already been amended once to fit the actual alignment of the raisebored shaft at the lower bend. Further amendments will be needed once the remaining part of the slashed out shaft is excavated and surveyed and the actual lining terminal point is determined.

5.5 Aqueduct Tunnel

The aqueduct tunnel is about 24 km long and so far has not been upset by any problems such as have been encountered in the pressure tunnels. It is a smaller tunnel and hence is not quite as susceptible to problems as the larger pressure tunnels. It also lies for the most part nearer to the surface in the type of material which presented little problems with the pressure tunnels. It is to be hoped that this good progress will continue.

The changes imposed by the Contractor have also had an effect on this tunnel. The alignment has been amended in several places to suit revised portal locations and to minimise adit lengths. The outlet structure has been redesigned several times and relocated once to suit the Contractors overall construction methods and at the time of writing is undergoing yet another redesign. At the inlet end through the first 25 m of soft ground the tunnel has been redesigned to be supported on minipiles in order to prevent settlement of the steel lined portion. The intermediate shafts and deaeration chambers at the foot of the shafts have all been redesigned to assist the Contractor's working methods. The gateshaft, for which an oval, rather than circular, excavated profile has been adopted is another example where the design has been adapted to suit the Contractors needs.

5.6 Tailrace Tunnel

The outlet portal of the tailrace tunnel is sited in an alluvial flood plain and tunnelling is particularly difficult over the first 100 m. The original outlet shaft was changed to a tunnel portal to aid the tunnelling. The outlet drive was advanced cautiously as a soft ground NATM tunnelling exercise employing lattice arches, multiple heading, shotcrete crown, walls and invert, rockbolts, dewatering holes together with a full range of convergence, settlement and load cell instrumentation. The excavation support through the soft ground section here and throughout the whole project was designed and supervised by the site design team.

5.7 Conclusion

The Pergau Hydroelectric Project is a major civil engineering undertaking involving the design of some 33 km of tunnels, shafts and underground caverns. Tunnel design is not static; tunnel designers expect to review and amend their designs as construction proceeds and more information is obtained. In the design and construct context another complete layer of review is added as the designer has to react to the ever changing requirements of the contractor as well. These reviews are sometimes challenging and sometimes irksome as well but there is no doubt that they increase the involvement of the designer in the project as a whole. The Pergau experience has shown that in the design and construct context, close cooperation between the contractor, the site based design team and the main design office is a vital factor in ensuring the smooth evolution of the design.

Acknowledgement: The authors would like to thank Tenaga Nasional Berhad (TNB) for permission to publish this paper. The views and opinions expressed therein are those of the authors and do not necessarily reflect those of TNB.

Serra da Mesa project, Brazil: excavation of shafts and penstocks

Guilherme Tibau
C. C. Camargo Corrêa S.A., São Paulo, Brazil
Erik Gutenwik
Skanska Raise Boring Team, Stockholm, Sweden
Maurício Santana
C. C. Camargo Corrêa S.A., São Paulo, Brazil
Pierre Jean Pellegri
C. C. Camargo Corrêa S.A., São Paulo, Brazil

This paper describes with full details the construction methods of the shafts and penstocks of the Serra da Mesa Power Plant, that is being constructed in Brazil. For the excavation of these structures special equipments have been assembled in the site workshop and proper construction methods have been used to excavate more than 1,000m of shafts without any accident.

To excavate the pilot holes of all shafts and penstocks it has been used the raise boring method.

The final rock surfaces obtained with these construction methods and equipments were considered very good.
Keywords: Serra da Mesa Project, shafts, penstocks, raise boring.

1 General description

1.1 The Serra da Mesa Project

The Serra da Mesa Project, shown in "Figure 1", is being constructed in the State of Goias, Brazil, 225 km from Brasflia. The installed capacity will be 1.200 mw, with three turbines of 400 mw each.

The hydraulic circuit of generation, shown in "Figure 2", is all subterranean and include the intake tunnel, the penstocks, the powerhouse, the surge tank and the outlet tunnel.

Complementing these structures there are three intake gate shafts, three busduct shafts, one ventilation shaft, one powerhouse access shaft, one powerhouse access tunnel and two diversion tunnels.

The work began in 1986 and energy generation is expected to begin in 1999.

"Photo 1"shows the general view of the **Serra da Mesa Project** and "Photo 2" the general view of the intake gate shafts.

The principal subject of this paper will be the shafts and penstocks, whose main caracteristics are described bellow:

Three penstocks:	length	-	126.0m
	diameter	-	10.0m
Three busduct shafts:	depth	-	123.0m
	diameter	-	5.5m
One ventilation shaft:	depth	-	138.0m
	diameter	-	from 5.7m to 8.5m
Three intake gate shafts:	depth	-	59.0m
	rectangular section	-	10.6m x 8.0m

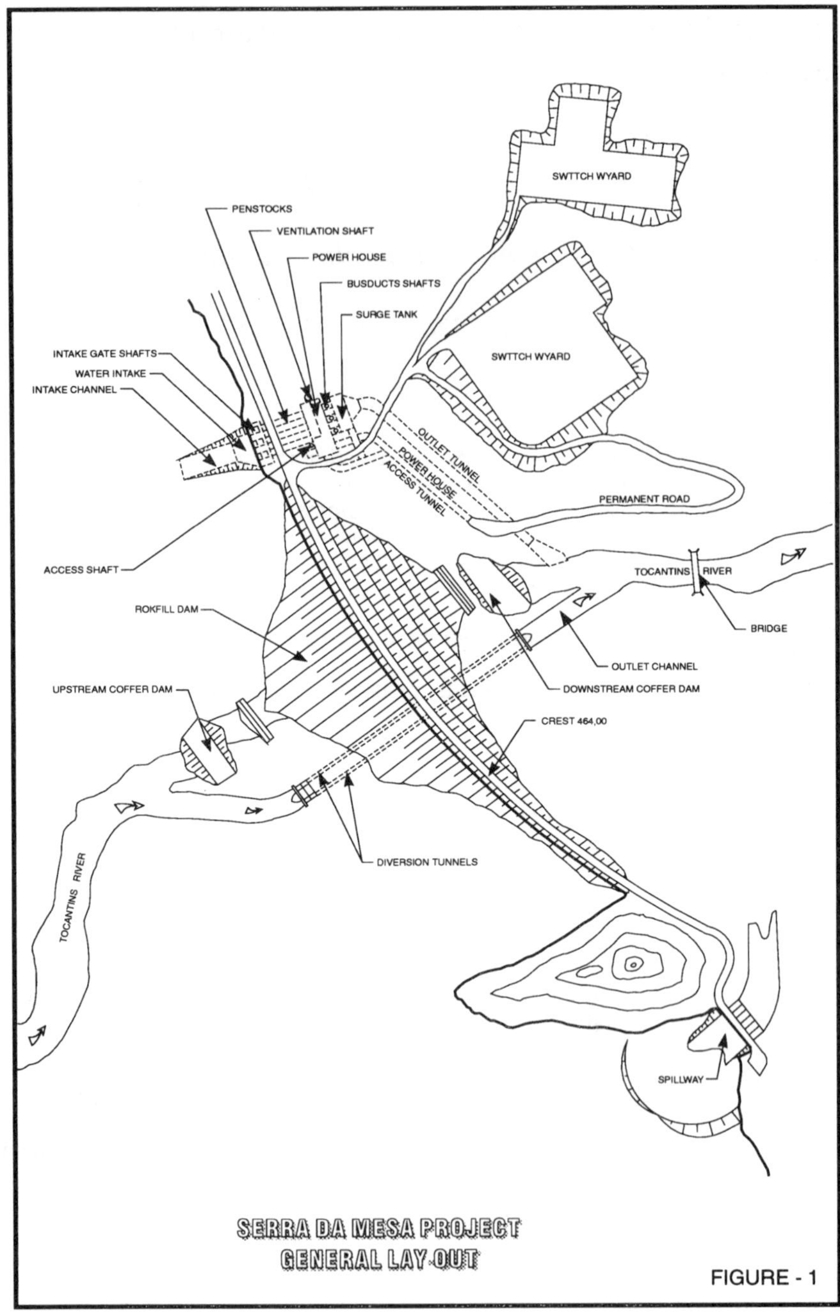

SERRA DA MESA PROJECT
GENERAL LAY-OUT

FIGURE - 1

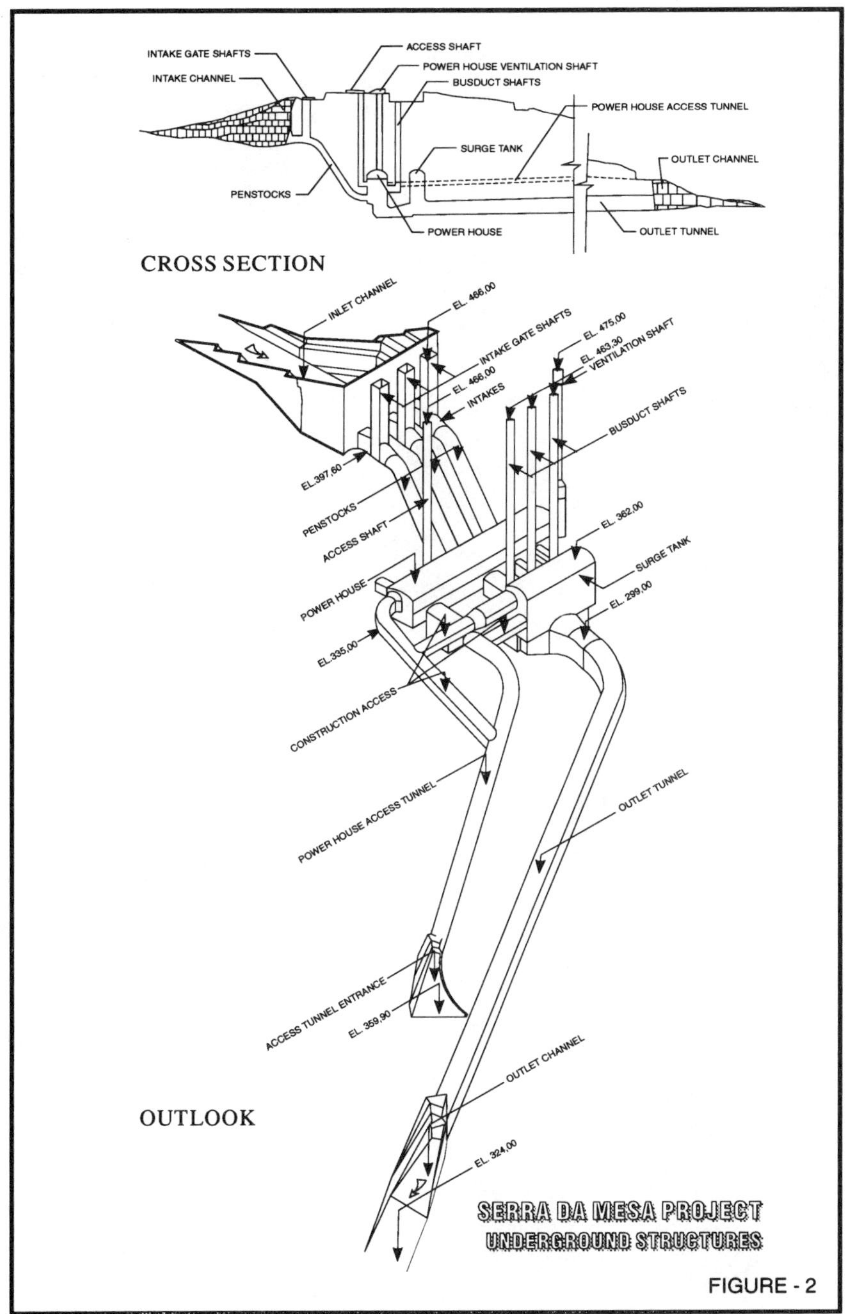
INTAKE GATE SHAFTS
INTAKE CHANNEL
ACCESS SHAFT
POWER HOUSE VENTILATION SHAFT
BUSDUCT SHAFTS
POWER HOUSE ACCESS TUNNEL
SURGE TANK
OUTLET CHANNEL
PENSTOCKS
POWER HOUSE
OUTLET TUNNEL
CROSS SECTION
INLET CHANNEL
EL. 466,00
INTAKE GATE SHAFTS
EL. 475,00
EL. 463,30
VENTILATION SHAFT
EL. 466,00
INTAKES
BUSDUCT SHAFTS
EL.397,60
PENSTOCKS
ACCESS SHAFT
EL. 362,00
SURGE TANK
POWER HOUSE
EL. 299,00
EL.335,00
CONSTRUCTION ACCESS
POWER HOUSE ACCESS TUNNEL
OUTLET TUNNEL
ACCESS TUNNEL ENTRANCE
EL. 359,90
OUTLET CHANNEL
OUTLOOK
EL. 324,00
SERRA DA MESA PROJECT
UNDERGROUND STRUCTURES

FIGURE - 2

Photo 1

Photo 2

Table 1. Quantities of the underground structures

Structure	Qt	Length Unit (m)	Length Total (m)	Volume Unit (m³)	Volume Total (m³)
Diversion tunnels	2	680	1.360	145.000	290.000
P. house access tunnel	1	690	690	47.000	47.000
Outlet tunnel	1	500	500	139.616	139.616
Inlet tunnel	3	24	72	3.918	11.754
Suction tunnels	3	36	108	5.400	16.200
Powerhouse	1	137	137	186.641	186.641
Surge tank	1	69	69	91.077	91.077
Busduct shafts	3	123	369	2.930	8.790
P. house access shaft	1	121	121	4.395	4.395
Ventilation shaft	1	138	138	3.855	3.855
Intake gate shafts	3	59	177	5.079	15.237
Penstocks	3	126	378	10.917	32.751
Total of tunnels			2.936		782.288
Total of shafts			1.183		64.028
TOTAL					847.312

1.2 Brief description of the main contractor

Construçoes e Comércio Camargo Corrêa S.A. came into being in 1938. As Brazil's railway system grew during the first half of the 1950's, Camargo Corrêa S.A. grew with it, contracting to build several important railway lines, as well as airports, and highways.

In 1952 Camargo Corrêa moved into the field of large hydraulic works and hydroeletric plants. Camargo Corrêa also moved fully into the task of building Brasília (1957/1961) and constructed the first roads leading there. In 1960 work started on the hydroelectric plant at Jupiá where final output is 1,400,000 kw.

During Brasil's quick-pace development over the last years, Camargo Corrêa has grown and expanded together with the Federal Government's important accomplishments, such as the Ilha Solteira Power Plant (1965/1974), where total ouput is 3,200,000 kw. Camargo Corrêa also constructed the Tucurui Power Plant with a final output of 4,000,000 kw and was one of the five contractors that constructed the Itaipu Power Plant, the biggest Hydroeletric Power Plant in the world.

Among its foreign contracts, the constrution of the second phase of the Guri Dam, in Venezuela, should be mentioned.

At this moment Camargo Corrêa has, among others, the following construction contracts:

Serra da Mesa Power Plant;
Porto Primavera Power Plant;
São Paulo City Sewerage System;
São Paulo Internacional Airport;
Brasília Internacional Airport;
Several highways and railways.

1.3 Brief description of the Skanska Raise Boring AB

Skanska Raise Boring AB is an independent member of the Skanska Group which is one of the leading scandinavian civil engineering contractors with worldwide activities.

Skanska was founded 1886 as a small local company in the southern part of Sweden. Today Skanska has about 25,000 employees and a turnover of approximately Us$ 5 billion (1992).

Skanska Raise Boring started their activities early 1974 in Finland with raise boring of a ventilation shaft for a finish mine. The first contract awarded outside Scandinavia was in Austria in 1977 and only two years later the first contract was awarded outside of Europe in Guatemala in 1979.

Up to now, Skanska Raise Boring has operated in 21 countries in four continents and has drilled close to 30,000 m of shafts. The shafts have been drilled in a variety of rock conditions, inclinations, diameters and lenghts. The longest single raise bored shaft is 550 m and the largest diameter raised is 4.13 m.

Skanska Raise Boring has equipment and the experienced personnel capable of carrying out shafts of lengths of up to 700 m and up to 5 m in diameter, depending on the rock conditions.

Ongoing or recently completed raise boring sites are at the Pergau H.E.P. in Malaysia, at the Uri H.E.P. in India, at the Zinkgruvan Mine in Sweden and at the Äspö Nuclear Waste Laboratory in Sweden.

2.0 Geomechanical characteristics of the rock and excavation specifications

2.1 Geomechanical characteristics of the rock

The massif at the excavations site is made of hard granitic rock, very dense and slightly fractured, giving the massif excellent geomechanical caracteristics.

A microscopic petrografic analisis showed the following mineralogical composition to the Serra da Mesa rocks:

Classification : Biotite granite

Composition : Quartz, feldspar and mica biotite

The laboratory tests done in accordance with ASTM 2938-79 "unconfined compressive strength of intact rock core specimens", gave the following results.

The strain status in the powerhouse area, fixed through hydraulic fracturing tests is the following:

vertical strain (sv):

$Sv = 0.265\ kg/cm^2/m^*$ (*) Depth

horizontal strain parallel to the longitudinal axis of the caves:

$SH = 155 \pm 30\ kg/cm^2$

horizontal strain, parallel to the transversal axis of the caves:

$Sh = 95 \pm 10\ kg/m^2$

Table 2. Laboratory tests in accordance with ASTM 2938-79

Hole (Nº)	Simple Strenght of Compression (kg/cm²)	Standard Deviation (kg/cm²)	Depth of Sample (m)	Quantity of Tests (nº)
1st Group				
Nº 1	1321	379	356-378	5
Nº 2	791	170	105-155	12
2nd Group				
Nº 1	2036	130	89-90	8
Nº 2	1017	211	120-130	3
Nº 2	2032	236	152	

2.2 Excavation specifications

The excavation specifications prepared by the "Furnas Centrais Elétricas S.A." for the underground excavation works in the Serra da Mesa Project were very strict concerning the remaining rock surfaces and crew safety.

Among the most important items of these specifications the following should be mentioned:

The maximun volume of rock to be excavated in just one round must not exceed 500 m³.

The maximun particle speed at any point of the rocky massif around the excavation, and 1.00 m away from the theoretical line of excavation, must not exceed 30 cm/sec.

The maximun particle speed at the places where the concrete structures are located must not exceed 5 cm/sec.

The spacing between the excavation contour holes must not exceed 0.60 m and the explosive charges must be light and well distributed in the drill holes.

The aligment and parallelism of the drill holes must be strictly controlled, especially the contour holes, obeying the following rules: the drill hole deviation in the hole entrance must not exceed 5 cm and the aligment deviation of the drill holes must not exceed 2 cm/m.

The remaining rocky massif must always be protected with the aplication of the presplitting, smooth blasting or cushion blasting techniques.

The excavated surfaces must be scaled in a methodical way.

Maximum mechanization of the excavation methods must be achieved in order to have the minimum number of crew involved in excavation activities.

Wherever established by the designer a wire mesh must be installed to prevent the falling of pieces of rock.

3 Excavation methods

The main points of the excavation methods used to excavate the shafts and penstocks were safety, time schedule and excavation costs.

As some shafts had a depth of more than 100m the following items were the ones that influinced most the excavation planning, the selection of the excavation methods and the selection of equipment:

a) Maximization of crew safety on the work front and in transit, adopting the following steps:

When the advance is in an upward direction, never do the drilling with the front of the rock in an overhead position.

Excavate the shaft walls very carefully, always doing a methodic scaling, in order to prevent the pieces of rock from falling on the crew.

Always fit the equipment with protection roofs to protect the crew.

Always fit the transportation equipment and winches with multiple breaks.

b) Minimization of costs in removing the detonated rock:

It is known that the costs of removing the detonated rock by mechanical means is very high and time - consuming. So the excavation was planned in such a way that the blasted rock could always fall by force of gravity into the lower access galleries or caves.

c) Diminishing the round cycle time by adopting the following steps:

Elimination of wasted time in rock removal, by adopting inclined surfaces in the intermediate platforms between two blasts.

d) Diminishing the overbreak by adopting the following steps:

Make the use of prespliting, smooth-blasting or cushion blasting an excavation routine practice.

Control the drilling in order to minimize the deviations.

3.1 General description of the excavation method

The shafts and penstocks were excavated in a similar way, although some details were diferent, as will be seen in the descriptions bellow.

3.1.1. Vertical shafts

a) 1st phase

Excavation of a pilot shaft using the **raise boring equipment** as shown in "Figure 3" and "Photos 3 and 4".

b) 2nd and 3rd phases

Excavation of the widenings with the drill and blast method, in a downward direction.

For the busduct, ventilation and powerhouse access shafts there was only one widening, as shown in "Figures 4 and 5".

As the intake gate shafts had a bigger cross section there were two widenings, as shown in "Figures 6 and 7".

The excavation of the shafts was always preceeded by the excavation of the upper and lower access galleries. All the blasted rock was removed through the lower access galleries.

3.1.2 Penstocks

a) 1st phase

Excavation of a pilot shaft using the **raise boring equipment**, as shown in "Figure 8". This pilot shaft was excavated with the same inclination as the penstock, in an upward direction, and was located touching the lower generatrix of the penstock.

b) 2nd phase

Excavation of the first widening in an upward direction, with the drill and blast method, as shown in "Figure 9".

This phase was excavated with the drilling done in a radial way as shown in "Figure 9" (section B-B)

c) 3rd phase

Excavation of the second widening, in a downward direction, also with the drill and blast method, as shown in "Figure 10".

The excavation of the penstocks was preceeded by the excavation of the upper and lower access galleries.

All the blasted rock was removed through the lower access gallery.

3.2 The raise boring method (1st excavation phase of all shafts and penstocks)

All the pilot shafts were excavated by the **raise boring equipment**.

3.2.1. Shaft excavation by raise boring

Raise boring is a method developed for shaft excavation between two levels. The method has been known since the early seventies but has through the years become more and more used instead of older and more traditional shaft excavation methods.

The machinery equipment consists of four major parts: the raise boring machine itself, an electric power pack, a hydraulic unit and the operators' console. For operating the equipment, electric power has to be supplied together with water or compressed air. The equipment is easily operated by two operators per shift and

THE RAISE BORING METHOD
HOW TO DO IT.
FIRST WE DRILL THE PILOT HOLE WITH A DIAMETER OF EITHER 279 OR 311mm. WE USE A DRILL STRING FITTED WITH STABILIZERS TO MINIMIZE THE DEVIATION.CUTTINGS ARE FLUSHED OUT WITH WATER.
THEN WE MOUNT THE REAMER.
WHILE THE MACHINE PULLS THE DRILL STRING BACK, THE ROCK IS GROUND INTO FRAGMENTS BY THE CUTTERS ON THE ROTATING REAMER.
RAISE BORING RIG
PILOT HOLE
REAMER
RAISE
FIGURE - 3

Photo 3

Photo 4

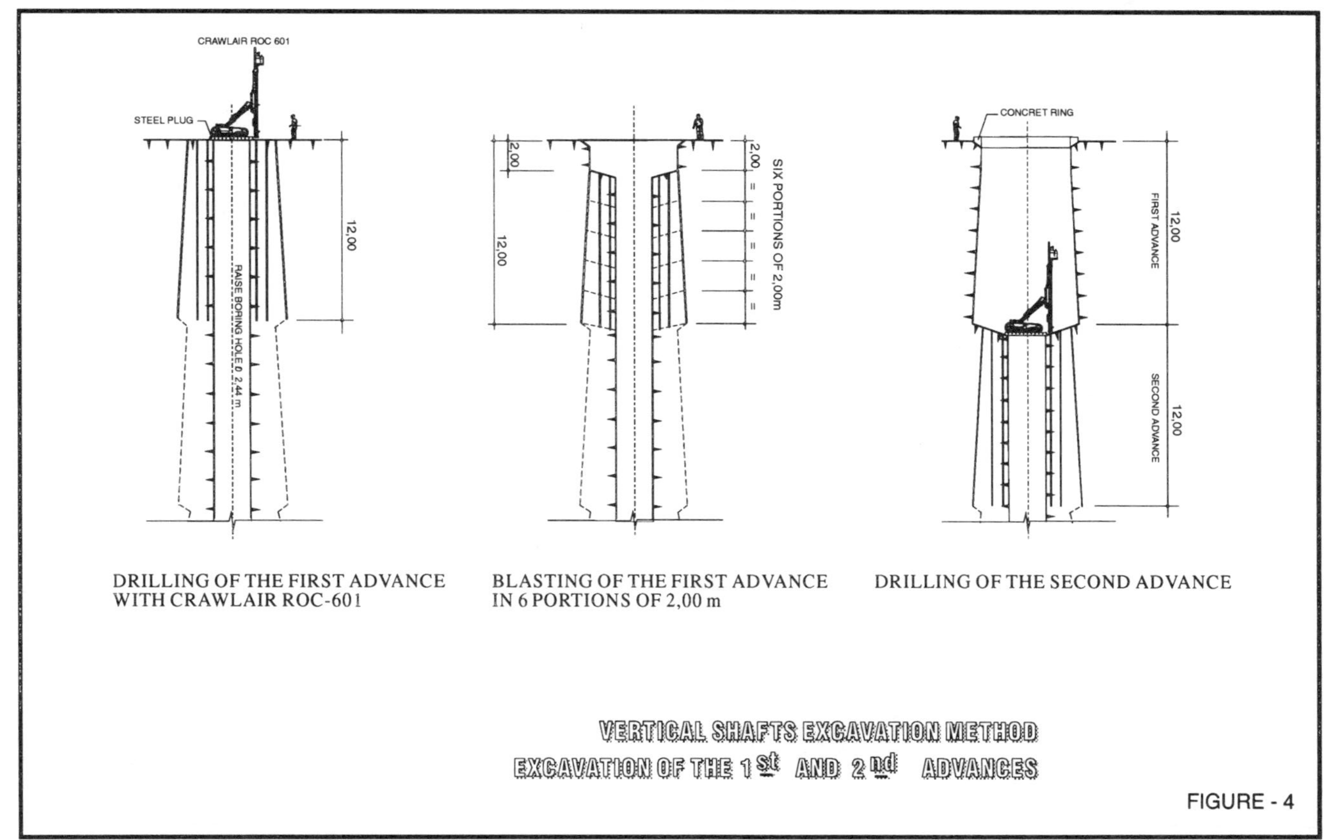
CRAWLAIR ROC 601
STEEL PLUG
12,00
RAISE BORING HOLE Ø 2,44 m
2,00
12,00
2,00
SIX PORTIONS OF 2,00m
CONCRET RING
12,00
FIRST ADVANCE
12,00
SECOND ADVANCE
DRILLING OF THE FIRST ADVANCE
WITH CRAWLAIR ROC-601
BLASTING OF THE FIRST ADVANCE
IN 6 PORTIONS OF 2,00 m
DRILLING OF THE SECOND ADVANCE
VERTICAL SHAFTS EXCAVATION METHOD
EXCAVATION OF THE 1st AND 2nd ADVANCES
FIGURE - 4

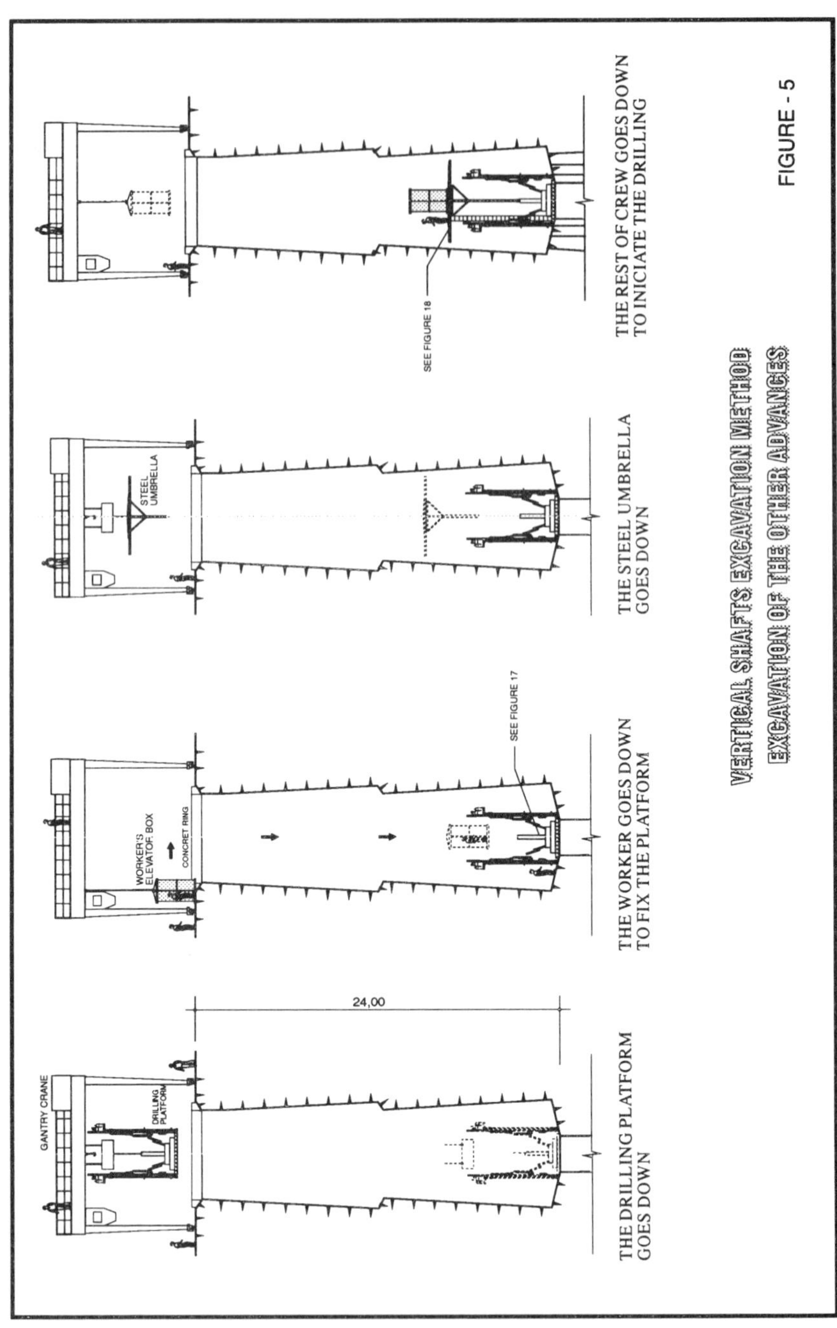

VERTICAL SHAFTS EXCAVATION METHOD

EXCAVATION OF THE OTHER ADVANCES

FIGURE - 5

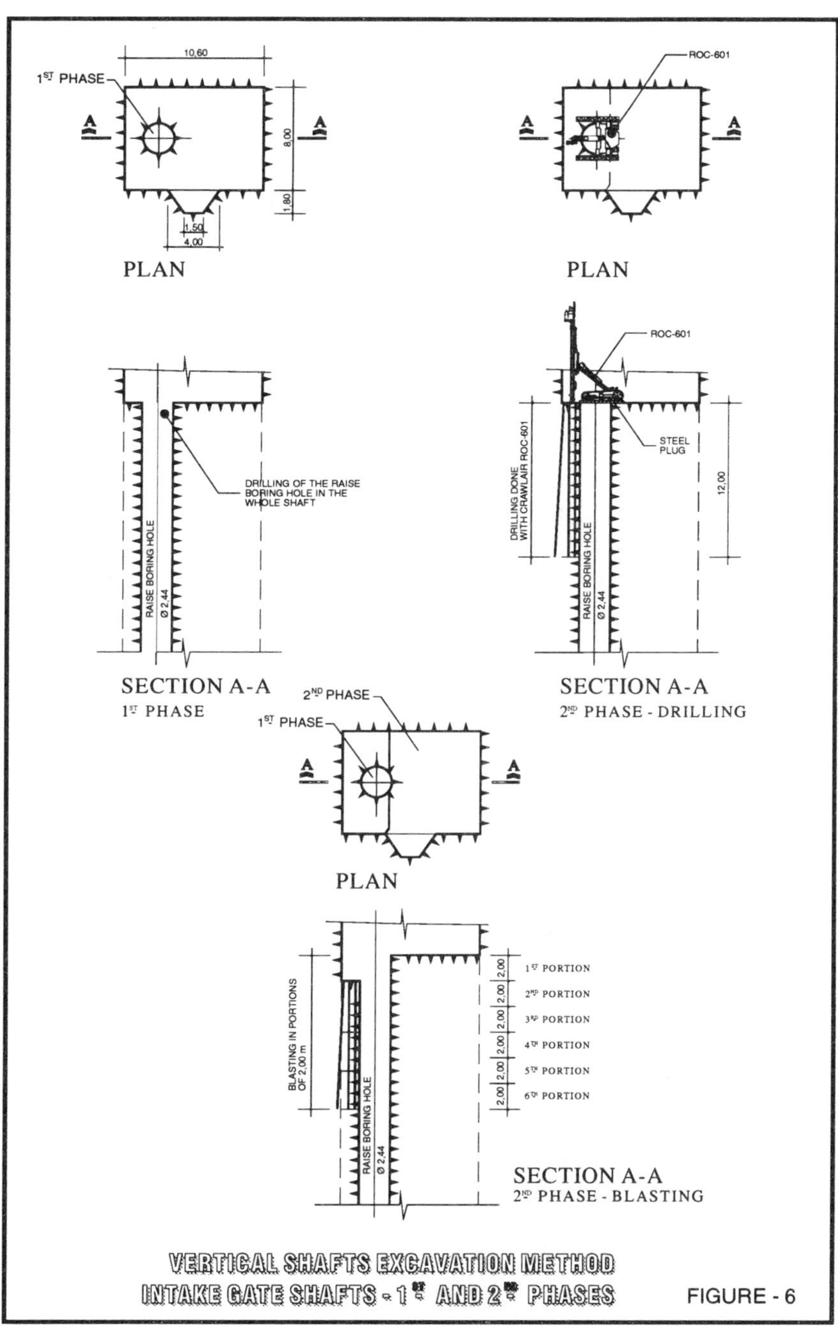

VERTICAL SHAFTS EXCAVATION METHOD
INTAKE GATE SHAFTS - 1ST AND 2ND PHASES

FIGURE - 6

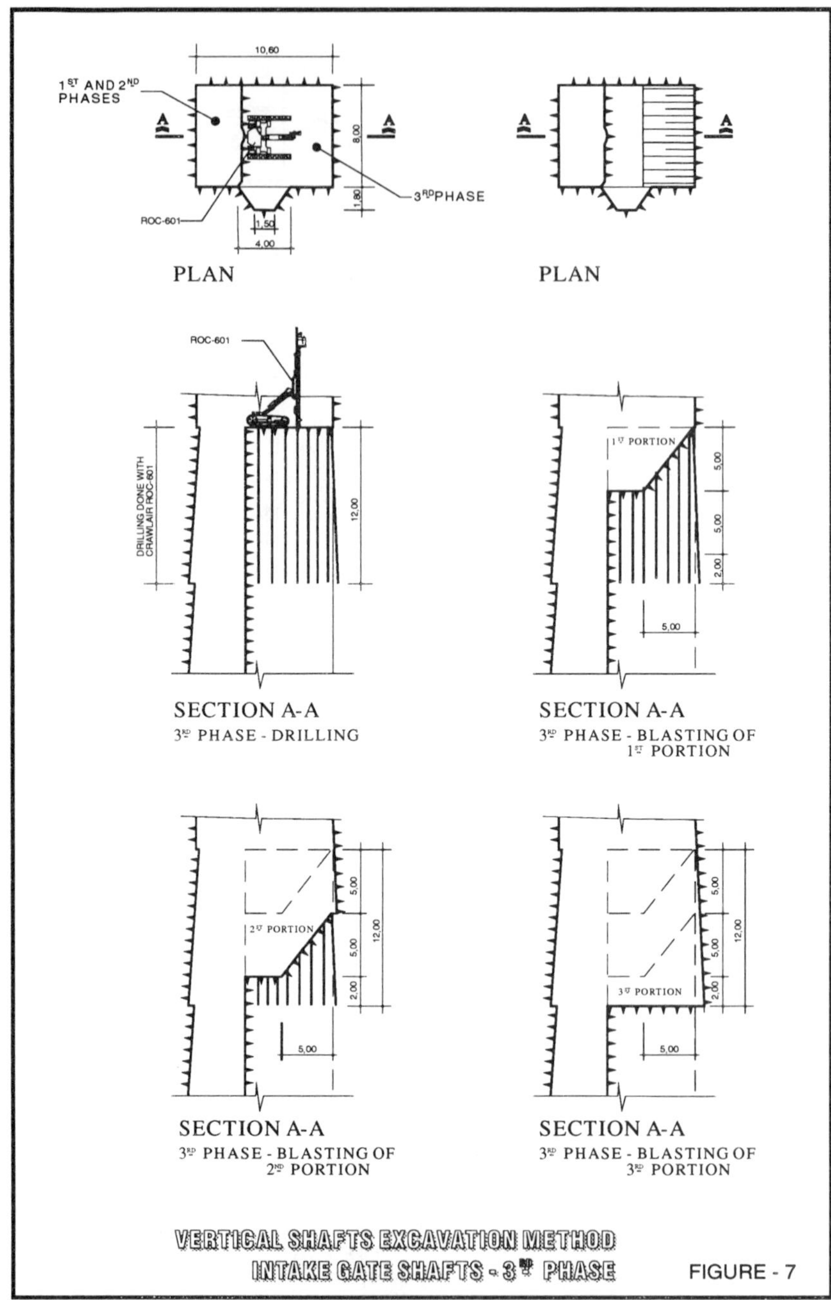
10,60
1ST AND 2ND PHASES
3RD PHASE
ROC-601
8,00
1,80
1,50
4,00
PLAN
PLAN
ROC-601
DRILLING DONE WITH CRAWLAIR ROC-601
12,00
1ST PORTION
5,00
5,00
2,00
5,00
SECTION A-A
3RD PHASE - DRILLING
SECTION A-A
3RD PHASE - BLASTING OF 1ST PORTION
2ND PORTION
3RD PORTION
12,00
SECTION A-A
3RD PHASE - BLASTING OF 2ND PORTION
SECTION A-A
3RD PHASE - BLASTING OF 3RD PORTION
VERTICAL SHAFTS EXCAVATION METHOD
INTAKE GATE SHAFTS - 3RD PHASE
FIGURE - 7

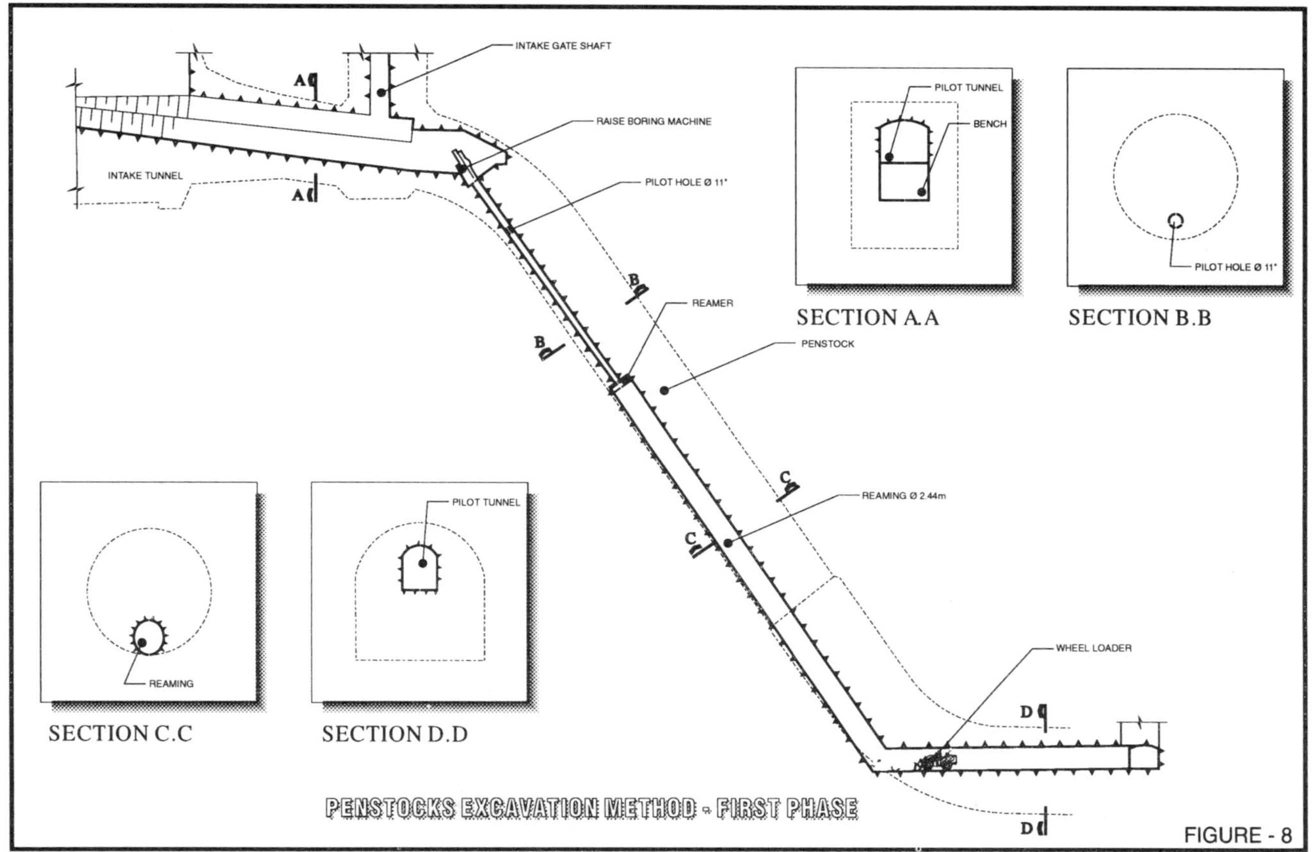

INTAKE GATE SHAFT
RAISE BORING MACHINE
PILOT HOLE Ø 11"
INTAKE TUNNEL
A
B
REAMER
PENSTOCK
C
REAMING Ø 2.44m
WHEEL LOADER
D
PILOT TUNNEL
BENCH
SECTION A.A
PILOT HOLE Ø 11"
SECTION B.B
REAMING
SECTION C.C
PILOT TUNNEL
SECTION D.D
PENSTOCKS EXCAVATION METHOD - FIRST PHASE
FIGURE - 8

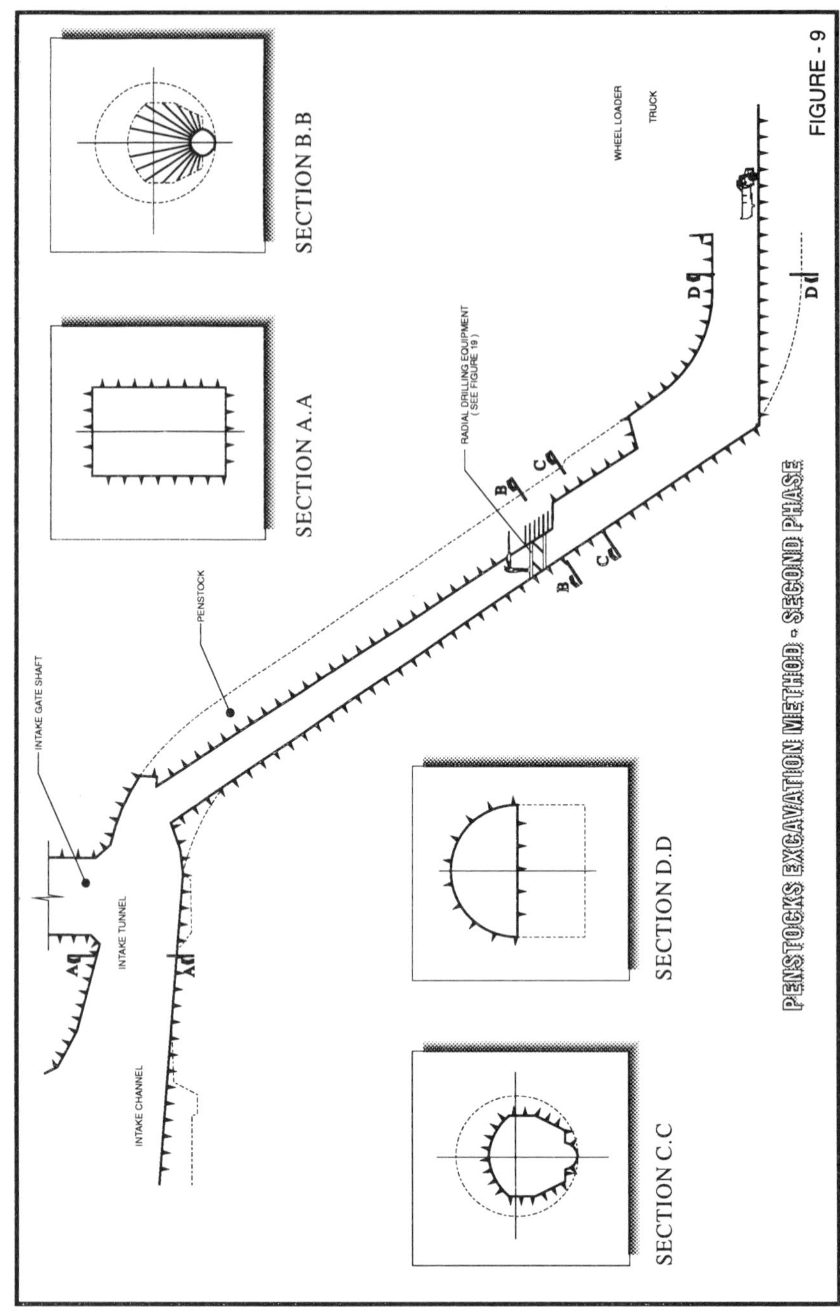
SECTION B.B
SECTION A.A
INTAKE GATE SHAFT
PENSTOCK
RADIAL DRILLING EQUIPMENT
(SEE FIGURE 19)
WHEEL LOADER
TRUCK
INTAKE TUNNEL
INTAKE CHANNEL
A
B
C
D
SECTION D.D
SECTION C.C
PENSTOCKS EXCAVATION METHOD - SECOND PHASE
FIGURE - 9

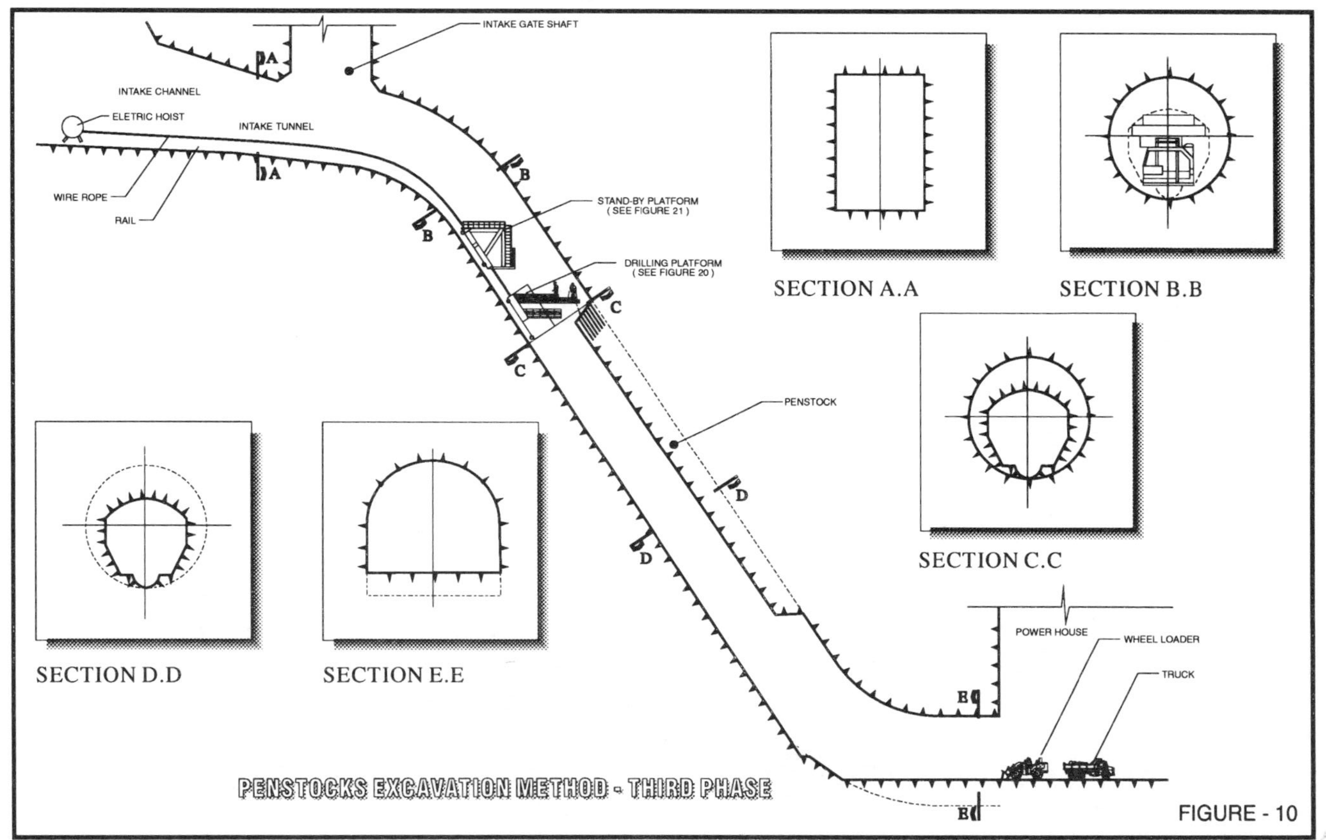
INTAKE GATE SHAFT
INTAKE CHANNEL
ELETRIC HOIST
INTAKE TUNNEL
WIRE ROPE
RAIL
STAND-BY PLATFORM (SEE FIGURE 21)
DRILLING PLATFORM (SEE FIGURE 20)
PENSTOCK
SECTION A.A
SECTION B.B
SECTION C.C
SECTION D.D
SECTION E.E
POWER HOUSE
WHEEL LOADER
TRUCK
PENSTOCKS EXCAVATION METHOD - THIRD PHASE

FIGURE - 10

the whole operation is controlled from the operators' console in the machine.

The method can be briefly described as follows:

The raise borer is erected at the upper part of the shaft and adjusted into the correct direction. Any direction between vertical and horizontal can be chosen.

A small hole, usually called the pilot hole, is drilled downwards by adding drillrods, each 1.5 m, attached to each other. The rock is excavated by a tricone drill bit by using thrust and rotation. The rock cuttings are flushed up with water or air. The pilot hole diameter is normally in the range of 250mm to 380mm.

After the break through into the tunnel the pilot bit is dismantled and a reamer head assembled onto the drill string. The reamer head can be chosen to fit requirements and are available in diameters between 0.7 m and 6.0 m.

The reamer head, which is equipped with cutters, is pulled and rotated upwards by the raise borer by deducting drillrods from the drill string. The cutters excavate the rock which falls down to the shaft bottom due to gravity.

When the reamer head reaches the top of the shaft the machine is dismantled and taken away. The shaft is completed.

The advantages of the method compared to more traditional methods are many.

a) Soft excavation:

Due to the soft excavation of the rock less reinforcement is needed compared to a drill and blasted shaft. Also the overbreak which normally occurs when using explosives for the shaft excavation is limited. This is important when the shaft has to be concrete lined, for example.

The method also gives a minimum of vibrations to the surrounding rock mass and this can be important when the shaft is located in urban areas where there might be sensitive equipment and so forth, in the vicinity.

b) Shaft section:

The method creates a perfectly cylindrical shaft which is strong due merely to its shape. The cylindrical shape is also preferable in the shaft walls with respect to friction loss when the shaft is used for ventilation or for water transport. The shaft section is also ideal when the shaft is enlarged by drill and blast.

c) Capacity:

This excavation method is normally 2-3 times faster than conventional methods and the availability of the equipment is also very high with few major break downs.

There are also some less favourable aspects with the raise boring method compared to other shaft excavation methods:

a) Accuracy:

Up to now it has been very difficult to control the drilling direction of the pilot hole, which also decides the direction of the final shaft. The result has depended on the experience and the skill of the drilling operators together with the specific geology of the rock mass. However there are now special tools under development which can control and adjust the drilling direction of the pilot hole while drilling.

b) Price:

The raise boring equipment is very cost intensive and requires highly skilled operators which results in the method being a little more expensive than other more personnel intensive methods, especially in countries where labour wages are low.

3.2.2 Raise Boring at Serra da Mesa Hydro Electric Project

In January 1989 Skanska Raise Boring AB was employed by Camargo Corrêa to carry out the raise boring works of 11 shafts of 2,44 m in diameter with a total lenght of approximately 1100 m at the Serra da Mesa Project in the state of Góias, Brazil. Camargo Corrêa's intention was to use the raise bored shafts as pilot shafts for the later enlargement works, by drill and blast, of the shafts to their final diameter. The equipment was shipped from Sweden in June 1989 and work started at the end of August 1989. Technically the work was not difficult. The shafts had a single length of between 140 m and 58 m and they were all vertical except for the three penstock shafts which were inclined at 55 degrees to the horizontal. The rock consisted mainly of granite with an RQD value close to 100 and there were no faults in the area.

One shaft was a little special and that was the raise boring of the ventilation shaft. Camargo Corrêa wanted an early access to the shaft in order to provide ventilation during the construction and excavation of the gallery in the powerhouse.

The final length of the shafts had to be 140 m and therefore the pilot hole was drilled to full depth. After the completion of the pilot hole the drillstring was pulled up to a depth of 115 m where a tunnel was excavated into the shaft location from the gallery of the powerhouse. The first stage of the upreaming started then from that level.

Later when the excavation in the powerhouse had reached the turbine level, a new tunnel was blasted into the shaft location where the pilot hole was situated.

The raise borer was mobilized a second time over the open shaft and the drillstring was lowered down, into the earlier drilled pilot hole to a depth of 140 m. The reamerhead was assembled and the last 22 m was completed.

The work was executed in accordance with the time schedule without any delays and the only surprise was the abrasiveness of the granite to the pilot bits and the reaming cutter. The life of the tools where about the half of the expected lifetime.

The explanation to the high wearing was probably due to the unusually high content of quartz in the granite together with a low frequency of micro fissures in the rock mass. Furthermore the hardness of the rock mass in the shaft location was probably more than the average rock hardness in the area.

The only problem which occurred during the project was a mechanical breakdown in the gearbox of the raise boring machine. Fortunately however, the forced stop was in any case close to a planned stop in the schedule. All details regarding the raise boring excavation is presented in item "6.0 Performance Indicators".

All the raise boring work was completed in early December 1990 and from Skankas point of view the raise boring work was a technical success.

The table 3 shows the hole deviations obtained in the several pilot holes drilled in the Serra da Mesa Project.

Table 3. Deviations of the pilot holes

Description	Total depht(m)	Deviation(m)	Desviat/m(%)
Ventilation shaft	138	0.47	0.34
Busduct shaft 1	123	0.46	0.37
Busduct shaft 2	123	0.41	0.33
Busduct shaft 3	123	0.42	0.34
P.house access shaft	121	0.45	0.34
Intake gate shaft 1	59	0.04	0.07
Intake gate shaft 2	59	0.12	0.20
Intake gate shaft 3	59	0.05	0.08
Penstock 1	93	0.27	0.29
Penstock 2	93	0.48	0.52
Penstock 3	93	0.49	0.53

3.3 The Shaft widening method (2nd and 3rd phases)

3.3.1 Busduct, ventilation and powerhouse access shafts

The busduct, ventilation and powerhouse access shafts, due to their small cross section area, were excavated with only one widening, and the excavation method is described below.

The first step consisted of drilling holes of 12.00m in depth, as shown in "Figure 3". To drill the first two advances, a crawlair Atlas Copco Roc-601 was used, placed directly on the rock surface, with the pilot hole closed by a steel plug made expecially for this purpose, as shown in "Figure 4", and in "Photo 5".

To drill the following 12.00m advance, a drilling platform with a circular base was made in the site workshop, where two booms of the Atlas Copco Roc-601 were erected, with all their hydraulic commands, as shown in "Figures 5, 11 and 12", and in "Photo 6".

The way of installing and using the drilling platform is demonstrated in the "Figure 5". As shown in this figure, the drilling platform has a steel umbrella to protect the crew from the accidental falling of pieces of rock, and this umbrella is moved away from the platform to diminish the weight of the kit in order to prevent accidents.

The crew can move through the vertical shafts in the workers' elevator box shown in "Figure 5".

The drilling of the shafts was done with an accurate control of positioning and alignment of the holes and after the conclusion of an entire 12.00m drill round, the platform was moved to another shaft to start a new cycle.

After the conclusion of the drill round, the holes were immediatly filled with rock dust, leaving unfilled the upper 2.00m space. Then, this space was filled with the explosive charge and the blast operations took place. The 12.00m holes were blasted in portions of 2.00m in order to prevent the obstruction of the pilot shaft with big pieces of rock, as shown in "Figures 4 and 5".

The holes next to the pilot shaft were always deeper than the others, in order to create an inclined surface in the direcion of the pilot shaft in order to facilitate the natural falling of the blasted rock. After blasting a portion of 2.00 m, the proccess begins again until the blasting of the entire 12.00m holes.

3.3.2 Intake gate shafts

As the intake gate shafts have a big cross section area, they had to be excavated in two widenings.

The first widening had the same excavation method as that used to excavate the busduct shafts, described above.

The drilling equipment used was a crawlair Atlas Copco Roc-60l, as shown in "Figure 6". The second widening was also drilled with a Roc-60l, as shown in "Figure 7", and the length of the drill holes was 12.00m.

As the first widening provided a larger area for the blasted rock to pass through, the portions blasted were bigger than those of the first widening, as shown in "Figure 7". Portions varying from 3.50m to 5.00m were used instead of 2.00 m portions.

Also in the second widening the surface between two portions was made inclined toward the opening, in order to facilitate the natural falling of the blasted rock.

3.4 The penstock widening method (2nd and 3rd phases)

The excavation of the penstock had three phases.

The first one was the execution of the pilot shaft with the raise boring method, already described in "Item 3.2", and shown in "Figure 8".

Photo 5

Photo 6

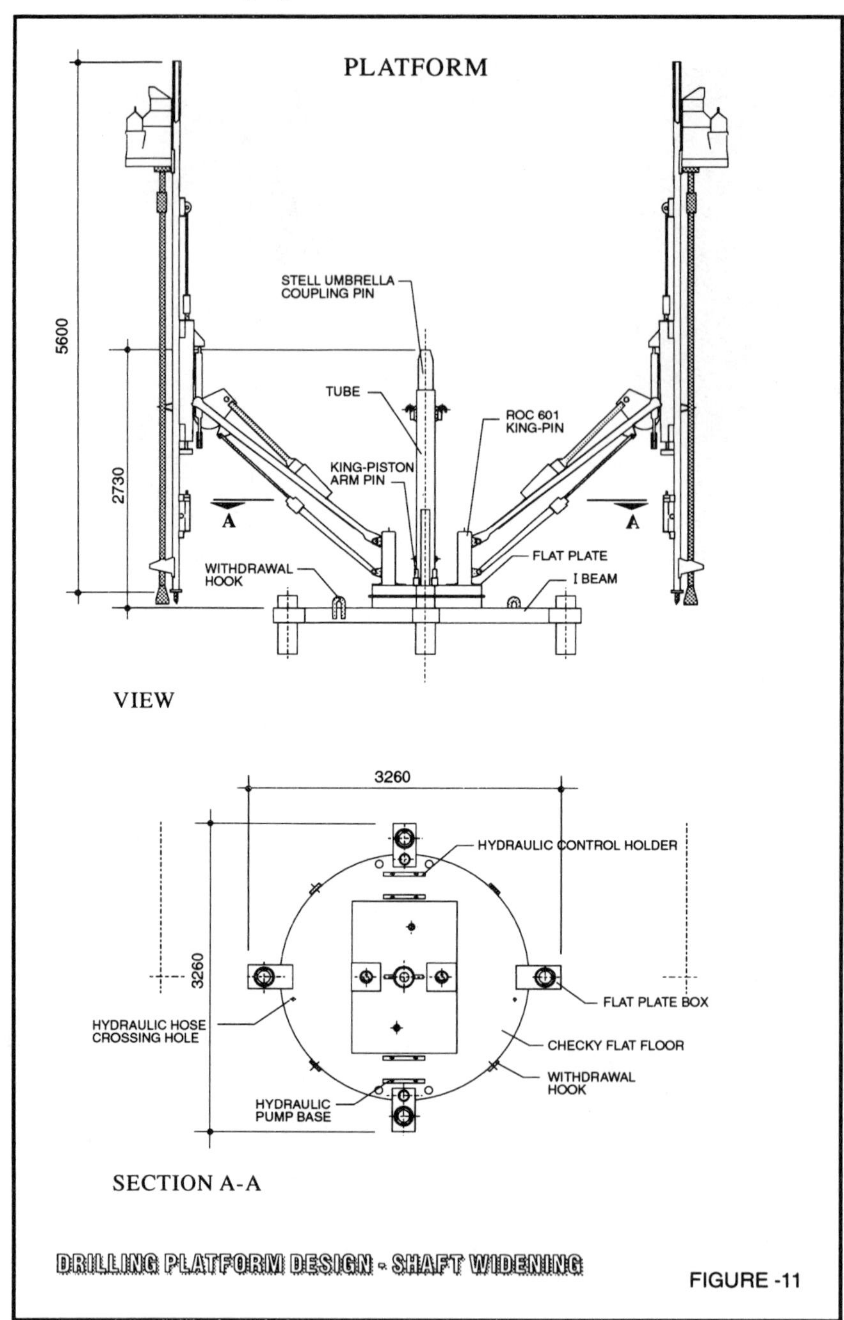

DRILLING PLATFORM DESIGN - SHAFT WIDENING

FIGURE -11

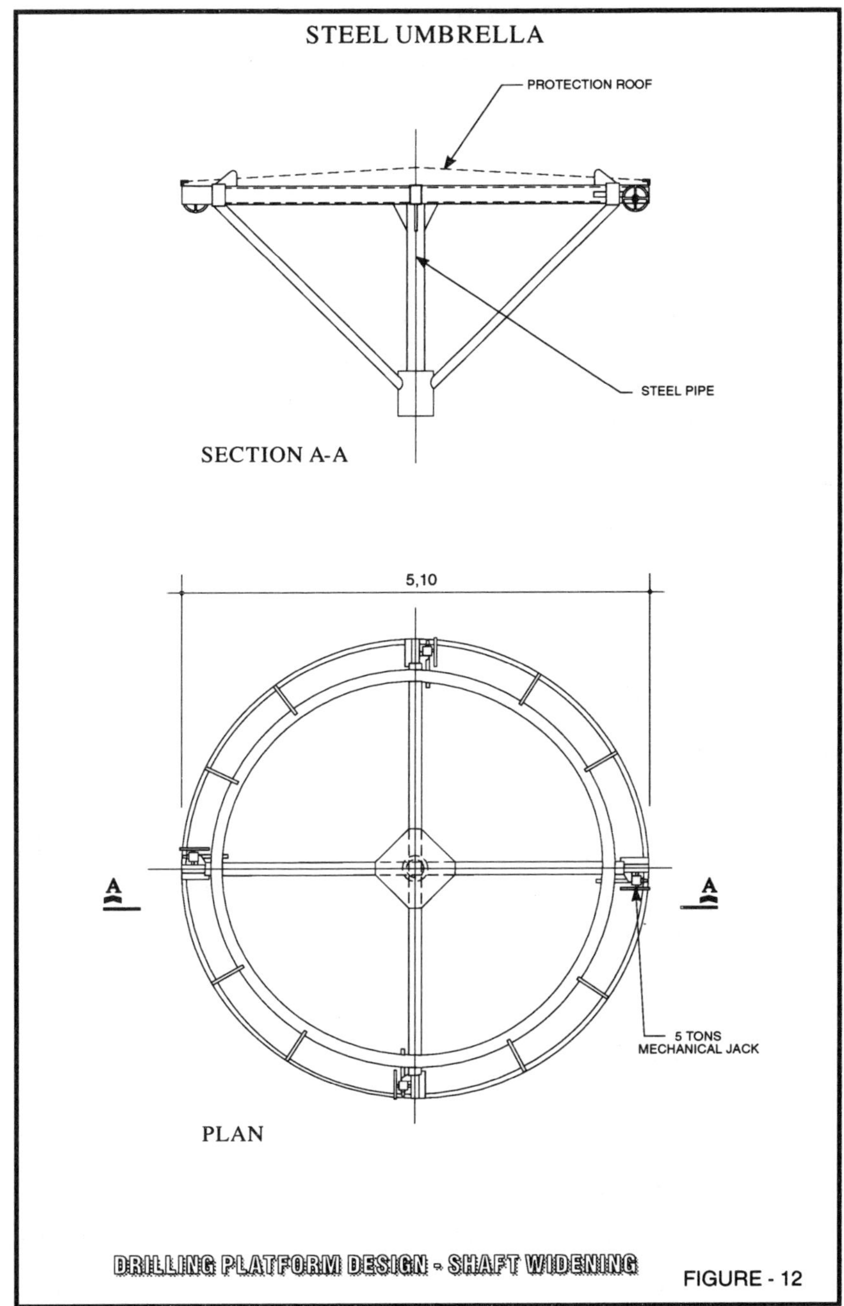
STEEL UMBRELLA
PROTECTION ROOF
STEEL PIPE
SECTION A-A
5,10
A
A
5 TONS
MECHANICAL JACK
PLAN
DRILLING PLATFORM DESIGN - SHAFT WIDENING

FIGURE - 12

The second phase, shown in "Figure 9"consisted of the widening of the pilot shaft.

The drill holes were drilled in a radial way and the excavation progressed in an upward direction. To drill the holes an equipment assembled in the site workshop was used consisting of a steel structure that moves through the pilot shaft gliding over ski shaped beams, and a boom of a Gardner-Denver drill AT-50 fixed onto that steel structure, as seen in "Figures 8 and 13"and in "Photo 7".

The holes were drilled from the inside of the pilot shaft, in a radial shape as shown in the "Figure 9" (section B-B).

The blasted rock falls directly into the lower access gallery where the frontal wheel loaders load the off-highway trucks.

The major advantage of this method is that the excavation cycle time is very short due to the blasted rock not having to be removed and the scaling not having to be done, and so the excavation cycle has only the drilling, charging and blasting activities. After the conclusion of the **2nd phase** of excavation in the whole shaft, the **3rd phase** of excavation begins in the downwards direction, as shown in "Figure 10". In this case, the holes were drilled longitudinally, from a platform (assembled in the site workshop), were the crew handled the manual drills, Atlas Copco BBC-17w and the design of the drilling platform is shown in "Figure 14" and in "Photo 8".

To move the crew, the equipment and the materials through the penstocks, the stand-by platform was used, whose design is shown in "Figure 15" and in "Photo 9".

4 Analysis of alternative excavation methods

Besides the excavation methods described in the above items, alternative excavation methods were analysed. However, after detailed studies these methods were abandoned.

The alternative methods were:

a) To excavate the pilot shaft:

The drill and blast method, done in an upward direction, with manual drills.

The drill and blast method, done in a downward direction, with manual drills and the blasted rock being removed by a gantry crane and a clamshell.

b) To excavate the widenings:

The drilling done by manual drills instead of the Roc-601 booms, in the vertical shafts.

The drilling done by a hydraulic jumbo instead of the drilling platform, in the penstocks.

These alternative methods were not used for reasons of safety, costs and time schedule, in that order of priority.

5 Equipment

5.1 Raise boring equipment

The equipment Skanska used for the raise boring works was:

Raise borer:	Robbins 73 RM DC (force ventilated) with a maximum pulling thrust of 3600 kN and with a maximum continuous reaming torque of 190 kNm.
Stabilizers:	Drilco high strength rods, diameter 254mm
Drillrods:	Drilco high strength rods, diameter 254mm

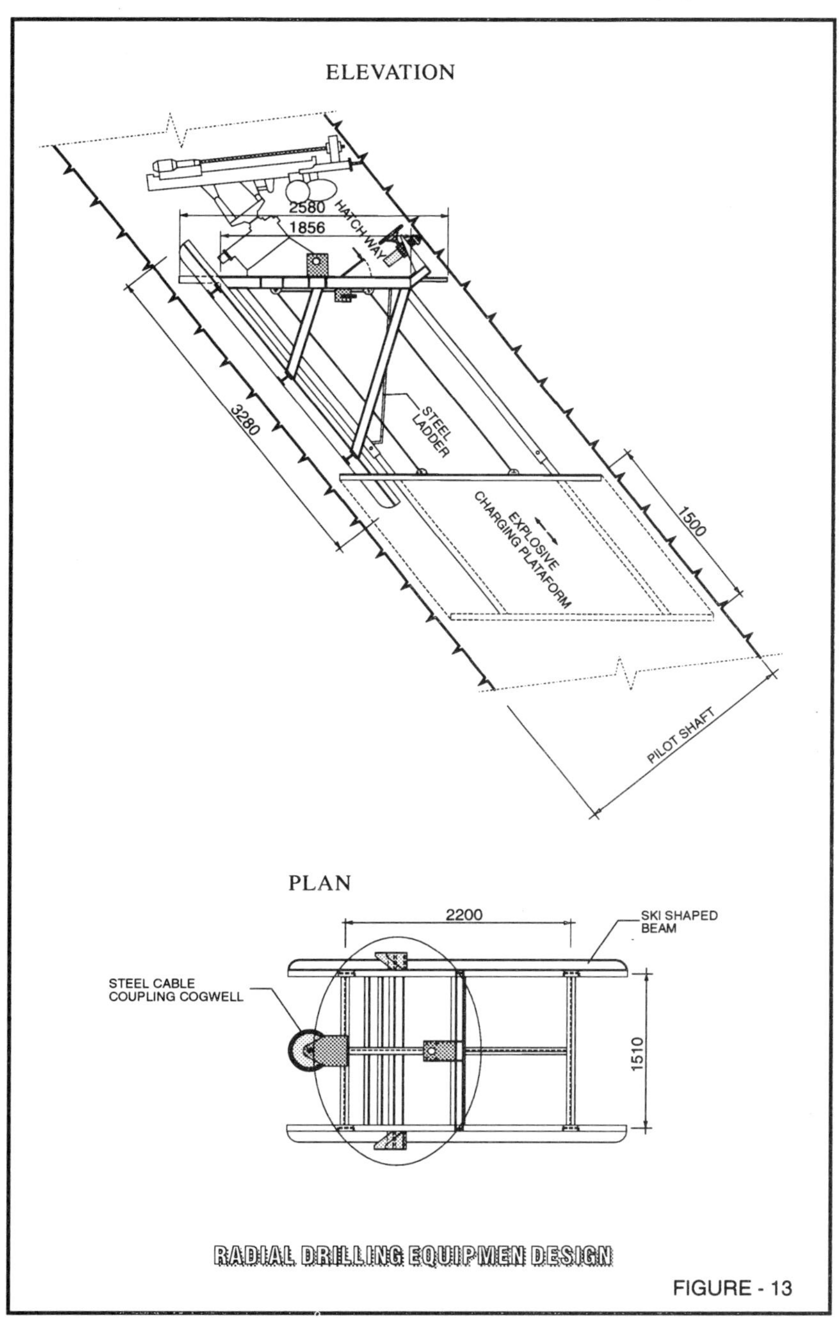

RADIAL DRILLING EQUIPMEN DESIGN

FIGURE - 13

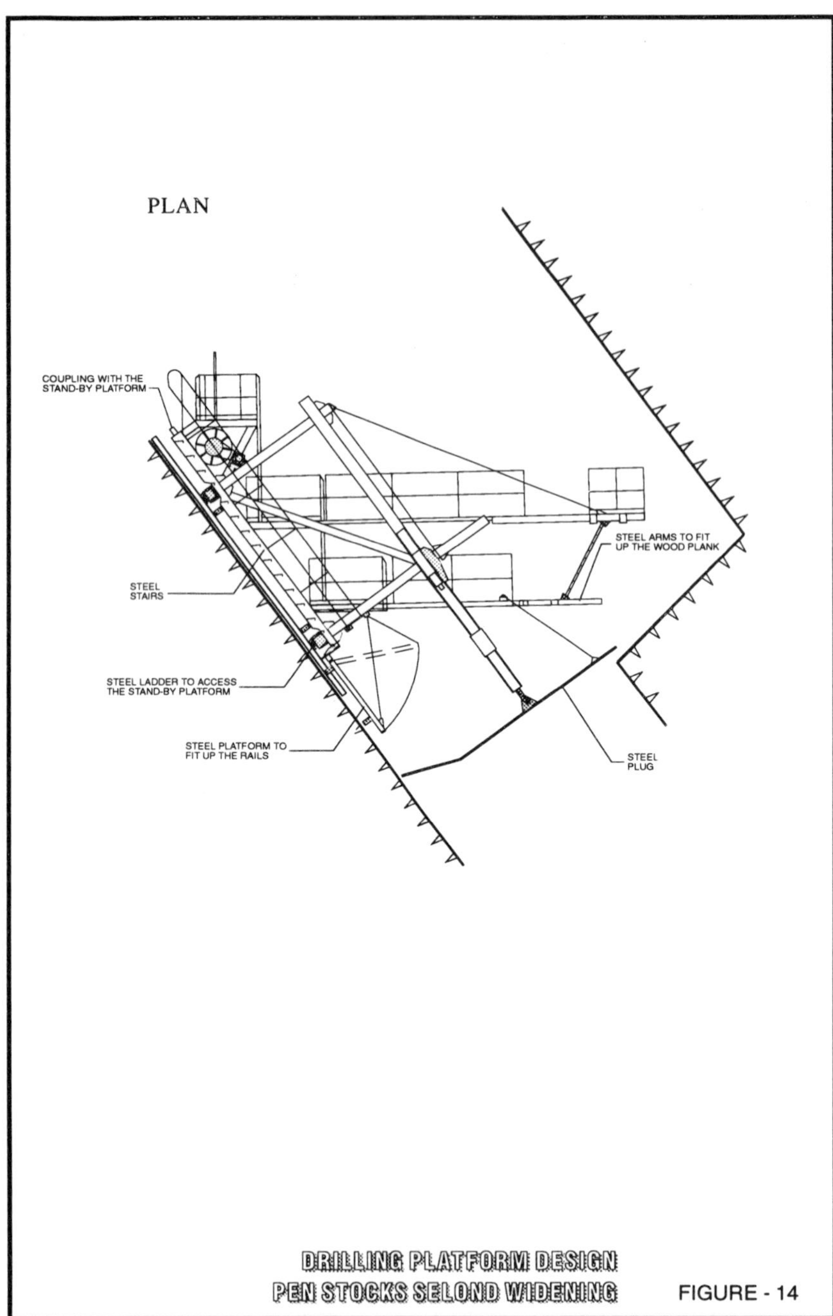
PLAN
COUPLING WITH THE STAND-BY PLATFORM
STEEL STAIRS
STEEL LADDER TO ACCESS THE STAND-BY PLATFORM
STEEL PLATFORM TO FIT UP THE RAILS
STEEL ARMS TO FIT UP THE WOOD PLANK
STEEL PLUG
DRILLING PLATFORM DESIGN
PEN STOCKS SELOND WIDENING

FIGURE - 14

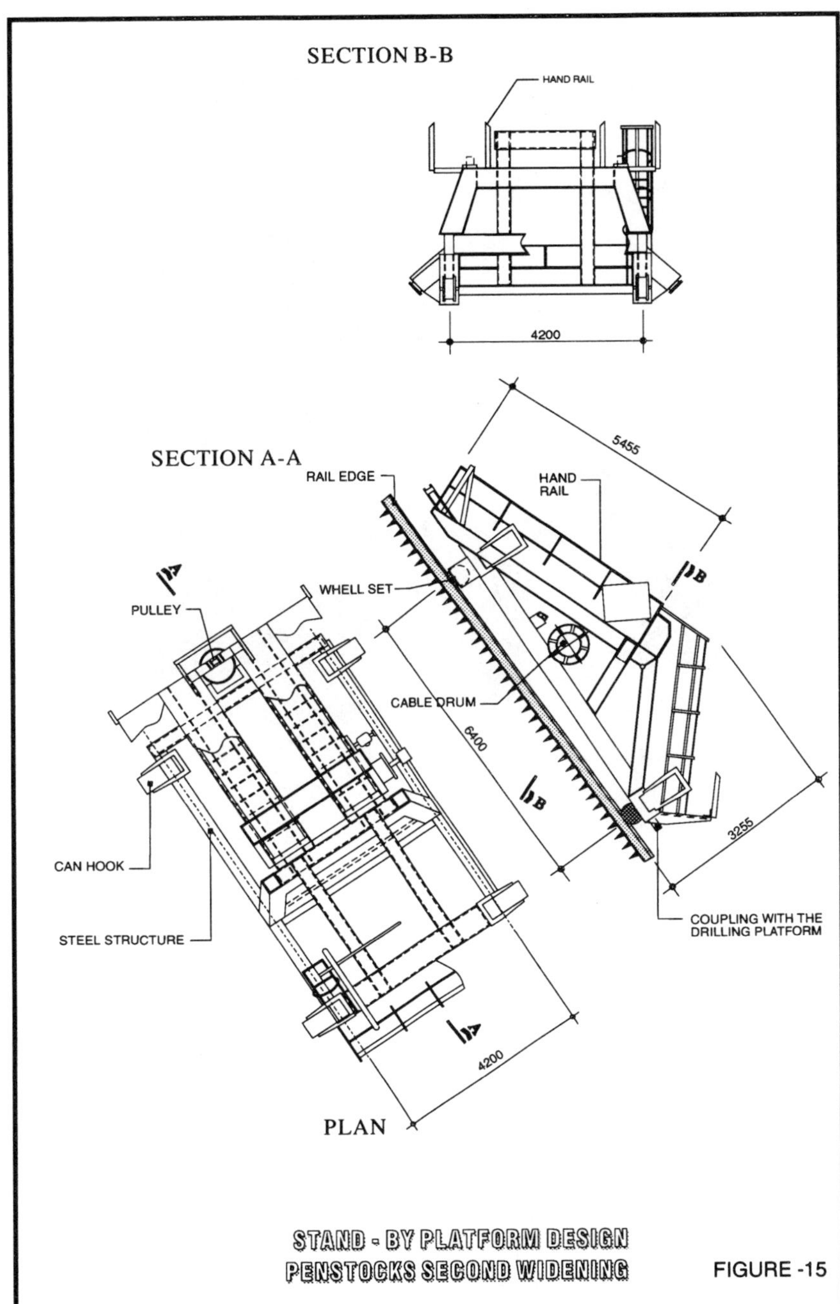
SECTION B-B
HAND RAIL
4200
SECTION A-A
RAIL EDGE
HAND RAIL
5455
WHELL SET
PULLEY
CABLE DRUM
6400
B
B
A
A
CAN HOOK
STEEL STRUCTURE
3255
COUPLING WITH THE DRILLING PLATFORM
4200
PLAN
STAND - BY PLATFORM DESIGN
PENSTOCKS SECOND WIDENING

FIGURE -15

Photo 7

Photo 8

Photo 9

Pilotbits: Baker Hughes BHS 80, diameter 280mm
Reamerhead: Sandvik CRH8, diameter 2.44m.
Cutters: Sandvik CMR 41 and CMR 52.

This equipment is shown in "Photos 3 and 4".

The crew consisted of one engineer and two technicians working in two shifts of ten hours each.

5.2 Vertical drilling platform

This platform was assembled in the site workshop and was used to drill the vertical shafts. Its design is shown in "Figures 11 and 12" and in "Photo 6". The principal component parts and auxiliary equipments of this platform are:

A steel plug that shuts the pilot shaft, preventing people or things from falling through it.

Two hidraulic controls for the Atlas Copco Roc-601 booms.

Two hydraulic comanded supports to level the platform.

A steel floor where the crew stay.

Two hydraulic comanded booms, model Atlas Copco BMM-36K.

Two compressed air hammers, model Atlas Copco Cop 131.

One steel umbrella to protect the crew from accidentally falling pieces of rock.

One elevator box to move the crew, the auxiliary equipment and materials.

One 18 ton gantry crane used to move the platform, the steel umbrella and the elevator box.

All this equipment is handled by two drill operators, one gantry crane operator and one production assistant.

5.3 Radial drilling equipment

This equipment was assembled in the site workshop and was used to drill the 1st widening of the penstocks. Its design is show in "Figure 13" and in "Photo 7".

The principal component parts and auxiliary equipment of this apparatus are:

A steel structure that glides though the pilot shafts on ski beams.

One Gardner-Denver boom AT-50, assembled on the steel structure, and which is able to make a manual radial movement.

One compressed air hammer, model Gardner Denver PRJ123.

One 10 ton Richier eletric hoist used to move the equipment, and placed in the upper part of the penstock.

This equipment is handled by one drill operator, one production assistant, one hoist operator and one electrician.

5.4 Penstocks stand-by and drilling platforms

These platforms were also assembled in the site workshop and were used to support and drill the 2nd widening of the penstocks. Their design is shown in "Figures 14 and 15" and in "Photos 8 and 9".

The principal component parts and auxiliary equipments of these platforms are:

One steel platform, assembled over steel wheel sets, able to move over rails, consisting of three decks where ten Atlas Copco BBC-17-W manual drills are handled.

One steel platform, assembled over steel wheel sets, able to move over rails, and used to transport the crew, the equipment and materials betwen the penstock top and the drilling platform.

One 50 ton Peiner eletric hoist used to move these platforms, and placed in the top of the penstock.

These platforms and the drilling and auxiliary equipment are handled by fourteen drill operators, one electrician, one plumber, one carpenter, one driver and one hoist operator.

5.5 Loading and hauling equipments

The blasted rock that falls into the lower access galleries is loaded and hauled by a front loader Caterpillar Cat-966 and hauled by 22 tons off-highway haul trucks Scania-Randon Kockum RK-425.

6 Performance Indicators

6.1 Raise boring

Table 4. Performance indicators of raise boring

Shaft	C.Times (Days)	Lenght (m)	Deviation (m)	Penetration Rate P.(m/h)	Penetration Rate R.(m/h)
V. Shaft I	29	140/115	0.47	1.46	0.78
B. Shaft I	28	125	0.46	1.81	0.54
B. Shaft III	22	125	0.42	1.70	0.48
B. Shaft II	25	125	0.41	1.68	0.58
V. Shaft II	4	0/25	-	-	0.41
P.House access	36	122	0.50	1.59	0.44
I.Gate shaft II	18	59	0.14	1.59	0.85
I.Gate shaft III	15	59	0.05	1.49	0.76
I.Gate shaft I	15	59	0.04	1.60	0.72
Penstock I	23	93	0.45	1.89	0.58
Penstock II	18	94	0.42	1.77	0.74
Penstock III	15	94	0.26	1.81	0.68

The times given do not include the time for moving the equipment between the different shaft locations. Normally one week was spent for dismantling, transporting and assembling.

6.2 Widening of vertical shafts

Shafts	Average excavation rate (m/month)
Busduct shafts	45
Ventilation shaft	45
Powerhouse acess shafts	45
Intake gate shafts - 1st widening	35
Intake gate shafts - 2nd widening	50

6.3 Widening of penstocks

Penstock	Average excavation rate (m/month)
1st widening	64
2nd widening	40

7 Safety

In the choice of the excavation methods and equipments for excavating the Serra da Mesa shafts and penstocks one of the main imposing factors was safety.

This care is reflected in the results because in more than 1,000m of shaft and penstock excavation, no serious accidents with the crew or the equipment were registered.

One of our precautions is to adopt the raise boring method.

The raise boring method does not require in-shaft personnel. All operations during the drilling are done from the operators' console at the top of the shaft. If the shaft has to be reinforced during the up reaming phase, the drilling can be stopped and the shaft can be entered from the shaft bottom. The shaft front is then secured by the reamer head which prevents bigger pieces of rock from falling down. Also, pilot drilling indicates what conditions which will be encountered during the upreaming phase.

Other precautions adopted are pointed out below:

Not using excavation methods where the holes had to be drilled in an overhead position.

Always excavating the lateral walls with controlled excavation methods, such as smooth blasting or cushion blasting.

Doing the scaling methodically.

Protecting the shaft top with a 0.50m high concrete ring.

Providing the drilling platform with a protection roof (steel umbrella).

Providing the hoists with multiple safety brakes.

Administering training courses on work safety.

Keeping a safety engineer involved in the excavation activities at all times.

Providing the crew with proper protection material.

Maintaining proper ventilation, illumination, signalling and comunication systems.

8 Excavation round cycles

The excavation of the shafts and penstocks was done together with the other underground excavations, such as the powerhouse, surge tank and tunnels. So the time schedule round cycles were affected by these other underground excavations.

Session 13: Tunnel Machinery

Rock cutting with roadheaders

R. J. Fowell
Department of Mining and Mineral Engineering, University of Leeds, England
G. Richardson
Dosco Overseas Engineering, Ltd., Tuxford, Nottinghamshire, England
M. J. Gollick
Dosco Overseas Engineering, Ltd., Tuxford, Nottinghamshire, England

The paper describes with practical case studies how some of the important factors influencing the excavation performance of boom tunnelling machines equipped with drag tools may be predicted. The pitfalls of basing performance predictions on compressive strength alone are discussed.

Use is made of results from a full scale instrumented laboratory boom tunnelling research rig, field trial data, and a standardized instrumented cutting test to provide estimates of instantaneous cutting rates for any given strata and machine.

The paper concludes with a review of options available should a machine encounter more difficult strata than originally anticipated.
Keywords: Performance prediction, roadheader, rock cutting, tunnels.

1 Introduction

When considering a boom tunnelling machine for a potential application the main initial requirements are the prediction of an excavation rate in cubic metres per hour and the likely number of cubic metres of rock cut per tool.

A decade ago this was a comparatively easy matter as roadheaders had similar specifications and applications were confined to the coal measures rocks, sandstone, siltstone and mudstone. Today the variety of rock materials to be excavated and the range of machines has grown considerably with a wide diversity of application. There is an obvious need to correctly match the machine to a specific duty, as in most cases, the end user is contractually obliged to complete on time.

This paper sets out a rational method of performance prediction based on field trial and fundamental research experience.

2 Performance Prediction

Using the method to be described not only can the production rate of a known machine be specified but it is also possible to determine the necessary increases in specification to give performance beyond this limit. This aspect highlights such factors as boom force and cutting head area having particular relevance when

TABLE 1 Comparison between Rock Salt and Coal Cutting Characteristics

	Salt	Coal
Unconfined compressive strength (MPa)	28.70	27.20
Cerchar abrasivity number	Nil	Nil
Density (t/m^3)	2.14	2.40
Mean Cutting Force (kN)	1.40	0.31
Mean Normal Force (kN)	0.46	0.19
Cutting Tool Wear (mm/m)	Nil	Nil
SPECIFIC ENERGY (MJ/m^3)	10.52	4.72

investigating potential applications and machine suitability.

It is important to clarify the term excavation or production rate in relation to the investigating potential applications and machine suitability. two terms in common usage,-Instantaneous Cutting Rate, (ICR), and Cutting and Loading Rate, (CLR). ICR reflects the ability of a machine to cut in an uninterrupted mode, for example when taking traversing cuts [1] and results in a maximum production rate. CLR on the other hand includes all the factors necessary to excavate a face; for example sumping, boom manoeuvring, loading, trimming etc., and gives overall machine performance. A factor dependant on the site situation will be required to establish the representative CLR estimate from the ICR value.

For many years unconfined compressive strength (UCS) has been used as a predictor of cutting performance. A few words are required to explain why UCS should not be taken in isolation to predict cutting performance

- It does not model the cutting process
- It is subject to so many influences that a variety of results can be obtained for the same rock material.
- It takes no account of the abrasivity of the rock and the rock mass characteristics.

For tunnel site investigations, borehole cores are used which have increased in diameter due to the need to run additional tests down the finished borehole. Most site investigations are conducted to provide information for lining design both permanent and temporary. The size effect on compressive strength is well known. To assess the cuttability of rock, the size of the UCS sample should model the amount of rock that an individual cutting tool would encounter when excavating the rock. Hence 30 to 40mm diameter cores are more appropriate than 80-100mm diameter core. Smaller cores also provide the added benefit of obtaining a sample free from imperfections or discontinuities which will give a better estimate of the intact strength of the rock.

Major departures from a UCS cuttability relationship are : anisotropic rocks, evaporites and rocks containing hard inclusions such as breccia, conglomerates, and chalk containing flints. Table 1 illustrates a comparison between the cutting characteristics of a hard coal and rock salt.The UCS values are similar but the energy to excavate a cubic metre is not.

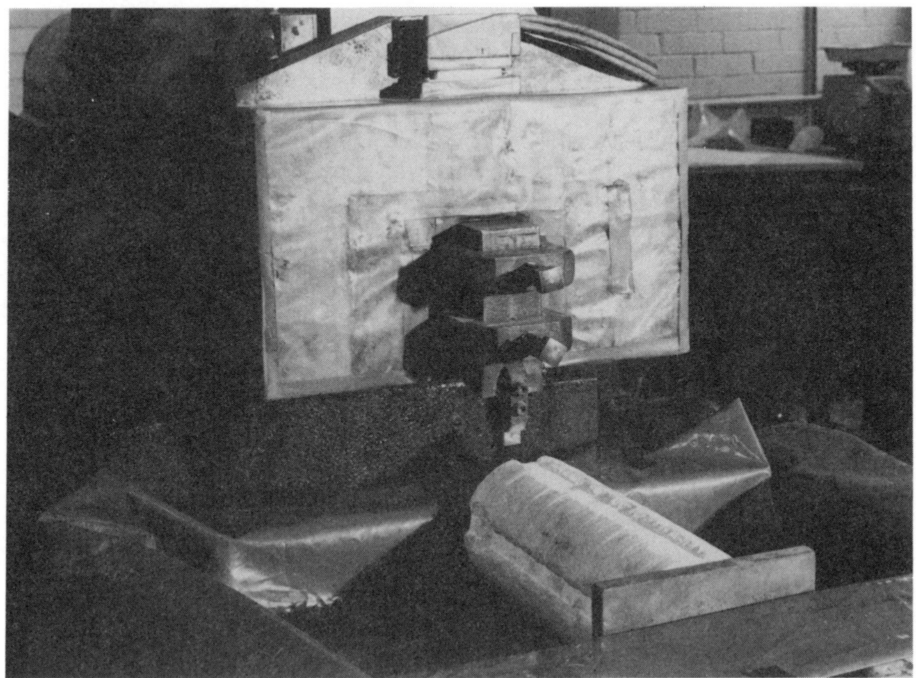

Figure 1. Instrumented cutting test.

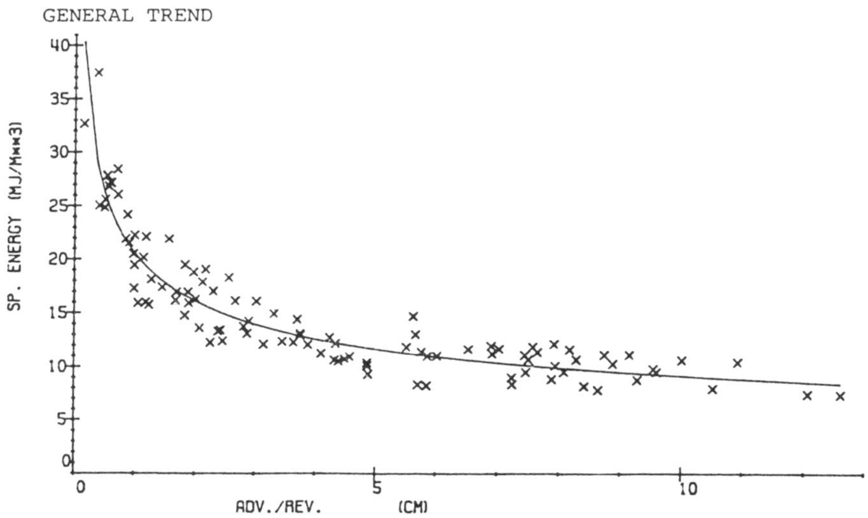

Figure 2. Typical specific energy versus advance per revolution relationship.

To overcome these problems the Instrumented Cutting Test (ICT) is used. This involves taking 5mm depth cuts along the outer periphery of a 75mm diameter, or greater, sample of borehole core with a standard geometry drag tool and measuring the force components acting on the tool [2], (figure 1). From the mean cutting force component and the yield a specific energy value is obtained which relates to the cuttability of the rock. Specific energy varies with depth of cut (figure 2) but under standardized conditions the only variable is the rock material under test.

Examination of the worn tungsten carbide insert under a travelling microscope gives a measure of the abrasivity of the rock and an indication of the other wear mechanisms likely to be operating under field conditions, such as impact chipping of the carbide or gross failure in service.

The ICT through the specific energy gives an indication of the cutting energy required to excavate unit volume of rock. The cutting motor power divided by the specific energy can give a general estimate of the excavation rate provided the cutting tool depth is similar to that used in the ICT. A measure of the boom force required for a particular rock is obtained from the mean normal force component obtained from the ICT.

The boom force is considered to be the most important parameter influencing production rate. For track mounted boom tunnelling machines (roadheaders) the weight of the machine is a deciding factor. If machine stability problems are to be encountered stelling can be used but this does not appear popular for certain operational reasons. Shield mounted boom tunnelling machines on the other hand have a natural in built stability provided by the near constant contact of shield and tunnel wall.

To obtain relationships between advance per revolution, head swept area and boom force a full scale instrumented laboratory boom tunnelling research rig was constructed,[3][4].

3 Instrumented Boom Tunnelling Research Rig

The basis of the rig was a boom from a Dosco Mk IIA roadheader powered by a 36 kW thyristor controlled motor, providing variable head rotational speeds. The pedestal and boom were mounted within a steel framework in which was mounted the sample box. (Figure 3).

The reinforced steel sample box was capable of holding a 20 tonne sample of artificial or natural rock, representing a mock tunnel approximate 3m wide by 2m high. Cuts could be made exactly simulating the operation of a boom tunnelling machine under production conditions. The sample was supported on rails and pulled towards the boom by hydraulic rams. The total weight of the rig was some 40 tonnes.

The rig functions were measured by 13 transducers which were monitored directly by a microcomputer based data storage and analysis system. Parameters measured included sumping, arcing and lifting pressures and displacements and cutter motor voltage and current.

Data reduction was carried out in a matter of minutes via interactive computer analysis programs. Parameters calculated include sumping, arcing and lifting forces and speeds, spatial position of the cutting head, cutting rate, cutting head power,

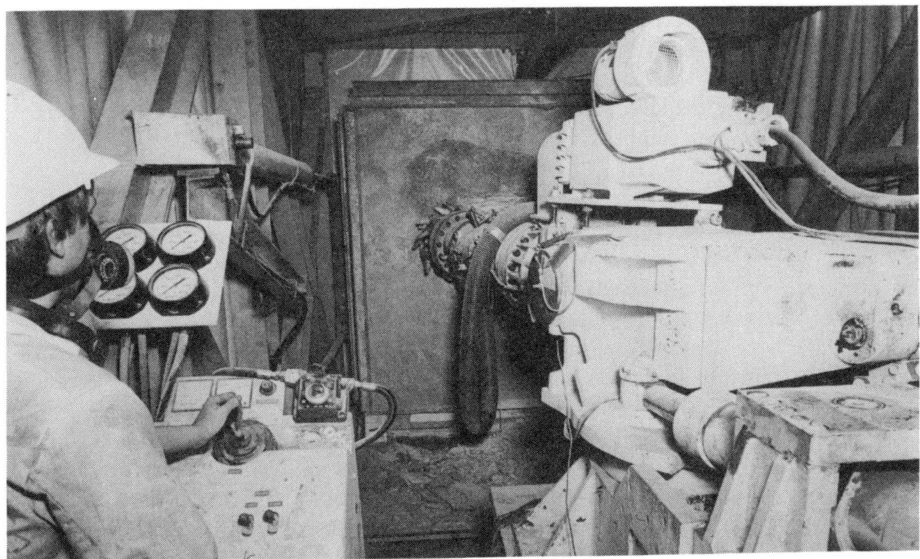

Figure 3. Instrumented boom tunnelling machine test rig.

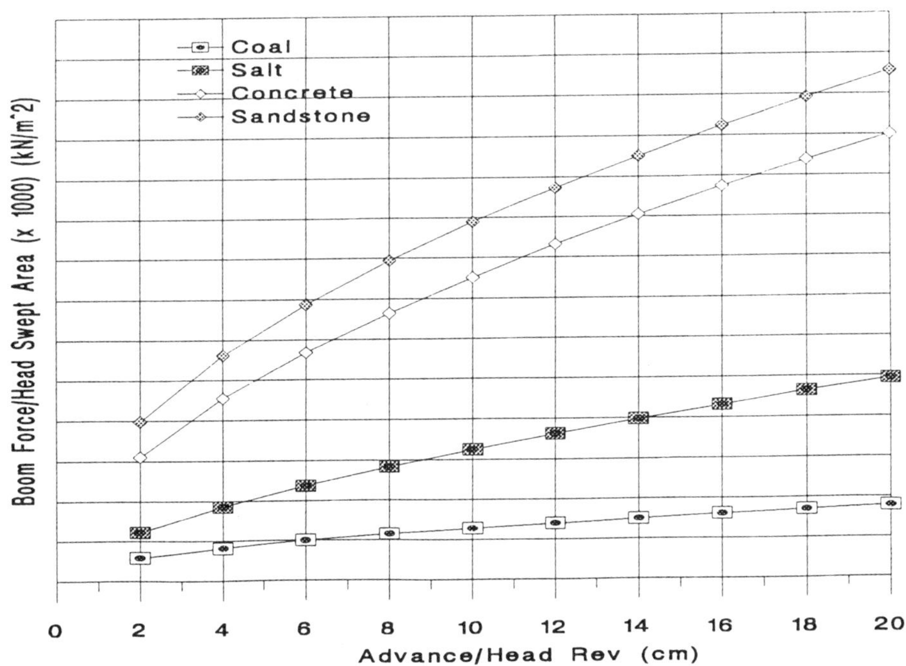

Figure 4. Typical boom force per unit area versus advance per revolution relationships.

torque and rotational speed and arcing rate per head revolution (Advance/Rev). Data analysis was carried out at the rig location.

Individual cuts and associated analysis took approximately five minutes with a similar period being required for cut set-up. Some 200 individual trial cuts were possible in a single sample, hence a large quantity of accurate roadheader performance data was rapidly obtained.

4 Experimental Programme

The principal aim of the trials was to investigate the relationships between the monitored parameters for a range of rock materials and cutting head designs. A complementary programme to investigate the benefits of high pressure water jet assistance was also undertaken. Waterjets were applied to individual tools in contact with the rock at pressures upto 140 MPa,[4][5].

For each cutting speed and advance per revolution the specific energy and boom force was determined. Figure 2 shows the relationship obtained for specific energy against advance per revolution. There is a rapid increase in cutting efficiency with increase in advance per revolution. The high specific energy at low tool penetrations is associated with the failure of the tool to cause the rock to break through into adjacent cuts.

Boom force is the arcing or lifting force at the cutting head, analogous to normal or hold in force for the instrumented cutting test. Boom force was normalised by expressing it per unit area swept by the cutting head during arcing or lifting. Thus experiments with different penetrations of the head were able to be compared. Such normalisation oversimplifies the actual geometric conditions during cutting, however it provides a simple and valuable means of comparison for cutting rate prediction purposes.

All parameters were related to advance per revolution of the head across or up the face. This is analogous to the depth of cut taken in single tool experiments.

Figure 4 illustrates the relationships between Boom Force per unit swept area (kN/m^2) and advance per head revolution for different materials. It can be seen that penetration is dependent on actual force applied and area of head in contact with the rock.

5 The prediction of instantaneous cutting rate

Three main parameters are used in the ICR prediction.

1. Cutting head contact or swept area (SA) in m^2. For a traverse cut this is the profile area of the head projected in the direction of boom movement. For undercutting this is the profile of that proportion of the head in contact with the rock projected in the direction of boom movement.
2. Head rotational speed in revolutions per minute. (RPM).
3. Pick penetration or advance per head revolution (A/R) (usually expressed in cm), determined by the available boom force (BF) in kN. (Figure 4)

These parameters are combined to give the ICR in m^3/h.

$$SA \times A/R \times RPM \times 60/100 = ICR\ m^3/h \qquad (1)$$

Cutting and loading rates (CLR) are determined by apply a site specific factor (normally between 0.5 and 0.9),

Example: To determine the excavation rate for a machine operating in hard coal.
Head contact swept area (SA) = 1.0 m^2
Head rotational speed = 52 rpm
To determine the advance/revolution of the head.
Using the advance/revolution characteristic for this coal (figure 4) which may be expressed as:

$$\frac{Boom\ Force\ (BF)}{Swept\ Area\ (SA)} = 41.0\ (A/R)^{0.49} \qquad (2)$$

where Boom Force is expressed in kN and A/R in centimetres per revolution.
For an available boomforce of 130 kN this gives an advance per revolution of 10.5 cm.

The excavation rates is given by equation (1) and using a value of 0.6 for the factor F_1 gives a cutting and loading rate of <u>196.6 m^3/h</u>.

Field trials subsequently showed close agreement with the predicted cutting and loading rate.

6 Difficult Cutting Conditions

Harder rock than foreseen is an obvious situation where difficulties arise. The options available are numerous and site specific conditions are usually an important factor in the solution.

1. Increased Boom Force.
Utilising higher boom force allied to a stable machine increases tool penetration force allowing harder rock to be cut.
2. Smaller Cutting Head.
Machines are often slewing or boom force limited therefore fitting a smaller head designed for hard rock cutting conditions is beneficial as it increases the boom force per unit area and the hold in force on each cutting tool.

3. Slower head speed.
Reducing the head speed can increase available torque,but relies on power consumption remaining constant.This reduces the wear rate of the cutting tools, but at the expense of production rate.

7 Prediction of Tool Consumption Rates

One of the difficult tasks in drag tool tunnelling performance assessment is the prediction of cutting tool consumption rates. This is because there are many factors influencing the rate that tools wear and the forms that wear may take. The principal factors may be grouped under three broad headings. The rock, the machine and the tunnel environment. A summary of factors is given in Table 2.

Wear may take several forms, abrasive wear, gross impact failure, thermal fatigue or spalling. Most drag tools are tipped with tungsten carbide. The composition of the tungsten carbide influences the wear rates. Several trends have been established. The higher the cobalt content the higher the toughness of the carbide but this is usually at the expense of abrasive resistance. The larger the grain size the tougher the carbide but in practice the composition is a compromise between abrasion resisting hardness and impact resisting toughness.

When to change tools is difficult as there is no firm definition of a worn tool. In hard rock even a small wearflat can quadruple the hold-in force required to maintain the required penetration for a given excavation rate. In these circumstances tool changing will be more frequent than for softer conditions where machines can give the required performance with highly worn tools.

Larger more powerful machines can take higher advances per head revolution and this results in lower tool consumption rates when compared with lower rated machines. The use of high pressure waterjet assistance reduces the rate of wearflat development due to cooling of the carbide insert and hence economic excavation rates can be maintained for far longer before production drops and tool replacement is required.

Figure 5 shows a relationship between tool consumption rate and a measure of strata abrasivity,obtained from a modified Cerchar abrasivity test [6]. Several points should be noted. Where an observer was present to record tool changes higher replacement rates were recorded. More rigid heavy weight roadheaders recorded lower consumption rates than light weight machines for the same rock abrasivity. High pressure waterjet assistance resulted in low tool consumption rates compared to unassisted machines. Broken ground reduced consumption rates dramatically. Boom tunnelling machines operating within a shield recorded lower consumption rates than track mounted machines. This is attributed to the more rigid cutting conditions for shield mounted machines.

To reduce tool consumption rates in abrasive materials the following options are available:

- Fit a slow speed cutter motor.
- Change head and tools to a conical self rotating geometry with larger tungsten carbide inserts at appropriate line spacing.
- Use water jet assistance.

TABLE 2 Factors Influencing drag tool consumption in Sedimentary rock

ROCK	VARIABLES	
Abrasive properties	Hard mineral content	e.g. quartz
	Grain size	
	Angularity	
	Cementing material	
Mechanical properties	Strength	ICT SE
	Cuttability	
Mass properties	Discontinuities	Volumetric intensity
	Mixed face conditions.	Orientation
Environment	Water	Within rock mass
		Dust suppression.
		H.P. water jets
	Tunnel geometry	Size
		Shape
		Gradient
	In-situ stresses	
Machine Cutting Head	Lacing pattern	
	Line spacing	
	H.P. Water Assistance	
	Dynamic balance	
	No. of tools/line	
Tool type	Radial/Conical	
Tool tip	Geometry	
	Material	Tungsten Carbide
		Cobalt %
		Grain size
		Special treatment
		Diamond
Machine	Slewing and lifting forces	
	Head speed	
	Head power	
	Rigidity of machine construction.	
Operational Features	Degree of automation	
	Profile guidance	
	Torque control	
	Speed control	

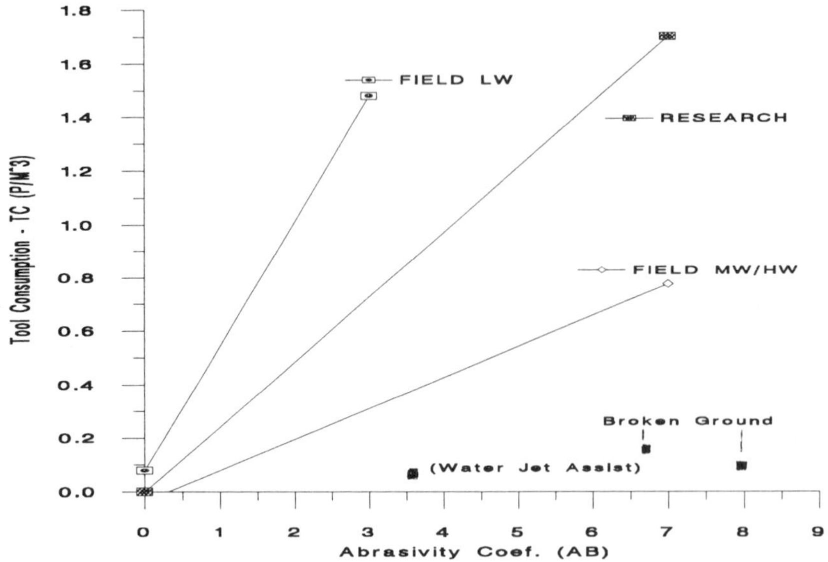

Figure 5. Abrasivity Coefficient vs Tool Consumption.

Figure 6. Dosco LH 1300H.

- Ensure rigid cutting system.
- Maintain as large an advance per revolution as possible by ensuring maximum boomforce with reduced head contact area.

8 Case Studies

Case Study A.
A heavy weight track mounted roadheader working within a British colliery encountered a dolerite dyke of UCS 380 MPa. To improve cutting performance the large 1 m diameter head was replaced by a much smaller 0.55 m diameter head and the head rotational speed reduced by fitting a 6 pole 87 kW motor giving a pick speed of 0.5 m/s compared with the original 1.4 m/s.

The tool consumption reduced from 7 tools per cubic metre to 2 tools per cubic metre and the excavation rate improved from less than 0.5 cubic metres per hour to around 2.0 cubic metres per hour.

Though the length of drivage through the dolerite was only 13.5m and assisted to some extent by veins within the dolerite, the use of the smaller head reducing the contact area by one third of the original, was considered beneficial, avoiding the need to change the whole tunnelling system to drill and blast. The use of high pressure waterjet assistance if available would have been of advantageous under these conditions.

Case Study B
In the early eighties a Dosco Mk2A machine was used in trials to investigate the potential benefits of high pressure waterjet assistance (WJA), the trial was conducted in a Derbyshire Limestone Mine. Without WJA severe vibrations and lack of penetration was experienced. With WJA assistance vibration was greatly reduced and production improved dramatically,[7].

Though WJA roadheaders were introduced into British collieries this was overtaken by the need to provide cooling to the track left by each tool when cutting in quartzitic rock. There is a danger of a potential methane ignition when excavating quartz bearing rock with drag tools in coal measures strata due to hot spots left by the cutting action. This has resulted in water sprays being placed behind each tool.If water jets were placed before each tool as well, the amount of water on the floor could cause floor breakup and slurrying problems. WJA is little used today but the potential benefits are there to be gained under the right conditions allowing harder stone to be cut economically with lower tool consumption rates.

Case Study C
This is a comprehensive study of the application of a boom tunnelling machine to rock salt excavation. At first sight it does not appear to be a difficult cutting situation but the toughness of the salt relative to its compressive strength, Table 1, and the product size restrictions, combined with high production and low tool consumption rates make it a valuable case study.

The benefits of conducting monitored trials in salt are its relative uniformity and extremely low abrasivity, which removes the influence of wear flat formation during the course of the trial.

TABLE 3

Dosco LH 1300 H Specification

Machine weight	52 t
Maximum cutting height	4.34 m
Maximum cutting width	6.26 m
Maximum sumping force	357 kN
Maximum lifting force	106 kN
Maximum arcing force	130 kN
Cutter motor at 60 H_z	170/270 kW
Cutting head speeds at 60 H_z	33/60 and 63 rpm

During the trial the amount of fine salt produced, excavation rate and tool consumption rates were monitored for changes in head rotational speed, tool line spacing on the head and tool reach and tip geometry.

The machine was a refurbished Dosco LH1300H roadheader. Figure 6. The main specifications are given in Table 3. Initial trials with a 30mm line spacing, 3 start head produced 70 m^3/h and too high a tool consumption. Wider spacing of the tools from the nose to the back of the head increased the hold in force per tool and advance per revolution. These changes resulted increased production rate, reduced fines and an increase in tool life.

Increase in the cutting head speed for the same penetration per revolution did not influence the amount of fines produced but did result in higher production rates.

The alterations to tool geometry and reach changed the forces required and the amount of fines produced. A compromise between service life and tool forces resulted in a 12mm diameter conical tungsten carbide insert being selected for the proving trials.

Final trials produced salt at a rate of 121 m^3/h and a tool consumption rate of 140 m^3/tool and the product had less than 10% - 1.5mm fines.

The boomforce per unit swept area against advance per revolution was used to predict the requirements for a larger twin boom machine which would give an ICR of 280 m^3/h.

9 Future Developments

Larger machines will be developed with the higher boomforces that are required for hard rock urban tunnels where their short length and restrictions on explosives make them more economical than full face tunnel boring machines.

The use of diamond surfaced drag tools will find increased application in harder rock with further improvements in their manufacture.

10 Conclusions

1. Providing adequate boom force is paramount to gaining a meaningful depth of cut for each tool and breakout between cutter lines. This must be achieved within the limits of machine stability.
2. Using the instrumented cutting test and boom force/swept area advance per rev characteristics production rates for existing machines can be provided. The general specification of new machines allied to a particular task can also be realistically considered.
3. Modifications to the mode of operation or machine specification can allow cutting of unforeseen harder ground conditions.
4. With the increased trend of guaranteed performance and penalties for not handing over tunnels on time, prior knowledge of excavation rates is of prime importance for competitive tendering.

11 Acknowledgements

The authors wish to express their gratitude to Dosco Overseas Engineering Ltd for support of the laboratory studies and permission to use information gained during field trials.

The opinions expressed in the paper are those of the authors' and not necessarily those of Dosco Overseas Engineering Ltd.

Acknowledgement is made to Drs H.E. Speight and A. Waggott former students of the University of Newcastle who built and developed the instrumented full scale boom tunnelling machine research rig.

12 References

1. Fowell, R.J. and McFeat-Smith, I.(1976)Factors influencing the cutting performance of a selective tunnelling machine. Proc. **Tunnelling '76**,ed M.J. Jones IMM, London 1976, pp 301-309.
2. McFeat-Smith, I and Fowell, R.J. 'The selection and application of roadheaders for rock tunnelling.' **Proc. 4th Rapid Excavation and Tunnelling Conf.** (Atlanta, USA) AIME 1979, pp261-279.
3. Speight, H.E. and Fowell, R.J. (1984) 'The influence of operational parameters on roadheader productivity and efficiency with particular reference to cutting pick wear.' Conf., on **Tribology in Mineral Extraction: War on Wear**, I.Mech.E., pp 175-182.
4. Speight, H.E. and Fowell, R.J. (1987) 'Roadheader studies using a full scale laboratory research facility.' **28th US Symp., on Rock Mechanics** (Tuscon, Arizona), pp 973-980.
5. Fowell, R.J., Waggott, A. and Anderson, I. (1988) 'Full scale boom tunnelling machine water jet trials at pressures upto 140 MPa', **9th Int., Symp., on Jet Cutting Technology** (Sedai, Japan), BHRA Cranfield, pp 323-339.
6. Johnson, S.T. and Fowell, R.J. (1986) 'Compressive strength is not enough - Assessing pick wear rates for drag tool equipped machines', Proc., **27th U.S. Rock Mechanics Symp.**, AIME, pp 840-845.

7. Barham, D.K. and Tomlin, M.G. (1986) 'High pressure water jet assisted rock and coal cutting with boom type roadheaders and shearers' Proc., **8th, Int., Symp., on Jet Cutting Technology** (Durham, UK) BHRA, Cranfield, UK, pp 57-71.

Increased productivity in construction of civil and mining tunnels through the use of high-capacity tunnel-boring machines and continuous belt conveyor muck haulage

J. G. Beatty
U.S. Heavy Division, Perini Corporation, Chicago, Illinois, U.S.A.
R. J. Ganey
Continental Conveyor and Equipment Company, Winfield, Alabama, U.S.A.
J. E. Killingsworth
Continental Conveyor and Equipment Company, Winfield, Alabama, U.S.A.

ABSTRACT

The use of a large diameter high production tunnel boring machine interfaced with a high capacity continuous belt conveyor system provides a highly productive and cost effective construction system for both civil and mining tunnels.

Continuous advance of the tunnel boring machine for a distance of 1,000 feet (305 m) allows for very efficient operation of the system. The available cost reductions will likely prove that this approach to muck handling will make marginally viable projects economically feasible. The contractor who elects to use this system will find that he is positioned to be much more competitive to owners.

The continuous advance system uses well proven technologies which are now being interfaced to provide new operational capabilities. Tunnel boring machines are well proven in many different types of materials and tunnel sizes. Belt conveyor systems of the type used in this application have a history of well over 40 years proven usage in underground mines such as coal mining, potash mining, salt mining, and copper mining. Overland material haulage has been very successful using the same conveyor concepts.

HAC® - High Angle Conveyor, Muck, TARP, TBMs - Tunnel Boring Machines, Trailing Conveyor,

This paper is demonstrated on a Metropolitan Water Reclamation District of Greater Chicago project. The total project, Tunnel and Reservoir Plan (TARP) was designed in the early 1970's with the intent to reduce pollution and provide flood relief for Chicago and many neighboring communities.

1. Tunnel and Reservoir Plan (TARP)

The TARP is designed to collect flows from combined sewers prior to discharge into local waterways, deposit the mixture of rain water and sewage in the tunnel and pump it to a treatment plant where it is purified then released into the waterways as clean water.

The TARP system, as a complete project, consists of four separate systems: Mainstream, Calumet, Des Plaines and O'Hare. These four systems all have many miles of Tunnel Boring Machine (TBM) mined tunnels, with connections to the surface called drop shafts. TARP will consist of 130 miles (210 km) of TBM mined tunnels, along with 243 drop shafts and 460 connecting structures when complete.

The TARP system consists of TBM mined tunnels over 300 feet (91 m) below ground level. This system has used various diameters of TBMs from the enormous 35'-4" (10.6 m) being the largest and 6'-6" (2 m) being the smallest.

This paper is written on what is called the TARP North Branch Tunnel.

The North Branch TARP project was bid on June 6, 1992, with award of the $167M contract a month later. The successful bidders were a joint venture composed of Perini Corporation, Framingham, Massachusetts (U.S.) as sponsor, ICA Civil Constructors in Mexico, and O&G Industries of Torrington, Connecticut (U.S.).

The project consists of 46,000 linear feet (14 km) of 32'-4" (9.86 m) diameter TBM bored tunnel in a single heading approximately 210 feet (64 m) deep. Also, 6,000 linear feet (1.8 km) of 10'-8" (3.25 m) TBM tunnel in four (4) separate headings.

These tunnels are connected to the surface structures by 20 drop shafts and two 30' (9 m) diameter main construction shafts. Drop shafts vary in diameter either 7' (2.1 m) or 9' (2.7 m) in diameter. In soft ground, which is about 70 feet (21 m) in depth, rings and lagging were used in a 20 foot (6 m) diameter shaft. Then we installed a steel casing of approximately the diameter of the drilled shaft and filled the shaft with water to begin our reverse circulation down boring to about spring line of the TBM bored tunnel. The boring in rock was approximately another 130 feet (40 m) in depth. Once the boring was complete, we moved in the drill and shoot operation to turn under and create the boot area. Every portion of exposed rock on the project gets a concrete lining. There is over 250,000 cubic yards (191,000 cubic meters) of concrete to be used on the overall project.

Each drop shaft area has to be complete prior to the TBM passing. The reason is to keep emergency access to the heading at a minimum distance and for emergency ventilation.

2. TBM's

For the projects 32'-4" (9.86 m) TBM we looked at various alternatives. There were four used TBMs in the diameter we needed for the large TBM. We did an evaluation on each TBM, also looking at the possibility of purchasing a new TBM from Robbins or Atlas Copco. We chose a used 32'-4" (9.86 m) diameter TBM (Robbins Serial No. 322-254) with approximately 25,000 linear feet (7.6 KM) of tunnel completed. We purchased the TBM and moved it to a local shop where we and a technical representative from Robbins Equipment Company began the seven month task of rebuilding.

With Robbins, all of the upgrades that they have developed since 1988, were used when 322-254 had been rebuilt. The main change was on the propel cylinders, changing them from pin connections to a ball and socket configuration. Robbins suggested to increase the horsepower and the rpm, but we thought it was not economically feasible for the production increase that the Robbins Company advertised.

The 32'-4" (9.86 m) TBM has 55 each of 17 inch (432 mm) disc cutters on the face. The TBM applies approximately 55,000 lbs of thrust per cutter. Using 3,900 horsepower (2900 kw) delivered by 13 each of a 300 horsepower (225 kw) motor. With over 3 million lbs of thrust, this can make the advance rate over 18 linear feet per hour (5.5 meters per hour) in a 25,000 to 30,000 psi rock.

The 32'-4" (9.86 m) TBM is scheduled to make the most muck out of a single heading of any machine in the U.S., almost 2 million cubic yards (1.53 million cubic meters).

For 10'-8" (3.25 m) drives, we elected to purchase from Harrison Western Company a Robbins Machine Serial No. 1011-198. This TBM has tunneled over 85,000 LF (25.9 km) on six different projects, and one of the most powerful TBMs ever built. The machine has had proven world record production just about every time it goes to work. This 10'-8" (3.25 m) TBM has 25 each of 17" (43.2 mm) disc cutter, giving it approximately 35,000 lbs of thrust per cutter. It develops 600 horsepower (450 kw) using four each of a 150 horsepower (118.5 kw) motor. We are skidding about 80 feet (24.4 m) of support backup behind this TBM with a Boretec/Long Airdox Conveyor System.

When purchased, this TBM required very little rebuilding which was accomplished in conjunction with the rebuild of Serial No. 322-254 in the same shop in Milwaukee, Wisconsin (U.S.).

Perini/ICA/O&G began the actual work on the project by dropping the main construction shaft to 150 feet (45.7 m) deep. There we established our first 10'-8" (3.25 m) TBM heading - this drive is over 3,200 linear feet (1 km). While shooting the starter tunnel, we installed our conveyor system.

This Boretec/Long Airdox Conveyor system has a belt storage unit installed on the surface. The belt ran under a set of bucket guide tubes down the shaft to the brow, then pass through a bucket loading tower, then was attached to the crown of the tunnel. The bucket loading tower system was built to load a set of 12 cubic yard (9 cubic meters) buckets in a stacked position.

While mining, a chute is loading the first bucket until full, then the chute flips up and stores the muck until the loaded bucket is hoisted past, then the chute is flipped back down and continues to load the second bucket. The system was designed to work in a 9'-6" (2.9 m) shaft. Two full buckets of muck equals the four foot advance of the TBM #1011-198.

The bucket guide tubes also carry the utilities needed for underground construction, including the ventilation. This has proven to be an extremely economical and simple system for the muck removal. The TBM averaged over 90 linear feet (27 m) per day throughout the first 3,200 foot (1 km) drive.

After the completion of the 10'-8" (3.25 m) drive, the shaft was sunk another 60 feet (18 m) to approximately 210 feet (64 m) deep, in which we began our starter and tail tunnels. We decided that our scheme is to minimize the length of the two tunnels because of the high cost and low production drill and shoot tunneling. A plan was developed to use 140 linear feet (42 m) of a 40 foot (12 m) X 40 foot (12 m) horseshoe shaped starter tunnel and a 25 linear feet (7.6 m) for a Tail Tunnel. In the preplanning of the 32'-4" (9.86 m) TBM installation, computer models were used in several schemes to try to eliminate the Tail Tunnel. Elimination of the Tail Tunnel through computer models was proven to be economically feasible.

During the drill and shoot development of the 32'-4" (9.86 m) starter tunnels, we were on the surface setting up our conveyor belt storage unit just as we did for the small tunnel while assembling our 100 foot (30 m) X 42" (1067 mm) belt piggy-back conveyor system and radial stacker supplied by Marco Conveyor. This system consists of six piggy-back conveyors and one 150 foot (45 m) radial stacker for handling the muck as it reaches the stockpile.

As the surface development continued, installation of the 32'-4" (9.86 m) TBM began. There are many extremely heavy loads to be lowered down the 210 foot (64 m) deep shaft, the heaviest being 110 short tons.

After completing assembly of the TBM, it was advanced 140 feet (42 m) to the heading. Then we installed the first three trailing decks which contains the transformers and electrical panels to operate the TBM. Upon completion, we began to mechanically mine another 70 linear feet (21 m) of tunnel, mucking with boxes, which gave the space for our final three trailing decks, the High Angle Conveyor (HAC®) and the main Continental Conveyor drive unit.

The trailing decks are a series of six 15 foot (4.6 m) long decks which contain all support for the 32'-4" (9.86 m) TBM. On the decks, over three stories in height, contain a generator to power the machine during power cable additions, maintenance shift and emergencies, and a mechanic container for repair of cutters and miscellaneous items. There are two 6' (2 m) diameter fans creating 60,000 cfm of air each, using a 72" (1829 mm) collapsible bagline attached to the crown of the tunnel for ventilation. Emergency ventilation is supplied at the nearest drop shaft to the heading, using a 6' (2 m) diameter fan at the surface. Also on the backup is office space to get out of the noise of mining, so supervision can discuss problems, and plan the work ahead. With the backup, we were concerned with the various curves in the tunnel, so we developed a rail system that "leapfrogs" from the rear of the backup to the heading by use of a trolley system. This prevents the backup from climbing the walls in the curves. Also on the lower deck is the advancing tail piece which allows the conveyor system to be installed while the TBM is mining. The conveyor system was proposed by three separate companies to supply the system for continuous belt for the 46,000 linear feet (14 km).

Continental Conveyor had the best proposed system considering engineering and installation procedures. They also had a very good track record of supply and engineering to several other tunneling and mining contractors. The difficult part of this project for a conveyor manufacturer was the 17 horizontal curves throughout the length of the tunnel.

All support of the TBM will be provided by rubber-tired vehicles. To respond to the problem of turning the vehicles around, a ramp and turntable system was designed that will advance once a week behind the large TBM. At several areas along the length of the tunnel, ramps were provided for the larger vehicles to pass, similar to a California switch. One of the advantages to a rubber-tired system is that the invert can be cleaned after the TBM passes, which gives the tunnel concrete operation very little clean-up to perform before setting the form. It also eliminates the high exposure of locomotives and track maintenance along with a reduction of labor. With the Continental Conveyor system working hand in hand with rubber-tired vehicle support, it has brought us a new and economical way to support the TBM heading.

3. Conveyor Utilization

The new TBM technology places greater emphasis on the method of removing tunnel muck from tunnel heading(s) with minimum delays. Conveyor systems are a quick, reliable, and economical means of removing muck. Conveyor systems and components have matured over the past ten years, reaching high technological levels due to advanced analytical techniques using computers. The result is that reputable conveyor companies can now use advanced analytical techniques to offer a wide variety of conveyor systems suitable for many haulage duties previously considered out-of-bounds to conveyors. U.S. tunnel development utilizing conveyors began in the late 1950's on the OAHE Tunnel Project in South Dakota, U.S.A.

TBM trailing conveyors were introduced into the tunneling industry in the early 1980's. The trailing (advancing) conveyors allow high production TBM's to continuously advance with minimum delays. High capacity conveyors allow contractors the ability to complete a mining project in less time than by conventional means of train/rail haulage. A conveyor will transport material for long distances with greater speed and capacity while negotiating multiple horizontal and vertical curves in single flights.

Perini/ICA/O&G considered muck removal options of train/rail haulage or conveyor. A careful study of train haulage revealed the TBM could advance at its maximum rate of 18 feet/hour (5.5 m/h) up to a tunnel length of 15,000 linear feet (4,572 m). The train haulage would require five double locomotives with fourteen (14) 20 cubic yard (15.3 cubic meter) muck cars per train with switches located at a maximum distance of 5,000 linear feet (1524 m) apart. As the length exceeded 15,000 linear feet (4,572 m), the TBM advance rate would decrease due to train haulage availability. The TBM advance rate would have been approximately 9 feet/hour (2.7 m/h) when the TBM approached the end of the tunnel. The conveyor availability would allow the TBM to mine at its maximum rate with minimum delays (Figure C-1).

Comparing train/rail haulage and conveyors based upon optimum number of mining shifts, the conveyors would permit the TBM to complete the project in approximately one half of the time of train/rail haulage over the entire tunnel (Figure C-2).

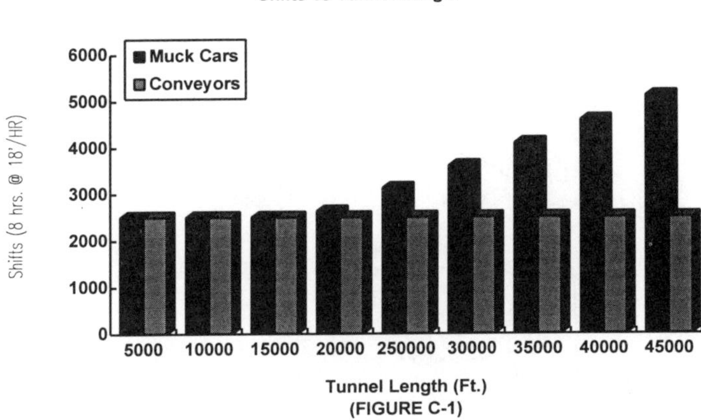

(FIGURE C-1)

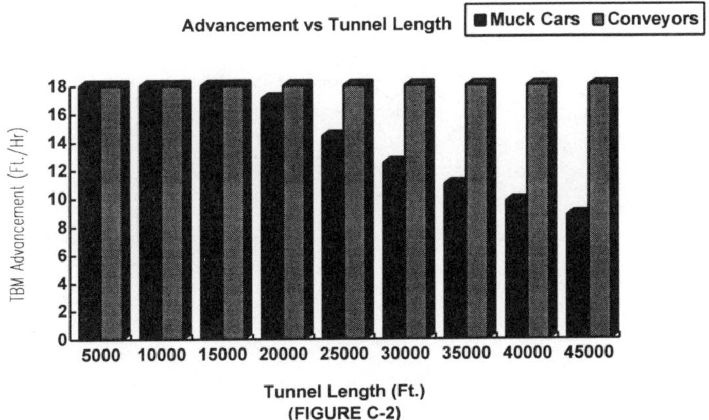

(FIGURE C-2)

The TARP North Branch Tunnel encompasses 8.7 miles (14 km) of tunnel including 17 horizontal curves with 1,000 linear feet (300 m) radii. Two conveyor flights remove the tunnel muck from the single heading tunnel. The first conveyor flight includes 3.2 miles (5.1 km) of tunnel and negotiates seven (7) horizontal curves with curve arc lengths of 248 linear feet (77 m) to 1,277 linear feet (374 m) with the second conveyor flight length of 5.5 miles (8.9 km) and maneuvering ten (10) horizontal curves with curve arc lengths varying from 240 linear feet (73 m) to 1,793 linear feet (546 m). The first conveyor (5.1 km) includes 1,250 hp (940 kw) total with one main drive and three tripper booster drives while the second conveyor flight includes 1,750 hp (1,310 kw) total with one main drive and five tripper booster drives (ref. Figure C-3). These conveyors are 36 inches (914 mm) wide, designed to handle 1,400 stph (1,270 mtph) at 700 FPM (3.6 mps).

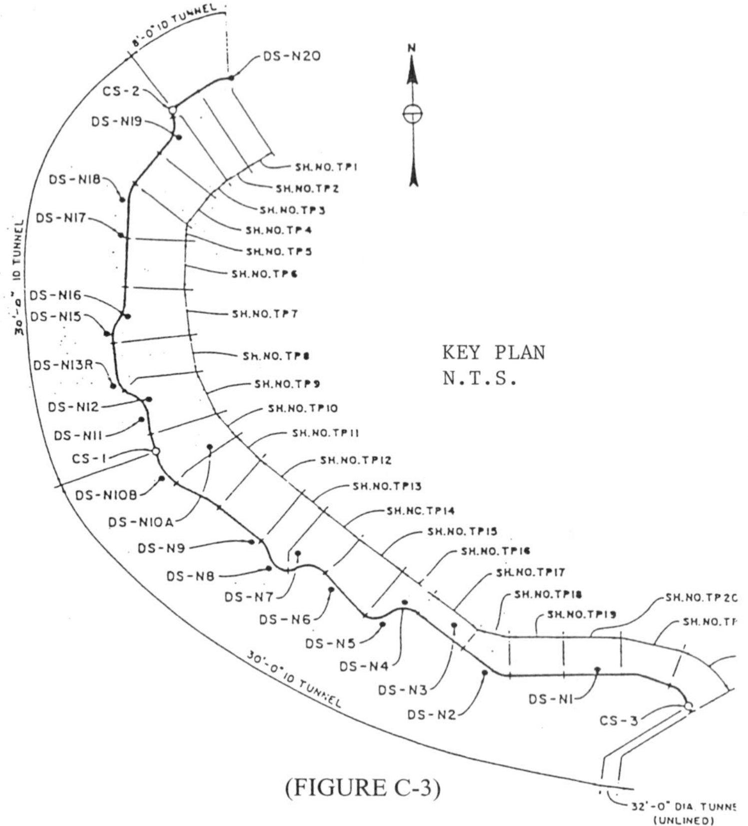

(FIGURE C-3)

Our experience has encountered tunnels with a constant curve radius in any one alignment with curves varying from 919 linear feet (280 m) to 3,200 linear feet (1,000 m). A horizontal curve analysis must be calculated to determine the lateral forces (edge pressure) the conveyor belting encounters under various conditions (loading, loaded, unloading or empty). These calculations determine the amount of intermediate super-elevation needed per curve to prevent the conveyor belt from pulling away from the conveyor intermediate structure.

The complexity of the North Branch Tunnel required the maximum amount of intermediate super-elevation and specially designed idlers in the curved areas to maintain the conveyor belt tracking under worst case conditions.

The tripper booster drives are strategically located to both optimize each unit's power contribution to the conveyor system and maintain proper belt tension in one or more horizontal curves. Each booster drive location is equipped with a specially designed electronic feedback device which provides the booster drive controller with an electrical process signal proportional to outbye belt tension (ref. Figure C-4).

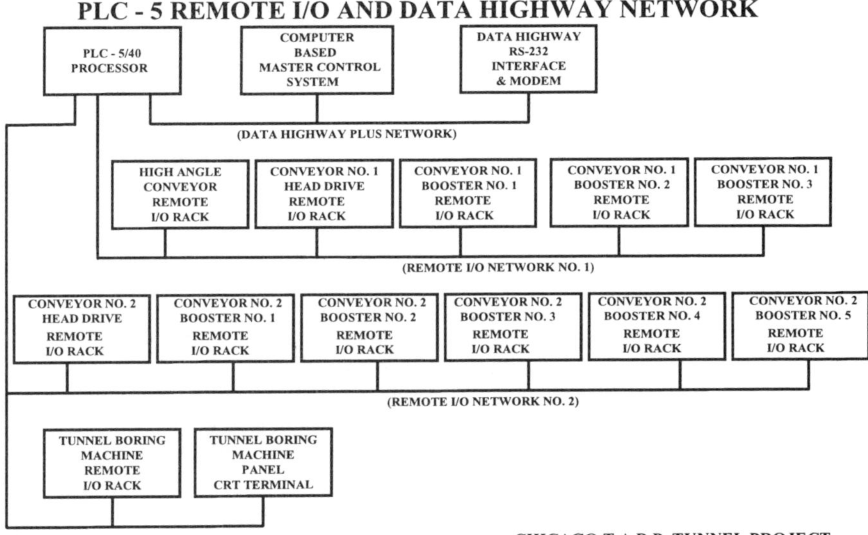

CHICAGO T.A.R.P. TUNNEL PROJECT
CONTRACTOR: PERINI / ICA / O&G
SUPPLIER : CONTINENTAL CONVEYOR & EQUIPMENT CO.
(FIGURE C-4)

The belt tension feedback is compared with a pre-programmed target tension which is calculated during the final conveyor design. The booster drive controller uses this tension feedback and preset information in a mathematical algorithm to control the amount of torque the booster drive motor produces. For example, if the belt tension is greater than the desired preset, the booster drive controller will increase motor torque thereby lowering the outbye tension value to the desired preset. If the belt tension is less than the desired preset, the booster drive controller will decrease motor torque thereby raising the outbye tension value to the desired preset.

Solid State Digital (SCR) controls combined with shunt wound direct current (DC) motors were selected to power the conveyor system. This electrical power train was selected based on performance, reliability, simplicity and economics. Technical selection considerations included the ability to vary conveyor speed to maximize loading capacity, creep speed operation for belt splicing and inspection, and torque control operation for booster drive tension regulation.

All conveyor control and monitoring functions are performed by a single Allen-Bradley PLC-5/40 based on programmable controller system. The PLC-5/40 processor interfaces with remote I/O racks located at each conveyor drive location. This includes one (1) High Angle Conveyor (HAC®) Controller, two (2) Head Drive Controllers and eight (8) Booster Drive Controllers.

In addition, the PLC-5/40 processor interfaces with an additional remote I/O rack and a PanelView color CRT operator interface located on board the TBM.

The PLC-5/40 processor also communicates via Allen-Bradley's Data Highway Plus network to a computer based master control system and a remote site support telephone modem located on the surface. Figure C-4 shows the fully developed remote I/O and data highway plus network for the control system.

The computer based master control system provides a windows based graphical operator interface for the site engineer. Functions of the master control system include centralized start/stop control of all conveyors, system alarming for all conveyors and TBM equipment, display of pertinent status information for both the conveyor system and the TBM equipment, trending analysis of trailing conveyor load demand and tension at each drive location, trending analysis of TBM cutter head motor loading, and printer datalogging of all selected status information and all alarms.

The master control system includes detailed status screens for each conveyor drive location. These detail status screens include comprehensive information about that particular location and include status for all safety and sequence devices, DC drive and motor status, auxiliary support equipment status, motor operating parameters such as armature voltage and amperage and belt tension operating levels.

The master control system provides an alarm message window which displays the most current alarm detected by the control system. All messages include the time and date of occurrences and must be acknowledged before they can be cleared from the display screen. In addition, an alarm history screen is provided which displays the twenty (20) most recent alarms detected by the system.

The master control system also provides trending screens to monitor belt loading and belt tension. Each conveyor has a separate trending screen showing loading and belt tension at all drive locations.

The master control system uses password protection to prevent unauthorized use of the system. Privilege levels are individually assigned for each password, thus allowing only key personnel full access to the system. Such things as parameter changes, tension presets and belt speed can each be assigned to be changed only if the operator logged on has the appropriate password privileges.

The panelview CRT operator interface located on the TBM provides the operator with the same conveyor start/stop control as the above ground master control system. In addition, it provides system alarming for all conveyor and TBM equipment, pertinent status information for both the conveyor system and the TBM equipment, conveyor load demand and tension values for all drive locations, and TBM cutter head motor load demand.

Once the conveyor system is started, the speed of the trailing conveyors may be varied by the TBM operator manually or automatically based on TBM production levels. This production level is monitored via the TBM primary conveyor which is equipped with a belt scale. The belt scale provides a process control signal proportional to operating tonnage. This tonnage information is used by the trailing conveyor PLC-5/40 to compute an optimum operational speed to maintain a constant profile of material.

Many of the diagnostic alarms for the conveyor system and the TBM equipment provide the TBM operator with early warning of a potential shutdown situation. This early warning information allows the TBM operator to take actions that will prevent the shutdown from occurring. A typical example would be to reduce the rate of TBM production based on an overload alarm warning for one of the cutterhead motors.

The Perini organization opted to located the belt storage unit on the surface at the main construction shaft to avoid installing it underground in adverse conditions and lengthening the starter tunnel an additional 400 linear feet (121 m). Surface mounting of the belt storage unit along with a surface mounted beltwinder/splice station allows the convenience of above ground handling of conveyor belt splicing and handling. This allows the maintenance crew the ease of installing conveyor belting into the belt storage during daily maintenance periods or non-production times. The belt storage capacity of 2,100 linear feet (640 m) allows TBM advancement of 1,000 linear feet (305 m) prior to adding additional conveyor belting. The belt storage was furnished with a hydraulic constant tension winch that maintains a preset tension thus allowing the payout of the belting as the TBM advances.

From the TBM, through the boosters, the belt tracks to the main drive. The main drive is located in front of the HAC®. Using a transition hopper, the muck is dumped onto the HAC® while the main underground belt tracks up the backside of the HAC® to the belt storage unit, then returns to the TBM (Figure C-5).

(FIGURE C-5)

4. High Angle Conveyor - HAC®

One of the most significant advances in conveyor technology, the High Angle Conveyor - HAC®, was introduced by Continental Conveyor in June of 1983.

The system has matured and proven its worth over the years with forty-five (45) units in various applications. Table 1 lists the main parameters of the various systems. Figure H-1 illustrates the first vertical HAC®, used to elevate muck from an underground excavation in Los Angeles, California,

Table 1. High Angle Conveyor - HAC® Installations - Continental Conveyor & Equipment Company

Company/Location	Material/ Rate (t/h)	Conveying Angle (°)	Elevating Height (m)	Length (m)	Belt Width (mm)	Belt Speed (m/s)	Drives (kw) Top/Bttm	In Operation
1. Demo Unit/ Winfield,AL,USA	Various/ To 2,903	30 to 60	7.9 to 19.5	35.0	1524	0 to 6.1	75/112	1983
2. Triton Coal Co./ WY,USA	Coal/2,540	60	32.9	56.7	1524	5.33	149/224	1984
3. Majdanpek Mine/Yugoslavia	Copper Ore/ 4,000	35.5	93.5	173.7	2000	2.67	450/900	1992
4. Coal Co./W.USA	Coal/2,903	35	29	61.9	1829	4.57	149/224	1987
5. Granite Constr./ CA, USA	Excavated Earth/272	90	31.7	39.9	914	1.6	22.4/22.4	1988
6. Waste Treatment Co./NY,USA	Sludge/.272	90	3.66	8.6	610	0.3	0.0/2.2	1989
7. Boise Cascade/ WA, USA	Wood Chips/ 173	53	32.6	49.3	1219	2.03	22.4/22.4	1989
8. Coal Prep Plant/ E. USA	Raw Coal/ 1,089	49	21.9	40.2	1372	2.79	56/56	1990
9. Bethenergy/ WV, USA	Clean Coal/ 726	90	76.2	90.2	1372	2.79	112/112	1991
10. Boise Cascade/ WA,USA	Wood Chips/ 65.3	90	15.5	31.4	914	2.03	7.5/7.5	1991
11. Valley Camp of Utah/UT,USA	Raw Coal/ 1,089	65	30.7	44.2	1372	3.56	93.2/93.2	1990
12. Island Creek/ VA,USA	Coal Refuse/ 454	To 41	174.8	454.2	914	2.34	186/186	1992
13. Steel Cement/ Australia	Gypsum,Slag/ 50	90	16.2	37.8	600	1.67	7.5/7.5	1991
14. Kimberly Clark/ Canada	Wood Chips/ 229	53	22.9	40.5	1219	2.03	18.6/18.6	1991
15. Cape May Co./ NJ, USA	Compost/ 40.3	90	9.0	17.5	762	1.27	0.0/11.2	1991
16. Cape May Co./ NJ, USA	Compost/ 40.3	90	13.0	31.8	762	1.27	0.0/11.2	1991
17. Shipping Co./ Mexico	Grain/584	90	18.9	27.4	1524	4.06	56/56	1991

18.	Shipping Co./ Mexico	Grain/1361	90	22.0	181.0	1829	4.06	112/112	1993
19.	Coal Co./ E.USA	Clean Coal/ 544	90	16.1	69.4	1372	2.79	37.3/75	1991
20.	Shipping Co./ Mexico	Grain/907	65	30.7	44.2	1372	3.73	75/75	1993
21.	Gleason-Pequiven/ Venezuela	Phosphate Rock/668	-35.5	Drop 34.0	113.0	914	2.29	0/93.2	1992
22.	Cementos Veracruz/ Mexico	Hot Clinker/ 715	35	41.3	198.9	1219	1.73	56/112	1992
23.	Mid-West Conv./ FL, USA	Coal/1,814	48	14.2	57.0	1829	3.56	75/112	1992
24.	U.S.Gypsum/ NY, USA	Gypsum Rock/ 363	90	36.6	48.5	1067	1.52	37.3/37.3	1992
25.	The Conveyor Co./WI, USA	Sludge/9.1	90	6.5	15.6	610	1.22	0.0/7.5	1992
26.	Mountain Coal Co./CO,USA	Raw Coal/1361	51	22.6	44.2	1524	3.56	75/93.2	1992
27.	Mountain Coal Co./CO,USA	Raw Coal/272	35	15.0	37.5	1219	1.27	11.2/14.9	1992
28.	Taulman Sys./ Canada	Compost/81.6	90	20.0	36.3	762	1.78	11.2/11.2	1992
29.	Montague Sys./ WY, USA	Coal/1950	57	59.4	90.8	1829	3.66	186/298	1993
30.	Turris Coal Co/ IL, USA	Coal/1361	90	102.0	113.0	1524	4.57	298/298	1993
31.	Sasol/So.Africa	Coal/400	90	13.3	39.3	1200	2.2	30/30	1993
32.	Sasol/So.Africa	Coal/400	90	13.3	40.3	1200	2.2	30/30	1993
33.	Sasol/So.Africa	Coal/400	90	13.3	43.4	1200	2.2	30/30	1993
34.	Sasol/So.Africa	Coal/600	90	13.3	40.3	1350	2.6	45/45	1993
35.	Sasol/So.Africa	Coal/600	90	13.3	40.3	1350	2.6	45/45	1993
36.	Bechtel/NV,USA	Gold Ore/689	60	28.9	58.4	1219	1.65	37.3/56	1993
37.	Perini/MA,USA	TBM Muck/ 1266	90	70.1	83.8	1372	3.56	186/186	1993
38.	Palm Beach Res./ FL, USA	RDF/45.3	45	23.8	40.2	1372	1.15	0/15	1993

39.	Colver Pwr Plnt/ PA, USA	Coal/260	55	28.3	60.4	762	2.29	22.4/22.4	1994
40.	Colver Pwr Plnt/ PA, USA	Coal/260	To 60	48.5	75.0	762	2.29	30/37.3	1994
41.	Butterley Engr./ Ripley, U.K.	Various/ To 49	90	9.0	11.3	500	2.5	3/3	1993

U.S.A. and Figure H-2 illustrates the twentieth (20th) vertical HAC, used to elevate muck from tunnel boring in Chicago, Illinois, U.S.A. by Perini/ICA/O&G.

(FIGURE H-1)

High Angle Conveyors - HAC®s, are proven versatile systems for elevating or lowering materials continuously at steep angles to 90°. Widely varying industries have exploited the benefits of HAC®s. Materials handled include coal, refuse, coarse copper ore, hot clinker, municipal sludge, wood chips, gypsum, slag, excavated TBM muck, various grains and RDF (refuse derived fuel - garbage). Material sizes can range from 10 inch (254 mm) maximum lump, and smaller. Tonnage rates vary from 0.27 to 4,000 stph. The conveying profiles vary widely and elevating heights range from 12 ft. (3.66 m) to 547 ft. (175 m).

While noting a wide variety of applications, this article focuses on the suitability of HACs for the elevating duties from underground TBM muck output, to muck loadout and disposal.

The Continental Conveyor HAC® represents logical evolution and development of the sandwich belt concept. The sandwich belt approach employs two ordinary rubber belts which sandwich the conveyed material. Additional force on the belt provides hugging pressure to the conveyed material in order to develop sufficient friction at the material-to-belt and material-to-material interface to prevent sliding back at the design conveying angle.

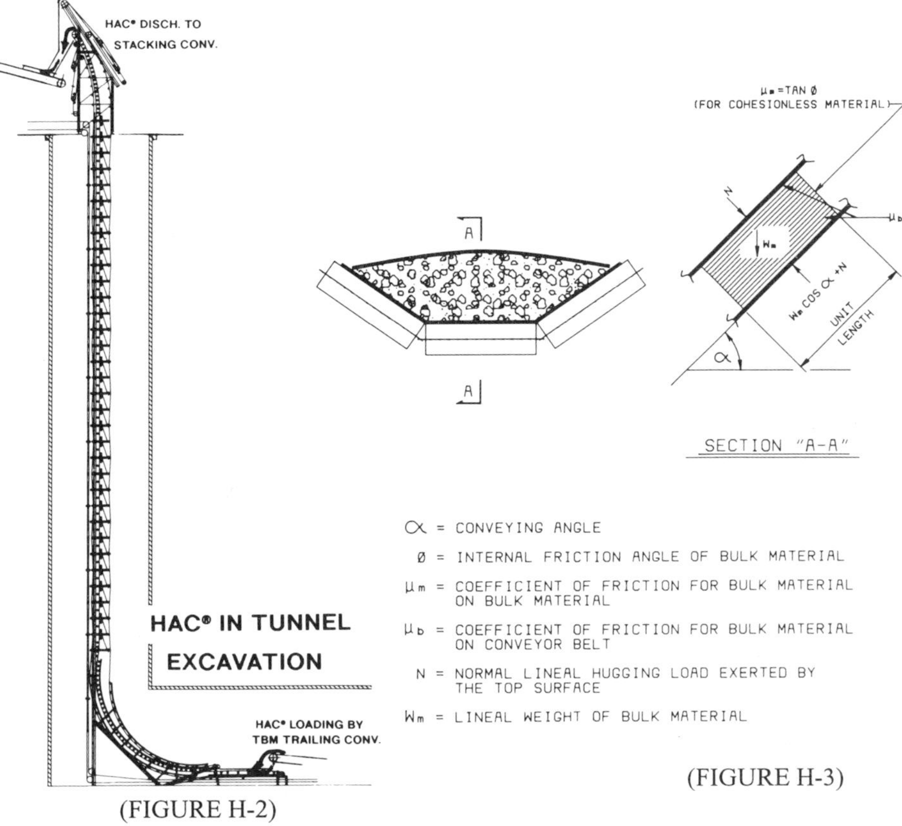

(FIGURE H-2)

(FIGURE H-3)

Figure H-3 is a simplified illustration of the interaction of forces. If the cover belt is not driven, then the lineal force N to provide the required hugging pressure at conveyor angle α is given the following equation:

$$N \geq Wm\left(\frac{\sin\alpha}{\mu} - \cos\alpha\right) \tag{1}$$

Where $\mu = \mu m$ or $\mu = \mu b$ whichever is the smaller.

A more realistic model is shown in Figure H-4. An ample belt edge distance assures a sealed material package during operation even when minor belt misalignment occurs. A more comprehensive treatment of force interaction for a complex model along with the implications of driving both belts is not within the present scope and can be found in [2] and [3].

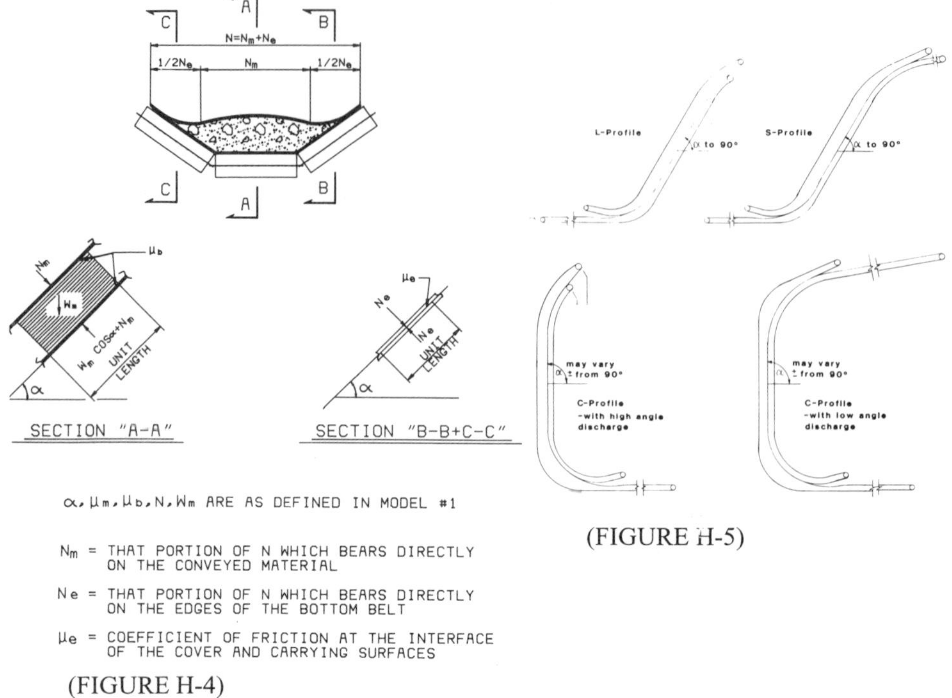

(FIGURE H-5)

(FIGURE H-4)

Per previously discussed studies by the Perini organization, the HAC® was chosen to formulate a continuous belt haulage system. The HAC® was chosen because it can take on various forms (Figure H-5) and offers many advantages over other conveying systems including:

Conveyor simplicity allows the use of all conventional conveyor hardware. Operating experience has revealed that HACs have very high availability and low maintenance costs.

Conventional conveyor components permit high conveying speeds with virtually unlimited capacity. Available belts and hardware up to 120 inch (3 m) wide make capacities greater than 15,000 stph possible.

High lifts beyond 350 ft. (107 m) are possible with standard fabric belts and single run lifts greater than 1,000 ft. (305 m) are possible with steel cord belts. High conveying angles to 90° are possible.

The Continental Conveyor sandwich belt lends itself to multi-flight conveying systems using self-contained units as well as to single-run systems using externally anchored, high angle conveyors which gives flexibility in planning and operation. In either case, the conveyor unit may be shortened or lengthened or the conveying angle may be altered according to the requirements of a new location.

Smooth surface belts are easily cleaned, allowing continuous cleaning by belt scrapers or plows. This is especially important in handling wet and sticky material. Smooth surface belts present no obstruction to quick repair of a damaged belt by hot or cold vulcanizing.

The material is sealed between the carrying and cover belts. Well-centered loading and ample belt edge distance result in spillage free operation along the conveyor length.

HAC®s are well established, with the first commercial unit in operation since June 1984. Thirty-five additional significant HACs have been delivered, and eight more units are presently in various stages of engineering and manufacturing, scheduled for operation in 1994 and 1995. Materials handled have included coarse ores at rates to 4,000 stph and lifts to 575 ft. (175 m).

Conveying angles vary from 35° to 90°. It is worth noting that twenty HACs elevate the conveyed materials at 90°; four (4) such units from underground. For the present project, HAC® specifications were derived from a study of the material and the maximum advancement of the TBM.

The TARP HAC® was designed for tunnel muck:

Material Density:	85 to 105 PCF
Material Size:	6 inch (152 mm) minus
Conveying Rate:	1400 STPH
Conveying Angle:	90° (Vertical)
Total Lift:	230 ft. (70.1 M)
Belt Width:	54 in. (1371 mm)
Belt Speed:	700 FPM (3.6 MPS)
Top Belt:	250 HP (187.5 kW)
Bottom Belt:	250 HP (187.5 kW)

4.1. Surface Conveyors

The surface muck handling system at the CS-2 stockpile utilizes one (1) 150' (45.7 m) radial stacker and eight (8) 100' (30.5 m) "piggy-back" (Marco conveyors as manufactured by E.F.Marsh Engineering Company), along with some type of conventional material handling equipment (a Cat D8 dozer). All conveyors are 42" (1067 mm) belt width designed to handle 1400 stph at 425 FPM (2.16 mps). The stacker conveyor is driven by 100 HP (75 kw) and the piggy-back is driven by 40 hp (30 kw). The "piggy-back" conveyors will transport the muck from the HAC® discharge to the stacker. Over a typical day, some 7500 cy (5734 cm) of material will be generated by the tunneling operation. By utilizing a 150' (45.7 m) stacker, a volume of over 9000 cy (6881 cm) could be placed without having to swing the unit. The D8 will swivel the stacker to a new position at the beginning of shift. The dozer would then distribute the 7500 cy piled during the previous day (3 shifts of tunneling) as muck is being stacked at the new location. This operation will continue until it becomes economical to relocate the stacker to a new set-up location. Each set-up will distribute approximately 400,000 cy (305,820 cm) of material in lifts of 40 vertical feet (12.2 m). There will be three set-ups on the first lift, two on the second, and one on the third.

5. Summary and Conclusions

Due to modern analytical methods suitable for high speed computers, it is now possible to accurately predict operating belt tensions and tension variations throughout the profile of any long conveyor. This, along with tension control by automatic take-up means and by controlled starting means, permits the design and operation of long high capacity conveyors with vertical and horizontal curves along the profile. In above ground applications, horizontally curving conveyors can permit long single flights around obstacles, such as rugged terrain, limited right-of-ways, structures, etc., that would otherwise require multiple conveyor flights and an equal number of transfer points. In underground mining, horizontally curving conveyors are used to follow tunnel boring machines (TBMs) to haul away the muck. In general, the tunnel path has many horizontal curves which the conveyor follows without transfer. After tunneling is complete, such conveyors may be operated in reverse to deliver fresh concrete, for tunnel lining as the conveyor is retracted.

Continuous haul conveyor systems are totally dependent upon the reliability and efficiency of each component. The high angle conveyor (HAC®) has added a new dimension in continuous haulage. Conveying systems have developed to the point of being extremely adaptable to TBM type excavation. With the introduction of horizontal and vertical conveying systems in the 1980's, combined with the advanced TBM technology, continuous belt conveyor haulage systems are more promising than ever in the future.

6. References

[1] Mevissen, E.A., Siminerio, A.C. and Dos Santos, J.A.: High Angle Conveyor Study; Dravo Corp. for the Bureau of Mines, U.S. Dept. of the Interior, under BuMine Contract No. J0295002, Vol. 1, 291 pp. and Vol. II, 276 pp. (1981).

[2] Dos Santos, J.A. and Frizzell, E.M.: Evolution of Sandwich Belt High Angle Conveyors; CIM bull. Vol. 576 (1983) No. 855, pp. 51-66.

[3] Dos Santos, J.A.; Sandwich Belt High Angle Conveyors - HAC - Evolution to Date; bulk solids handling, Vol. 6 (1986) No. 2, pp. 299-314.

[4] Belt Conveyors for Bulk Materials; Conveyor Equipment Manufacturers Association (CEMA), CBI Publishing Co., Inc., Boston, USA, 2nd Ed., 1979.

[5] Dos Santos, J.A.: Sandwich Belt High Angle Conveyors - Broad Horizons; bulk solids handling, Vol. 12 (1992) No. 3, pp. 425-432.

[6] Dos Santos, J.A.: HACs Continuous Vertical Haulage; presentation at Mine Hoisting '93, Second International Conference, Royal School of Mines, London, U.K., June 28-30, 1993.

[7] Smith, Bob: EM-Kayan, Tunneling in South Dakota, April 1960, pp 6-8.

[8] Petersen, Tom: Perini Corporation (unpublished), Muck Car vs. Conveyor Haulage, Study.

[9] Shupp, Craig: South Atlantic Controls (unpublished), PLC Controls.

Subject index